AF736038

Geometric and Spectral Analysis

CONTEMPORARY MATHEMATICS

630

Centre de Recherches Mathématiques Proceedings

CRM

Geometric and Spectral Analysis

CRM Workshops on:
Geometry of Eigenvalues and Eigenfunctions
June 4–8, 2012

Manifolds of Metrics and Probabilistic Methods
in Geometry and Analysis
July 2–6, 2012

Spectral Invariants on Non-compact and Singular Spaces
July 23–27, 2012
Centre de Recherches Mathématiques,
Université de Montréal, Quebec, Canada

Pierre Albin
Dmitry Jakobson
Frédéric Rochon
Editors

American Mathematical Society
Providence, Rhode Island

Centre de Recherches Mathématiques
Montréal, Québec, Canada

2010 *Mathematics Subject Classification.* Primary 58J05, 58J40, 58J50, 58J52, 53C55, 60D05, 49R05, 46E35, 53A30, 35Q41.

Library of Congress Cataloging-in-Publication Data

Geometric and spectral analysis / Pierre Albin, Dmitry Jakobson, Frédéric Rochon, editors.
p. cm. – (Contemporary mathematics ; volume 630)
Centre de Recherches Mathématiques Proceedings.
CRM Workshops on: Geometry of Eigenvalues and Eigenfunctions, June 4–8, 2012, Manifolds of Metrics and Probabilistic Methods in Geometry and Analysis, July 2–6, 2012, Spectral Invariants on Non-compact and Singular Spaces, July 23–27, 2012, Centre de Recherches Mathématiques, Université de Montreal, Montréal, Quebec, Canada.
Includes bibliographical references.
ISBN 978-1-4704-1043-8 (alk. paper)
1. Spectrum analysis–Congresses. 2. Geometric analysis–Congresses. I. Albin, Pierre, 1976– II. Jakobson, Dmitry, 1970– III. Rochon, Frédéric, 1978–.
QC20.7.S64G46 2014
516–dc23 2014021478

Contemporary Mathematics ISSN: 0271-4132 (print); ISSN: 1098-3627 (online)

DOI: `http://dx.doi.org/10.1090/conm/630`

∞ The paper used in this book is acid-free and falls within the guidelines
established to ensure permanence and durability.
Visit the AMS home page at `http://www.ams.org/`

10 9 8 7 6 5 4 3 2 1 19 18 17 16 15 14

Contents

Preface

In 2012, thematic programs on *Geometric Analysis and Spectral Theory* and on *Moduli spaces, Extremality and Global Invariants* were held at the CRM. There were several workshops on cutting-edge topics in geometric analysis and spectral theory, notably on the *Geometry of Eigenvalues and Eigenfunctions*, on *Manifolds of Metrics and Probabilistic Methods in Geometry and Analysis* and on *Spectral invariants on non-compact and singular spaces.* This was the occasion to learn about the most recent developments in various interconnected fields and also the starting point for new investigations and fruitful collaborations.

The 2012 Spring Semester on Geometric Analysis and Spectral Theory at the CRM brought together researchers working in these fields interpreted in a broad sense, so as to include applications to fundamental problems in geometry, PDE, dynamical systems and mathematical physics. The present CRM Proceedings includes articles written by participants of the workshops in *Geometry of Eigenvalues and Eigenfunctions* and on *Manifolds of Metrics and Probabilistic Methods in Geometry and Analysis.*

The workshop on Geometry of Eigenvalues and Eigenfunctions was held at the CRM on June 4–8, 2012. It was organized by D. Jakobson (McGill) and I. Polterovich (Montréal). The workshop brought together the leading researchers and young mathematicians working in various areas of geometric spectral theory. Many problems in the field are motivated by questions originating in the study of real-life phenomena: quantum-mechanical effects, vibration of membranes and plates, oscillations of fluids, etc. One of the main highlights of the meeting was a series of Aisenstadt lectures on Quantum Unique Ergodicity by Elon Lindenstrauss. Quantum Ergodicity and properties of eigenfunctions in the semiclassical limit were among the central themes of the workshop. In recent work, Jakobson, Safarov and Strohmaier analyzed the semiclassical limit of spectral theory on manifolds whose metrics have jump-like discontinuities; the corresponding semiclassical limit does not relate to a classical flow but rather to branching (ray-splitting) billiard dynamics. A Quantum Ergodicity theorem was established for discontinous systems, and a new notion of ergodicity for the ray-splitting dynamics was introduced, and a class of examples of such ergodic systems was provided by Yves Colin de Verdière. Very important technical results in that work were established in the paper by Y. Safarov appearing in these proceedings. His paper discusses Fourier integral operators (FIOs) associated with canonical transformations. It is well known that the composition of two such operators is a FIO associated with the composition of the corresponding transformations. It is also known that the principal symbol of a FIO is a half-density with values in a linear bundle over a Lagrangian manifold. The latter makes it impossible to write down a simple formula for the principal symbol of the composition: such a formula must involve the product of principal symbols but it is not a half-density. It will be shown that, under a slightly different definition of the principal symbol, this problem can be avoided, and then the formulae for principal symbols of compositions and adjoint operators become almost as simple as for pseudodifferential operators.

Another topic considered at the conference was the behavior of eigenfunctions of the Laplacian on Riemannian manifolds and Euclidean domains. In their paper appearing in these proceedings, B. Nguyen, A. Delytsin, and D. Grebenkov discuss the eigenvalue problem for the Laplace operator in a planar domain which can be decomposed into a bounded domain of arbitrary shape and elongated "branches" of variable cross-sectional profiles. They show that when the eigenvalue is smaller than a prescribed threshold, the corresponding eigenfunction decays exponentially along each branch.

Two papers appearing in these proceedings concern the behavior of eigenvalues of differential operators. In the paper by V. Guillemin, E. Legendre, and R. Sena-Dias, the authors calculate the variation of the Rayleigh quotient under conformal variations and under variations within a Kähler class. They use the conformal formula to reprove Uhlenbeck's result: a generic metric can be perturbed by a conformal factor into a metric whose eigenspaces for the Laplace operator have dimension 1. In the case of a Kähler manifold, they write an explicit variational formula for the Rayleigh quotient under Kähler deformations and discuss this relationship with the following question: Is the spectrum of a generic Kähler metric simple?

G. Poliquin's paper considers lower bounds for the principal frequency of the p-Laplacian on n-dimensional Euclidean domains. In particular, he extends the classical results involving the inner radius of a domain and the first eigenvalue of the Laplace operator to the case $p \neq 2$. As a by-product, he obtains a lower bound on the size of the nodal set of an eigenfunction of the p-Laplacian on planar domains.

The workshop on Manifolds of Metrics and Probabilistic Methods in Geometry and Analysis was held at CRM on July 2–6, 2012. It was organized by D. Jakobson (McGill), S. Klevtsov (ULB), and S. Zelditch (Northwestern). The workshop brought together mathematicians and physicists working on questions related to random geometry in a broad sense.

The study of geometry of the infinite-dimensional space of Kähler metrics was the subject of many talks at the conference, as well as two papers in these proceedings. The main result in the paper of R. Berman provides a new formula for the distribution of the exponentially small eigenvalues of Dolbeault Laplacians associated to high tensor powers of a line bundle over a compact complex manifold, which in physics terminology is a measure of "tunneling" of the Dolbeault complex. Along the way a new proof of the asymptotics of the induced Quillen metric on the corresponding determinant line is obtained. A brief comparison with the tunneling effect for Witten Laplacians and large deviation principles for fermions is also made.

A long paper by Y. Rubinstein provides a survey on Kähler-Einstein metrics in the smooth and singular (edge) setting, presented from a unifed point of view. The paper takes the reader from the very beginning of Kähler geometry and the Einstein equations, through the edge calculus for Kähler-Einstein edge metrics as developed in the work of Rubinstein, Jeffres and Mazzeo, over the geometry of the space of Kähler metrics and its associated functionals, to the very recent work of

Rubinstein-Cheltsov on asymptotic log Fano varieties. The survey is concluded with a conjectural picture concerning a generalization of the Calabi conjecture in the asympotic log setting.

Several sessions at the workshop were devoted to the lectures on Liouville 2d quantum gravity and the physical definition of random metrics. L. Chen outlined a construction (jointly with D. Jakobson) of random measures in dimension four, inspired by the construction of Duplantier and Sheffield in dimension two, and described an approach to derive a KPZ-type relation for spherical averages of those measures. The construction used a conformally covariant differential operator (Paneitz operator) in dimension four, in place of the Laplacian used in dimension two. R. Ponge gave an overview of Fefferman's program, conformally covariant differential operators (including the Yamabe and the Paneitz operators). In his paper appearing in these proceedings, Ponge studied the logarithmic singularities of the Green functions of conformally covariant operators, including the Yamabe and Paneitz operators, as well as the conformal fractional powers of the Laplacian arising from scattering theory for Poincaré-Einstein metrics. He explains how to compute explicitly the logarithmic singularities of the Green functions of the conformal powers of the Laplacian. The results are formulated in terms of Weyl conformal invariants arising from the ambient metric of Fefferman-Graham. As applications we obtain characterizations in terms of Green functions of locally conformally flat manifolds and a spectral theoretic characterization of the conformal class of the round sphere.

Several talks at the conference concerned the study of random functions. L. Nicolaescu's talk described his results on critical values of random functions on a given compact Riemannian manifold. In his paper appearing in these proceedings, Nicolaescu surveys his work on the distribution of critical points and critical values of those functions, defined as a Fourier-like eigenseries whose coefficients are independent Gaussian random variables.

Nicolas Burq gave a talk on probabilistic Sobolev embeddings, showing that, from a PDE point of view, randomly chosen functions may behave much more nicely than what the deterministic theory would predict. In the paper by N. Burq and G. Lebeau appearing in these proceedings, the authors present some probabilistic versions of Sobolev embeddings and an application to the growth rate of the L^p norms of spherical harmonics, as well as eigenfunctions on manifolds with transitive isometry group actions.

Shortly after these workshops, the CRM thematic year on *Moduli spaces, extremality and global invariant* started with its first workshop on *Spectral invariants on non-compact and singular spaces.* It was held on July 23–27 at the CRM and was organized by P. Albin (UIUC) and F. Rochon (UQAM). Since it involved spectral theory and geometric analysis, as well as global invariants and moduli spaces (which are a good source of examples of singular spaces), this workshop was a natural transition between the two thematic programs. A good part of it was concerned with spectral invariants arising in index theory, like the eta invariant, analytic torsion or the determinant of Laplacian. The latter also arises naturally in scattering theory and more generally in spectral theory, which was the other important theme of the workshop.

The workshop was the starting point of an interesting collaboration between Dean Baskin, Jeremy Marzuola, and Jared Wunsch, who obtained global Strichartz estimates for exterior polygonal domains as well as some smoothing estimates. These results have natural applications to the study of some non-linear Schrödinger equations on exterior polygonal domains. Their results appear in the present proceedings as one of the five articles related to this workshop.

There is also the collaboration of Zhiqin Lu and Julie Rowlett, which provides a careful analysis of the gap between the first two positive eigenvalues of the Laplacian of a convex domain in $\mathbb{R}^n$ that degenerates to a domain in $\mathbb{R}^{n-1}$. One of the important conclusions is that this gap detects the geometry of a one-dimensional collapse. In fact, depending on the geometry of the collapse, the gap can diverge or remain bounded.

In another direction, Maxim Braverman and Boris Vertman give a new proof of the Bismut-Zhang formula for the ratio of the analytic torsion and the Reidemeister torsion. Their proof works when the representation of the fundamental group lies in a connected component of a unitary representation and has the advantage of being significantly simpler. The idea is to use the Cheeger-Müller theorem for the unitary representations and extend the result to the whole connected component using analytic continuation.

Gerardo A. Mendoza contributes to these proceedings with a nice paper on b-complex manifolds, which are complex manifolds arising from a complex structure of the b-tangent bundle of Melrose. This is a subject he has developed over the years. In the paper published in these proceedings, Mendoza studies the rich structure of the cohomology of the indicial complex obtained by restricting the Dolbeault complex to the boundary.

Finally, B.-W. Schulze gives a valuable and useful overview of the iterative approach, an approach he and his collaborators developed to construct and study a pseudodifferential calculus suitable for doing analysis on stratified spaces.

Pierre Albin
Dmitry Jakobson
Frédéric Rochon

Contemporary Mathematics
Volume **630**, 2014
http://dx.doi.org/10.1090/conm/630/12662

A New Proof of a Bismut-Zhang Formula for Some Class of Representations

Maxim Braverman and Boris Vertman

Abstract. Bismut and Zhang computed the ratio of the Ray-Singer and the combinatorial torsions corresponding to non-unitary representations of the fundamental group. In this note we show that for representations which belong to a connected component containing a unitary representation the Bismut-Zhang formula follows rather easily from the Cheeger-Müller theorem, i.e. from the equality of the two torsions on the set of unitary representations. The proof uses the fact that the refined analytic torsion is a holomorphic function on the space of representations.

1. Introduction

Let M be a closed oriented odd-dimensional manifold and let $\mathrm{Rep}(\pi_1(M),\mathbb{C}^n)$ denote the space of representations of the fundamental group $\pi_1(M)$ of M. For each $\alpha\in\mathrm{Rep}(\pi_1(M),\mathbb{C}^n)$, let (E_α,∇_α) be a flat vector bundle over M, whose monodromy representation is equal to α. We denote by $H^\bullet(M,E_\alpha)$ the cohomology of M with coefficients in E_α. Let $\mathrm{Det}(H^\bullet(M,E_\alpha))$ denote the determinant line of $H^\bullet(M,E_\alpha)$.

Reidemeister [**21**] and Franz [**10**] used a cell decomposition of M to construct a combinatorial invariant of the representation $\alpha\in\mathrm{Rep}(\pi_1(M),\mathbb{C}^n)$, called the *Reidemeister torsion*. In modern language it is a metric on the determinant line $\mathrm{Det}(H^\bullet(M,E_\alpha))$, cf. [**2, 19**]. If α is unitary, then this metric is independent of the cell decomposition and other choices. In general to define the Reidemeister metric one needs to make some choices. One of such choices is a Morse function $F:M\to\mathbb{R}$. Bismut and Zhang [**2**] call the metric obtain using the Morse function F the *Milnor metric* and denote it by $\|\cdot\|^{\mathrm{M}}_F$.

Ray and Singer [**20**] used the de Rham complex to give a different construction of a metric on $\mathrm{Det}(H^\bullet(M,E_\alpha))$. This metric is called the *Ray-Singer metric* and is denoted by $\|\cdot\|^{\mathrm{RS}}$. Ray and Singer conjectured that the Ray-Singer and the Milnor metrics coincide for unitary representation of the fundamental group. This conjecture was proven by Cheeger [**8**] and Müller [**16**] and extended by Müller [**17**] to unimodular representations. For non-unitary representations the two metrics are not equal in general. In the seminal paper [**2**] Bismut and Zhang computed the ratio of the two metrics using very non-trivial analytic arguments.

2010 *Mathematics Subject Classification*. Primary 58J52.
The first author was supported in part by the NSF grant DMS-1005888.

In this note we show that for a large class of representations the Bismut-Zhang formula follows quite easily from the original Ray-Singer conjecture. More precisely, let $\alpha_0 \in \mathrm{Rep}(\pi_1(M), \mathbb{C}^n)$ be a unitary representation which is a regular point of the complex analytic set $\mathrm{Rep}(\pi_1(M), \mathbb{C}^n)$ and let $\mathcal{C} \subset \mathrm{Rep}(\pi_1(M), \mathbb{C}^n)$ denote the connected component of $\mathrm{Rep}(\pi_1(M), \mathbb{C}^n)$ which contains α_0. We derive the Bismut-Zhang formula for all representations in $\mathcal{C}$ from the Cheeger-Müller theorem. In other words, we show that knowing that the Milnor and the Ray-Singer metrics coincide on unitary representations one can derive the formula for the ratio of those metrics for all representations in the connected component $\mathcal{C}$.

The proof uses the properties of the refined analytic torsion $\rho_{\mathrm{an}}(\alpha)$ introduced in [**3, 5, 6**] and of the refined combinatorial torsion $\rho_{\varepsilon,\mathfrak{o}}(\alpha)$ introduced in [**9, 27**]. Both refined torsions are non-vanishing elements of the determinant line $\mathrm{Det}(H^\bullet(M, E_\alpha))$ which depend holomorphically on $\alpha \in \mathrm{Rep}(\pi_1(M), \mathbb{C}^n)$. The ratio of these sections is a holomorphic function

$$\alpha \;\mapsto\; \frac{\rho_{\mathrm{an}}(\alpha)}{\rho_{\varepsilon,\mathfrak{o}}(\alpha)}$$

on $\mathrm{Rep}(\pi_1(M), \mathbb{C}^n)$. We first use the Cheeger-Müller theorem to compute this function for unitary α. Let now $\mathcal{C} \subset \mathrm{Rep}(\pi_1(M), \mathbb{C}^n)$ be a connected component and suppose that a unitary representation α_0 is a regular point of $\mathcal{C}$. The set of unitary representations can be viewed as the real locus of the connected complex analytic set $\mathcal{C}$. As we know $\frac{\rho_{\mathrm{an}}(\alpha)}{\rho_{\varepsilon,\mathfrak{o}}(\alpha)}$ for all points of the real locus, we can compute it for all $\alpha \in \mathcal{C}$ by analytic continuation. Since the Ray-Singer norm of $\rho_{\varepsilon,\mathfrak{o}}$ and the Milnor norm of ρ_{an} are easy to compute, we obtain the Bismut-Zhang formula for all $\alpha \in \mathcal{C}$.

The paper is organized as follows. In Section 2, we briefly outline the main steps of the proof. In Subsection 3.5 and Section 3 we recall the construction and some properties of the Milnor metric and of the Farber-Turaev torsion. In Section 4 we recall some properties of the refined analytic torsion. In Section 5 we recall the construction of the holomorphic structure on the determinant line bundle and show that the ratio of the refined analytic and the Farber-Turaev torsions is a holomorphic function on $\mathrm{Rep}(\pi_1(M), \mathbb{C}^n)$. Finally, in Section 6 we present our new proof of the Bismut-Zhang theorem for representations in the connected component $\mathcal{C}$.

2. The idea of the proof

Our proof of the Bismut-Zhang theorem for representations in the connected component $\mathcal{C}$ consists of several steps. In this section we briefly outline these steps.

Step 1. In [**25, 26**], Turaev constructed a refined version of the combinatorial torsion associated to an acyclic representation α. Turaev's construction depends on additional combinatorial data, denoted by ε and called the *Euler structure*, as well as on the *cohomological orientation* of M, i.e., on the orientation $\mathfrak{o}$ of the determinant line of the cohomology $H^\bullet(M, \mathbb{R})$ of M. In [**9**], Farber and Turaev extended the definition of the Turaev torsion to non-acyclic representations. The Farber-Turaev torsion associated to a representation α, an Euler structure ε, and a cohomological orientation $\mathfrak{o}$ is a non-zero element $\rho_{\varepsilon,\mathfrak{o}}(\alpha)$ of the determinant line $\mathrm{Det}\big(H^\bullet(M, E_\alpha)\big)$.

Let us fix a Hermitian metric h^{E_α} on E_α. This scalar product induces a norm $\|\cdot\|^{\rm RS}$ on $\mathrm{Det}\big(H^\bullet(M,E_\alpha)\big)$, called the *Ray-Singer metric.* In Subsection 3.5 we use the Cheeger-Müller theorem to show that for unitary α

$$\|\rho_{\varepsilon,\mathfrak{o}}(\alpha)\|^{\rm RS} \;=\; 1. \tag{2.1}$$

REMARK 2.1. Theorem 10.2 of [**9**] computes the Ray-Singer norm of $\rho_{\varepsilon,\mathfrak{o}}$ for arbitrary representation $\alpha\in\mathrm{Rep}(\pi_1(M),\mathbb{C}^n)$; however, the proof uses the result of Bismut and Zhang, which we want to prove here for $\alpha\in\mathcal{C}$.

Theorem 1.9 of [**5**] computes the Ray-Singer metric of $\rho_{\rm an}(\alpha)$. Combining this result with (2.1) we conclude, cf. Subsection 5.7, that if α is a unitary representation, then

$$\left|\frac{\rho_{\rm an}(\alpha)}{\rho_{\varepsilon,\mathfrak{o}}(\alpha)}\right| \;=\; \frac{\|\rho_{\rm an}(\alpha)\|^{\rm RS}}{\|\rho_{\varepsilon,\mathfrak{o}}(\alpha)\|^{\rm RS}} \;=\; 1. \tag{2.2}$$

Step 2. The Farber-Turaev torsion $\rho_{\varepsilon,\mathfrak{o}}(\alpha)$ is a holomorphic section of the determinant line bundle

$$\mathcal{D}et \;:=\; \bigsqcup_{\alpha\in\mathrm{Rep}(\pi_1(M),\mathbb{C}^n)} \mathrm{Det}\big(H^\bullet(M,E_\alpha)\big)$$

over $\mathrm{Rep}(\pi_1(M),\mathbb{C}^n)$. We denote by $\rho_{\rm an}(\alpha)/\rho_{\varepsilon,\mathfrak{o}}(\alpha)$ the unique complex number such that

$$\rho_{\rm an}(\alpha) \;=\; \frac{\rho_{\rm an}(\alpha)}{\rho_{\varepsilon,\mathfrak{o}}(\alpha)}\cdot\rho_{\varepsilon,\mathfrak{o}}(\alpha) \;\in\; \mathrm{Det}\big(H^\bullet(M,E_\alpha)\big).$$

Since both $\rho_{\varepsilon,\mathfrak{o}}$ and $\rho_{\rm an}$ are holomorphic sections of $\mathcal{D}et$,

$$\alpha \;\mapsto\; \frac{\rho_{\rm an}(\alpha)}{\rho_{\varepsilon,\mathfrak{o}}(\alpha)}$$

is a holomorphic function on $\mathrm{Rep}(\pi_1(M),\mathbb{C}^n)$.

Step 3. Let α' denote the representation dual to α with respect to a Hermitian scalar product on $\mathbb{C}^n$. Then the Poincaré duality induces, cf. [**9**, §2.5] and [**5**, §10.1], an anti-linear isomorphism[1]

$$D:\,\mathrm{Det}\big(H^\bullet(M,E_\alpha)\big) \;\longrightarrow\; \mathrm{Det}\big(H^\bullet(M,E_{\alpha'})\big).$$

In particular, when α is a unitary representation, D is an anti-linear automorphism of $\mathrm{Det}\big(H^\bullet(M,E_\alpha)\big)$. Hence,

$$\frac{D(\rho_{\rm an}(\alpha))}{D(\rho_{\varepsilon,\mathfrak{o}}(\alpha))} \;=\; \overline{\frac{\rho_{\rm an}(\alpha)}{\rho_{\varepsilon,\mathfrak{o}}(\alpha)}}. \tag{2.3}$$

Using Theorem 7.2 and formula (9.4) of [**9**] we compute the ratio $D(\rho_{\varepsilon,\mathfrak{o}}(\alpha))/\rho_{\varepsilon,\mathfrak{o}}(\alpha)$, cf. (6.6) (here α is a unitary representation). On the analytic side Theorem 10.3 of [**5**] computes the ratio $D(\rho_{\rm an}(\alpha))/\rho_{\rm an}(\alpha)$. Combining these two results we get

$$\overline{\frac{\rho_{\rm an}(\alpha)}{\rho_{\varepsilon,\mathfrak{o}}(\alpha)}} \;=\; f_2(\alpha)\cdot\frac{\rho_{\rm an}(\alpha)}{\rho_{\varepsilon,\mathfrak{o}}(\alpha)}, \tag{2.4}$$

where f_2 is a function on $\mathrm{Rep}(\pi_1(M),\mathbb{C}^n)$ computed explicitly in (6.7).

[1]There is a sign difference in the definition of the duality operator in [**9**] and [**5**], which is not essential for the discussion in this paper.

From (2.3) and (2.4) we conclude that

$$\left(\frac{\rho_{\rm an}(\alpha)}{\rho_{\varepsilon,\mathfrak{o}}(\alpha)}\right)^2 \;=\; f_1(\alpha)^2\cdot f_2(\alpha) \tag{2.5}$$

for any unitary representation α, cf. (6.9), where $f_1(\alpha)=\rho_{\rm an}(\alpha)/\rho_{\varepsilon,\mathfrak{o}}(\alpha)$.

Step 4. The right hand side of (2.5) is an explicit function of a unitary representation α. It turns out that it is a restriction of a holomorphic function $f(\alpha)$ on $\operatorname{Rep}(\pi_1(M),\mathbb{C}^n)$ to the set of unitary representations. Recall that the connected component $\mathcal{C}$ contains a regular point which is a unitary representation. The set of unitary representations can be viewed as the *real locus* of the complex analytic set $\mathcal{C}$. Hence any two holomorphic functions which coincide on the set of unitary representations, coincide on $\mathcal{C}$. We conclude now from (2.5) that

$$\left(\frac{\rho_{\rm an}(\alpha)}{\rho_{\varepsilon,\mathfrak{o}}(\alpha)}\right)^2 \;=\; f(\alpha), \qquad \text{for all}\quad \alpha\in\mathcal{C}. \tag{2.6}$$

Step 5. Recall that we denote by $\|\cdot\|_F^M$ the Milnor metric associated to the Morse function F. In Section 3 we compute the Milnor metric

$$\|\rho_{\varepsilon,\mathfrak{o}}(\alpha)\|_F^M \;=\; h_1(\alpha), \tag{2.7}$$

where $h(\alpha)$ is a real valued function on $\operatorname{Rep}(\pi_1(M),\mathbb{C}^n)$ given explicitly by the right hand side of (3.14)

Theorem 1.9 of [**5**] computes the Ray-Singer norm of the refined analytic torsion:

$$\|\rho_{\rm an}(\alpha)\|^{\rm RS} \;=\; h_2(\alpha), \tag{2.8}$$

where $h_2(\alpha)$ is a real valued function on $\operatorname{Rep}(\pi_1(M),\mathbb{C}^n)$ given explicitly by the right hand side of (4.5). Combining (2.6) with (2.8), we get

$$\frac{\|\cdot\|^{\rm RS}}{\|\cdot\|_F^{\rm M}} \;=\; \frac{\|\rho_{\rm an}(\alpha)\|^{\rm RS}}{\|\rho_{\varepsilon,\mathfrak{o}}(\alpha)\|_F^{\rm M}}\cdot\left|\frac{\rho_{\varepsilon,\mathfrak{o}}(\alpha)}{\rho_{\rm an}(\alpha)}\right| \;=\; \frac{h_2(\alpha)}{h_1(\alpha)\cdot|f(\alpha)|}. \tag{2.9}$$

This is exactly the Bizmut-Zhang formula [**2**, Theorem 0.2].

The rest of the paper is occupied with the details of the proof outlined above.

3. The Milnor metric and the Farber-Turaev torsion

In this section we briefly recall the definitions and the main properties of the Milnor metric and the Farber-Turaev refined combinatorial torsion. We also compute the Milnor norm of the Farber-Turaev torsion.

3.1. The Thom-Smale complex. Set

$$C^k(K,E_\alpha) \;=\; \bigoplus_{\substack{x\in Cr(F)\\ {\rm ind}_F(x)=k}} E_{\alpha,x}, \qquad k=1,\dots,n,$$

where $E_{\alpha,x}$ denotes the fiber of E_α over x and the direct sum is over the critical points $x\in Cr(F)$ of the Morse function F with Morse-index ${\rm ind}_F(x)=k$. If the Morse function is F generic, then using the gradient flow of F one can define the *Thom-Smale complex* $(C^\bullet(K,E_\alpha),\partial)$ whose cohomology is canonically isomorphic to $H^\bullet(M,E_\alpha)$, cf. for example [**2**, §I c].

3.2. The Euler structure. The *Euler structure* ε on M can be described as (an equivalence class of) a pair (F,c) where c is a 1-chain in M such that

$$\partial c = \sum_{x\in Cr(F)} (-1)^{\mathrm{ind}_F(x)}\cdot x, \tag{3.1}$$

cf. [**7**, §3.1]. We denote the set of Euler structures on M by $\mathrm{Eul}(M)$.

REMARK 3.3. The Euler structure was introduced by Turaev [**26**]. Turaev presented several equivalent definitions and the equivalence of these definitions is a nontrivial result. Burghelea and Haller [**7**] found a very nice way to unify these definitions. They suggested a new definition which is obviously equivalent to the two definitions of Turaev. In this paper we use the definition introduced by Burghelea and Haller.

3.4. The Kamber-Tondeur form. To define the Milnor and the Ray-Singer metrics on $\mathrm{Det}(H^\bullet(M,E_\alpha))$ we fix a Hermitian metric h^{E_α} on E_α. This metric is not necessarily flat and the measure of non-flatness is given by taking the trace of $(h^{E_\alpha})^{-1}\nabla_\alpha h^{E_\alpha}\in\Omega^1(M,\mathrm{End}E_\alpha)$ which defines the *Kamber-Tondeur form*

$$\theta(h^{E_\alpha}) := \mathrm{Tr}\left[(h^{E_\alpha})^{-1}\nabla_\alpha h^{E_\alpha}\right] \in \Omega^1(M), \tag{3.2}$$

cf. [**14**] (see also [**2**, Ch. IV]).

Let $\mathrm{Det}(E_\alpha)\to M$ denote the determinant line bundle of E_α, i.e. the line bundle whose fiber over $x\in M$ is equal to the determinant line $\mathrm{Det}(E_{\alpha,x})$ of the fiber $E_{\alpha,x}$ of E_α. The connection ∇_α and the metric h^{E_α} induce a flat connection $\nabla_\alpha^{\mathrm{Det}}$ and a metric $h^{\mathrm{Det}(E_\alpha)}$ on $\mathrm{Det}(E_\alpha)$. Then

$$\theta(h^{\mathrm{Det}(E_\alpha)}) = \theta(h^{E_\alpha}). \tag{3.3}$$

For a curve $\gamma:[a,b]\to M$ let

$$\alpha(\gamma): E_{\alpha,\gamma(a)} \to E_{\alpha,\gamma(b)}, \quad \alpha^{\mathrm{Det}}(\gamma): \mathrm{Det}(E_{\alpha,\gamma(a)}) \to \mathrm{Det}(E_{\alpha,\gamma(b)}) \tag{3.4}$$

denote the parallel transports along γ. Then

$$\mathrm{Det}\big(\alpha(\gamma)\big) = \alpha^{\mathrm{Det}}(\gamma). \tag{3.5}$$

Let $\widetilde{\gamma}(t)\in\mathrm{Det}(E_{\alpha,\gamma(t)})$ denote the horizontal lift of the curve γ. By the definition of the Kamber-Tondeur form we have

$$\log\frac{h^{\mathrm{Det}(E_\alpha)}(\widetilde{\gamma}(b),\widetilde{\gamma}(b))}{h^{\mathrm{Det}(E_\alpha)}(\widetilde{\gamma}(a),\widetilde{\gamma}(a))} = \int_\gamma \theta(h^{\mathrm{Det}(E_\alpha)}) = \int_\gamma \theta(h^{E_\alpha}), \tag{3.6}$$

where in the last equality we used (3.3).

If γ is a closed curve, $\gamma(a)=\gamma(b)$, we obtain

$$\frac{h^{\mathrm{Det}(E_\alpha)}(\widetilde{\gamma}(b),\widetilde{\gamma}(b))}{h^{\mathrm{Det}(E_\alpha)}(\widetilde{\gamma}(a),\widetilde{\gamma}(a))} = \left|\alpha^{\mathrm{Det}}(\gamma)\right|^2 = \left|\mathrm{Det}\big(\alpha(\gamma)\big)\right|^2.$$

Hence from (3.6) we obtain

$$\left|\mathrm{Det}\big(\alpha(\gamma)\big)\right| = e^{\frac{1}{2}\int_\gamma\theta(h^{E_\alpha})}. \tag{3.7}$$

3.5. The Milnor metric. The Hermitian metric h^{E_α} on E_α defines a scalar product on the spaces $C^\bullet(K,E_\alpha)$ and, hence, a metric $\|\cdot\|_{\mathrm{Det}(C^\bullet(K,E_\alpha))}$ on the determinant line of $C^\bullet(K,E_\alpha)$. Using the isomorphism

$$\phi: \mathrm{Det}\big(C^\bullet(K,E_\alpha)\big) \longrightarrow \mathrm{Det}\big(H^\bullet(M,E_\alpha)\big), \tag{3.8}$$

cf. formula (2.13) of [5], we thus obtain a metric on $\mathrm{Det}\big(H^\bullet(M,E_\alpha)\big)$, called the *Milnor metric* associated with the Morse function F and denoted by $\|\cdot\|_F^{\mathrm{M}}$.

3.6. The Farber-Turaev torsion. Turaev [26] showed that if an Euler structure is fixed, then the scalar product on the spaces $C^k(K,E_\alpha)$ allows one to construct not only a metric on the determinant line $\mathrm{Det}\big(C^\bullet(K,E_\alpha)\big)$ but also an element of this line, defined modulo sign.

We recall briefly Turaev's construction. Fix a base point $x_* \in M$. Then every Euler structure ε can be represented by a pair (F,c) such that

$$c = \sum_{x\in Cr(F)} (-1)^{\mathrm{ind}_F(x)}\gamma_x,$$

with $\gamma_x : [0,1] \to M$ being a smooth curve such that $\gamma_x(0) = x_*$ and $\gamma_x(1) = x$. The chain c is often referred to as a *Turaev spider*.

We need to construct an element of the the determinant line $\mathrm{Det}\big(C^\bullet(K,E_\alpha)\big)$ of the cochain complex $C^\bullet(K,E_\alpha)$. It is easier to start with constructing an element in the determinant line of the *chain* complex. Since the cochain complex is dual to the chain complex of the bundle $E_{\alpha'}$, where α' denote the representation dual to α, we construct an element in the determinant line $\mathrm{Det}\big(C_\bullet(K,E_{\alpha'})\big)$. This is done as follows:

Fix an element $v_* \in \mathrm{Det}(E_{\alpha',x_*})$ whose norm with respect to the Hermitian metric $h^{\mathrm{Det}(E_{\alpha'})}$ is equal to 1 and set

$$v_x := {\alpha'}^{\mathrm{Det}}(\gamma_x)(v_*) \in \mathrm{Det}(E_{\alpha',x}),$$

where ${\alpha'}^{\mathrm{Det}}$ is the monodromy of the induced connection on the determinant line bundle $\mathrm{Det}(E_{\alpha'})$, cf. (3.4). Let

$$|v|^{\mathrm{Det}(E_{\alpha'})} := \sqrt{h^{\mathrm{Det}(E_{\alpha'})}(v,v)}$$

denote the norm induced on $\mathrm{Det}(E_{\alpha'})$ by the Hermitian metric $h^{\mathrm{Det}(E_{\alpha'})}$. Then from (3.6) we obtain

$$|v_x|^{\mathrm{Det}(E_{\alpha'})} = \frac{|v_x|^{\mathrm{Det}(E_{\alpha'})}}{|v_*|^{\mathrm{Det}(E_{\alpha'})}} = e^{\frac12\int_{\gamma_x}\theta(h^{\mathrm{Det}(E_{\alpha'})})} = e^{-\frac12\int_{\gamma_x}\theta(h^{\mathrm{Det}(E_\alpha)})}. \tag{3.9}$$

Let

$$v = \prod_{x\in Cr(F)} v_x^{(-1)^{\mathrm{ind}_F(x)}} \in \mathrm{Det}\big(C_\bullet(K,E_{\alpha'})\big)/\pm.$$

(The sign indeterminacy comes form the choice of the order of the critical points of F.) From (3.9) we conclude that

$$\|v\|_{\mathrm{Det}(C_\bullet(K,E_{\alpha'}))} = e^{-\frac12\int_c\theta(h^{\mathrm{Det}(E_\alpha)})}. \tag{3.10}$$

Let $\langle\cdot,\cdot\rangle$ denote the natural pairing

$$\mathrm{Det}\big(C^\bullet(K,E_\alpha)\big)\times\mathrm{Det}\big(C_\bullet(K,E_{\alpha'})\big) \to \mathbb{C}$$

and let $\nu \in \operatorname{Det}\big(C^\bullet(K,E_\alpha)\big)/\pm$ be the unique element such that $\langle \nu, v\rangle = 1$. From (3.10) we now obtain

$$\|\nu\|_{\operatorname{Det}(C^\bullet(K,E_\alpha))} \;=\; e^{\frac{1}{2}\int_c \theta(h^{\operatorname{Det}(E_\alpha)})}. \tag{3.11}$$

Using the isomorphism (3.8) we obtain an element

$$\phi(\nu) \;\in\; \operatorname{Det}\big(H^\bullet(M,E_\alpha)\big)/\pm. \tag{3.12}$$

To fix the sign one can choose a *cohomological orientation* $\mathfrak{o}$, i.e. an orientation of the determinant line $\operatorname{Det}(H^\bullet(M,\mathbb{R}))$. Thus, given the Euler structure ε and the cohomological orientation $\mathfrak{o}$ we obtain a sign refined version of $\phi(\nu)$ which we call the *Farber-Turaev torsion* and denote by

$$\rho_{\varepsilon,\mathfrak{o}}(\alpha) \in \operatorname{Det}\big(H^\bullet(M,E_\alpha)\big). \tag{3.13}$$

3.7. The Milnor norm of the Farber-Turaev torsion. From (3.11) we immediately get

$$\|\rho_{\varepsilon,\mathfrak{o}}(\alpha)\|^{\mathrm{M}}_F \;=\; e^{\frac{1}{2}\int_c \theta(h^{E_\alpha})}. \tag{3.14}$$

In particular, if α is a unitary representation, then h^{E_α} is a flat Hermitian metric and $\theta(h^{E_\alpha})=0$. Hence, if α is unitary, then

$$\|\rho_{\varepsilon,\mathfrak{o}}(\alpha)\|^{\mathrm{M}}_F \;=\; 1. \tag{3.15}$$

We now use the Cheeger-Müller theorem to conclude that

$$\|\rho_{\varepsilon,\mathfrak{o}}(\alpha)\|^{\mathrm{RS}} \;=\; 1, \qquad \text{if } \alpha \text{ is unitary}. \tag{3.16}$$

3.8. Dependence of the Farber-Turaev torsion on the Euler structure. For a homology class $h \in H_1(M,\mathbb{Z})$ and an Euler structure $\varepsilon = (F,c) \in \operatorname{Eul}(M)$ we set

$$h\varepsilon \;:=\; (F, c+h) \;\in\; \operatorname{Eul}(M). \tag{3.17}$$

This defines a free and transitive action of $H_1(M,\mathbb{Z})$ on $\operatorname{Eul}(M)$, cf. [**9**, §5] or [**7**, §3.1].

One easily checks, cf. [**9**, page 211], that

$$\rho_{h\varepsilon,\mathfrak{o}}(\alpha) \;=\; \operatorname{Det}\big(\alpha(h)\big)\cdot \rho_{\varepsilon,\mathfrak{o}}(\alpha). \tag{3.18}$$

From (3.7) and (3.14) we now obtain

$$\|\rho_{h\varepsilon,\mathfrak{o}}(\alpha)\|^{\mathrm{M}}_F \;=\; e^{-\frac{1}{2}\int_{c+h} \theta(h^{E_\alpha})}. \tag{3.19}$$

4. The Ray-Singer norm of the Refined Analytic Torsion

In [**5**] Braverman and Kappeler defined an element of $\operatorname{Det}(H^\bullet(M,E_\alpha))$ called the *refined analytic torsion* and denoted by $\rho_{\mathrm{an}}(\alpha)$. They also computed the Ray-Singer norm $\|\rho_{\mathrm{an}}(\alpha)\|^{\mathrm{RS}}$ of the refined analytic torsion. In this section we recall the result of this computation.

4.1. The odd signature operator. Fix a Riemannian metric g^M on M and let $* : \Omega^\bullet(M, E_\alpha) \to \Omega^{m-\bullet}(M, E_\alpha)$ denote the Hodge $*$-operator, where $m = \dim M$. Define the *chirality operator*

$$\Gamma = \Gamma(g^M) : \Omega^\bullet(M, E_\alpha) \to \Omega^\bullet(M, E_\alpha)$$

by the formula

$$\Gamma\,\omega := i^r\,(-1)^{\frac{k(k+1)}{2}} * \omega, \qquad \omega \in \Omega^k(M, E), \tag{4.1}$$

where $r = \frac{m+1}{2}$. The numerical factor in (4.1) has been chosen so that $\Gamma^2 = 1$, cf. Proposition 3.58 of [**1**].

The *odd signature operator* is the operator

$$\mathcal{B} = \mathcal{B}(\nabla_\alpha, g^M) := \Gamma\,\nabla_\alpha + \nabla_\alpha\,\Gamma : \Omega^\bullet(M, E_\alpha) \longrightarrow \Omega^\bullet(M, E_\alpha). \tag{4.2}$$

4.2. The eta invariant. We recall from [**5**, §3] the definition of the sign-refined η-invariant $\eta(\nabla_\alpha, g^M)$ of the (not necessarily unitary) connection ∇_α.

Let $\Pi_>$ (resp. $\Pi_<$) be the projection whose image contains the span of all generalized eigenvectors of $\mathcal{B}$ corresponding to eigenvalues λ with $\operatorname{Re}\lambda > 0$ (resp. with $\operatorname{Re}\lambda < 0$) and whose kernel contains the span of all generalized eigenvectors of $\mathcal{B}$ corresponding to eigenvalues λ with $\operatorname{Re}\lambda \leq 0$ (resp. with $\operatorname{Re}\lambda \geq 0$), cf. [**18**, Appendix B]. We define the η-function of $\mathcal{B}$ by the formula

$$\eta_\theta(s, \mathcal{B}) = \operatorname{Tr}\left[\Pi_> \mathcal{B}_\theta^s\right] - \operatorname{Tr}\left[\Pi_<(-\mathcal{B})_\theta^s\right], \tag{4.3}$$

where θ is an Agmon angle for both operators $\mathcal{B}$ and $-\mathcal{B}$ and $\mathcal{B}_\theta^s$ denotes the complex power of $\mathcal{B}$ defined relative to the spectral cut along the ray $\{re^{i\theta} : r > 0\}$, cf. [**22**, **24**]. It was shown by Gilkey, [**11**], that $\eta_\theta(s, \mathcal{B})$ has a meromorphic extension to the whole complex plane $\mathbb{C}$ with isolated simple poles, and that it is regular at $s = 0$. Moreover, the number $\eta_\theta(0, \mathcal{B})$ is independent of the Agmon angle θ.

Let $m_+(\mathcal{B})$ (resp., $m_-(\mathcal{B})$) denote the number of eigenvalues of $\mathcal{B}$, counted with their algebraic multiplicities, on the positive (resp., negative) part of the imaginary axis. Let $m_0(\mathcal{B})$ denote algebraic multiplicity of 0 as an eigenvalue of $\mathcal{B}$.

DEFINITION 4.3. *The η-invariant $\eta(\nabla_\alpha, g^M)$ of the pair (∇_α, g^M) is defined by the formula*

$$\eta(\nabla_\alpha, g^M) = \frac{\eta_\theta(0, \mathcal{B}) + m_+(\mathcal{B}) - m_-(\mathcal{B}) + m_0(\mathcal{B})}{2}. \tag{4.4}$$

If the representation α is unitary, then the operator $\mathcal{B}$ is self-adjoint and $\eta(\nabla_\alpha, g^M)$ is real. If α is not unitary then, in general, $\eta(\nabla_\alpha, g^M)$ is a complex number. Notice, however, that while the real part of $\eta(\nabla_\alpha, g^M)$ is a non-local spectral invariant, the imaginary part $\operatorname{Im}\eta(\nabla_\alpha, g^M)$ of $\eta(\nabla_\alpha, g^M)$ is local and relatively easy to compute, cf. [**11**, **15**].

We also note that the imaginary part of the η-invariant is independent of the Riemannian metric g^M.

4.4. The Ray-Singer norm of the refine analytic torsion. Let $\eta(\nabla_\alpha, g^M)$ denote the η-invariant of the odd signature operator corresponding to the connection ∇_α. By Theorem 1.9 of [**5**]

$$\|\rho_{\rm an}(\alpha)\|^{\rm RS} = e^{\pi\,\operatorname{Im}(\eta(\nabla_\alpha, g^M))}. \tag{4.5}$$

In particular, when α is a unitary representation, $\eta(\nabla_\alpha, g^M)$ is real and we get

$$\|\rho_{\rm an}(\alpha)\|^{\rm RS} = 1. \tag{4.6}$$

5. The Determinant Line Bundle over the Space of Representations

The space $\mathrm{Rep}(\pi_1(M), \mathbb{C}^n)$ of complex n-dimensional representations of $\pi_1(M)$ has a natural structure of a complex analytic space, cf., for example, [**6**, §13.6]. The disjoint union

$$\mathcal{D}et := \bigsqcup_{\alpha \in \mathrm{Rep}(\pi_1(M), \mathbb{C}^n)} \mathrm{Det}\left(H^\bullet(M, E)\right) \tag{5.1}$$

is a line bundle over $\mathrm{Rep}(\pi_1(M), \mathbb{C}^n)$, called the *determinant line bundle.* In [**4**, §3], Braverman and Kappeler constructed a natural holomorphic structure on $\mathcal{D}et$, with respect to which both the refined analytic torsion $\rho_{\mathrm{an}}(\alpha)$ and the Farber-Tureav torsion $\rho_{\varepsilon,\mathfrak{o}}(\alpha)$ are holomorphic sections. In this section we first recall this construction and then consider the ratio $\rho_{\mathrm{an}}/\rho_{\varepsilon,\mathfrak{o}}$ of these two sections as a holomorphic function on $\mathrm{Rep}(\pi_1(M), \mathbb{C}^n)$.

5.1. The flat vector bundle induced by a representation. Denote by $\pi : \widetilde{M} \to M$ the universal cover of M. For $\alpha \in \mathrm{Rep}(\pi_1(M), \mathbb{C}^n)$, we denote by

$$E_\alpha := \widetilde{M} \times_\alpha \mathbb{C}^n \longrightarrow M \tag{5.2}$$

the flat vector bundle induced by α. Let ∇_α be the flat connection on E_α induced from the trivial connection on $\widetilde{M} \times \mathbb{C}^n$.

For each connected component (in classical topology) $\mathcal{C}$ of $\mathrm{Rep}(\pi_1(M), \mathbb{C}^n)$, all the bundles E_α, $\alpha \in \mathcal{C}$, are isomorphic, see e.g. [**12**].

5.2. The combinatorial cochain complex. Fix a CW-decomposition $K = \{e_1, \ldots, e_N\}$ of M. For each $j = 1, \ldots, N$, fix a lift $\widetilde{e}_j$, i.e., a cell of the CW-decomposition of $\widetilde{M}$, such that $\tau(\widetilde{e}_j) = e_j$. By (5.2), the pull-back of the bundle E_α to $\widetilde{M}$ is the trivial bundle $\widetilde{M} \times \mathbb{C}^n \to \widetilde{M}$. Hence, the choice of the cells $\widetilde{e}_1, \ldots, \widetilde{e}_N$ identifies the cochain complex $C^\bullet(K, \alpha)$ of the CW-complex K with coefficients in E_α with the complex

$$0 \to \mathbb{C}^{n\cdot k_0} \xrightarrow{\partial_0(\alpha)} \mathbb{C}^{n\cdot k_1} \xrightarrow{\partial_1(\alpha)} \cdots \xrightarrow{\partial_{m-1}(\alpha)} \mathbb{C}^{n\cdot k_m} \to 0, \tag{5.3}$$

where $k_j \in \mathbb{Z}_{\geq 0}$ $(j = 0, \ldots, m)$ is equal to the number of j-dimensional cells of K and the differentials $\partial_j(\alpha)$ are $(nk_j \times nk_{j-1})$-matrices depending analytically on $\alpha \in \mathrm{Rep}(\pi_1(M), \mathbb{C}^n)$.

The cohomology of the complex (5.3) is canonically isomorphic to $H^\bullet(M, E_\alpha)$. Let

$$\phi_{C^\bullet(K,\alpha)} : \mathrm{Det}\left(C^\bullet(K, \alpha)\right) \longrightarrow \mathrm{Det}\left(H^\bullet(M, E_\alpha)\right) \tag{5.4}$$

denote the natural isomorphism between the determinant line of the complex and the determinant line of its cohomology, cf. [**5**, §2.4]

5.3. The holomorphic structure on $\mathcal{D}et$. The standard bases of $\mathbb{C}^{n\cdot k_j}$ $(j = 0, \ldots, m)$ define an element $c \in \mathrm{Det}\left(C^\bullet(K, \alpha)\right)$, and, hence, an isomorphism

$$\psi_\alpha : \mathbb{C} \longrightarrow \mathrm{Det}\left(C^\bullet(K, \alpha)\right), \qquad z \mapsto z \cdot c.$$

Then the map

$$\sigma : \alpha \mapsto \phi_{C^\bullet(K,\alpha)}\left(\psi_\alpha(1)\right) \in \mathrm{Det}\left(H^\bullet(M, E_\alpha)\right), \tag{5.5}$$

where $\alpha \in \mathrm{Rep}(\pi_1(M), \mathbb{C}^n)$ is a nowhere vanishing section of the determinant line bundle $\mathcal{D}et$ over $\mathrm{Rep}(\pi_1(M), \mathbb{C}^n)$.

DEFINITION 5.4. *We say that a section* $s(\alpha)$ *of* $\mathcal{D}et$ *is* holomorphic *if there exists a holomorphic function* $f(\alpha)$ *on* $\mathrm{Rep}(\pi_1(M),\mathbb{C}^n)$, *such that* $s(\alpha) = f(\alpha)\cdot\sigma(\alpha)$.

This defines a holomorphic structure on $\mathcal{D}et$, which is independent of the choice of the lifts $\tilde{e}_1,\dots,\tilde{e}_N$ of $e_1,\dots,e_N$, since for a different choice of lifts the section $\sigma(\alpha)$ will be multiplied by a constant. It is shown in [**4**, §3.5] that this holomorphic structure is also independent of the CW-decomposition K of M.

THEOREM 5.5. *Both the refined analytic torsion* $\rho_{\mathrm{an}}(\alpha)$ *and the Farber-Turaev torsion* $\rho_{\varepsilon,\mathfrak{o}}(\alpha)$ *are holomorphic sections of* $\mathcal{D}et$ *with respect to the holomorphic structure described above.*

PROOF. The fact that the Farber-Turaev torsion is holomorphic is established in Proposition 3.7 of [**4**]. The fact that the refined analytic torsion is holomorphic is proven in Theorem 4.1 of [**4**]. □

5.6. The ratio of the torsions as a holomorphic function. Since both $\rho_{\varepsilon,\mathfrak{o}}$ and ρ_{an} are holomorphic nowhere vanishing sections of the same line bundle there exists a holomorphic function

$$\kappa : \mathrm{Rep}(\pi_1(M),\mathbb{C}^n) \to \mathbb{C}\backslash\{0\}$$

such that

$$\rho_{\mathrm{an}}(\alpha) = \kappa(\alpha)\cdot\rho_{\varepsilon,\mathfrak{o}}(\alpha).$$

We shall denote this function by

$$\kappa(\alpha) = \frac{\rho_{\mathrm{an}}(\alpha)}{\rho_{\varepsilon,\mathfrak{o}}(\alpha)}. \tag{5.6}$$

5.7. The absolute value of $\frac{\rho_{\mathrm{an}}(\alpha)}{\rho_{\varepsilon,\mathfrak{o}}(\alpha)}$ **for unitary representations.** Combining (4.6) with (3.16) we obtain

$$\left|\frac{\rho_{\mathrm{an}}(\alpha)}{\rho_{\varepsilon,\mathfrak{o}}(\alpha)}\right| = \frac{\|\rho_{\mathrm{an}}(\alpha)\|^{\mathrm{RS}}}{\|\rho_{\varepsilon,\mathfrak{o}}(\alpha)\|^{\mathrm{RS}}} = 1, \qquad \text{if } \alpha \text{ is unitary.} \tag{5.7}$$

6. The Bismut-Zhang theorem for some non-unitary representations

We now present our proof of the Bismut-Zhang theorem [**2**, Theorem 0.2] for representations in the connected component $\mathcal{C}$.

6.1. The duality operator. Let α' denotes the representation dual to α. The Poincaré duality defines a non-degenerate pairing

$$\mathrm{Det}\left(H^k(M,E_\alpha)\right)\times\mathrm{Det}\left(H^{m-k}(M,E_{\alpha'})\right) \to \mathbb{C}, \qquad k=0,\dots,m,$$

and, hence, an anti-linear map

$$D : \mathrm{Det}\left(H^\bullet(M,E_\alpha)\right) \to \mathrm{Det}\left(H^\bullet(M,E_{\alpha'})\right) \tag{6.1}$$

see [**9**, §2.5] and [**5**, §10.1] for details.

By Theorem 10.3 of [**5**] we have

$$D\,\rho_{\mathrm{an}}(\alpha) = \rho_{\mathrm{an}}(\alpha')\cdot e^{2i\pi\left(\eta(\nabla_\alpha,g^M)-(\mathrm{rank}\,E)\,\eta_{\mathrm{trivial}}(g^M)\right)}, \tag{6.2}$$

where $\eta(\nabla_\alpha,g^M)$ is defined in Definition 4.3 and η_{trivial} is the η-invariant corresponding to the standard connection on the trivial line bundle $M\times\mathbb{C}\to M$.

6.2. The dual of the Farber-Turaev torsion. By Theorem 7.2 of [**9**]

$$D\,\rho_{\varepsilon,\mathfrak{o}}(\alpha) \;=\; \pm\rho_{\varepsilon^*,\mathfrak{o}}(\alpha'), \tag{6.3}$$

where $\varepsilon^* := (-F,-c)$ is the *dual Euler structure* on M.

We shall use formula (3.18) in the following situation: if $\varepsilon = (F,c) \in \mathrm{Eul}(M)$ then the Euler structure $\varepsilon^* := (-F,-c)$ is called *dual* to ε. Since $H_1(M,\mathbb{Z})$ acts freely and transitively on $\mathrm{Eul}(M)$ there exists $c_\varepsilon \in H_1(M,\mathbb{Z})$ such that

$$\varepsilon \;=\; c_\varepsilon\,\varepsilon^*. \tag{6.4}$$

The homology class c_ε was introduced by Turaev [**26**] and is called the *characteristic class of the Euler structure.* From (3.18) and (6.3) we now conclude that

$$D\,\rho_{\varepsilon,\mathfrak{o}}(\alpha) \;=\; \pm\rho_{\varepsilon^*,\mathfrak{o}}(\alpha') \;=\; \pm\,\mathrm{Det}\,\big(\alpha'(c_\varepsilon)\big)\cdot\rho_{\varepsilon,\mathfrak{o}}(\alpha'). \tag{6.5}$$

If α is a unitary representation, then $\alpha = \alpha'$. Hence, it follows from (6.5) that

$$\rho_{\varepsilon,\mathfrak{o}}(\alpha') \;=\; \pm\big(\mathrm{Det}\,\big(\alpha(c_\varepsilon)\big)\big)^{-1}\cdot D\,\rho_{\varepsilon,\mathfrak{o}}(\alpha). \tag{6.6}$$

6.3. The ratio of torsions for unitary representations. Combining (6.2) and (6.6) we conclude that for unitary α

$$\frac{D\rho_{\mathrm{an}}(\alpha)}{D\rho_{\varepsilon,\mathfrak{o}}(\alpha)} \;=\; \pm\,\mathrm{Det}\,\big(\alpha(c_\varepsilon)\big)\cdot e^{2i\pi\,\left(\eta(\nabla_\alpha,g^M)-(\mathrm{rank}\,E)\,\eta_{\mathrm{trivial}}(g^M)\right)}\cdot\frac{\rho_{\mathrm{an}}(\alpha)}{\rho_{\varepsilon,\mathfrak{o}}(\alpha)}. \tag{6.7}$$

Since D is an anti-linear involution we have

$$\frac{D\rho_{\mathrm{an}}(\alpha)}{D\rho_{\varepsilon,\mathfrak{o}}(\alpha)} \;=\; \overline{\frac{\rho_{\mathrm{an}}(\alpha)}{\rho_{\varepsilon,\mathfrak{o}}(\alpha)}}.$$

Hence, it follows from (6.7) that
(6.8)

$$\left(\frac{\rho_{\mathrm{an}}(\alpha)}{\rho_{\varepsilon,\mathfrak{o}}(\alpha)}\right)^2 \;=\; \pm\,\mathrm{Det}\,\big(\alpha(c_\varepsilon)\big)\cdot e^{-2i\pi\,\left(\eta(\nabla_\alpha,g^M)-(\mathrm{rank}\,E)\,\eta_{\mathrm{trivial}}(g^M)\right)}\cdot\left|\frac{\rho_{\mathrm{an}}(\alpha)}{\rho_{\varepsilon,\mathfrak{o}}(\alpha)}\right|^2.$$

Combining this equality with (5.7) we obtain for unitary α

$$\left(\frac{\rho_{\mathrm{an}}(\alpha)}{\rho_{\varepsilon,\mathfrak{o}}(\alpha)}\right)^2 \;=\; \pm\,\mathrm{Det}\,\big(\alpha(c_\varepsilon)\cdot e^{-2i\pi\,\left(\eta(\nabla_\alpha,g^M)-(\mathrm{rank}\,E)\,\eta_{\mathrm{trivial}}(g^M)\right)}. \tag{6.9}$$

6.4. The ratio of torsions for non-unitary representations. Suppose now that $\mathcal{C} \subset \mathrm{Rep}(\pi_1(M),\mathbb{C}^n)$ is a connected component and $\alpha_0 \subset \mathcal{C}$ is a unitary representation which is a regular point of the complex analytic set $\mathcal{C}$. The set of unitary representations is the fixed point set of the anti-holomorphic involution

$$\tau:\,\mathrm{Rep}(\pi_1(M),\mathbb{C}^n)\;\to\mathrm{Rep}(\pi_1(M),\mathbb{C}^n), \qquad \tau:\,\alpha\;\mapsto\;\alpha'.$$

Hence it is a totally real submanifold of $\mathrm{Rep}(\pi_1(M),\mathbb{C}^n)$ whose real dimension is equal to $\dim_{\mathbb{C}}\mathcal{C}$, see for example [**13**, Proposition 3]. In particular there is a holomorphic coordinates system $(z_1,\dots,z_r)$ near α_0 such that the unitary representations form a *real neighborhood* of α_0, i.e. the set $\mathrm{Im}\,z_1 = \dots = \mathrm{Im}\,z_r = 0$. Therefore, cf. [**23**, p. 21], if two holomorphic functions coincide on the set of unitary representations they also coincide on $\mathcal{C}$. We conclude that the equation (6.9) holds for all representations $\alpha \in \mathcal{C}$. Hence, using (4.5) and (3.14) we obtain for every $\alpha \in \mathcal{C}$

$$\frac{\|\cdot\|^{\mathrm{RS}}}{\|\cdot\|^{\mathrm{M}}_F} \;=\; \frac{\|\rho_{\mathrm{an}}(\alpha)\|^{\mathrm{RS}}}{\|\rho_{\varepsilon,\mathfrak{o}}(\alpha)\|^{\mathrm{M}}_F}\cdot\left|\frac{\rho_{\varepsilon,\mathfrak{o}}(\alpha)}{\rho_{\mathrm{an}}(\alpha)}\right| \;=\; \big|\,\mathrm{Det}\,\big(\alpha(c_\varepsilon)\big)\big|^{-1/2}\cdot e^{-\frac12\int_c\theta(h^{E_\alpha})}. \tag{6.10}$$

6.5. The absolute value of the determinant of $\alpha(c_\varepsilon)$. Let

$$\mathrm{PD}:\ H_1(M,\mathbb{R}) \ \to\ H^{n-1}(M,\mathbb{R})$$

denote the Poincaré isomorphism. By Proposition 3.9 of [**7**] there exists a map

$$P:\mathrm{Eul}(M)\ \to\ \Omega^{n-1}(M,\mathbb{R})$$

such that

$$\begin{aligned} P(h\varepsilon) &= P(\varepsilon) + \mathrm{PD}(h),\\ P(\varepsilon^*) &= -P(\varepsilon), \end{aligned} \tag{6.11}$$

and if $\varepsilon=(X,c)$ then for every $\omega\in\Omega^1(M,\mathbb{R})$

$$\int_c \omega \ =\ \int_M \omega\wedge X^*\Psi(g)\ -\ \int_M \omega\wedge P(\varepsilon). \tag{6.12}$$

Here $\Psi(g)$ is the Mathai-Quillen current on TM, cf. [**2**, §III c] and $X^*\Psi(g)$ denotes the pull-back of this current by $X: M\to TM$.

Combining (6.4) with (6.11) we obtain

$$P(\varepsilon)\ =\ P(\varepsilon^*)\ +\ \mathrm{PD}(c_\varepsilon)\ =\ -P(\varepsilon)\ +\ \mathrm{PD}(c_\varepsilon).$$

Thus

$$\mathrm{PD}(c_\varepsilon)\ =\ 2\,P(\varepsilon). \tag{6.13}$$

Combining this equality with (6.12) we get

$$\int_c \omega\ =\ \int_M \omega\wedge X^*\Psi(g)\ -\ \frac{1}{2}\int_M \omega\wedge \mathrm{PD}(c_\varepsilon). \tag{6.14}$$

Notice now that

$$\int_M \omega\wedge\mathrm{PD}(c_\varepsilon)\ =\ \int_{c_\varepsilon}\omega.$$

Hence, from (6.14) we obtain

$$\int_{c_\varepsilon}\omega\ =\ -2\int_c\omega\ +\ 2\int_M \omega\wedge X^*\Psi(g). \tag{6.15}$$

In particular, setting $\omega=\theta(h^{E_\alpha})$ and using (3.7) we obtain

$$\big|\,\mathrm{Det}\,\big(\alpha(c_\varepsilon)\big)\big|\ =\ e^{-\int_c\theta(h^{E_\alpha})+\int_M\theta(h^{E_\alpha})\wedge X^*\Psi(g)}. \tag{6.16}$$

Combining this equality with (6.10)

$$\frac{\|\cdot\|^{\mathrm{RS}}}{\|\cdot\|^{\mathrm{M}}_F}\ =\ e^{-\frac{1}{2}\int_M\theta(h^{E_\alpha})\wedge X^*\Psi(g)}, \tag{6.17}$$

which is exactly the Bizmut-Zhang formula [**2**, Theorem 0.2].

References

[1] N. Berline, E. Getzler, and M. Vergne, *Heat kernels and Dirac operators*, Grundlehren der Mathematischen Wissenschaften [Fundamental Principles of Mathematical Sciences], vol. 298, Springer-Verlag, Berlin, 1992. MR1215720 (94e:58130)

[2] J.-M. Bismut and W. Zhang, *An extension of a theorem by Cheeger and Müller* (English, with French summary), Astérisque **205** (1992), 235 pp. With an appendix by François Laudenbach. MR1185803 (93j:58138)

[3] M. Braverman and T. Kappeler, *A refinement of the Ray-Singer torsion* (English, with English and French summaries), C. R. Math. Acad. Sci. Paris **341** (2005), no. 8, 497–502, DOI 10.1016/j.crma.2005.09.015. MR2180817 (2006g:58065)

[4] M. Braverman and T. Kappeler, *Ray-Singer type theorem for the refined analytic torsion*, J. Funct. Anal. **243** (2007), no. 1, 232–256, DOI 10.1016/j.jfa.2006.10.008. MR2291437 (2008e:58042)

[5] M. Braverman and T. Kappeler, *Refined analytic torsion as an element of the determinant line*, Geom. Topol. **11** (2007), 139–213, DOI 10.2140/gt.2007.11.139. MR2302591 (2008a:58031)

[6] M. Braverman and T. Kappeler, *Refined analytic torsion*, J. Differential Geom. **78** (2008), no. 2, 193–267. MR2394022 (2009a:58041)

[7] D. Burghelea and S. Haller, *Euler structures, the variety of representations and the Milnor-Turaev torsion*, Geom. Topol. **10** (2006), 1185–1238, DOI 10.2140/gt.2006.10.1185. MR2255496 (2007i:58043)

[8] J. Cheeger, *Analytic torsion and the heat equation*, Ann. of Math. (2) **109** (1979), no. 2, 259–322, DOI 10.2307/1971113. MR528965 (80j:58065a)

[9] M. Farber and V. Turaev, *Poincaré-Reidemeister metric, Euler structures, and torsion*, J. Reine Angew. Math. **520** (2000), 195–225, DOI 10.1515/crll.2000.021. MR1748274 (2001m:58069)

[10] W. Franz, *Über die torsion einer überdeckung.*, Journal für die reine und angewandte Mathematik **173** (1935), 245–254.

[11] P. B. Gilkey, *The eta invariant and secondary characteristic classes of locally flat bundles*, Algebraic and differential topology—global differential geometry, Teubner-Texte Math., vol. 70, Teubner, Leipzig, 1984, pp. 49–87. MR792686 (88a:58190)

[12] W. M. Goldman and J. J. Millson, *The deformation theory of representations of fundamental groups of compact Kähler manifolds*, Inst. Hautes Études Sci. Publ. Math. **67** (1988), 43–96. MR972343 (90b:32041)

[13] N.-K. Ho, *The real locus of an involution map on the moduli space of flat connections on a Riemann surface*, Int. Math. Res. Not. **61** (2004), 3263–3285, DOI 10.1155/S107379280414004X. MR2096257 (2005g:53167)

[14] F. Kamber and Ph. Tondeur, *Flat bundles and characteristic classes of group-representations*, Amer. J. Math. **89** (1967), 857–886. MR0220309 (36 #3374)

[15] X. Ma and W. Zhang, *η-invariant and flat vector bundles*, Chinese Ann. Math. Ser. B **27** (2006), no. 1, 67–72, DOI 10.1007/s11401-005-0320-8. MR2209952 (2007a:58024)

[16] W. Müller, *Analytic torsion and R-torsion of Riemannian manifolds*, Adv. in Math. **28** (1978), no. 3, 233–305, DOI 10.1016/0001-8708(78)90116-0. MR498252 (80j:58065b)

[17] W. Müller, *Analytic torsion and R-torsion for unimodular representations*, J. Amer. Math. Soc. **6** (1993), no. 3, 721–753, DOI 10.2307/2152781. MR1189689 (93m:58119)

[18] R. Ponge, *Spectral asymmetry, zeta functions and the noncommutative residue*, Internat. J. Math. **17** (2006), no. 9, 1065–1090, DOI 10.1142/S0129167X06003825. MR2274011 (2007h:58048)

[19] D. Quillen, *Determinants of Cauchy-Riemann operators over a Riemann surface*, Funct. Anal. Appl. **14** (1985), 31–34.

[20] D. B. Ray and I. M. Singer, *R-torsion and the Laplacian on Riemannian manifolds*, Advances in Math. **7** (1971), 145–210. MR0295381 (45 #4447)

[21] K. Reidemeister, *Homotopieringe und Linsenräume* (German), Abh. Math. Sem. Univ. Hamburg **11** (1935), no. 1, 102–109, DOI 10.1007/BF02940717. MR3069647

[22] R. T. Seeley, *Complex powers of an elliptic operator*, Singular Integrals (Proc. Sympos. Pure Math., Chicago, Ill., 1966), Amer. Math. Soc., Providence, R.I., 1967, pp. 288–307. MR0237943 (38 #6220)

[23] B. V. Shabat, *Introduction to complex analysis. Part II*, Translations of Mathematical Monographs, vol. 110, American Mathematical Society, Providence, RI, 1992. Functions of several variables; Translated from the third (1985) Russian edition by J. S. Joel. MR1192135 (93g:32001)

[24] M. A. Shubin, *Pseudodifferential operators and spectral theory*, Springer Series in Soviet Mathematics, Springer-Verlag, Berlin, 1987. Translated from the Russian by Stig I. Andersson. MR883081 (88c:47105)

[25] V. G. Turaev, *Reidemeister torsion in knot theory* (Russian), Uspekhi Mat. Nauk **41** (1986), no. 1 (247), 97–147, 240. MR832411 (87i:57009)

[26] V. G. Turaev, *Euler structures, nonsingular vector fields, and Reidemeister-type torsions* (Russian), Izv. Akad. Nauk SSSR Ser. Mat. **53** (1989), no. 3, 607–643, 672; English transl., Math. USSR-Izv. **34** (1990), no. 3, 627–662. MR1013714 (90m:57021)

[27] V. G. Turaev, *Introduction to combinatorial torsions*, Lectures in Mathematics ETH Zürich, Birkhäuser Verlag, Basel, 2001. Notes taken by Felix Schlenk. MR1809561 (2001m:57042)

DEPARTMENT OF MATHEMATICS, NORTHEASTERN UNIVERSITY, BOSTON, MASSACHUSETTS 02115
E-mail address: `maximbraverman@neu.edu`
URL: `www.math.neu.edu/~braverman/`

MATHEMATISCHES INSTITUT, UNIVERSITÄT BONN, 53115 BONN, GERMANY
E-mail address: `vertman@math.uni-bonn.de`
URL: `www.math.uni-bonn.de/people/vertman`

Contemporary Mathematics
Volume **630**, 2014
http://dx.doi.org/10.1090/conm/630/12663

Tunneling, the Quillen Metric and Analytic Torsion for High Powers of a Holomorphic Line Bundle

Robert J. Berman

ABSTRACT. Let L be a line bundle over a compact complex manifold X and denote by h_L and h_X fixed Hermitian metrics on L and TX, respectively. Our main result provides a formula for the average distribution of the exponentially small eigenvalues of the corresponding Dolbeault Laplacians associated to high tensor powers of L, which in physics terminology is a measure of "tunneling" of the Dolbeault complex. Along the way a new proof of the asymptotics of the induced Quillen metric on the corresponding determinant line is obtained. A brief comparison with the tunneling effect for Witten Laplacians and large deviation principles for fermions is also made.

CONTENTS

1. Introduction

1.1. Setup. Let $L \to X$ be a holomorphic line bundle over a compact complex manifold X and let h_L be a smooth metric on the line bundle L with normalized curvature form ω (the most interesting case will be when ω is not semi-positive). The normalization is made so that ω is real and defines an integer cohomology class: $[\omega] \in H^2(X, \mathbb{Z})$. It will be convenient to use the weight notation for h_L, i.e. fixing a local trivialization s of L we may locally write $\|s\|_{h_L}^2 := e^{-\phi}$ so that $\omega = dd^c\phi := \frac{i}{2\pi}\partial\bar{\partial}\phi$ where ϕ will be called a *weight* on L. We also fix a metric h_X on X and denote its volume form by dV. These metrics induce, in the standard way, Hermitian products on the space $\Omega^{0,q}(X, L)$ of smooth $(0, q)-$forms with values in L and we will write

$$\Delta_{\bar{\partial}}^{0,q} := \bar{\partial}\bar{\partial}^* + \bar{\partial}^*\bar{\partial},$$

for the corresponding Dolbeault-Kodaira Laplacian acting on $\Omega^{0,q}(X, L)$.

2010 *Mathematics Subject Classification.* Primary 32L05, 58J52, 81Q60.
Key words and phrases. Holomorphic line bundle, determinant bundle and analytic torsion, supersymmetry and quantum mechanics.

We will be concerned with the asymptotic situation when L is replaced by a large tensor power, written in our additive notation as kL. As is well-known, in terms of semi-classical analysis, $1/k$ plays the role of Planck's constant and the main motivation for the present paper is to understand the notion of "tunneling" for the Dolbeault complex associated to the line bundle kL equipped with a non-positively curved metric, i.e. the distribution of the exponentially small eigenvalues of the corresponding Dolbeault-Kodaira Laplacians $\Delta_{\bar\partial}^{0,q}$. To be more precise, a sequence λ_k of positive numbers will be said to be *exponentially small* if

$$\liminf_{k\to\infty}(\log\lambda_k)/k<0$$

(this is sometimes written as $\lambda_k=\mathcal{O}(e^{-\epsilon k})$ in the literature). As recalled in Section 3.1 the corresponding situation for the De Rham complex has been studied extensively, motivated by the seminal work of Witten relating the corresponding tunneling to supersymmetric quantum mechanics and Morse theory and very precise results have been obtained. However, the situation for the (asymptotic) Dolbeault complex appears to be somewhat mysterious and there seem to be essentially no prior results available concerning the corresponding tunneling. On the other hand, as shown by Demailly [**14**], who was inspired by Witten's approach to Morse theory, the local weight ϕ of the given metric on L plays the role of a Morse function and this leads to asymptotic estimates on the corresponding Dolbeault cohomology groups (the so called weak and strong holomorphic Morse inequalities of Demailly). The main analytical ingredient in Demailly's work is a formula for the leading asymptotics of the number of lower-lying eigenvalues of the Dolbeault Laplacian (see section 3.1 for related results). However, the formula in question gives no information on the distribution of the corresponding exponentially small eigenvalues λ_k or more precisely the distribution of the corresponding "phases" $(\log\lambda_k)/k$. One of the main points of the present paper is to show that the Bergman kernel type asymptotics obtained in [**3**] can be used to give an explicit asymptotic formula for a certain average of $(\log\lambda_k)/k$ (see the concluding Section 3.3 for a discussion about much more precise conjectural results). 3.3 To this end a truncated version of the analytic torsion of the Dolbeault complex will be introduced. Recall that the analytic torsion, introduced by Ray-Singer, is defined as

$$T(h_L,h_X):=\prod_{q=1}^{n}(\det\Delta_{\bar\partial}^{0,q})^{(-1)^{q+1}q},$$

using zeta function regularization, i.e. $\log\det\Delta_{\bar\partial}^{0,1}:=-\frac{\partial\zeta^{0,q}}{\partial s}{}_{s=0}$, where $\zeta^{0,q}(s)=\sum_i(\lambda_i^{0,q})^{-s}$ is the meromorphic continuation to $\mathbb{C}$ of the zeta function for the *non-zero* eigenvalues $\{\lambda_i^{0,q}\}$ of $\Delta_{\bar\partial}^{0,q}$ (see [**9, 12, 20, 22**] and references therein). In the particular case of a Riemann surface

$$T(h_L,h_X):=\det\Delta_{\bar\partial}^{0,1}=\det\Delta_{\bar\partial}^{0,0}.$$

The "truncated" version $T_{]0,\lambda[}(h,\omega_0)$ of the analytic torsion $T(h_L,h_X)$ is obtained by simply replacing the regularized determinants with the product of all positive eigenvalues strictly smaller than a given positive number λ. Using functional calculus we may hence write

$$\log T_{]0,\lambda[}(h,\omega_0):=-\sum_{q=1}^{n}q(-1)^q\log\det(1_{]0,\lambda[}(\Delta_{\bar\partial}^{0,q}).\tag{1.1}$$

The asymptotics of the corresponding analytic torsions was studied, using heat kernel asymptotics, in the seminal work of Bismut-Vasserot [**1**] and is closely related to the Quillen metric on the determinant of cohomology whose definition we briefly recall. Given a holomorphic line bundle L over X the corresponding *determinant line* $DET(L)$ is the one-dimensional complex vector space defined as the following tensor product of $\bar{\partial}-$cohomology groups:

$$\mathrm{DET}(L) := \bigotimes_{q=1}^{n} \det(H^q(X,L))^{(-1)^q},$$

where $\det(W)$ denotes the top exterior power of a complex vector space W. Given Hermitian metrics $h_L(= e^{-\phi_L})$ and h_X on L and X, respectively, the corresponding *Quillen metric* on $\mathrm{DET}(kL)$, that we will denote by $\mathcal{Q}(h_L^{\otimes k}, h_X)$, is defined as the L^2-metric multiplied by the analytic torsion, where the L^2-metric is the one induced from the isomorphisms between the vector spaces $H^q(X,L)$ and the kernel of $\Delta_{\bar{\partial}}^{0,q}$. Similarly, we will denote by $\mathcal{Q}_{]0,\lambda_k[}(h_L^{\otimes k}, h_X)$, the "truncated" Quillen metrics obtained by replacing the analytic torsion with its truncation, defined above.

Before stating our main results we also recall the following well-known bi-functional on the space of metrics (weights) on L :

$$\mathcal{E}(\phi,\phi') := \frac{1}{(n+1)!}\int_X \sum_{j=0}^{n} (\phi-\phi')(dd^c\phi)^{n-j}\wedge(dd^c\phi')^j \left(= \tilde{ch}(h,h')\right),$$

where $\tilde{ch}(h,h')$ is (up to a multiplicative constant) the Bott-Chern class attached to the first Chern class [**9, 22**]. Moreover, in the case when L is an ample line bundle and ϕ is a given smooth weight on L we will write $P\phi$ for the semi-positively curved weight defined as the following upper envelope:

$$P\phi = \sup\{\psi : \psi \leq \phi\},$$

where ψ ranges over all smooth weights on L with positive curvature form. Then the functional $\mathcal{E}(P\phi,\phi)$ is still well-defined by basic pluripotential theory, only using that $P\phi$ is semi-positively curved and locally bounded (see [**6**] and references therein). Alternatively, by the regularity result in [**4**] $P\phi$ is locally $\mathcal{C}^{1,1}-$smooth and hence the current $dd^c(P\phi)$ has locally bounded coefficients and exists point-wise almost everywhere on X. As a consequence $\mathcal{E}(P\phi,\phi)$ can be defined as a standard Lebesgue integral over X.

1.2. Statement of the main results.

THEOREM 1.1. *Let $L \to X$ be an ample line bundle equipped with a smooth Hermitian metric h_L and let h_X be a fixed smooth Hermitian metric on X (which is not assumed to be Kähler). Let λ_k be a sequence such that $\lambda_k = o(k)$ and such that λ_k is not exponentially small. Then the large $k-$limits of the corresponding analytic torsions $\frac{1}{k^{n+1}}\log T(h_L^{\otimes k}, h_X)$ and their "truncations" $\frac{1}{k^{n+1}}\log T_{]0,\lambda_k[}(h_L^{\otimes k}, h_X)$ exist and both coincide with $\mathcal{E}(P\phi_L,\phi_L)$*

In fact, it will be clear that the only contribution to the truncated analytic torsions in the previous theorem comes from the exponentially small eigenvalues (in an appropriate sense). In the case when h_X is a Kähler metric the asymptotics of the (non-truncated) analytic torsion in the previous theorem were deduced in [**6**] from the exact anomaly formula in [**9**] for the Quillen metric (referred to above) combined with the asymptotics for the L^2-part of the Quillen metric proved in [**6**]. Of course, since L is ample the manifold X is automatically Kähler, but the point is that one does not need to assume that the Hermitian metric h_X is Kähler.

As a consequence of the previous theorem we get the following geometric criterion for the existence of exponentially small eigenvalues of $\bar{\partial}-$Laplacians:

COROLLARY 1.2. *Let L be a line bundle and h_L and h_X metrics as in the previous theorem. Suppose that*

$$\mathcal{E}(P\phi_L, \phi_L) \neq 0.$$

Then, for some q, there is a sequence of exponentially small eigenvalues of the corresponding $\bar{\partial}-$Laplacian acting on $(0,q)-$forms with values in kL. In the particular case of a line bundle of positive degree over a Riemann surface equipped with a metric h_L whose curvature is not semi-positive everywhere, the condition above is always satisfied. More precisely,

$$\frac{1}{k^2}\log\sum_i \lambda_i^{(k)} \to -\frac{1}{2}\left\|d(P\phi-\phi)\right\|_X^2 (<0), \tag{1.2}$$

where $(\lambda_i^{(k)})$ are the "small" positive eigenvalues of the $\bar{\partial}-$Laplacian on the space of smooth sections with values in kL (i.e the eigenvalues are in $]0, k^{1-\epsilon}[$)

One of the main ingredients in the proof of the previous theorem is an asymptotic anomaly formula for a truncated version of the Quillen metric, which leads to the following

THEOREM 1.3. *Let $L \to X$ be a holomorphic line bundle over a compact complex manifold (not necessarily Kähler). Fix a Hermitian metric h_X on X and two metrics h_L and h'_L on L. Then the following asymptotic anomaly formula for the corresponding Quillen metrics on the determinant lines DET(kL), as well as its truncations $\mathcal{Q}_{]0,\lambda_k[}(h_L^{\otimes k}, h_X)$, holds*

$$\lim_{k\to\infty}\frac{1}{k^{n+1}}\log(\frac{\mathcal{Q}(h_L^{\otimes k}, h_X)}{\mathcal{Q}(h_{L'}^{\otimes k}, h_X))}) = \tilde{ch}(h_L, h'_L),$$

for any sequence λ_k as in the previous theorem.

In the case when h_X is a Kähler metric the previous theorem is a direct consequence of the deep *exact* anomaly formula for the Quillen metric of Bismut-Gillet-Soula for the determinant line applied to any fixed tensor power of L (see [**9**], Theorem 0.3 stated in part I). Even though the exact formula is not known in the non-Kähler case it should be stressed that the previous theorem can be deduced from the leading heat-kernel asymptotics in [**8, 10, 20**], as they are independent of any Kähler assumptions (see Remark 2.6). As explained below the key point of the present proof, where the role of the heat-kernel is played by Bergman type kernels [**3**], is to first obtain the asymptotics for the "truncated" Quillen metric obtained by replacing the analytic torsions with suitable truncations.

By standard arguments the formula in the previous theorem can be used to obtain an asymptotic version of Theorem 0.1 in [**9**] in the case when $\pi : X \to S$ is a submersion of a base S and a pair (h_L, h_X) as above is fixed. Then the curvature of the Quillen metric on DET(kL), seen as a line bundle over S, converges when divided by k^{n+1} to the push-forward to S of the top exterior power of the curvature form of h_L.

The starting point of the proof of Theorem 1.3 is the observation that differential of the normalized logarithm $\mathcal{Q}_{]0,\lambda_k[}$ of the "truncated" Quillen metric on $\mathrm{DET}(kL)$, seen as a functional on the space of all Hermitian metrics ϕ (or rather weights) on L, is represented by the alternating sum of the one-point measures (i.e. Bergman type measures) associated to the spaces of "low-energy" $(0,q)-$forms. Using the large k asymptotics of the latter measures obtained in [**3**] gives

$$\frac{1}{k^{n+1}} d_\phi \mathcal{Q}_{]0,\lambda_k[}(k\phi, h_X) \to (dd^c\phi)^n/n!$$

and integrating between the line segment connecting ϕ_0 and ϕ_1 in the space of all weights on L then concludes the proof of Theorem 1.3 for the "truncated" Quillen metrics. Adapting arguments in [**10**] to the present setting we also show that asymptotically the analytic torsions may be replaced by truncations as above, concluding the proof of Theorem 1.3. Theorem 1.1 is then obtained by using the asymptotics for the $H^0(X, kL)-$part of the Quillen metrics for a general smooth (but possible non-positively curved) metric in [**6**] (or alternatively the asymptotics for the corresponding one-point (Bergman) measures in [**4**]).

Acknowledgments. The author is grateful to Johannes Sjöstrand and Frédéric Faure for their interest in this work and for stimulating discussions related to the topic of the present paper, as well as to the referees whose comments helped to improve the exposition of the paper. This work has been supported by grants from the Swedish and European Research Councils and the Wallenberg Foundation.

2. Proofs of the main results

In the following we will fix the metric h_X on TX. Then any given weight ϕ (which we will eventually vary) induces Hermitian products $\langle \cdot, \cdot \rangle_\phi$ on the space $\Omega^{0,q}(X, L)$ using the metrics $(e^{-\phi}, h_X)$ on L and TX, respectively. Given ϕ we denote the associated $\bar{\partial}-$Laplacians on $\Omega^{0,q}(X, L)$ by $\Delta^{0,q}$ (and sometimes by $\Delta^{0,q}_\phi$ to indicate the dependence on ϕ). We will denote by $\mathcal{H}^{0,q}_{]0,\lambda[}$ the subspace of $\Omega^{0,q}(X, L)$ spanned by the eigenforms of $\Delta^{0,q}$ with eigenvalues in $]0, \lambda[$ (sometimes called the *space* of *low-energy* $(0,q)-$*forms*). To this space we associate its (non-normalized) one-point measure, i.e.

$$\mathbb{B}^{0,q}_{]0,\lambda[} := \sum_i |\Psi^{0,q}_i|^2_\phi dV, \tag{2.1}$$

where $(\Psi^{0,q}_i)$ is any orthonormal base for $\mathcal{H}^{0,q}_{]0,\lambda[}$. We will also use the notation $\mathcal{H}^{0,q} := \ker \Delta^{0,q}$ and write $\mathbb{B}^{0,q}$ for the corresponding one-point measure.

2.1. Variational formulae.

If $\mathcal{F}(\phi)$ is a (Gateaux differentiable) functional on the affine space of all weights on L we will write $d\mathcal{F}$ for its differential, which is a one-form on the space of all weights. We will identify the linear functional $d\mathcal{F}_{|\phi}$ on $\mathcal{C}^\infty(X)$ with a measure in the usual way. Concretely, this means that

$$\frac{d}{dt}\mathcal{F}(\phi_t) = \int_X (d\mathcal{F}_{|\phi_t})\frac{d\phi_t}{dt}. \tag{2.2}$$

Given a pair of weights ϕ and ϕ' on L we now let

$$\mathcal{L}^{0,q}(\phi, \phi') := -\log\det(\left\langle \Psi^{0,q}_i, \Psi^{0,q}_j \right\rangle_\phi)$$

where $(\Psi_i^{0,q})_i$ is a basis in $\ker \Delta_{\phi'}^{0,q}$ which is orthonormal wrt $\langle \cdot, \cdot \rangle_{\phi'}$. We then have the following basic

LEMMA 2.1. *Fix a weight ϕ' on L and write $\mathcal{L}^{0,q}(\phi) := \mathcal{L}^{0,q}(\phi, \phi')$ and $\mathcal{E}(\phi) := \mathcal{E}(\phi, \phi')$. Then*

$$(i)\, d\mathcal{E}_{|\phi} = \frac{(dd^c\phi)^n}{n!}, \quad (ii)\, d\mathcal{L}^{0,q}{}_{|\phi} = \mathbb{B}^{0,q}.$$

PROOF. The first point was first shown by Mabuchi [**21**, Theorem 2.3] (see also [**6**] and references therein for a more general setting). As for the second point it holds in a general setting where Π_ϕ is the orthogonal projection (wrt the Hermitian product $\langle \cdot, \cdot \rangle_\phi$) on a given $N-$dimensional subspace of $\mathcal{C}^\infty(X, L \otimes E)$ where E is a given Hermitian complex vector bundle (here $E = \Lambda^{0,q}(X, h_X)$). Indeed, we can then compute the time derivative of the corresponding functional $\mathcal{L}(\phi_t, \phi)$ (here $\mathcal{L}^{0,q}(\phi, \phi')$) using the basic formula

$$\frac{d}{dt}_{t=0} \log\det(H(t)) = \operatorname{Tr}(H(0)^{-1} \frac{d}{dt}_{t=0} H(t)) \tag{2.3}$$

By the cocycle property $\mathcal{L}(\phi_0, \phi_1)+\mathcal{L}(\phi_1, \phi_2)+\mathcal{L}(\phi_2, \phi_0) = 0$ we have that $\frac{d}{dt}\mathcal{L}(\phi_t, \phi_{'})$ is independent of ϕ' and hence we can set $\phi' = \phi_0$ above so that $H(0) = I$, which gives the desired formula. □

We will also have great use for the following lemma, whose proof is inspired by some arguments in [**22**]:

LEMMA 2.2. *The differential of the functional $\phi \mapsto \tau_\lambda(\phi) := \log T_{]0,\lambda[}(e^{-\phi}, h_X)$ given by the truncated analytic torsion (formula* 1.1*) is given by*

$$d\tau_{|\phi} = \sum_{q=1}^{n} (-1)^q \mathbb{B}_{]0,\lambda[}^{0,q}$$

for a generic number λ. In particular, the differential of the truncated Quillen metric wrt ϕ satisfies

$$d_\phi \mathcal{Q}_{]0,\lambda_k[}(e^{-\phi}, h_X) = \sum_{q=1}^{n} (-1)^q \mathbb{B}_{[0,\lambda[}^{0,q}. \tag{2.4}$$

PROOF. In the proof we will repeatedly use that $\bar\partial$ commutes with $\Delta (:= \Delta_{\bar\partial})$ and hence if $\Psi^{0,q}$ is an eigenform of $\Delta^{0,q}$ then $\bar\partial\Psi^{0,q}$ is an eigenform of $\Delta^{0,q+1}$ unless it vanishes identically. To fix ideas we start with

The case $n = 1$:

Since, $\Delta = \bar\partial^*\bar\partial$ on $\Omega^{0,q}(X, L)$ it follows immediately from the definition of the determinant that

$$\det(1_{]0,\lambda[}(\Delta_{\bar\partial}^{(0)}) = \det(\langle \Psi_i, \Psi_j \rangle)^{-1} \det(\langle \bar\partial\Psi_i, \bar\partial\Psi_j \rangle)$$

for any given base in $\mathcal{H}_{]0,\lambda[}^{0,q}$ (with $q = 0$). Next we note that, given the path ϕ_t, we may find a path $(\Psi_i^{0,q})(t)$ of bases in $\mathcal{H}_{]0,\lambda[}^{0,q}(t)$ such that $(\Psi_i^{0,q})(t)$ is orthonormal wrt ϕ_0 for $t = 0$ and for any fixed index i $(\Psi_i^{0,q})(0)$ is an eigenform for Δ_{ϕ_0} and

$$\left\langle \frac{d}{dt}_{t=0} (\Psi_i^{0,q})(t), \mathcal{H}_{]0,\lambda[}^{0,q}(0) \right\rangle_{\phi_0} = 0. \tag{2.5}$$

To see this we first recall that, as a well-known consequence of the ellipticity of Laplacians, for a generic λ the family $\mathcal{H}^{0,q}_{[0,\lambda[}(t)$ defines a vector bundle over $\{t\} :=]-\epsilon,\epsilon[$ (see Proposition 1 on p. 123 in [**22**]). Moreover, since, for any t, $\ker\Delta^{0,q}_{\bar\partial}(t)$ is isomorphic to the Dolbeault cohomology group $H^{0,q}_{\bar\partial}(X,L)$ it follows that $F := \mathcal{H}^{0,q}_{]0,\lambda[}(t)$ is also a vector bundle. Hence, we can start with an arbitrary smooth curve $(s_i)(t)$ of bases in F_t satisfying the first requirements above at $t=0$. Then, for t sufficiently small, $(\Psi^{0,q}_i)(t) := s_i(t) - t\Pi_t(\frac{d}{dt}_{t=0}(s_i)(t))$ has the desired properties, since $\Pi_0(\frac{d}{dt}_{t=0}(\Psi^{0,q}_i)(t)) = 0$, where Π_t denotes the orthogonal projection onto $\mathcal{H}^{0,q}_{]0,\lambda[}(t)$.

Now we can decompose

$$\text{(2.6)}\qquad \frac{d}{dt}_{t=0}\log\det(1_{]0,\lambda[}(\Delta^{(0)}_{\bar\partial}) = -\frac{d}{dt}_{t=0}\log\det(\langle\Psi_i(t),\Psi_j(t)\rangle + \frac{d}{dt}_{t=0}\log\det(\langle\bar\partial\Psi_i(t),\bar\partial\Psi_j(t)\rangle).$$

Using formula formula 2.3 and the fact that

$$\frac{d}{dt}_{t=0}(\langle\Psi_i(t),\Psi_j(t)\rangle_{\phi_t} = \left\langle -\frac{d\phi_t}{dt}\Psi_i(t),\Psi_j(t)\right\rangle + 0$$

(by Leibniz rule and 2.5) the first term in 2.6 above gives (also using that $H(0)=I$)

$$-\frac{d}{dt}_{t=0}\log\det(\langle\Psi_i(t),\Psi_j(t)\rangle = \int_X \mathbb{B}^{0,0}_{]0,\lambda[}\frac{d\phi_t}{dt}.$$

The second term is computed similarly, using that $\alpha_i := \bar\partial\Psi_i/\left\|\bar\partial\Psi_i\right\|_{\phi_0}$ is an orthonormal base in $\mathcal{H}^{0,1}_{]0,\lambda[}(t)$ for $t=0$ and that the relation 2.5 still holds with Ψ_i replaced by α_i (indeed, since $\bar\partial$ commutes with $\frac{d}{dt}$ and $\Delta_{\bar\partial}$ we get $\Pi^{0,q+1}_{]0,\lambda[}(\frac{d}{dt}\bar\partial\Psi) = \bar\partial\Pi^{0,q}_{]0,\lambda[}(\frac{d}{dt}\Psi) = 0$).

General n

By Hodge theory we have a decomposition

$$\Omega^0(X,L) = \ker\Delta_{\bar\partial}\oplus\mathrm{Im}\bar\partial\oplus\bar\partial^*,$$

which is orthogonal wrt the corresponding Hermitian product. Restricting to $\mathcal{H}^{0,q}_{]0,\lambda[}$ (which by definition is in the orthogonal complement of $\ker\Delta_{\bar\partial}$ in $\Omega^0(X,L)$) gives an induced orthogonal decomposition

$$\mathcal{H}^{0,q}_{]0,\lambda[} = M_q\oplus N_q.$$

Next, we note that $\bar\partial$ induces a bijection, intertwining the corresponding restricted Laplacians, such that

$$\text{(2.7)}\qquad \bar\partial:\ N_q\to M_{q+1},\ \ \Delta_{N_q} = \bar\partial^*\bar\partial$$

and hence $\det(\Delta_{M_q}) = \det(\Delta_{N_{q-1}})$ so that $\det(1_{]0,\lambda[}(\Delta^{(q)}_{\bar\partial}) = \det(\Delta_{N_{q-1}})\det(\Delta_{N_q})$. The latter relation implies, since $\det(\Delta_{N_{-1}}) := 1 = \det(\Delta_{N_n})$ that

$$\text{(2.8)}\qquad -\sum_{q=1}^n q(-1)^q\log\det(1_{]0,\lambda[}(\Delta^{(q)}_{\bar\partial}) = \sum_{q=1}^n(-1)^q\log\det(\Delta_{N_q}).$$

Using the bijection 2.7 we can now repeat the arguments used above (when $n = 1$) to deduce that

$$\frac{d}{dt}_{t=0}\log(\det(\Delta_{N_q}) = \int_X (\mathbb{B}^{0,q}_{]0,\lambda[\cap N_q} - \sum_{q=1}^{n} \mathbb{B}^{0,q+1}_{]0,\lambda[\cap M_{q+1}})\frac{d\phi_t}{dt},$$

where the intersection with N_q indicates that we have replaced $\mathcal{H}^{0,q}_{]0,\lambda[}$ with its subspace N_q in the definition 2.1 (and similarly for M_{q+1}). Hence, taking the alternating sum over q and using 2.8 proves the lemma in general dimensions. Finally, formula 2.4 now follows by invoking the formula (ii) in the previous lemma. □

2.2. Asymptotics. Next we recall the following asymptotics from [**3**] (which can be seen as a local version of Demailly's strong holomorphic Morse inequalities):

PROPOSITION 2.3. *Let λ_k be a sequence of positive numbers such that $\lambda_k = O(k^{1-\epsilon})$ for some $\epsilon < 1/2$. Then the sequence $k^{-n}\mathbb{B}^{0,q}_{[0,\lambda_k[}/dV$ is uniformly bounded and*

$$k^{-n}\mathbb{B}^{0,q}_{[0,\lambda_k[} \to (-1)^q 1_{X(q)}(dd^c\phi)^n/n!$$

weakly as $k \to \infty$, where $X(q)$ is the subset of X where $dd^c\phi$ has exactly q negative eigenvalues. In particular,

$$\dim \mathcal{H}^{0,q}_{[0,\lambda_k[} = k^n \int_X (-1)^q 1_{X(q)}(dd^c\phi)^n/n! + o(k^n).$$

PROOF. As shown in [**3**] (Proposition 5.1) the upper bound in the convergence above holds for any sequence $\lambda_k = k\mu_k$ with $\mu_k \to 0$. As for the lower bound it was was shown to hold as long as $\delta_k/\mu_k \to 0$ for δ_k a certain non-explicit sequence tending to zero (Proposition 5.3). This latter sequence appears in Lemma 5.2 in [**3**] and the proof given there actually shows that δ_k can be taken as $\delta_k = 1/k^{1/2-\delta}$ for any $\delta > 0$ (since the radius R_k appearing in that proof is equal to $\log k$). A further refinement of this argument will be considered in Proposition 3.1. □

The previous asymptotics will allow us to obtain the asymptotics of truncated analytic torsions (and hence of exponentially small eigenvalues). Using the next proposition we will then deduce the corresponding asymptotics for the usual analytic torsions.

PROPOSITION 2.4. *The following asymptotics hold for any line bundle $L \to X$ and smooth Hermitian metrics on L and X :*

$$\log(\det(\Delta^{(q)}_k) = \log(\det(1_{]0,\lambda_k]}(\Delta^{(q)}_k)) + O(k^{n+\epsilon})$$

if $\lambda_k = O(k^{1-\epsilon})$ for a given $\epsilon \in]0,1[$.

PROOF. We will adapt to our setting some arguments from [**10**], where it was among other things shown that $\log(\det(\Delta^{(q)}_k) = O(k^n)\log k$ if the metric on L is positively curved (the point being that the smallest positive eigenvalue of $\Delta^{(q)}_k$ is then of the order k). As follows immediately from the definition (see below) the statement of the proposition to be proved is equivalent to

$$\zeta'_{1_{[\lambda_k,\infty[}(\Delta^{(q)})}(0) = O(k^{n+\epsilon}). \tag{2.9}$$

To prove the latter asymptotics we first recall some general facts about spectral zeta functions (following chapter V in [**22**]). If D is an Hermitian non-negative operator on a Hilbert space then its spectral zeta function is defined by

$$\zeta_D(s) = \frac{1}{\Gamma(s)} \int_0^\infty (t^s(\mathrm{Tr}(e^{-tD}) - \Pi_D)\frac{dt}{t}$$

where Π_D denotes the orthogonal projection on the kernel of D (we assume that $\zeta_D(s)$ exists for s a complex number of sufficiently negative real part and is then analytically contained to other values for s). In other words $\zeta_D(s)$ is the Mellin transform of the spectral theta function

$$\Theta_D(t) := (\mathrm{Tr}(e^{-tD}) - \Pi_D) = \sum_{\nu_i > 0} e^{-t\nu_i}$$

summing over the positive eigenvalues of D. Next we assume that

- $\Theta_D(t)$ converges for $t > 0$
- For every positive integer M there are real numbers $a_i = 0$ with $a_i = 0$ for $i < -n$ such that

$$\Theta_D(t) = \sum_{j=-n}^{j=M} a_j t^j + O(t^{M+1}) \tag{2.10}$$

  uniformly when $t \to 0$.

Under these assumptions one obtains (see formula 11 on p. 99 in [**22**])

$$\zeta_D(0) = a_0, \quad \zeta'_D(0) = \int_1^\infty \Theta_D(t)\frac{dt}{t} + \left(\gamma a_0 + \sum_{j<0} \frac{a_j}{j} + \int_0^1 \rho_0(t)\frac{dt}{t}\right) \tag{2.11}$$

where $\rho_0(t)$ is the $O(t^{M+1})$ term in 2.10 for $M = 0$ and γ denotes Euler's constant. We also define a scaled version of ζ_D by setting $\widetilde{\zeta}_D := k^{-n}\zeta_{k^{-1}D}$ so that the derivative $\zeta'_D(0)$ at $s = 0$ satisfies

$$k^{-n}\zeta'_D(0) = -\log k\widetilde{\zeta}_D(0) + \widetilde{\zeta}'_D(0) \tag{2.12}$$

(compare formula 40 in [**10**]).

We now come back to the present complex geometric setting. As shown in [**10**] (Theorem 2), for any positive integer M, there is an asymptotic expansion

$$k^{-n}\mathrm{Tr}(e^{-tk^{-1}\Delta^{(q)}})\frac{dt}{t} = \sum_{j=-n}^{j=M} a_j^q(k)t^j + O(t^{M+1}), \tag{2.13}$$

when $t \to 0$ uniformly for $t \in [0, 1]$ and moreover

$$a_j^q(k) = a_j^q + O(k^{-1/2}), \tag{2.14}$$

as $k \to \infty$. Next, we note that, for $t \in [0, 1]$

$$k^{-n}\mathrm{Tr}(e^{-tk^{-1}\Delta^{(q)}}) - k^{-n}\mathrm{Tr}(e^{-tk^{-1}1_{[k^{1-\epsilon},\infty[}(\Delta^{(q)})}) = \int (-1)^q 1_{X(q)}(dd^c\phi)^n/n! + o(1) \tag{2.15}$$

uniformly when $k \to \infty$ (strictly speaking these asymptotics only hold when $\epsilon < 1/2$ but in general the argument will show that the lhs above is uniformly bounded which is all that will be used to deduce the lemma). Indeed, the left hand side above may be written and estimated from below and above as

$$e^{-tk^{-\epsilon}}k^{-n}\dim \mathcal{H}^{0,q}_{]0,\lambda_k[} \leq k^{-n}\sum_i e^{-tk^{-1}\nu_{i,k}} \leq k^{-n}\dim \mathcal{H}^{0,q}_{]0,\lambda_k[},$$

which combined with Proposition 2.3 proves the previous formula. Now combining 2.13, 2.14 and 2.16 gives an expansion

$$k^{-n}\mathrm{Tr}(e^{-tk^{-1}1_{[k^{1-\epsilon},\infty[}(\Delta^{(q)})}) = \sum_{j=-n}^{j=M} b_j^q t^j + O(t^{M+1}) + o(1) \tag{2.16}$$

uniformly for $t \in [0,1]$ as $k \to \infty$ (where $b_j^q = a_j^q$ for $j \neq 0$). Accordingly, it follows from 2.11 that

$$\widetilde{\zeta}_{\Delta^{(q)}}(0) = -b_0^q(k) = -b_0^q + o(1) \tag{2.17}$$

and

$$\widetilde{\zeta}_{1_{[\lambda_k,\infty[}(\Delta^{(q)})}(0) = -(b_0^q - \int_{X(q)} (-1)^q (dd^c\phi)^n/n!) + o(1).$$

Next, we apply the derivative formula in 2.11 to $D = k^{-1}1_{[k^{1-\epsilon},\infty[}(\Delta^{(q)}$ to get

$$\widetilde{\zeta}'_{1_{[k^{1-\epsilon},\infty[}(\Delta^{(q)})}(0) = \int_1^\infty k^{-n}\mathrm{Tr}(e^{-tk^{-1}1_{[k^{1-\epsilon},,\infty[}(\Delta^{(q)})})\frac{dt}{t} + O(1),$$

also using 2.16. From the basic estimate

$$k^{-n}\mathrm{Tr}(e^{-tk^{-1}1_{[\lambda_k,\infty[}(\Delta^{(q)})}) \leq e^{-(t-1)k^{-1}\lambda_k}(k^{-n}\mathrm{Tr}(e^{-k^{-1}(\Delta^{(q)})}))$$

we deduce, since the second factor is uniformly bounded according to 2.13 applied to $t = 1$ and since we have assumed $\lambda_k = O(k^{1-\epsilon})$ the bound

$$\int_1^\infty k^{-n}\mathrm{Tr}(e^{-tk^{-1}1_{[k^{1-\epsilon},,\infty[}(\Delta^{(q)})})\frac{dt}{t} \leq C\int_1^\infty e^{-(t-1)k^{-\epsilon}}\frac{dt}{t} \leq C'k^\epsilon.$$

All in all this means that $\widetilde{\zeta}'_{1_{[k^{1-\epsilon},\infty[}(\Delta^{(q)})}(0) = O(k^\epsilon)$ which combined with 2.12 and 2.17 proves 2.9 and hence finishes the proof of the proposition. □

2.3. Proof of Theorem 1.3. We start with a given ϕ and take a smooth path ϕ_t such that $\phi_1 = \phi$ and ϕ_0 has positive curvature. Setting

$$f_k(t) := k^{-(n+1)}(\log T_{]0,\lambda_k[}(e^{-k\phi_t}, h_X) + \sum_{q=0}^n (-1)^q \mathcal{L}^{0,q}(k\phi_t) \tag{2.18}$$

we have to prove that

$$\frac{1}{V}\lim_{k\to\infty}(f_k(1) - f_k(0)) = \mathcal{E}(\phi_1) - \mathcal{E}(\phi_0)(:= \mathcal{E}(\phi_1,\phi_0)). \tag{2.19}$$

To this end note that combining Lemma 2.1 and Lemma 2.2 gives $\frac{df_k(t)}{dt}$

$$= k^{-n}\int_X (\sum_{q=1}^n (-1)^q (\mathbb{B}^{0,q}_{\phi_t,]0,\lambda_k[} + \mathbb{B}^{0,1}_{\phi_t})\frac{d\phi_t}{dt} = k^{-n}\int_X (\sum_{q=1}^n (-1)^q \mathbb{B}^{0,q}_{\phi,[0,\lambda_k[})\frac{d\phi_t}{dt}.$$

Using Proposition 2.3 together with the dominated convergence theorem hence gives

$$\frac{1}{V}\lim_{k\to\infty}(f_k(1) - f_k(0)) = \lim_{k\to\infty}\int_0^1 \frac{df_k(t)}{dt}dt = \frac{1}{Vn!}\int_0^1\int_X (dd^c\phi_t)^n \frac{d\phi_t}{dt}dt.$$

According to the variational characterization of $\mathcal{E}$ in Lemma 2.1 (or by directly computing the integral) this finishes the proof of 2.19. Finally, by the asymptotics 2.4 this finishes the proof of the theorem.

2.4. Proof of Theorem 1.1. By the Kodaira vanishing theorem we have that $H^{0,q}(X,kL)=\{0\}$ for $k>>1$ (since L is assumed ample) and hence $\mathcal{L}^{0,q}(k\phi_t)=0$ for $q>0$. By Theorem A in [6] $(\mathcal{L}(\phi_1)-\mathcal{L}_k(\phi_0)/k^{(n+1)}$ converges to $(\mathcal{E}(P\phi_1)-\mathcal{E}(P\phi_0)$ and hence the previous theorem (or rather its proof) gives

$$\lim_{k\to\infty}\log\left(T_{]0,\lambda_k[}(e^{-k\phi_1},\omega_0)/T_{]0,\lambda_k[}(e^{-k\phi_0},\omega_0)\right)=(\mathcal{E}(\phi_1)-\mathcal{E}(\phi_0))-(\mathcal{E}(P\phi_1)-\mathcal{E}(P\phi_0))$$

Since, ϕ_0 is assumed positively curved we have on one hand that $P\phi_0=\phi_0$ so that the terms involving ϕ_1 in the rhs above cancel. On the other hand, the smallest positive eigenvalue $\lambda_k^{0,q}$of $\Delta_{k\phi_0}^{0,q}$ satisfies $\lambda_k^{0,q}\geq Ck$ (see below) and hence the second term in the lhs above also vanishes. All in all this gives the convergence of the truncated analytic torsion in Theorem 1.1 and by Proposition 2.4 the same asymptotics then hold for the analytic torsions. As for the eigenvalue estimate used above it is a standard consequence of the Kodaira-Nakano identity [**15**] in the case when ω is Kähler. In the non-Kähler case it was shown in [**10**] (Theorem 1).

REMARK 2.5. The eigenvalue estimate referred to above can also be deduced from the Kähler case as follows: first one notes that the first positive eigenvalue on the subspace $\ker\bar{\partial}$ of the $\bar{\partial}-$Laplacian associated to $(k\phi,h_X)$ is the supremum over all constants $C_k^{(q)}(h_X)$ such that for any f, a $\bar{\partial}-$closed $(0,1)-$form with values in L, the inhomogeneous $\bar{\partial}-$equation $\bar{\partial}u=f$ for can be solved with an estimate

$$\|u\|^2_{(k\phi,h_X)}\leq\frac{1}{C_k^{(q)}(\omega)}\|f\|^2_{(k\phi,h_X)}.$$

But since X is compact there is a constant A such that $A^{-1}\omega_0\leq h_X\leq A\omega_0$ for ω_0 a fixed Kähler metric and hence replacing h_X with ω_0 only distorts $C_k^{(q)}(h_X)$ with a multiplicative constant independent of k. Finally, using the bijection 2.7 (and since $H^{0,q}(X,kL)=\{0\}$ for $k>>1$) and the fact that $\bar{\partial}$ commutes with Δ this gives the desired lower bound for the first positive eigenvalue on all of $\Omega^{0,q}(X,kL)$ for all q.

2.5. Proof of Corollary 1.2. Let $\lambda_k=k^{1-\epsilon}$ for some $\epsilon\in]1/2,1[$.By Proposition 2.3 the number of terms in the sum

$$k^{-(n+1)}\log\det(1_{]0,\lambda_k[}(\Delta_{\bar{\partial}}^{(q)})=\frac{1}{k^n}\sum_i\frac{1}{k}\log\lambda_{i,k}^{(q)} \tag{2.20}$$

grows as constant times k^n. Hence, if there were no sequence of eigenvalues as in the statement of the corollary then the previous sum would converge to zero for any q contradicting the positivity assumption in the corollary (according to the previous theorem). As for the final statement we first note that, when $n=1$:

$$\mathcal{E}(P\phi,\phi)=\frac{1}{2}\|d(P\phi-\phi)\|_X^2,$$

which follows from integration by parts, using the "orthogonality relation" $\int(P\phi-\phi)dd^c(P\phi)=0$ [**6**]. Hence $\mathcal{E}(P\phi,\phi)\geq0$ with equality iff $P\phi=\phi+c$ for some constant c. But then it follows from the definition of $P\phi$ that $c=0$, i.e. ϕ has semi-positive curvature.

REMARK 2.6. By way of comparison let us briefly explain how to deduce the asymptotics for the (non-truncated) Quillen metric Theorem 1.3 from the results in [**1**, **9**, **20**] (following the suggestions of a referee). First, by [**9**, III, Theorem 1.18] (or [**20**, Theorem 5.5.6]) the differential of the Quillen metric associated to $(kL, e^{-k\phi}, h_X)$ can be expressed as the constant term $C_{k,0}$ in the $t-$expansion $C_{k,t}$ of the super trace of the heat-kernel associated to the Kodaira-Dolbeault Laplacian (this is the "non-truncated" version of Lemma 2.2). By [**1**, Theorem 2] the corresponding heat-kernel asymptotics are uniform in $t \in [0, T]$ as $k \to \infty$ and calculating $C_{k,0}$, by comparing with a constant coefficient model operator, gives that $C_{k,0} = k^{n+1}(dd^c\phi)^n + o(k^{n+1})$ (this is the "non-truncated" heat-kernel analog of Proposition 2.3). Finally, the proof is concluded by integrating along line segments in the space of metrics on L, just as above (see also [**20**, Theorem 5.5.9] for a detailed proof in the non-Kähler case).

3. Further remarks on tunneling and outlook

3.1. Comparison with the tunneling effect for Witten Laplacians and large deviations for fermions. It may be illuminating to compare the asymptotics of the truncated analytic torsions in Theorem 1.1 with the well-known results concerning the tunneling effect for Witten's deformation of the De Rham complex, which appeared in Witten's heuristic approach to Morse theory based on supersymmetric quantum mechanics [**23**] (see [**16**] for rigorous results based on semi-classical analysis). Geometrically this latter setting corresponds to letting $L \to X$ be the *trivial* line bundle over a *real* manifold X and replacing the operator $\bar{\partial}$ with the exterior derivative d. Any given *global* function ϕ on X induces a Hermitian metric h_{kL} on kL represented as $h_{kL} = e^{-k\phi}$ in the standard global trivialization (i.e. $s = 1$) of L, where now k makes sense for any positive number and where the semi-classical parameter $\hbar := 1/k$ corresponds to Planck's constant in the quantum mechanical picture. To this setting one associates as before a Laplacian on $(0, q)-$forms, $\Delta_k^{(q)}$, depending on $k\phi$, which may be identified with the *Witten Laplacian* (in a unitary frame). Assuming that ϕ is a Morse function we denote by $X(q)$ the finite set of all critical points of ϕ where the Hessian of ϕ has index q. As shown in [**16**] the set of all "small" eigenvalues $(\lambda_{i,k}^{(q)})$ of the Witten Laplacian $\Delta_k^{(q)}$ is in a one to one correspondence with the finite set $X(q)$. Moreover, all the non-zero eigenvalues are actually *exponentially* small:

$$\lim_{k\to\infty} \frac{1}{k} \log \lambda_{i,q}^{(k)} = -c_{i,q}(\phi), \tag{3.1}$$

where the positive number $c_{i,q}(\phi)$ may be expressed in terms of an Agmond distance betweeen critical points. In the present complex geometric setting of the $\bar{\partial}-$Laplacian the number of "small" eigenvalues $N_q^{(k)}$ depends on k and is of the order $O(k^n)$, more precisely:

$$N_q^{(k)} = k^n \int_{X(q)} (-1)^q (dd^c\phi)^n/n! + o(k^n),$$

where now $X(q)$ is the subset of X where the curvature form $dd^c\phi$ (i.e. the *complex* Hessian of ϕ) has index q (as shown by Demailly in his proof of his holomorphic Morse inequalities [**14**]; compare the discussion in the beginning of Section 3.2 below). Hence the asymptotics of the truncated analytic torsions in Theorem 1.1

may be interpreted as an averaged version of 3.1. This analogy becomes particularly striking in the Riemann surface case (compare the asymptotics 1.2).

The approach in [**16**] to study the asymptotics of the exponentially small eigenvalues, i.e. of the quantum mechanical *tunneling effect*, uses that the Witten Laplacians are *elliptic* when viewed as semi-classical elliptic differential operators. In the particular case when $q = 0$ there is also a probabilistic approach to the corresponding Witten Laplacian $\Delta_k^{(0)}$ which refines the information in 3.1 with asymptotics for pre-factors under certain genericity assumptions (see [**11**] and references therein for the case of Euclidean domains). The point is that the operator $\Delta_k^{(0)}$ is the Feller semigroup generator for the stochastic differential equation on the Riemannian manifold X describing a Brownian particle in the gradient field of $-\phi$ at temperature $1/k$. The invariant probability measure of the process is

$$\mu_{k\phi} := \frac{e^{-k\phi}dV}{Z_{k\phi}} \tag{3.2}$$

and its relation to the exponentially small eigenvalues was studied in [**18**] in connection to simulated annealing. As explained in [**11**] each local minima of ϕ corresponds to a metastable state and the corresponding eigenvalue to the *inverse life time* of the state (i.e. the inverse mean exit time). See also the very recent work [**19**] for refinements of 3.1 for general degrees q.

There are also quantum mechanical, as well as probabilistic motivations for considering the present complex geometric setting. For example in the Riemann surface case the $\bar{\partial}-$Laplacian is the Pauli Hamiltonian for a single fermion with spin up ($q = 0$) and spin down ($q = 1$) subject to the magnet field with two-form $F_{kA} := kdd^c\phi$. Moreover, from a probabilistic point of view the asymptotics of the corresponding truncated analytic torsions appear in the large deviation principle for the corresponding determinantal random point process on X with N_k particles defined by the following probability measure on X^{N_k} :

$$\mu_{k\phi}^{(N_k)} := \frac{|\det\Psi|^2(x_1, ...x_{N_k})e^{-k(\phi_1(x_1)+\cdots+\phi_1(x_N))}dV^{\otimes N_k}}{\mathcal{Z}_{k\phi}},$$

where N_k is the dimension of the space $H^0(X, kL)$ of global holomorphic sections of kL and $\det\Psi$ is any generator of the complex line $\det H^0(X, kL)$, which we identify with a holomorphic section over X^{N_k} [**5**]. Physically, this is the maximally filled fermionic many-particle ground state and $\det\Psi$ is the corresponding Slater determinant. The large deviation principle referred to above may be symbolically formulated as follows:

$$\text{Prob}\left(\frac{1}{N_k}\sum_i \delta_{x_i} \in \mathcal{B}_\epsilon(\mu)\right) \sim \prod_{q=1}^{n}(\det\Delta_{\bar{\partial},k}^{0,q})^{-(-1)^{q+1}q}e^{-kN_kE(\mu)}, \tag{3.3}$$

where, by Theorem 1.1 above, the weighted determinant above may by replaced the weighted product of the corresponding exponentially small eigenvalues of the Kodaira-Dolbeault Laplacian. In the left hand side above $\mathcal{B}_\epsilon(\mu)$ denotes a ball of radius ϵ centered at a given probability measure μ in the space of all probability measures on X (equipped with a fixed metric) and $E(\mu)$ appearing in the right hand side denotes the pluricomplex energy of the measure μ, defined with respect to the curvature form ω. The relation 3.3 holds asymptotically as $N \to \infty$ and $\epsilon \to 0$ and (see [**5**] for the precise statement). Loosely speaking the left hand side in formula 3.3 gives the probability of finding a "cloud" of particles at the points $x_1, x_2, ..., x_N$ in X representing a macroscopic state μ, in the sense that $\frac{1}{N_k}\sum_i \delta_{x_i}$ approximates μ (the small parameter ϵ can physically be seen as a measure of the coarse-graining). As explained in [**5**] the formula 3.3 may also be interpreted as an (asymptotic) higher dimensional bosonization formula, generalizing the well-known exact bosonization formula [**1**] on a Riemann surface.

Note that in the setting of the Witten Laplacian the role of $H^0(X, kL)$ is played by the *one*-dimensional kernel of d acting on the space smooth functions, inducing the *one* particle point-process 3.2. The main analytical difference here is that the $\bar{\partial}-$Laplacian on $(0,q)-$forms with values in kL is *not* an elliptic operator, when viewed as a semi-classical differential operator (see [**7**]). There appears to be very few results concerning the tunneling effect for non-elliptic semi-classical operators (see however [**17**] and references therein for the study of Kramers Fokker-Planck type operators generalizing the Witten Laplacian for $q = 0$).

3.2. "Small" vs. "exponentially small" eigenvalues for the Dolbeault complex. Fix a number $\epsilon \in]0, 1/2[$. For a given positive integer k we say that an eigenvalue ν_k of the $\bar{\partial}-$Laplacian $\Delta_{\bar{\partial}}$ associated to kL is "small" if $\nu_k \leq k^{1-\epsilon}$. By Proposition 2.3 the number of "small" eigenvalues is of the order $O(k^n)$. Hence, as explained above the "small" eigenvalues which are not "exponentially small" make no contribution to the large k limit of the sum 2.20. But it is tempting to ask whether *all* "small" eigenvalues are actually "exponentially small"? As recalled in section 3.1 this is indeed the case in the setting of the Witten Laplacian under the assumption that ϕ be a Morse function. However in the present setting there are no non-degeneracy assumptions on the weight ϕ of the metric and hence it seems unlikely that the answer to the question above is yes, in general. Still when $q = 0$ the number of "small" and "exponentially" small eigenvalues are the same to the leading order according to the following proposition (that should be compared with Theorem 3.14 in [**14**] concerning eigenvalues of the order $k\lambda$ for λ fixed):

PROPOSITION 3.1. *Given any $\delta > 0$ there is a constant C_δ such that the number $N_{\delta,k}$ of eigenvalues in $[0, e^{-C_\delta K}]$ of the $\bar{\partial}-$Laplacian $\Delta_{\bar{\partial}}$ acting on smooth sections of kL satisfies*

$$\int_{X(0)} (dd^c\phi)^n/n! - \delta \leq N_{\delta,k} \leq N_k \leq \int_{X(0)} (dd^c\phi)^n/n! + \delta$$

for k sufficiently large, where N_k is the number of "small" eigenvalues.

PROOF. Given $\delta > 0$ take K_δ a compact subset of the open set $X(0) := \{dd^c\phi > 0\}$ such that

$$\int_{X(0)} (dd^c\phi)^n/n! - \delta \leq \int_{K_\delta} (dd^c\phi)^n/n!.$$

By Fatou's lemma it will be enough to prove that

$$\liminf_{k\to\infty} k^{-n}\mathbb{B}^{0,q}_{[0,e^{-C_\delta K}[}(x) \geq (dd^c\phi)^n(x)/n!$$

for any given $x \in K_\delta$. As shown in [**3**] (Proposition 5.3) it is enough to, given a point $x \in K_\delta$, find a smooth section α_k of kL such that

$$(i) \liminf_{k\to\infty} k^{-n}\frac{(|\alpha_k|^2e^{-k\phi})(x)}{\int_X |\alpha_k|^2e^{-k\phi}dV} \geq (dd^c\phi)^n(x)/n!, \quad (ii)\, k^{-1}\left\langle \Delta_{\bar{\partial}}\alpha_k, \alpha_k\right\rangle_{k\phi} \leq \delta_k$$

for $\delta_k = e^{-2C_\delta k}$. To this end we simply take $\alpha_k = \chi s$ where χ is a smooth cut-off function which is equal to 1 close to x and s is local holomorphic frame for L on a neighborhood x such that the corresponding local weight $\phi(z)$ is given by $\phi(z) = \sum_{i=1}^n \mu_i |z_i|^2 + O(|z|^3)$ (where $\mu_i > 0$ since $dd^c\phi > 0$ at x). Then (i) above follows from computing a Gaussian integral. Moreover,

$$\langle \Delta_{\bar{\partial}} \alpha_k, \alpha_k \rangle_{k\phi} = \langle \bar{\partial}\alpha_k, \bar{\partial}\alpha_k \rangle_{k\phi} = \int |(\partial\chi)|^2 e^{-k\phi} dV \leq e^{-2C_\delta k}.$$

By the compactness of K_δ the constant C_δ can be taken to be independent of x. This finishes the proof of the lower bound on $N_{\delta,k}$ and the upper bound is a special case of Proposition 2.3. □

It would be interesting to know if the analog for $q > 0$ of the previous proposition is also valid? If one replaces e^{-kC_δ} with a sequence δ_k of the form $\delta_k = O(k^{-\infty})$, i.e. $\delta_k \leq C_m k^{-m}$ for any $m > 0$ then this is indeed case. This follows from taking α_k in the previous proof to be the given by $\Pi_k(x, \cdot)$ where $\Pi_k(x, y)$ is the local projector for $(0, q)-$forms constructed in [**7**], defining a local Fourier integral operator with a complex phase (in fact, only the phase function constructed in [**7**] is needed together with the leading term in the symbol expansion).

3.3. Discussion and outlook. This paper gives a first small step towards understanding the tunneling effect for the asymptotic Dolbeault complex associated to high powers of a line bundle $L \to X$ equipped with a (non-positively curved) Hermitian metric. Ideally, one could hope to give an asymptotic description, as $k \to \infty$, of the following measures on $\mathbb{R}$

$$\nu_q^{(k)} := -\frac{1}{k^{n+1}} \sum_{i=1}^{N_q^{(k)}} \log \lambda_{i,q}^{(k)},$$

where $\{\lambda_{i,q}^{(k)}\}_{i=1,}^{N_q^{(k)}}$ ranges over the "small" non-zero eigenvalues of the Kodaira-Dolbeault Laplacian acting on $\Omega^{0,q}(X, L^{\otimes k})$. Conjecturally, $\nu_q^{(k)}$ converges weakly as $k \to \infty$ to a measure ν_q only depending on the fixed metric on L. If such a measure exists, say when $n = 1$, Corollary 1.2 shows that its first moment is given by a Dirichlet norm:

$$\int_{\mathbb{R}} t d\nu_1(t) = \frac{1}{2} \|d(P\phi - \phi)\|_X^2 \,.$$

Furthermore, it would be very interesting to give an asymptotic description of the corresponding eigenforms. In the De Rham case it is well-known that the eigenforms are, for a generic Morse function, localized along the gradient paths of the Morse function connecting critical points of index q and index $q + 1$ [**16**] (in physics terminology these are the "tunneling paths" or "instanton paths"[**23**]). It seems likely that, in the asymptotic Dolbeault case, the role of the gradient paths (at least for $q = 0$) are played by the leaves of the Monge-Ampére foliation attached to the weight $\psi := P\phi$ on the open subset Ω of X where $P\phi < \phi$. Recall that the corresponding leaves are defined as the holomorphic curves along which ψ is locally harmonic. Since, $(dd^c\psi)^n = 0$ such leaves are well-known to exist under appropriate (but rather strong) regularity assumptions, but in the present case when ψ is merely known to be $\mathcal{C}^{1,1}-$smooth the existence of the leaves is a very delicate issue. Still, one could hope to be work with a suitable approximate notion of leaves.

References

[1] L. Alvarez-Gaumé, J.-B. Bost, G. Moore, P. Nelson, and C. Vafa, *Bosonization on higher genus Riemann surfaces*, Comm. Math. Phys. **112** (1987), no. 3, 503–552. MR908551 (88j:81061)

[2] N. Berline, E. Getzler, and M. Vergne, *Heat kernels and Dirac operators*, Grundlehren Text Editions, Springer-Verlag, Berlin, 2004. Corrected reprint of the 1992 original. MR2273508 (2007m:58033)

[3] R. J. Berman, *Bergman kernels and local holomorphic Morse inequalities*, Math. Z. **248** (2004), no. 2, 325–344, DOI 10.1007/s00209-003-0630-z. MR2088931 (2005g:32024)

[4] R. J. Berman, *Bergman kernels and equilibrium measures for line bundles over projective manifolds*, Amer. J. Math. **131** (2009), no. 5, 1485–1524, DOI 10.1353/ajm.0.0077. MR2559862 (2010g:32030)

[5] R. J. Berman, *Determinantal point processes and fermions on complex manifolds: large deviations and bosonization*, Comm. Math. Phys. **327** (2014), no. 1, 1–47, DOI 10.1007/s00220-014-1891-6. MR3177931

[6] R. J. Berman and S. Boucksom, *Growth of balls of holomorphic sections and energy at equilibrium*, Invent. Math. **181** (2010), no. 2, 337–394, DOI 10.1007/s00222-010-0248-9. MR2657428 (2011h:32021)

[7] R. J. Berman and J. Sjöstrand, *Asymptotics for Bergman-Hodge kernels for high powers of complex line bundles* (English, with English and French summaries), Ann. Fac. Sci. Toulouse Math. (6) **16** (2007), no. 4, 719–771. MR2789717 (2012e:32034)

[8] J.-M. Bismut, *Demailly's asymptotic Morse inequalities: a heat equation proof*, J. Funct. Anal. **72** (1987), no. 2, 263–278, DOI 10.1016/0022-1236(87)90089-9. MR886814 (88j:58131)

[9] J.-M. Bismut, H. Gillet, and C. Soul, *Analytic torsion and holomorphic determinant bundles.* I, II, III. Comm. Math. Phys. **115** (1988), 49–78, 79–126, 301–351

[10] J.-M. Bismut and É. Vasserot, *The asymptotics of the Ray-Singer analytic torsion associated with high powers of a positive line bundle*, Comm. Math. Phys. **125** (1989), no. 2, 355–367. MR1016875 (91c:58141)

[11] A. Bovier, V. Gayrard, and M. Klein, *Metastability in reversible diffusion processes. II. Precise asymptotics for small eigenvalues*, J. Eur. Math. Soc. (JEMS) **7** (2005), no. 1, 69–99, DOI 10.4171/JEMS/22. MR2120991 (2006b:82113)

[12] J. I. Burgos Gil, G. Freixas i Montplet, and R. Litcanu, *Generalized holomorphic analytic torsion.* arXiv:1011.3702

[13] J.-P Demailly, *Complex analytic and algebraic geometry.* Available at www-fourier.ujf-grenoble.fr/~demailly/books.html

[14] J.-P. Demailly, *Champs magnétiques et inégalités de Morse pour la d''-cohomologie* (French, with English summary), C. R. Acad. Sci. Paris Sér. I Math. **301** (1985), no. 4, 119–122. MR799607 (86k:32028)

[15] P. Griffiths and J. Harris, *Principles of algebraic geometry*, Wiley Classics Library, John Wiley & Sons, Inc., New York, 1994. Reprint of the 1978 original. MR1288523 (95d:14001)

[16] B. Helffer and J. Sjöstrand, *Puits multiples en mécanique semi-classique. IV. Étude du complexe de Witten* (French), Comm. Partial Differential Equations **10** (1985), no. 3, 245–340, DOI 10.1080/03605308508820379. MR780068 (87i:35162)

[17] F. Hérau, M. Hitrik, and J. Sjöstrand, *Tunnel effect and symmetries for Kramers-Fokker-Planck type operators*, J. Inst. Math. Jussieu **10** (2011), no. 3, 567–634, DOI 10.1017/S1474748011000028. MR2806463 (2012h:35249)

[18] R. A. Holley, S. Kusuoka, and D. W. Stroock, *Asymptotics of the spectral gap with applications to the theory of simulated annealing*, J. Funct. Anal. **83** (1989), no. 2, 333–347, DOI 10.1016/0022-1236(89)90023-2. MR995752 (92d:60081)

[19] D. Le Peutrec, F. Nier, and C. Viterbo, *Precise Arrhenius law for p-forms: the Witten Laplacian and Morse-Barannikov complex*, Ann. Henri Poincaré **14** (2013), no. 3, 567–610, DOI 10.1007/s00023-012-0193-9. MR3035640

[20] X. Ma and G. Marinescu, *Holomorphic Morse inequalities and Bergman kernels*, Progress in Mathematics, vol. 254, Birkhäuser Verlag, Basel, 2007. MR2339952 (2008g:32030)

[21] T. Mabuchi, *Some symplectic geometry on compact Kähler manifolds. I*, Osaka J. Math. **24** (1987), no. 2, 227–252. MR909015 (88m:53126)

[22] C. Soulé, *Lectures on Arakelov geometry*, Cambridge Studies in Advanced Mathematics, vol. 33, Cambridge University Press, Cambridge, 1992. With the collaboration of D. Abramovich, J.-F. Burnol and J. Kramer. MR1208731 (94e:14031)
[23] E. Witten, *Supersymmetry and Morse theory*, J. Differential Geom. **17** (1982), no. 4, 661–692 (1983). MR683171 (84b:58111)

Mathematical Sciences - Chalmers University of Technology and University of Gothenburg - SE-412 96 Gothenburg, Sweden
E-mail address: robertb@chalmers.se

Contemporary Mathematics
Volume **630**, 2014
http://dx.doi.org/10.1090/conm/630/12664

Simple Spectrum and Rayleigh Quotients

Victor Guillemin, Eveline Legendre, and Rosa Sena-Dias

ABSTRACT. In this note we calculate the variation of the Rayleigh quotient under specific variations of the metric. We consider two cases. First we study conformal variations and secondly variations within a Kähler class. We use the first variational formula to reprove an important result of Uhlenbeck's: a generic metric can be perturbed by a conformal factor into a metric whose eigenspaces for the Laplace operator have dimension 1. In the case of a Kähler manifold, we write an explicit variational formula for the Rayleigh quotient under Kähler deformations. We relate our calculation with the following question: Is the spectrum of a generic Kähler metric simple?

1. Introduction

In the 1970's, Uhlenbeck proved that given a compact manifold M, "most" Riemannian metrics g on M are *simple* meaning that each eigenspace of the Laplace operator Δ^g is one-dimensional. Here, "most" means that the set of simple metrics is residual in the space of smooth Riemannian metrics for an appropriate topology. Uhlenbeck also proved that the set of simple Riemannian metrics is pathwise connected (in a residual sense: the set of paths of simple metrics is residual the set of paths of metrics). Uhlenbeck's approach is very general: her idea is to study the following operator

$$\begin{array}{rccl} F: & \mathcal{M}\times\mathcal{L}^2\times\mathbb{R} & \rightarrow & \mathcal{L}^2 \\ & (g,u,\lambda) & \mapsto & \Delta^g u-\lambda u \end{array}$$

where $\mathcal{M}$ stands for the space of metrics and $\mathcal{L}^2$ stand for the space of square integrable function. Using general theory of operators and transversality theorems for Fredholm operators on Banach spaces, she proved that λ is simple if and only if 0 is a regular value of $F(g,\cdot,\lambda)$ and that the set of metrics $g\in\mathcal{M}$ such that 0 is a regular value for $F(g,\cdot,\lambda)$ is residual in $\mathcal{M}$. For the last step she uses specific variations (conformal variations) in the space of Riemannian metrics.

In this very short note, we study the variation of Rayleigh quotients and by elementary means we prove a refinement of Uhlenbeck's famous result in a conformal class of Riemannian metrics. We also discuss similar issues on the space, $\mathcal{M}_\Omega$, of Kähler metrics in a fixed Kähler class Ω on a fixed compact Kähler manifold.[1]

2010 *Mathematics Subject Classification.* Primary 58E99.

[1]After this work was completed Professor Dmitry Jakobson mentioned [**5**] to the second author. Teytel's work is very close in spirit to our own work: he studies variations of Rayleigh quotients of a family of self-adjoint operators and introduces two conditions namely the strong Arnold hypothesis 2 (SAH2) and the strong Arnold hypothesis 1 (SAH1) which ensure that the

1.1. Conformal case. In the conformal case, we prove the following Theorem.

THEOREM 1.1. *Let X be a compact manifold (or orbifold) of dimension greater than* 1 *and* $\mathcal{C}$ *a conformal class of Riemannian metrics of fixed volume on* X. *Given* $k \geq 1$ *and* $N \geq 2$, *the subset of* $\mathcal{C}$ *of metrics for which the* k*-th eigenspace is of dimension* N *is a submanifold of codimension at least* 1. *In particular, the subset of* $\mathcal{C}$ *of metrics admitting a non-simple eigenvalue of the Laplacian is a countable union of submanifolds of codimension at least* 1.

This theorem was previously known. It is implicit in Uhlenbeck's paper [**6**] and proved by Bleecker and Wilson in [**1**].

There are examples of non-simple metrics g, on the 2–torus for instance, whose tangent space in $\mathcal{C}$ has a codimension 1 subspace tangent to a set of non-simple metric (see Remark 2.9).

1.2. Kähler case. To explain our result concerning Kähler manifolds, let us recall that a Kähler manifold is a triple (X,g,J) where (X,g) is a Riemannian manifold, J is an integrable almost complex structure such that the 2-form ω defined by the formula $\omega(\cdot,\cdot) = g(J\cdot,\cdot)$ defines a symplectic form on X (that is ω is closed) and g is J–invariant (that is $g(J\cdot,J\cdot) = g(\cdot,\cdot)$). We assume X to be compact, without boundary and connected.

The symplectic form ω determines a cohomology class called the Kähler class, say $[\omega] = \Omega$, in the second group of de Rham cohomology $H^2(X,\mathbb{R})$. We take the following approach: we fix the complex structure J on X and consider the space of compatible Kähler metrics

$$\mathcal{M}_\Omega := \{g\,|\,(X,g,J) \text{ is Kähler and } [\omega] \in \Omega\}.$$

Note that any two of the structures ω, g, J determine the third and once J fixed, the *Kähler metric* refers to g or ω as well.

The endomorphism $J : M \to \mathrm{End}(TM)$ acts on tensors and in particular on p–forms as $J\psi(X_1,\dots,X_p) = \psi(J^{-1}(X_1),\dots,J^{-1}(X_p))$. One can define the twisted differential $d^c = J^{-1}\circ d\circ J^{-1}$ and we have

$$dd^c = -d^c d = 2\partial\bar{\partial}.$$

It follows from the $\partial\bar{\partial}$–lemma (or dd^c-lemma) that any two J–invariant 2–forms in the same cohomology class are related by $\omega' - \omega = dd^c\varphi$ for some smooth function $\phi \in C^\infty(X)$. Consequently, $\mathcal{M}_\Omega$ is naturally locally modeled on $C^\infty(X)/\mathbb{R}$ and, thus, endowed with the structure of a Fréchet manifold. Precisely, denoting $\mathbf{v}_g$ the volume form induced by g, we make the following identification

$$T_g\mathcal{M}_\Omega = \left\{ f \in C^\infty(X) \,\middle|\, \int_X f\,\mathbf{v}_g = 0 \right\}.$$

The $\mathcal{L}^2$–inner product defines a metric on $\mathcal{M}_\Omega$ sometimes called the Mabuchi metric.

Finally, we denote the $\delta : \Omega^p(X) \to \Omega^{p-1}(X)$ and $\delta^c : \Omega^p(X) \to \Omega^{p-1}(X)$ the adjoint (with respect to the $\mathcal{L}^2$–inner product on $\Omega^p(X)$) of d and d^c. We prove the following.

family of non-simple operators is of positive codimension in the full family of operators. Our work can be reformulated in terms of Teytel's terminology: we are making explicit SAH1 and SAH2 for families of Laplacians corresponding to metrics in a fixed conformal class in one case and Kähler metrics in a fixed Kähler class in the other.

THEOREM 1.2. *Let X be a complex compact manifold (or orbifold) X of Kähler type with a fixed Kähler class Ω. Given $k \geq 1$ and $N \geq 2$, the subset $\mathcal{U}_{k,N}$ of $\mathcal{M}_\Omega$ of metrics for which the k-th eigenvalue is of dimension N is a Fréchet submanifold of $\mathcal{M}_\Omega$. Moreover at $g \in \mathcal{U}_{k,N}$ the codimension of $T_g\mathcal{U}_{k,N}$ in $T_g\mathcal{M}_\Omega$ is at least 2 if there exist two eigenfunctions f, h of $\lambda_k(g)$ such that $\delta\delta^c(fdd^c h + hdd^c f)$ and $\delta\delta^c(fdd^c f - hdd^c h)$ are linearly independent.*

2. The Conformal class

Given a compact manifold (or orbifold) X, we fix a conformal class of Riemannian metrics on X with fixed volume, say $\mathcal{C}$. The tangent space of $\mathcal{C}$ at any point $g \in \mathcal{C}$ is identified with $C_0^\infty(X) := \{f \in C^\infty(X) \,|\, \int_X f\mathbf{v}_g = 0\}$ via the exponential map

$$\phi \mapsto e^{t\phi}g.$$

The normalization $\int_X \phi\mathbf{v}_g = 0$ (that is $\phi \in C_0^\infty(X)$) ensures that the volume of X is preserved along the path. Via this identification, the Hilbert product defines a metric on $\mathcal{C}$: for $\phi, \psi \in T_g\mathcal{C}$,

$$(\psi, \phi)_g = \int_X \phi\psi v_g$$

where $v = v_g$ denotes the volume form of the metric g on X.

We consider a family of Riemannian metrics g_t on X parameterized by $t \in (-\varepsilon, \varepsilon)$ for some positive constant ε. Assume that these are all in the same conformal class $\mathcal{C}$. For each t we have a Laplace operator and we consider a family of eigenfunctions for the eigenvalues λ_t. It is known (see [**3**]) that λ_t is a continuous function of t. In fact, one has the following lemma

LEMMA 2.1. *Let g_t be a smooth path, $t \in (-\varepsilon, \varepsilon)$, of Riemannian metrics on a compact orbifold X. Consider λ, an eigenvalue of the Laplacian of $g = g_0$ with eigenspace V_λ, and take $\delta > 0$ such that λ is the only eigenvalue of Δ^g in $(\lambda - \delta, \lambda + \delta)$. Define $V_t^\lambda(\delta) = \operatorname{span}_{\mathbb{R}}\{f \in C^\infty(X) \,|\, \exists \mu \in (\lambda - \delta, \lambda + \delta),\ \Delta^{g_t} f = \mu f\}$. Then, there exists $\varepsilon_o \in (0, \varepsilon)$ such that for each $t \in (0, \varepsilon_o)$ there is an isomorphism $\Phi_t : V_t^\lambda(\delta) \to V_\lambda$ which depends smoothly on t.*

For the proof, see Theorems XII.12 and XII.13 in [**3**]. Note that using the notation above we have

$$V_t^\lambda(\delta) = \bigoplus_{\{\lambda_t | \lambda_t - \lambda| \leq \delta\}} V_{\lambda_t}$$

Recall that, given a Riemannian metric, the Rayleigh quotient of a complex valued non-zero function $f \in C^2(X, \mathbb{C})$ is

$$R^g(f) = \operatorname{Re}\frac{\int_X f\Delta^g \bar{f} v}{\int_X |f|^2 v} = \operatorname{Re}\frac{\int_X g(\nabla f, \nabla \bar{f})v}{\int_X |f|^2 v}.$$

By the max-min characterization of eigenvalue, we know that Rayleigh quotient R^g is constant on any eigenspace of the Laplacian and that this is a way to characterize an eigenspace of the Laplacian.

2.1. The variation of the Rayleigh quotient. In coordinates $(x_1, \cdots, x_n)$ on X one can easily write down an expression for Δ in terms of the entries of the metric. Set

$$g_{ij} = \left(\frac{\partial}{\partial x_i}, \frac{\partial}{\partial x_i}\right)$$

to be the inner product of coordinate tangent vectors. Let $|g| = \det(g_{ij})$ be the volume of the metric in these coordinates i.e. $v = |g| dx$, where v is the volume form on X given by g. We also write g^{ij} for the entries of the inverse of the matrix (g_{ij}). Given f in $\mathcal{C}^\infty(X)$

$$\Delta f = -\frac{1}{|g|}\sum_{i,j}\frac{\partial}{\partial x_i}|g|g^{ij}\frac{\partial f}{\partial x_i}. \tag{1}$$

Assume we have a family of metrics g_t on X we would like to study how the Rayleigh quotient varies in the family. In this section we will consider a family of metrics which are all conformal to the initial metric g_0 i.e.

$$g_t = e^{\phi_t} g_0,\ t \in]-\epsilon, \epsilon[$$

with $\epsilon > 0$. Assume $\phi_0 = 0$. We denote by Δ_t the Laplace operators associated to g_t. We write Δ for Δ_0.

LEMMA 2.2. *Let f_t be a smooth family of smooth functions on X such that f_0 is an eigenfunction of Δ of eigenvalue λ. We have the following variational formula for the Rayleigh quotient of f_t*

$$\frac{dR_t(f_t)}{dt}_{|t=0} = \int_X \left(\lambda - \frac{\Delta}{2}\right)\dot{\psi}|f|^2 v, \tag{2}$$

where $\psi_t = e^{(n-1)\phi_t}$ and $\psi = \psi_0 = 1$.

REMARK 2.3. *Note that $\dot{\psi} = (\frac{d}{dt}(e^{\phi_t})^{n-1}))_{t=0} = (n-1)\dot{\phi}_0(e^{\phi_0})^{n-2} = (n-1)\dot{\phi}_0$ since $\phi_0 = 0$. For the rest of this note we assume $n > 1$.*

PROOF. Assume that $||f_t||_{\mathcal{L}^2(g_t)} = 1$. Cover X by coordinates patches $\mathcal{U}$ where one can choose coordinates $(x_1, \cdots, x_n)$ such that $|g_0| = 1$ and let $\rho_{\mathcal{U}}$ be a partition of unity subordinate to this cover.

For a coordinate patch $\mathcal{U}$ define

$$R^{g_t}_\rho(f) = \mathrm{Re}\int_X \rho f \,\Delta_{g_t} \bar{f} v_t.$$

We have $|g_t| = e^{n\phi_t}$ in the coordinates $(x_1, \cdots, x_n)$ and $(g^{ij}) = e^{-\phi_t}(g_0^{ij})$. Using 1 we have that this quantity is given by

$$R^{g_t}_\rho(f) = -\mathrm{Re}\int_X \rho e^{-n\phi_t}\sum_{i,j}\frac{\partial}{\partial x_i}\left(e^{(n-1)\phi_t}g_0^{ij}\frac{\partial \bar{f}_t}{\partial x_j}\right) f_t v_t.$$

Using the fact that $v_t = e^{n\phi_t}dx$ and writting $\psi_t = e^{(n-1)\phi_t}$, we have

$$R^{g_t}_\rho(f) = -\mathrm{Re}\int_X \rho \sum_{i,j}\frac{\partial}{\partial x_i}\left(\psi_t g_0^{ij}\frac{\partial \bar{f}_t}{\partial x_j}\right) f_t dx.$$

Thus, it is

$$-\mathrm{Re}\int_X \rho\psi_t \sum_{i,j}\frac{\partial}{\partial x_i}\left(g_0^{ij}\frac{\partial \bar{f}_t}{\partial x_i}\right) f_t dx - \mathrm{Re}\int_X \rho \sum_{i,j}\frac{\partial \psi_t}{\partial x_i}g_0^{ij}\frac{\partial \bar{f}_t}{\partial x_j} f_t dx.$$

Integrating by parts we see that

$$-\mathrm{Re}\int_X \rho\sum_{i,j}\frac{\partial\psi_t}{\partial x_i}g_0^{ij}\frac{\partial\bar{f}_t}{\partial x_j}f_t dx = \frac{1}{2}\int_X\sum_{i,j}\frac{\partial}{\partial x_j}\left(\rho\frac{\partial\psi_t}{\partial x_i}g_0^{ij}\right)|f_t|^2 dx.$$

Because $\phi_0 = 0$ we see that

$$-\mathrm{Re}\int_X \rho\sum_{i,j}\frac{\partial\psi_t}{\partial x_i}g_0^{ij}\frac{\partial\bar{f}_t}{\partial x_j}f_t dx = -\frac{1}{2}\int_X \triangle\rho\psi_t|f_t|^2 v,$$

and

$$-\mathrm{Re}\int_X \rho\psi_t\sum_{i,j}\frac{\partial}{\partial x_i}\left(g_0^{ij}\frac{\partial\bar{f}_t}{\partial x_i}\right)f_t dx = \mathrm{Re}\int_X \rho\psi_t\,\triangle\, f_t\bar{f}_t v,$$

so that

$$R_\rho^{g_t}(f_t) = \mathrm{Re}\int_X \rho\psi_t\,\triangle\, f_t\bar{f}_t v - \frac{1}{2}\int_X \triangle\rho\psi_t|f_t|^2 v,$$

Summing over ρ we see that

$$R^{g_t}(f_t) = \mathrm{Re}\int_X \psi_t\,\triangle\, f_t\bar{f}_t - \frac{1}{2}\int_X \triangle\psi_t|f_t|^2 v.$$

Taking derivatives with respect to t of the above expressions we see that $\frac{dR_t(f_t)}{dt}$ is given by

$$\mathrm{Re}\int_X \dot{\psi}_t\triangle\bar{f}_t f_t v+\mathrm{Re}\int_X \psi_t\triangle\dot{\bar{f}}_t f_t v+\mathrm{Re}\int_X \psi_t\triangle\bar{f}_t\dot{f}_t v-\frac{1}{2}\int_X \triangle\dot{\psi}_t|f_t|^2 v-\frac{1}{2}\int_X \triangle\psi_t\frac{d|f_t|^2}{dt}v.$$

Because

$$\mathrm{Re}\int_X \psi_t\,\triangle\,\dot{\bar{f}}_t f_t v + \mathrm{Re}\int_X \psi_t\,\triangle\,\bar{f}_t\dot{f}_t v - \frac{1}{2}\int_X \triangle\psi_t\frac{d|f_t|^2}{dt}v = 0,$$

it follows that

$$\frac{dR_t(f_t)}{dt} = \mathrm{Re}\int_X \dot{\psi}_t\,\triangle\,\bar{f}_t f_t v - \frac{1}{2}\int_X \triangle\dot{\psi}_t|f_t|^2 v.$$

so that at $t = 0$ we have

$$\frac{dR_t(f_t)}{dt} = \mathrm{Re}\int_X\left(\lambda\dot{\psi}_t - \frac{\Delta\dot{\psi}_t}{2}\right)|f_t|^2 v.$$

□

2.2. The codimension of the space of bad metrics.

DEFINITION 2.4. *Given a Riemannian manifold, we say that an eigenvalue of the Laplace operator is simple if its eigenspace is one dimensional. We say that that a metric is simple if all the eigenvalues of the corresponding Laplace operator are simple. We say a metric is bad if it is not simple.*

Next, we set up some notation.

- We denote the subset of simple metrics in $\mathcal{C}$ by $\mathcal{S}$,
- We write

$$0 = \lambda_0(g) < \lambda_1(g) \le \lambda_2(g) \le \cdots \nearrow \infty$$

  for the ordered spectrum of g,
- We write V_λ for the real eigenspace associated to an eigenvalue λ,
- For $k \ge 1$ we denote $\mathcal{S}_k = \{(g,\omega) \in \mathcal{C} \mid \lambda_l \text{ simple if } l \le k\}$,

- For $N \geq 1$ we define

$$\mathcal{U}_{k,N} = \{(g,\omega) \in \mathcal{C} \mid \dim V_{\lambda_k} = N\}.$$

- For $N \geq 1$ we define

$$\mathcal{Q}_{k,N} = \left\{(g,\omega) \in \mathcal{C} \;\middle|\; \max_{l\in\{1,\dots,k\}} \{\dim V_{\lambda_l}\} = N\right\}.$$

These spaces depend on the conformal class but we assume it is fixed throughout and do not make the dependence explicit in the notation for the sake of simplicity. We have the following relations among these spaces

(i) $\mathcal{Q}_{k,1} = \mathcal{S}_k$;
(ii) $\mathcal{C}\backslash\mathcal{S}_k = \bigsqcup_{N=2}^{\infty} \mathcal{Q}_{k,N}$;
(iii) $\mathcal{S} = \bigcap_{k=1}^{\infty} \mathcal{S}_k$;
(iv) $\mathcal{Q}_{k,N} = \bigcup_{l\leq k} \mathcal{U}_{l,N}$.

Theorem 1.1 follows from items (ii) and (iv) above once we show that, for each $k \geq 1$ and $N \geq 2$, $\mathcal{U}_{k,N}$ is a submanifold of $\mathcal{C}$ of codimension at least 1. The proof of this last claim is broken into the following few lemmas.

LEMMA 2.5. *The space $\mathcal{U}_{k,N}$ is locally modeled around g by*

$$\{\psi \in C_0^\infty(X,g) \mid \exists c \in \mathbb{R},\ \forall f \in V_{\lambda_k}\ ((\Delta^g - 2\lambda)\psi, f^2)_g = c\|f\|^2\}.$$

In particular, it is a submanifold of $\mathcal{M}$.

PROOF. Let g_t be a path in $\mathcal{U}_{k,N}$ starting at g and set $\lambda = \lambda_k$. This path of metrics remains in $\mathcal{U}_{k,N}$ if and only if $\dim V_{\lambda_k(g_t)} = N$. Thanks to Lemma 2.1, this happens if and only if for t and $\delta > 0$ small enough, $V_{\lambda_k(g_t)} = V_t^\lambda(\delta)$.

We claim there exists $\varepsilon > 0$ such that $\dim V_{\lambda_k(t)} = N$ for each $t \in (0,\varepsilon)$ if and only if

$$\left(\frac{d}{dt} R^{g_t}(f)\right)_{|t=0}$$

remains independent of f for all $f \in V_{\lambda_k}$. To prove this using the notation of Lemma 2.1, we note that for ε small enough, there is a path of isomorphisms $\Phi_t : V_t^\lambda(\delta) \to V_{\lambda_k(g_0)}$. We define real-valued functions

$$F(t) = R^{g_t}\big(\Phi_t^{-1}(f)\big), \quad \text{and} \quad H(t) = R^{g_t}\big(\Phi_t^{-1}(h)\big)$$

for $f, h \in V_{\lambda_k}$. We know that the multiplicity of $\lambda_k(g_t)$ cannot be greater than N for small t and, if it is equal to N then $V_{\lambda_k(g_t)} = V_\lambda(\delta)$ and thus $F(t) = H(t)$. And the claim follows.

Now by using formula (2) it follows that we can identify a neighborhood of g in $\mathcal{U}_{k,N}$ with the set

$$\{\varphi \in C_0^\infty(X,g) \mid \exists c \in \mathbb{R},\ \forall f \in V_{\lambda_k}\ ((\Delta^g - 2\lambda)\psi, f^2)_g = c\|f\|^2,\ \}.$$

□

Assume that $\lambda = \lambda_k$ is non-simple eigenvalue of Δ^g. Let ψ_o be a function with zero mean value, for which there exists $c_o \in \mathbb{R}$ satisfying $((\Delta^g - 2\lambda)\psi_o, f^2)_g = c_o\|f\|^2$, $\forall f \in V_\lambda$. The constant function equal to 0 is always a solution with $c_o = 0$ but we assume that, if a non trivial solution exists with $c_o \neq 0$, then ψ_o is such a solution, otherwise let ψ_o be the zero function. We have

$$\begin{aligned} T_g\mathcal{U}_{k,N} &= \{\psi \,|\, ((\Delta^g - 2\lambda)\psi, f^2)_g = c\|f\|^2, \forall\, f \in V_\lambda\} \\ &= \mathbb{R}\psi_o \oplus \{\psi \,|\, ((\Delta^g - 2\lambda)\psi, f^2)_g = 0,\ \forall f \in V_\lambda\}. \end{aligned} \tag{3}$$

We would like to find the codimension of $\mathcal{U}_{k,N}$ in $T_g\mathcal{C} \simeq C_0^\infty(X)$. Set $F_\lambda = \operatorname{span}\{2\lambda f^2 - \Delta f^2 \,|\, f \in V_\lambda\}$, so

$$\begin{aligned} T_g\mathcal{U}_{k,N} &= \mathbb{R}\psi_o \oplus \{\psi \,|\, ((\Delta^g - 2\lambda)\psi, f^2)_g = 0\ \forall f \in V_\lambda\} \\ &= \mathbb{R}\psi_o \oplus (\mathbb{R}1 \oplus F_\lambda)^\perp \end{aligned} \tag{4}$$

Therefore, the orthogonal complement of $\mathcal{U}_{k,N}$ in $T_g\mathcal{C} \simeq C_0^\infty(X)$ is

$$(\mathbb{R}\psi_o)^\perp \cap (\mathbb{R}1 \oplus F_\lambda) \cap C_0^\infty(X)$$

which is just $(\mathbb{R}1 \oplus F_\lambda) \cap C_0^\infty(X)$ whenever $\psi_o = 0$. Otherwise, an orthogonal complement of $\mathcal{U}_{k,N}$ in $T_g\mathcal{C} \simeq C_0^\infty(X)$ is given by the space

$$F_\lambda \cap C_0^\infty(X) = \left\{ \sum_{f\in V_\lambda} a_f(2\lambda f^2 - \Delta f^2) \,\middle|\, \sum_{f\in V_\lambda} a_f \int_X f^2 \mathbf{v} = 0 \right\}.$$

In any case, we get that an orthogonal complement of $\mathcal{U}_{k,N}$ in $T_g\mathcal{C}$ contains the space $F_\lambda \cap C_0^\infty(X)$.

We denote $G_\lambda = \operatorname{span}\{f^2 \,|\, f \in V_\lambda\}$.

LEMMA 2.6. *The dimension of the space $F_\lambda \cap C_0^\infty(X)$ coincides with the codimension of $V_{2\lambda} \cap G_\lambda$ in $G_\lambda \cap C_0(X)$.*

PROOF. The map $(2\lambda - \Delta) : G_\lambda \cap C_0(X) \to F_\lambda \cap C_0^\infty(X)$ is surjective and its kernel coincides with $V_{2\lambda} \cap G_\lambda$. □

LEMMA 2.7. *Let $f, h \in V_\lambda$ be two orthonormal eigenfunctions. The dimension of the space $\operatorname{span}_{\mathbb{R}}\{f^2, h^2, fh\} \cap V_{2\lambda}$ is strictly smaller than 2.*

PROOF. Since $V_{2\lambda} \subset C_0(X)$, the dimension of $\operatorname{span}_{\mathbb{R}}\{f^2, h^2, fh\} \cap V_{2\lambda}$ is strictly smaller than 3 and if $a, b, \in \mathbb{R}$ are such $af^2 + bh^2 + cfh \in V_{2\lambda}$ then $a = -b$. Thus, if there exist 2 linearly independent elements in $\operatorname{span}_{\mathbb{R}}\{f^2, h^2, fh\} \cap V_{2\lambda}$, both $f^2 - h^2$ and fh lie in it. Equivalently, using the formula $\Delta fh = 2\lambda fh - 2g(\nabla f, \nabla h)$, both relations $g(\nabla f, \nabla h) \equiv 0$ and $|\nabla f|^2 - |\nabla h|^2 \equiv 0$ hold which is impossible due to Lemma 2.8. □

LEMMA 2.8. *On a compact Riemannian manifold (X, g), there are no non-constant functions f, $h \in C^\infty(X)$ such that $g(\nabla f, \nabla h) \equiv 0$ and $|\nabla f|^2 - |\nabla h|^2 \equiv 0$.*

PROOF. The first condition implies that h is constant along the gradient flow of f while the second condition implies they have common critical points. Each point p of X is either a common critical point of f and h or is sent by the gradient flow of f to another a critical point of f, say q. Thus, $h(p) = h(q)$ is also critical value of h. This contradicts Sard's Theorem unless h is constant. □

Lemma 2.7 implies there is a 2–dimensional plane lying in G_λ which meets $V_{2\lambda}$ only at 0. Intersecting with $C_0(X)$ and mapping to $F_\lambda \cap C_0^\infty(X)$ we get that $\dim(F_\lambda \cap C_0^\infty(X)) \geq 1$. Therefore, the set of non-simple metrics in $\mathcal{C}$ is a countable union of sets codimension at least 1.

REMARK 2.9. *There exist Riemannian manifolds for which there are two linearly independent eigenfunctions of the same eigenvalue and satisfying* $g(\nabla f, \nabla h) \equiv 0$. *Indeed, one can think of product of a Riemannian manifolds with itself, the flat 2–torus is such an example.*

3. The Kähler class

3.1. A variational formula for the Rayleigh quotient in the Kähler setting. We consider a family of Kähler metrics g_t on X parameterized by $t \in (-\varepsilon, \varepsilon)$ for some positive constant ε. Assume that these are all in the same cohomology class. It follows from the $\partial\bar{\partial}$ lemma that the associated Kähler forms are then given by $\omega_t = \omega + dd^c\varphi_t$. For each t we have a Laplace operator and we consider a family of eigenfunctions for the eigenvalues λ_t. It is know (see [**3**]) that λ_t is a continuous function of t. As before we set $g = g_0$ and $\lambda = \lambda_0$ and we define $V_t^\lambda(\delta) = \mathrm{span}_{\mathbb{R}}\{f \in C^\infty(X) \,|\, \exists \mu \in (\lambda - \delta, \lambda + \delta),\ \Delta^{g_t} f = \mu f\}$. From lemma 2.1 there is an isomorphism $\Phi_t : V_t^\lambda(\delta) \to V_\lambda$ which depends smoothly on t. For the proof, see Theorems XII.12 and XII.13 in [**3**].

We would like to obtain a formula for the variation of the Rayleigh quotient $R^{g_t}(f_t)$. We denote $\Omega^{1,1}X$ the space of $(1,1)$–form, an element $\alpha \in \Omega^{1,1}X$ is a real two form $\alpha \in \Lambda^2 X$ such that at each point $p \in X$, $\alpha_p(Ju, Jv) = \alpha(u, v)$ for all vectors $u, v \in T_pX$.

LEMMA 3.1. *Let* $f_t \in C^\infty(X, \mathbb{R})$ *be a smooth path of functions such that* $f_0 = f$ *satisfies* $\Delta^g f = \lambda f$ *and* $\|f\|_g^2 = 1$. *Then, denoting the function* $\dot{\varphi} = (\partial_t \varphi_t)_{t=0}$,

$$\left(\frac{d}{dt} R^{g_t}(f_t)\right)_{|t=0} = \int_X (dd^c\dot{\varphi}, f dd^c f)\mathbf{v} \tag{5}$$

In particular, the variation of $R^{g_t}(f_t)$ *at* $t = 0$ *does not depend on the path* f_t *when* $f = f_0$ *is an eigenfunction of the Laplacian.*

PROOF. Recall that the volume form associated to a Kähler metric is $\mathbf{v}_{g_t} = \omega_t^n / n!$ and, a well-known formula, see for example [**4**], is

$$\frac{d}{dt}\mathbf{v}_{g_t} = \sum_{i=1}^{n} \dot{\omega} \wedge \frac{\omega^{n-1}}{n!} = (dd^c\dot{\varphi}) \wedge \frac{\omega^{n-1}}{(n-1)!} = -\Delta^{g_t}(\dot{\varphi})\mathbf{v}_{g_t} \tag{6}$$

for the last equality see [**2**, page 31].

REMARK 3.2. *We take the convention that the Laplace operator (acting on forms) is* $\Delta^g = \delta^g d + d\delta^g$ *where* $\delta = \delta^g$ *is the adjoint of* d *with respect the global inner product on forms. Recall that, with respect to this convention, a smooth function* f *is an eigenfunction of* Δ^g, *for the eigenvalue* $\lambda \geq 0$, *if* $\Delta^g f = \lambda f$. *Moreover, for any smooth functions* f *and* p*–form* α, *we have* $\delta f\alpha = f\delta\alpha - \nabla f \lrcorner\, \alpha$ *and, in particular,* $\Delta^g(fh) = f\Delta^g h + h\Delta^g f - 2g(\nabla f, \nabla h)$.

Observe that $(-\partial_t g_t)(\nabla^{g_t} f, \nabla^{g_t} f) = g_t((\partial_t \nabla^{g_t})f, \nabla^{g_t} f)$ since, in coordinates, if $g = \sum_{i,j} g_{ij} dx_i \otimes dx_j$, $\nabla^g f = \sum_{i,j} g^{ij} \frac{\partial f}{\partial x_i}\frac{\partial}{\partial x_j}$ where $(g^{ij}) = (g_{ij})^{-1}$. Hence, using (6) for any smooth path of functions $f_t \in C^\infty(X)$, we have

$$\begin{aligned}\frac{d}{dt}R^{g_t}[f_t] =& \frac{1}{\|f_t\|^2_{g_t}}\int_X ((-\partial_t g_t)(\nabla f_t, \nabla f_t) - g_t(\nabla f_t, \nabla f_t)\Delta^{g_t}\dot\varphi)\mathbf{v}_t \\ &+ \frac{1}{\|f_t\|^2_{g_t}}\int_X 2g_t(\nabla\partial_t f_t, \nabla f_t)\mathbf{v}_t - \frac{\|df_t\|^2_{g_t}}{\|f_t\|^4_{g_t}}\int_X 2f_t\partial_t\bar f_t\mathbf{v}_t \\ &+ \frac{\|df_t\|^2_{g_t}}{\|f_t\|^4_{g_t}}\int_X f_t^2\Delta^{g_t}\dot\varphi\mathbf{v}_t.\end{aligned} \tag{7}$$

At $t = 0$, since f is an eigenfunction, $R^g[f] = \lambda$ and the second line of (7) vanishes. Thus, the variation (7) at $t = 0$ is independent of the path f_t and reads

$$\begin{aligned}\left(\frac{d}{dt}R^{g_t}[f_t]\right)_{t=0} &= \frac{1}{\|f\|^2_g}\int_X (-\partial_t g_t)_{t=0}(\nabla f, \nabla f)\mathbf{v} - g(\nabla f, \nabla f)\Delta^g\dot\varphi\mathbf{v}_g \\ &+ \frac{\lambda}{\|f\|^2_g}\int_X f^2\Delta^g\dot\varphi\mathbf{v}.\end{aligned} \tag{8}$$

Now, since J does not depend on t, we have

$$\begin{aligned}(-\partial_t g_t)(\nabla f, \nabla f) &= (-\partial_t\omega_t)(\nabla f, J\nabla f) \\ &= -dd^c\dot\varphi(\nabla f, J\nabla f) \\ &= -g_t(dd^c\dot\varphi, df \wedge d^c f)\end{aligned} \tag{9}$$

(recall that $g(\alpha_1 \wedge \cdots \wedge \alpha_k, \beta_1 \wedge \cdots \wedge \beta_k) = \det(g(\alpha_i, \beta_j)))$. We get

$$\left(\frac{d}{dt}R^{g_t}[f_t]\right)_{t=0} = \frac{1}{\|f\|^2_g}\int_M (-(dd^c\dot\varphi, df \wedge d^c f) - |df|^2\Delta^g\dot\varphi + \lambda f^2\Delta^g\dot\varphi)\mathbf{v}. \tag{10}$$

Writing $\lambda f^2 = \frac{1}{2}\Delta^g(f^2) + g(\nabla f, \nabla f)$, we have

$$\begin{aligned}\left(\frac{d}{dt}R^{g_t}[f_t]\right)_{t=0} &= \frac{1}{\|f\|^2_g}\int_X (-(dd^c\dot\varphi, df \wedge d^c f) + \frac{1}{2}\Delta^g(f^2)\Delta^g\dot\varphi)\mathbf{v} \\ &= \frac{1}{\|f\|^2_g}\int_X (-(dd^c\dot\varphi, \frac{1}{2}dd^c(f^2) - f dd^c f) + \frac{1}{2}\Delta^g(f^2)\Delta^g\dot\varphi)\mathbf{v} \\ &= \frac{1}{\|f\|^2_g}\int_X ((dd^c\dot\varphi, f dd^c f) + \frac{1}{2}(-(dd^c\dot\varphi, dd^c(f^2)) + \Delta^g(f^2)\Delta^g\dot\varphi))\mathbf{v}.\end{aligned}$$

The last two terms of this equation cancel each other since, using the Kähler identities $dd^c = -d^c d$, $d\delta^c = -\delta^c d$ and $\delta^c d^c = \delta d$, see for example [**2**], we get

$$\int_X (dd^c\dot\varphi, dd^c(f^2))\mathbf{v} = \int_X (d^c\Delta\dot\varphi, d^c(f^2))\mathbf{v} = \int_X \Delta^g(f^2)\Delta^g\dot\varphi)\mathbf{v}.$$

□

3.2. The structure of the space of bad metrics. As explained in the introduction, we fix the complex structure J and the cohomology class Ω of the Kähler structure and consider the space $\mathcal{M} = \mathcal{M}_\Omega$ of Kähler structure (g, ω, J) on X with $\omega \in \Omega$.

The space $\mathcal{M}$ has the structure of a Fréchet space since the neighborhood of any point $(g,\omega) \in \mathcal{M}$ is parameterized by an open subset of

$$C_0^\infty(X,g) := \left\{\varphi \in C^\infty(X) \;\middle|\; \int_X \varphi\omega^n = 0\right\}$$

via $\omega + dd^c\varphi$. More precisely, the condition that $\omega + dd^c\varphi$ is non-degenerate is fulfilled for at least a small multiple of φ since X is compact. Thus, the map $\varphi \mapsto \omega + dd^c\varphi$ is well-defined on a small neighborhood of $0 \in C_0^\infty(X,g)$. The fact that this map is surjective is due to the dd^c–lemma. The fact that it is injective comes from the fact that the Laplacian Δ_g is injective on $C_0^\infty(X,g)$ and $g(\omega, dd^c\varphi) = -\Delta\varphi$. This gives a natural identification $T_g\mathcal{M} \simeq C_0^\infty(X,g)$ for which the exponential is explicit. Moreover, the Green operator can be thought as an inverse to this map locally defined in a neighborhood of a metric g, providing a Fréchet manifold chart around g. To learn more about this space, see for example [**2**, **4**].

We use the same notation as in the previous section (i.e the conformal class) namely

- We denote the subset of simple metrics in $\mathcal{M}$ by $\mathcal{S}$, for $k \geq 1$.
- We denote $\mathcal{S}_k = \{(g,\omega) \in \mathcal{M} \mid \lambda_l \text{ simple if } l \leq k\}$, for $N \geq 1$.
- We define $\mathcal{U}_{k,N} = \{(g,\omega) \in \mathcal{M} \mid \dim V_{\lambda_k} = N\}$.

These spaces depend on the cohomology class Ω but we assume it is fixed throughout and do not make the dependence explicit in the notation for the sake of simplicity.

LEMMA 3.3. *Via the standard diffeomorphism $T_g\mathcal{M} \simeq C_0^\infty(X,g)$, the tangent space of $\mathcal{U}_{k,N}$ at g is identified with*

$$\left\{\psi \in C_0^\infty(X,g) \;\middle|\; \exists c \in \mathbb{R},\ \int_X (dd^c\psi, f dd^c f)\mathbf{v} = c\|f\|^2,\ \ \forall f \in V_{\lambda_k}\right\}.$$

In particular, $\mathcal{U}_{k,N}$ is a submanifold of $\mathcal{M}$.

PROOF. As in the proof of 2.5, a path of Kähler metrics (g_t, ω_t) starting at (g,ω) remains in $\mathcal{U}_{k,N}$ if and only if

$$\left(\frac{d}{dt}R^{g_t}(f)\right)_{|t=0}$$

remains independent of f for all $f \in V_{\lambda_k}$.

Now, using Lemma 3.1, for a given path of metrics $(g_t, \omega_t = \omega_0 + dd^c\phi_t)$,

$$\left(\frac{d}{dt}R^{g_t}(f)\right)_{|t=0}$$

remains independent of the choice $f \in V_{\lambda_k}$ if and only if there exists $c \in \mathbb{R}$ such that

$$\int_X (dd^c\dot{\phi}_0, f dd^c f)\mathbf{v} = c\|f\|^2, \qquad \forall f \in V_{\lambda_k}.$$

This concludes the proof. □

Assume that $\lambda = \lambda_k(g)$ is a (non-simple) eigenvalue of multiplicity N. The metric is fixed and we write $C_0^\infty(X) = C_0^\infty(X,g)$.

The variational formula (5) can be rewritten as

$$\left(\frac{d}{dt}R^{g_t}(f_t)\right)_{|t=0} = \int_X \dot{\varphi}\,\delta^c\delta(f dd^c f)\mathbf{v}.$$

Define $E_\lambda = \operatorname{span}\{\delta^c\delta(fdd^cf) \mid f \in V_\lambda\} \subset C_0^\infty(X)$. We will compute the dimension of E_λ. Let ψ_o be a function with zero mean value, for which there exists $c_o \in \mathbb{R}$ satisfying $\int_X (dd^c\psi_o, dd^cf)_g \mathbf{v}_g = c_o\|f\|^2$, $\forall f \in V_\lambda$. The constant function equal to 0 is always a solution with $c_o = 0$ but we assume that, if a non trivial solution exists with $c_o \neq 0$, then ψ_o is such a solution. Otherwise we set ψ_o to be the zero function. Thanks to Lemma 2.5, we have

$$\begin{aligned} T_g\mathcal{U}_{k,N} &\simeq \left\{\psi \in C_0^\infty(X) \,\middle|\, \exists c \in \mathbb{R},\ \forall f \in V_\lambda \int_X (dd^c\psi, fdd^cf)\mathbf{v} = c\|f\|^2\right\} \\ &\simeq \mathbb{R}\psi_o \oplus (\mathbb{R}1 \oplus E_\lambda)^\perp, \end{aligned}$$

where $\mathbb{R}\psi_o$ and $\mathbb{R}1$ denote respectively the real span of ψ_o and 1 in $\mathcal{C}^\infty(X)$ and $\perp$ refers to the orthogonal complement in $\mathcal{C}^\infty(X)$ with respect to the global inner product induced by the Kähler metric g. Therefore, since $E_\lambda \subset \mathcal{C}_0^\infty(X)$,

$$(T_g\mathcal{U}_{k,N})^\perp \simeq (\mathbb{R}\psi_o)^\perp \cap (\mathbb{R}1 \oplus E_\lambda) \cap \mathcal{C}_0^\infty(X) = (\mathbb{R}\psi_o)^\perp \cap E_\lambda. \tag{11}$$

If $\psi_o \equiv 0$, this means $(T_g\mathcal{U}_{k,N})^\perp \simeq E_\lambda$. Suppose ψ_o is not the zero function that is there exist $c_0 \in \mathbb{R}\backslash\{0\}$ and $\psi_o \in C_0^\infty(X)$ such that $\int_X (dd^c\psi_o, fdd^cf)\mathbf{v} = c_0\|f\|^2$, $\forall f \in V_\lambda$. A general element of E_λ, say $\sum_{f\in V_\lambda} a_f\delta^c\delta(fdd^cf)$ lies in $(\mathbb{R}\psi_o)^\perp$, if and only if

$$\sum_{f\in V_\lambda} a_f \int_X f^2\mathbf{v} = \frac{1}{c_o}\int_X \left(\sum_{f\in V_{\lambda_k}} a_f\delta^c\delta(fdd^cf)\right)\psi_o\mathbf{v} = 0.$$

Thus, if ψ_o does not vanish identically, the orthogonal complement of $T_g\mathcal{U}_{k,N}$ is the the image of the following subspace of $(1,1)$–forms

$$W_\lambda = \operatorname{span}\left\{\sum_{f\in V_\lambda} a_f fdd^cf \,\middle|\, \sum_{f\in V_\lambda} a_f \int f^2 = 0\right\}$$

by the map $\delta^c\delta : \Omega^{1,1} \to C_0^\infty(X)$. We have just proved the following lemma.

LEMMA 3.4. *For any $g \in \mathcal{U}_{k,N}$, the orthogonal complement of $T_g\mathcal{U}_{k,N}$ in $C_0^\infty(X)$ contains the image of W_λ under the map $\delta^c\delta$.*

We are now in a position to prove our main Theorem 1.2.

PROOF. Assume that $\dim V_\lambda \geq 2$. Any pair of orthonormal functions $f, h \in V_\lambda$ define an element of W_λ namely $fdd^cf - hdd^ch$ because $\int f^2 - \int h^2 = 0$. Now if $\{f,h\}$ is an orthonormal set in V_λ so is $\{\frac{f+h}{\sqrt{2}}, \frac{f-h}{\sqrt{2}}\}$ so that another element in W_λ is

$$\frac{1}{2}[(f+h)dd^c(f+h) - (f-h)dd^c(f-h)]$$

i.e. $fdd^ch + hdd^cf \in W_\lambda$. In particular, Lemma 3.4 tells us that the span of $\delta\delta^c(fdd^ch + hdd^cf)$ and $\delta\delta^c(fdd^cf - hdd^ch)$ lies in the orthogonal complement of $T_g\mathcal{U}_{k,N}$ in $C_0^\infty(X)$. This concludes the proof of Theorem 1.2. □

Acknowledgments

The first author was partially supported by an NSF grant DMS-1005696 and the third author was partially supported by an FCT grant PTDC/MAT/098936/2008. The second and the third authors would like to thank the Mathematics Department at MIT for its warm hospitality.

References

[1] D. D. Bleecker and L. C. Wilson, *Splitting the spectrum of a Riemannian manifold*, SIAM J. Math. Anal. **11** (1980), no. 5, 813–818, DOI 10.1137/0511072. MR586909 (81j:58077)

[2] P. Gauduchon, *Calabi's extremal Kähler metrics: An elementary introduction*, note.

[3] M. Reed and B. Simon, *Methods of modern mathematical physics. IV. Analysis of operators*, Academic Press [Harcourt Brace Jovanovich, Publishers], New York-London, 1978. MR0493421 (58 #12429c)

[4] G. Tian, *Canonical metrics in Kähler geometry*, Lectures in Mathematics ETH Zürich, Birkhäuser Verlag, Basel, 2000. Notes taken by Meike Akveld. MR1787650 (2001j:32024)

[5] M. Teytel, *How rare are multiple eigenvalues?*, Comm. Pure Appl. Math. **52** (1999), no. 8, 917–934, DOI 10.1002/(SICI)1097-0312(199908)52:8⟨917::AID-CPA1⟩3.3.CO;2-J. MR1686977 (2000b:47065)

[6] K. Uhlenbeck, *Generic properties of eigenfunctions*, Amer. J. Math. **98** (1976), no. 4, 1059–1078. MR0464332 (57 #4264)

Department of Mathematics, Massachusetts Institute of Technology, Cambridge, Massachusetts 02139

E-mail address: `vwg@math.mit.edu`

Université Paul Sabatier, Institut Mathématique de Toulouse, 118 route de Narbonne, Toulouse, France

E-mail address: `eveline.legendre@math.univ-toulouse.fr`

Centro de Análise Matemática, Geometria e Sistemas Dinâmicos, Departamento de Matemática, Instituto Superior Técnico, Av. Rovisco Pais, 1049-001 Lisboa, Portugal

E-mail address: `rsenadias@math.ist.utl.pt`

Contemporary Mathematics
Volume **630**, 2014
http://dx.doi.org/10.1090/conm/630/12665

Smooth and Singular Kähler–Einstein Metrics

Yanir A. Rubinstein

Dedicated to Eugenio Calabi on the occassion of his 90th birthday

ABSTRACT. Smooth Kähler–Einstein metrics have been studied for the past 80 years. More recently, singular Kähler–Einstein metrics have emerged as objects of intrinsic interest, both in differential and algebraic geometry, as well as a powerful tool in better understanding their smooth counterparts. This article is mostly a survey of some of these developments.

CONTENTS

1. Introduction

The Kähler–Einstein (KE) equation is among the oldest fully nonlinear equations in modern geometry. A wide array of tools have been developed or applied towards its understanding, ranging from Riemannian geometry, PDE, pluripotential theory, several complex variables, microlocal analysis, algebraic geometry, probability, convex analysis, and more. The interested reader is referred to the numerous existing surveys on related topics [**10**, **28**, **33**, **34**, **40**, **41**, **62**, **99**, **115**–**120**, **125**, **199**, **200**, **227**, **235**, **237**, **246**–**248**].

In this article, which is largely a survey, we do not attempt a comprehensive overview, but instead focus on a rather subjective bird's eye view of the subject. Mainly, we aim to survey some recent developments on the study of *singular* Kähler–Einstein metrics, more specifically, of *Kähler–Einstein edge (KEE)* metrics, and while doing so to provide a unified introduction also to smooth Kähler–Einstein metrics. In particular, the new tools that are needed to construct KEE metrics give a new perspective on construction of smooth KE metrics, and so it seemed worthwhile to use this opportunity to survey the existence and non-existence theory of both smooth and singular KE metrics. We also put an emphasis on understanding concrete new geometries that can be constructed using the general theorems, concentrating on the complex 2-dimensional picture, and touch on some relations to algebraic geometry and non-compact Calabi–Yau spaces.

As noted above, most of the material in this article is a survey of existing results and techniques, although, of course, sometimes the presentation may be different than in the original sources. For the reader's convenience, let us also point out the sections that contain results that were not published elsewhere. First, some of the treatment of energy functionals in §5.3 and §5.6 extends some previous results of the author from the smooth setting to the edge setting, although the generalization is quite straightforward and is presented here for the sake of unity. The section on Bott–Chern forms §5.4 is essentially taken from the author's thesis, again with minor modifications to the singular setting. Second, some of the treatment of the a priori estimates in §7 is new. In particular, the reverse Chern–Lu inequality introduced in §7.5 is new, as is the proof in §7.6 that the classical Aubin–Yau Laplacian estimate follows from it. Finally, we note the minor observation made in §7.7 that some of the classical Laplacian estimates can be phrased also for Kähler metrics that need not satisfy a complex Monge–Ampère equation. Third, Proposition 8.14 is a very slight extension of the original result of Di Cerbo.

Many worthy vistas are not surveyed here, including: toric geometry; quantization; log canonical thresholds, Tian's α-invariant, and Nadel multiplier ideal sheaves [**62**, **189**, **238**]; noncompact Kähler–Einstein metrics; Berman's probabilistic approach to the KE problem [**19**]; Kähler–Einstein metrics on singular varieties, as, e.g., in [**24**, **108**]; recent work on algebraic obstruction to deforming singular KE metrics to smooth ones [**63**, **249**]; the GIT related aspects of the Kähler–Einstein problem, that we do not discuss in any detail in this article, instead referring to the survey of Thomas [**237**] and the articles of Paul [**194**, **195**].

Organization. Historically, the Kähler–Einstein equation was first phrased locally as a complex Monge–Ampère equation. This, and the corresponding global formulation, are discussed in Section 2. Section 3 is a condensed introduction to Kähler edge geometry, describing aspects of the theory that are absent from the study of smooth Kähler manifolds: new function spaces, a different notion of smoothness (polyhomogeneity), a theory of (partial) elliptic regularity, and new features of the reference geometry (e.g., unbounded curvature). We use these tools to describe the structure of the Green kernel of Kähler edge metrics, and the proof of higher regularity of KEE metrics [**141**]. Section 4 summarizes the main existence and non-existence results on KE(E) metrics. First, it describes the classical obstructions due to Futaki and Mastushima coming from the Lie algebra of holomorphic vector fields (and their edge counterparts), and their more recent generalization to various notions of "degenerations", that capture more of the complexity of Mabuchi K-energy, the functional underlying the KE problem. Second, it states the main existence and regularity theorems for KE(E) metrics. The strongest form of these results appears in [**141**, **175**], generalizing the classical results of Aubin, Tian, and Yau from the smooth setting, and those of Troyanov from the conical Riemann surface setting, to the edge setting. We also describe other approaches to this problem. The main objective of §6–§7 is to describe the analytic tools needed to carry out the proof of these theorems, in a unified manner, both in the smooth and the edge settings, and regardless of the sign of the curvature. We describe this in more detail below; before that, however, Section 5 reviews the variational theory underlying the Monge–Ampère equation in this setting. In particular, it reviews the basic properties of the Mabuchi K-energy and the functionals introduced by Aubin, Mabuchi, and others. The alternative definition of these via Bott–Chern forms is described in detail. A relation to the Legendre transform due to Berman is also described. Finally, we describe the properness and coercivity properties of these functionals. The former is needed for the actual statement of the existence theorem in the positive case from §4.

Section 6 describes a new approach, the *Ricci continuity method*, developed by Mazzeo and the author in [**141**] to prove existence of KE(E) metrics in a unified manner, and with only one-sided curvature bounds. This approach works in a unified manner for all cases (negative, zero, and positive Einstein constant) and for both smooth and singular metrics. Section 7 describes the a priori estimates needed to carry out the Ricci continuity method. In doing so, this section also provides a unified reference for a priori estimates for a wide class of singular Monge–Ampère equations. Section 8 describes work of Cheltsov and the author [**60**] on classification problems in algebraic geometry related to the KEE problem. Here, the notion of *asymptotically log Fano varieties* is introduced. This is a class of varieties much larger than Fano varieties, but where we believe there is still hope of classification and a complete picture of existence and non-existence of KEE metrics. Section 9 then builds on these classification results to phrase a logarithmic version of Calabi's conjecture. We then briefly describe a program to relate this conjecture concerning KEE metrics to Calabi–Yau fibrations and global log canonical thresholds. Finally, some progress towards this conjecture is described, in particular giving new KEE metrics on some explicit pairs, and proving non-existence on others.

2. The Kähler–Einstein equation

A Kähler manifold is a complex manifold (M, J) equipped with a closed positive 2-form ω that is J-invariant, namely $\omega(J\,\cdot\,, J\,\cdot\,) = \omega(\,\cdot\,,\,\cdot\,)$. Here M is a differentiable manifold, and J is a complex structure on M, namely an endomorphism of the tangent bundle TM satisfying $J^2 = -I$ and $[T^{1,0}M, T^{1,0}M] \subseteq T^{1,0}M$, where $TM \otimes_{\mathbb{R}} \mathbb{C} = T^{1,0}M \oplus T^{0,1}M$, with $T^{1,0}M$ the $\sqrt{-1}$-eigenspace of J, and $T^{0,1}M$ the $-\sqrt{-1}$-eigenspace of J. Associated to (M, J, ω) is a Riemannian metric g, defined by $g(\,\cdot\,,\,\cdot\,) = \omega(\,\cdot\,, J\,\cdot\,)$, whose Levi-Civita connection we denote by ∇. Equivalently, one may also define (M, J, g) to be a Kähler manifold when g is J-invariant, $J^2 = -I$, and $\nabla J = 0$ (i.e., g-parallel transport preserves $T^{1,0}M$).

Schouten called this new type of geometry "unitary," [**220**] and this can be understood from the fact that not only does one obtain a reduction of the structure group to $U(n)$ (this merely characterizes almost-Hermitian manifolds) but also the holonomy is reduced from $O(2n)$ to $U(n)$. In retrospect this name seems quite fitting, on a somewhat similar footing to "symplectic geometry," a name first suggested by Ehresmann. However, only a few authors in the 1940's and 1950's were familiar with Schouten-van Dantzig's work. The first few accounts of Hermitian geometry, mainly by Bochner, Eckmann and Guckenheimer referred only to Kähler's works, and the name "Kähler geometry" became rooted (for more on this topic, see [**216**, §2.1.4]).

One of Schouten–van Dantzig's and Kähler's discoveries was that on the eponymous manifold the Einstein equation is implied, locally, by the complex Monge–Ampère equation

$$\det[u_{i\bar{j}}] := \det\left[\frac{\partial^2 u}{\partial z^i \partial \bar{z}^j}\right] = e^{-\mu u}, \quad \text{on } U, \tag{2.1}$$

where $\mu \in \mathbb{R}$ is the Einstein constant [**144, 220**]. Here u is a plurisubharmonic function defined on some coordinate chart $U \subset M$, with holomorphic coordinates $z_1, \ldots, z_n : U \to \mathbb{C}^n$; Indeed, if we let $g_{\mathrm{H}}|_U = \sqrt{-1}u_{i\bar{j}}dz^i \otimes \overline{dz^j}$ denote the Hermitian metric associated to u (with summation over repeated indices), then the Riemannian metric $g = 2\mathrm{Re}\, g_{\mathrm{H}}$ is Einstein if (2.1) holds because Ric $g = 2\mathrm{Im}((\log\det[u_{i\bar{j}}])_{k\bar{l}}dz^k \otimes \overline{dz^l})$. Thus, in the "unitary gauge", the Einstein system of equations reduces to a single, albeit fully nonlinear, equation.

For some time though, it was not clear how to patch up these equations globally. The crucial observation is that both sides of (2.1) are the local expressions of global Hermitian metrics on two different line bundles: the anticanonical line bundle $\Lambda^n T^{1,0}M$ on the left hand side, and the line bundle associated to $\mu[\omega]$ on the right. In particular this implies that these line bundles must be isomorphic, and

$$2\pi c_1(M, J) - \mu[\omega] = 0. \tag{2.2}$$

This of course puts a serious cohomological restriction on the problem. But assuming this, the local Monge–Ampère equation can be converted to a global one. Let $\mathbf{dz} := dz^1 \wedge \cdots \wedge dz^n$. The expressions

$$\sqrt{-1}^{n^2}\det[u_{i\bar{j}}]\mathbf{dz} \wedge \overline{\mathbf{dz}} = e^{-\mu u}\sqrt{-1}^{n^2}\mathbf{dz} \wedge \overline{\mathbf{dz}}$$

represent globally defined volume forms on M, appropriately interpreted. We choose a representative Kähler form ω of $[\omega]$. By definition, $\sqrt{-1}\partial\bar{\partial}u$ is the curvature of e^{-u}, hence it lies in $[\omega]$. On each U, $\sqrt{-1}\partial\bar{\partial}u = \omega + \sqrt{-1}\partial\bar{\partial}\varphi$ for a globally defined $\varphi \in C^\infty(M)$: this follows from the Hodge identities [**124**, p. 111] by setting $\varphi = \mathrm{tr}_\omega G_\omega(\sqrt{-1}\partial\bar{\partial}u - \omega)$, where G_ω denotes the Green operator associated to the Laplacian (on the exterior algebra).

Thus,

$$(\omega + \sqrt{-1}\partial\bar{\partial}\varphi)^n = \omega^n e^{-\log\det[v_{i\bar{j}}] - \mu v - \mu\varphi}$$

where locally $\omega = \sqrt{-1}v_{i\bar{j}}dz^i \wedge \overline{dz^j}$. But by (2.2) $e^{-\log\det[v_{i\bar{j}}] - \mu v}$ is the quotient of two Hermitian metrics on the same bundle, hence a globally defined positive function that we denote by e^{f_ω}. We thus obtain the Kähler–Einstein equation,

$$\omega_\varphi^n = \omega^n e^{f_\omega - \mu\varphi}, \quad \text{on } M \tag{2.3}$$

for a global smooth function φ (called the Kähler potential of ω_φ relative to ω). The function f_ω, in turn, is given in terms of the reference geometry and is thus known. It is called the Ricci potential of ω, and satisfies $\sqrt{-1}\partial\bar{\partial}f_\omega = \text{Ric}\,\omega - \mu\omega$, where it is convenient to require the normalization $\int e^{f_\omega}\omega^n = \int \omega^n$. Equivalently, (2.3) says that $f_{\omega_\varphi} = 0$. Thus, in the language of the introduction to §4, the KE problem has a solution precisely when the vector field $f : \phi \mapsto f_{\omega_\phi}$ on the space of all Kähler potentials has a zero.

3. Kähler edge geometry

This section introduces the basics of Kähler edge geometry. First, we describe some general motivation for introducing these more general geometries in §3.1. The one-dimensional geometry of a cone is the topic of §3.2. It is fundamental, since a Kähler edge manifold looks like a cone transverse to its 'boundary' divisor. Subsection 3.3 phrases the KEE problem as a singular complex Monge–Ampère equation, generalizing the discussion in §2. What is the appropriate smooth structure on a Kähler edge space? This is discussed in §3.4 where the notion of polyhomogeneity is introduced. The edge and wedge scales of Hölder function space are defined in §3.5. Next, the various Hölder domains relevant to the study of the *complex* Monge–Ampère equation are introducted in §3.6.

3.1. A generalization of Kähler geometry. The cohomological obstruction (2.2) is necessary for the local KE geometries to patch up to a global one. However, the condition (2.2) is very restrictive. Is there a way of constructing a KE metric at least on a large subset of a general Kähler manifold? Of course, such a question makes sense and is very interesting also in the Riemannian context. The interpretation of the local KE equation in terms of Hermitian metrics on line bundles, though, distinguishes between these two settings.

Thus, suppose that $c_1(M,J)-\mu[\omega]$ is not zero (nor torsion) but that this difference, or 'excess curvature' can be decomposed as follows: there exist divisors $D_1,\ldots,D_N$ and numbers $\beta_i>0$ such that

$$c_1(M,J)-\frac{\mu}{2\pi}[\omega]=\sum_{i=1}^{N}(1-\beta_i)c_1(L_{D_i}),$$

with L_F denoting the line bundle associated to a divisor F. At least when M is projective, by the Lefschetz theorem on (1,1)-classes [**124**, p. 163] this can always be done when the left hand side belongs to $H^2(M,\mathbb{Z})\hookrightarrow H^2(M,\mathbb{R})$, and therefore also when it is merely a rational class, or a real class. Of course this means that we might need to take a non-reduced divisor on the right hand side, and limits the usefulness of such a generalization to the case when the divisor $D=\sum D_i$ is not too singular, and the β_i are not too large.

Thus, necessarily any KE potential u must satisfy locally,

$$\det[u_{i\bar{j}}]=e^{-\mu u}\prod_{i=1}^{N}|e_i|^{2\beta_i-2},\quad \text{on } U, \tag{3.1}$$

where e_i denotes a local holomorphic section of D_i. This equation is then quite similar to the local KE equation (2.1) on neighborhoods U contained in the complement of the D_i. However, if U intersects any of the D_i nontrivially, the equation becomes singular or degenerate.

To understand this equation better, it is helpful to consider the model case with $N=1$, $M=\mathbb{C}^n$ and $D=\{e_1=z_1=0\}=\{0\}\times\mathbb{C}^{n-1}$. Then, $u=\frac{1}{2}(\frac{1}{\beta^2}|z_1|^{2\beta}+|z_2|^2+\ldots+|z_n|^2)$ is a solution of (3.1) with $\mu=0$. Note that u corresponds to a singular, but continuous, Hermitian metric e^{-u}. We call the associated curvature form the *model edge form on* $(\mathbb{C}^n,\mathbb{C}^{n-1})$,

$$\omega_\beta=-\sqrt{-1}\partial\bar{\partial}\log e^{-u}=\frac{1}{2}\sqrt{-1}\Big(|z_1|^{2\beta-2}dz^1\wedge\overline{dz^1}+\sum_{j=2}^{n}dz^j\wedge\overline{dz^j}\Big), \tag{3.2}$$

and also denote by

$$g_\beta = |z_1|^{2\beta-2}|dz^1|^2 + \sum_{j=2}^{n} |dz^j|^2, \tag{3.3}$$

the *model edge metric on* $(\mathbb{C}^n, \mathbb{C}^{n-1})$. More generally, if $D = D_1 + \ldots + D_N$ is the union of N coordinate hyperplanes in $\mathbb{C}^n$ which intersect simply and normally at the origin, we set $\beta = (\beta_1, \ldots, \beta_N)$ and denote the *model edge form on* $(\mathbb{C}^n, D)$ by

$$\omega_\beta = \frac{1}{2}\sqrt{-1}\Big(\sum_{i=1}^{N} \lambda_i |z_{\epsilon(i)}|^{2\beta_i - 2} dz^{\epsilon(i)} \wedge \overline{dz^{\epsilon(i)}} + \sum_{j\in\{1,\ldots,n\}\setminus\{\epsilon(i)\}_{i=1}^N} dz^j \wedge \overline{dz^j}\Big),$$

where $\lambda_i = 1$ if $0 \in D_i$ and $\lambda_i = 0$ otherwise, and where $\epsilon(i) = 0$ if $\lambda_i = 0$, and otherwise $\epsilon(i) \in \{1, \ldots, n\}$ is such that $\{z_{\epsilon(i)} = 0\} = D_i$.

This model case motivates the following generalization of a Kähler manifold. Some visual references are given in Figure 5 in §4.4 below.

DEFINITION 3.1. *A Kähler edge manifold is a quadruple*

$$(M, D, \beta = (\beta_1, \ldots, \beta_N), \omega),$$

with M a smooth Kähler manifold, $D = D_1 + \ldots D_N \subset M$ a simple normal crossing (snc) divisor, $\beta_i : D_i \to \mathbb{R}_+$ a function for each $i = 1, \ldots, N$, and, finally, ω a Kähler current on M that is smooth on $M \setminus D$ and asymptotically equivalent to $\omega_{\beta(p)}$ near each point p of D.

For the notion of a Kähler current (also called a positive $(1,1)$ current) we refer to [**124**, Chapter 3]. One could make the definition more general, e.g., by allowing D to be more singular, but we do not explore that here. Furthermore, in our discussion below, we will always assume that β_i is constant on each component D_i. (This assumption is present in essentially all works on the subject so far.) Lastly, for the moment, we are deliberately vague on the meaning of "asymptotically equivalent." Several working definitions are given in §3.7 (see in particular Lemma 3.11).

The study of Kähler edge metrics was initiated by Tian [**243**], motivated in part by the possibility of endowing more Kähler manifolds with a generalized KE metric, when the obstruction (2.2) does not vanish. Of course, the possibility of uniformizing more Kähler manifolds is exciting in itself, and there are many possible applications of such metrics in algebraic geometry (see, e.g., [**243, 261**]). However, as we will see later, this generalization sheds considerable light also on the theory of smooth KE metrics. Finally, Kähler edge manifolds are also a natural generalization of conical Riemann surfaces, who were first systematically studied by Troyanov [**260**]; see §4.4 for more on this topic.

3.2. The geometry of a cone. In the previous subsection we arrived at a generalization of Kähler geometry by seeking a generalization of the Kähler–Einstein equation. Before going back to the latter, let us first say a few words on the geometry described by (3.2).

The basic observation is that

$$C_\beta := (\mathbb{C}, |z|^{2\beta-2}\sqrt{-1}dz \wedge \overline{dz}) \tag{3.4}$$

is a cone with tip at the origin. For instance, when $\beta = 1/k$ with $k \in \mathbb{N}$, we obtain an orbifold. We emphasize that here we equip the smooth manifold $\mathbb{C}$ with a singular

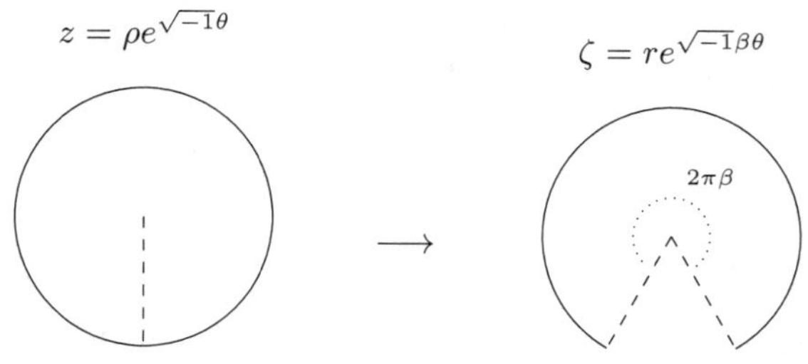

FIGURE 1. The map $z \mapsto z^\beta =: \zeta$ maps the disc to a wedge of angle $2\pi\beta$ (the inverse map identifies the two 'sides' of the wedge). This map pulls-back the flat metric $|d\zeta|^2$ on the wedge to the singular metric $\beta^2|z|^{2\beta-2}|dz|^2$ on the disc.

metric, and we are claiming that thus equipped it is isometric to a singular metric space, a cone. Perhaps the simplest way to see this is to recall the construction of a cone out of a wedge. Starting with a wedge of angle $2\pi\beta$, one identifies the two sides of the boundary. This can of course be done in many ways, in other words there are many maps of the form $(r, \theta) \mapsto (f(r, \theta), \theta/\beta)$ (so the angle in the target indeed varies between 0 and 2π, or in other words, the target indeed 'closes up' and is homeomorphic to $\mathbb{R}_+ \times S^1(2\pi)$). However, there is an essentially unique such map that preserves angles, or in other words is holomorphic. By the Cauchy–Riemann equations (i.e.,

$$r\partial_r(\mathrm{Re}F, \mathrm{Im}F) = \partial_\theta(\mathrm{Im}F, -\mathrm{Re}F)$$

for a holomorphic function F) it must be of the form $f(r, \theta) = Cr^{1/\beta}$, for some constant C. The inverse of this map when $C = 1$, from the cone to the wedge, is given by $(\rho, \phi) \mapsto (\rho^\beta, \beta\phi)$, or simply $z = \rho e^{\sqrt{-1}\phi} \mapsto z^\beta =: \zeta$ (see Figure 1). Declaring this map to be an isometry determines the geometry of the cone; pulling back the Euclidean metric $|d\zeta|^2 = \sqrt{-1}\partial\bar{\partial}|\zeta|^2$ endows the cone with the metric $\sqrt{-1}\partial\bar{\partial}|z|^{2\beta} = \beta^2|z|^{2\beta-2}\sqrt{-1}dz \wedge \overline{dz}$.

3.3. The Kähler–Einstein edge equation. We now return to our discussion of the generalized *local* Kähler–Einstein equation (3.1).

To turn (3.1) into a global equation we seek (at least formally in the case we are not dealing with $\mathbb{Q}$-line bundles), an equality of two continuous Hermitian metrics: one on $\mu\Omega$ and the other on $-K_M + \sum_i(1 - \beta_i)L_{D_i}$. In analogy with the discussion of §2, we now choose a reference metric that is locally asymptotic to the model edge geometry, instead of the Euclidean geometry on $\mathbb{C}^n$ that was implicitly the model there. Suppose that h_i is a smooth Hermitian metric on L_{D_i}, and that s_i is a global holomorphic section of L_{D_i}, so that $D_i = s_i^{-1}(0)$. We let

$$\omega := \omega_0 + \sqrt{-1}\partial\bar{\partial}\phi_0, \tag{3.5}$$

with

$$\phi_0 := c\sum(|s_i|^2_{h_i})^{\beta_i}. \tag{3.6}$$

An easy computation shows that ω is locally equivalent to ω_β and that for small enough $c > 0$ it defines a Kähler metric away from D [**141**, Lemma 2.2]. Moreover, in the coordinates above, near a smooth point of D_1, for example, $|z_1|^{2\beta_1-2}\det\psi_{i\bar{j}}$

is continuous, as desired. Similar properties hold near crossing points of D. Thus, we may proceed exactly as in §2 to obtain an (seemingly!) identical equation,

$$\omega_\varphi^n = \omega^n e^{f_\omega - \mu\varphi}, \qquad \text{on } M \setminus D, \tag{3.7}$$

only now away from D, and with respect to the reference form ω, where φ is required to lie in the space of Kähler edge potentials

$$\begin{aligned}\mathcal{H}_\omega := \{\varphi \in C^\infty(M\backslash D) \cap C^0(M) \,:\, \omega_\varphi := \omega + \sqrt{-1}\partial\bar\partial\varphi > 0 \text{ on } M \\ \text{and } \omega_\varphi \text{ is asymptotically equivalent to } \omega_\beta\}.\end{aligned} \tag{3.8}$$

Equation (3.7) is the *Kähler–Einstein edge (KEE) equation.*

By construction, the twisted Ricci potential of ω, f_ω, satisfies

$$\sqrt{-1}\partial\bar\partial f_\omega = \text{Ric}\,\omega - 2\pi(1-\beta)[D] - \mu\omega, \tag{3.9}$$

where it is again convenient to require the normalization $\int e^{f_\omega}\omega^n = \int \omega^n$. Differentiating (3.7) leads to the following equivalent formulation of the KEE equation.

DEFINITION 3.2. *With all notation as above, a Kähler current $\omega_{\rm KE}$ is called a Kähler–Einstein edge current with angle $2\pi\beta$ along D and Ricci curvature μ if $\omega_{\rm KE} \in \mathcal{H}_\omega$ (see (3.8)), and if*

$$\text{Ric}\,\omega_{\rm KE} - 2\pi(1-\beta)[D] = \mu\omega_{\rm KE}, \tag{3.10}$$

where $[D]$ is the current associated to integration along D, and where $\text{Ric}\,\omega$ *denotes the Ricci current (on M) associated to ω, namely, in local coordinates* $\text{Ric}\,\omega = -\sqrt{-1}\partial\bar\partial \log\det[g_{i\bar j}]$ *if $\omega = \sqrt{-1}g_{i\bar j}dz^i \wedge \overline{dz^j}$.*

The KEE equation may also be rewritten in terms of the background smooth geometry. We carry this out for pedagogical purposes, since it allows to unravel the difference between equations (3.7) and (2.3), that are formally the same but involve different geometric objects.

Let e be a local holomorphic frame for L_D valid in a neighborhood intersecting D, such that $s = z_1 e$ on that neighborhood, and denote by

$$a_i := |e_i|^2_{h_i}$$

a smooth positive function on that neighborhood. Define F_{ω_0} (up to a constant, for the moment) by $\sqrt{-1}\partial\bar\partial F_{\omega_0} = \text{Ric}\,\omega_0 - \mu\omega_0 + \sum_i (1-\beta_i)\sqrt{-1}\partial\bar\partial \log a_i$. Setting $\tilde\varphi := \phi_0 + \varphi$, it easy to see that

$$(\omega_0 + \sqrt{-1}\partial\bar\partial\tilde\varphi)^n = \prod_i |s_i|^{2\beta_i - 2}_{h_i}\omega_0^n e^{F_{\omega_0} - \mu\tilde\varphi}, \tag{3.11}$$

where we think of this equation as determining a normalization of F_{ω_0} such that both sides have equal integrals (cf. [**141**, (53)]).

For much of the rest of this article we will be concerned with solving this equation with optimal estimates on the solution. In this regard, we note that global solutions, smooth away from D, of this (when $\mu \le 0$) and quite more general Monge–Ampère equations were constructed already in much earlier work of Yau [**268**, §7–9]. Our goal though, is to explain how to go beyond such weak solutions and obtain estimates that show the metric in fact has edge singularities near the 'boundary' D. To start, we define the appropriate function spaces to prove such estimates.

REMARK 3.3. A point we glossed over in the discussion is the fact that a necessary condition for the existence of a KEE metric is that the first Chern class of the adjoint bundle,

$$-K_M + \sum_i (1-\beta_i)L_{D_i}, \quad \text{is } \tfrac{\mu}{2\pi} \text{ times an ample class.} \tag{3.12}$$

This is obvious in the smooth case ($\beta_i = 1$) from the KE equation. In general, the KEE equation only guarantees that this class is represented by ($\frac{\mu}{2\pi}$ times) a Kähler current. But the edge singularities are sufficiently mild that the divisor D has zero volume [**141**, §6], and the Lelong numbers of this current along D are zero. As observed by Di Cerbo [**87**] this implies that the class is actually ($\frac{\mu}{2\pi}$ times) an ample class, i.e., represented by a smooth Kähler metric.

3.4. Three smooth structures, one polyhomogeneous structure. Naturally, if one's goal is to obtain existence and regularity of certain geometric objects (such as KEE metrics) on a Kähler edge manifold, then understanding the differentiable structure underlying such a space should play a key rôle.

First, let us consider the simplest compact closed example, that of $M = \mathbb{P}^1$ with a single cone point $D = p$. Then M of course has its natural conformal structure, and there is a corresponding smooth structure. The former is represented locally near p by the holomorphic coordinate z, and the latter can be represented by the associated polar coordinates (ρ, θ), where of course all the points $(0, \theta)$ are identified with p.

Recall from §3.2 that an alternative conformal structure is represented locally by the coordinate $\zeta := z^\beta$. Of course, ζ is multi-valued, and one must choose a branch of the Riemann surface associated to $z \mapsto z^\beta$. Whenever we work with ζ, we assume such a choice was made. More specifically we slit the disc, and work with the associated polar coordinates denoted by (r, θ), where now $\theta \in [0, 2\pi\beta)$, and these represent the associated smooth structure. That is, smooth functions are smooth functions of r, θ. Clearly, this smooth structure is incompatible with the one of the previous paragraph.

Which smooth structure should we work with? The point of view stressed in [**141**], following Melrose's general framework [**181**], is that it is rather the *polyhomogeneous structure* that is central to the problem, and not either of these smooth structures. The purpose of this section is to explain what is meant by this. To carry this out, it will be preferable to work with a bona fide *singular space* associated to M (recall M itself is smooth). That is, we consider the singular metric space M_{sing} obtained as the metric completion from the underlying distance function associated to M equipped with its Kähler edge metric. It is readily seen that this space is homeomorphic to M (since the divisor is at finite distance from any point by (3.3)). What is important to the development of the theory described below is that the smooth locus of this singular space M_{sing} is precisely the complement of a simple normal crossing divisor in a nonsingular space. Thus, as we describe below, it will admit an *edge structure*.

The first step we take is actually to *desingularize* M_{sing}, i.e., resolve its singularities to obtain a manifold with boundary. The edge structure we will define shortly, will, by definition, be on this desingularized space. The desingularization is done, as usual, by a series of resolutions by real blow-ups; when D is smooth and connected a single such blow-up suffices. In the simplest example of $(\mathbb{P}^1, p)$

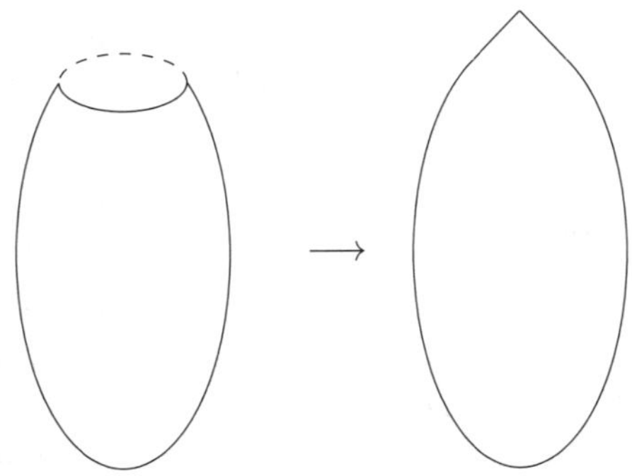

FIGURE 2. The real blow-up of the tear-drop space

considered just above this amounts to a single real blow-up at p, resulting in the manifold with boundary X consisting of the disjoint union of $\mathbb{P}^1 \setminus \{p\}$ and S^1 (see Figure 2). In other words, the smallest manifold with boundary on which the polar coordinates are well-defined, without identifying the points $\{(0,\theta) : \theta \in [0,2\pi)\}$. In general, the manifold X is the real blow-up of M_{sing} at D, i.e., the disjoint union $M \setminus D$ and the circle normal bundle of D in M, endowed with the unique smallest topological and differential structure so that the lifts of smooth functions on M and polar coordinates around D are smooth.

The advantage of working with functions on X rather than on M or M_{sing} is convenience. For instance, it is much easier to keep track of singularities of distributions (such as the Green kernel) on the desingularization—this is explained in detail in §3.9. When D has crossings this desingularization also serves to give a reasonable description for distributions near crossing points [**175**, **176**].

The relevant coordinates on X are (r,θ,y) where $y = (y_1,\ldots,y_{2n-2})$ denotes the 'conormal' coordinates, i.e., coordinates on D. The relevant smooth structure to our problem turns out to be the smooth structure of X as a manifold with boundary. Namely, smooth functions are smooth functions of r,θ, and $y_1,\ldots,y_{2n-2}$.

Next, we defined the associated 'edge structure' of X in the sense of Mazzeo [**172**, §3].

DEFINITION 3.4. *The edge structure on the compact manifold with boundary X is the Lie algebra of vector fields $\mathcal{V}_e(X)$ generated by*

$$\mathcal{V}_e(X) = \text{span}_{\mathbb{R}}\Big\{r\frac{\partial}{\partial r}, \frac{\partial}{\partial\theta}, r\frac{\partial}{\partial y_1},\ldots, r\frac{\partial}{\partial y_{2n-2}}\Big\}.$$

The Lie algebra $\mathcal{V}_e$ consists of vector fields on X tangent to the S^1 fibers on ∂X. The space C_e^∞ is defined to be the space of C^∞ functions on X with respect to this edge structure, i.e., functions that are infinitely-differentiable with respect to the vector fields in $\mathcal{V}_e(X)$; see §3.5 for an alternative, somewhat more geometric, definition of this space. Note that a function on X can be pushed-forward, using the blow-down map, to a function on $M \setminus D$, and, by abuse of notation, we often make this identification without mention. Much care is needed here, however, since a function in C_e^∞ need not correspond even to a continuous function on M (consider, e.g., the function $\sin(\log r)$)!

In Kähler edge geometry there is a clear distinction between differentiation in the direction normal to the edge and in the complementary directions. Thus, it is convenient to define the Lie algebra $\mathcal{V}_b(X)$ consisting of vector fields on X that are tangent to ∂X, i.e.,

$$\mathcal{V}_b(X) = \operatorname{span}_{\mathbb{R}}\Big\{r\frac{\partial}{\partial r}, \frac{\partial}{\partial \theta}, \frac{\partial}{\partial y_1}, \ldots, \frac{\partial}{\partial y_{2n-2}}\Big\},$$

and introduce the following terminology.

DEFINITION 3.5. *The space of bounded conormal functions on X is*

$$\mathcal{A}^0 = \mathcal{A}^0(X) := \{f \in L^\infty(X) \,:\, \textit{for all } k \in \mathbb{N} \textit{ and } V_1, \ldots, V_k \in \mathcal{V}_b(X),\ V_1 \cdots V_k f \in L^\infty(X)\}.$$

In other words, these are bounded functions on X that are infinitely differentiable with respect to vector fields in $\mathcal{V}_b(X)$. Thus, they are infinitely-differentiable in directions tangent to D (the 'conormal directions'), but still potentially rather badly behaved with respect to the vector field $\frac{\partial}{\partial r}$. Another geometric description of $\mathcal{A}^0$ is given in §3.5 below.

Now, let us, finally, define the *polyhomogeneous structure* of X. For that we first set

$$\mathcal{A}^{\gamma,p} := r^\gamma (\log r)^p \mathcal{A}^0.$$

DEFINITION 3.6. *The space of bounded polyhomogeneous functions on X is*

$$\mathcal{A}^0_{\mathrm{phg}} := \Big\{f \in \mathcal{A}^0(X) \,:\, f \sim \sum_{j=0}^{\infty}\sum_{p=0}^{N_j} a_{jp}(\theta, y) r^{\sigma_j} (\log r)^p\Big\},$$

with $\operatorname{Re}\sigma_j$ *increasing to* ∞, *and* $\operatorname{Re}\sigma_j \ge 0$ *with* $N_j = 0$ *if* $\operatorname{Re}\sigma_j = 0$.

Similarly, one defines $\mathcal{A}^{\gamma,p}_{\mathrm{phg}} := r^\gamma(\log r)^p \mathcal{A}^0_{\mathrm{phg}}$, and the *polyhomogeneous structure of X* is defined as the union

$$\mathcal{A}^*_{\mathrm{phg}} := \bigcup_{\gamma,p} \mathcal{A}^{\gamma,p}_{\mathrm{phg}}.$$

For later use, we define the index set associated to a function $u \in \mathcal{A}^0_{\mathrm{phg}}$ to be the set

$$E_u := \{(\sigma_j, p) \,:\, a_{jp} \not\equiv 0\}. \tag{3.13}$$

In Definition 3.6, the $\sim$ symbol means that f admits an asymptotic expansion in powers of r and $\log r$. By definition, this means that for every $k \in \mathbb{N}$ and $m \in \{0, \ldots, N_k\}$,

$$f - \sum_{j=0}^{k-1}\sum_{p=0}^{N_j} a_{jp}(\theta,y) r^{\sigma_j}(\log r)^p - \sum_{p=0}^{N_k - m} a_{kp}(\theta,y) r^{\sigma_k}(\log r)^p \in \begin{cases} \mathcal{A}^{\sigma_k, N_k+1-m}_{\mathrm{phg}} & \text{if } m \in \mathbb{N}, \\ \mathcal{A}^{\sigma_{k+1}, N_{k+1}}_{\mathrm{phg}} & \text{if } m = 0, \end{cases} \tag{3.14}$$

and moreover that corresponding remainder estimates hold whenever any number of vector fields in $\mathcal{V}_b(X)$ are applied to the left hand side of (3.14). These are referred to as remainder estimates since if a function u lies in $\mathcal{A}^{\gamma,p}$, it satisfies $|u| \le C r^\gamma (\log r)^p$ on X.

Several remarks are in order. First, on a smooth space, Taylor's theorem implies that any smooth function admits a Taylor series expansion. In our setting, however, there is a certain 'gap' between the space C^∞_e (or even $\mathcal{A}^0$) and the space of functions admitting a bi-graded expansion as in Definition 3.6. Second, the push-forward of a function in $\mathcal{A}^0_{\mathrm{phg}}$ (unlike a function in C^∞_e) can be considered as a continuous function on M, provided its leading term is independent of θ (and not just on $M \setminus D$). Third, expansions of polyhomogeneous functions are rarely convergent, but only give 'order of vanishing' type estimates, as described above. Fourth, the remainder estimate (3.14) can be written as an equality

$$f - \sum_{j=0}^{k-1} \sum_{p=0}^{N_j} a_{jp}(\theta, y) r^{\sigma_j} (\log r)^p - \sum_{p=0}^{N_k - m} a_{kp}(\theta, y) r^{\sigma_k} (\log r)^p = u, \tag{3.15}$$

with u belonging to one of the two spaces in the right hand side of (3.14), and this equality is understood to hold on some ball (in the coordinates r, y, θ) centered at a point on D. It is usually difficult to control the size of this ball, i.e., to give lower bounds for its radius. However, since such a positive radius exists for each $p \in D$ and D is compact, the equality (3.15) actually holds on some tubular neighborhood (of positive distance—as r is uniformly equivalent to the distance function close enough to D) of D. Finally, note that the polyhomogeneous structure associated to either the (ρ, θ, y) or the (r, θ, y) coordinates is the same, precisely because we are allowing fractional powers of r and $\beta \log \rho = \log r$.

3.5. Edge and wedge function spaces. There are two distinct scales of function spaces naturally associated to a Kähler edge metric, that we will denote by $C_s^{k,\alpha}$

with s equal to either w or e.

We consider both of these as subspaces of $L^\infty(M)$.

The first, the wedge spaces $C_w^{k,\alpha}$, are the usual $C^{k,\alpha}$ spaces on $M \setminus D$ with respect to a(ny) Kähler edge metric, intersected with $L^\infty(M)$ (and, in fact, are contained in $C^0(M)$). That is, we consider $M \setminus D$ as an *incomplete* Riemannian manifold with the usual distance to the edge being a distance function.

The second, the edge spaces $C_e^{k,\alpha}$ [**172**], are the intersection of $L^\infty(M)$ with the usual $C^{k,\alpha}$ spaces on $M \setminus D$ with respect to a conformal deformation of a Kähler edge metric for which the logarithm of the usual distance to the edge is now a distance function. Thus, we consider $M \setminus D$ as a *complete* Riemannian manifold. Equivalently, $C_e^{k,\alpha}$ consists of functions on $M \setminus D$ that are push-forwards of functions on X that are $C^{k,\alpha}$ with respect to the edge differentials $\mathcal{V}_e(X)$ of §3.4.

Let us now explain how to define these function spaces explicitly in the model edge $C_\beta \times \mathbb{R}^{2n-2}$. In the notation of §3.2, the flat metric on a cone C_β takes the form $dr^2 + r^2 d\theta^2$, with $\theta \in [0, 2\pi\beta)$. Thus the flat model edge metric ω_β is given by $dr^2 + r^2 d\theta^2 + dy^2$, with $y = (y_1, \ldots, y_{2n-2})$ coordinates on $\mathbb{R}^{2n-2}$, and it is this metric that defines the spaces $C_w^{k,\alpha}$. The conformally rescaled metric $(d \log r)^2 + d\theta^2 + r^{-2} dy^2$ defines the spaces $C_e^{k,\alpha}$. It follows that the defining vector fields for the spaces $C_s^{k,\alpha}$ are $r^{\delta_e(s)} \{\frac{\partial}{\partial r}, \frac{1}{r}\frac{\partial}{\partial \theta}, \frac{\partial}{\partial y_j}\}$, where $\delta_e(e) = 1, \delta_e(w) = 0$. We will say a bit more about these function spaces below, but for a thorough discussion of these function spaces as well as the different coordinate choices involved we refer to [**141**, §2]. For the moment we observe the obvious inclusion $C_w^{k,\gamma} \subset C_e^{k,\gamma}$; the wedge spaces are in fact much smaller than their edge counterparts. E.g., as noted earlier, $\sin \log r \in C_e^\infty$ shows that $C_e^\infty \not\subset C^0(M)$ (though it is contained in $L^\infty(M)$), and $r^{k+\epsilon} \in C_w^{k,\epsilon} \cap C_e^\infty$ but is not contained in any higher wedge space.

Note, finally, that the space $\mathcal{A}^0$ defined in §3.4 admits a similar geometric description. Namely it is the space of bounded functions that are C^∞ space with respect to the metric $d(\log r)^2 + d\theta^2 + dy^2$, namely the product metric obtained by conformally rescaling the flat metric on the cone together with the standard (non-rescaled) Euclidean metric on $\mathbb{R}^{2n-2}$.

3.6. Various Hölder domains. Perhaps surprisingly, it turns out that the complex Monge–Ampère equation, and, as a special case, the Poisson equation, cannot be solved in general in $C^{2,\alpha}_w$. In fact, as already noted, $\operatorname{Re} z_1$ is pluriharmonic and belongs merely to $C^{1,\frac{1}{\beta}-1}_w$, which fails to lie in $C^{2,\alpha}_w$ when $\beta > 1/2$. On the other hand, $C^{2,\alpha}_e$ is certainly a large enough space to find solutions; however, it is not even contained in C^0 and so seems to give too little control/regularity to develop a reasonably strong existence theory for the Monge–Ampère equation.

First, we introduce the *maximal Hölder domains*

$$\mathcal{D}^{0,\gamma}_s(\Delta_\omega) = \mathcal{D}^{0,\gamma}_s := \{u \in C^{0,\gamma}_s \,:\, \Delta_\omega u \in C^{0,\gamma}_s\}.$$

These are Banach spaces with associated norm

$$||u||_{\mathcal{D}^{0,\gamma}_s(\Delta_\omega)} := ||\Delta_\omega u||_{C^{0,\gamma}_s} + ||u||_{C^{0,\gamma}_s}.$$

We also define the *little Hölder domains*

$$\widetilde{\mathcal{D}}^{0,\gamma}_s(\Delta_\omega),$$

as the closure of the space $\mathcal{A}^0_{\mathrm{phg}}$ of polyhomogeneous functions in the $\mathcal{D}^{0,\gamma}_s(\Delta_\omega)$ norm. The name 'maximal' comes from the analogy with the usual definition of the maximal domain of the Laplacian in L^2, namely $\mathcal{D}_{\max}(\Delta) := \{u \in L^2 \,:\, \Delta u \in L^2\}$. On the other hand, the 'little' spaces are defined similarly to the usual little Hölder spaces (see, e.g., [**164**]). The latter are separable, unlike the former, and of course $\widetilde{\mathcal{D}}^{0,\gamma}_s(\Delta_\omega) \subset \mathcal{D}^{0,\gamma}_s(\Delta_\omega)$.

The space $\mathcal{D}^{0,\gamma}_w$ was introduced by Donaldson [**102**] (where it is denoted $C^{2,\gamma,\beta}$); it gives wedge Hölder control of the wedge Laplacian, which is a sum of *certain* second wedge derivatives of type (1,1). Motivated by this, the space $\mathcal{D}^{0,\gamma}_e$ was introduced in [**141**], and gives edge Hölder control of the wedge Laplacian. Thus, unlike $\mathcal{D}^{0,\gamma}_w$, it is a sort of hybrid space. In fact, $\mathcal{D}^{0,\gamma}_s$ can be characterized by requiring the (1,1) wedge Hessian (and not only its trace!) to be Hölder (in fact, a slightly stronger characterization holds, see Theorem 3.7 below). When $\beta \le 1/2$ an even sharper characterization holds for $s = e$: the full (real) wedge Hessian is Hölder. Let

$$P_{1\bar{1}} := \partial_r^2 + \frac{1}{r}\partial_r + \frac{1}{\beta^2 r^2}\partial_\theta^2,$$

and define

$$\mathcal{Q} := \{\partial_r, r^{-1}\partial_\theta, \partial_{y_a}, \partial^2_{y_a y_b}, \partial_r\partial_{y_a}, \partial_r\partial_\theta, r^{-1}\partial_\theta\partial_{y_a}, P_{1\bar{1}}, a, b = 1, \ldots 2n-2\}.$$

THEOREM 3.7. *(i) There exists a constant $C > 0$ independent of u such that*

$$||Tu||_{s;0,\gamma} \le C(||\Delta_\omega u||_{s;0,\gamma} + ||u||_{s;0,\gamma}),$$

for all $T \in \mathcal{Q}$. Thus, $u \in \mathcal{D}^{0,\gamma}_s$ if and only if $Tu \in C^{0,\gamma}_s$ for all $T \in \mathcal{Q}$.
(ii) If $\beta \in (0, 1/2]$, the previous statement holds with $\mathcal{Q}$ replaced by

$$\mathcal{Q} \cup \{\partial_r^2, r^{-1}\partial_r, r^{-2}\partial_\theta^2, r^{-1}\partial_r\partial_\theta, r^{-1}\partial_r\partial_{y_a}, a, b = 1, \ldots 2n-2\}.$$

Also, $\mathcal{D}^{0,\gamma}_w \subset C^{2,\min\{\frac{1}{\beta}-2,\gamma\}}_w$.

REMARK 3.8. Part (ii) quantifies the difference between the "orbifold" regime $\beta \in (0, 1/2]$ and the harder $\beta \in (1/2, 1)$ regime. In fact, one can write down stronger and stronger characterizations of $\mathcal{D}_e^{0,\gamma}$ under mild assumptions when $\beta \in (0, \beta_0)$ as $\beta_0 \le 1/2$ approaches 0.

For the proof of (i) and (ii) with $s = e$, we refer to [**141**, Proposition 3.3], and for (i) and (ii) with $s = w$ to [**141**, Proposition 3.8]. The key ingredient in the proofs is a precise description of the singularities (or, in other words, of the the polyhomogeneous structure) of the Green kernel of the Laplacian of the (curved) reference metric ω (3.5), considered as a distribution on certain blow-up of $X \times X$. This is explained in §3.9. For the proof of (i) with $s = w$ for most of the operators in $\mathcal{Q}$, we also refer to Donaldson [**102**, Theorem 1]; the verification for the remaining operators is straightforward from his arguments. Donaldson's approach is more elementary in that he only obtains the polyhomogeneous structure of the Laplacian of the flat model metric ω_β (3.2) (which can be done by separation of variables arguments, and also goes back to the work of Mooers [**183**]); using the Schauder estimate for this flat model he then obtains (i) with $s = w$ by a partition of unity argument. The advantage of the extra work done in [**141**] is that the polyhomogeneous structure of the Green kernel of the curved metric itself allows much more refined information, e.g., (ii), and also leads eventually to the higher regularity of solutions of the Monge–Ampère equation—see §3.10—that does not seem to be accessible using the arguments of [**102**].

Next, similarly to the definition of the spaces $C_s^{k,\alpha}$, also the spaces $\mathcal{D}_s^{0,\gamma}$ can be defined with respect to *any* Kähler edge metric of the form ω_φ with $\varphi \in \widetilde{\mathcal{D}}_s^{0,\gamma}$. This property is absolutely crucial in applications to 'openness' along the continuity method (§6.4), and to higher regularity (§3.10).

THEOREM 3.9. *Suppose that* $u \in \widetilde{\mathcal{D}}_s^{0,\gamma}$. *Then*

$$\mathcal{D}_s^{0,\gamma}(\Delta_\omega) = \mathcal{D}_s^{0,\gamma}(\Delta_{\omega_u}) := \{v \in C_s^{0,\gamma} \, : \, \Delta_{\omega_u} v \in C_s^{0,\gamma}\}.$$

This is proved in [**141**, Corollary 3.5]; Donaldson [**102**] does not state any such result explicitly, but briefly sketches related ideas in the case $s = w$ in [**102**, p. 64]. This result is crucial in our approach and we describe the proof in some detail below.

The goal is to show that $\mathcal{D}_s^{0,\gamma}(\Delta_\omega) = \mathcal{D}_s^{0,\gamma}(\Delta_{\omega_u})$. First, one observes that this is true when u is polyhomogeneous, by the explicit polyhomogeneous structure of the Green kernel of polyhomogeneous elliptic edge operators (see Theorem 3.15 below). Next, note that

$$\Delta_{\omega_u} f = \frac{\omega^n}{\omega_u^n} \frac{n\sqrt{-1}\partial\bar\partial f \wedge \omega_u^{n-1}}{\omega^n}.$$

Thus, if $u \in \mathcal{D}_s^{0,\gamma}$ then $|\Delta_{\omega_u} f| \le C_1(|\Delta f| + (\sqrt{-1}\partial\bar\partial f, \alpha)_\omega)$, where $|\alpha|_\omega \le C_2$, and C_1, C_2 are both controlled by a polynomial function of $\sum_{i,j} [u_{i\bar j}]_{s;0,\gamma}$. By Theorem 3.7 (i) these constants are then also controlled by $||u||_{\mathcal{D}_s^{0,\gamma}}$. Thus, $\mathcal{D}_s^{0,\gamma}(\Delta_\omega) \subset \mathcal{D}_s^{0,\gamma}(\Delta_{\omega_u})$.

Since Δ_{ω_u} is injective on the L^2-orthogonal complement to the constants in $\mathcal{D}_s^{0,\gamma}(\Delta_{\omega_u})$, that we denote by $\mathcal{D}_s^{0,\gamma}(\Delta_{\omega_u})'$, it is also injective on $\mathcal{D}_s^{0,\gamma}(\Delta_\omega)'$. The main point is that it is also surjective when restricted to the latter space; since $\Delta_{\omega_u} : \mathcal{D}_s^{0,\gamma}(\Delta_{\omega_u})' \to C_s^{0,\gamma}$ is by definition a bijection, this immediately gives the reverse inclusion and concludes the proof. We now explain the proof of the surjectivity statement.

The surjectivity follows in several steps. First, given any nonzero $f \in C^{0,\gamma}_s$ there is a unique solution $\phi \in \mathcal{D}^{0,\gamma}_s(\Delta_{\omega_u})'$ to $\Delta_{\omega_u}\phi = f$. Approximate u in $\mathcal{D}^{0,\gamma}_s(\Delta_\omega)$ by polyhomogeneous $u^{(k)}$, and let $\phi^{(k)}$ be the associated solution of $\Delta_{\omega_{u^{(k)}}}\phi^{(k)} = f$. Since $u^{(k)}$ are polyhomogeneous, there is a unique constant c_k such that $\phi^{(k)} - c_k \in \mathcal{D}^{0,\gamma}_s(\Delta_\omega)'$, and by Theorem 3.7 (i) (which, as remarked in the beginning of the proof, applies to ω replaced by any polyhomogeneous metric) this implies an estimate

$$\sum_{Q\in\mathcal{Q}} ||Q\phi^{(k)}||_{s;0,\gamma} \le C^{(k)}(||f||_{e;0,\gamma} + ||\phi^{(k)}||_{e;0,\gamma}). \tag{3.16}$$

We claim that the constant $C^{(k)}$ in (3.16) is locally uniform in the $C^{0,\gamma}_s$ norm of $u^{(k)}_{i\bar{j}}$ (i.e., that if $u^{(k)}$ were supported in some small ball then the constant on the right hand side of (3.16) would be controlled uniformly in terms of the $C^{0,\gamma}_s$ norm of $u^{(k)}_{i\bar{j}}$). Of course, since $||u^{(k)}_{i\bar{j}} - u_{i\bar{j}}||_{s;0,\gamma}$ is uniform in k, this would imply also that the $C^{(k)}$ are locally uniformly controlled, independently of k, by the $C^{0,\gamma}_s$ norm of $u_{i\bar{j}}$. The claim is true for polyhomogeneous u by the Green kernel construction of Theorem 3.15. For a more general $u \in \widetilde{\mathcal{D}}^{0,\gamma}_s$, we freeze coefficients of Δ_{ω_u} only in a small neighborhood of any point in M and approximate this operator by a global polyhomogeneous operator. The estimate above follows for such a concatenated operator by the standard method of proof that makes clear that the constant in the estimate depends only on the local $C^{0,\gamma}_s$ norm of $u_{i\bar{j}}$. Using a partition of unity to paste these estimates then concludes the proof of the claim (since both D and M are compact).

Thus, the $\phi^{(k)}$ are uniformly in $\mathcal{D}^{0,\gamma}_s(\Delta_\omega)'$ (note that the constants c_k are uniformly controlled). When $s = w$ we can now take a subsequence converging in $\mathcal{D}^{0,\gamma'}_w$ (for any $\gamma' \in (0,\gamma)$) to a function $\phi \in \mathcal{D}^{0,\gamma}_w(\Delta_\omega)'$ that solves $\Delta_{\omega_u}\phi = f$. When $s = e$, we cover $M \setminus D$ by a countable collection of Whitney cubes so that on each such cube $T\phi^{(k)}$ converges in $C^{0,\gamma'}_e$ to $T\phi$, for all $\gamma' \in (0,\gamma)$. Taking a diagonal sequence then produces a solution ϕ to $\Delta_{\omega_u}\phi = f$ that belongs to $\mathcal{D}^{0,\gamma}_s(\Delta_\omega)'$, by Theorem 3.7 (i). In either case ($s = e$ or $s = w$), (iii) follows.

Finally, we mention several further basic regularity properties of the Hölder domains.

THEOREM 3.10. *(i) For any $T \in \mathcal{Q} \setminus P_{1\bar{1}}$, T maps $\mathcal{D}^{0,\gamma}_e$ into $C^0(M)$. In particular, if $u \in \mathcal{D}^{0,\gamma}_e$, then Tu is continuous up to D, and has a well-defined restriction to D, independent of θ.*
(ii) Let $\gamma \ge 0$. Then, $\mathcal{D}^{0,\gamma}_e \subset C^{1,\alpha}_w$ for $\alpha \in (0, \frac{1}{\beta} - 1] \cap (0,1)$.
(iii) If $u \in \widetilde{\mathcal{D}}^{0,\gamma}_e$ then u has the partial expansion near D

$$u = a_0(y) + (a_{01}(y)\sin\theta + b_{01}(y)\cos\theta)r^{\frac{1}{\beta}} + O(r^2).$$

For (i) refer to [**141**, Corollary 3.6], for some of the operators mentioned, while the proof for the remaining ones follows from (ii). Part (ii) is the singular analogue for the usual $C^{1,\alpha}$ estimates for a function in $W^{2,\infty}$, and can be proved using [**141**, Proposition 3.8]—we sketch the proof in Lemma 3.12. In fact, (ii) implies the statement of (i) holds for a larger class of operators. Finally, (iii) is a corollary of Theorem 3.9 and the polyhomogeneous structure of the Green kernel stated in Theorem 3.15.

3.7. Kähler edge metrics. We start with a completely elementary, yet important, lemma.

LEMMA 3.11. *Let g be a continuous Kähler metric on $M \setminus D$, and such that in any local holomorphic coordinate system near D where $D = \{z_1 = 0\}$, and $z_1 = \rho e^{\sqrt{-1}\theta}$,*

$$g_{1\bar{1}} = F\rho^{2\beta-2},\ g_{1\bar{j}} = g_{i\bar{1}} = O(\rho^{\beta-1+\epsilon}),\ \text{and all other } g_{i\bar{j}} = O(1), \tag{3.17}$$

for some $\epsilon \geq 0$, where F is a bounded nonvanishing function which is continuous at D. Then there exists some $C > 0$ such that in any such coordinate chart

$$\frac{1}{C} g_\beta \leq g \leq C g_\beta. \tag{3.18}$$

Moreover, the converse implication holds.

PROOF. Assume first that (3.17) holds. This amounts to showing that $\frac{1}{C} I_\beta \leq [g_{i\bar{j}}] \leq C I_\beta$, where $I_\beta := \operatorname{diag}(\rho^{2\beta-2}, 1, \ldots, 1)$. Let $v = (v_1, \ldots, v_n) \in \mathbb{C}^n$ be any vector. Then,

$$\begin{aligned} \sum_{i,j=1}^n g_{i\bar{j}} v_i \overline{v_j} &\leq g_{1\bar{1}} |v_1|^2 + \sum_{i,j=2}^n g_{i\bar{j}} v_i \overline{v_j} + \sum_{j=2}^n |g_{1\bar{j}}|^2 |v_1|^2 + \sum_{j=2}^n |v_j|^2 \\ &\leq C\rho^{2\beta-2}|v_1|^2 + C\sum_{j=2}^n |v_j|^2, \end{aligned}$$

for some $C > 0$, proving the second inequality (since ρ is small and $\beta \in (0,1)$). The first inequality follows similarly, since (3.17) implies that

$$g^{1\bar{1}} = F\rho^{2-2\beta},\ g^{1\bar{j}}, g^{i\bar{1}} = O(\rho^{1-\beta+\epsilon}),\ \text{and all other } g^{i\bar{j}} = O(1). \tag{3.19}$$

Conversely, choosing $v = (0, v')$ shows that $CI \leq g' := [g_{i\bar{j}}]_{i,j=2}^n \leq C'I$, and thus $g_{i\bar{j}} = O(1)$, for all $i, j \geq 2$. Similarly, it follows that $g_{1\bar{1}} = O(\rho^{2\beta-2})$. Equation (3.18) implies $C^{-n}\omega_\beta^n \leq \omega^n \leq C^n\omega_\beta^n$, but

$$\det[g_{i\bar{j}}] = g_{1\bar{1}} \det[g_{i\bar{j}}]_{i,j=2}^n - |(g_{1\bar{2}}, \ldots, g_{1\bar{n}})|_{g'}^2,$$

where $|v|_{g'}^2 := v^H [g_{i\bar{j}}]_{i,j=2}^n v$. It follows that $|g_{1\bar{j}}| \leq C\rho^{\beta-1}$, for all $j \geq 2$. □

We define

$$\mathcal{H}_\omega^\epsilon := \{\varphi \in C^\infty(M \backslash D) \cap C^0(M) : \omega_\varphi > 0 \text{ on } M, \text{ and } \omega_\varphi \text{ satisfies (3.17)}\}. \tag{3.20}$$

We call

$$\mathcal{H}_\omega^e := \cup_{\epsilon>0} \mathcal{H}_\omega^\epsilon,$$

the *space of Kähler edge metrics*.

We note that another, simpler, way of deriving the preceeding lemma, but which conceals some of what is going on, is by working in the singular coordinate chart $(\zeta, z_2, \ldots, z_n)$. In a nutshell, in terms of the singular coordinates, the definition of a Kähler edge metric simply means that the cross terms in the matrix $[g_{i\bar{j}}]$ (i.e., terms for which precisely one of i and j equals 1) have a fixed rate of decay $O(r^{\epsilon/\beta})$ near D, so that the metric is asymptotically a product. In particular, this means the corresponding Laplacian is also approximated in a certain precise sense by the

Laplacian of the product of a flat cone and $\mathbb{C}^{n-1}$ (i.e., the Laplacian of ω_β). This is very important in proving existence of asymptotic expansions, and structure results for the Green kernel, as we discuss later. In fact, as will follow from the general theory we explain later, in essentially all of the discussion below one may take concretely $\epsilon = \min\{1-\beta, \beta\}$. To be more precise, we will show that solutions to essentially any reasonable complex Monge–Ampère equation always lie in $\mathcal{H}^e_\omega$, and KEE metrics, in fact, even lie in $\mathcal{H}^\beta_\omega$.

We end this subsection with a lemma that shows that $\mathcal{H}^e_\omega$ is a natural choice of space of Kähler metrics in this context. Indeed, once we have found one reference Kähler edge metric, we may produce many other such metrics by adding a Kähler potential with *merely bounded* complex Hessian (with respect to the singular coordinates). This fact may seem counterintuitive at first. It is absolutely crucial for all that follows. Its proof relies crucially on the fact that $\beta < 1$, or, ultimately, on the fact that no nonzero $2\pi\beta$-periodic function b can satisfy $b_{\theta\theta} + b = 0$, i.e., -1 is not in the spectrum of the Laplacian on $S^1(2\pi\beta)$, which equals $-\mathbb{N}_0/\beta^2$.

LEMMA 3.12. *Suppose that $\eta \in \mathcal{H}^\epsilon_\omega$ and that $[u_{i\bar{j}}]$ is bounded (with respect to the coordinates $(\zeta, z_2, \ldots, z_n)$, i.e., $u \in \mathcal{D}^{0,0}_s$). Then $\eta + \sqrt{-1}\partial\bar{\partial}u \in \mathcal{H}^{\tilde{\epsilon}}_\omega$, where $\tilde{\epsilon} = \min\{\epsilon, \beta, 1-\beta\}$.*

This is a corollary of Theorem 3.10 (ii). The proof of this result uses the basic Schauder type estimates in the edge spaces. Indeed, the assumption implies of course that $\Delta_\omega u = f \in L^\infty$. Thus $u = Gf + k$ with k a constant. On the other hand, it is not difficult to show that both $\partial_r \circ G$ and $\frac{1}{r}\partial_\theta \circ G$ map L^∞ to itself (see Theorem 3.7 (i)). Thus, $\frac{1}{r}\partial_\theta u \in L^\infty$ and $\partial_r u \in L^\infty$. We will use both of these facts in a moment. Now, by our assumption, $\partial_y\partial_r u = O(1)$. Integrating twice we find that $u = v(y,\theta) + a(r,\theta) + O(r)$ (with a and v having suitable differentiability). It follows that $\partial_\theta(v + a) = O(r)$. Plugging $r = 0$ this implies that $\partial_\theta v(y,\theta)$ is a function of θ alone, but plugging that back in we see that necessarily then $v + a$ must be a equal to $v_1(y) + a_1(r)$, with $a_1(r) = O(r)$, so $u = O(r) + v_1(y)$. We fix a value of y and regard u as a function on the flat cone $\{y\} \times C_\beta$. We know that $\Delta_\beta u = P_{1\bar{1}}u = F^{(y)} \in L^\infty$, and so $u = G_\beta F + k$ (here $\Delta_\beta = P_{1\bar{1}}$ denotes the Laplacian on the flat cone C_β and G_β its Green function, while $k \in \ker\Delta_\beta$). But now, by the properties of G_β and Δ_β we conclude that $u - v_1(y)$ must vanish to order $O(r^{\min\{1/\beta,2\}})$, and so it follows that $\partial_r\partial_y u = O(r^{\min\{1/\beta-1,1\}})$, from which the lemma follows.

REMARK 3.13. It is a very interesting problem to find regularity properties enjoyed by functions belonging to $\widetilde{\mathcal{D}}^{0,\gamma}_e$. In particular, to find natural conditions that guarantee when a function in $\mathcal{D}^{0,\gamma}_e$ actually belongs to $\widetilde{\mathcal{D}}^{0,\gamma'}_e$ for some $\gamma' > 0$.

3.8. The reference geometry. *"[A] Hermitian metric has the peculiarity of favoring negative curvature over positive curvature."*

— Solomon Bochner [**35**, p. 179].

One of the novelties of [**141**] was to prove a priori second order estimates for a fully nonlinear PDE without curvature bounds on the reference geometry, but only with a one-sided curvature bound. We are not aware of other situations where this is possible.

The next lemma shows that there is no uniform bound, in general, for the curvature of the reference geometry. But, it provides a one-sided bound, which turns out to be very useful for the Laplacian estimates (see §7.3).

LEMMA 3.14. *The bisectional curvature of the reference metric ω (recall 3.5) is bounded from above. In general, it is not bounded from below.*

The proof, due to Li and the author, appears in [**141**, Appendix] and its generalization to the case of normal crossings in [**175, 176**]. It relies on a careful computation, using an adapted normal coordinate system appearing in the work of Tian–Yau [**253**]. The lack of the lower bound can be seen directly from the computations in [**141**, Appendix] by considering the $R_{1\bar{1}1\bar{1}}$ component of the bisectional curvature, and observing that the upper bound can be made as negative as one wishes. If one is only interested in proving the upper bound (but without proving that no lower bound exists), the proof of Lemma 3.14 can be simplified considerably, as pointed out to the author by J. Sturm in November 2013. Indeed, working locally on a ball $B \subset M$ centered at a point in D, one considers the holomorphic map $F : (z_1, \ldots, z_n) \mapsto (z_1, \ldots, z_n, z_1^{\frac{1}{\beta}})$. Let $\pi : \mathbb{C}^{n+1} \to \mathbb{C}^n$ denote the projection to the first n components. The Kähler form $\pi^\star\omega_0 + \sqrt{-1}\partial\bar{\partial}(H(z_1, \ldots, z_n)|z_{n+1}|^2)$ is a smooth metric on a ball in $\mathbb{C}^{n+1}$ whose bisectional curvature is uniformly bounded. Pulling this form back from the complement of $\{z_1 = 0\} \cup \{z_{n+1} = 0\}$ under the holomorphic map F yields the Kähler form $\omega_0 + \sqrt{-1}\partial\bar{\partial}(H|z_1|^{2\beta})$ on $\mathbb{C}^n$ whose bisectional curvature can only decrease by the classical expression for the second fundamental form of a complex submanifold [**124**, p. 79]. Even though the map F is only multivalued, the pull-back of the Kähler form above is continuous single-valued since H is a smooth function of z_1 and the expression $|z_1|^{2\beta}$ is single-valued under this pullback.

A corollary of both statements in the Lemma is that the Ricci curvature of ω is unbounded from below: if it were not, the upper bound on the bisectional curvature would force a lower one as well. As alluded to above, this fact motivates using the Ricci continuity method that starts out, morally, with a metric whose Ricci curvature bound is $-\infty$, and gradually produces a better lower bound until, eventually, arriving at the Kähler–Einstein metric.

In the case of the model edge $C_\beta \times \mathbb{C}^{n-1}$ the curvature is identically zero outside the divisor. In general, it is natural to expect that the holomorphic submanifold geometry of the divisor D in M should be related to whether a reference metric exists with bounded bisectional curvature. A first step in this direction was taken by Arezzo–Della Vedova–La Nave [**4**].

3.9. The structure of the Green kernel. In this subsection we describe in detail the structure of the Green kernel in the case M is S^2 with D a single point. The general case is not much more complicated, but we prefer concreteness over generality in this discussion.

Thus, we start with the singular manifold (S^2, p) and blow-up the point p as in Figure 2. What this means is that we 'separate' the different directions in which one may approach the point p. Each of these directions is now a separate point in the blow-up. As in §3.4, what this real blow-up amounts to is to introduce polar coordinates (r, θ) around p, with $r(p) = 0$. Then $X = \mathrm{Bl}_p^{\mathbb{R}_+} S^2$ is the disjoint union of $S^2 \setminus \{p\}$ and a circle $S^1_{2\pi\beta}$ of radius $2\pi\beta$. It comes equipped with a blow-down map $X \to S^2$ and the inverse image of p is $S^1_{2\pi\beta}$. (In the real setting a real blow-up

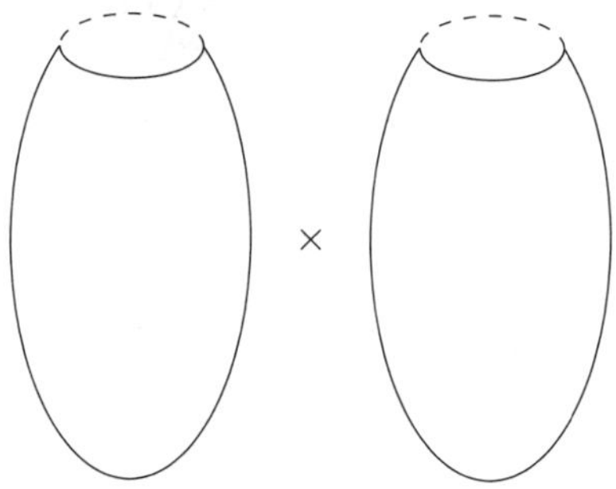

FIGURE 3. The space $X \times X$.

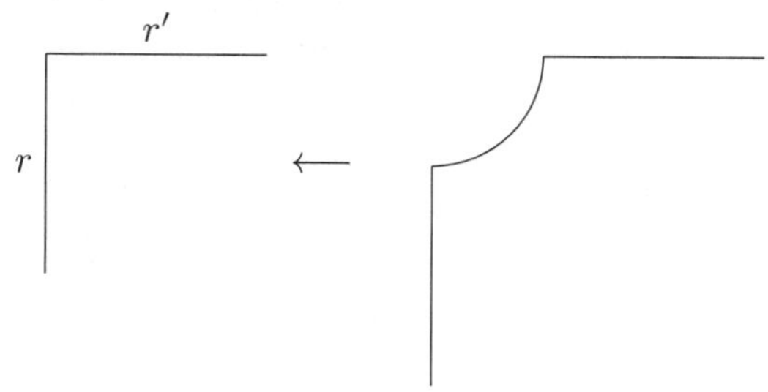

FIGURE 4. The real blow-up of the origin in the quadrant; the origin represents any point in the diagonal of $\partial X \times \partial X$.

sometimes also refers to an $\mathbb{R}$ blow-up where the resulting fiber over p is a half circle or $\mathbb{RP}^1$; therefore we used the superscript $\mathbb{R}_+$ for our 'oriented' blow-up.)

We do not describe the Green kernel G_ω associated to ω (or any phg edge metric) on $X \times X$ (see Figure 3), but instead pull it back under one more blow-up map. The purpose of this is to 'separate' certain directions near the boundary diagonal. In other words, if $\{(0,\theta)\} \times \{(0,\theta')\}$ is a point on $\partial X \times \partial X$ there are many different ways to approach it from the interior: if r, r' are the radial variables on each of the two copies of X, we may let r approach zero faster than r' or vice versa. Thus we consider r, r' as coordinates on the positive orthant $\mathbb{R}^2_+ := \mathbb{R}_+ \times \mathbb{R}_+$ and blow-up the origin (see Figure 4). The result $\mathrm{Bl}^{\mathbb{R}_+}_{(0,0)}\mathbb{R}^2_+$ is the disjoint union of $\mathbb{R}^2_+ \setminus \{(0,0)\}$ and a quarter circle S^1_{++} of radius $\pi/2$. The quarter circle parametrizes various values of the ratio r/r', via the map $\tan^{-1} : [0,\infty) \to S^1_{++}$. This blow-up is a purely local construction; we may thus perform it on $X \times X$, which amounts to blowing-up each point on the boundary diagonal $\partial X \times \partial X$, or in other words blowing-up that whole submanifold. We denote the resulting space by the edge double space

$$X^2_e := \mathrm{Bl}^{\mathbb{R}}_{\partial X \times \partial X} X \times X$$

(here the $\mathbb{R}$ and $\mathbb{R}_+$ blow-ups coincide). In higher dimensions, there are $2n-2$ additional conormal directions y. Then we only blow-up the fiber diagonal $\{r = r' = 0, y = y'\}$ which is a strict submanifold of $\partial X \times \partial X$. We denote by

$$\pi : X^2_e \to X^2$$

the blow-down map. Observe that $\pi^{-1}(\partial X \times \partial X)$ is the union of three hypersurfaces:

$$\mathrm{rf} := \pi^{-1}\big(\{r = 0\}\big), \quad \mathrm{lf} := \pi^{-1}\big(\{r' = 0\}\big),$$

called the right and left faces, and the new hypersurface

$$\mathrm{ff} := \pi^{-1}\big(\{r = r' = 0\}\big),$$

diffeomorphic to $S^1 \times S^1 \times S^1_{++}$, called the front face. These hypersurfaces all have coordinate descriptions in terms of polar coordinates about the corner $\{r = r' = 0\}$. Namely, denote $R := \sqrt{r^2 + r'^2}$ and set $(\psi, \psi') := (r/R, r'/R) \in S^1_{++}$. The double edge space then is parametrized near its boundary by $(R, \psi, \psi', \theta, \theta')$ (while, by comparison, X^2 was parametrized by (r, r', θ, θ')), and $\mathrm{rf} = \{\psi = 0\}, \mathrm{lf} = \{\psi' = 0\}, \mathrm{ff} = \{R = 0\}$. Away from its boundary X^2_e is locally diffeomorphic to $M \times M$, of course, and is parametrized there by coordinates on the latter.

Now, we return to the general setting of a Kähler edge manifold, and describe G_ω as the push-down of a distribution on X^2_e under the blow-down map π. For concreteness, the reader may focus on the example given above, even though the result below is stated in the general setting. The first main reason for doing so is that G_ω is of course singular along the diagonal of X^2, however the diagonal intersects the boundary nontransversally, while the lifted diagonal

$$\mathrm{diag}_e := \{(R, 1/\sqrt{2}, 1/\sqrt{2}, \theta, \theta) \,:\, R \in \mathbb{R}_+, \theta \in S^1\}$$

intersects ff transversally at $\{(0, 1/\sqrt{2}, 1/\sqrt{2}, \theta, \theta)\} \cong S^1$, and the intersection points lie in the interior of ff. Another reason for working on X^2_e is the dilation invariance structure. Finally, here is the description of G_ω, a corollary of a general result of Mazzeo [**172**, Theorem 6.1], see [**141**, Proposition 3.8].

THEOREM 3.15. *Let g be a polyhomogeneous edge metric with index set E_g with angle β along D and denote by G the generalized inverse to the Friedrichs extension of $-\Delta_g$. Then the Schwartz kernel K_G of G is a distribution on X^2_e that can be decomposed as $K_G = K_1 + K_2$ with the following properties:*
K_1 is supported on a neighborhood of diag_e disjoint from lf and rf; $R^{2n-2}K_1$ has a classical pseudodifferential singularity of order -2 on diag_e that remains conormal when extended to ff.
K_2 is polyhomogeneous on X^2_e with index sets $2 - 2n$ at ff, and E at both rf and lf, where

$$E \subset \{(j/\beta + k, \ell) : j, k, \ell \in \mathbb{Z}_+ \text{ and } \ell = 0 \text{ for } j + k \le 1,\ (j, k, \ell) \ne (0, 1, 0)\},$$

when $E_g = \{0\}$, and otherwise E is contained in a larger set that is determined by E_g and the above set. In particular,

$$T \circ G : C^{0,\gamma}_e \to C^{0,\gamma}_e,$$

is a bounded operator where $T \in \mathcal{Q}$ is as in Theorem 3.7.

The proof of this occupies a good part of [**172**] and we will leave a detailed exposition of it in our complex setting to a separate exposition.

3.10. Higher regularity for solutions of the Monge–Ampère equation. The standard theory of elliptic regularity applies directly to the Monge–Ampère equation, despite it being fully nonlinear [**123**, §17]. Indeed, the linearization of the Monge–Ampère equation is simply the Poisson equation (with a potential)—we review this shortly. It implies that any $C^{2,\alpha}$ solution of (2.3) is automatically smooth (for an alternative approach see [**83**]). This assumes, of course, that the reference form ω is smooth. A penetrating feature of Kähler edge geometry is that although ω (see (3.5)) is no longer smooth, it is possible to develop an analytical theory that isolates, so to speak, the only direction in which the geometry is singular, or, equivalently, in which the associated Laplacian is degenerate (fails to be elliptic). Thus, KEE metrics turn out to be smooth in all conormal directions. One could stop here and argue that this is already the right analogue of higher regularity in the singular setting. However, as already indicated in §§3.4, Taylor's theorem does not apply to a merely conormal function. The analogue of smoothness is thus the space of polyhomogeneous conormal distributions. This is more than an academic difference. One of the thrusts of this approach is that polyhomogeneity yields basic geometric information. Also, the proofs of these facts are rather standard by now—essentially a matter of high-level bookkeeping, somewhat analogously to tools in algebraic geometry. These proofs have their origin in the fundamental work of Melrose, developed in-depth in the case of real edges by Mazzeo, and further developed in the case of complex codimension one edges in [**141**] and in any codimension for crossing complex edges of codimension one [**175, 176**]. For the rest of this subsection we describe, in broad strokes, how higher regularity is proved in the edge setting and mention some of its basic analytic and geometric consequences.

A basic fact is that solutions of a general class of complex Monge–Ampère equations are automatically polyhomogeneous as soon as they lie in the maximal Hölder domain $\mathcal{D}_e^{0,\gamma}$. For concreteness, we concentrate on the special class of equations that we will consider in §6. We denote by $\mathrm{PSH}(M,\omega)$ the space of ω-plurisubharmonic functions on M, namely upper semi-continuous functions u from M to $[-\infty,\infty)$ such that $\omega_u=\omega+\sqrt{-1}\partial\bar\partial u$ is a nonnegative current on M.

THEOREM 3.16. *Let ω be a polyhomogeneous Kähler edge metric. Suppose that $\varphi\in\widetilde{\mathcal{D}}_s^{0,\gamma}\cap\mathrm{PSH}(M,\omega)$, satisfies*

$$\omega_\varphi^n=\omega^n e^{f+c\varphi},\qquad \text{on } M\setminus D, \tag{3.21}$$

where $c\in\mathbb{R}$ and $f\in\mathcal{A}^0_{\mathrm{phg}}$. Then $\varphi\in\mathcal{A}^0_{\mathrm{phg}}$.

The proof, that we now outline, relies mostly on the linear theory for $\mathcal{D}_e^{0,\gamma}$. The detailed proof appears in [**141**, §4].

PROOF. Differentiating (the logarithm of (3.21)) in a conormal direction y_a (the same discussion applies to differentiation in θ), the Monge–Ampère equation becomes a seemingly linear equation,

$$\Delta_{\omega_\varphi}\partial_{y_a}\varphi=c\partial_{y_a}\varphi+\partial_{y_a}f\in C_e^{0,\gamma}, \tag{3.22}$$

where we used Theorem 3.7 (i) for the inclusion. The nonlinearity comes, of course, from the fact that the operator Δ_{ω_φ} itself depends on φ. However, by Theorem 3.9 this is irrelevant, as we assume that $\varphi\in\widetilde{\mathcal{D}}_s^{0,\gamma}\subseteq\widetilde{\mathcal{D}}_e^{0,\gamma}$! Thus, we conclude that $\partial_{y_a}\varphi,\partial_\theta\varphi\in\mathcal{D}_e^{0,\gamma}$. Again, because $\varphi\in\mathcal{D}_e^{0,\gamma}$ this implies that $\partial_y(\Delta_{\omega_\varphi}\varphi)\in C_e^{0,\gamma}$. By induction, $\partial_\theta^l\partial_y^k(\Delta_{\omega_\varphi}\varphi),\Delta_{\omega_\varphi}(\partial_\theta^l\partial_y^k\varphi)\in C_e^{0,\gamma}$ for all $l,k\in\mathbb{N}\cup\{0\}$ with

$l + k > 0$, where ∂_y^k denotes any operator of the form $\partial_{y_{a(1)}} \circ \cdots \circ \partial_{y_{a(k)}}$, with $a(i) \in \{1, \ldots, 2n-2\}$. Thus, by composing with G_{ω_φ}, we see that $\partial_y \varphi, \partial_\theta \varphi$ are infinitely differentiable with respect to ∂_θ and ∂_y. Moreover, by (3.22) and standard elliptic regularity in the edge spaces [**172**] it follows that $\partial_y \varphi, \partial_\theta \varphi$ are infinitely differentiable with respect to $r\partial_r$. And the inductive argument above then gives the same also for $\partial_\theta^l \partial_y^k \varphi$. Namely, $\partial_y \varphi, \partial_\theta \varphi \in \mathcal{A}^0$.

Next, we write $\partial_{y_a} \varphi = G_{\omega_\varphi}(\partial_{y_a} f + c\partial_{y_a} \varphi) + \kappa$, where $\kappa \in \ker \Delta_{\omega_\varphi}$ is simply a constant. We show now that $\partial_{y_a} \varphi \in \mathcal{A}^0_{\text{phg}}$. We claim that the Green operator G_{ω_φ} maps a conormal function to a function in $r^2\mathcal{A}^0$, at least modulo something polyhomogoneous (phg). This is an improvement on the general fact (cf. [**172**, Proposition 3.28]) that such an operator (we do not go into the details of what "such" means here, but refer to [**141**, Lemma 4.2] for a precise statement) maps $\mathcal{A}^0$ to itself. By induction, this is all we need to conclude the proof, as we have already showed that $\Delta_{\omega_\varphi} \partial_{y_a} \varphi = \partial_{y_a} f + \partial_{y_a} \varphi \in \mathcal{A}^0$; applying G_{ω_φ} to both sides would then show $\partial_{y_a} \varphi \in r^2\mathcal{A}^0 + \mathcal{A}^0_{\text{phg}}$. But, since f is phg it then follows that $\partial_{y_a} f + \partial_{y_a} \varphi \in r^2\mathcal{A}^0 + \mathcal{A}^0_{\text{phg}}$. And so, applying the previous claim again we conclude $\partial_{y_a} \varphi \in r^4\mathcal{A}^0 + \mathcal{A}^0_{\text{phg}}$, and thus by induction $\partial_{y_a} \varphi \in \mathcal{A}^0_{\text{phg}}$.

Here is the idea behind the proof of the claim: by a theorem of Mazzeo [**172**, Theorem 6.1] we know that G_β, the Green kernel of the reference metric ω, has a polyhomogeneous kernel. If all terms in its expansion, with one variable frozen, are $O(r^2)$ or better we would be done. In general though there will be finitely many (positive, of course) exponents $\gamma_1, \ldots, \gamma_N$ in the polyhomogeneous expansion of G that are smaller than 2. (In our setting, there is only one such, $1/\beta$.) But then post-composing G_β with $\Pi_{i=1}^N (r\partial_r - \gamma_i)$ eliminates these terms. Integrating the equation $\Pi_{i=1}^N (r\partial_r - \gamma_i) \circ G_\beta v = O(r^2)$ in r thus yields that $G_\beta v = O(r^2) + \sum u_i(y, \theta) r^{\gamma_i}$. When $v \in \mathcal{A}^0$ this yields therefore $G_\beta v \in r^2\mathcal{A}^0 + \mathcal{A}^0_{\text{phg}}$. However, we are dealing with the Green kernel G_{ω_φ} which does not necessarily have a phg expansion (since φ is not phg as of yet). However, by Theorem 3.9 the domain $\widetilde{\mathcal{D}}^{0,\gamma}_e(\Delta_{\omega_\varphi})$ is independent of $\varphi \in \widetilde{\mathcal{H}}^e_\omega$, and simply equals $\widetilde{\mathcal{D}}^{0,\gamma}_e$ (see Remark 3.13). In particular, G_{ω_φ} has the same asymptotic behavior near the lf and rf of X. In other words, while it is not phg, it has a partial expansion (the regularity implied by belonging to $\mathcal{D}^{0,\gamma}_e$), of the form $a_0(y) + (a_1(y)\cos\theta + a_2(y)\sin\theta) r^{1/\beta}) + O(r^2)$. Thus, the same argument as above implies that for $v \in \mathcal{A}^0$ one has $G_{\omega_\varphi} v \in r^2\mathcal{A}^0 + \mathcal{A}^0_{\text{phg}}$.

Thus, $\nabla_y \varphi \in \mathcal{A}^0_{\text{phg}}$, and by the same reasoning also $\partial_\theta \varphi \in \mathcal{A}^0_{\text{phg}}$. Thus, integrating, $\varphi = \varphi_0 + \varphi_1$ with φ_0 a function of r alone, and $\varphi_1(r, \theta, y) \in \mathcal{A}^0_{\text{phg}}$. But we already know that $\varphi \in C^\infty_e$. Thus, φ_0 lies in C^∞_e, but then necessarily also in $\mathcal{A}^0$ since it is independent of y, θ. Thus, $\varphi_0(r)$ and hence φ lie in $\mathcal{A}^0$ and, moreover, $\varphi - \varphi_0(r) \in \mathcal{A}^0_{\text{phg}}$. Thus, to obtain Theorem 3.16 it remains to prove that $\partial_r \varphi \in \mathcal{A}^0_{\text{phg}}$.

To that end, we apply ∂_r to the logarithm of (3.21). This yields $\partial_r \varphi = G_{\omega_\varphi}(\partial_r f + c\partial_r \varphi) + \kappa_2$, where now the term in paranthesis belongs to $r^{-1}\mathcal{A}^0$, and $\kappa_2 \in \mathbb{R}$. Since the volume form with respect to which G_{ω_φ} is defined is asymptotically equivalent to $rdrd\theta dy$ we may still apply G_{ω_φ} to a distribution in $r^{-1}\mathcal{A}^0$ and obtain a distribution that lies in $r^{-1}\mathcal{A}^0$. But now we know that $G_{\omega_\varphi} \partial_r f$ is actually bounded, since by Theorem 3.7 (i) $\partial_r \varphi$ is as $\varphi \in \mathcal{D}^{0,\gamma}_e$. Thus, again, we may treat the equation as an ODE in r, and show that, in fact, $\partial_r\varphi$ lies in $r\mathcal{A}^0+\mathcal{A}^0_{\mathrm{phg}}$. As before, an induction results in $\partial_r\varphi\in\mathcal{A}^0_{\mathrm{phg}}$. In conclusion then $\varphi\in\mathcal{A}^0_{\mathrm{phg}}$, completing the proof. □

REMARK 3.17. A precursor to this sort of argument is the work of Lee–Melrose [**154**] on the complex Monge–Ampère equation on a domain, and related work of Mazzeo on the singular Yamabe problem [**173**]. We also refer to Rochon–Zhang [**209**] for recent work. All of these articles deal with several quite different situations, but are all on complete spaces, as opposed to our incomplete setting.

We end this subsection by discussing the significance of the terms that appear in the asymptotic expansion. When the function on the right hand side f has a certain index set E_f (see (3.13)) the solution then has an index set that will depend on E, E_ω, and E_f where E is the index set of G_β, and E_ω that of ω (see (3.5)), by which we mean the index set of $\psi_0+\phi_0$ where ψ_0 is any Kähler potential for ω_0 and ϕ_0 is defined in (3.6). In the setting of the Kähler–Einstein equation, one may determine the index set of the solution from this observation. This is a somewhat tedious, albeit completely inductive routine—one simply treats the Monge–Ampère equation as an ODE in r and computes the terms that can occur in the expansion, relying on our alternative characterization of $\mathcal{D}^{0,\gamma}_e$ (Theorems 3.7 and 3.9). It follows [**141**, Proposition 4.3] that solutions to essentially all the complex Monge–Ampère equations considered in the setting of the Kähler–Einstein equation have the following expansion:

$$\varphi(r,\theta,y)=a_{00}(y)+(a_{01}(y)\sin\theta+b_{01}(y)\cos\theta)r^{\frac{1}{\beta}}+a_{20}(y)r^2+O(r^{2+\epsilon}) \tag{3.23}$$

for some $\epsilon=\epsilon(\beta)>0$.

REMARK 3.18. When $\beta\in(1/3,1/2]$, one may readily show that the term in the expansion after $O(r^{1/\beta})$ is $O(r^{1/\beta+1})$. This is roughly equivalent to the divisor D being totally geodesic for all $\beta\in(0,1/2)$ (for $\beta\in(0,1/3)$ this is related to Atiyah–LeBrun [**5**]). This is another justification for calling the regime $\beta\in(0,1/2]$ the orbifold regime.

Note that the term of order $r^{1/\beta}$ in (3.23) is annihiliated by $\partial^2/\partial\zeta\overline{\partial\zeta}$. From this, and the explicit formulas for the reference metric ω and its derivatives [**141**, §2.2] one immediately deduces the following geometric information [**141**, Theorem 2].

COROLLARY 3.19. *Let $\varphi\in\widetilde{\mathcal{H}}^e_\omega\cap\mathcal{D}^{0,\gamma}_e$, and suppose that ω_φ is KEE or else is a solution of a complex Monge–Ampère equation of the type (3.21) with E_F (see (3.13)) suitably small. Then $\omega_\varphi\in C^{0,\min\{1,\frac{1}{\beta}-1\}}_w$, but in general its Christoffel symbols and curvature tensor are unbounded. More precisely, in the coordinates (r,θ,y), $g_{i\bar{j}}=O(1+r+r^{\frac{1}{\beta}-1}+r^{\frac{2}{\beta}-2})$, $R_{i\bar{j}k\bar{l}}=O(1+r^{\frac{1}{\beta}-2}+r^{\frac{2}{\beta}-4})$, while $\Gamma^k_{ij}=g^{k\bar{l}}g_{i\bar{l},j}=O(1+r^{\frac{1}{\beta}-2}+r^{\frac{2}{\beta}-3})$. Moreover, $\omega_\varphi|_D$ is a smooth Kähler metric.*

As another corollary, Song–Wang observed that the expansion (3.23) directly implies that the curvature tensor of a KEE metric is in L^2 with respect to the metric [**230**, §4.1].

4. Existence and non-existence

In this section we survey necessary and sufficient conditions for existence of KE(E) metrics. We start in §4.1 with the easier nonpositive curvature regime, where the cohomological criterion (3.12) is necessary and sufficient. We then move on to obstructions (§4.2–4.3), and finally, an essentially optimal sufficient condition for existence in the positive case (§4.5). In §4.4 we pause to discuss the Riemann surface case. The existence theorems described in §4.1 and §4.5 are the main existence results for KE(E) metrics, and their proof uses results from Sections 3, 6, and 7.

We now review some background that is useful in describing some of the obstructions below. Denote by

$$\mathcal{H}_\omega = \{\varphi \in C^\infty(M) \,:\, \omega_\varphi := \omega + \sqrt{-1}\partial\bar\partial\varphi > 0\}$$

the (moduli) space of Kähler potentials representing Kähler forms (equivalently, metrics) in a fixed cohomology class $\Omega = [\omega]$. When we discuss Kähler edge metrics, the natural replacement is $\mathcal{H}^\epsilon_\omega$ defined in (3.20). The corresponding space of Kähler forms is denoted by

$$\mathcal{H}_\Omega = \{\alpha \text{ is a Kähler form with } [\alpha] = \Omega\}.$$

These are infinite-dimensional Fréchet manifolds, whose transformations, functions, forms, and vector fields control different aspects of the KE problem. We first introduce some basic objects on this space.

The tangent bundle of $\mathcal{H}_\omega$ is isomorphic to $\mathcal{H}_\omega \times C^\infty(M)$, and similarly $T^\star\mathcal{H}_\omega \cong \mathcal{H}_\omega \times \Gamma(M, \Lambda^{2n}T^\star M)$, the latter factor denoting the space of top degree forms on M, with the fiberwise pairing given by integration over M. The Mabuchi metric on $\mathcal{H}_\omega$ is defined by [**94, 168, 221**]

$$g_{\mathrm{M}}(\nu,\eta)|_\varphi := \frac{1}{V}\int_M \nu\eta\,\omega_\varphi^n, \quad \nu,\eta \in T_\varphi\mathcal{H}_\omega \cong C^\infty(M), \tag{4.1}$$

where $V = [\omega]^n/n!$. Note that the constant vector field $\mathbf{1}(\varphi) = 1$ is of unit norm. This induces a Riemannian splitting $\mathcal{H}_\omega = \iota_\omega(\mathcal{H}_\Omega) \times \mathbb{R}$, with $d\iota_\omega(T\mathcal{H}_\Omega) \perp \mathbf{1}$, $\mathcal{H}_\Omega$ thus identified with a totally geodesic submanifold of $\mathcal{H}_\omega$ passing through $0 \in \mathcal{H}_\omega$. Thus, with some abuse of notation, we may speak of geodesics in $\mathcal{H}_\omega$ or in $\mathcal{H}_\Omega$ interchangeably, with the latter meaning geodesics in $\iota_\omega(\mathcal{H}_\Omega)$. Remarkably, $\mathbf{1}$ is a gradient vector field for g_{M} [**168**, Theorem 2.3], and the corresponding potential

$$L : \mathcal{H}_\omega \to \mathbb{R} \tag{4.2}$$

is thus a distance function for g_{M} (in the sense that the norm of its gradient is one; as an aside we remark that it would be interesting to find an interpretation of this fact in terms of the Mabuchi distance), as first observed by Mabuchi [**167**, Theorem 2.3],[**168**, Remark 3.3]. It is known as the Aubin–Mabuchi or Monge–Ampère energy since $dL|_\varphi = {\omega_\varphi}^n$ (see also §5.1), and sometimes also referred to as the Aubin–Yau functional, and has been studied by many authors, see, e.g., [**18, 21–23, 200**]. Consequently, since $\mathbf{1}$ is constant, for any C^2 curve $\gamma(s)$ in $\mathcal{H}_\omega$,

$$\ddot{L}(\gamma(s)) = g_{\mathrm{M}}(1, \nabla^{g_{\mathrm{M}}}_{\dot\gamma}\dot\gamma), \tag{4.3}$$

with the notation $\dot f := df/ds, \ddot f := d^2f/ds^2$. That is, if $\nabla^{g_{\mathrm{M}}}_{\dot\gamma}\dot\gamma \geq 0$ then $\gamma(s)$ is a geodesic iff $L(\gamma(s))$ is linear in s. Note also that $\iota_\omega(\mathcal{H}_\Omega) = L^{-1}(0)$.

4.1. Nonpositive curvature: the Calabi–Tian conjectures. As reviewed in §2 the early history of Kähler–Einstein metrics revolved around the local version of this equation. Two decades later, Calabi first formulated an ambitious program for constructing KE metrics on *closed* manifolds. He explicitly formulated in writing the case of zero Ricci curvature, but in analogy with the uniformization theorem, he expected that the negative case should then follow as well. The following statement was first formulated as a theorem in [**44**], before the existence part quickly became a conjecture thanks to discussions between Calabi and Nirenberg on the need of a priori estimates (private communication of E. Calabi to the author, 2011).

CONJECTURE 4.1. (Calabi's conjecture 1953 [**44–46**]) *Let (M, J) denote a closed Kähler manifold with $c_1(M, J) < 0$ or $c_1(M, J) = 0$. Then (M, J) admits a Kähler–Einstein metrics that are unique up to homothety in the former case, and unique in each Kähler class in the latter.*

When $\mu < 0$, the uniqueness is immediate from the maximum principle: if u, v satisfy $\omega_u^n/\omega_v^n = e^{|\mu|(u-v)}$, then, if $u - v$ is maximized in $p \in M$, the form $\sqrt{-1}\partial\bar{\partial}(u - v)$ has non-positive eigenvalue with respect to ω_v, and so $u - v \leq 0$; by symmetry $v - u \leq 0$, so $u = v$. When $\mu = 0$ the uniqueness part was proved by Calabi [**46**] by exploiting the algebraic structure of the Monge–Ampère equation (more specifically, properties of determinants and integration by parts). Indeed, if $\omega_u^n = \omega_v^n$ then $0 = \omega_u^n - \omega_v^n = \sqrt{-1}\partial\bar{\partial}(u-v)\wedge T$ with T a positive $(n-1, n-1)$-form. Multiplying by $u - v$ and integrating by parts shows $u - v$ is constant.

The existence part of Calabi's conjecture was established by Aubin in the negative case and by Yau in both cases [**8, 268**] (the zero case under the restrictive assumption that the manifold admits a reference Kähler metric with nonnegative bisectional curvature was established earlier by Aubin [**6**]). The main innovation was to establish the a priori C^0 and Laplacian estimates conjectured by Calabi and Nirenberg. The higher derivative estimates then followed by work of Calabi described in §7.8, and by elliptic bootstrapping.

Four decades later, motivated by application to algebraic geometry, Tian formulated a generalization for pairs of Calabi's conjecture.

CONJECTURE 4.2. (Tian's conjecture 1994 [**243**]) *Let (M, J) denote a closed Kähler manifold and $D = D_1 + \ldots + D_r \subset M$ a divisor with simple normal crossing support. Suppose that $c_1(M, J) - \sum_{i=1}^{r}(1-\beta_j)[D_j]$ is negative or zero. Then (M, J) admits a KEE metric with angle $2\pi\beta_j$ along D_j that is unique in the former case, and unique in each Kähler class in the latter.*

This conjecture, similarly to Calabi's original conjecture, was also first stated as a theorem in Jeffres' Ph.D. thesis [**142, 243**], but subsequently only the uniqueness was published [**143**] due to the need of a priori estimates, as well as a good linear theory. A program outlining a collaboration between Jeffres and Mazzeo toward such a linear theory was announced by Mazzeo [**174**]. As described in Theorem 3.7, two independent sets of linear estimates, in the case of a smooth divisor, were obtained, finally, first by Donaldson [**102**], and subsequently in [**141**]. What was lacking from the approach suggested in [**174**] was the simple, but crucial, observation of [**102**] that if one simply considers Hölder control only of the complex $(1, 1)$-type derivatives, then a Schauder theory can be established; if one considers the full Hessian then this is not the case, due to the harmonic function $z^{1/\beta}$. In the case of a smooth divisor, this conjecture was resolved by Mazzeo and the author [**141**], that also established higher regularity of the metric. The key here was, first, to establish the a priori estimates in the absence of curvature bounds and, second, to develop a good linear theory in the edge spaces that, among other things, allows to obtain higher regularity (polyhomogeneity). A different proof of existence was later given by Guenancia–Păun [**127**] relying on work of Guenancia–Păun–Campana [**52**] that appeared around the same time as [**141**] (also around the same time, Brendle [**42**] obtained the special case of a smooth divisor and angle $\beta \in (0, 1/2]$ using Theorem 3.7 (i) with $s = w$ whose proof, in turn, appeared subsequently in [**141**]). As observed by Datar–Song [**81**], the existence in the general snc case actually follows easily by combining the Chern–Lu approach of [**141**] and a standard regularization argument à la Demailly; a more computationally involved proof by Guenancia–Păun of the general case appeared slightly earlier than [**81**], and around the same time also another approach based on regularization was developed by Yao [**266**]. Finally, the general snc case has also been settled by [**175, 176**] (independently of [**127**]), where, in addition, the higher regularity of solutions is proved by establishing a linear theory also in the snc case; such a theory is considerably more complicated than in the smooth divisor case.

Much of Sections 3, 6, and 7 are devoted to describing the proof of Conjecture 4.2 due to Mazzeo and the author [**141, 175**]. In fact, the proof of Conjecture 4.2 is, in fact, a corollary of the proof of the parallel result in the harder positive case—Theorem 4.13. The strategy of proof, following [**141**], is described in §4.5 below, where the latter result is stated. We also survey other approaches in §6.6.

Fulfilling one of the original motivations for introducing edges—see §3.1, the following result was first obtained in 2011 in [**141**] (an alternative proof was given in 2013 by Guenancia–Păun [**127**]):

COROLLARY 4.3. (KEE metrics on all projective manifolds) *Every projective manifold admits a KEE metric of negative curvature for any angle* $2\pi\beta \in (0, 2\pi)$.

In fact, on any projective manifold M there exists an ample class H. Therefore, given $\beta \in (0, 1)$, for large enough $m = m(\beta) \in \mathbb{N}$, both $|mH|$ admits a smooth representative by Bertini's theorem, and the class $K_M + (1-\beta)m_\beta H$ is positive. Note that if M is *minimal*, i.e., K_M is nef, then one may take $m_\beta = 1$ independently of β by Kleiman's criterion, so that the fixed pair (M, H) admits KEE metrics for all $\beta \in (0, 1)$.

PROBLEM 4.4. *Let M be minimal and H a smooth ample divisor. Describe the limiting behavior of the KEE metrics on (M, H) as β tends to 1.*

EXAMPLE 4.5. To illustrate some of the new metrics arising from Corollary 4.3 consider the projective plane $\mathbb{P}^2$ which is typically thought of as 'positively curved'. By choosing a smooth plane curve D of degree $d \geq 4$ and any $\beta \in (0, (d-3)/d)$, one can thus construct a KEE metric of negative Ricci curvature. As $d \to \infty$ the angle $2\pi\beta$ can be made arbitrarily close to 2π. Of course, this example can be generalized to any $\mathbb{P}^n$, with D a smooth hypersurface of degree d, and $\beta \in (0, (d-1-n)/d)$.

PROBLEM 4.6. *As $d \to \infty$, does the divisor D becomes dense in $\mathbb{P}^n$ with respect to the KEE metric? How do the diameters of D and M behave? What is a Gromov–Hausdorff limit of such a metric space (for different limiting values of β)?*

The existence problem in the positive Ricci curvature realm is more complicated, as the cohomological assumption is not sufficient. In the rest of this section we first describe obstructions to existence, and then, finally, formulate a general existence criterion.

4.2. Reductivity of the automorphism group. The (connected) isometry group of the round 2-sphere complexifies to give the (identity component of the) Möbius group of its conformal transformations. A similar fact also holds for the isometry group of a football: it complexifies to the conformal transformations that fix the poles, or cone points. Matsushima's theorem states that the same is true for any KEE manifold.

THEOREM 4.7. *Let (M, J, D, g) be a KEE manifold. Let $\mathrm{Isom}_0(M, g)$ denote the identity component of the isometry group and denote by $\mathrm{Aut}_0(M, D)$ the identity component of the Lie group of automorphisms preserving D. Then $\mathrm{Aut}_0(M, D) = \mathrm{Isom}_0(M, g)^{\mathbb{C}}$. In particular, $\mathrm{Aut}_0(M, D)$ is reductive.*

Matsushima's original result was proved for KE manifolds and extended to constant scalar curvature manifolds by Lichnerowicz [**160, 171**]. The singular version appearing in [**60**] follows the same lines, by using regularity results from [**141**]. A more general result (assuming the manifold only has a weak KE metric in the sense of [**108**]) but in the special case M is Fano and $D \in |-K_M|$ can be found in [**63, 249**].

Instead of reviewing the proofs above, we describe a formal proof due to Donaldson in the smooth setting, in the spirit of infinite-dimensional geometry [**98**].

First, let $\mathrm{diff}(M, D)$ denote the Lie algebra of vector fields on M that vanish on D, and denote by $\mathrm{ham}(M, D, \omega)$ the Lie subalgebra of fields X satisfying $[\iota_X\omega] = 0$. The corresponding Lie subgroup, $\mathrm{Ham}(M, D, \omega)$, consists of Hamiltonian diffeomorphisms of $(M \setminus D, \omega)$ that preserve D. The group $\mathrm{Ham}(M, D, \omega)$ acts in a natural way on the space $\mathcal{J}$ of ω-compatible complex structures. Furthermore, $\mathcal{J}$ comes equipped with a natural symplectic structure. A result of Donaldson in the smooth setting shows that the $\mathrm{Ham}(M, D, \omega)$-action on $\mathcal{J}$ is Hamiltonian, with a moment map given by the scalar curvature $s(\omega, J)$ minus its average. As long as we restrict to diffeomorphisms F that preserve polyhomogeneity and for which further $F^\star\omega - \omega = \sqrt{-1}\partial\bar{\partial}u_F$, with $u_F \in \mathcal{D}^{0,\gamma}_w$, it is not hard to show that this result extends to the edge setting.

Now, suppose a holomorphic Hamiltonian group action $\mathcal{G}$ can be complexified to $\mathcal{G}^{\mathbb{C}}$. A basic fact in finite-dimensional moment map geometry says that for an element belonging to the zero of the moment map, the isotropy group of the $\mathcal{G}^{\mathbb{C}}$-action is the complexification of the isotropy group of the $\mathcal{G}$-action. Now for any $J \in \mathcal{J}$ the isotropy group of $\mathrm{Ham}(M, D, \omega)$ are those diffeomorphisms of $M \setminus D$ that preserve both J and ω, i.e., isometries. Similarly, if one is willing to think of $\mathrm{Ham}(M, D, \omega)^{\mathbb{C}}$ as all diffeomorphisms preserving D and the (1,1)-type of ω, then the isotropy group of J can be identified with $\mathrm{Aut}(M, D)$ (some motivation for such a formal identification is discussed in [**94, 221**]). Thus, if $J \in \mathcal{J}$ is a constant scalar curvature structure, the Matsushima–Lichnerowicz criterion follows.

EXAMPLE 4.8. The automorphism group of all strongly asymptotically log del Pezzo pairs (these pairs are defined in Definition 8.6) are computed in [**60**]. Here are some explicit examples. The pair $(\mathbb{P}^n, H)$ with H a hyperplane in $\mathbb{P}^n$, satisfies

$$\mathrm{Aut}(\mathbb{P}^n, H) \cong \mathrm{Aut}(\mathbb{P}^n, p) \cong \mathrm{Aut}(\mathrm{Bl}_p\mathbb{P}^n) \cong \mathbb{G}_a^n \rtimes \mathrm{GL}_n(\mathbb{C}),$$

for a point $p \in \mathbb{P}^n$, where $\mathrm{Bl}_p\mathbb{P}^n$ denotes the blow-up of $\mathbb{P}^n$ at p, and where $\mathbb{G}_a$ the additive group $(\mathbb{C}, +)$. The latter group is not reductive. This generalizes the well-known obstruction to the existence of a constant curvature metric on the teardrop (S^2 with one cone point), see §4.4 below (another way to see this pair is obstructed is to use the Bogomolov–Miyaoka–Yau inequality [**230, 243**]). For the pair $(\mathbb{P}^2, D)$, with $D \in |2H|$ a smooth quadric, any automorphism of D lifts to an automorphism of $\mathbb{P}^2$, since $\mathbb{P}^2(H^0(D, -K_D)) = \mathbb{P}^2$ (again, note that $\mathrm{Aut}(\mathbb{P}^2) \to \mathrm{Aut}(D)$ is injective, since an automorphism of $\mathbb{P}^2$ can only fix a linear subspace). Thus $\mathrm{Aut}(M, D)$ equals $\mathrm{PGL}_2(\mathbb{C})$ (and is reductive). This pair, however, turns out to be obstructed only when $\beta \in (0, 1/4]$ [**157**].

4.3. Mabuchi energy, Futaki character, and their relatives. When (2.2) holds, the KE problem reduces to finding a constant scalar curvature metric in $\mathcal{H}_\omega$. Indeed, $\bar\partial s_\omega = \partial^\star \mathrm{Ric}\,\omega$, where $s_\omega = \mathrm{tr}_\omega \mathrm{Ric}\,\omega$, denotes the scalar curvature (more precisely, half of the standard convention in Riemannian geometry). Or simpler, $s_\omega \omega^n = n\mathrm{Ric}\,\omega \wedge \omega^{n-1}$ and thus s_ω is constant iff $\mathrm{Ric}\,\omega$ is harmonic, i.e., a multiple of ω. This can be considered as an easy Kählerian analogue of the Obata theorem in conformal geometry.

Thus, it is natural to consider the following vector field on $\mathcal{H}_\omega$:

$$\mathbf{s} : \omega \mapsto s_\omega - g_{\mathrm{M}}(s_\omega, 1) = s_\omega - nc_1.[\omega]^{n-1}/[\omega]^n. \tag{4.4}$$

4.3.1. *Mabuchi K-energy.* The zeros of the vector field $\mathbf{s}$ are the constant scalar curvature (csc) metrics in $\mathcal{H}$, and its integral curves are the trajectories of the Calabi flow [**48**]. A remarkable fact is that $\mathbf{s}$ is in fact a gradient vector field $\nabla^M E_0 = \mathbf{s}$ [**167**]. Its potential $E_0 : \mathcal{H}_\omega \times \mathcal{H}_\omega \to \mathbb{R}$

$$E_0(\omega_0, \omega_1) = \int_\gamma \mathbf{s}^\flat,$$

is known as the Mabuchi K-energy, where $\mathbf{s}^\flat$ denotes the 1-form associated to $\mathbf{s}$ via g_{M}, and γ is any (sufficiently regular) path between ω_0 and ω_1.

We now mention several fundamental properties of Mabuchi's K-energy due to Bando–Mabuchi [**14**]. Any critical point of E_0 is KE as already remarked; and in fact all such critical points lie in a connected finite-dimensional totally geodesic submanifold of $\mathcal{H}_\Omega$ parametrizing all KE metrics in $\mathcal{H}_\Omega$ and isometric to the symmetric space $\mathrm{Aut}_0(M, J)/\mathrm{Isom}_0(M, \omega_{KE})$ (the zero subscript stands for the identity components of these groups)—the orbit of a single KE metric $\omega_{\mathrm{KE}} \in \mathcal{H}_\Omega$ under the action of the identity component of $\mathrm{Aut}(M, J)$ on $\mathcal{H}_\Omega$ by pull-back. Moreover, a computation shows that the second variation of E_0 at a critical point is nonnegative, so KE metrics are local minima. A more difficult result is that KE metrics are in fact global minima of E_0. Thus, the KE problem is equivalent to determining when E_0 attains its absolute minimum. In particular, for a solution to exist, E_0 must be bounded from below. In practice, it is easier to use this criterion to prove non-existence, as we discuss in the next paragraph. A stronger criterion introduced by Tian [**244**], "properness of E_0", does turn out to be equivalent to existence; see §5.

4.3.2. *Futaki character.* A naïve way to show that E_0 is unbounded from below is to find a path parametrized by $\mathbb{R}_+$ along which the derivative of E_0 is uniformly negative. The simplest kind of path arises from a holomorphic vector field, and is the pull-back of a fixed metric by a one-paramter group of automorphisms. To spell this out, denote by ψ^X the vector field on $\mathcal{H}_\omega$ associated to the holomorphic vector field X,

$$\psi^X : \omega \mapsto \psi^X_\omega \in C^\infty(M),$$

with $\mathcal{L}_X\omega = d\iota_X\omega = \sqrt{-1}\partial\bar\partial\psi^X_\omega$, and $\int \psi^X_\omega\omega^n = 0$. Thus, ψ^X is tangent to $\iota_\omega(\mathcal{H}_\Omega)$. Moreover, if $\gamma(s)$ is an integral curve of ψ^X in $\iota_\omega(\mathcal{H}_\Omega)$, then $\omega(s) := \iota_\omega^{-1}\gamma(t) = (\exp sX)^\star\omega(0)$ and $\dot L(\gamma(s)) = g_{\mathrm{M}}(\mathbf{1}, \psi^X)|_{\omega(s)} = \int \psi^X_{\omega(0)}\omega(0)^n = g_{\mathrm{M}}(\mathbf{1}, \psi^X)|_{\omega(0)}$ since $\psi^X_{\omega(s)} = (\exp sX)^\star\psi^X_\omega(0)$ from the definition of ψ^X and the fact that the pull-back by an automorphism commutes with $\partial\bar\partial$. Thus, $\gamma(s)$ is a geodesic by (4.3).

Along such geodesics $\dot E_0(\omega(0), \omega(s)) = g_{\mathrm{M}}(\mathbf{s}, \psi^X)(\gamma(s))$ is manifestly constant in s. What is less obvious is that this latter constant does not depend on the initial metric $\omega(0)$. To see this, note that $\mathbf{s}^\flat = dE_0$ is evidently a closed form (here $\flat$ and $\sharp$ denote the "musical operators" associating to a vector field a one-form via the metric g_{M} and vice versa). Denote by $\exp t\psi^X$ the flow on $\mathcal{H}_\omega$ associated to ψ^X, given by pull-back by $\exp tX$. Since the scalar curvature is natural then $(\exp t\psi^X)^\star\mathbf{s}^\flat = \mathbf{s}^\flat$. Combining these facts, $\mathcal{L}_{\psi^X}\mathbf{s}^\flat = d\iota_{\psi^X}\mathbf{s}^\flat = 0$. Thus, $\psi^X(E_0)$ is constant on $\mathcal{H}_\omega$, as claimed. It is thus denoted by $F(X)$, and is called the Futaki invariant of X [**39, 49, 114**]. Since $F(-X) = -F(X)$, it follows that $F : \mathrm{aut}(M, J) \to \mathbb{R}$ must vanish identically or else E_0 is unbounded from below.

Like most objects in Kähler geometry, also the above discussion generalizes in a straightforward manner to include edges when (3.12) holds. Considering the 1-form

$$\mathbf{s}^\flat_\beta : \omega \mapsto (s_\omega - nc_1.[\omega]^{n-1}/[\omega]^n)\omega^n - (1-\beta)\omega^n|_D,$$

proofs conceptually similar to the ones in the smooth setting show that this 1-form is closed, admits a potential E_0^β, and an associated Futaki character on the Lie algebra $\mathrm{aut}(M, D)$ of holomorphic vector fields vanishing on D, and that E_0^β must be bounded from below when a KEE metric exists. To obtain these results one works on the space $\mathcal{H}^\epsilon_\omega$ (3.20), where integration by parts arguments, as in the smooth setting, are justified, thanks to the higher regularity results of [**141**] discussed in §3.10.

We refer to [**89, 115, 156, 163, 191, 246**] for explicit computations of Futaki invariants.

REMARK 4.9. In the smooth 2-dimensional Fano setting Futaki's and Matshusima's obstructions coincide. But in higher dimensions, or else in the singular two-dimensional setting, there exist examples where only one of the obstructions appears. For the latter see §9. We now give an example of the former. For instance, $\mathbb{P}(E)$, where $E = \mathcal{O}_{\mathbb{P}^1}(-1) \oplus \mathcal{O}_{\mathbb{P}^2}(-1)$ is the rank 2 bundle over $\mathbb{P}^1 \times \mathbb{P}^2$, is a Fano 4-fold with reductive automorphism group and nonvanishing Futaki character [**117**, pp. 24–26] (see also [**114, 262**]). On the other hand, there exists a Fano 3-fold whose connected automorphism group is $\mathbb{G}_a = (\mathbb{C}, +)$, see case 2) in the main result of Prokhorov [**207**], and so its Futaki character must vanish by a theorem of Mabuchi [**169**, Theorem 0.1]. We also remark that for some time it was believed that these two obstructions should be also sufficient [**114**, §4], [**167**, p. 575], [**138**, Conjecture E]. This was verified by Tian for del Pezzo surfaces [**240**] (see §9), but disproved for 3-folds, as we discuss next.

4.3.3. *Degenerations and geodesic rays.* One-parameter subgroups of automorphisms are particularly amenable to computations, as we saw in the previous subsection. Yet, generic complex manifolds do not admit such automorphisms, and the complexity of E_0 goes beyond automorphisms: Tian constructed Fano 3-folds with no nontrivial one-parameter subgroups of automorphisms that admit no KE metrics [**244**]. As a replacement, Tian suggested to consider M as embedded in a one-parameter family of complex manifolds and consider a $\mathbb{C}^\star$ action on this whole family. He observed that one may still associate a Futaki type invariant to this family, often referred to as a *special degeneration* or *test configuration*, and this reduces to the ordinary invariant when the family is a product. (Of course, we are glossing over many technical details here, among them that the action should lift to an action on a polarizing line bundle, and that the invariant is computed by considering high powers of this bundle. As noted in the Introduction, we refer to Thomas [**237**] for GIT aspects of KE theory.) Furthermore, he conjectured, in what became later known as the Yau–Tian–Donaldson conjecture, that if the sign of this invariant is identical for all special degenerations, and is zero only for product configurations, then the manifold should admit a KE metric [**244**]. This criterion is called *K-polystability*. In Tian's original definition only certain singularities were allowed on the "central fiber" of such a degeneration. Donaldson extended this to much more general ones [**97**]. Li–Xu finally showed that the original definition suffices [**158**]. A different definition of stability has been introduced by Paul [**194, 195, 251**], who also conjectured its equivalence to the existence of KE metrics. Very recently, a solution to these conjectures has been announced by Chen–Donaldson–Sun and Tian [**63, 249**], crucially building upon the theory of KEE metrics described in this article, and in particular on [**102, 141**]. Shortly after the appearance of [**63, 249**], Székelyhidi observed that those may be adapted to give similar conclusions without using KEE metrics [**234**].

Recently, Ross–Witt-Nyström introduced the notion of an analytic test configuration [**212**]. Very roughly, in their approach the singularities of the central fiber are replaced by a "singularity type" of a curve of ω-psh functions, and a (usually singular) generalized geodesic ray is constructed out of this data. This generalized previous constructions of Arezzo–Tian, Phong–Sturm, Song–Zelditch, that constructed such rays out a degeneration in the sense of the previous paragraph [**2, 201, 229**]. A generalized, or weak, geodesic existing for time $s \in [0, T]$, is a solution to the homogeneous complex Monge–Ampère equation on the product $S_T \times M$, where M is a strip $[0, T] \times \mathbb{R}$ of width T. It need not be a path in $\mathcal{H}$, but only in $\mathrm{PSH}(M, \omega)$. These can be regarded as bona fide geodesics if one enlarges the space of Kähler metrics appropriately, as considered by Darvas [**78**]. A different approach to construction of geodesic rays is suggested by Zelditch and the author via the Cauchy problem for the Monge–Ampère equation [**217–219**].

Coming full circle, a conjecture of Donaldson states that the existence problem should actually be completely characterized by generalizing Futaki's criterion to all geodesic rays. Namely, the non-existence of a csc Kähler metric is conjectured to be equivalent to the existence of a geodesic ray (whose regularity is not specified in the conjecture) along which the derivative of the K-energy is negative [**94**].

4.3.4. *Flow paths and metric completions.* Another conceivable way to "destabilize" a Fano manifold is to use Hamilton's (volume normalized) Ricci flow [**129**],

$$\frac{\partial\omega(t)}{\partial t} = -\mathrm{Ric}\,\omega(t) + \omega(t), \quad \omega(0) = \omega \in \mathcal{H}_{c_1}, \tag{4.5}$$

which preserves the space $\mathcal{H}_{c_1}$ of Kähler forms cohomologous to c_1 and exists for all time [**54**]. A theorem of Perleman asserts that the flow will converge to a Kähler–Einstein metric if and only if such a metric exists [**256**]. Thus, it is tempting to hope that Ricci flow trajectories could actually give another way of constructing geodesic rays. The following conjecture was suggested by La Nave–Tian [**152**, §5.4], based on a description of the Ricci flow as a Monge–Ampère equation in one extra dimension, that should bear a relationship to the homogeneous complex Monge–Ampère equation governing geodesics in the Mabuchi metric (4.1).

CONJECTURE 4.10. *The Kähler–Ricci flow is asymptotic to a g_{M}-geodesic in a suitable sense.*

The first attempt to understand the metric completion of $\mathcal{H}_\omega$ is due to Clarke and the author [**75**, Theorem 5.6] where the metric completion with respect to the Calabi metric is computed, motivated by an old research announcement of Calabi [**44**] that speculated a different answer. Calabi's metric is the L^2 metric on the level of Kähler forms, that also takes the form

$$g_{\mathrm{C}}(\nu,\eta)|_\varphi := \int_M \Delta_\varphi\nu\Delta_\varphi\eta\,\frac{\omega_\varphi^n}{n!}. \tag{4.6}$$

Interestingly, Calabi's metric was his original motivation for introducing the Calabi conjecture [**44**], see [**75**, Remark 4.1]. The following result relates degenerations arising from the Ricci flow, the existence problem of KE metrics, and the metric geometry of $\mathcal{H}_\omega$.

THEOREM 4.11. *Let (M,J) denote a Fano manifold. The following are equivalent:*
(i) (M,J) admits a KE metric;
(ii) Any Ricci flow trajectory in $\mathcal{H}_{c_1}$ converges in the Calabi metric g_{C} (4.6).

Note that the derivative of the K-energy is negative along the Ricci flow. Also, any divergent path must have infinite length in the corresponding metric. Thus, this result stands in precise analogy to Donaldson's conjecture on "geodesic stability," with Ricci flow paths taking the place of g_{M}-geodesic rays. This gives some intuitive motivation both for Donaldson's conjecture and for Conjecture 4.10. We conjecture that these statements are also equivalent to the statement that any Ricci flow trajectory in $\mathcal{H}_{c_1}$ converges in the Mabuchi metric g_{M} (4.1). A weaker statement was obtained by McFeron [**178**] where "converges" is replaced by "has finite length".

Convergence in the metric sense is rather weak from a PDE point of view. Thus, the main point in proving this theorem is to show that such convergence implies strong convergence. A sequence of Kähler metrics ω_i converges in the Calabi metric precisely when the volume forms ω_i^n converge in $L^1(M,\omega_0^n)$ [**75**, Corollary 5.5], and in particular do not charge Lesbegue null sets. Now, the limit along the Ricci flow converges to a metric that charges an analytic subvariety [**75**, Lemma 6.5] (this builds on results of Nadel [**189, 190**] and the author [**215**]), which proves that (ii) implies (i). The converse is an immediate corollary of the exponential convergence of the flow when a KE metric exists [**203, 256**].

Finally, we remark that both Donaldson's conjecture and Conjecture 4.10 should be related to the Hamilton–Tian conjecture, stipulating that the Kähler–Ricci flow on Fano manifolds should converge in a suitable sense to a Kähler–Ricci soliton (with respect to a possibly different complex structure) with mild singularities [**244**, Conjecture 9.1]. This relation could be related to the following problem.

PROBLEM 4.12. *Determine which g_{M}-geodesic rays (possibly non-smooth) come from automorphisms of a manifold with a nearby complex structure.*

For smooth Riemann surfaces the Ricci flow converges to a constant scalar curvature by results of Hamilton and Chow [**72, 130**], and to a soliton in the case of orbifold Riemann surfaces (i.e., angles β_i of the form $1/m_i$ with $m_i \in \mathbb{N}$) [**73**]. The Hamilton–Tian conjecture was also recently established in the setting of conical Riemann surfaces [**177**] and in the smooth 3-dimensional setting by Tian–Zhang [**254**]. The smooth 2-dimensional case was previously known by Tian's uniformization of del Pezzo surfaces relying on their classification (see Remark 4.9 and the introduction of §9) and Tian–Zhu's generalization of Perleman's convergence result for the Kähler–Ricci flow [**256**].

Finally, we mention that much of the discussion above has an analogue for the Calabi flow [**48**]. Calabi conjectured that the eponymous flow exists for all time and converges to a csc or, in an appropriate sense, to an extremal metric. The most general long time existence result known is due to Streets [**231**] though the question of regularity of such weak flows is open (assuming curvature bounds along the flow a smooth solution exists for all time by Chen–He [**64**]). An analogue of Conjecture 4.10 in this setting is shown by Chen–Sun [**66**]. It would be interesting to understand an analogue of Theorem 4.11. We mention in this context that according to [**64**, Theorem 4.1] (see [**55**, §3] for a recent alternative and conceptual proof) the Calabi flow converges exponentially fast as soon as it converges smoothly, say. In addition, a result of Berman states that, when the Kähler class is a multiple of the canonical class, the Calabi flow, when it exists, converges to a Kähler–Einstein metric when one exists [**18**, Theorem 1.4]. Finally, results of Chen–He, Feng–Huang, He, Huang, and Tosatti–Weinkove, among others, give various conditions for convergence of the Calabi flow [**65, 111, 133, 137, 258**].

4.4. Conical Riemann surfaces. It is interesting to contrast the general results surveyed so far with the 1-dimensional picture: when can one uniformize a conical Riemann surface? By uniformization we refer to the construction, in a given conformal class, of a constant scalar curvature metric with prescribed conical singularities. This question was studied, in the negative case, already in Picard's work more than a century ago [**205, 206**] and was first treated definitively in Troyanov's thesis [**259, 260**] (see also McOwen [**179, 180**]) where a sufficient condition for existence was established; uniqueness and necessity of this condition were addressed by Luo–Tian [**165**]. We refer to [**177**] for background and an alternative approach via the Ricci flow (see also [**269, 270**] for the nonpositive cases).

Fix a smooth compact surface M, along with a conformal, or equivalently, complex structure J. A divisor D is now a collection of distinct points $\{p_1, \ldots, p_N\} \subset M$ and the associated class $c_1(M, J) - \sum(1-\beta_j)p_j$, that can be thought of as a modified

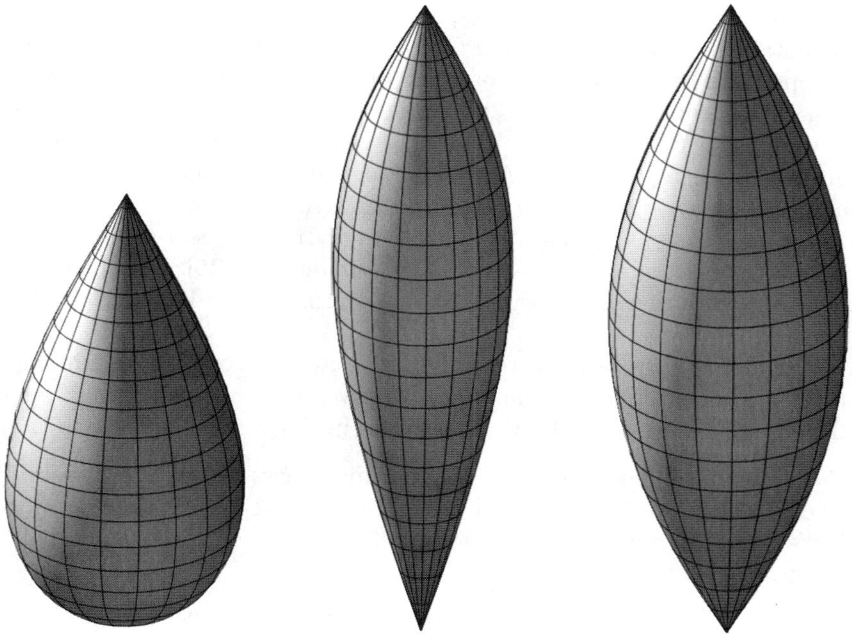

FIGURE 5. The teardrop and football (with different angles) solitons with non-constant curvature, and a constant curvature football with equal angles (courtesy of D. Ramos [**208**]).

Euler characteristic, is

$$2 - 2g(M) - \sum_{j=1}^{N}(1 - \beta_j). \tag{4.7}$$

In particular, this is now a number, and so the cohomological condition (3.12) is always satisfied. In a local conformal coordinate chart near each p_j, $g = \sqrt{-1}\gamma_j dz \otimes \overline{dz} = \sqrt{-1}\gamma_j |dz|^2$, with $\gamma = |z|^{2\beta_j - 2}F_j$ and F_j bounded. Suppose that g has constant curvature away from the cone points. The Poincaré–Lelong formula asserts that $-\Delta_g \log |z|$ is a multiple of the delta function at $\{z = 0\}$ (this can be seen by excising a small neighborhood near the cone point and using Stokes' formula). Together with the standard formula for the scalar curvature, $K_g = -\Delta_g \log \gamma$ (up to a constant factor), it follows that

$$K_g - 2\pi \sum (1 - \beta_i)\delta_{p_i} = \text{const}, \tag{4.8}$$

with the constant given by (4.7). The existence in the nonpositive regime uses standard methods as in the smooth setting, e.g., the variational method of Berger [**17**], or the method of sub- and super-solutions used by Kazdan–Warner [**145**]. A somewhat surprising discovery of Troyanov was the sufficient condition

$$\beta_j - 1 > \sum_{i \neq j} \beta_i - 1, \quad \text{for each } j = 1, \ldots, N, \tag{4.9}$$

for existence in the positive case, that followed by generalizing Moser's inequality [**185, 186**] to the singular setting. For instance, if $N = 2$, this is violated when

$\beta_1 \neq \beta_2$. As observed by Ross–Thomas, this condition can be rephrased by saying that the Futaki invariant of the pair $(M, \sum(1-\beta_i)p_i)$ has the right sign, or that the pair is slope (poly)stable [**211**, Theorem 8.1] (see [**211**, Remark 8.4] for a careful treatment of the equality case in (4.9)). The only additional unobstructed pair is $N = 2$ with $\beta_1 = \beta_2$, and such a csc metric can be constructed explicitly using ODE methods; when $\beta_1 \neq \beta_2$ or $N = 1$ one can still construct a shrinking Ricci soliton [**130**] (see also [**26, 208**]). These are depicted in Figure 5. We also remark that [**141**] and Berman's work gave a new proof of Troyanov's original results [**18**], and Berndtsson's work gave a new approach to uniqueness [**25**]. Finally, the higher regularity of such a metric was only obtained much later [**141**], as a corollary of Theorem 3.16.

The variational approach has recently been extended considerably through the work of Malchiodi et al. to allow angles $\beta > 1$, even when coercivity fails, see, e.g., [**15, 56**]. For some higher regularity results also in this regime we refer to [**177**].

4.5. Existence theorem for positive curvature. The following essentially optimal existence result in the positive case is due to [**141**]. It parallels and generalizes Tian's theorem from the smooth setting [**246**, §6]. We postpone the definition of properness to §5.2.

THEOREM 4.13. (Kähler–Einstein edge metrics with positive Ricci curvature) *Let (M, ω_0) be a compact Kähler manifold, $D \subset M$ a smooth divisor, and suppose that $\beta \in (0, 1]$ and $\mu > 0$ are such that*

$$c_1(M) - (1-\beta)[D] = \frac{\mu}{2\pi}[\omega_0],$$

and that the twisted K-energy E_0^β is proper. Then, there exists a Kähler–Einstein edge metric $\omega_{\varphi_{\mathrm{KE}}}$ with Ricci curvature μ and with angle $2\pi\beta$ along D, that is unique up to automorphisms that preserve D. This metric is polyhomogeneous, namely, φ_{KE} admits a complete asymptotic expansion with smooth coefficients as $r \to 0$ of the form

$$\varphi_{\mathrm{KE}}(r, \theta, z_2, \ldots, z_n) \sim \sum_{j,k \geq 0} \sum_{\ell=0}^{N_{j,k}} a_{jk\ell}(\theta, z_2, \ldots, z_n) r^{j+k/\beta} (\log r)^\ell, \tag{4.10}$$

where locally D is cut-out by z_1, $r = |z_1|^\beta/\beta$ and $\theta = \arg z_1$, and with each $a_{jk\ell} \in C^\infty$. There are no terms of the form $r^\zeta(\log r)^\ell$ with $\ell > 0$ if $\zeta \leq 2$. In particular, $\varphi_{\mathrm{KE}} \in \mathcal{A}^0 \cap \mathcal{D}_w^{0, \frac{1}{\beta}-1}$, i.e., $\omega_{\varphi_{\mathrm{KE}}}$ has infinite conormal regularity, and is $(\frac{1}{\beta} - 1)$-Hölder continuous with respect to the reference edge metric ω.

A parallel result in the snc case can be stated [**175, 176**]. In fact, the a priori estimates of [**141**] apply to the snc case without any change. The essential new difficulty, however, as compared to the smooth divisor case, is to extend the linear theory to one with crossing edges. This is new even in the real setting, and goes beyond the original methods of Mazzeo [**172**]. A different approach to the existence, avoiding such linear theory, but producing only $\mathcal{D}_w^{0,0}$ solutions without Hölder estimates on the metric (or even continuity of the metric up to D) nor higher regularity, and under the assumption that a C^0 (or even a weaker) solution exists, has been developed by Guenancia–Păun [**127**] (a similar result is obtained by Yao [**266**] who derives the C^0 estimate based on the approximation scheme of [**63, 249**]), as described in §4.1. Their approach starts from the C^0 solution constructed by

Berman [**18**] and [**141**], and produces a $\mathcal{D}_w^{0,0}$ estimate by a careful approximation by smooth metrics[1]. Finally, the uniqueness is due to Berndtsson [**25**].

STRATEGY OF PROOF OF CONJECTURE 4.2 AND THEOREM 4.13. The proof uses results from Sections 3, 6, and 7. We now describe how these results piece together.

Solving the KE equation is equivalent to solving the Monge–Ampère equation (3.7). We embed this equation in a one-parameter family of equations (6.3), called the Ricci continuity path (Ricci CP), and define as usual the set

$$I := \{s \in (-\infty, \mu] \,:\, (6.3) \text{ with parameter value } s \text{ admits a solution in } \mathcal{D}_w^{0,\gamma} \cap C^4(M \setminus D)\}.$$

Equation (6.3) with parameter value μ is precisely (3.7). By Proposition 6.1, there exists some $S > -\infty$ such that $(-\infty, S) \subset I$, so that I is not empty. Furthermore, I is open. As a matter of fact, the linearization of (6.3) at the parameter value s is given by $\Delta_{\omega_{\varphi(s)}} + s$. This is clearly invertible when $s < 0$, and is also invertible when $s > 0$ since, as shown in §7.1, the first eigenvalue of $\Delta_{\omega_{\varphi(s)}} + s$ is positive whenever $s \in (0, \mu)$. Invertibility at $s = 0$ follows by working on the orthogonal complement of the constants [**9**]. Thus, Theorem 3.7 (i) yields the openness. The fact that I is closed (and hence equal to $(-\infty, \mu]$) follows from Theorem 6.3. Thus, (3.7) admits a solution in $\mathcal{D}_w^{0,\gamma} \cap C^4(M \setminus D)$. Theorem 3.16 implies that the solution is polyhomogeneous and belongs to $\mathcal{A}^0_{\text{phg}}$; Equation (3.23) further implies the precise regularity statements about the solution. □

Note that the proof gives a new and unified proof for the classical results of Aubin, Tian and Yau, on the existence of smooth KE metrics, in that it uses a single continuity method for all signs of μ. This point is discussed in detail in §6.2.

Finally, consider the special case that M is Fano and D is a smooth anticanonical divisor (the existence of such a divisor is highly nontrivial, we refer to Problem 8.9 and the discussion there). Then, as noted by Berman [**18**], the twisted K-energy is proper for small $\mu = \beta > 0$. Theorem 4.13 thus gives the following corollary conjectured by Donaldson [**101**].

COROLLARY 4.14. *Let M be a Fano manifold, and suppose that there exists a smooth anticanonical divisor $D \subset M$. Then there exists some $\beta_0 \in (0, 1]$ such that for all $\beta \in (0, \beta_0)$ there exists a KEE metric with angle $2\pi\beta$ along D and with positive Ricci curvature equal to β.*

Li–Sun [**157**] observed that the exact same arguments actually prove existence also in the plurianticanonical setting: supposing there exists a smooth divisor D in $|-mK_M|$ (this always holds if M is Fano and $m \in \mathbb{N}$ is sufficiently large by Kodaira's theorem), there exists a small $\beta_0 > 0$ such that there exists a KEE metric with $\beta \in (1 - \frac{1}{m}, 1 - \frac{1}{m} + \beta_0)$ along D and with $\mu = 1 - (1 - \beta)m$.

[1]Just before posting this article, Guenancia–Păun have considerably revised their article [**127**]. In this version of their article they also develop, among other things, a new alternative approach for the $\mathcal{D}_w^{0,\gamma}$ estimate.

5. Energy functionals

The Mabuchi energy was crucial in stating the main existence theorem (Theorem 4.13). The purpose of this section is to describe several energy functionals, including the Mabuchi energy, that cast the Kähler–Einstein (edge) problem as a variational one. (The variational approach can be applied to much more general settings, and we do not seek full generality, for which we refer to Berman et al. [**18**, **22**, **23**].) First, in §5.1, we describe the Aubin energy functionals, that are nonlinear generalizations of the $W^{1,2}$-seminorm. Using these functionals it is easy to construct the Ding functional whose Euler–Lagrange equation is the inhomogeneous Monge–Ampère equation. Using the Aubin energies, one can then define a notion of relative compactness of level sets of an energy functional, and this leads to a calculus of variations formulation of the Kähler–Einstein problem in §5.2. Both the Mabuchi and Ding energies can be defined in a more conceptual way, using Bott–Chern forms. We describe this in detail in §5.4 since we are not aware of a single easy-to-read reference for this (we refer to [**198**, **228**] for the related Deligne pairing formalism). In §5.3 we also describe other natural functionals, the Kähler–Ricci energies, that lend themselves to a similar description, mainly to illustrate the richness of the theory, but also to show that there are many more-or-less equivalent functionals whose variational theory underlies the KE problem. Subsection 5.5 describes a relation between the Ding energy and the Mabuchi energy, involving the Legendre transform. Building on the preceeding subsections, in §5.6 we prove the equivalence, in suitable senses, of the Ding, Mabuchi, and Kähler–Ricci energies.

5.1. Nonlinear Dirichlet energies and the Berger–Moser–Ding functional. The most basic functionals, going back to the work of Aubin [**9**], are defined by

$$
\begin{aligned}
I(\eta,\eta_\varphi) &= \frac{1}{V}\int_M \sqrt{-1}\partial\varphi\wedge\bar\partial\varphi\wedge\sum_{l=0}^{n-1}\eta^{n-1-l}\wedge\eta_\varphi^l = \frac{1}{V}\int_M \varphi(\eta^n-\eta_\varphi^n),\\
J(\eta,\eta_\varphi) &= \frac{V^{-1}}{n+1}\int_M \sqrt{-1}\partial\varphi\wedge\bar\partial\varphi\wedge\sum_{l=0}^{n-1}(n-l)\eta^{n-l-1}\wedge\eta_\varphi^l.
\end{aligned}
\tag{5.1}
$$

The functionals I, J, and $I-J$ are all equivalent (and hence the latter is nonnegative), in the sense that $\frac{1}{n}J\le I-J\le\frac{n}{n+1}I\le nJ$. Granted, these might not look so 'basic' at a first glance. Let us give some motivation to these definitions.

The first motivation comes from the calculus of variations: is there a sort of nonlinear Dirichlet energy whose Euler–Lagrange equation is (2.3)? In the case $n=1$ such an energy was studied by Berger and Moser [**17**, **186**]

$$
F(\eta,\eta_\varphi)=\begin{cases}\dfrac{1}{V}\displaystyle\int\frac{1}{2}\sqrt{-1}\partial\varphi\wedge\bar\partial\varphi-\frac{1}{V}\int\varphi\eta-\frac{1}{\mu}\log\frac{1}{V}\int e^{f_\eta-\mu\varphi}\eta, & \text{for } \mu\neq 0,\\[2ex] \dfrac{1}{V}\displaystyle\int\frac{1}{2}\sqrt{-1}\partial\varphi\wedge\bar\partial\varphi-\frac{1}{V}\int\varphi\eta+\frac{1}{V}\int\varphi e^{f_\eta}\eta, & \text{for } \mu=0.\end{cases}
\tag{5.2}
$$

(Recall the definition of f_η (3.9).) Its critical points are precisely constant scalar curvature metrics in the conformal class. It is straightforward to generalize in higher dimensions the last term to $\frac{1}{\mu}\log\frac{1}{V}\int e^{f_\eta-\mu\varphi}\eta^n$ (respectively, $\frac{1}{V}\int\varphi e^{f_\eta}\eta^n$) so that its variational derivative comes out to be the right hand side of (2.3) (up to the implicit normalization that the integral of the right hand side is 1). How to replace the first two terms so that the resulting Euler–Lagrange equation comes out right? This amounts to finding a functional whose differential is exactly minus the Monge–Ampère operator, i.e., the left hand side of (2.3). This is precisely the functional $-L$ (recall (4.2) and the line following it). Decomposing $-L$ into two terms, one of which is $-\frac{1}{V}\int \varphi\eta^n$, yields precisely J as the second term! (Incidentally, this also explains why L, defined in (4.2), is called the Aubin–Mabuchi functional.)

Thus, the Euler–Lagrange equation for the Berger–Moser–Ding energy (or Ding energy for short) [**88**]

$$F^\beta(\eta,\eta_\varphi)=\begin{cases} J(\eta,\eta_\varphi)-\dfrac{1}{V}\displaystyle\int\varphi\eta^n-\dfrac{1}{\mu}\log\dfrac{1}{V}\displaystyle\int e^{f_\eta-\mu\varphi}\eta^n, & \text{for } \mu\neq 0,\\ J(\eta,\eta_\varphi)-\dfrac{1}{V}\displaystyle\int\varphi\eta^n+\dfrac{1}{V}\displaystyle\int e^{f_\eta}\eta^n, & \text{for } \mu=0,\end{cases}\tag{5.3}$$

is precisely (3.7).

5.2. A nonlinear variational problem. This brings us to a second motivation for J. As we just noticed, $-L(\varphi)=J(\eta,\eta_\varphi)-\int\varphi\eta^n$. Recall that $\pm L$ (4.2) is a distance function for the Mabuchi metric. Thus, it is tempting to think of J as a sort of approximate distance function. This is of course not quite true, since the submanifold $\{\varphi\in\mathcal{H}^e_\omega : \int\varphi\eta^n=0\}$ of $\mathcal{H}^e_\omega$ is not totally geodesic. But ignoring this subtlety, one is then tempted to think of J as a good way to define coercivity for our nonlinear problem on $\mathcal{H}^e_\omega$; of course this temptation also arises from the first motivation discussed earlier. Following Tian [**244**], one says a functional E on $\mathcal{H}_\omega\times\mathcal{H}_\omega$ is proper provided that it dominates J (or, by their equivalence, I or $I-J$) on each $\mathcal{H}_\omega$ slice. This is an analogue of the standard assumption in the direct method in the calculus of variations, namely that sublevel sets (of E) are compact. In particular, if E is proper, a sublevel set of E is contained in some sublevel set of J. Theorem 4.13 can be recast as a nonlinear analogue of the fundamental theorem of the direct method (cf. [**232**, Theorem 1.1]) in this setting, justifying the definition of properness.

THEOREM 5.1. *Suppose that $F^\beta(\omega,\cdot)$ is proper on $\mathcal{H}^e_\omega$. Then it is bounded from below, and its infimum is attained at a solution of (2.3).*

For the proof we refer to [**141**, Theorem 2] (that proof assumes that the Mabuchi functional E^β_0 is proper, however the arguments are identical; a conceptual proof of the equivalence coercivity of F^β and that of E^β_0 is given in Theorem 5.11 (ii), although a direct proof (i.e., one that does not use the existence of a minimizer) of equivalence of the properness of these functionals seems to be unknown). The special case when $\beta=1$ goes back to Ding–Tian [**90**] and Tian [**244**] (see also [**248**, §2.2] for a generalization that allows for automorphisms). On the other hand, Berman showed that under the assumption in the theorem the infimum is attained within the larger set $\mathrm{PSH}(M,\omega)\cap C^0(M)$, and while it is easy to see that the minimizer is contained in $C^\infty(M\setminus D)$ by local ellipticity and the usual arguments as in the smooth case, the proof that the minimizer lies in $\mathcal{H}^e_\omega$ [**141**] relies on the Ricci continuity method, as described in §6–7.

5.3. The staircase energies and Kähler–Ricci energies. A generalization of the Aubin energy J (5.1) was introduced in [**213**, (6)],

$$I_k(\omega,\omega_\varphi)=\frac{V^{-1}}{k+1}\int_M \varphi(k\omega^n-\sum_{l=1}^{k}\omega^{n-l}\wedge{\omega_\varphi}^l). \tag{5.4}$$

Note that $I_n=J$, $I_{n-1}=((n+1)J-I)/n$. These functionals are nonnegative (see (5.5) below) and are related to, but different from, functionals defined by Chen–Tian [**67**]—see [**216**, p. 133]. The functionals I_k can be thought of as gradual nonlinear generalizations of the Dirichlet energy that interpolate between the latter and J. In fact, I_1 is simply a multiple of the Dirichlet energy $\int\sqrt{-1}\partial\varphi\wedge\bar\partial\varphi\wedge\omega^{n-1}$, and increasing k is analogous to climbing a staircase (see [**216**, p. 132] for a pictorial description) where I_k incorporates one additional 'mixed' term proportional to $\int\sqrt{-1}\partial\varphi\wedge\bar\partial\varphi\wedge\omega^{n-k}\wedge{\omega_\varphi}^{k-1}$; indeed, by integrating by parts,

$$I_k(\omega,\omega_\varphi)=\frac{1}{V}\int_M\sqrt{-1}\partial\varphi\wedge\bar\partial\varphi\wedge\sum_{l=0}^{k-1}\frac{k-l}{k+1}\omega^{n-1-l}\wedge{\omega_\varphi}^l. \tag{5.5}$$

The definition (5.4) certainly makes sense for pairs of smooth Kähler forms, and by Theorem 3.10 (i) and the continuity of the mixed Monge–Ampère operators on $\mathrm{PSH}(M,\omega_0)\cap C^0(M)$ [**16**, Proposition 2.3], these functionals can be uniquely extended to pairs $(\omega_{\varphi_1},\omega_{\varphi_2})$, with $\varphi_1,\varphi_2\in\mathcal{H}^e_\omega$ with ω the reference edge metric as in (3.5). Thus, these functionals are well-defined on $\mathcal{H}^e_\omega\times\mathcal{H}^e_\omega$. Moreover, (5.5) is still justified for edge metrics: working on a tubular neighborhood of D the boundary term one obtains from integrating by part in (5.4) tends to zero with the radius of the neighborhood about D.

As we just saw, the functionals I_k are a natural generalization of the Dirichlet energy on the one hand, and Aubin's J functional on the other. It turns out that I_k play a role in describing a scale of energy functionals E_k that similarly depend gradually on additional 'mixed terms' as k increases, loosely speaking. This can be made precise using the language of Bott–Chern forms that we review in §5.4. Another, more down to earth way, of defining this functional is as follows.

Denote the normalized elementary symmetric polynomials of the eigenvalues of the twisted Ricci form $\mathrm{Ric}\,\eta-(1-\beta)[D]$ (with respect to η) by

$$\sigma_k^\beta(\eta)=\frac{(\mathrm{Ric}\,\eta-(1-\beta)[D])^k\wedge\eta^{n-k}}{\eta^n},\qquad k=1,\ldots,n,$$

and their average (which is independent of the representative of $[\eta]$) over (M,ω^n) by

$$\mu_k^\beta:=\frac{(c_1-(1-\beta)c_1(L_D))^k\cup[\eta]^{n-k}([M])}{[\eta]^n([M])}. \tag{5.6}$$

When $k=1$, $\sigma_1^\beta(\eta)=s_\eta/n=\mathrm{tr}_\eta(\mathrm{Ric}\,\eta-(1-\beta)[D])/n$, where $s_\beta(\eta)$ denotes the twisted scalar curvature of η. When $k=n$ this is $\det_\eta(\mathrm{Ric}\,\eta-(1-\beta)[D])$, the determinant of the twisted Ricci curvature with respect to the metric.

A straightforward generalization of a theorem of the author from the case $\beta=1$ [**213**, Proposition 2.6] yields the following.

PROPOSITION 5.2. *Let $k \in \{0,\dots,n\}$. Suppose $\frac{\mu}{2\pi}[\eta] = c_1 - (1-\beta)c_1(L_D)$. The 1-form*

$$\eta \mapsto \Big[\Delta_\eta \sigma_k^\beta(\eta) - \frac{n-k}{k+1}\Big(\sigma_{k+1}^\beta(\eta) - \mu_{k+1}^\beta\Big)\Big]\eta^n$$

is exact. Its potential, considered as a function on $\mathcal{H}_\eta^e \times \mathcal{H}_\eta^e$, is given by

$$\begin{aligned}\mu^{-k}E_k^\beta(\eta,\eta_\varphi) = E_0^\beta(\eta,\eta_\varphi) &+ \mu I_k(\eta_\varphi, \mu^{-1}\mathrm{Ric}\,\eta_\varphi - \mu^{-1}(1-\beta)[D]) \\ &- \mu I_k(\eta, \mu^{-1}\mathrm{Ric}\,\eta - \mu^{-1}(1-\beta)[D]),\end{aligned} \tag{5.7}$$

uniquely determined by the normalization $E_k^\beta(\eta,\eta) = 0$.

PROOF. As remarked earlier, I_k is continuous in the topology of uniform convergence (on the level of Kähler potentials) by a result of Bedford–Taylor. By definition (see the second part of Definition 3.2, (3.20), and (3.17)), $\mathrm{Ric}\,\omega_\varphi - (1-\beta)[D]$ admits a continuous Kähler potential with respect to ω, and moreover this potential can be suitably uniformly approximated. Smoothly approximate (ω,ω_φ) uniformly while approximating $\mathrm{Ric}\,\omega_\varphi - (1-\beta)[D]$ uniformly on the level of potentials. Thus, it suffices to verify the proposition when the Kähler forms are smooth, when integration by parts, precisely those in the proof in the smooth case [**213**], are justified. But then the proof in the smooth case can be applied verbatim, with the definition of the twisted Ricci potential as in (3.9) (recall f_ω is continuous for all β) making the proof consistent with the introduction of the terms $(1-\beta)[D]$ in the right hand side of (5.7). □

Similar arguments allow to generalize the following formula of Tian [**242**] from the smooth case.

LEMMA 5.3. *Let $\eta, \eta_\varphi \in \mathcal{H}_\omega^e$. One has,*

$$E_0^\beta(\eta,\eta_\varphi) = \frac{1}{V}\int_M \log\frac{\eta_\varphi^n}{\eta^n}\eta_\varphi^n - \mu(I-J)(\eta,\eta_\varphi) + \frac{1}{V}\int_M f_\eta(\eta^n - \eta_\varphi^n). \tag{5.8}$$

The functionals E_0 and E_n were introduced by Mabuchi [**167**] and Bando–Mabuchi [**14**], while the remaining ones by Chen–Tian [**67**]. Formula (5.7) shows that the E_k interpolate between E_0 and E_n, to wit [**213**, (17)]

$$\begin{aligned}E_0^\beta(\eta,\eta_\varphi) &= ((1-\tfrac{l}{k+1})E_0 + \tfrac{l}{k+1}E_n)(\eta,\eta_\varphi) \\ &\quad + (I_k - \tfrac{l}{k+1}J)(\eta_\varphi, \mu^{-1}\mathrm{Ric}\,\eta_\varphi - (1-\beta)[D]/\mu) \\ &\quad - (I_k - \tfrac{l}{k+1}J)(\eta, \mathrm{Ric}\,\eta), \quad \forall l \in \{0,\dots,k+1\}.\end{aligned}$$

Since E_0 is known as the K-energy or Kähler energy, and E_n as the Ricci energy, it is natural, following [**213**], to refer to E_k as the *Kähler–Ricci energies.*

The Ricci energy is special. It lends a geometric interpretation to the Ding functional, via the inverse Ricci operator, in the sense of [**214**, §9]. That is, suppose that (3.12) holds with $\mu = 1$. Define by $\mathrm{Ric}_\beta^{-1} : \mathcal{H}_\omega^\epsilon \to \mathcal{H}_\omega^\epsilon$ the twisted inverse Ricci operator, by letting $\mathrm{Ric}_\beta^{-1}\eta := \chi$ be the unique Kähler form cohomologous to η satisfying $\mathrm{Ric}\,\chi - (1-\beta)[D] = \eta$. Such a unique form exists by Conjecture 4.2, since this equation can be written as a Monge–Ampère equation of identical form to that of the equation for a Ricci flat Kähler edge metric.

PROPOSITION 5.4. *One has $(\mathrm{Ric}_\beta^{-1})^\star E_n^\beta = F^\beta$.*

Again, the proof is a simple adaptation of the proof in the smooth case [**214**, Proposition 10.4], following the arguments in the proof of Proposition 5.2 above.

5.4. Bott–Chern forms. In this subsection we will describe the theory of Bott–Chern forms and energy functionals, inspired by work of Bott and Chern [**37**], and developed by Donaldson and Bismut–Gillet–Soulé [**29, 92**]. This will be applied to showing that the functionals E_k^β have an expression in terms of Bott–Chern forms, slightly generalizing but very closely following the discussion in [**216**, §4.4.5].

The main idea of Bott–Chern forms is that given a moduli space of Hermitian metrics on a bundle one may construct canonically defined "universal" functions on the moduli space associated to curvature. These functions arise via a "potential" for the curvature form of a "universal" bundle over the whole moduli space.

Let $E \to M$ be a holomorphic vector bundle of rank r.[2] A vector bundle represents a Čech cohomology class in $H^1(M, \mathcal{O}_M(GL(r,\mathbb{C})))$. Here by $\mathcal{O}_M(GL(r,\mathbb{C}))$ we mean the sheaf of germs of holomorphic functions to $GL(r,\mathbb{C})$. When $r = 1$ it is denoted by $\mathcal{O}_M^\star$. Let us identify E with its Čech class representative, i.e., by a collection of transition functions $g = \{g_{\alpha\beta}\}$ that are holomorphic maps from the intersection of any two coordinate neighborhoods $U_\alpha, U_\beta \subset M$ to $GL(r,\mathbb{C})$, $g_{\alpha\beta} : U_\alpha \cap U_\beta \to GL(r,\mathbb{C})$, satisfying the Čech cocycle conditions [**124**, p. 66]

$$(\delta g)_{\alpha\beta\gamma} := g_{\alpha\beta} \cdot g_{\beta\gamma} \cdot g_{\gamma\alpha} = \mathrm{I}.$$

Note that here the groups comprising the sheaf have a multiplicative structure (and not an additive one) and hence $\delta g = \mathrm{I}$ expresses closedness, i.e., $[g]$ represents a Čech cohomology class.

Denote by $\mathcal{H}_E$ the space of all Hermitian metrics on E. Let $\mathrm{Herm}(r)$ denote the space of positive Hermitian $r \times r$ matrices. Any Hermitian metric $H \in \mathcal{H}_E$ can be represented by smooth maps $H_\alpha : U_\alpha \to \mathrm{Herm}(r)$ such that with respect to local bases of sections one has $(H_\alpha)_{i\bar{j}} = (g_{\alpha\beta})_{ik}(H_\beta)_{k\bar{l}}\overline{(g_{\alpha\beta})_{jl}}$, or simply $H_\alpha = g_{\alpha\beta}^\star H_\beta g_{\alpha\beta}$. This is summarized in the notation $H = \{H_\alpha\}$.

To every $H \in \mathcal{H}_E$ there is associated a unique complex connection D_H. For a general holomorphic vector bundle the connection D_H is a 1-form on M with values in the bundle $\mathrm{End}(E)$ of endomorphisms of E. One may write globally

$$D_H = H^{-1} \circ \partial \circ H.$$

The derivation of this formula follows the same argument as in the line bundle case. The exact meaning of how to understand this and other similar expressions involving compositions of endomorphisms and differential operators will be explained in detail below in several computations. With respect to a local holomorphic frame $e_1, \ldots, e_r$ over $U_\alpha \subset M$ the endomorphism H may be represented by a matrix and one has the special expression $(D_H|_{U_\alpha})^i_j = \partial h_{j\bar{l}} \cdot h^{\bar{l}i}$. This expression is not valid with respect to an arbitrary frame.

The global expression for the curvature is then

$$F_H := D_H \circ D_H = \bar{\partial} \circ H^{-1} \circ \partial \circ H. \tag{5.9}$$

Implicit in this notation as well as in the sequel is the convention of working with endomorphisms with values in the exterior algebra of differential forms on M. Locally with respect to a holomorphic frame one has the special expression $F_H|_{U_\alpha} = \bar{\partial}(\partial h_{j\bar{l}} \cdot h^{\bar{l}i})$. This expression is not valid with respect to an arbitrary frame.

---

[2] For the general discussion of Bott–Chern forms we will only make use of the fact that M is complex (rather than Kähler). In our applications we will always work with line bundles, i.e., $r = 1$. If one were interested only in that case the discussion below could be slightly simplified. However we have chosen to maintain this level of generality.

However it demonstrates that F_H is $(1,1)$-form on M (with values in $\mathrm{End}(E)$), since the type is independent of the choice of frame.

Let ϕ denote an elementary symmetric polynomial on $gl(r,\mathbb{C}) \times \cdots \times gl(r,\mathbb{C})$ (p times) that is invariant under the adjoint action of $GL(r,\mathbb{C})$ (conjugation). The idea behind Chern classes is that one may plug into such polynomials matrices that have differential forms as their entries, for example F_H. Since the polynomials are $GL(r,\mathbb{C})$-invariant one obtains a differential form that is invariant under a change of local trivializations for the bundle, hence is intrinsically defined. Moreover it turns out that such forms are closed, hence define intrinsic cohomology classes that depend only on the complex structure of $E \to M$.

Now we come back to our original task of constructing functions on $\mathcal{H}_E$. One may show that $\phi(F_H) := \phi(F_H, \ldots, F_H)$ is a closed $2p$-form. It certainly depends on the metric H, however its cohomology class in $H^{2p}(M,\mathbb{Z})$ does not. This means that the difference

$$\phi(F_{H_0}) - \phi(F_{H_1})$$

is exact. Moreover, and here we arrive at the main point of Bott–Chern theory, one may find a $(p-1,p-1)$-form $\mathbf{BC}(\phi; H_0, H_1)$, well-defined up to ∂- and $\bar\partial$-exact forms, such that

$$\bar\partial\partial \mathbf{BC}(\phi; H_0, H_1) = \phi(F_{H_1}) - \phi(F_{H_0}).$$

The form $\mathbf{BC}(\phi; H_0, H_1)$ may then be integrated against a $(n-p+1, n-p+1)$-form on M to give a number. Fixing H_0 and letting H_1 vary we therefore obtain a function on $\mathcal{H}_E$ as desired. And, if $p-1 = n$, we even do not need to make a choice of such a form in order to integrate (this will be the case in our applications). We now show how to construct the Bott–Chern form $\mathbf{BC}(\phi; H_0, H_1)$. The proof below is a slow pitch version of the orignal one.

PROPOSITION 5.5. (See [**92**, Proposition 6].) *Let ϕ be a $GL(r,\mathbb{C})$-invariant elementary symmetric polynomial. Given $H_0, H_1 \in \mathcal{H}_E$ and any path $\{H_t\}_{t\in[0,1]}$ in $\mathcal{H}_E$ connecting them, the $(p-1,p-1)$-form*

(5.10)
$$\mathbf{BC}(\phi; H_0, H_1) := p(\sqrt{-1})^{p-1} \int_{[0,1]} \phi(H_t^{-1}\dot H_t, F_{H_t}, \ldots, F_{H_t})dt \mod \mathrm{Im}\partial + \mathrm{Im}\bar\partial,$$

is well-defined, namely does not depend on the choice of path. In addition,

$$\mathbf{BC}(\phi; H_0, H_1) + \mathbf{BC}(\phi; H_1, H_2) + \mathbf{BC}(\phi; H_2, H_0) = 0, \tag{5.11}$$

and

$$\bar\partial\partial \mathbf{BC}(\phi; H_0, H_1) = \phi(F_{H_1}) - \phi(F_{H_0}). \tag{5.12}$$

Notice that the first argument of ϕ is an endomorphism while the rest of its arguments are endomorphism-valued 2-forms.

PROOF. Note that $\mathbf{BC}(\phi; H_0, H_1)$ is given by integration over a path connecting H_0 and H_1 of a globally defined 1-form on $\mathcal{H}_E$ with values in $(p-1,p-1)$-forms on M. We call this form θ. To show independence of path we show that this form is closed modulo ∂- and $\bar\partial$-exact terms. Let $H \in \mathcal{H}_E$. Let $h, k \in T_H\mathcal{H}_E$ and extend them to constant vector fields near H. Then,

$$d\theta(h,k) = k\,\theta|_H(h) - h\,\theta|_H(k). \tag{5.13}$$

First we obtain an expression for the infinitesimal change of the curvature under a variation of a Hermitian metric. Write $H+tk=H\circ(\mathrm{I}+tH^{-1}k)=: H\circ f$. We write $\circ$ to emphasize that composition of endomorphisms is taking place (in coordinates: multiplication of matrices). According to (5.9) we have

$$\begin{aligned} F_{H\circ f} &= \bar\partial\big((H\circ f)^{-1}\circ\partial(H\circ f)\big)\\ &= \bar\partial\big(f^{-1}\circ(H^{-1}\circ\partial H)\circ f+f^{-1}\circ\partial f\big) \qquad (5.14)\\ &= \bar\partial\circ f^{-1}\circ D_H^{1,0}\circ f. \end{aligned}$$

Here it should be emphasized that D_H decomposes according to type (of its 1-form part) into $D_H^{1,0}$ and $D_H^{0,1}$ and, that while originally the connection D_H was defined on E, it may be extended naturally to $\mathrm{End}(E)$ and it is this extension that we use in the equation above (f is a section of $\mathrm{End}(E)$ and not of E). The same applies to the operator $\bar\partial$ that we also extend to act on $\mathrm{End}(E)$. To understand the last equation better, we let the endomorphism it defines act on a holomorphic section s of E, and compute:

$$\begin{aligned} F_{H\circ f}s &= \bar\partial\circ f^{-1}\circ D_H^{1,0}(fs)\\ &= \bar\partial\circ f^{-1}\circ\big((D_H^{1,0}f)s+fD_H^{1,0}s\big)\\ &= \Big(\bar\partial\big(f^{-1}(D_H^{1,0}f)\big)+\bar\partial\circ D_H^{1,0}\Big)s. \end{aligned}$$

Therefore we write

$$F_{H\circ f}=F_H+\bar\partial\big(f^{-1}(D_H^{1,0}f)\big),$$

and the second term should be understood to be distinct from (5.14). This subtle notational issue can be a cause for great confusion when consulting the literature on vector bundles and Yang-Mills theory. Putting now $f=\mathrm{I}+tH^{-1}k$ we obtain

$$F_{H+tk}=F_H+t\bar\partial\big(D_H^{1,0}(H^{-1}k)\big)+O(t^2).$$

Hence the first term in (5.13) is given by

(5.15)
$$\begin{aligned} \frac{1}{p(\sqrt{-1})^{p-1}}k\theta|_H(h) &= \frac{d}{dt}\Big|_0\phi((H+tk)^{-1}h,F_{H+tk},\ldots,F_{H+tk})\\ &= \phi(-\sqrt{-1}Hk\sqrt{-1}Hh,F_H,\ldots,F_H)\\ &\quad+\textstyle\sum\phi(\sqrt{-1}Hh,F_H,\ldots,\bar\partial D_H^{1,0}(\sqrt{-1}Hk),\ldots,F_H). \end{aligned}$$

Therefore one has, putting $\sigma=\sqrt{-1}Hh,\ \tau=\sqrt{-1}Hk$,

(5.16)
$$\begin{aligned} \frac{1}{p(\sqrt{-1})^{p-1}}d\theta(h,k) &= \phi([\sigma,\tau],F_H,\ldots,F_H)+\textstyle\sum\phi(\sigma,F_H,\ldots,\bar\partial D_H^{1,0}\tau,\ldots,F_H)\\ &\quad-\textstyle\sum\phi(\tau,F_H,\ldots,\bar\partial D_H^{1,0}\sigma,\ldots,F_H). \end{aligned}$$

σ and τ are sections of the endomorphism bundle of E. Note that as operators on this bundle one has

$$\bar\partial\circ D_H^{1,0}+D_H^{1,0}\circ\bar\partial=(\bar\partial+D_H^{1,0})\circ(\bar\partial+D_H^{1,0})=D_H^2=F_H,$$

since we already saw that the curvature is of type $(1,1)$. Hence for example

$$\bar\partial\circ D_H^{1,0}\sigma=-D_H^{1,0}\circ\bar\partial\sigma+[F_H,\sigma]. \qquad (5.17)$$

Note $[F_H, \sigma] \equiv F_H\sigma$, and the bracket notation simply emphasizes that we have extended F_H to act on the endomorphism bundle and so the endomorphism part of F_H will actually act by bracket on the endomorphism σ (the 2-form part will simply be multiplied along). By the Bianchi identity $D_H F_H = 0$ and so $D_H^{1,0} F_H = 0$, $\bar\partial F_H = 0$. Now,

$$
\begin{aligned}
\phi(\sigma, \bar\partial D_H^{1,0}\tau, F_H, \dots, F_H) &= \bar\partial\phi(\sigma, D_H^{1,0}\tau, F_H, \dots, F_H) \\
&\quad - \phi(\bar\partial\sigma, D_H^{1,0}\tau, F_H, \dots, F_H) \\
&\quad - \textstyle\sum \phi(\sigma, D_H^{1,0}\tau, F_H, \dots, \bar\partial F_H, \dots, F_H) \\
&= \bar\partial\phi(\sigma, D_H^{1,0}\tau, F_H, \dots, F_H) \\
&\quad - \phi(\bar\partial\sigma, D_H^{1,0}\tau, F_H, \dots, F_H).
\end{aligned} \tag{5.18}
$$

Using (5.17) the corresponding term in the second sum of (5.16) is

(5.19)

$$
\begin{aligned}
-\phi(\tau, \bar\partial D_H^{1,0}\sigma, F_H, \dots, F_H) &= -\phi(\tau, -D_H^{1,0}\bar\partial\sigma + [F_H, \sigma], F_H, \dots, F_H) \\
&= -\phi(\tau, [F_H, \sigma], F_H, \dots, F_H) + \partial\phi(\tau, \bar\partial\sigma, F_H, \dots, F_H) \\
&\quad - \phi(D_H^{1,0}\tau, \bar\partial\sigma, F_H, \dots, F_H) \\
&\quad - \textstyle\sum \phi(\tau, \bar\partial\sigma, F_H, \dots, D_H^{1,0} F_H, \dots, F_H). \\
&= \phi(\tau, [\sigma, F_H], F_H, \dots, F_H) + \partial\phi(\tau, \bar\partial\sigma, F_H, \dots, F_H) \\
&\quad - \phi(D_H^{1,0}\tau, \bar\partial\sigma, F_H, \dots, F_H).
\end{aligned}
$$

Note that it is not necessarily true that

$$-\phi(D_H^{1,0}\tau, \bar\partial\sigma, F_H, \dots, F_H) \tag{5.20}$$

cancels with

$$-\phi(\bar\partial\sigma, D_H^{1,0}\tau, F_H, \dots, F_H).$$

But, eventually taking the equations (5.18) and (5.19) for all pairs appearing in the sums (5.16) then, e.g., the term (5.20) will cancel with the term

$$-\phi(\bar\partial\sigma, F_H, \dots, F_H, D_H^{1,0}\tau).$$

Indeed we are only allowed to permute the arguments of ϕ cyclically (e.g., for three matrices A, B, C one has $\mathrm{tr}(ABC) = \mathrm{tr}(CAB)$, but in general $\mathrm{tr}(ABC)$ is different from $\mathrm{tr}(BAC)$). Note also that while ϕ does not change when permuting matrices cyclically, when we permute cyclically matrix valued 1-forms a sign appears, as usual. This explains the cancellation above.

Hence, modulo ∂- and $\bar\partial$-exact terms, we are left with

$$\sum \phi(\tau, F_H, \dots, [\sigma, F_H], F_H, \dots, F_H)$$

which cancels with the first term in (5.16); this can be seen by using the invariance of ϕ under the action of $GL(r, \mathbb{C})$ by conjugation

$$\frac{d}{dt}\Big|_0 \phi(e^{-tB}A_1 e^{tB}, \dots, e^{-tB}A_p e^{tB}) = \sum \phi(A_1, \dots, [A_j, B], \dots, A_p),$$

concluding the proof. □

EXAMPLE 5.6. Let E denote an ample line bundle polarizing a Kähler class $\Omega = c_1(E)$. We identify $\mathcal{H}_E$ with $\mathcal{H}_\omega$ where $\omega = -\sqrt{-1}\partial\bar\partial \log h = \sqrt{-1}\bar\partial(h^{-1}\partial h) = \sqrt{-1}F_h$ is a Kähler form with $h \in \mathcal{H}_E$ (and hence $[\omega] = \Omega$). Now $r = 1$ and so no traces are needed (the matrices are all one-dimensional). Put

$$\phi(A_1, \ldots, A_{n+1}) := A_1 \cdots A_{n+1}. \tag{5.21}$$

Take a path of Hermitian metrics $h_t = e^{-\varphi_t}h$. Then $F_{h_t} = F_h + \partial\bar\partial\varphi_t = \omega_{\varphi_t}/\sqrt{-1}$. Then the Bott–Chern form is
(5.22)

$$\mathbf{BC}(\phi; h_0, h_1) = (n+1)\int_0^1 (\sqrt{-1})^n \phi(-\dot\varphi_t, F_{h_t}, \ldots, F_{h_t})dt = -(n+1)\int_0^1 \dot\varphi_t\,\omega_{\varphi_t}^n \wedge dt.$$

This expresses a $2n$-form on M. Integrating this over M gives the function $-(n+1)L(\varphi)$, on $\mathcal{H}_\omega$. By 5.5 this is independent of the choice of path since any two choices differ by ∂- and $\bar\partial$-exact terms and hence by d-exact terms since our expressions are real.

Now we turn to the setting of a Kähler manifold with an integral Kähler class $[\omega]$. Let L be a line bundle polarizing the Kähler class, namely $c_1(L) = [\omega]$. Let K_M^{-1} denote the anticanonical bundle polarizing the class c_1, and L_D the line bundle associated to D. Given a Hermitian metric h on L of positive curvature, and a global holomorphic section s of L_D one obtains a metric $\det F_h|s|^{2-2\beta} = \det(\omega/\sqrt{-1})$ on $K_M^{-1} \otimes L_D^{\beta-1}$ (we mean that locally $\det F_h = \det(g_{i\bar j})$ if $\omega = \sqrt{-1}g_{i\bar j}dz^i \wedge \overline{dz^j}$) where $\omega = -\sqrt{-1}\partial\bar\partial\log h$. We write $h \in \mathcal{H}_\omega =: \mathcal{H}_L^+$. Note that the Bott–Chern forms defined below are defined on $\mathcal{H}_L^+$ rather than on all of $\mathcal{H}_L$ (or to be more precise on a set isomorphic to $\mathcal{H}_L^+$, see [**245**, p. 214]).

The main result of this subsection is the following expression for the Kähler–Ricci functionals in terms of Bott–Chern forms. While this generalizes a theorem of Tian for the K-energy (the case $k = 0$) [**245**, §2] and its extension by the author to all k [**216**, §4.4.5], the computations are almost identical. In essence, the (twisted) Kähler–Ricci functionals are realized as a linear combination of Bott–Chern forms, one for each of the $\mathbb{R}$-line bundles $E_j = K_M^{-1} \otimes L_D^{\beta-1} \otimes L^{n-2j}$ for $j = 0, \ldots, n$. One of these terms (the simplest contribution) is a multiple of the form appearing in Example 5.6.

THEOREM 5.7. *Let $k \in \{0, \ldots, n\}$ and let ϕ be defined by (5.21). Let (M, J, ω) be a projective Kähler manifold, and let L be a line bundle with $c_1(L) = [\omega]$. Let μ_k^β be given by (5.6). For each $k \in \{0, \ldots, n\}$,*
(5.23)

$$\begin{aligned}
&\Big[\Delta_\omega \sigma_k^\beta - \frac{n-k}{k+1}\Big(\sigma_{k+1}^\beta - \mu_{k+1}^\beta\Big)\Big]\omega^n = \\
&\quad \frac{2^{-n}}{(n+1)!}\binom{n}{k}^{-1}\frac{1}{V}\sum_{j=0}^n (-1)^j\binom{n}{j}(n-2j)^k\,\mathbf{BC}(\phi; h_0^{n-2j}\det F_{h_0}|s|^{2-2\beta}, \\
&\qquad\qquad h_1^{n-2j}\det F_{h_1}|s|^{2-2\beta}) \\
&\quad - \frac{1}{V}\frac{\mu_{k+1}^\beta}{n+1}\frac{n-k}{k+1}\mathbf{BC}(\phi; h_0, h_1).
\end{aligned}$$

PROOF. The proof in the case $\beta = 1$ is given in [**216**, Proposition 4.22]. The general case follows from the same computations by subtracting the *fixed* current $(1-\beta)[D]$ from each Ricci current that appears in the computation. □

Remark that from the proof it follows that similar "twisted" functionals can thus be defined by replacing $(1-\beta)[D]$ with some other fixed curvature current of a line bundle.

REMARK 5.8. Tian [**242**, p. 255–257] gave an interpretation of the "complex Hessian" of the K-energy E_0 in terms of a certain "universal" Hermitian metric $\mathbf{h}$ (see the references above for the notation and definitions):

$$-\sqrt{-1}\partial\bar{\partial}E_0 = \frac{1}{V}\int_M (\sqrt{-1})^{n+1}\Big(F_{\det F_{\mathbf{h}}} - \frac{n\mu_0}{n+1}F_{\mathbf{h}}\Big)\wedge (F_{\mathbf{h}})^n. \tag{5.24}$$

We note in passing the following generalization of this formula to the Kähler–Ricci functionals
(5.25)

$$-\sqrt{-1}\partial\bar{\partial}E_k = \frac{1}{V}\int_M (\sqrt{-1})^{n+1}\Big((F_{\det F_{\mathbf{h}}})^{k+1} - \frac{(n-k)\mu_k}{(k+1)(n+1)}(F_{\mathbf{h}})^{k+1}\Big)\wedge (F_{\mathbf{h}})^{n-k}.$$

We also remark that using the techniques of [**213**] one may generalize appropriately Theorem 5.1 in terms of properness of the functional E_k^β on the space of Kähler forms η for which $\operatorname{Ric}\eta - (1-\beta)[D]$ is positive and cohomologous to $[\eta]$. Finally, also L has such a formula due to Tian [**245**, p. 214],

$$\sqrt{-1}\partial\bar{\partial}L = \frac{1}{V}\int_M (\sqrt{-1}F_{\mathbf{h}})^{n+1},$$

that has been extended by Berman–Boucksom [**21**, (4.1)] to Kähler potentials with low regularity, and can be used to characterize weak geodesics in the Mabuchi metric.

5.5. Legendre transform. In this section we restrict to the case $\mu > 0$ to simplify the notation, and describe work of Berman [**18**] and Berman–Boucksom [**21**] that ties F^β with E_0^β via the Legendre transform.

Defining the probability measure (recall the normalization (3.9))

$$\nu_\eta := (\operatorname{Ric}^{-1}\eta)^n = e^{f_\eta}\eta^n, \tag{5.26}$$

we rewrite (5.3) and (5.8) as

$$\begin{aligned} F^\beta(\eta,\eta_\varphi) &= -L_\eta(\varphi) - \frac{1}{\mu}\log\frac{1}{V}\int e^{-\mu\varphi}\nu_\eta, \\ E_0^\beta(\eta,\eta_\varphi) &= \frac{1}{V}\int_M \log\frac{\eta_\varphi^n}{\nu_\eta}\eta_\varphi^n - \mu(I-J)(\eta,\eta_\varphi) + \frac{1}{V}\int_M f_\eta\eta^n. \end{aligned} \tag{5.27}$$

The last term in E_0^β is a constant. It can be eliminated if we had normalized $\int_M f_\eta\eta^n = 0$, and then set $\nu_\eta := \frac{e^{f_\eta}\eta^n}{\frac{1}{V}\int e^{f_\eta}\eta^n}$. The first term in E_0^β is the entropy of η_φ^n relative to the measure ν_η considered as a functional on the space of measures,

$$\operatorname{Ent}(\nu,\chi) = \frac{1}{V}\int_M \log\frac{\chi}{\nu}\chi, \tag{5.28}$$

which, in terms of the density $d = \chi/\nu$, takes the familiar form for the entropy $\int d \log d\, \nu$. On the other hand, it is classical that the last term in F^β is precisely the Legendre transform of the entropy, in the sense that [**85**, p. 264],

(5.29)
$$\Lambda_\nu(-\mu\varphi) = \log \frac{1}{V} \int e^{-\mu\varphi}\nu = \mathrm{Ent}(\nu, \cdot)^\star(-\mu\varphi) = \sup_{\chi \in \mathcal{V}_V} \{\langle -\mu\varphi, \chi\rangle - \mathrm{Ent}(\nu, \chi)\},$$

where $\mathcal{V}_V = \{\nu : 0 \le \nu/\omega^n \in C^0(M, \omega^n), \int \nu = V\}$. Conversely, by convexity,

$$\mathrm{Ent}(\nu, \mu) = \sup_{\psi \in C^0(M)} \{\langle \mu, \psi\rangle - \log \frac{1}{V} \int e^\psi \nu\} = \Lambda_\nu^\star(\mu).$$

One of Berman's insights was that the remaining terms in E_0^β and F^β are similarly related by the Legendre transform [**18**, §2] (cf. [**23**, Theorem 5.3]).

LEMMA 5.9. *Let* $\mu > 0$. *Then,* $\sup_{\psi \in \mathrm{PSH}(M,\omega)\cap C^0}\{\langle -\mu\psi, \omega_\varphi^n\rangle + \mu L_\omega(\psi)\} = \mu(I - J)(\omega, \omega_\varphi)$.

We could have introduced a minus sign into the usual inner product between functions and measures in order to obtain that $(-L)^\star = I - J$. Instead of doing that, we kept the usual inner product but then the left hand side in Lemma 5.9 is not precisely the Legendre transform.

PROOF. The proof would be easier if we knew that the supremum over all functions coincides with that over ω-psh ones. Indeed, in that case, $F(\psi) := \langle -\mu\psi, \omega_\varphi^n\rangle + \mu L_\omega(\psi)$ is concave in ψ in the sense that

$$-\frac{d^2}{dt^2}\bigg|_{t=0} F(\psi + t\phi) = \mu \int n\sqrt{-1}\partial\phi \wedge \bar\partial\phi\omega_\psi^n \ge 0$$

(recall that $dL_\omega|_\psi = \omega_\psi^n$). Differentiating, one sees a critical point ψ, necessarily a maximum, must satisfy $\omega_\psi^n = \omega_\varphi^n$ (here we are being a bit loose since $C^0(M)$ is infinite-dimensional; however the same reasoning as for the Legendre transform in finite-dimensions applies), hence by uniqueness $\varphi = \psi + C$ [**31**]. Plugging back in, and using the formula $L(\varphi) = (I - J)(\omega, \omega_\varphi) + \frac{1}{V}\int \varphi\omega_\varphi^n$ then yields the statement.

To make the reduction to the subset $\mathrm{PSH}(M, \omega) \cap C^0(M) \subset C^0(M)$, recall the definition of the ω-psh envelope operator $P_\omega : \varphi \mapsto \sup\{\phi \in \mathrm{PSH}(M, \omega) \cap C^0(M) : \phi \le \varphi\}$. By a result of Berman–Boucksom [**21**], $L_\omega \circ P_\omega$ is concave on $C^0(M)$, Gateaux differentiable, and $dL_\omega \circ P_\omega|_\varphi = \omega_{P_\omega\varphi}^n$. Thus, $\sup_{\psi \in C^0}\{\langle -\psi, \omega_\varphi^n\rangle + L_\omega \circ P_\omega(\psi)\} = (I - J)(\omega, \omega_{P\varphi})$. This concludes the proof, since $P\varphi \le \varphi$ and so the supremum must actually be attained at $\varphi \in \mathrm{PSH}(M, \omega)$, as $Lu \le Lv$ if $u \le v \le 0$, and we can normalize ψ so that $\sup\psi = 0$. □

REMARK 5.10. As pointed out by Berman, one can, in fact, avoid using the result of Berman–Boucksom to prove the preceeding Lemma, since simpler convexity arguments already show that the supremum must be attained at φ. However, that result is useful to show that a maximizer ψ must satisfy $\omega_\psi^n = \omega_\varphi^n$ (and hence $\psi \in L^\infty$ [**149**] and so equal to φ up to a constant [**31**]). In addition, the result of Berman–Boucksom is essentially needed if one replaces ω_φ^n by a more general measure or even volume form ν; the result can be seen as the starting point of the variational approach to constructing a weak solution of the equation $\omega_\varphi^n = \nu$.

5.6. Equivalence of functionals. Motivated by variational calculus, one expects that the KEE problem is solvable if and only if it can be cast as a variational problem with a coercive functional. One calls E coercive if there exist uniform positive constants A, B such that the c-sublevel set of E is contained in the $A(c+B)$-sublevel set of J, i.e., $E \geq \frac{1}{A}J - B$. In particular, E is then proper, thus bounded from below. We have seen a number of functionals that have KEE metrics as critical points. The following basic result says that they are all more or less equivalent as far as boundedness, coercivity, and existence of KEE metrics is concerned. To state it we introduce some notation. Suppose that $\mu > 0$, and define

$$\mathcal{H}^{e,+}_\omega = \{\varphi \in \mathcal{H}^e_\omega \, : \, \mathrm{Ric}\,\omega_\varphi - (1-\beta)[D] \text{ is a positive current}\}. \tag{5.30}$$

This space is nonempty as a corollary of the existence theorem for the case $\mu = 0$. Let $l(\omega) = \inf_{\varphi\in\mathcal{H}^e_\omega} F^\beta(\omega,\omega_\varphi)$ and

$$l_k(\omega) = \begin{cases} \inf_{\varphi\in\mathcal{H}^e_\omega} E_k(\omega,\omega_\varphi), & \text{for } k = 0, 1, \\ \inf_{\varphi\in\mathcal{H}^{e,+}_\omega} E_k(\omega,\omega_\varphi), & \text{for } k = 2, \dots, n. \end{cases}$$

THEOREM 5.11. *Let $\mu > 0$. (i) The lower bounds of E^β_k and that of F^β are related by*

$$\mu l(\omega) + \frac{1}{V}\int f_\omega\omega^n = l_0(\omega) = \mu^{-k} l_k(\omega) - \mu I_k(\omega, \mu^{-1}\mathrm{Ric}\,\omega - \mu^{-1}(1-\beta)[D]). \tag{5.31}$$

In particular, F^β, E^β_0 and E^β_1 are simultaneously bounded or unbounded from below on $\mathcal{H}^e_\omega$.
(ii) The coercivity of E^β_0 is equivalent to that of F^β.

PROOF. (i) First, by Proposition 5.2 and the fact that I_k is nonnegative on $\mathcal{H}^e_\omega \times \mathcal{H}^e_\omega$, it follows that if E^β_0 is bounded below on $\mathcal{H}^e_\omega$ then E^β_k, $k = 1 \dots n$ is bounded below on $\mathcal{H}^{e,+}_\omega$. In particular, this is true for E^β_n, but then, by Proposition (5.4) and the edge version of the Calabi–Yau theorem (Conjecture 4.2, that guarantees that $\mathrm{Ric}^{-1}_\beta : \mathcal{H}^e_\omega \to \mathcal{H}^{e,+}_\omega$ is an isomorphism) it follows that F^β is bounded below on $\mathcal{H}^e_\omega \times \mathcal{H}^e_\omega$. But, by a formula of Ding–Tian [**90**]

$$(E^\beta_0 - \mu F^\beta)(\omega,\omega_\varphi) = \frac{1}{V}\int f_\omega\omega^n - f_{\omega_\varphi}\omega^n_\varphi \geq \int f_\omega\omega^n, \tag{5.32}$$

where we used the normalization (3.9) and Jensen's inequality

$$\frac{1}{V}\int f_{\omega_\varphi}\omega^n_\varphi \leq \log\frac{1}{V}\int e^{f_{\omega_\varphi}}\omega^n_\varphi = 0.$$

This proves (i), since the precise lower bounds (5.31) can be deduced from the proof and a theorem of Ding–Tian—see [**213**, Remark 4.5].

We now give a second derivation, due to Berman, of a special case of (i), namely, the equivalence of the lower bounds of F^β and E^β_0. First, $F^\beta(\omega,\,\cdot\,) \geq -C$, is equivalent to $-\mu L_\omega \geq \log\frac{1}{V}\int e^{-\mu\varphi}\nu - \mu C$. The Legendre transform, in the sense of the previous subsection, is order-reversing. Thus, according to (5.29) and Lemma 5.9, $\mu(I-J)(\omega,\omega_\varphi) \leq \mathrm{Ent}(\nu,\omega^n_\varphi) + \mu C$ i.e., $E^\beta_0(\omega,\,\cdot\,) \geq \frac{1}{V}\int f_\omega\omega^n - \mu C$. This concludes this derivation since the Legendre transform is an involution.

(ii) Suppose that E^β_0 is coercive. Then, by definition, $E^\beta_0 - \epsilon(I-J)$ is bounded from below. It follows from (i) that so is F^β, but now with μ replaced by $\mu+\epsilon$, in other words

$$-L_\omega(\varphi) \geq \frac{1}{\mu+\epsilon}\log\frac{1}{V}\int e^{-(\mu+\epsilon)\varphi}\nu - C,$$

for all $\varphi \in \mathrm{PSH}(M,\omega)\cap C^0$. Normalize φ so that $\int\varphi\omega^n = 0$, and substitute $\frac{\mu}{\mu+\epsilon}\varphi \in \mathrm{PSH}(M,\omega)$ in this inequality to obtain

$$J(\omega,\omega_{\mu\varphi/(\mu+\epsilon)}) = -L_\omega\Big(\frac{\mu}{\mu+\epsilon}\varphi\Big) \geq \frac{1}{\mu+\epsilon}\log\frac{1}{V}\int e^{-\mu\varphi}\nu - C,$$

so

$$\begin{aligned}\frac{1}{\mu}\log\frac{1}{V}\int e^{-\mu\varphi}\nu - C' &\leq \frac{\mu+\epsilon}{\mu}J(\omega,\omega_{\mu\varphi/(\mu+\epsilon)})\\ &\leq \Big(\frac{\mu}{\mu+\epsilon}\Big)^{1/n}J(\omega,\omega_\varphi) =: -(1-\epsilon')L(\omega,\omega_\varphi),\end{aligned}$$

where the last inequality is due to Ding [**88**, Remark 2]. Thus, $F^\beta \geq \epsilon' J - C'$. The converse follows from (5.32). □

REMARK 5.12. Part (i) and its proof above is due to [**213**]. The special case of the equivalence of F^β and E_0^β being bounded below was obtained independently by H. Li [**159**] using results of Perelman [**222**] on the Ricci flow, and a third proof was later given by Berman [**18**] using the Legendre transform, as presented above. Part (ii) is a special case of a result of Berman (which allows to replace ν by a rather general probability measure), which generalized a result of Tian and its subsequent refinement by Phong et al. (in the smooth case) [**202, 244, 246**]. For a comparison of properness and coercivity in the KE setting we refer to [**248**, §2].

6. The Ricci continuity method

This section describes a new input that goes into the proof of Conjecture 4.2 and Theorem 4.13 in §4.5. It is a new continuity method that is essentially the only one that can be used to prove existence of KEE metrics, since the classical continuity methods that were previously used to construct KE metrics break down when β belongs to the more challenging regime $(1/2, 1)$ where, among other things, the curvature of the reference geometry is no longer bounded (Lemma 3.14). We describe all of this in detail in §6.1–§6.4. Subsection 6.5 is an interlude about the Ricci iteration that origingally motivated the Ricci continuity method. Finally, §6.6 describes other approaches to existence.

6.1. Ricci flow meets the continuity method. The Ricci continuity method was introduced in [**214**] and was further developed and first used systematically to construct KE(E) metrics in [**141, 175**]. The idea is to prove existence of a continuity path, or, in other words, a one-parameter family of Monge–Ampère equations and solutions thereto, in a canonical geometric manner. To that end, we start with the Ricci flow

$$\frac{\partial\omega(t)}{\partial t} = -\mathrm{Ric}\,\omega(t) + (1-\beta)[D] + \mu\omega(t), \quad \omega(0) = \omega \in \mathcal{H}, \tag{6.1}$$

Fix $\tau \in (0,\infty)$. The (time τ) Ricci iteration is the sequence $\{\omega_{k\tau}\}_{k\in\mathbb{N}} \subset \mathcal{H}_\omega$, satisfying the equations

$$\omega_{k\tau} = \omega_{(k-1)\tau} + \tau\mu\omega_{k\tau} - \tau\mathrm{Ric}\,\omega_{k\tau} + \tau(1-\beta)[D], \qquad \omega_{0\tau} = \omega,$$

for each $k \in \mathbb{N}$ for which a solution exists in $\mathcal{H}_\omega$. This is the backwards Euler discretization of the flow (4.5) [**214**]. Equivalently, let $\omega_{k\tau} = \omega_{\psi_{k\tau}}$, with $\psi_{k\tau} = \sum_{l=1}^{k} \varphi_{l\tau}$. Then,

$$\omega^n_{\psi_{k\tau}} = \omega^n e^{f_\omega - \mu\psi_{k\tau} + \frac{1}{\tau}\varphi_{k\tau}}. \tag{6.2}$$

We now change slightly our point of view by fixing $k = 1$ but instead varying τ in $(0, \infty)$. This yields $\omega^n_{\varphi_\tau} = \omega^n e^{f_\omega + (\frac{1}{\tau} - \mu)\varphi_\tau}$, and setting $s := \mu - \frac{1}{\tau}$, we obtain

$$\omega^n_\varphi = \omega^n e^{f_\omega - s\varphi}, \quad s \in (-\infty, \mu], \tag{6.3}$$

where $\varphi(-\infty) = 0$, and $\omega_{\varphi(-\infty)} = \omega$. We call this the *Ricci continuity path (Ricci CP)*.

Aside from the formal derivation that relates (6.3) to the Ricci flow, it turns out that the two equations share several key analytic and geometric properties:
(i) Short-time existence: the Ricci CP exists for all $0 < \tau << 1$, i.e., for all $s << -1$.
(ii) Monotonicity: $\dot{E}_0(\omega(0), \omega_{\varphi_s}) \le 0$ with equality iff $\omega(0)$ is KEE.
Moreover, the Ricci CP inherits one additional property that is not satisfied by the Ricci flow, and which provides an important advantage:
(iii) Ricci lower bound: Along the Ricci CP $\operatorname{Ric} \omega_{\varphi(s)} > s\omega_{\varphi(s)}$. Moreover, this holds even if the initial metric has unbounded Ricci curvature! Note that properties (i) and (iii) were first noticed by Wu and Tian–Yau, respectively, in their study of non-compact KE metrics with negative Ricci curvature [**252, 265**], while (ii) goes back to [**14**].

In the rest of this section we will explain how to obtain a unified proof of existence of both smooth and edge KE metrics (the proof of Conjecture 4.2 and Theorem 4.13) using the Ricci CP. In §6.6 we also review other approaches to existence. But first, we make a comparison to some other CPs and explain where each of them would break down in the edge setting.

6.2. Other continuity methods. The continuity path (6.3) has several useful properties, some already noted above, which are necessary for the proof of Conjecture 4.2 and Theorem 4.13 when $\beta > 1/2$. In other words, one could use various CPs (including the Ricci CP) when β is in the "orbifold regime" $\beta \in (0, 1/2]$, but it seems that the CPs we discuss below break down in the regime $\beta \in (1/2, 1)$. To illustrate this, we now describe these CPs and where they fail.

When $\mu \le 0$, Calabi suggested the following path [**46**, (11)] that was later used by Aubin and Yau [**8, 268**]

$$\omega^n_{\varphi_t} = \omega^n e^{tf_\omega + c_t - \mu\varphi_t}, \quad t \in [0, 1]. \tag{6.4}$$

In the case $\mu > 0$, Aubin suggested the following extension of Calabi's path [**9**]

$$\omega^n_{\varphi_t} = \begin{cases} \omega^n e^{tf_\omega + c_t}, & t \in [0, 1], \\ \omega^n e^{f_\omega - (t-1)\varphi_t}, & t \in [1, 1 + \mu]. \end{cases} \tag{6.5}$$

Still when $\mu > 0$, an alternative path was considered by Demailly and Kollár [**84**, (6.2.3)], given by

$$\omega^n_{\varphi_t} = \omega^n e^{t(f_\omega/\mu - \varphi_t)}, \quad t \in [0, \mu]. \tag{6.6}$$

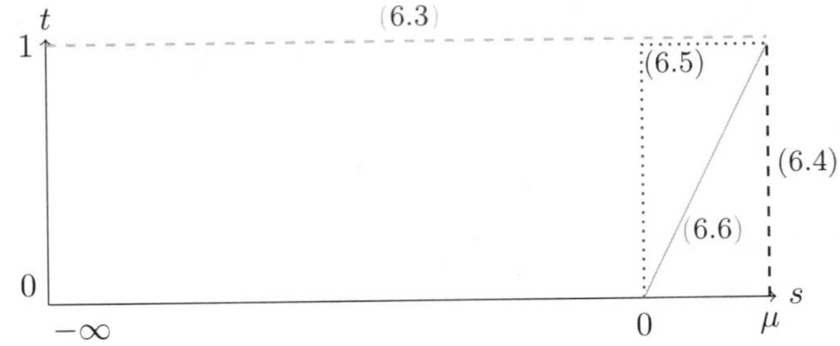

FIGURE 6. Continuity paths of the form (6.7) with $(s,t) \in A$ (assuming $\mu > 0$). The Ricci continuity path is (6.3).

All of these paths, as well as the Ricci CP, correspond to different curves within the two-parameter family of equations

$$\omega_\varphi^n = e^{tf_\omega + c_t - s\varphi}\omega^n, \quad c_t := -\log \frac{1}{V}\int_M e^{tf_\omega}\omega^n, \quad (s,t) \in A, \tag{6.7}$$

where $A := (-\infty, 0] \times [0,1] \cup [0,\mu] \times \{1\}$ (see Figure 6).

In the smooth setting, any one of these paths may be used to prove existence of a KE metric, assuming the K-energy is proper. Note that different paths have been used to prove existence, depending on μ: (6.4) when $\mu \le 0$, and (6.5) or (6.6) when $\mu > 0$. Below, we will prove existence in a unified manner, i.e., regardless of the sign of μ or whether $\beta = 1$ or $\beta \in (0,1)$. In fact, we show that when $\beta \in (0,1/2] \cup \{1\}$ then (6.7) has a solution for each $(s,t) \in A$. On the other hand, when $\beta \in (0,1/2)$ the Ricci curvature of the reference metric ω is unbounded from below as a corollary of Lemma 3.14. Thus, for each $(s,t) \in A$,

$$\operatorname{Ric}\omega_{\varphi(s,t)} = (1-t)\operatorname{Ric}\omega + s\omega_{\varphi(s,t)} + (\mu t - s)\omega + (1-\beta)[D],$$

and this has a lower bound *only if* $t = 1$. Thus, on the one hand, the Chern–Lu inequality, which requires such a lower bound is inapplicable; on the other hand, the Aubin–Yau inequality which requires a lower bound on the bisectional curvature of the reference geometry is inapplicable once again due to Lemma 3.14. This reasoning sifts out naturally the Ricci CP among all other possible curves in A.

6.3. Short time existence. Intuitively, the Ricci continuity path (6.3) has the trivial solution $\omega(-\infty) = \omega$ at $s = -\infty$. Producing solutions for very negative s can be considered as the continuity method analogue of showing short-time existence for the Ricci flow. However, it is not possible to apply the implicit function theorem directly to obtain solutions for large negative finite values of s (this observation is due to Wu, who noted that the last displayed equation on [**252**, p. 589] is valid for $s_0 > 0$ but not for $s_0 = 0$). Indeed, reparametrizing (6.3) by setting $\sigma = -1/s$, then the linearization of the Monge-Ampère equation at σ equals $\sigma\Delta_{\varphi(-1/\sigma)} - 1$, which degenerates at $\sigma = 0$. More concretely, $L_\sigma := \sigma\Delta_{\varphi(-1/\sigma)} - 1$ has bounded operator norm when considered as acting from $\mathcal{D}^{0,\gamma}_w$ to $C^{0,\gamma}_w$ only when $\sigma > 0$: there is no constant $C > 0$ such that $||L_0 v||_{\mathcal{D}^{0,\gamma}_w} = ||v||_{\mathcal{D}^{0,\gamma}_w} \le C||v||_{C^{0,\gamma}_w}$, of course.

Thus a different method is needed to produce a solution of (6.3) for sufficiently negative, but finite, values of s. We present two arguments. The first, described in

§6.3.1, works only when $\beta \in (0, 1/2] \cup \{1\}$. Wu's original argument also only works when β is in that range; in §6.3.2 we present a generalization of Wu's argument that does not require lower curvature bounds on ω and thus is applicable for all $\beta \in (0, 1]$. Finally, we remark that an interpretation of the $s \to -\infty$ limit in terms of thermodynamics together with a variational approach has been given by Berman [**20**].

6.3.1. *The two-parameter family trick.* When $\beta \in (0, 1/2] \cup \{1\}$, the difficulty with applying the implicit function theorem can be circumvented as follows [**141**]. Indeed, the original continuity path (6.3) embeds into the two-parameter family (6.7), and it is trivial that solutions exist for the finite parameter values $(s, 0)$. Thus, while this does not show directly that our original equation has solutions for all $(s, 1)$ with s sufficiently negative, it yields that result eventually, *provided* we have a priori estimates for all values (s, t). This is somewhat reminiscent of adding variables or symmetries to a given equation in order to solve it.

The reason this trick does not seem to work when $\beta > 1/2$ is that it is not clear how to obtain the a priori estimates needed to carry out the rest of the continuity argument for the two-parameter family, unless $\beta \in (0, 1/2] \cup \{1\}$. In essence then, the classical continuity path with parameter values $\{(-\mu, t),\ t \in [0, 1]\}$ may simply fail to exist within the space of $\widetilde{\mathcal{D}}^{0,\gamma}_s$-regular Kähler edge potentials. It would be interesting to understand the maximal set of values (s, t) for which a solution exists in (6.7),

$$M := \{(s, t) \in A : (6.7) \text{ admits a solution } \varphi(s, t) \in \mathrm{PSH}(M, \omega) \cap \mathcal{D}^{0,\gamma}_w\}, \tag{6.8}$$

as well as analogues of this set for lower regularity classes.

6.3.2. *Newton iteration arguments.* Thus, to handle the general case, another method must be used to obtain a solution of (6.3) for some very negative value of s. Wu used a Newton iteration argument to obtain such a solution in a different setting [**265**, Proposition 7.3]. However, his argument requires a Ricci curvature bound on the reference metric (see [**265**, p. 431] where the expression $\Delta_\omega f_\omega$, that on $M \setminus D$ equals the scalar curvature up to a constant, enters), which we lack. What follows is an adaptation of Wu's argument that requires no curvature control on the reference metric, and thus requires more delicate estimates. We compare our approach to Wu's in Remark 6.2.

Define

$$N_\sigma : \mathcal{D}^{0,\gamma}_w \to C^{0,\gamma}_w, \quad N_\sigma(\Phi) := \log(\omega^n_{\sigma\Phi}/e^{f_\omega}\omega^n) - \Phi.$$

This is equivalent to the original Monge–Ampère equation (6.3) upon substituting $\sigma = -1/s$ and $\Phi = -s\varphi$. Note that $DN_\sigma|_\Phi = \sigma\Delta_{\sigma\Phi} - \mathrm{Id}$. Now, suppose that $\sigma\Phi \in \mathcal{H}_\omega$, and, say, $s < -1$. By the maximum principle (adapted to the edge setting by adding a barrier function, see §6.4), the nullspace of DN_σ is trivial provided $s < 0$. Thus, Theorem 3.7 (i) implies that $DN_\sigma|_\Phi : \mathcal{D}^{0,\gamma}_w \to C^{0,\gamma}_w$ is an isomorphism, with

$$||u||_{\mathcal{D}^{0,\gamma}_w} \le C||DN_\sigma u||_{C^{0,\gamma}_w}. \tag{6.9}$$

Denote by $DN_\sigma|_\Phi^{-1}$ the inverse of this map on $C^{0,\gamma}_w$.

PROPOSITION 6.1. *Define,* $\Phi_0 = 0$, $\Phi_k = (Id - DN_\sigma|_{\Phi_{k-1}}^{-1} \circ N_\sigma)(\Phi_{k-1})$, $\quad k \in \mathbb{N}$. *There exists* $0 < \sigma_0 \ll 1$ *and* $\gamma' > 0$, *such that if* $\sigma \in (0, \sigma_0)$ *then* $\sigma \lim_{k\to\infty} \Phi_k \in \mathcal{D}^{0,\gamma'}_w \cap \mathrm{PSH}(M, \omega)$ *solves* (6.3) *with* $s = -1/\sigma$.

The proof appears in [**141**, §9]. The crucial step is showing that $\sigma\Phi_1$ has *small* $\mathcal{D}_w^{0,\gamma'}$ norm, and therefore it is still a Kähler edge potential. We now sketch some of the details.

When $k=1$, $N_\sigma(0)=-f_\omega$ and $\Phi_1=(\sigma\Delta_\omega-\mathrm{Id})^{-1}f_\omega$. To prove that $\sigma\Phi_1\in\mathcal{H}_\omega$, it suffices to show that the pointwise norm $|\partial\bar\partial\,\sigma\Phi_1|_\omega$ is small, for then $\omega+\sqrt{-1}\partial\bar\partial\,\sigma\Phi>0$. By Theorem 3.7 (i), it is enough to prove that $\Delta_\omega(\sigma\Phi_1)$ is small in $C_w^{0,\gamma}$. However, by definition,

$$\sigma\Delta_\omega\Phi_1=\sigma\Delta_\omega(\sigma\Delta_\omega-1)^{-1}f_\omega=\Delta_\omega(\Delta_\omega+s)^{-1}f_\omega.$$

The difficulty is that $f_\omega\in C_w^{0,\gamma}$ for $\gamma\in(0,1/\beta-1)$, but not higher in the wedge Hölder scale. To overcome this, we consider f as varying over a range of function spaces (some of which f_ω does not belong to!), and estimate the norm of the map $C_w^{\ell_1,\gamma_1}\ni f\mapsto\Delta_\omega(\Delta_\omega+s)^{-1}f\in C_w^{\ell_2,\gamma_2}$, for different values of (ℓ_j,γ_j), and interpolate. This eventually leads to the estimate $||\Delta_\omega(\Delta_\omega+s)^{-1}f_\omega||_{w;0,\gamma''}\le C|s|^{-\eta}$ for some $\eta>0$ and $\gamma''\in(0,\gamma)$, proving that $\sigma\Phi_1$ is a Kähler edge potential for small enough σ. The rest of the proof of Proposition 6.1 then follows by induction.

REMARK 6.2. It is worth comparing the above approach to Wu's original argument. Appropriately translating Wu's argument into our setting one would have considered the operator $N'_\sigma(\Phi):=\log(\tilde\omega^n_{\sigma\Phi}/\omega^n)-\Phi$, with $\tilde\omega:=\omega-\sigma\sqrt{-1}\partial\bar\partial f_\omega$. This is clearly equivalent to the original complex Monge–Ampère equation. With this definition,

$$\Phi_1=(\sigma\Delta_{\tilde\omega}-1)^{-1}\log\frac{\tilde\omega^n}{\omega^n}.$$

When a small multiple of f_ω is a Kähler edge potential, equivalently when the Ricci curvature of the reference ω is uniformly bounded, then ω and $\tilde\omega$ are uniformly equivalent for all small σ. Then it is straightforward to show $\sigma\Phi_1$ is small in $\mathcal{D}_w^{0,\gamma}$, essentially from its definition and by computing D^2N'. The approach we described before was devised precisely to circumvent this lack of differentiability of f_ω.

6.4. Convergence. To show the convergence of the Ricci CP, in particular implying the existence of a KE(E) metric, one must prove, as usual, openness and closedness of the set $M\cap(-\infty,\mu]\times\{1\}$ in $(-\infty,\mu]\times\{1\}$ (recall (6.8)); indeed, since this set is nonempty by §6.3, this implies that it is equal to $(-\infty,\mu]\times\{1\}$. Openness follows as described in §4.5. Closedness follows from the following a priori estimate.

THEOREM 6.3. *Along the Ricci continuity path* (6.3),

$$||\varphi(s)||_{\mathcal{D}_w^{0,\gamma}}\le C,\tag{6.10}$$

where $C=C(||\varphi(s)||_{L^\infty(M)},M,\omega,\beta,n)$. *When* $\mu\le0$ *or the twisted Mabuchi energy is proper then*

$$||\varphi(s)||_{L^\infty(M)}\le c,\tag{6.11}$$

with c *depending only on* M,ω,β,n.

For very negative values of s this is a consequence of §6.3. Thus, (6.11) follows from Proposition §7.1, and (6.10) is a corollary of the a priori estimates (7.8) for the Laplacian of $\varphi(s)$ together with the maximum principle, and (7.24) for the Hölder semi-norm of the Laplacian of $\varphi(s)$. These estimates are described in detail in §7 below. There is, however, one caveat in applying the maximum principle in the edge setting: the maximum could be attained on D and then, as the metric blows up there, one cannot make sense of its Laplacian. A trick due to Jeffres [**143**] is to add the barrier function $c|s|_h^\epsilon$ with $c, \epsilon > 0$ small. This is easily seen to "push" the maximum away from D, while not changing the value of the function being maximized by a whole lot: the latter fact is obvious, while the maximum is pushed away from D precisely because the gradient of the barrier function blows up near D. One can even let c tend to zero to see that the same exact estimates as in the smooth case hold. An improved version of this maximum principle is proved in [**141**, Lemma 5.1].

6.5. The Ricci iteration. As explained in §6.1, the Ricci continuity method is motivated by the Ricci iteration, introduced in [**214**] (cf. Keller [**146**]). It is then natural to go back and prove convergence of the Ricci iteration $\{\omega_{k\tau}\}_{k\in\mathbb{N}}$. When $\tau = 1$, this leads to a particularly natural result:

$$\lim_{k\to\infty} \operatorname{Ric}_{\beta}^{-k}\omega = \omega_{\mathrm{KE}},$$

where ω_{KE} is a KEE metric of angle β, $\operatorname{Ric}^{-k} := (\operatorname{Ric}^{-1})^k$, and Ric^{-1} is the twisted inverse Ricci operator defined in §5.3. It is interesting to note that when $\beta = 1$, results of Donaldson [**100**] show that Ric^{-1} can be approximated by certain finite-dimensional approximations, and this was further studied by Keller [**146**] yielding Bergman type approximations to KE metrics.

The convergence of the Ricci iteration when $\tau > 1$, or when $\tau = 1$ and the α-invariant is bigger than one was proved in [**214**] and adapts to the edge setting once the a priori estimates needed for the Ricci CP are established. The weak convergence when $\tau \leq 1$ in general was first established in [**22**] (and the strong convergence then can be deduced from arguments of [**141**, **214**]) using a new pluripotential estimate from [**18**, **23**] that can be stated as follows:

LEMMA 6.4. *Suppose $J(\omega, \omega_\varphi) \leq C$. Then for each $t > 0$ there exists $C' = C'(C, M, \omega, t)$ such that $\int_M e^{-t(\varphi - \sup\varphi)}\omega^n \leq C'$.*

In particular, since the K-energy decreases along the Ricci iteration, the properness assumption means that $J(\omega, \omega_{k\tau})$ is uniformly bounded, independently of k. Thus, rewriting (6.2) as

$$\omega_{\psi_{k\tau}}^n = \omega^n e^{f_\omega - (1-\frac{1}{\tau})\psi_{k\tau} - \frac{1}{\tau}\psi_{(k-1)\tau}},$$

choosing p sufficiently large depending only on τ, say $p/3 = \max\{1 - \frac{1}{\tau}, \frac{1}{\tau}\}$. Using Kołodziej's estimate and the Hölder inequality this yields the uniform estimate $\operatorname{osc}\psi_{k\tau} \leq C$. Unlike for solutions of (6.3), the functions $\psi_{k\tau}$ need not be changing signs. But an inductive argument shows that $|(1-\frac{1}{\tau})\psi_{k\tau} - \frac{1}{\tau}\psi_{(k-1)\tau}| \leq C$ [**214**, p. 1543]. The higher derivative estimates follow as in the Ricci CP since the Ricci curvature is uniformly bounded from below along the iteration.

It is natural to also hope for similar results for a suitable Kähler–Ricci edge flow (4.5). For Riemann surfaces, a rather complete understanding is given by [**177**], as described in §4.4. Different approaches to short time existence in higher dimensions are developed by Chen and Wang [**68**, **69**, **264**], Liu–Zhang [**161**], as well as by Mazzeo and the author [**176**].

6.6. Other approaches to existence. The Ricci continuity method gives a unified proof of the classical results of Aubin, Tian, and Yau on existence of KE metrics in the smooth setting, and naturally generalizes to give new and optimal existence results for KEE metrics. This was the main contribution of [**141**, **175**], in addition to the the linear theory and higher regularity. Later, two other alternative approaches to existence were brought to fruition.

The first is a combination of a variational approach of Berman, and an approximation technique of Campana, Guenancia, and Păun. Guenancia–Păun, building on work of Campana–Guenancia–Păun, developed a smooth approximation method [**52**, **127**] to prove existence assuming a weak, say C^0, solution exists. In their scheme, such a solution is obtained by using the variational approach of Berman [**18**], under a properness assumption on the K-energy (when $\mu \leq 0$ such a C^0 solution exists automatically by Kołodziej's estimate [**149**]). One first approximates the reference form ω (3.5) by a particular sequence of smooth cohomologous Kähler forms $\omega_0 + \sqrt{-1}\partial\bar\partial\psi_\epsilon$ (with $\lim_{\epsilon\to 0}\psi_\epsilon = (|s|_h^2)^\beta$), and then solves the regularized Monge–Ampère equation

$$(\omega + \sqrt{-1}\partial\bar\partial(\psi_\epsilon + \phi_\epsilon))^n = \omega_0^n e^{f+\mu(\psi_\epsilon+\phi_\epsilon)}(|s|^2+\epsilon^2)^{\beta-1}. \tag{6.12}$$

These equations can be solved by standard results in the smooth setting, i.e., when $\epsilon > 0$. It thus suffices to prove a Laplacian estimate. When $\beta \in (0, 1/2]$ this follows directly from the classical Aubin–Yau estimate (7.16), once a careful and tedious computation establishes that the bisectional curvature of $\omega_0 + \sqrt{-1}\partial\bar\partial\psi_\epsilon$ is bounded from below independently of ϵ [**52**]. When $\beta \in (0,1)$ this follows by additional clever and lengthy computations showing that the negative contribution of the bisectional curvature of $\omega_0 + \sqrt{-1}\partial\bar\partial\psi_\epsilon$ in the right hand side of (7.16), can be cancelled by adding terms of the form $\Delta_\omega\chi(\epsilon^2 + |s|_h^2)$ on the left hand side, where $\chi : \mathbb{R}_+ \to \mathbb{R}_+$ is a certain auxiliary function [**127**]. As remarked in [**127**], a somewhat similar trick appears in [**42**] to deal with C^3 estimates, and is reviewed in §7.8. Later, Datar–Song observed that relying on Lemma 3.14 and the Chern–Lu inequality, as developed in [**141**, §7] (Corollary 7.2), one can avoid the aforementioned lengthy computations. Either way, the main advantage of this method over a continuity method is that, as first observed by Berman [**18**], no openness argument is needed.

The second is an angle-deforming continuity method that applies in the special case of a smooth plurianticanonical divisor $D \in |-mK_M|$ in a Fano manifold. It was introduced by Donaldson in lectures at Northwestern University in 2009 and later published in [**102**] in the case D is anticanonical, and the immediate, yet useful, extension to the case of a plurianticanonical divisor, was noted by Li–Sun [**157**]. Here, one constructs first a KEE of angle β along D for some small $\beta_0 \in (0,1)$. Equation (3.10) reads, $\operatorname{Ric}\omega_{\mathrm{KE}\beta_0} = \beta_0\omega_{\mathrm{KE}\beta_0} + (1-\beta_0)[D]/m$, and now we may consider β_0 as a *parameter* and try to deform it to a given β. This was first achieved for *all* small β_0 by the combined results of Berman [**18**] and [**141**]: the former shows that for all small $\beta > 0$ the twisted K-energy is proper, while the latter shows that properness implies existence. Alternatively, Berman also observes that when $\beta_0 = 1/k$ for $k \in \mathbb{N}$ sufficiently large, an orbifold KE metric can be constructed using Demailly–Kollár's orbifold version of Tian's α-invariant existence criterion [**84**, **238**]. Next, Donaldson's openness result implies that the KEE metric of angle β_0 can be deformed to a KEE of slightly larger angle, as long as the Lie algebra $\mathrm{aut}(M, D)$ is trivial. This always holds in this Fano setting (but not in general [**60**]) as first observed by Berman [**18**, p. 1291] (an algebraic proof of this was later given by Song–Wang [**230**]). Finally, the recently announced results of [**63**, **249**], together with Berman's observation that properness of the twisted K-energy is an open property (in β), can be combined to prove existence.

7. A priori estimates for Monge–Ampère equations

This section surveys the a priori estimates pertinent for the study of the (possibly degenerate) complex Monge–Ampère equations (6.7), both in the smooth setting ($\beta = 1$) and the edge setting ($\beta \in (0,1)$). The L^∞ estimate can be proved in at least three different ways, as discussed in §7.1 with little dependence on the (possibly unbounded) curvature of the background geometry. The Laplacian estimate, on the other hand, is quite sensitive to the latter, and more care is needed here. We take the opportunity to give a rather self-contained introduction to the Laplacian estimate for the complex Monge–Ampère equation in §7.2–7.7. The Chern–Lu inequality was used first by Bando–Kobayashi in the 80's to obtain a Laplacian estimate with bounded reference geometry [**13**], but fell into disuse since and was first systematically put to use for a general class of Monge–Ampère equations (more specifically, whenever the solution metric has Ricci curvature uniformly bounded from below) in the author's work on the Ricci iteration [**214**] and fully exploited in [**141**] to obtain estimates under a one-sided curvature bound on the reference geometry. This is described in §7.3–7.4. Traditionally, except in those three articles, the Laplacian estimate was essentially always derived using the Aubin–Yau estimate (with the exception of [**18**] that used the estimate from [**214**]). The latter estimate depends on a lower bound on the bisectional curvature of the reference metric, and is therefore not directly applicable for the Ricci continuity method. However, it always seemed curious to the author that the Chern–Lu inequality can be derived as a corollary of a general statement about holomorphic mappings, while the Aubin–Yau estimate is classically derived using a lengthy and rather un-enlightening computation. In §7.5 we describe a new inequality on holomorphic *embeddings*, that we call the *reverse Chern–Lu inequality* that yields the Aubin–Yau estimate as a corollary (§7.6–7.7). The remainder of this section then describes approaches to Hölder continuity of the metric. When $\beta \in (0, 1/2] \cup \{1\}$, the asymptotic expansion of (3.23) proves that third mixed derivatives of the type $\varphi_{i\bar{j}k}$ are bounded. Subsection 7.8 indicates how to obtain a uniform estimate for such derivatives by slightly modifying the original approach of Calabi in the smooth setting. In general, the expansion (3.23) shows that $\varphi_{i\bar{j}k} \notin L^\infty(M)$ but that $\varphi_{i\bar{j}k} \in L^2$. Tian's approach to proving a uniform *local* $W^{3,2}$ estimate on φ is the topic of §7.9; Campanato's characterization of Hölder spaces implies a uniform $\mathcal{D}_w^{0,\gamma}$ estimate on φ. Finally, §7.10 describes three other approaches to $\mathcal{D}_s^{0,\gamma}$ estimates.

7.1. Uniformity of the potential. Kołodziej's estimate gives a uniform bound on the oscillation of the solution u of $(\omega_0 + \sqrt{-1}\partial\bar{\partial}u)^n = F\omega_0^n$ in terms of $||F||_{L^{1+\epsilon}(M,\omega_0^n)}, \omega_0$, and $\epsilon > 0$ [**149**]. By (3.11) it suffices to take any ϵ in the range $(0, \frac{\beta}{1-\beta})$. Thus, Kołodziej's estimate (together with the normalization along (6.3)) directly provides the L^∞ bound on the Kähler potential along the Ricci CP for all $s \le 0$, and this is of course enough when $\mu \le 0$. A different approach is to use Moser iteration, as in Yau's work in the smooth setting, which directly adapts to this setting without change. The real challenge is then to obtain the estimate when $s > 0$.

The method used in [**141**] is to prove uniform bounds on the Sobolev and Poincaré constants along the Ricci CP. Then, a standard Moser iteration argument [**246**] gives a uniform control on the C^0 norm of $\varphi(s)$ in terms of $I(\omega, \omega_{\varphi(s)})$, which is in turn controlled under the properness assumption.

The proof of the uniformity of the Poincaré inequality is a quick consequence of the asymptotic expansion of the solutions $\varphi(s)$. This expansion precisely shows that $\varphi(s) \in W^{3,2}$ and so the integration by parts in the Bochner–Weitzenböck formula is justified, and readily implies $\lambda_1(\omega_{\varphi(s)}) \geq s$ with strict inequality for all $s < \mu$, and equality when $s = \mu$ iff there exist holomorphic vector fields on M tangent to D [**102**, Proposition 8],[**141**, Lemma 6.1].

The Sobolev inequality is trickier, but still the key is to use the validity of the integrated form of the Bochner–Weitzenböck formula. More precisely, standard results of Bakry and others on diffusive semigroups imply both the existence and the uniformity of a Sobolev inequality under a general curvature-dimension condition (that precisely corresponds to the Bochner–Weitzenböck inequality holding on a class of functions) as well as some assumptions on the algebra of functions on which the (uniform) curvature-dimension condition holds [**12**]. In our setting, the existence of a (possibly non-uniform in s) Sobolev inequality is easily verified by the change of coordinate $z \mapsto \zeta$ and a covering argument. Moreover, one can verify, using basic results on polyhomogeneity of solutions to quasilinear elliptic equations, that the class $\mathcal{D}_w^{0,\gamma}$ satisfies the conditions necessary for Bakry's approach to be carried out [**141**, §6]. Thus, this approach furnishes a *uniform* in s Sobolev inequality. This approach also gives a uniform diameter estimate along the Ricci continuity method. It is interesting to note that one could also use classical Riemannian geometry arguments (e.g., Croke's approach for the isoperimetric inequality [**77**], and Myers' approach for the diameter bound [**188**]) provided one knew that between every two points in $M \setminus D$ there exists a minimizing geodesic entirely contained in $M \setminus D$. This was shown very recently by Datar [**79**] building on a result on Colding–Naber [**76**].

A completely different approach is to regularize the equation and prove that solutions $\varphi(s)$ can be approximated by smooth Kähler metrics whose Ricci curvature is also bounded from below by s, and then use the standard results on Sobolev bounds [**63, 249**].

Finally, as in [**18**], one may use the pluripotential estimate of Lemma 5.9 to obtain a C^0 estimate via Kołodziej's result. In fact, more recent and sophisticated methods yield a Hölder estimate in this setting (see, e.g., [**105, 150**]). Furthermore, under stronger regularity assumptions on the right hand side there is also a Lipschitz estimate due to Błocki [**32**].

7.2. Uniformity of the metric, I. We say that ω, ω_φ are uniformly equivalent if $C_1\omega \leq \omega_\varphi \leq C_2\omega$, for some (possibly non-constant) $C_2 \geq C_1 > 0$. This is implied by either

$$n + \Delta_\omega \varphi = \mathrm{tr}_\omega \omega_\varphi \leq C_2 \text{ and } \det_\omega \omega_\varphi \geq C_1 C_2^{n-1}/(n-1)^{n-1}, \tag{7.1}$$

or,

$$n - \Delta_{\omega_\varphi} \varphi = \mathrm{tr}_{\omega_\varphi} \omega \leq 1/C_1 \text{ and } \det_\omega \omega_\varphi \leq C_1 C_2^{n-1}(n-1)^{n-1}; \tag{7.2}$$

conversely, it implies $\mathrm{tr}_\omega\omega_\varphi \le nC_2$ and $\det_\omega \omega_\varphi \ge C_1^n$, as well as $\mathrm{tr}_{\omega_\varphi}\omega \le n/C_1$ and $\det_\omega \omega_\varphi \le C_2^n$. Indeed, $\sum(1+\lambda_j) \le A$, and $\Pi(1+\lambda_j) \ge B$ implies $1+\lambda_j \ge (n-1)^{n-1}B/A^{n-1}$; conversely, $\Pi(1+\lambda_j) \ge \left(\frac{1}{n}\sum \frac{1}{1+\lambda_j}\right)^{-n} \ge C_1^n$.

Let $\iota : (M,\omega_\varphi) \to (M,\omega)$ denote the identity map. Consider $\partial\iota^{-1}$ either as a map from $T^{1,0}M$ to itself, or as a map from $\Lambda^n T^{1,0}M$ to itself. Alternatively, it is section of $T^{1,0\star}M \otimes T^{1,0}M$, or of $K_M \otimes K_M^{-1}$, and we may endow these product bundles with the product metric induced by ω on the first factor, and by ω_φ on the second factor. Then, (7.1) means that the norm squared of $\partial\iota^{-1}$, in its two guises above, is bounded from above by C_2, respectively bounded from below by $C_1C_2^{n-1}/(n-1)^{n-1}$. Similarly, (7.2) can be interpreted in terms of $\partial\iota$.

Now, $\det_\omega \omega_\varphi = {\omega_\varphi}^n/\omega^n = F(z,\varphi)$ is given solely in terms of φ (without derivatives), it thus suffices to find an upper bound for either $|\partial\iota^{-1}|^2$ or $|\partial\iota|^2$ (from now on we just consider maps on $T^{1,0}M$).

The standard way to approach this is by using the maximum principle, and thus involves computing the Laplacian of either one of these two quantities. The classical approach, due to Aubin [**7**, **8**] and Yau [**268**], is to estimate the first, while a more recent approach is to estimate the second [**141**, **176**, **214**], and this builds on using and finessing older work of Lu [**162**] and Bando–Kobayashi [**13**].

We take the opportunity to explain here both of these approaches in a unified manner since such a unified treatment seems to be missing in the literature. In particular, essentially all known Laplacian estimates are seen to be a direct corollary of the Chern–Lu inequality or its reverse form. We now explain this in detail.

7.3. Chern–Lu inequality. Let $f : (M,\omega) \to (N,\eta)$ be a holomorphic map between Kähler manifolds. We choose two holomorphic coordinate charts $(z_1,\dots,z_n)$ and $(w_1,\dots,w_n)$ centered at a point $z_0 \in M$ and at a point $f(z_0) \in N$, respectively, such that the first is normal for ω while the second is normal for η. In those coordinates, we consider the map

$$f : z = (z^1,\dots,z^n) \mapsto f(z) = (f^1(z),\dots,f^n(z)),$$

and write $\omega = \sqrt{-1}g_{i\bar{j}}(z)dz^i \wedge \overline{dz^j}$, $\eta = \sqrt{-1}h_{i\bar{j}}(w)dw^i \wedge \overline{dw^j}$, and

$$\partial f|_{T^{1,0}M} = \frac{\partial f^j(z)}{\partial z^i}dz^i|_z \otimes \frac{\partial}{\partial w^j}\Big|_{w(z)} = f_i^j dz^i|_z \otimes \frac{\partial}{\partial w^j}\Big|_{w(z)},$$

so

$$|\partial f_{T^{1,0}M}|^2 = g^{i\bar{l}}(z)h_{j\bar{k}}(f(z))f_i^j(z)\overline{f_l^k(z)}. \tag{7.3}$$

Thus, at z_0,

$$\begin{aligned}\Delta_\omega|\partial f_{T^{1,0}M}|^2(z_0) &= \sum_{p,q} g^{p\bar{q}}\frac{\partial^2(g^{i\bar{l}}h_{j\bar{k}}f_i^j\overline{f_l^k})}{\partial z^p\overline{\partial z^p}} \\ &= \sum_p g^{p\bar{q}}\Big[g^{i\bar{l}}h_{j\bar{k},d\bar{e}}f_i^j\overline{f_l^k}f_p^d\overline{f_q^e} - h_{j\bar{k}}g^{i\bar{t}}g^{s\bar{l}}g_{s\bar{t},p\bar{q}}f_i^j\overline{f_l^k} + g^{i\bar{l}}h_{j\bar{k}}f_{ip}^j\overline{f_{lq}^k}\Big] \\ &= -\omega^\# \otimes \omega^\# \otimes R_\eta(\partial f,\bar\partial f,\partial f,\bar\partial f) + (\mathrm{Ric}\,\omega)^\# \otimes \eta(\partial f,\bar\partial f) \\ &\qquad + g^{p\bar{q}}g^{i\bar{l}}h_{j\bar{k}}f_{ip}^j\overline{f_{lq}^k},\end{aligned} \tag{7.4}$$

where the last line is now coordinate independent. Here R_η denotes the curvature tensor of η (of type $(0,4)$), while $\omega^\#$ denotes the metric g^{-1} on $T^{1,0\star}M$ (i.e., of type $(2,0)$), and similarly $(\mathrm{Ric}\,\omega)^\#$ denotes the $(2,0)$-type tensor obtained from $\mathrm{Ric}\,\omega$ by raising indices using g. The last term in (7.4) is equal to $|\nabla\partial f|^2$, the covariant differential of ∂f, a section of $T^{1,0}M\otimes T^{1,0}M\otimes f^\star T^{1,0}N$. Note, finally, that $\eta(\partial f,\bar\partial f)=f^\star\eta$, and similarly other terms above can be expressed as pull-backs from N to M.

PROPOSITION 7.1. (Chern–Lu inequality) *Let $f:(M,\omega)\to(N,\eta)$ be a holomorphic map between Kähler manifolds. Then,*

(7.5)
$$\begin{aligned}|\partial f|^2\Delta_\omega\log|\partial f|^2 &= ((\mathrm{Ric}\,\omega)^\#\otimes\eta)(\partial f,\bar\partial f)-(\omega^\#\otimes\omega^\#\otimes R_\eta)(\partial f,\bar\partial f,\partial f,\bar\partial f)+e(f)\\ &\geq(\mathrm{Ric}\,\omega)^\#\otimes\eta(\partial f,\bar\partial f)-\omega^\#\otimes\omega^\#\otimes R_\eta(\partial f,\bar\partial f,\partial f,\bar\partial f),\end{aligned}$$

where $e(f)=|\nabla\partial f|^2-|\partial f|^2|\partial\log|\partial f|^2|^2$.

The proof follows from the previous paragraph, the identity $u\Delta_\omega\log u=\Delta_\omega u-u|\partial\log u|^2$, and the Cauchy–Schwarz inequality: indeed, since f is holomorphic,

$$|\partial f|^2|\nabla\partial f|^2\geq\langle\nabla\partial f,\bar\partial f\rangle^2=|\partial|\partial f|^2|^2, \tag{7.6}$$

therefore $e(f)\geq 0$. Note that $\partial^2 f$ and ∂f are of course sections of different bundles, and so we are abusing notation a bit when we write $\langle\partial\partial f,\bar\partial f\rangle$, however the meaning should be clear from (7.3). Note that the right hand side of (7.5) can be thought of as the Ricci curvature of the bundle $T^{1,0\,\star}M\otimes f^\star T^{1,0}N$ equipped with the metric $g^\#\otimes f^\star h$.

A few words about the history of inequality (7.5). It was shown by Lu (more generally on Hermitian manifolds) [**162**] (though "=" should be replaced by "$\geq$" in [**162**, (4.13)]). Chern [**71**] carried out a similar computation earlier for $\Delta_\omega|\partial f|^2$ for ∂f considered as a map from $\Lambda^nT^{1,0}M$ to $\Lambda^nT^{1,0}N$. As pointed out to us by Donaldson, (7.4) is in fact a special case of the computation of Eells–Sampson (with a slightly different sign convention) [**103**, (16)] on the Laplacian of the energy density of a harmonic map (holomorphic maps are harmonic by op. cit., pp. 116–118). In fact, as Eells–Sampson observe, when f is an immersion, (7.4) can also be proved using the Gauss equations.

7.4. A corollary of the Chern–Lu inequality. The next result shows how to estimate $\Delta_{\omega_\varphi}\log|\partial\iota|_{T^{1,0}M}|^2$ solely under an upper bisectional curvature bound on the target, and a generalized lower Ricci curvature bound on the domain.

PROPOSITION 7.2. *Let $f:(M,\omega)\to(N,\eta)$ be a holomorphic map between Kähler manifolds. Assume that* $\mathrm{Ric}\,\omega\geq-C_1\omega-C_2f^\star\eta$ *and that* $\mathrm{Bisec}_\eta\leq C_3$, *for some $C_1,C_2,C_3\in\mathbb{R}$. Then,*

$$\Delta_\omega\log|\partial f|^2\geq-C_1-(C_2+2C_3)|\partial f|^2. \tag{7.7}$$

In particular, if $f=\iota:(M,\omega)\to(M,\eta)$ is the identity map, and $\omega=\eta+\sqrt{-1}\partial\bar\partial\varphi$ then

$$\Delta_\omega\big(\log\mathrm{tr}_\omega\eta-(C_2+2C_3+1)\varphi\big)\geq-C_1-(C_2+2C_3+1)n+\mathrm{tr}_\omega\eta. \tag{7.8}$$

Using the Chern–Lu inequality to prove a Laplacian estimate for complex Monge–Ampère equations seems to go back to Bando–Kobayashi [**13**], who considered the case $\mathrm{Ric}\,\omega\geq-C_2\eta$. Next, the case $\mathrm{Ric}\,\omega\geq-C_1\omega$ first appeared in proving a priori Laplacian estimate for the Ricci iteration [**214**], where both the Bando–Kobayashi estimate and the Aubin–Yau estimate do not work directly. Proposition 7.2 combines both cases, and first appeared in [**141**, Proposition 7.1].

7.5. The reverse Chern–Lu inequality. The following is a new, reverse form of the Chern–Lu inequality.

PROPOSITION 7.3. (Reverse Chern–Lu inequality) *Let $f : (M, \omega) \to (N, \eta)$ be a holomorphic map between Kähler manifolds that is a biholomorphism onto its image. Then,*

(7.9)

$$\begin{aligned}|\partial f|^2 \Delta_\eta \log |\partial f|^2 \circ f^{-1} =& - (\omega^\# \otimes \operatorname{Ric} \eta)(\partial f, \bar\partial f)\\ &+ ((R_\omega)^\# \otimes \eta \otimes \eta^\#)(\partial f, \bar\partial f, \partial f^{-1}, \bar\partial f^{-1}) + e(f)\\ \geq& - \omega^\# \otimes \operatorname{Ric} \eta(\partial f, \bar\partial f) + (R_\omega)^\# \otimes \eta \otimes \eta^\#(\partial f, \bar\partial f, \partial f^{-1}, \bar\partial f^{-1}),\end{aligned}$$

where $e(f) = |\nabla^{1,0}\partial f|^2 - |\partial f|^2 |\nabla^{1,0}_\eta \log |\partial f|^2|^2$.

PROOF. Mostly keeping the notation of §7.3 we compute $\Delta_\eta \operatorname{tr}_\omega \eta$ with respect to two holomorphic coordinate charts, but now we only assume $z = (z^1, \ldots, z^n)$ is normal for g, and let $w = (w^1, \ldots, w^n) = f(z)$. By our assumption on f, w is a holomorphic coordinate on $f(M)$. Then,

(7.10)

$$\begin{aligned}\Delta_\eta |\partial f_{T^{1,0}M}|^2 &= \sum_{p,q} h^{p\bar q} \frac{\partial^2}{\partial w^p \overline{\partial w^q}} \Big[g^{i\bar l}(z) h_{j\bar k}(f(z)) f^j_i(z) \overline{f^k_l(z)} \Big]\\ &= \sum_{p,q} h^{p\bar q} \Big[g^{i\bar l} h_{j\bar k, p\bar q} f^j_i(z) \overline{f^k_l(z)} - g^{i\bar t} g^{s\bar l} g_{s\bar t, d\bar e} h_{j\bar k} (f^{-1})^d_p \overline{(f^{-1})^e_q} f^j_i \overline{f^k_l} \Big]\\ &= -(\omega)^\# \otimes \operatorname{Ric} \eta(\partial f, \bar\partial f) + (R_\omega)^\# \otimes \eta(\partial f, \bar\partial f, \partial f^{-1}, \bar\partial f^{-1}) + e_1(f).\end{aligned}$$

Here,

$$h^{p\bar q} h_{j\bar k, p\bar q} = h^{p\bar q}(-R_{j\bar k p\bar q} + h^{s\bar t} h_{j\bar t, p} h_{s\bar k, \bar q}),$$

so

$$e_1(f) := g^{i\bar l}(z) h^{p\bar q} h^{s\bar t} h_{j\bar t, p} h_{s\bar k, \bar q}(w) f^j_i(z) \overline{f^k_l(z)} = |\nabla^{1,0}\partial f|^2. \tag{7.11}$$

Also, by $(R_\omega)^\#$ we denote the $(2, 2)$-type version of the curvature tensor. Next,

$$e_2(f) := \Delta_\eta |\partial f|^2 - |\partial f|^2 \Delta_\eta \log |\partial f|^2 = |\partial f|^2 |\nabla^{1,0}_\eta \log |\partial f|^2|^2.$$

This time,

$$e_1(f)|\partial f|^2 = |\nabla^{1,0}\partial f|^2 |\partial f|^2 \geq \langle \nabla^{1,0}\partial f, \bar\partial f \rangle^2 = |\nabla^{1,0}_\eta |\partial f|^2|^2 = e_2(f)|\partial f|^2, \tag{7.12}$$

using the Kähler condition. Therefore,

$$|\partial f|^2 \Delta_\eta \log |\partial f|^2 \geq -(\omega)^\# \otimes \operatorname{Ric} \eta(\partial f, \bar\partial f) + (R_\omega)^\# \otimes \eta(\partial f, \bar\partial f, \partial f^{-1}, \bar\partial f^{-1}),$$

as desired. □

Note that (7.12) seems to simplify, or at least cast invariantly, Aubin–Yau's derivation, done in coordinates, of a similar inequality, cf. [**268**, (2.15)], [**227**, p. 99].

REMARK 7.4. The reverse Chern–Lu inequality is *not* the same inequality one would obtain from the Chern–Lu inequality by considering the inverse of the identity map. In fact, the latter would yield

$$|\partial f^{-1}|^2 \Delta_\eta \log |\partial f^{-1}|^2 \geq (\operatorname{Ric} \eta)^\# \otimes \omega(\partial f^{-1}, \bar\partial f^{-1}) - \eta^\# \otimes R_\omega(\partial f^{-1}, \bar\partial f^{-1}, \partial f^{-1}, \bar\partial f^{-1}),$$

or specifically, considering the map $\iota^{-1}:(M,\omega)\to(M,\omega_\varphi)$,

$$\Delta_\omega \log(n+\Delta_\omega\varphi) \ge (\mathrm{Ric}\,\omega)^\# \otimes \omega_\varphi(\partial\iota,\bar\partial\iota) - \omega^\# \otimes R_{\omega_\varphi}(\partial\iota,\bar\partial\iota,\partial\iota,\bar\partial\iota), \tag{7.13}$$

which is less useful, since it is hard to estimate the full bisectional curvature of a solution to a complex Monge–Ampère equation (which only controls the Ricci curvature).

7.6. The Aubin–Yau inequality as a corollary. We now demonstrate how the classical Aubin–Yau inequality [**7**, **268**] (see Siu [**227**, p. 114] for a comparison between the approaches of Aubin and Yau) can be deduced using the reverse Chern–Lu inequality. It allows to work under somewhat complementary curvature assumptions to those in Proposition 7.2.

PROPOSITION 7.5. (Aubin–Yau Laplacian estimate) *In the above, let* $f=\mathrm{id}:(M,\omega)\to(M,\eta)$ *be the identity map, and assume that* $\mathrm{Ric}\,\eta\le C_1\omega+C_2\eta$ *and that* $\mathrm{Bisec}_\omega\ge -C_3$*, for some* $C_1,C_2,C_3\in\mathbb{R}$*. Then,*

$$\mathrm{tr}_\omega\eta\Delta_\eta\log\mathrm{tr}_\omega\eta \ge -n(C_1+C_3)-C_2\mathrm{tr}_\omega\eta - C_3\mathrm{tr}_\omega\eta\,\mathrm{tr}_\eta\omega. \tag{7.14}$$

In particular, if $\eta=\omega+\sqrt{-1}\partial\bar\partial\varphi$ *then*

$$\Delta_{\omega_\varphi}\big(\log\mathrm{tr}_\omega\omega_\varphi-(C_3+1)\varphi\big)\ge -n\frac{C_1+C_3}{\mathrm{tr}_\omega\omega_\varphi}-(C_2+n(C_3+1))+\mathrm{tr}_{\omega_\varphi}\omega, \tag{7.15}$$

hence

$$\Delta_{\omega_\varphi}\big(\log\mathrm{tr}_\omega\omega_\varphi-(2C_3+C_1+1)\varphi\big)\ge -(C_2+n(2C_3+C_1+1))+\mathrm{tr}_{\omega_\varphi}\omega. \tag{7.16}$$

PROOF. Proposition 7.3 implies (7.14) by direct computation. Since $\mathrm{tr}_{\omega_\varphi}\omega = n-\Delta_{\omega_\varphi}\varphi$, this inequality is equivalent to (7.15). Since $\mathrm{tr}_\omega\omega_\varphi\,\mathrm{tr}_{\omega_\varphi}\omega\ge n$ (since $\mathrm{tr}A\,\mathrm{tr}A^{-1}\ge n$ for every positive matrix A), this last inequality implies (7.16). □

REMARK 7.6. An older inequality of Aubin [**6**, p. 408] also follows from (7.10) when ω has nonnegative bisectional curvature. Aubin used this inequality to prove the Calabi conjecture under this curvature assumption.

REMARK 7.7. The reverse Chern–Lu inequality also yields, by considering the inverse of the identity map $\iota^{-1}:(M,\omega_\varphi)\to(M,\omega)$,

$$\Delta_\omega\log(n-\Delta_{\omega_\varphi}\varphi)\ge -(\omega_\varphi)^\#\otimes\mathrm{Ric}\,\omega(\partial f,\bar\partial f)+(R_{\omega_\varphi})^\#\otimes\omega(\partial f,\bar\partial f,\partial f^{-1},\bar\partial f^{-1}), \tag{7.17}$$

which is not very useful for the Monge–Ampère equations we consider here for the same reasons as in Remark 7.4. In summary, there are four quantities one can estimate, and the corresponding four inequalities are (7.8) (Chern–Lu), (7.13) (Chern–Lu backwards), (7.15) (reverse Chern–Lu, i.e., Aubin–Yau), and (7.17) (reverse Chern–Lu backwards), and it is the first and the third which are most useful. One may also easily derive four corresponding parabolic versions of the above inequalities: we leave the details to the reader.

7.7. Uniformity of the metric, II. The estimates of the previous paragraphs imply uniformity of the metric under various curvature assumptions coupled with uniform estimates on the potential and/or the volume form. Let us now state these consequences carefully. Part (i) in the next result is a corollary of the Chern–Lu inequality, and seems to have first been formulated in this generality in [**141**]. Part (ii) is a corollary of the reverse Chern–Lu inequality, and seems to be phrased in this generality for the first time here. Part

(iii) in the case $\mathrm{Ric}\,\omega_\varphi = \mathrm{Ric}\,\omega + \sqrt{-1}\partial\bar\partial\psi_2 - \sqrt{-1}\partial\bar\partial\psi_1$ is due to Păun [**196**] (whose more quantitative formulation appears in [**22, 24**]) together with Campana–Guenancia–Păun [**52**] that allows to assume only a lower bound on the bisectional curvature of the reference metric (without an upper bound on the scalar curvature). Here we explain how it (or rather its slight generalization to the case $\mathrm{Ric}\,\omega_\varphi \le \mathrm{Ric}\,\omega + \sqrt{-1}\partial\bar\partial\psi_2 - \sqrt{-1}\partial\bar\partial\psi_1$), too, follows from the reverse Chern–Lu inequality.

COROLLARY 7.8. *Let* $\varphi \in \mathcal{D}^{0,\gamma}_w \cap C^4(M\setminus D) \cap \mathrm{PSH}(M,\omega)$.
(i) Suppose that $\mathrm{Ric}\,\omega_\varphi \ge -C_1\omega - C_2\omega_\varphi$ *and* $\mathrm{Bisec}_\omega \le C_3$ *on* $M\backslash D$. *Then*

$$-n < \Delta_\omega\varphi \le (C_1 + n(C_2+2C_3+1))e^{(C_2+2C_3+1)\,\mathrm{osc}\,\varphi} - n. \tag{7.18}$$

(ii) Suppose that $\mathrm{Ric}\,\omega_\varphi \le C_1\omega + C_2\omega_\varphi$ *and* $\mathrm{Bisec}_\omega \ge -C_3$ *on* $M\backslash D$. *Then*
(7.19)

$$-n < \Delta_\omega\varphi \le \frac{(C_2+n(2C_3+C_1+1))^{n-1}}{n-1}\Big|\Big|e^{(2C_3+C_1+1)(\varphi-\min\varphi)}\frac{{\omega_\varphi}^n}{\omega^n}\Big|\Big|_{L^\infty} - n.$$

(iii) Let $\psi_1,\psi_2 \in \mathcal{D}^{0,\gamma}_w$ *with* $\psi_2/C_4 \in \mathrm{PSH}(M,\omega)$. *Suppose that* $\mathrm{Ric}\,\omega_\varphi \le \mathrm{Ric}\,\omega + \sqrt{-1}\partial\bar\partial\psi_2 - \sqrt{-1}\partial\bar\partial\psi_1$ *and* $\mathrm{Bisec}_\omega \ge -C_3$ *on* $M\backslash D$. *Then,*

$$-n < \Delta_\omega\varphi \le \frac{(nA)^{n-1}}{n-1}\Big|\Big|e^{(A\varphi+\psi_2-\min(A\varphi+\psi_2))}\frac{{\omega_\varphi}^n}{\omega^n}\Big|\Big|_{L^\infty} - n, \tag{7.20}$$

where $A := 1 + C_4 + C_3 + \frac{|\inf\Delta_\omega\psi_1|}{n}$.

PROOF. (i) By (7.8), $\mathrm{tr}_\omega\omega_\varphi(p) \le C_1 + n(C_2+2C_3+1)$, where $\log\mathrm{tr}_\omega\omega_\varphi - (C_2+2C_3+1)\varphi$ is maximized at $p \in M\setminus D$, proving (7.18). Notice that here we used that the maximum is always attained on $M\setminus D$, as explained in §6.4 (see [**141**, §7] for details) by using barrier functions of the form $\epsilon_1|s|^{\epsilon_2}$.
(ii) By (7.16), $\mathrm{tr}_{\omega_\varphi}\omega(p) \le C_2 + n(2C_3+C_1+1)$, where $\log\mathrm{tr}_\omega\omega_\varphi - (2C_3+C_1+1)\varphi$ is maximized at $p\in M\setminus D$. But, $\mathrm{tr}_\omega\omega_\varphi\frac{\omega^n}{{\omega_\varphi}^n} \le \frac{1}{n-1}(\mathrm{tr}_{\omega_\varphi}\omega)^{n-1}$. Thus,

$$\begin{aligned}\max\nolimits_M \mathrm{tr}_\omega\omega_\varphi &\le \mathrm{tr}_\omega\omega_\varphi(p)e^{(2C_3+C_1+1)(\varphi(p)-\min\varphi)}\\ &\le e^{(2C_3+C_1+1)(\varphi(p)-\min\varphi)}\frac{{\omega_\varphi}^n}{\omega^n}(p)\frac{1}{n-1}(C_2+n(2C_3+C_1+1))^{n-1}.\end{aligned}$$

(iii) Proposition 7.3 implies

$$\mathrm{tr}_\omega\omega_\varphi\Delta_{\omega_\varphi}\log\mathrm{tr}_\omega\omega_\varphi \ge -nC_3 - C_3\mathrm{tr}_\omega\omega_\varphi\mathrm{tr}_{\omega_\varphi}\omega - s_\omega + \Delta_\omega\psi_1 - \Delta_\omega\psi_2,$$

where s_ω is the scalar curvature of ω. However, since we do not want to assume an upper bound for the scalar curvature (e.g., if $\mathrm{Bisec}_\omega \le C_5$ then $-s_\omega \ge -n(n+1)C_5$ [**148**, pp. 168–169]), we observe that Proposition 7.3 also implies

$$\mathrm{tr}_\omega\omega_\varphi\Delta_{\omega_\varphi}\log\mathrm{tr}_\omega\omega_\varphi \ge R^{\#}_\omega \otimes \omega_\varphi \otimes {\omega_\varphi}^{\#} - s_\omega + \Delta_\omega\psi_1 - \Delta_\omega\psi_2. \tag{7.21}$$

Recall that here the first term on the right hand side is defined as the contraction of the curvature tensor of ω that appears in (7.10). As in [**52**, Lemma 2.2], one can then combine the terms depending on the curvature of ω, as we now explain. Indeed, this is most easily seen upon diagonalization (choosing normal holomorphic coordinates so that at a given point p, ω is represented by the identity matrix, while

ω_φ is represented by a diagonal matrix $\mathrm{diag}\{l_1,\dots,l_n\}$ while $\omega_\varphi{}^\#$ is represented by $\mathrm{diag}\{1/l_1,\dots,1/l_n\}$). Then,

$$\begin{aligned} R_\omega^\# \otimes \omega_\varphi \otimes \omega_\varphi{}^\# - s_\omega &= \sum_{j,k}\Big(\frac{l_j}{l_k}-1\Big)R_{j\bar{j},k\bar{k}} = \sum_{j\le k}\Big(\frac{l_j}{l_k}+\frac{l_k}{l_j}-2\Big)R_{j\bar{j},k\bar{k}} \\ &\ge -C_3\sum_{j\le k}\Big(\frac{l_j}{l_k}+\frac{l_k}{l_j}-2\Big), \end{aligned}$$

where, by assumption, the numbers $R_{j\bar{j},k\bar{k}}$ are bounded from below $-C_3$ and $R_{j\bar{j},k\bar{k}} = R_{k\bar{k},j\bar{j}}$ by the symmetries of the curvature tensor. But now,

$$\sum_{j\le k}\Big(\frac{l_j}{l_k}+\frac{l_k}{l_j}-2\Big) = \sum_{j,k}\Big(\frac{l_j}{l_k}-1\Big) = -n(n+1)+\sum_{j,k}\frac{l_j}{l_k} \le -n(n+1)+\sum_j\frac{1}{l_j}\sum_k l_k,$$

and this last expression equals $-n(n+1)+\mathrm{tr}_{\omega_\varphi}\omega\,\mathrm{tr}_\omega\omega_\varphi(p)$. Thus, returning to (7.21), we now deduce,

$$\Delta_{\omega_\varphi}\log\mathrm{tr}_\omega\omega_\varphi \ge (n(n+1)C_3+\Delta_\omega\psi_1-\Delta_\omega\psi_2)/\mathrm{tr}_\omega\omega_\varphi - C_3\mathrm{tr}_{\omega_\varphi}\omega.$$

Since $\mathrm{tr}_\omega\omega_\varphi\,\mathrm{tr}_{\omega_\varphi}\omega \ge n$, $\Delta_{\omega_\varphi}\log\mathrm{tr}_\omega\omega_\varphi \ge -[C_3+\frac{1}{n}|\inf\Delta_\omega\psi_1|]\mathrm{tr}_{\omega_\varphi}\omega - \Delta_\omega\psi_2/\mathrm{tr}_\omega\omega_\varphi$, hence,

$$\Delta_{\omega_\varphi}\Big(\log\mathrm{tr}_\omega\omega_\varphi - \Big[1+C_3+\frac{|\inf\Delta_\omega\psi_1|}{n}\Big]\varphi\Big) \ge \mathrm{tr}_{\omega_\varphi}\omega - n\Big[1+C_3+\frac{|\inf\Delta_\omega\psi_1|}{n}\Big] - \frac{\Delta_\omega\psi_2}{\mathrm{tr}_\omega\omega_\varphi}.$$

Păun's additional observation is that the term $\Delta_\omega\psi_2/\mathrm{tr}_\omega\omega_\varphi$ can be controlled even though it has the 'wrong' sign, under the plurisubharmonicity assumption. Indeed, since $0 \le C_4\omega+\sqrt{-1}\partial\bar\partial\psi_2$ then each eigenvalue of this nonnegative form (that we take with respect to ω_φ) is controlled by the sum of the eigenvalues, i.e., $C_4\omega+\sqrt{-1}\partial\bar\partial\psi_2 \le (C_4\mathrm{tr}_{\omega_\varphi}\omega+\Delta_{\omega_\varphi}\psi_2)\omega_\varphi$. Taking the trace of this inequality with respect to ω,

$$-\frac{\Delta_\omega\psi_2}{\mathrm{tr}_\omega\omega_\varphi} \ge \frac{nC_4}{\mathrm{tr}_\omega\omega_\varphi} - C_4\mathrm{tr}_{\omega_\varphi}\omega - \Delta_{\omega_\varphi}\psi_2 \ge -C_4\mathrm{tr}_{\omega_\varphi}\omega - \Delta_{\omega_\varphi}\psi_2.$$

Thus,

$$\begin{aligned} \Delta_{\omega_\varphi}\Big(\log\mathrm{tr}_\omega\omega_\varphi - \Big[1+C_4+C_3+\frac{|\inf\Delta_\omega\psi_1|}{n}\Big]\varphi+\psi_2\Big) &\ge \mathrm{tr}_{\omega_\varphi}\omega \\ &\quad - n\Big[1+C_4+C_3+\frac{|\inf\Delta_\omega\psi_1|}{n}\Big]. \end{aligned} \tag{7.22}$$

Arguing as in (ii) proves (7.20). □

In all three cases, it follows that $\frac{1}{C}\omega \le \omega_{\varphi(s)} \le C\omega$, with C depending only on the constants appearing in (7.18)–(7.20) and in (7.1)–(7.2), with the precise dependence computed in §7.2. Of course, if one assumes equalities in the expressions for $\mathrm{Ric}\,\omega_\varphi$ instead of inequalities, then one can express the ratio of the volume forms more explicitly to make the estimates (7.19)–(7.20) more explicit.

7.8. Uniformity of the connection: Calabi's third derivative estimate. The term $e_1(f) = |\nabla\partial f|^2$ of (7.11) is the norm squared of the connection associated to $g^\#\otimes f^\star h$. on $T^{1,0\,\star}M\otimes f^\star T^{1,0}N$. When $f=\mathrm{id}$ it equals $|\nabla^{1,0}\partial\bar\partial\varphi|^2$, with the norm taken with respect to $\omega_\varphi\otimes\omega_\varphi\otimes\omega$. But assuming that the metrics ω_φ and ω are uniformly equivalent, this term is uniformly equivalent to a term obtained by using the norm associated to ω_φ alone, namely $S := |\nabla^{1,0}\partial f|^2_{\omega_\varphi}$. Under appropriate bounds on R_ω and $\mathrm{Ric}\,\omega_\varphi$ it follows from (7.10) that $\Delta_{\omega_\varphi}\mathrm{tr}_\omega\omega_\varphi \geq C_1S - C_2$, with C_i depending also on the equivalence between ω and ω_φ. Standard computations going back to Calabi [**47**] also show that $\Delta_{\omega_\varphi}S \geq -C_3'|R_\omega|S - C_4'|DR_\omega|S^{1/2} \geq -C_3S - C_4$, with constants depending on bounds on R_ω and its covariant derivative. Thus, $\Delta_{\omega_\varphi}(S + \frac{C_3+1}{C_1}\mathrm{tr}_\omega\omega_\varphi) \geq S - \frac{C_3+1}{C_1}C_2 - C_4$. Using the maximum principle now yields a bound on S. This summarizes the proof in the smooth setting [**6**, p. 410], [**268**, §3],[**204**].

In the edge setting, R_ω and its covariant derivative are no longer bounded, and so this approach has somewhat limited applicability. For instance, when $\beta \in (0, 1/2)$, Brendle observed (and this also follows from the asymptotic expansion of [**141**, Proposition 4.3]) that $|\nabla R_\omega| \leq \frac{C}{|s|_h^{\epsilon-\beta}}$ for some $\epsilon > 0$. Thus, an inequality from the previous paragraph implies $\Delta_{\omega_\varphi}S \geq -C_3S - \frac{C_4'}{2}(|DR_\omega|^2 + S) \geq -(C_3 + \frac{C_4'}{2})S - \frac{C_4'}{2}\frac{C^2}{|s|_h^{2\epsilon-2\beta}}$. But since $\Delta_{\omega_\varphi}|s|_h^{2\epsilon} \geq C_5|s|_h^{2\epsilon-2\beta} - C_6$, the maximum principle can be applied, this time to $S + \frac{C_3+1}{C_1}\mathrm{tr}_\omega\omega_\varphi + |s|_h^{2\epsilon}$, to conclude [**42**].

7.9. Tian's $W^{3,2}$-estimate. A general result due to Tian [**239**], proved in his M.Sc. thesis, gives a local a priori estimate in $W^{3,2}$ for solutions of both real and complex Monge–Ampère equations under the assumption that the solution has bounded real or complex Hessian and the right hand side is at least Hölder. By Campanato's classical integral characterization of Hölder spaces [**53, 131**] this implies a uniform Hölder estimate on the Laplacian. This result can be seen as an alternative to the Evans–Krylov theorem (and, in fact, appeared independently around the same time). Let $B_1 \subset \mathbb{C}^n$ be the unit ball. Consider the equation

$$\det[u_{i\bar{j}}] = e^{F-cu}, \quad \text{on } B_1. \tag{7.23}$$

The following is due to Tian. The proof in [**239**] is written for the real Monge–Ampère, but applies equally to the complex Monge–Ampère.

THEOREM 7.9. *Suppose that $u \in C^4 \cap \mathrm{PSH}(B_1)$ satisfies (7.23). For any $\gamma \in (0, 1)$, there exists and $C > 0$ such that for any $0 < a < 1/2$,*

$$\int_{B_a} |u_{i\bar{j}k}|^2 \leq Ca^{2n-2+2\gamma}. \tag{7.24}$$

Thus, $||\varphi_{i\bar{j}}||_{C^{0,\gamma}(B_{1/4})} \leq C'$. The constants C, C' depend only on γ, β, ω, n, $||u_{i\bar{j}}||_{L^\infty(B_1)}$, $||F||_{C^{0,\gamma}(B_1)}$, and $||u||_{L^\infty(B_1)}$.

The main ideas are as follows. First, an easy computation shows that $(\det[u_{k\bar{l}}]u^{i\bar{j}})_i = 0$ [**239, 263**]. Consider the Monge–Ampère equation $\det[u_{i\bar{j}}] = h$. Taking the logarithm and differentiating twice, multiplying by h, and using the previous identity, yields

$$-hu^{i\bar{s}}u^{t\bar{j}}u_{t\bar{s},\bar{l}}u_{i\bar{j},k} + (hu^{i\bar{j}}u_{k\bar{l},i})_{\bar{j}} = h_{k\bar{l}} - h_kh_{\bar{l}}/h, \quad \text{for each } k, l \tag{7.25}$$

Thus, the Monge–Ampère equation roughly becomes a second order *system* of equations in divergence form for the Hessian, with a quadratic nonlinearity, resembling the harmonic map equation. In the harmonic map setting, a result of Giaquinta–Giusti shows that a bounded weak $W^{1,2}$-solution is necessarily Hölder [**122**]. Since we already have bounds on the real/complex Hessian, the situation is quite analogous. As shown by Tian, the Monge–Ampère equation can be treated in just the same way, proving Theorem 7.9. A key difference between the two settings is dealing with a system, so some extra algebra facts are needed. This method makes clear that no additional curvature assumptions on the reference geometry are needed for this estimate.

The discussion so far was in the absence of edge singularities. However, a very nice feature of the above result is that since its proof involves *integral* quantities, they carry over verbatim to the edge situation $C_\beta \times \mathbb{C}^{n-1}$ (recall (3.4)), modulo only one very minor difference [**141**, Theorem B.1]: since the vector field $V_1 = z_1^{1-\beta}\partial_{z_1} = \partial_\zeta$ is multivalued, then if f is a smooth function in $(z_1,\ldots,z_n)=(z_1,Z)$ then choosing any branch and considering the *function* $f_1 = V_1 f$ on the model wedge, f_1 satisfies the boundary condition

$$f_1(re^{\sqrt{-1}2\pi\beta}, Z) = e^{\sqrt{-1}2\pi(1-\beta)} f_1(r,Z) \tag{7.26}$$

(as well as a corresponding boundary condition on the first derivatives). Thus, the Sobolev inequality satisfied by such functions deteriorates as β approaches 1 (but this is harmless if β is fixed, for instance). Moreover, harmonic functions with such boundary conditions satisfy the estimate

$$||dh||^2_{L^2(C_\beta(a),\omega_\beta)} \le Ca^{2n-4+2\beta^{-1}} ||dh||^2_{L^2(C_\beta(1),\omega_\beta)}, \tag{7.27}$$

instead of the usual one with a^{2n}. Again, this is harmless, and the only effect it has is to restrict the range of possible Hölder exponents in the following immediate corollary of Theorem 7.9 [**141**, **239**]. Consider the singular equation

$$\det[u_{i\bar{j}}] = e^{F-cu}|z_1|^{2\beta-2}, \quad \text{on } B_1 \setminus \{z_1=0\}. \tag{7.28}$$

COROLLARY 7.10. *Suppose that* $u \in C^4(\{z_1 \neq 0\}) \cap \mathrm{PSH}(B_1)$ *satisfies* (7.28). *For any* $\gamma \in (0, \frac{1}{\beta}-1) \cap (0,1)$, *there exists a* $C>0$ *such that for any* $0<a<1/2$,

$$\int_{B_a} |u_{i\bar{j}k}|^2 \le Ca^{2n-2+2\gamma}. \tag{7.29}$$

Here $f_i := V_i f$, *where* $V_1 = z_1^{1-\beta}\partial_{z_1} = \partial_\zeta$ *(a choice of one branch),* $V_2 = \partial_{z_2}, \ldots, V_n = \partial_{z_n}$. *Thus,* $||u_{i\bar{j}}||_{C^{0,\gamma}} \le C'$. *The constants* C, C' *depend only on* γ, β, ω, n, $||u_{i\bar{j}}||_{L^\infty(B_1)}$, $||F||_{C^{0,\gamma}(B_1)}$, *and* $||u||_{L^\infty(B_1)}$.

7.10. Other Hölder estimates for the Laplacian. Next, we describe several other Hölder estimates on the Laplacian of a solution of a complex Monge–Ampère equation. The first is a weaker $\mathcal{D}^{0,\gamma}_e$ estimate that follows the Evans–Krylov method. The other estimates are alternative approaches to the $\mathcal{D}^{0,\gamma}_w$ estimate descried in §7.9 (we also mention the approach in [**51**] under the restriction $\beta \in (0,2/3)$).

7.10.1. *The complex Evans–Krylov edge estimate.* Compared with the approach presented in §7.9, perhaps a more well-known approach to the Hölder estimate on the Laplacian of a solution of a complex Monge–Ampère equation is an adaptation of the Evans–Krylov estimate to the complex setting. This has been carried out for the original formulation [**227**] and in divergence form [**263**]; the latter requires slightly less control on the right hand side of the equation than the former (see also [**34**]). Let us then concentrate on the relevant adaptation to the singular setting.

The standard complex Evans–Krylov estimate adapts rather easily to the edge setting to give a uniform $\mathcal{D}^{0,\gamma}_e$ estimate in terms of a Laplacian estimate. The key point is to use properties of the edge Hölder spaces under rescaling, and a Lipschitz estimate on the right hand side that is valid for the Ricci continuity path (and fails for some other paths). We present the result, closely following [**141**, §8].

The Evans–Krylov technique is local; we may thus concentrate on the neighborhood of D where $r \leq 1$. We cover this region with *Whitney cubes*

$$W = W_R(y_0) := \{p \in M \setminus D \,:\, |y(p) - y_0| < R, \quad \theta(p) \in I_{\pi\beta}, \\ r(p) \in (R, 2R)\} \subset M \setminus D,$$

where $R > 0$ and $y_0 \in D$. Here $I_{\pi\beta}$ denotes any interval in $S^1_{2\pi\beta}$ of length $\pi\beta$, so each $W_R(y_0)$ is simply connected. Clearly $\{r \leq 1\} \setminus D$ is covered by the union of such cubes. Our goal is to show that there exists a fixed $\gamma' \in (0,1)$ and a uniform $C > 0$ such that $||P_{ij}\varphi||_{C_e^{0,\gamma'}(W)} \leq C$ for all such Whitney cubes W (here P_{ij} are the special second order operators discussed in §2.1, cf. the discussion in §3.4). Taking the supremum over W gives the uniform estimate $[\varphi]_{\mathcal{D}_e^{0,\gamma'}} \leq C'$.

The key point here is that $\log r$ and y_i/r are distance functions for the complete metrics $(d\log r)^2$ and $|dZ|^2/r^2$, respectively, and the model metric is the product of these two, i.e., $\omega_\beta/r^2 = (d\log r)^2 + d\theta^2 + |dZ|^2/r^2$. Thus, we may in fact restrict to such cubes when computing the Hölder norm. Indeed, if the supremum in the definition of the Hölder seminorm was nearly attained for two points $p, q \in M\setminus D$ not contained in one such cube then either $r(p)/r(q) > 2$ or $r(q)/r(p) > 2$. But then the distance between p and q with respect to ω_β/r^2 would be bigger than some fixed constant, and the Hölder seminorm would then be uniformly controlled by the C^0 norm, which is, by assumption, already bounded. Similarly we see that $|y(p)/r(p) - y(q)/r(q)|$ must be uniformly bounded.

This property of ω_β/r^2 manifests itself in another way in the proof. Namely, denoting $S_\lambda(r,\theta,y) = (\lambda r, \theta, \lambda y)$ the dilation map then

$$||S_\lambda^\star f||_{C_e^{0,\gamma}(W_1(y_0))} = ||f||_{C_e^{0,\gamma}(W_\lambda(y_0))}.$$

Thus, we can perform all estimates on a cube of fixed size.

Finally, the last crucial observation is that for any $\varphi \in \mathcal{H}^e_\omega$, the metric ω_φ when viewed in a "microscope" looks (up to perhaps a linear transformation of the original coordinates) very close to the model (product) metric ω_β. In particular, given $\epsilon > 0$, there exists $\lambda_0 = \lambda_0(\epsilon, \varphi)$ such that for all $\lambda > \lambda_0$,

$$(1-\epsilon)|d\vec{Z}|^2 \leq \lambda^{-2} S_\lambda^\star \omega(\vec{Z}) \leq (1+\epsilon)|d\vec{Z}|^2, \tag{7.30}$$

where $\vec{Z} = (Z_1, \ldots, Z_n)$ are holomorphic coordinates on W_1, where $\lambda\vec{Z} = \vec{\zeta} = (\zeta, z_2, \ldots, z_n)$, are the original coordinates on W_λ.

To summarize, when we work in the rescaled cubes (that we may assume are of fixed size) and use the coordinates $\vec{Z}$, the "rescaled pulled-back" metric is essentially equivalent to the model Euclidean metric (7.30). Fortunately, also the complex Monge–Ampère equation we are trying to solve transforms very nicely under this same rescaling coupled with pull-back under the dilation map. So, at the end of the day, one may simply apply the standard Evans–Krylov argument on this cube of fixed size (that is disjoint from D). There is one small caveat, however. In the (divergence form of the) Evans–Krylov argument one differentiates the Monge–Ampère equation twice to obtain a differential inequality and one must control the right hand side in $C^{0,1}$. More precisely, provided then that we can estimate the Lipschitz norm of the right hand side $e^{f_\omega - s\varphi}\omega^n$, we can carefully put all these observations together to prove a uniform $\mathcal{D}_e^{0,\gamma}$ estimate by directly applying the divergence form complex Evans–Krylov estimate to the (uniformly elliptic) metric $\lambda^{-2}S_\lambda^\star\omega$ on the (fixed size) cube $W_1(y_0)$, and to the local Kähler potential $\lambda^{-2}(\psi+\varphi)\circ S_\lambda$ (here ψ is a local potential for ω). The required aforementioned Lipschitz estimate is proved in [**141**, Lemmas 4.4,8.3] and can be summarized as follows (note that it proves a wedge Lipschitz estimate which is more than we need).

LEMMA 7.11. *Let* $\log h(s,t) := \log F + \log\det[\psi_{i\bar j}] = tf_\omega + c_t - s\varphi + \log\det[\psi_{i\bar j}]$, *with* $s > S$. *Then the following estimates hold with constants independent of* t, s:
(i) For $\beta \le 2/3$, $||h(s,t)||_{w;0,1} \le C = C(S, M, \omega, \beta, ||\varphi(s,t)||_{C_w^{0,1}})$.
(ii) For $\beta \le 1$, $||h(s,1)||_{w;0,1} \le C = C(S, M, \omega, \beta, ||\varphi(s,1)||_{C_w^{0,1}})$.

The point of the proof of this lemma is that f_ω and $\det[\psi_{i\bar j}]$ are in $C_w^{0,\frac{2}{\beta}-2}$, and therefore so is h. This is however in $C_w^{0,1}$ only when $\beta \le 2/3$. Fortunately, the combination $f_\omega + \log\det[\psi_{i\bar j}]$ is nevertheless always in $C_w^{0,1}$. On the other hand, this crucial cancellation is false for $tf_\omega + \log\det[\psi_{i\bar j}]$ when $t < 1$! Thus, we must use the Ricci continuity path in this juncture.

To summarize, we have:

THEOREM 7.12. *Let* $\varphi(s) \in \mathcal{D}_e^{0,\gamma} \cap C^4(M\setminus D) \cap \mathrm{PSH}(M,\omega)$ *be a solution to* (6.3) *with* $s > S$ *and* $0 < \beta \le 1$. *Then*

$$||\varphi(s)||_{\mathcal{D}_e^{0,\gamma}} \le C, \tag{7.31}$$

where $\gamma > 0$ *and* C *depend only on* M, ω, β, S *and* $||\Delta_\omega\varphi(s)||_{C^0}, ||\varphi(s)||_{C^0}$.

7.10.2. *A harmonic map type argument vs. a Schauder type argument.* Equation (7.30) was simply a consequence of the definition of a Kähler edge metric. However, it turns out that under some geometric assumptions it is possible to obtain a priori control on $\lambda(\epsilon,\varphi)$. Intuitively, of course, if along a family $\{\omega_{\varphi_j}\}$ of Kähler edge metrics $\sup_j \lambda(\epsilon,\varphi_j) = \infty$ then there exists $p \in D$ such that the family of metrics, while being bounded, fails to be uniformly continuous up to the boundary at $p \in D$. This kind of behavior can be easily ruled out by the existence of a unique tangent cone at p. In [**63**, II] this is proved when ω_{φ_j} are KEE metrics of angle $\beta_j < \beta_\infty < 1$. This can be generalized to Kähler edge metrics with a uniform lower bound on the Ricci curvature [**250**]. One idea is to notice that a rescaled limit will be Ricci flat and then use results of [**57, 63, 249**].

Be it as it may, this improved control on the metric allows to upgrade to Hölder bounds. We explain two approaches.

The first is similar in spirit to the proof of Theorem 7.9 and is due to Tian [**250**]. In the following paragraph all norms, covariant derivatives and Laplacian are with respect to ω_β. Thus, let v be the unique ω_β-harmonic (1,1)-form equal to ω_φ on $\partial B_a(y) \subset U$, where U is a neighborhood of D (this neighborhood is the same for the whole family of potentials φ we are considering) on which $(1-\epsilon)\omega_\beta < \omega_\varphi < (1+\epsilon)\omega_\beta$. Since ω_β is ω_β-harmonic, then $\Delta|v-\omega_\beta|^2 = |\nabla(v-\omega_\beta)|^2 \ge 0$. The maximum principle then gives that $-\epsilon\omega_\beta < v - \omega_\beta < \epsilon\omega_\beta$ on $B_a(y)$, and thus also $-\epsilon\omega_\beta < v - \omega_\varphi < \epsilon\omega_\beta$ on $B_a(y)$. Multiplying (7.25) by $\hat\omega := v - \omega_\varphi$ and integrating directly implies that $||\nabla\hat\omega||^2_{L^2(B_a(y))}$ is controlled from above by $C\epsilon||\nabla\omega_\varphi||^2_{L^2(B_a(y))} + Cr^{2n}$. Now, since $||\nabla v||^2_{L^2(B_\sigma(y))} \le C\left(\frac{\sigma}{a}\right)^{2n-4+\frac{2}{\beta}}||\nabla v||^2_{L^2(B_a(y))}$, and using Dirichlet's principle $||\nabla v||_{L^2(B_a(y))} \le ||\nabla\omega_\varphi||_{L^2(B_a(y))}$, we conclude that

$$||\nabla\omega_\varphi||^2_{L^2(B_\sigma(y))} \le C\left(\epsilon + \left(\frac{\sigma}{a}\right)^{2n-4+\frac{2}{\beta}}\right)||\nabla\omega_\varphi||^2_{L^2(B_a(y))} + Ca^{2n}.$$

In fact, these arguments are very similar to the ones that go into the proof of Theorem 7.9; the only difference is in showing the smallness of $||\nabla\hat{\omega}||^2_{L^2(B_a(y))}$. The latter proof uses completely elementary tools (Moser iteration, essentially). At any rate, given the inequality above, it is standard to show the estimate (7.29). The arguments above assume that $F_{i\bar{j}} \in L^\infty$, but a close examination shows that had F only been assumed Hölder than by elementary arguments one must replace the term Ca^{2n} by a slightly more singular term $Ca^{2n-\delta}$, however still sub-critical. We omit the elementary details.

The second approach is to cleverly use the Schauder estimate for Δ. A key point is that given any $\delta \in (0,1)$ there exists a sufficiently small ball $B_a(y)$, with a uniformly positive, such that

$$|| \det[u_{i\bar{j}}] - \Delta u||_{C^{0,\gamma}_w} \leq \delta[u]_{\mathcal{D}^{0,\gamma}_w}.$$

This can be proved by elementary properties of the function $\det(\,\cdot\,) - \operatorname{tr}(\,\cdot\,)$ near the identity on the space of positive Hermitian matrices [**239**, Lemma 2.2],[**63**, II]. By the Monge–Ampère equation we know that $\det[u_{i\bar{j}}]$ is uniformly controlled. Thus, by the triangle inequality and Schauder estimate of Theorem 3.7 (i),

$$||\Delta u||_{C^{0,\gamma}_w} \leq \delta[u]_{\mathcal{D}^{0,\gamma}_w} + C \leq C\delta(||\Delta u||_{C^{0,\gamma}_w} + ||u||_{C^{0,\gamma}_w} + 1).$$

Choosing δ small enough gives a uniform bound on $||\Delta u||_{C^{0,\gamma}_w}$. In the argument above we have been particularly sloppy in keeping track of the scaling; we refer to [**63**, II] for details.

7.10.3. *An approximation by orbifolds argument.* In the very recent revision of the paper of Guenancia–Păun (that appeared during the final revision of the present article) one finds a yet different approach to the $\mathcal{D}^{0,\gamma}_w$ estimate, that we only attempt to sketch briefly, restricting for simplicity to the case D is smooth. First, the authors assume that β is rational, namely $\beta = p/q$ for $p, q \in \mathbb{N}$ with no common nontrivial divisors. Then, working on the ramified q cover essentially reduces one to the situation of edge metrics of angles $2\pi p$. In fact, under this cover, that amounts to the map $z \mapsto z^{1/q}$, the metric $|z|^{2\beta-2}|dz|^2$ pulls-back to $q^2|w|^{2p-2}|dw|^2$ (substitute $z = w^q$). When $p = 1$ this solves the problem, of course (the orbifold case). When $p \in \mathbb{N}$, the authors show that all the tools from the standard complex Evans–Krylov theorem have appropriate analogues in this large angle regime. Namely, a Sobolev inequality with an appropriate constant and a Harnack inequality (for such degenerate (vanishing along D to possibly high order) metrics). The key point here is that these inequalities do not break down when p tends to infinity. Then, the authors express the Monge–Ampère equation in terms possibly singular vector fields $\frac{1}{q}w^{1-p}\partial_w, \partial_{z_2}, \ldots, \partial_{z_n}$, to obtain the desired conclusion. Finally, the authors approximate an arbitrary $\beta \in (0,1)$ by rational numbers, and approximate the initial Monge–Ampère equation by Monge–Ampère equations with β replaced by those rational numbers. By stability results for the complex Monge–Ampère operator the solutions of these approximate equations will converge to the solution of the original equation. Thus, taking a limit, the $\mathcal{D}^{0,\gamma}_w$ estimate carries over to the solution of the original equation.

8. The asymptotically logarithmic world

As discussed in §3.3, a basic obstruction to existence of KEE metrics is the cohomological requirement that the $\mathbb{R}$-divisor

$$K^\beta = K_M^\beta := K_M + D_\beta := K_M + \sum_{i=1}^r (1-\beta_i)D_i \tag{8.1}$$

satisfy

$$-K_M^\beta \text{ equals } \mu \text{ times an ample class, with } \mu \in \mathbb{R}. \tag{8.2}$$

Here, $\beta := (\beta_1, \ldots, \beta_r) \in (0,1]^r$, M is smooth, $D \neq 0$ has simple normal crossings (as we will assume throughtout this whole section), and K_M^β is sometimes referred to as the twisted canonical bundle associated to the triple (M, D, β). In this section we will be interested in *classification questions*, perhaps the most basic of which is:

QUESTION 8.1. *What are all triples* (M, D, β) *for which* (8.2) *holds?*

When $\mu \leq 0$, according to Theorem 4.13, such a classification is tantamount to a classification of KEE manifolds of nonpositive Ricci curvature. When $\mu > 0$, such a classification would yield a class of manifolds containing the KEE manifolds. Narrowing this class down then of course depends on notions of stability that are a further challenging obstacle—more on that in §9.

Question 8.1 is, of course, too ambitious in the sense that even when all $\beta_i = 1$ and M is smooth there is no complete classification, or list, of projective manifolds satisfying (8.2), unless $\mu > 0$ and n is small. In particular, when $\mu < 0$, which is a subset of the world of "general type" varieties, a classification is quite hopeless; in §8.5 we will review what can still be said when $n = 2$. Thus, we will largely concentrate on the case $\mu > 0$ and further restrict to the small angle regime where some classification can be achieved, that furthermore has interesting geometric consequences. The small angle, or asymptotically logarithmic, regime, can be thought of as the other extreme from the smooth regime ($\beta_i = 1$). In the next few subsections we discuss some of its interesting properties.

8.1. Warm-up: classification of del Pezzo surfaces. Which compact complex surfaces admit a Kähler metric of positive Ricci curvature? The following basic classification result, together with the Calabi–Yau theorem gives a complete list.

THEOREM 8.2. *Let S be a compact complex surface. Then $c_1(S)$ is ample if and only if S is either $\mathbb{P}^1 \times \mathbb{P}^1$, or otherwise $\mathbb{P}^2$ blown-up at at most 8 distinct points, of which no three are collinear, no six lie on a conic, and no eight lie on a cubic with one of the points being a double point.*

Del Pezzo first described some of the eponymous surfaces in the ninteenth century [**82**]; more precisely, he described the ones with up to six blown-up points, or, in his language, surfaces of degree $d = c_1^2$ embedded in $\mathbb{P}^d$. In other words, the ones for which $-K_M$ is very ample, i.e., for which the linear series $|-K_M|$ gives a projective embedding; indeed, since $-K_M > 0$, by Riemann–Roch $\dim H^0(M, \mathcal{O}_M(-K_M)) = \chi(-K_M) = \chi(\mathcal{O}_M) + c_1^2 = 1 + c_1^2$. What are now known as del Pezzo surfaces are the surfaces for which $-K_M$ is ample, i.e., those in the statement of Theorem 8.2. By the Kodaira Embedding Theorem, those are the surfaces for which $|-mK_M|$ gives a projective embedding for some $m \in \mathbb{N}$. It is hard to trace precisely the original discoverers of those remaining del Pezzo surfaces, let alone the first time Theorem 8.2 was stated in this form in the literature, but the contributions of Clebsch, Segre, Enriques, Nagata, among others, played a crucial role. We refer to [**11**, **50**, **74**, **91**, **104**] for more references and historical notes.

This result is used, and generalized, in Theorem 8.10. We describe a proof, closely following Hitchin [**135**] (see also, e.g., [**91**, **112**, **267**]). The detailed proof serves to motivate later classification results, as well as to establish notation and basic results that are useful later.

PROOF. To start, one checks that indeed the surfaces in the statement are del Pezzo. We concentrate on the converse.

Step 1. Let $\Omega \in H^2(M,\mathbb{R}) \cap H^{1,1}_{\bar\partial}$. By Nakai's criterion [**43**, **153**, **192**]

$$\Omega > 0 \text{ if and only if } \Omega^2 > 0 \text{ and } \Omega.C > 0 \text{ for every curve } C \text{ in } S. \tag{8.3}$$

Thus, any blow-down $\pi : \tilde S \to S$ of $\tilde S$ with $c_1(\tilde S) > 0$ will also satisfy $c_1(S) > 0$: indeed, if E is the exceptional divisor of π, then $E^2 = \deg L_E|_E = -1$ and [**124**, p. 185,187]

$$K_{\tilde S} = \pi^\star K_S + E. \tag{8.4}$$

Thus, $c_1^2(S) = c_1^2(\tilde S) + 1 > 0$. Additionally, if Σ is any holomorphic curve in S then the associated cohomology class $[\Sigma]$ (represented by the current of integration along it), that by abuse of notation we still denote by Σ, satisfies $\tilde\Sigma + mE = \pi^\star\Sigma$, where $\tilde\Sigma := \overline{\pi^{-1}(\Sigma \setminus E)}$ denotes the proper transform of Σ, and where m is the multiplicity of $\tilde\Sigma$ at the blow-up point. Since cup product is invariant under birational transformations, $K_S.\Sigma = \pi^\star K_S.\pi^\star\Sigma = (K_{\tilde S} - E).\pi^\star\Sigma = K_{\tilde S}.(\tilde\Sigma + mE) \le 0$. Here, we used the fact that any pulled-back class has zero intersection number with the exceptional divisor.

It thus remains to classify all del Pezzos with no -1-curves, since by a classical theorem of Castelnuovo–Enriques [**124**, p. 476] there always exists such a birational blow-down π contracting any given -1-curve. The goal is to show that these "minimal del Pezzos" are precisely $\mathbb{P}^2$ and $\mathbb{P}^1 \times \mathbb{P}^1$. To that end, one first observes that S is rational, i.e., birational to $\mathbb{P}^2$. Indeed, by the Kodaira Vanishing Theorem [**124**, p. 154], $H^k(S, \mathcal{O}(mK_S)) = 0, k = 0, 1, m \in \mathbb{N}$. In particular, $H^0(S, \mathcal{O}_S(2K_S)) = 0$, and by Dolbeault's theorem and Kodaira–Serre duality $H^{0,1}_{\bar\partial} \cong H^1(S, \mathcal{O}_S) \cong H^1(S, \mathcal{O}_S(K_S))$. Consequently, by the Castelnuovo–Enriques characterization S is rational [**124**, p. 536]. The classification of minimal rational surfaces implies that these are precisely $\mathbb{P}^2$ and the Hirzebruch surfaces,

$$\mathbb{F}_m := \mathbb{P}(\mathcal{O} \oplus \mathcal{O}(m)), \tag{8.5}$$

with $m \in \mathbb{N}_0 \setminus \{1\}$, the projectivization of the rank 2 bundle over $\mathbb{P}^1$ that is obtained by the direct sum of the trivial bundle and the degree m bundle. (Note that $\mathbb{F}_0 = \mathbb{P}^1 \times \mathbb{P}^1$, while $\mathbb{F}_1$ is the blow-up of $\mathbb{P}^2$ at one point, hence is not minimal.) Finally, observe that $\mathbb{F}_m, m \ge 2$ are not del Pezzo: indeed they contain a rational non-singular curve of self-intersection $-m$, while by adjunction, any curve C on a del Pezzo surface S satisfies $C^2 = 2g_C - 2 - K_S.C > -2$.

Step 2. Next, we classify the admissible blow-ups of $\mathbb{P}^2$ and $\mathbb{P}^1 \times \mathbb{P}^1$. Since the two-point blow-up of the former equals the one-point blow-up of the latter, we may concentrate on $\mathbb{P}^2$. Since $c_1^2(\mathbb{P}^2) = 9$, by (8.4) and the relation following it, at most 8 blow-ups are allowed, according to (8.3). Next, it remains to determine the allowable configurations of blow-ups. Suppose that $k \leq 8$ points have been blown up and that the resulting surface is not del Pezzo. Then by (8.3) this means that there exists a curve $C \subset S$ with $C.K_S \geq 0$. We may assume C is irreducible, since at least one of its components will have nonnegative intersection number with K_S. Denote by $\pi : S \to \mathbb{P}^2$ the blow-down map, by $E \subset S$ the exceptional divisor, and by $\{p_1, \ldots, p_k\} \subset \mathbb{P}^2$ the blown-up points. We denote by $q_1, \ldots, q_l$ the singular points of C, and let $p_{k+i} := \pi(q_i)$, $i = 1, \ldots, l$. We also let $\Sigma = \pi(C)$ and write $E = \sum_{i=1}^k E_i$, with $\pi^{-1}(p_i) = E_i$. Since π is an isomorphism outside E, Σ will have multiplicity one everywhere except, possibly, at the points $\{p_i\} \cap \Sigma$, and we denote each of these multiplicities by m_i. Denote by d the degree of Σ. By the genus formula for planar irreducible projective curves [**124**, p. 220, 505] the genus of Σ equals $g_\Sigma = (d-1)(d-2)/2$ precisely when Σ is smooth, and in general

$$(8.6)\quad 2g_\Sigma \leq (d-1)(d-2) - \sum_{i=1}^{k+l} m_i(m_i - 1) \leq (d-1)(d-2) - \sum_{i=1}^{k} m_i(m_i - 1).$$

Here the genus of a possibly singular irreducible curve is defined either as $\dim H^1(\Sigma, \mathcal{O}_\Sigma)$ [**124**, p. 494], or as the genus of its unique desingularization [**124**, p. 500]; in particular, it is nonnegative. Letting $m := \sum_{i=1}^k m_i$, convexity of $f(x) = x^2$ gives $\sum m_i^2/k \geq (m/k)^2$. Combined with $g_\Sigma \geq 0$ this yields $m^2/k - m \leq (d-1)(d-2)$, hence

$$(8.7)\qquad 2m \leq k + \sqrt{k^2 + 4k(d-1)(d-2)}.$$

Additionally, $C = \pi^\star\Sigma - \sum_{i=1}^k m_i E_i$, and so by (8.4),

$$(8.8)\qquad 0 \leq K_S.C = (-\pi^\star 3H + E).(\pi^\star\Sigma - \sum_{i=1}^{k} m_i E_i) = -3d + m,$$

since $\Sigma \in |dH|$, where $H \to \mathbb{P}^2$ is the hyperplane bundle. Here we are implicitly assuming that the blown-up points are distinct, or in other words that we have not blown up any point on the exceptional divisor of a previous blow-up. This is justified by the observation that had we blown-up a point on a -1-curve, we would obtain a -2-curve, contradicting the Fano assumption (recall the end of the previous paragraph).

Thus, $3d \leq m$. Plugging this back into (8.7) and expanding the resulting inequality yields

$$(8.9)\qquad (9-k)d^2 \leq 2k.$$

It follows that $(k,d) \in \{(3,1),(4,1),(5,1),(6,1),(6,2),(7,1),(7,2),(8,1),(8,2),(8,3),(8,4)\}$. When $d \in \{1,2\}$ by (8.6) all m_i must equal 1, thus equality holds in (8.6). Thus, the cases $\{(k,1) : k = 3,\ldots,8\}$ correspond to Σ being a line passing through three or more of the $\{p_i\}$, and the cases $\{(k,2) : k = 6,\ldots,8\}$ correspond to Σ being a smooth conic passing through six or more of the points. If $d = 3$ then by (8.8) at least one m_i equals 2, thus by (8.6) exactly one such m_i exists. Thus, $(k,d) = (8,3)$ and Σ is a singular rational cubic with a double point passing through one of the blow-up points. Finally, the case $(8,4)$ is the only one that is excluded, since it forces $m \geq 12$, while the equation following (8.6), namely, $m^2/8 - m \leq 6$ implies $m \leq 12$, thus $m = 12$. Now equality in the latter means precisely that equality also occurs in $\sum m_i^2/k \geq (m/k)^2$. By strict convexity of $f(x) = x^2$ this means that all the m_i are equal; but since they are also all at least 2, this implies $m \geq 16$, a contradiction. □

8.2. Log Fano manifolds. The definition of log Fano manifolds goes back to work of Maeda [**166**].

DEFINITION 8.3. *We say that the pair $(M, D = \sum D_i)$ is* log Fano *if $-K_M - D$ is ample.*

In dimension 2, these are also called log del Pezzo surfaces (to avoid confusion, we remark that some authors use this terminology to refer to rather different objects). The motivation for the adjective "logarithmic", according to Maeda, is from the work of Iitaka on classification of open algebraic varieties where logarithmic differential forms are used to define invariants of the pair. The open variety associated to (M, D) is the Zariski open set $M \setminus D$.

Maeda posed the following problem.

PROBLEM 8.4. *Classify log Fano manifolds.*

This problem has a beautiful inductive structure. Indeed, by the adjunction formula [**124**, p. 147], any component D_i of D, or more precisely the pair $(D_i, \sum_{j\neq i} D_i \cap D_j)$, is itself a log Fano manifold of one dimension lower, to wit

$$K_{D_i} + \sum_{j\neq i} D_j|_{D_i} = (K_M + D)|_{D_i}.$$

When $n = 1$, log Fanos consist precisely of $(\mathbb{P}^1, \{\text{point}\})$ (we always omit the case of empty boundary, that in this dimension corresponds to $(\mathbb{P}^1, \emptyset)$). Thus, the first step in Problem 8.4 should be a classification for $n = 2$. This was provided by Maeda [**166**, §3.4].

THEOREM 8.5. *Log del Pezzo surfaces (S, C) are classified as follows:*
(i) $S \cong \mathbb{P}^2$, and C is a line in S,
(ii) $S \cong \mathbb{P}^2$, and $C = C_1 + C_2$, where each C_i is a line in S.
(iii) $S \cong \mathbb{P}^2$, and C is a smooth conic in S.
(iv) $S \cong \mathbb{F}_n$ for any $n \geqslant 0$, and C is a smooth rational curve in S such that $C^2 = -n$ (such curve is unique if $n \geqslant 1$).
(v) $S \cong \mathbb{F}_n$ for any $n \geqslant 0$, and $C = C_1 + C_2$ where C_1 is as in (iv) and C_2 is a smooth fiber (i.e., a smooth rational curve such that $C_2^2 = 0, C_2.C_1 = 1$).
(vi) $S \cong \mathbb{F}_1$, and C is a smooth rational curve such that $C \in |C_1 + C_2|$, with C_1, C_2 as in (v).
(vii) $S \cong \mathbb{P}^1 \times \mathbb{P}^1$, and C is a smooth rational curve in $|H_1 + H_2|$ where H_1, H_2 are lines in each copy of $\mathbb{P}^1$.

Building on this result and considerable more work, Maeda then tackles the case $n = 3$. Much more recently, Fujita was able to obtain some results in higher dimensions, especially for pairs with high log Fano index [**113**].

8.3. Asymptotically log Fano manifolds. In this section we finally get to the case $0 < \beta_i \ll 1$, that generalizes both extremal cases $\beta_i = 1$ and $\beta_i = 0$ studied in the last two sections.

DEFINITION 8.6. *We say that a pair (M, D) is (strongly) asymptotically log Fano if the divisor $-K_M^\beta = -K_M - \sum_{i=1}^r (1 - \beta_i) D_i$ is ample for (all) sufficiently small $(\beta_1, \dots, \beta_r) \in (0, 1]^r$.*

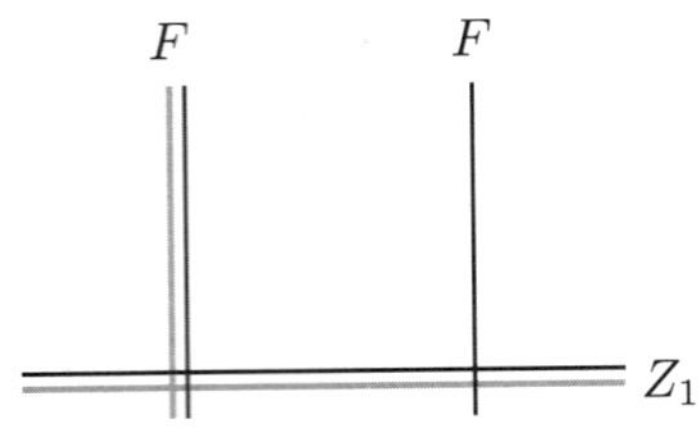

FIGURE 7. A non-strongly asymptotically log del Pezzo pair (see Example 8.7): $\mathbb{F}_1$ with boundary consisting of two fibers and the -1-curve (the fibration structure of the surface $\mathbb{F}_1$ is indicated in green, the boundary C (consisting of those three curves) in red).

Both of these classes (strongly asymptotically log Fano and asymptoticaly log Fano) generalize the class of log Fanos since ampleness (of $-K_X - D$) is an open property. They also generalize the class of Fanos (at least in small dimensions, see the discussion surrounding Problem 8.9), since if D is a smooth anticanonical divisor in a Fano M, then (M, D) is strongly asymptotically log Fano.

The notion of strongly asymptotically log Fano coincides with that of asymptoticaly log Fano in the case $r = 1$, i.e., when D consists of a single smooth component. However, they differ in general, as the following example demonstrates.

EXAMPLE 8.7. (See Figure 7.) Let $S = \mathbb{F}_1$ (recall (8.5)) and $C = C_1 + C_2 + C_3$ where $C_1, C_2 \in |F|$ are both fibers and $C_3 = Z$ is the -1-curve. Note that $-K_S = 2Z + 3F$. Then $-K_S^\beta.Z = \beta_1 + \beta_2 - \beta_3$ and so (S, C) is not strongly asymptotically log del Pezzo. However, one may verify that it is asymptotically log del Pezzo: when $\beta_1 + \beta_2 > \beta_3$, the class $-K_S - (1-\beta_1)C_1 - (1-\beta_2)C_2 - (1-\beta_3)C_3$ is positive. Note this pair is the blow-up of the pair $\mathbb{P}^2$ with two lines at their intersection (the pair II.1B in Figure 9 below).

We pose the following problem.

PROBLEM 8.8. *Classify the (strongly) asymptotically log del Pezzo surfaces and Fano 3-folds.*

In §8.4 below we explain the solution to this problem in the case of a smooth divisor [**60**], where Problem 8.8 is solved more generally for strongly asymptotically log del Pezzos. This generalizes Maeda's result and the classical classification of del Pezzo surfaces (Theorems 8.5 and 8.2). The general (i.e., not necessarily strongly) case of surfaces, as well as the three dimensional case are open and challenging.

8.3.1. *Comparison between the asymptotic and classical logarithmic regimes.* We end this subsection with a few comparisons between the log Fano and asymptotically log Fano regimes, emphasizing the flexibility in the asymptotic classes as opposed to the rigidity of the class of log Fanos.

One point of similarity between both classes is that unlike Fano manifolds, for neither logarithmic classes is the degree of the logarithmic anticanonical bundle bounded uniformly for fixed dimension. E.g., when $n = 2$, $K_M^2 \leq 9$ for del Pezzos, while following the notation of Example 8.7, $(\mathbb{F}_m, Z)$ (see Theorem 8.5 (iv)) satisfies $(-K_M - Z)^2 = ((m+2)F + Z)^2 = m+4$. Thus, already in the log Fano class there are infinitely many non-diffeomorphic pairs.

Aside from these properties though, these classes are quite different.

First, the asymptotic notion is no longer inductive, in the sense that $(D_i, D_i \cap \cup_{j\neq i} D_j)$, is not necessarily itself asymptotically log Fano. In fact,

$$K_{D_i} + \sum_{j\neq i}(1-\beta_j)D_j|_{D_i} = K_M^\beta|_{D_i} + \beta_i D_i|_{D_i},$$

and the right hand side may fail to be negative. Perhaps the simplest example is the pair ($\mathbb{P}^2$, smooth cubic curve), where the boundary is an elliptic curve, hence not Fano. Thus, in every dimension one may encounter boundaries that were absent from the classification in lower dimensions.

Second, while D is always connected in the classical setting [**166**, Lemma 2.4], this is certainly not so in the asymptotic regime. As an example, consider $M = \mathbb{P}^1 \times \mathbb{P}^1$ and $D = D_1 + D_2$ a union of two disjoint lines in the same linear series, say, $D_i \in |H_1|$. However, there is an upper bound on the number of disjoint components D may have. The reason D is always connected in the non-asymptotic regime is the standard logarithmic short exact sequence

$$0 \to \mathcal{O}_M(-D) \to \mathcal{O}_M \to \mathcal{O}_D \to 0.$$

Note that $H^0(M, \mathcal{O}_M(-D)) = \{0\}$ since holomorphic functions on M vanishing on D must be identically zero, as M is connected. Also, $H^1(M, \mathcal{O}_M(-D)) = \{0\}$ since by Serre duality this vector space is isomorphic to $H^1(M, \mathcal{O}_M(K_M + D)) = \{0\}$, by Kodaira Vanishing. Therefore, $H^0(M, \mathcal{O}_M) \cong H^0(D, \mathcal{O}_D)$, and thus the connectivity of D is 'inherited' from that of M.

Other more refined connectivity properties are also interesting to compare. According to Maeda (op. cit.), when (M, D) is log Fano, D is always "strongly connected," meaning that any two components of D intersect. This follows immediately from the inductive structure already mentioned. Indeed, this certainly holds for $n = 1$. Suppose now that D_1 intersects both D_2 and D_3. Then since $(D_1, \sum_{j=2}^r D_1 \cap D_j)$, is itself log Fano, then by induction $(D_2 \cap D_1) \cap (D_3 \cap D_1) \neq \emptyset$, therefore also $D_2 \cap D_3 \neq \emptyset$, as desired. Such strong connectivity again fails in the asymptotic world. In fact, Example 8.7, or even simpler, the disconnected example or the previous paragraph, or even ($\mathbb{P}^1$, 2 distinct points) (which are both strongly asymptotically log Fano) provide instances of that.

Moreover, the number of components in the boundary of log Fanos is bounded from above by the dimension (op. cit.): by strong connectivity any two components intersect and thus all components have a common point. But the components cross normally! There is no analogue for this property in the asymptotic regime. As we will see, the number of boundary components can be arbitrary.

Finally, the class of asymptotically log Fano manifolds seems like a more natural generalization of the class of Fano manifolds than the class of log Fanos. Indeed the latter do contain the Fano manifolds as a subclass if one allows the case of empty boundary. However the class of Fanos actually can be considered as a subset of the asymptotically log Fanos, if one considers pairs (M, D) with M Fano and $D \in |-K_M|$ a snc divisor, when such a divisor exists. As an aside, we mention that this last existence problem is known to hold for all smooth Fano up to dimension three (then even a smooth anticanonical divisor exists by the classification and a theorem Shokurov [**109, 139, 140, 184, 187, 224, 225**]), and in general it falls under the world of the Elephant conjectures going back to Iskovskih [**36**]. In fact, even the existence of such a divisor is open, although examples show that, in general, one needs to allow for worse singularities than snc [**136**].

PROBLEM 8.9. *Determine whether an anticanonical divisor exists on a smooth Fano manifold, and whether it has some regularity, at least in sufficiently low dimensions.*

One approach to this problem that does not seem to have been tried so far would be to use Geometric Measure Theory. Indeed, any (holomorphic) divisor is automatically a minimal submanifold, in fact area minimizing in its homology class by Wirtinger's inequality [**110**, §5.4.19]. In other words, the Kähler form provides for a calibration in the sense of Harvey–Lawson [**132**]. The question is then whether an area minimizing representative of the homology class $[-K_M]$ can be found that is also a complex subvariety, and if so whether it has some regularity beyond that provided by general results of GMT. In view of [**182**] this seems to be a delicate question. In the real setting, a famous result says that hypersurfaces can have singularities only in codimension 7 or higher [**226**, Theorem 37.7]. Perhaps one approach to Problem 8.9 would be to develop a regularity theory for complex hypersurfaces. The rigidity of the holomorphic setting might just be enough for such a theory, which in the general real codimension greater than one setting breaks down, of course aside from Almgren's fundamental result saying that singularities then occur in real codimension two or higher [**1**].

8.4. Classification of strongly asymptotically log del Pezzo surfaces. The following result gives a complete classification of strongly asymptotically log del Pezzo surfaces with smooth connected boundary.

THEOREM 8.10. *Let S be a smooth surface (the surface), and let C be an irreducible smooth curve on S (the boundary curve). Then $-K_S-(1-\beta)C$ is ample for all sufficiently small $\beta>0$ if and only if S and C can be described as follows:*

(I.1A) *$S\cong\mathbb{P}^2$, and C is a smooth cubic elliptic curve,*
(I.1B) *$S\cong\mathbb{P}^2$, and C is a smooth conic,*
(I.1C) *$S\cong\mathbb{P}^2$, and C is a line,*
(I.2.n) *$S\cong\mathbb{F}_n$ for any $n\geq 0$, and $C=Z_n$,*
(I.3A) *$S\cong\mathbb{F}_1$, and $C\in|2(Z_1+F)|$,*
(I.3B) *$S\cong\mathbb{F}_1$, and $C\in|Z_1+F|$,*
(I.4A) *$S\cong\mathbb{P}^1\times\mathbb{P}^1$, and C is a smooth elliptic curve of bi-degree $(2,2)$,*
(I.4B) *$S\cong\mathbb{P}^1\times\mathbb{P}^1$, and C is a smooth rational curve of bi-degree $(2,1)$,*
(I.4C) *$S\cong\mathbb{P}^1\times\mathbb{P}^1$, and C is a smooth rational curve of bi-degree $(1,1)$,*
(I.5.m) *S is a blow-up of the surface in (I.1A) at $m\leq 8$ distinct points on the boundary curve such that $-K_S$ is ample, i.e., S is a del Pezzo surface, and C is the proper transform of the boundary curve in (I.1A), i.e., $C\in|-K_S|$,*
(I.6B.m) *S is a blow-up of the surface in (I.1B) at $m\geq 1$ distinct points on the boundary curve, and C is the proper transform of the boundary curve in (I.1B),*
(I.6C.m) *S is a blow-up of the surface in (I.1C) at $m\geq 1$ distinct points on the boundary curve, and C is the proper transform of the boundary curve in (I.1C),*
(I.7.n.m) *S is a blow-up of the surface in (I.2.n) at $m\geq 1$ distinct points on the boundary curve, and C is the proper transform of the boundary curve in (I.2),*

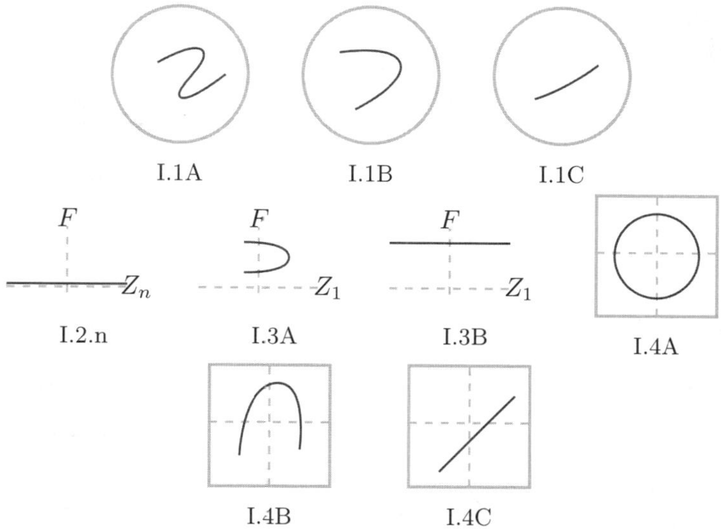

FIGURE 8. Strongly asymptotically log del Pezzo surfaces with smooth connected boundary: the minimal pairs (the surface is indicated in green, the boundary in red, the fibration structure, when one exists, is indicated by the dashed green lines). The remaining pairs listed in Theorem 8.10 are obtained by blowing-up along the boundary curves as follows: I.1A as described in Theorem 8.2; I.1B, I.1C, I.2.n, I.3B, and I.4C at any number of distinct points; I.4B at any number of distinct point with no two on a single $(0,1)$-fiber. Note that I.4A and I.3A may also be blown-up but these cases are covered by blow-ups of I.1A and I.4B, respectively.

(I.8B.m) *S is a blow-up of the surface in* (I.3B) *at* $m \geq 1$ *distinct points on the boundary curve, and C is the proper transform of the boundary curve in* (I.3B),

(I.9B.m) *S is a blow-up of the surface in* (I.4B) *at* $m \geq 1$ *distinct points on the boundary curve with no two of them on a single curve of bi-degree* $(0,1)$, *and C is the proper transform of the boundary curve in* (I.4B),

(I.9C.m) *S is a blow-up of the surface in* (I.4C) *at* $m \geq 1$ *distinct points on the boundary curve, and C is the proper transform of the boundary curve in* (I.4C).

The proof appears in [**60**]. We sketch the main steps. Figure 1 illustrates the pairs graphically.

One starts by checking directly that indeed all the pairs in the list above are asymptotically log Fano. Let us concentrate on the reverse implication.

First, it follows from the asymptotic assumption that $-K_S - C$ is nef. Also, $-K_S$ is big and nef, since it is a linear combination of an ample class $-K_S-(1-\beta)C$ and an effective class $(1-\beta)C$. The former implies that the genus of C is at most one. The latter, together with a version of Nadel Vanishing Theorem and a theorem of Castelnuovo, imply that S is rational. This further implies that if C is elliptic then $C \in |-K_S|$, and S is del Pezzo, i.e., is one of (1A),(4A), or (5_m). On

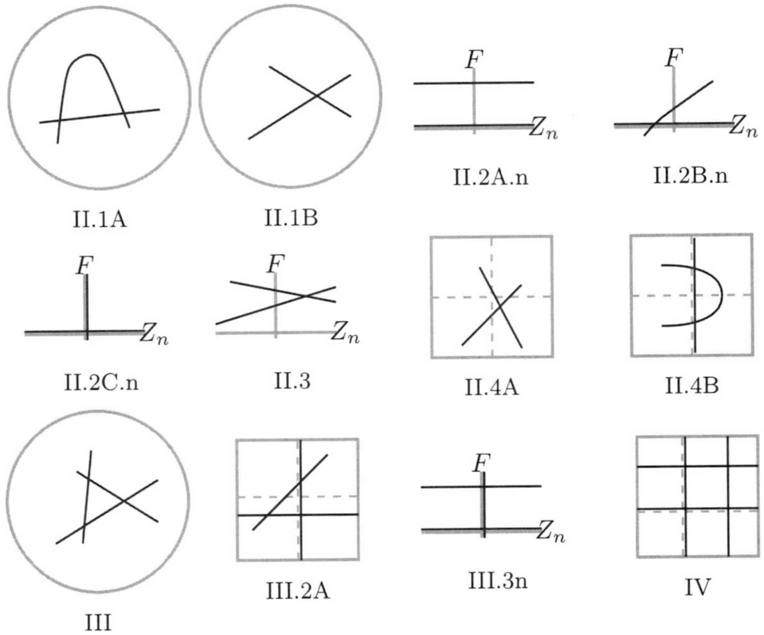

FIGURE 9. Strongly asymptotically log del Pezzo surfaces with general snc boundary: the minimal pairs. Essentially, these pairs are obtained by "degenerating" the elliptic/rational boundaries of Figure 8 into cycles/chains of rational curves.

the other hand, if C is rational, then it must "trap" all the negative curvature of $-K_S$. More precisely, the only curve that can intersect $-K_S$ nonpositively is C, and that happens if and only if $C^2 \leq -2$ (compare to the end of Step 1 in the proof of Theorem 8.2). Thus, all other negative self-intersection curves must be -1-curves. Furthermore, these curves must be either disjoint from C, or intersect it transversally at exactly one point. This motivates the following definition.

DEFINITION 8.11. *We say that the pair (S, C) is* minimal *if there exist no smooth irreducible rational -1-curve $E \neq C$ on the surface S such that $E \cap C \neq \emptyset$.*

The importance of this definition is in the following.

LEMMA 8.12. *Suppose that (S, C) is non-minimal asymptotically log del Pezzo and let E be as in Definition 8.11. Then there exists a birational morphism $\pi\colon S \to s$ such that s is a smooth surface, $\pi(E)$ is a point, the morphism π induces an isomorphism $S\backslash E \cong s\backslash\pi(E)$, the curve $\pi(C)$ is smooth, and $(s, \pi(C))$ is asymptotically log del Pezzo.*

Thus, it remains to classify all minimal pairs. First, one proves that minimality implies the rank of the Picard group of S is at most two. Second, one shows, using the classical theory of rational surfaces, that a minimal pair with this rank restriction must be (I.1B), (I.1C), (I.2.n), (I.3A), (I.3B), (I.4B), or (I.4C), and a non-minimal one must equal (I.6B.1) or (I.6C.1). This concludes the proof since (I.6B.m), (I.6C.m), with $m \geq 2$, and (I.7.n.m),(I.8B.m),(I.9B.m), (I.9C.m), with

$m \geq 1$, are precisely the only blow-ups of minimal pairs that are still asymptotically log del Pezzo.

Building on this, the case of a snc boundary is handled in [**60**]. Essentially, aside from a few cases of disconnected boundary, the only new boundaries allowed beyond the smooth connected boundary case are boundaries that can be considered as "degenerations" of smooth ones. For instance, the smooth elliptic boundary of (I.1A) can be replaced by a triangle of lines, or a conic and a line; the elliptic boundary of (I.4A) can similarly break up to no more than 4 components. However, an additional complication in the snc case is that a -1-curve in a non-minimal pair could be a component of the boundary. Luckily, one can show that such a curve must be at the 'tail': it cannot intersect two boundary components. Thanks to this, paying attention to the combinatorical structure of the boundary, the main idea from the proof of Theorem 8.10 carries over to give a classification of strongly asymptotically log del Pezzos [**60**]. We list these pairs in Figures 9–10, and refer to [**60**] for their precise construction.

8.5. The negative case. In this subsection we assume that $\mu < 0$. We fix M and seek necessary and sufficient restrictions on a snc divisor $D \subset M$ in order to be an admissible boundary for all small enough β.

DEFINITION 8.13. *We say that a pair (M, D) is (strongly) asymptotically log general type if the divisor $K_M^\beta = K_M + \sum_{i=1}^r (1-\beta_i)D_i$ is ample for (all) sufficiently small $(\beta_1, \ldots, \beta_r) \in (0,1]^r$.*

The following theorem comes close to describing the (strongly) asymptotically log general type surfaces as a subclass of the class of log general type minimal surfaces. The proof we describe is due to Di Cerbo [**87**] (who showed *(ii)* $\Leftrightarrow$ *(iv)*), with slight modifications to include also the asymptotic classes defined above.

PROPOSITION 8.14. *Let S be projective surface and $C \subset S$ a snc curve such that $K_S + C$ is big and nef. Consider the following statements:*
(i) (S, C) is strongly asymptotically log general type.
(ii) $K_S^\beta > 0$ for $\beta = \beta_1(1, \ldots, 1)$ for $0 < \beta_1 \ll 1$.
(iii) (S, C) is asymptotically log general type.
(iv) Every rational -1-curve not contained in C intersects C at least in two points, and any rational -2-curve F satisfies $F \not\subset S \setminus C$. Every rational component C_i of C intersects $\cup_{j \neq i} C_j$ at least in two points, and, if $C_i^2 \in \{-1, -2\}$ then at least in three points.
(v) Every -1-curve not contained in C intersects C at least in two points, and any rational -2-curve F satisfies $F \not\subset S \setminus C$. Every rational component C_i of C intersects $\cup_{j \neq i} C_j$ at least in two points.
Then (i) $\Rightarrow$ (ii) $\Leftrightarrow$ (iv) $\Rightarrow$ (iii) $\Rightarrow$ (v).

Note that, similarly to the last section, the assumption that $K_S + C$ is big and nef is of course a consequence of (S, C) being asymptotically log general type.

PROOF. Suppose first that (S, C) is (strongly) asymptotically log canonical. Let F be a holomorphic rational curve. Then, $K_S^\beta . F > 0$, i.e.,
(8.10)
$$C.F > -K_S.F + \sum \beta_i C_i.F = 2 - 2g_F + F^2 + \sum \beta_i C_i.F = 2 + F^2 + \sum \beta_i C_i.F.$$

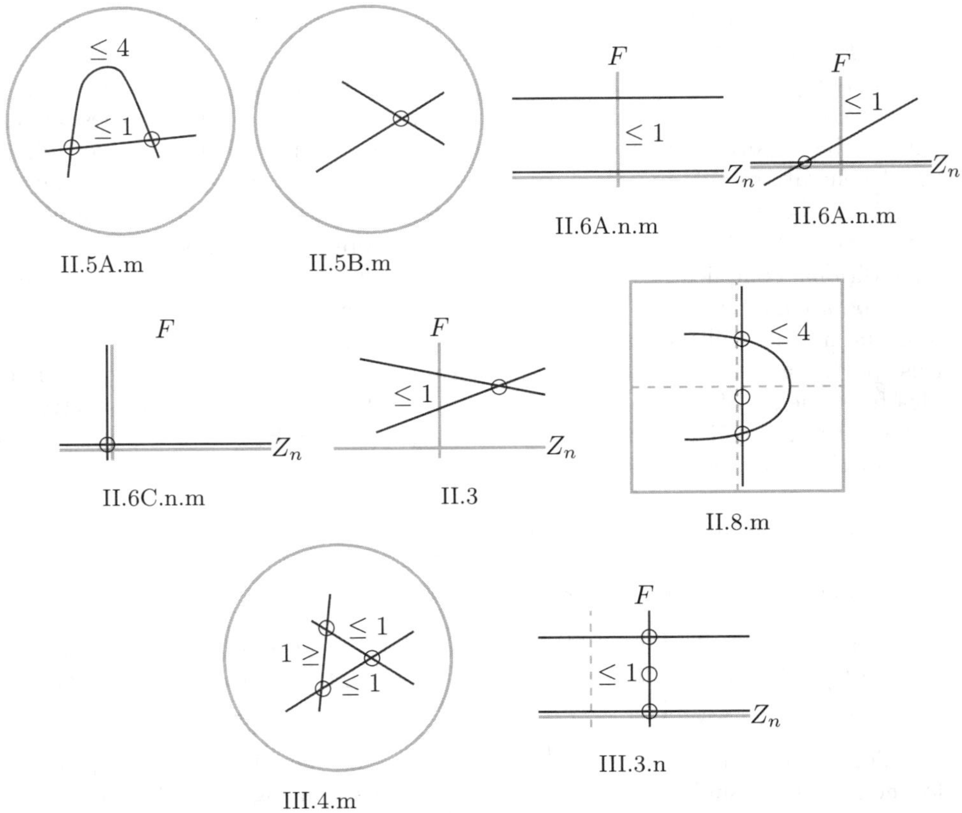

FIGURE 10. Strongly asymptotically log del Pezzo surfaces with general snc boundary: the remaining cases obtained by blowing-up the minimal pairs in Figure 9. A circle corresponds to a point that may not be blown-up. An indication "$\leq k$" next to a curve means that no more than k distinct points may be blown-up on that curve. In III.3.n no more than one point may be blown up on any single fiber and none on the fiber belonging to the boundary.

Thus $(C-F).F \geq 2$ which proves the second part of (v) by letting $F = C_i$. To prove the second part of (iv), suppose now that we are in the strong regime. By putting $\beta_i = \beta_1 \ll 1$, then $(1-\beta_1)C.F > 2 + F^2$, i.e., $(1-\beta_1)F.(C-F) > 2 + \beta_1 F^2$. If $F^2 \in \{-1,-2\}$ this then majorizes $2 - 2\beta_1$, so $F.(C-F) > 2$ proving the second part of (iv). Finally, suppose now that $F^2 = -2$ (but not necessarily in the strong regime). Then, $F^2 < -2 + F.\sum(1-\beta_i)C_i$. Thus, if $F \subset S \setminus C$ so that $F.C_i = 0$ for each i, then necessarily $F^2 < -2$, contradicting $F^2 = -2$. To prove the remainder of the first part of (iv) and (v), let $F^2 = -1$. If $F \neq C_i$ for each i, then $F.C_i \geq 0$, thus (8.10) implies $C.F > 1$. In the strong regime the same inequality gives $C.F > (1-\beta_1)^{-1} > 1$ without further assumption on F.

Suppose now that (iv) hold. As $(K_S + C)^2 > 0$ also $(K_S^\beta)^2 > 0$ for all small enough $|\beta|$. By Nakai's criterion (8.3), it remains to show that K_S^β intersects positively with every irreducible curve in S. By taking $|\beta|$ sufficiently small, this is certainly the case for every curve Z such that $(K_S+C).Z>0$. Thus, suppose that $(K_S+C).Z=0$. Note that the cup product Q on $H^{1,1}_{\bar\partial}$ has exactly one positive eigenvalue [**124**, p. 126]. Thus if $Q(x,x)>0$, $Q(x,y)=0$ for some $x,y\in H^{1,1}_{\bar\partial}$ then $Q(y,y)=Q(x\pm y,x\pm y)-Q(x,x)\mp 2Q(x,y)=Q(x+y,x+y)-Q(x,x)$, so necessarily $Q(y,y)<0$, otherwise $Q(ax+by,ax+by)>0$, for all $a,b\in\mathbb{R}$, and Q would have at least two positive eigenvalues. Thus, $Z^2<0$. Now, by our assumption on Z, $(C-Z).Z=2-2g_Z$ (here g_Z denotes the genus of the desingularization of Z). Therefore, $g_Z\geq 1$ implies $C.Z\leq Z^2<0$. But then $K_S^\beta.Z=(K_S+C).Z-\sum\beta_iC_i.Z=-\sum\beta_iC_i.Z$, which is necessarily positive for certain β in any neighborhood of $0\in\mathbb{R}^r_+\setminus\{0\}$ (for instance, if $\beta_i=\beta_1$). On the other hand, suppose $g_Z=0$. As we just saw, we may suppose that $C.Z\geq 0$; since $C.Z=2+Z^2$, this implies $Z^2\in\{-1,-2\}$. If $Z^2=-2$, so $C.Z=0$, then either $Z\subset S\setminus C$—but this is precluded by the first part of (iv)—or, $Z\not\subset S\setminus C$, so necessarily $Z=C_i$ for some i, but then $C.Z=(Z+\sum_{j\neq i}C_j).Z\geq -2+3\geq 1$, contradicting the second part of (iv). If $Z^2=-1$ then $C.Z=1$. Thus, $Z\not\subset S\setminus C$. If $Z\neq C_i$ for each i we obtain a contradiction to the first part of (iv). If $Z=C_i$ for some i, then $C.Z=(Z+\sum_{j\neq i}C_j).Z\geq -1+3\geq 2$, a contradiction. □

8.6. Uniform bounds. A natural question is whether there exist uniform bounds on the asymptotic range of β; and if so, what do they depend on? This was first addressed by Di Cerbo–Di Cerbo [**86**] in the case $\beta=\beta_1(1,\ldots,1)$, and this subsection is mostly a review of these results. As can be expected, the results are more complete in the negative regime.

8.6.1. *Strongly asymptotically log general type regime.* Perhaps the simplest example of an asymptotically log general type pair is $(\mathbb{P}^n,D)$ with $D\in|\mathcal{O}(n+2)|$. Then for every β in the range $(0,\frac{1}{n+2})$ K_M^β is still positive. As shown by Di Cerbo–Di Cerbo [**86**], this is always the case when restricting to the ray $\beta=\beta_1(1,\ldots,1)$.

PROPOSITION 8.15. *Suppose that (M,D) is such that $K_M^\beta>0$ for $\beta=\beta_1(1,\ldots,1)$ for some $0<\beta_1\ll 1$ (recall (8.1)). Then the same is true for $0<\beta_1<\frac{1}{n+2}$.*

PROOF. First, recall the following fact:

(8.11) if C is an irreducible curve such that $(K_M+D).C=0$ then $K_M.C>0$.

In fact, for some $t<1$, $(K_M+tD).C>0$, and since $D.C=-K_M.C$ we conclude $0<(K_M+tD).C=(1-t)K_M.C$.

Second, K_M+tD is nef for every $t\in[\frac{n+1}{n+2},1]$. This implies the Proposition, since then for any $t\in(\frac{n+1}{n+2},1]$, K_M+tD is a convex combination of an ample divisor and a nef divisor, hence positive by Kleiman's criterion. We now prove the nefness claim. It suffices to show that if C is an irreducible curve with $(K_M+D).C>0$ then $(K_M+tD).C\geq 0$; indeed, this is already true by the first paragraph if $(K_M+D).C=0$ (and since K_M+D is nef as a limit of ample divisors, it is always true that $(K_M+D).C\geq 0$). Now, we decompose C according to the cone theorem (see, e.g., [**86**])

$$C\sim_{\mathbb{Z}}\sum_{i=1}^r a_iC_i+F,\qquad a_i>0,\quad F.K_X\geq 0,\quad C_i.K_X\in(-n-1,0). \tag{8.12}$$

Thus, once again using that $K_M + D$ is nef,

$$(K_M + (n+1)(K_M + D)).C = (K_M + (n+1)(K_M + D)).\Big(\sum_{i=1}^{r} a_i C_i + F\Big)$$
$$\geq -(n+1)\sum a_i + (n+1)\sum a_i C_i.(K_M + D) \geq 0,$$

since $C_i.(K_M + D) \geq 1$ as otherwise, by nefness of $K_M + D$, $C_i.(K_M + D) = 0$ which would imply $K_M.C_i = 0$ by (8.11), contradicting (8.12). Thus, $K_M + tD$ is nef for every $t \in [\frac{n+1}{n+2}, 1]$. □

REMARK 8.16. In [**86**] it is shown that (8.11) also implies that $K_M + (1-\beta_1)D$ is ample for some $0 < \beta_1 \ll 1$ in the case $K_M + D$ is big.

8.6.2. *Log Fano regime.* Theorem 8.10 implies that an analogue of Proposition 8.15 in the asymptotically log Fano regime is false, and the correct analogue remains to be found. In the more restrictive log Fano regime (see §8.2), Di Cerbo–Di Cerbo prove an interesting first result in this direction [**86**], based on deep results from algebraic geometry. It is an a priori bound on the asymptotic regime depending only on the degree of $-K_M - D$ and n.

PROPOSITION 8.17. *Suppose (M, D) is log Fano. Then $-K_M - (1-\beta_1)D$ is positive for every $\beta_1 \in [0, \beta_{\max})$ with $\beta_{\max}$ depending only on n and $(-K_M - D)^n$.*

9. The logarithmic Calabi problem

For simplicity, in what follows we always suppose the boundary is smooth and connected and that the dimension is two. We refer to [**60**] for more general considerations.

The preceding section sets the stage for the asymptotic logarithmic Calabi problem:

PROBLEM 9.1. *Determine which (strongly) asymptotically logarithmic Fano manifolds admit KEE metrics for sufficiently small β.*

In dimension two, the smooth version of Calabi's problem was solved by Tian in 1990 who showed that among the list of Theorem 8.2, only $\mathbb{P}^2$ blown-up at one or two distinct points do not admit KE metrics [**240**]. In light of Theorem 8.10 it is very natural and tempting to hope for a counterpart for strongly asymptotically log del Pezzo surfaces. The formulation conjectured in [**60**] is the following:

CONJECTURE 9.2. *Suppose that (S, C) is strongly asymptotically log del Pezzo with C smooth and irreducible. Then S admits KEE metrics with angle β along C for all sufficiently small β if and only if $(K_S + C)^2 = 0$.*

In Tian's solution of the smooth case the vanishing of the Futaki invariant provided a necessary and sufficient condition for existence. More generally, since the work of Hitchin, Kobayashi, and many others, a standard condition for the existence of canonical metrics that can be described as zeros of an infinite-dimensional moment map is some sort of 'stability' condition. How, then, does Conjecture 9.2 fit into this scheme?

9.1. First motivation: positivity classification and Calabi–Yau fibrations. It turns out to be quite useful to re-classify the pairs appearing in Theorem 8.10 according to the positivity of the logarithmic anticanonical bundle $-K_S - C$. We distinguish between four mutually exclusive classes. Class ($\aleph$): S is del Pezzo and $C \sim -K_S$; class ($\beth$): $C \not\sim -K_S$ and $(K_S + C)^2 = 0$; class ($\gimel$): $-K_S - C$ is big but not ample; class ($\daleth$): $-K_S - C$ is ample.

THEOREM 9.3. *The asymptotically log del Pezzo pairs appearing in Theorem* 8.10 *are classified according to the positivity properties* ($\aleph$), ($\beth$), ($\gimel$), *and* ($\daleth$) *as follows:*

($\aleph$) (S, C) *is one of* (I.1A), (I.4A), *or* (I.5.m).
($\beth$) (S, C) *is one of* (I.3A), (I.4B), *or* (I.9B.m).
($\gimel$) (S, C) *is one of* (I.6B.m), (I.6C.m), (I.7.n.m), (I.8B.m), *or* (I.9C.m).
($\daleth$) (S, C) *is one of* (I.1B), (I.1C), (I.3B), (I.2.n), *or* (I.4C).

This list nicely puts the discussion of §8.1–8.2 in perspective. Class ($\daleth$) is Maeda's class of log del Pezzo surfaces [**166**], while class ($\aleph$) is the classical class of del Pezzo surfaces together with the information of a simple normal crossing anticanonical curve. The classes ($\beth$) and ($\gimel$) are new.

The next result is a structure result for surfaces of class ($\beth$) [**60**]. It is slightly stronger than what Kawamata–Shokurov basepoint freeness would give: there the relevant linear system giving a morphism is $| - lK_S - lC|$, for some $l \in \mathbb{N}$.

PROPOSITION 9.4. *If* (S, C) *is of class* ($\beth$), *then the linear system* $| - K_S - C|$ *is free from base points and gives a morphism* $S \to \mathbb{P}^1$ *whose general fiber is* $\mathbb{P}^1$, *and every reducible fiber consists of exactly two components, each a* $\mathbb{P}^1$.

Thus, these surfaces are conic bundles, and the boundary C intersects each generic fiber at two points, whose fiber complement is a cylinder! It is therefore tempting to conjecture:

CONJECTURE 9.5. *Let* (S, C, ω_β) *be KEE pairs of class* ($\aleph$) *or* ($\beth$). *Then* (S, C, ω_β) *converges in an appropriate sense to a a generalized KE metric* ω_∞ *on* $S \setminus C$ *as* β *tends to zero. In particular,* ω_∞ *is a Calabi–Yau metric in case* ($\aleph$), *and a cylinder along each generic fiber in case* ($\beth$).

This conjecture is itself a generalization of a folklore conjecture in Kähler geometry saying that $S \setminus C$ equipped with the Tian–Yau metric [**253**] should be a limit of KEE metrics on (S, C) when S is of class ($\aleph$) (see, e.g., [**174**, p. 9], [**102**, p. 76]).

This gives strong motivation for the 'if' part of Conjecture 9.2 because it suggests what the small-angle KEE metrics could be considered as a perturbation of the complete Calabi–Yau metrics on the complement of C. It also motivates the 'only if' part: then there is no good limit, as the limit class is 'too' positive, which should morally preclude the existence of a smooth non-compact complete metric on it (having Myers' theorem in mind).

9.2. Second motivation: asymptotic log canonical thresholds. Perhaps further evidence for Conjecture 9.2 is given by the following result.

THEOREM 9.6. *Assume (S, C) is asymptotically log del Pezzo with C smooth and irreducible. Then*

$$\lim_{\beta\to 0^+} \alpha(S,(1-\beta)C) = \begin{cases} 1 & \textit{class } (\aleph), \\ 1/2 & \textit{class } (\beth), \\ 0 & \textit{class } (\gimel) \textit{ or } (\daleth) \end{cases}$$

The result for class $(\aleph)$ is shown by Berman [**18**], and the remaining cases are shown in [**60**]. Note that $0, 1/2$ and 1 are the Tian invariants of $\mathbb{P}^n, n \to \infty, \mathbb{P}^1$, and $\mathbb{P}^0$, respectively. It is then tempting to think of $1/2$ as the Tian invariant of the generic rational fiber of Proposition 9.4, thus suggesting existence of approximate conic metrics on the football fibers, who should tend to cylinders in the limit. On the other hand, the smallness of the log canonical threshold for classes $(\gimel)$ and $(\daleth)$ suggests non-existence.

9.3. Third motivation: explicit computations. Tian's 1990 result in the smooth regime mentioned earlier can also be phrased equivalently by saying that a del Pezzo surface admits a KE metric if and only if its automorphism group is reductive (a simplification of Tian's original proof has been obtained by the work of Cheltsov [**58**] and Shi [**223**], see also Odaka–Spotti–Sun [**193**], and the expository article [**257**]). Given the logarithmic version of Matsushima's criterion (Theorem 4.7), it is tempting to check how far reductivity gets us in the asymptotic regime. Some explicit computations give [**60**]:

PROPOSITION 9.7. *The automorphism groups of all pairs of class $(\aleph)$ and $(\beth)$ are reductive. The pairs of classes $(\gimel)$ and $(\daleth)$ that have non-reductive automorphism groups, and hence admit no KEE metrics, are:* (I.1C), (I.2.n) *with any* $n \geq 0$, (I.6C.m) *with any* $m \geq 1$, (I.7.n.m) *with any* $n \geq 0$ *and* $m \geq 1$, (I.6B.1), (I.8B.1) *and* (I.9C.1).

Thus, Matsushima's criterion supports Conjecture 9.2 but does not solve the problem in the singular setting.

Another tool is Tian's criterion for existence of KEE metrics, which involves calculating log canonical threshold, and is especially useful in the presence of large finite symmetries. Using such tools, the following KEE metrics are constructed [**60**] on surfaces of class $(\beth)$:

THEOREM 9.8. *There exist strongly asymptotically log del Pezzo pairs of type* (I.3A), (I.4B), *and* (I.9B.5) *that admit KEE metrics for all sufficiently small β.*

An in-depth study of log canonical thresholds on pairs of class $(\aleph)$ was carried out by Martínez-García [**170**, §4] and Cheltsov–Martínez-García [**59**] whose results give lower bounds on how large β can be taken (see also [**60**] for some weaker bounds that hold in all dimensions). In fact, their results show that the threshold depends on the representative chosen in $|-K_S|$, and, similarly, examples of Székelyhidi [**233**] show the maximal allowed β might as well. Conjecture 9.2 predicts that such dependence does not appear in the asymptotic regime.

In work in progress [**61**], the 'only if' part of Conjecture 9.2 is verified by using techniques adapted from work of Ross–Thomas on slope stability, motivated in part by work of Li–Sun [**157**] that proved non-existence in the small angle regime for (I.1B), (I.3B), and (I.4C).

Acknowledgements

This article is an expanded version of a talk delivered at the CRM, Montréal, in July 2012. I am grateful to D. Jakobson for the invitation, the hospitality, and the encouragement to write this article, and to the city of Montréal for its inspiring Jazz Festival. I have benefited from many discussions on these topics with many people over the years, and in particular R. Berman, S. Donaldson, and C. Li. I thank L. Di Cerbo and J. Martínez-García for corrections, C. Ciliberto and I. Dolgachev for historical references on classical algebraic geometry, R. Conlon, T. Darvas, S. Dinew, H. Guenancia, T. Murphy, S. Zelditch, and the referees for comments improving the exposition, and O. Chodosh and D. Ramos for creating many of the figures. Sections 3, 6, and 7 are largely based on joint work with R. Mazzeo and I am grateful to him for everything he has taught me during our collaboration. Similarly, Sections 8–9 survey joint work with I. Cheltsov and I am grateful to him for an inspiring collaboration. I am indebted to Gang Tian for introducing me to this subject, for countless discussions and lectures over many years, encouragement, and inspiration. Finally, this survey is dedicated to Eugenio Calabi on the occasion of his ninetieth birthday. I feel privileged to have known him since my graduate school days. He continues to be an inspiring rôle model exemplifying the spirit of free and open exchange of ideas in a (mathematical) world that has become increasingly competitive. This research was supported by NSF grants DMS-0802923,1206284, and a Sloan Research Fellowship.

References

[1] F. J. Almgren Jr., *Almgren's big regularity paper*, World Scientific Monograph Series in Mathematics, vol. 1, World Scientific Publishing Co., Inc., River Edge, NJ, 2000. Q-valued functions minimizing Dirichlet's integral and the regularity of area-minimizing rectifiable currents up to codimension 2; With a preface by Jean E. Taylor and Vladimir Scheffer. MR1777737 (2003d:49001)

[2] C. Arezzo and G. Tian, *Infinite geodesic rays in the space of Kähler potentials*, Ann. Sc. Norm. Super. Pisa Cl. Sci. (5) **2** (2003), no. 4, 617–630. MR2040638 (2005c:32027)

[3] C. Arezzo, A. della Vedova, and G. la Nave, *Geometric flows and Kähler reduction*, preprint, arxiv:1304.5728.

[4] C. Arezzo, A. della Vedova, and G. la Nave, *On the curvature of conic Kähler–Einstein metrics*, preprint, 2014, available at http://www.math.uiuc.edu/~lanave/research.html.

[5] M. Atiyah and C. Lebrun, *Curvature, cones and characteristic numbers*, Math. Proc. Cambridge Philos. Soc. **155** (2013), no. 1, 13–37, DOI 10.1017/S0305004113000169. MR3065256

[6] T. Aubin, *Métriques riemanniennes et courbure* (French), J. Differential Geometry **4** (1970), 383–424. MR0279731 (43 #5452)

[7] T. Aubin, Équations du type Monge-Ampère sur les variétés kählériennes compactes, C. Rendus Acad. Sci. Paris 283 (1976), 119–121.

[8] T. Aubin, *Équations du type Monge-Ampère sur les variétés kählériennes compactes* (French, with English summary), Bull. Sci. Math. (2) **102** (1978), no. 1, 63–95. MR494932 (81d:53047)

[9] T. Aubin, *Réduction du cas positif de l'équation de Monge-Ampère sur les variétés kählériennes compactes à la démonstration d'une inégalité* (French), J. Funct. Anal. **57** (1984), no. 2, 143–153, DOI 10.1016/0022-1236(84)90093-4. MR749521 (85k:58084)

[10] T. Aubin, *Some nonlinear problems in Riemannian geometry*, Springer Monographs in Mathematics, Springer-Verlag, Berlin, 1998. MR1636569 (99i:58001)

[11] D. Babbitt and J. Goodstein, *Federigo Enriques's quest to prove the "completeness theorem"*, Notices Amer. Math. Soc. **58** (2011), no. 2, 240–249. MR2768117 (2012a:14001)

[12] D. Bakry, *L'hypercontractivité et son utilisation en théorie des semigroupes* (French), Lectures on probability theory (Saint-Flour, 1992), Lecture Notes in Math., vol. 1581, Springer, Berlin, 1994, pp. 1–114, DOI 10.1007/BFb0073872. MR1307413 (95m:47075)

[13] S. Bando and R. Kobayashi, *Ricci-flat Kähler metrics on affine algebraic manifolds*, Geometry and analysis on manifolds (Katata/Kyoto, 1987), Lecture Notes in Math., vol. 1339, Springer, Berlin, 1988, pp. 20–31, DOI 10.1007/BFb0083044. MR961470 (89k:32055)

[14] S. Bando and T. Mabuchi, *Uniqueness of Einstein Kähler metrics modulo connected group actions*, Algebraic geometry, Sendai, 1985, Adv. Stud. Pure Math., vol. 10, North-Holland, Amsterdam, 1987, pp. 11–40. MR946233 (89c:53029)

[15] D. Bartolucci, F. De Marchis, and A. Malchiodi, *Supercritical conformal metrics on surfaces with conical singularities*, Int. Math. Res. Not. IMRN **24** (2011), 5625–5643, DOI 10.1093/imrn/rnq285. MR2863376

[16] E. Bedford and B. A. Taylor, *The Dirichlet problem for a complex Monge-Ampère equation*, Invent. Math. **37** (1976), no. 1, 1–44. MR0445006 (56 #3351)

[17] M. S. Berger, *Riemannian structures of prescribed Gaussian curvature for compact 2-manifolds*, J. Differential Geometry **5** (1971), 325–332. MR0295261 (45 #4329)

[18] R. J. Berman, *A thermodynamical formalism for Monge-Ampère equations, Moser-Trudinger inequalities and Kähler-Einstein metrics*, Adv. Math. **248** (2013), 1254–1297, DOI 10.1016/j.aim.2013.08.024. MR3107540

[19] R. J. Berman, *Kähler–Einstein metrics, canonical random point processes and birational geometry*, preprint, arxiv:1307.3634.

[20] R. J. Berman, *From Monge–Ampère equations to envelopes and geodesic rays in the zero temperature limit*, preprint, arxiv:1307.3008.

[21] R. J. Berman and S. Boucksom, *Growth of balls of holomorphic sections and energy at equilibrium*, Invent. Math. **181** (2010), no. 2, 337–394, DOI 10.1007/s00222-010-0248-9. MR2657428 (2011h:32021)

[22] R. J. Berman, S. Boucksom, P. Eyssidieux, V. Guedj, and A. Zeriahi, *Kähler–Ricci flow and Ricci iteration on log Fano varieties*, preprint, arxiv:1111.7158.

[23] R. J. Berman, S. Boucksom, V. Guedj, and A. Zeriahi, *A variational approach to complex Monge-Ampère equations*, Publ. Math. Inst. Hautes Études Sci. **117** (2013), 179–245, DOI 10.1007/s10240-012-0046-6. MR3090260

[24] R. J. Berman and H. Guenancia, *Kähler-Einstein metrics on stable varieties and log canonical pairs*, preprint, arxiv:1304.2087.

[25] B. Berndtsson, *A Brunn–Minkowski type inequality for Fano manifolds and the Bando–Mabuchi uniqueness theorem*, preprint, arxiv:1103.0923.

[26] J. Bernstein and T. Mettler, *Two-dimensional gradient Ricci solitons revisited*, preprint, arxiv:1303.6854.

[27] A. L. Besse, *Einstein manifolds*, Ergebnisse der Mathematik und ihrer Grenzgebiete (3) [Results in Mathematics and Related Areas (3)], vol. 10, Springer-Verlag, Berlin, 1987. MR867684 (88f:53087)

[28] O. Biquard, *Métriques kählériennes extrémales sur les surfaces toriques (d'après S. Donaldson)* (French, with French summary), Astérisque **339** (2011), Exp. No. 1018, viii, 181–201. Séminaire Bourbaki. Vol. 2009/2010. Exposés 1012–1026. MR2906354

[29] J.-M. Bismut, H. Gillet, and C. Soulé, *Analytic torsion and holomorphic determinant bundles. I. Bott-Chern forms and analytic torsion*, Comm. Math. Phys. **115** (1988), no. 1, 49–78. MR929146 (89g:58192a)

[30] Z. Błocki, *Interior regularity of the complex Monge-Ampère equation in convex domains*, Duke Math. J. **105** (2000), no. 1, 167–181, DOI 10.1215/S0012-7094-00-10518-2. MR1788046 (2001i:32060)

[31] Z. Błocki, *Uniqueness and stability for the complex Monge-Ampère equation on compact Kähler manifolds*, Indiana Univ. Math. J. **52** (2003), no. 6, 1697–1701, DOI 10.1512/iumj.2003.52.2346. MR2021054 (2004m:32073)

[32] Z. Błocki, *A gradient estimate in the Calabi-Yau theorem*, Math. Ann. **344** (2009), no. 2, 317–327, DOI 10.1007/s00208-008-0307-3. MR2495772 (2010d:32046)

[33] Z. Błocki, *The Calabi-Yau theorem*, Complex Monge-Ampère equations and geodesics in the space of Kähler metrics, Lecture Notes in Math., vol. 2038, Springer, Heidelberg, 2012, pp. 201–227, DOI 10.1007/978-3-642-23669-3_5. MR2932444

[34] Z. Błocki, *The complex Monge-Ampère equation in Kähler geometry*, Pluripotential theory, Lecture Notes in Math., vol. 2075, Springer, Heidelberg, 2013, pp. 95–141, DOI 10.1007/978-3-642-36421-1_2. MR3089069

[35] S. Bochner, *Curvature in Hermitian metric*, Bull. Amer. Math. Soc. **53** (1947), 179–195. MR0019983 (8,490d)

[36] F. A. Bogomolov, Yu. G. Zarkhin, and Vik. S. Kulikov, et al., *Vyacheslav Vladimirovich Shokurov (on the occasion of his sixtieth birthday)* (Russian), Uspekhi Mat. Nauk **65** (2010), no. 6 (396), 195–199, DOI 10.1070/RM2010v065n06ABEH004724; English transl., Russian Math. Surveys **65** (2010), no. 6 1193–1198. MR2779369

[37] R. Bott and S.-S. Chern, *Hermitian vector bundles and the equidistribution of the zeroes of their holomorphic sections*, Acta Math. **114** (1965), 71–112. MR0185607 (32 #3070)

[38] S. Boucksom, P. Eyssidieux, V. Guedj, and A. Zeriahi, *Monge-Ampère equations in big cohomology classes*, Acta Math. **205** (2010), no. 2, 199–262, DOI 10.1007/s11511-010-0054-7. MR2746347 (2011k:32049)

[39] J.-P. Bourguignon, *Invariants intégraux fonctionnels pour des équations aux dérivées partielles d'origine géométrique* (French), Differential geometry, Peniscola 1985, Lecture Notes in Math., vol. 1209, Springer, Berlin, 1986, pp. 100–108, DOI 10.1007/BFb0076623. MR863748 (88b:58145)

[40] J.-P. Bourguignon, *Métriques d'Einstein–Kähler sur les variétés de Fano: obstructions et existence* (d'après Y. Matsushima, A. Futaki, S. T. Yau, A. Nadel et G. Tian), Astérisque No. 245 (1997), 277–305.

[41] J.-P. Bourguignon, *The unabated vitality of Kählerian geometry*, in: Erich Kähler, Mathematical works (R. Berndt et al., Eds.), Walter de Gruyter, 2003, pp. 737–766.

[42] S. Brendle, *Ricci flat Kähler metrics with edge singularities*, Int. Math. Res. Not. IMRN **24** (2013), 5727–5766. MR3144178

[43] N. Buchdahl, *On compact Kähler surfaces* (English, with English and French summaries), Ann. Inst. Fourier (Grenoble) **49** (1999), no. 1, 287–302. MR1688136 (2000f:32029)

[44] E. Calabi, *Extremal isosystolic metrics for compact surfaces* (English, with English and French summaries), Actes de la Table Ronde de Géométrie Différentielle (Luminy, 1992), Sémin. Congr., vol. 1, Soc. Math. France, Paris, 1996, pp. 167–204. MR1427758 (97k:53037)

[45] E. Calabi, *The space of Kähler metrics*, Proceedings of the International Congress of Mathematicians, 1954, pp. 206–207.

[46] E. Calabi, *On Kähler manifolds with vanishing canonical class*, Algebraic geometry and topology. A symposium in honor of S. Lefschetz, Princeton University Press, Princeton, N. J., 1957, pp. 78–89. MR0085583 (19,62b)

[47] E. Calabi, *Improper affine hyperspheres of convex type and a generalization of a theorem by K. Jörgens*, Michigan Math. J. **5** (1958), 105–126. MR0106487 (21 #5219)

[48] E. Calabi, *Extremal Kähler metrics*, Seminar on Differential Geometry, Ann. of Math. Stud., vol. 102, Princeton Univ. Press, Princeton, N.J., 1982, pp. 259–290. MR645743 (83i:53088)

[49] E. Calabi, *Extremal Kähler metrics. II*, Differential geometry and complex analysis, Springer, Berlin, 1985, pp. 95–114. MR780039 (86h:53067)

[50] A. Calabri and C. Ciliberto, *Birational classification of curves on rational surfaces*, Nagoya Math. J. **199** (2010), 43–93. MR2730411 (2012b:14022)

[51] S. Calamai and K. Zheng, *Geodesics in the space of Kähler cone metrics*, preprint, arxiv:1205.0056.

[52] F. Campana, H. Guenancia, and M. Păun, *Metrics with cone singularities along normal crossing divisors and holomorphic tensor fields* (English, with English and French summaries), Ann. Sci. Éc. Norm. Supér. (4) **46** (2013), no. 6, 879–916. MR3134683

[53] S. Campanato, *Proprietà di hölderianità di alcune classi di funzioni* (Italian), Ann. Scuola Norm. Sup. Pisa (3) **17** (1963), 175–188. MR0156188 (27 #6119)

[54] H. D. Cao, *Deformation of Kähler metrics to Kähler-Einstein metrics on compact Kähler manifolds*, Invent. Math. **81** (1985), no. 2, 359–372, DOI 10.1007/BF01389058. MR799272 (87d:58051)

[55] A. Carlotto, O. Chodosh, and Y.A. Rubinstein, *Slowly converging Yamabe flows*, preprint, arxiv:1401.3738. To appear in Geom. & Top.

[56] A. Carlotto and A. Malchiodi, *Weighted barycentric sets and singular Liouville equations on compact surfaces*, J. Funct. Anal. **262** (2012), no. 2, 409–450, DOI 10.1016/j.jfa.2011.09.012. MR2854708

[57] J. Cheeger, T. H. Colding, and G. Tian, *On the singularities of spaces with bounded Ricci curvature*, Geom. Funct. Anal. **12** (2002), no. 5, 873–914, DOI 10.1007/PL00012649. MR1937830 (2003m:53053)

[58] I. A. Chel′tsov, *Log canonical thresholds on hypersurfaces* (Russian, with Russian summary), Mat. Sb. **192** (2001), no. 8, 155–172, DOI 10.1070/SM2001v192n08ABEH000591; English transl., Sb. Math. **192** (2001), no. 7-8, 1241–1257. MR1862249 (2002j:14018)
[59] I. Cheltsov and J. Martínez-García, *Dynamic alpha-invariants of del Pezzo surfaces*, preprint, 2014.
[60] I. A. Cheltsov and Y. A. Rubinstein, *Asymptotically log Fano varieties*, preprint, 2013, arxiv:1308.2503.
[61] I. A. Cheltsov and Y. A. Rubinstein, *Flops and canonical metrics*, preprint, 2014.
[62] I. A. Chel′tsov and K. A. Shramov, *Log-canonical thresholds for nonsingular Fano threefolds* (Russian, with Russian summary), Uspekhi Mat. Nauk **63** (2008), no. 5(383), 73–180, DOI 10.1070/RM2008v063n05ABEH004561; English transl., Russian Math. Surveys **63** (2008), no. 5, 859–958. MR2484031 (2010b:14082)
[63] X. Chen, S. Donaldson, and S. Song, *Kähler–Einstein metrics on Fano manifolds, I, II, III*, preprints, arxiv:1211.4566,1212.4714,1302.0282. To appear in J. Amer. Math. Soc.
[64] X. X. Chen and W. Y. He, *On the Calabi flow*, Amer. J. Math. **130** (2008), no. 2, 539–570, DOI 10.1353/ajm.2008.0018. MR2405167 (2009c:53086)
[65] X. Chen and W. He, *The Calabi flow on Kähler surfaces with bounded Sobolev constant (I)*, Math. Ann. **354** (2012), no. 1, 227–261, DOI 10.1007/s00208-011-0723-7. MR2957626
[66] X.-X. Chen and S. Sun, *Calabi flow, Geodesic rays, and uniqueness of constant scalar curvature Kähler metrics*, preprint, arxiv:1004.2012.
[67] X. X. Chen and G. Tian, *Ricci flow on Kähler-Einstein surfaces*, Invent. Math. **147** (2002), no. 3, 487–544, DOI 10.1007/s002220100181. MR1893004 (2003c:53095)
[68] X.-X. Chen and Y.-Q. Wang, *Bessel Functions, heat kernel and the conical Kähler–Ricci flow*, preprint, arxiv:1305.0255.
[69] X.-X. Chen and Y.-Q. Wang, *On the long time behaviour of the conical Kähler–Ricci flows*, preprint, arxiv:1402.6689.
[70] S. Y. Cheng and P. Li, *Heat kernel estimates and lower bound of eigenvalues*, Comment. Math. Helv. **56** (1981), no. 3, 327–338, DOI 10.1007/BF02566216. MR639355 (83b:58076)
[71] S.-s. Chern, *On holomorphic mappings of hermitian manifolds of the same dimension.*, Entire Functions and Related Parts of Analysis (Proc. Sympos. Pure Math., La Jolla, Calif., 1966), Amer. Math. Soc., Providence, R.I., 1968, pp. 157–170. MR0234397 (38 #2714)
[72] B. Chow, *The Ricci flow on the 2-sphere*, J. Differential Geom. **33** (1991), no. 2, 325–334. MR1094458 (92d:53036)
[73] B. Chow and L.-F. Wu, *The Ricci flow on compact 2-orbifolds with curvature negative somewhere*, Comm. Pure Appl. Math. **44** (1991), no. 3, 275–286, DOI 10.1002/cpa.3160440302. MR1090433 (92g:53035)
[74] C. Ciliberto and E. Sallent Del Colombo, *Pasquale del Pezzo, Duke of Caianello, Neapolitan mathematician*, Arch. Hist. Exact Sci. **67** (2013), no. 2, 171–214, DOI 10.1007/s00407-012-0110-0. MR3021553
[75] B. Clarke and Y. A. Rubinstein, *Ricci flow and the metric completion of the space of Kähler metrics*, Amer. J. Math. **135** (2013), no. 6, 1477–1505, DOI 10.1353/ajm.2013.0051. MR3145001
[76] T. H. Colding and A. Naber, *Sharp Hölder continuity of tangent cones for spaces with a lower Ricci curvature bound and applications*, Ann. of Math. (2) **176** (2012), no. 2, 1173–1229, DOI 10.4007/annals.2012.176.2.10. MR2950772
[77] C. B. Croke, *Some isoperimetric inequalities and eigenvalue estimates*, Ann. Sci. École Norm. Sup. (4) **13** (1980), no. 4, 419–435. MR608287 (83d:58068)
[78] T. Darvas, *Envelopes and geodesics in the space of Kähler potentials*, preprint, arxiv:1401.7318.
[79] V. Datar, *On convexity of the regular set of conical Kähler–Einstein metrics*, preprint, arxiv:1403.6219.
[80] V. Datar, B. Guo, J. Song, and X.-W. Wang, *Connecting toric manifolds by conical Kähler–Einstein metrics*, preprint, arxiv:1308.6781.
[81] V. Datar and J. Song, *A remark on Kähler metrics with conical singularities along a simple normal crossing divisor*, preprint, arxiv:1309.5013.
[82] P. del Pezzo, *Sulle superficie delle n^{mo} ordine immerse nello spazio di n dimensioni*, Rend. del circolo matematico di Palermo 1 (1887), 241–271.

[83] Ph. Delanoë and A. Hirschowitz, *About the proofs of Calabi's conjectures on compact Kähler manifolds*, Enseign. Math. (2) **34** (1988), no. 1-2, 107–122. MR960195 (90g:32033)

[84] J.-P. Demailly and J. Kollár, *Semi-continuity of complex singularity exponents and Kähler-Einstein metrics on Fano orbifolds* (English, with English and French summaries), Ann. Sci. École Norm. Sup. (4) **34** (2001), no. 4, 525–556, DOI 10.1016/S0012-9593(01)01069-2. MR1852009 (2002e:32032)

[85] A. Dembo and O. Zeitouni, *Large deviations techniques and applications*, 2nd ed., Applications of Mathematics (New York), vol. 38, Springer-Verlag, New York, 1998. MR1619036 (99d:60030)

[86] G. Di Cerbo and L. Di Cerbo, *Positivity questions in Kähler–Einstein theory*, preprint, arxiv:1210.0218.

[87] L. Di Cerbo, *On Kähler–Einstein surfaces with edge singularities*, preprint, arxiv:1205.3203.

[88] W. Y. Ding, *Remarks on the existence problem of positive Kähler-Einstein metrics*, Math. Ann. **282** (1988), no. 3, 463–471, DOI 10.1007/BF01460045. MR967024 (90a:58186)

[89] W. Y. Ding and G. Tian, *Kähler-Einstein metrics and the generalized Futaki invariant*, Invent. Math. **110** (1992), no. 2, 315–335, DOI 10.1007/BF01231335. MR1185586 (93m:53039)

[90] W. Y. Ding and G. Tian, *The generalized Moser-Trudinger inequality*, in: Nonlinear Analysis and Microlocal Analysis: Proceedings of the International Conference at Nankai Institute of Mathematics (K.-C. Chang et al., Eds.), World Scientific, 1992, pp. 57–70. ISBN 9810209134.

[91] I. V. Dolgachev, *Classical algebraic geometry. A modern view*, Cambridge University Press, Cambridge, 2012. MR2964027

[92] S. K. Donaldson, *Anti self-dual Yang-Mills connections over complex algebraic surfaces and stable vector bundles*, Proc. London Math. Soc. (3) **50** (1985), no. 1, 1–26, DOI 10.1112/plms/s3-50.1.1. MR765366 (86h:58038)

[93] S. K. Donaldson, *Remarks on gauge theory, complex geometry and 4-manifold topology*, Fields Medallists' lectures, World Sci. Ser. 20th Century Math., vol. 5, World Sci. Publ., River Edge, NJ, 1997, pp. 384–403, DOI 10.1142/9789812385215_0042. MR1622931 (99i:57050)

[94] S. K. Donaldson, *Symmetric spaces, Kähler geometry and Hamiltonian dynamics*, Northern California Symplectic Geometry Seminar, Amer. Math. Soc. Transl. Ser. 2, vol. 196, Amer. Math. Soc., Providence, RI, 1999, pp. 13–33. MR1736211 (2002b:58008)

[95] S. K. Donaldson, *Scalar curvature and projective embeddings. I*, J. Differential Geom. **59** (2001), no. 3, 479–522. MR1916953 (2003j:32030)

[96] S. K. Donaldson, *Holomorphic discs and the complex Monge-Ampère equation*, J. Symplectic Geom. **1** (2002), no. 2, 171–196. MR1959581 (2003m:32037)

[97] S. K. Donaldson, *Scalar curvature and stability of toric varieties*, J. Differential Geom. **62** (2002), no. 2, 289–349. MR1988506 (2005c:32028)

[98] S. K. Donaldson, Kähler geometry and diffeomorphism groups, notes for lectures given in CIMAT, August 2004.

[99] S. K. Donaldson, *Kähler geometry on toric manifolds, and some other manifolds with large symmetry*, Handbook of geometric analysis. No. 1, Adv. Lect. Math. (ALM), vol. 7, Int. Press, Somerville, MA, 2008, pp. 29–75. MR2483362 (2010h:32025)

[100] S. K. Donaldson, *Some numerical results in complex differential geometry*, Pure Appl. Math. Q. **5** (2009), no. 2, Special Issue: In honor of Friedrich Hirzebruch., 571–618, DOI 10.4310/PAMQ.2009.v5.n2.a2. MR2508897 (2010d:32020)

[101] S. K. Donaldson, *Discussion of the Kähler–Einstein problem*, preprint, 2009, available at: http://www2.imperial.ac.uk/~skdona/KENOTES.PDF

[102] S. K. Donaldson, *Kähler metrics with cone singularities along a divisor*, Essays in mathematics and its applications, Springer, Heidelberg, 2012, pp. 49–79, DOI 10.1007/978-3-642-28821-0_4. MR2975584

[103] J. Eells Jr. and J. H. Sampson, *Harmonic mappings of Riemannian manifolds*, Amer. J. Math. **86** (1964), 109–160. MR0164306 (29 #1603)

[104] F. Enriques and F. Conforto, *Le superficie razionali*, Zanichelli, 1939.

[105] J.-P. Demailly, S. Dinew, V. Guedj, P. H. Hiep, S. Kołodziej, and A. Zeriahi, *Hölder continuous solutions to Monge–Ampère equations*, J. Eur. Math. Soc. (JEMS) **16** (2014), no. 4, 619–647, DOI 10.4171/JEMS/442. MR3191972

[106] C. L. Epstein, R. B. Melrose, and G. A. Mendoza, *Resolvent of the Laplacian on strictly pseudoconvex domains*, Acta Math. **167** (1991), no. 1-2, 1–106, DOI 10.1007/BF02392446. MR1111745 (92i:32016)

[107] L. C. Evans, *Classical solutions of fully nonlinear, convex, second-order elliptic equations*, Comm. Pure Appl. Math. **35** (1982), no. 3, 333–363, DOI 10.1002/cpa.3160350303. MR649348 (83g:35038)

[108] P. Eyssidieux, V. Guedj, and A. Zeriahi, *Singular Kähler-Einstein metrics*, J. Amer. Math. Soc. **22** (2009), no. 3, 607–639, DOI 10.1090/S0894-0347-09-00629-8. MR2505296 (2010k:32031)

[109] G. Fano, *Sopra alcune varietà algebriche a tre dimensione aventi tutti i generi nulli*, Atti Act. Torino 43 (1908), 973–984.

[110] H. Federer, *Geometric measure theory*, Springer, 1969.

[111] R. Feng and H. Huang, *The global existence and convergence of the Calabi flow on* $\mathbb{C}^n/\mathbb{Z}^n + i\mathbb{Z}^n$, J. Funct. Anal. **263** (2012), no. 4, 1129–1146, DOI 10.1016/j.jfa.2012.05.017. MR2927407

[112] R. Friedman, *Algebraic surfaces and holomorphic vector bundles*, Universitext, Springer-Verlag, New York, 1998. MR1600388 (99c:14056)

[113] K. Fujita, *Simple normal crossing Fano varieties and log Fano manifolds*, Nagoya Math. J. **214** (2014), 95–123, DOI 10.1215/00277630-2430136. MR3211820

[114] A. Futaki, *An obstruction to the existence of Einstein Kähler metrics*, Invent. Math. **73** (1983), no. 3, 437–443, DOI 10.1007/BF01388438. MR718940 (84j:53072)

[115] A. Futaki, *Kähler–Einstein metrics and integral invariants*, Lecture Notes in Mathematics 1314, Springer, 1988.

[116] A. Futaki, *Stability, integral invariants and canonical Kähler metrics*, Differential geometry and its applications, Matfyzpress, Prague, 2005, pp. 45–58. MR2268920 (2007h:32034)

[117] A. Futaki, T. Mabuchi, and Y. Sakane, *Einstein-Kähler metrics with positive Ricci curvature*, Kähler metric and moduli spaces, Adv. Stud. Pure Math., vol. 18, Academic Press, Boston, MA, 1990, pp. 11–83. MR1145246 (94c:32019)

[118] A. Futaki and H. Ono, *Einstein metrics and GIT stability [translation of MR 2422701]*, Sugaku Expositions **24** (2011), no. 1, 93–122. MR2866661

[119] A. Futaki and Y. Sano, *Multiplier ideal sheaves and geometric problems*, Variational problems in differential geometry, London Math. Soc. Lecture Note Ser., vol. 394, Cambridge Univ. Press, Cambridge, 2012, pp. 68–93. MR2882771

[120] P. Gauduchon, Calabi's extremal Kähler metrics: An elementary introduction, manuscript, 2010.

[121] F. W. Gehring, *The* L^p*-integrability of the partial derivatives of a quasiconformal mapping*, Acta Math. **130** (1973), 265–277. MR0402038 (53 #5861)

[122] M. Giaquinta and E. Giusti, *Nonlinear elliptic systems with quadratic growth*, Manuscripta Math. **24** (1978), no. 3, 323–349. MR0481490 (58 #1606)

[123] D. Gilbarg and N. S. Trudinger, *Elliptic partial differential equations of second order*, Classics in Mathematics, Springer-Verlag, Berlin, 2001. Reprint of the 1998 edition. MR1814364 (2001k:35004)

[124] P. Griffiths and J. Harris, *Principles of algebraic geometry*, Wiley-Interscience [John Wiley & Sons], New York, 1978. Pure and Applied Mathematics. MR507725 (80b:14001)

[125] V. Guedj, Degenerate complex Monge–Ampère equations and singular Kähler–Einstein metrics, lecture notes, June 2012.

[126] H. Guenancia, *Kähler-Einstein metrics with cone singularities on klt pairs*, Internat. J. Math. **24** (2013), no. 5, 1350035, 19, DOI 10.1142/S0129167X13500353. MR3070804

[127] H. Guenancia and M. Păun, *Conic singularities metrics with prescribed Ricci curvature: The case of general cone angles along normal crossing divisors*, preprint, arxiv:1307.6375.

[128] P. Hajłasz and P. Koskela, *Sobolev met Poincaré*, Mem. Amer. Math. Soc. **145** (2000), no. 688, 101 pp., DOI 10.1090/memo/0688. MR1683160 (2000j:46063)

[129] R. S. Hamilton, *Three-manifolds with positive Ricci curvature*, J. Differential Geom. **17** (1982), no. 2, 255–306. MR664497 (84a:53050)

[130] R. S. Hamilton, *The Ricci flow on surfaces*, Mathematics and general relativity (Santa Cruz, CA, 1986), Contemp. Math., vol. 71, Amer. Math. Soc., Providence, RI, 1988, pp. 237–262, DOI 10.1090/conm/071/954419. MR954419 (89i:53029)

[131] Q. Han and F. Lin, *Elliptic partial differential equations*, Courant Lecture Notes in Mathematics, vol. 1, New York University, Courant Institute of Mathematical Sciences, New York; American Mathematical Society, Providence, RI, 1997. MR1669352 (2001d:35035)
[132] R. Harvey and H. B. Lawson Jr., *Calibrated geometries*, Acta Math. **148** (1982), 47–157, DOI 10.1007/BF02392726. MR666108 (85i:53058)
[133] W. He, *Local solution and extension to the Calabi flow*, J. Geom. Anal. **23** (2013), no. 1, 270–282, DOI 10.1007/s12220-011-9247-3. MR3010280
[134] C. Hinde and P. Petersen, *Generalized doubling meets Poincaré*, J. Geom. Anal. **17** (2007), no. 3, 485–494, DOI 10.1007/BF02922093. MR2359976 (2009h:53084)
[135] N. Hitchin, *On the curvature of rational surfaces*, Differential geometry (Proc. Sympos. Pure Math., Vol. XXVII, Part 2, Stanford Univ., Stanford, Calif., 1973), Amer. Math. Soc., Providence, R. I., 1975, pp. 65–80. MR0400127 (53 #3962)
[136] A. Höring and C. Voisin, *Anticanonical divisors and curve classes on Fano manifolds*, Pure Appl. Math. Q. **7** (2011), no. 4, Special Issue: In memory of Eckart Viehweg, 1371–1393, DOI 10.4310/PAMQ.2011.v7.n4.a13. MR2918165
[137] H. Huang, *Convergence of the Calabi flow on toric varieties and related Kähler manifolds*, preprint, arxiv:1207.5969.
[138] A. D. Hwang and T. Mabuchi, *A conjecture on the group of biholomorphisms of a nonsingular Fano variety*, Internat. J. Math. **4** (1993), no. 5, 833–839, DOI 10.1142/S0129167X93000376. MR1245352 (94k:14036)
[139] V. A. Iskovskih, *Fano threefolds. I* (Russian), Izv. Akad. Nauk SSSR Ser. Mat. **41** (1977), no. 3, 516–562, 717. MR463151 (80c:14023a)
[140] V. A. Iskovskikh and Yu. G. Prokhorov, *Fano varieties*, Algebraic geometry, V, Encyclopaedia Math. Sci., vol. 47, Springer, Berlin, 1999, pp. 1–247. MR1668579 (2000b:14051b)
[141] T. D. Jeffres, R. Mazzeo, and Y. A. Rubinstein, *Kähler–Einstein metrics with edge singularities*, (with an appendix by C. Li and Y.A. Rubinstein), preprint, arxiv:1105.5216.
[142] T. D. Jeffres, *Kahler-Einstein cone metrics*, ProQuest LLC, Ann Arbor, MI, 1996. Thesis (Ph.D.)–State University of New York at Stony Brook. MR2695191
[143] T. D. Jeffres, *Uniqueness of Kähler-Einstein cone metrics*, Publ. Mat. **44** (2000), no. 2, 437–448, DOI 10.5565/PUBLMAT_44200_04. MR1800816 (2002c:32040)
[144] E. Kähler, *Über eine bemerkenswerte Hermitesche Metrik* (German), Abh. Math. Sem. Univ. Hamburg **9** (1933), no. 1, 173–186, DOI 10.1007/BF02940642. MR3069594
[145] J. L. Kazdan and F. W. Warner, *Curvature functions for compact 2-manifolds*, Ann. of Math. (2) **99** (1974), 14–47. MR0343205 (49 #7949)
[146] J. Keller, *Ricci iterations on Kähler classes*, J. Inst. Math. Jussieu **8** (2009), no. 4, 743–768, DOI 10.1017/S1474748009000103. MR2540879 (2010h:32026)
[147] S. Kobayashi, *On compact Kähler manifolds with positive definite Ricci tensor*, Ann. of Math. (2) **74** (1961), 570–574. MR0133086 (24 #A2922)
[148] S. Kobayashi and K. Nomizu, *Foundations of differential geometry. Vol. II*, Interscience Tracts in Pure and Applied Mathematics, No. 15 Vol. II, Interscience Publishers John Wiley & Sons, Inc., New York-London-Sydney, 1969. MR0238225 (38 #6501)
[149] S. Kołodziej, *The complex Monge-Ampère equation and pluripotential theory*, Mem. Amer. Math. Soc. **178** (2005), no. 840, x+64, DOI 10.1090/memo/0840. MR2172891 (2006d:32054)
[150] S. Kołodziej, *Hölder continuity of solutions to the complex Monge-Ampère equation with the right-hand side in L^p: the case of compact Kähler manifolds*, Math. Ann. **342** (2008), no. 2, 379–386, DOI 10.1007/s00208-008-0239-y. MR2425147 (2009g:32079)
[151] N. V. Krylov, *Boundedly inhomogeneous elliptic and parabolic equations* (Russian), Izv. Akad. Nauk SSSR Ser. Mat. **46** (1982), no. 3, 487–523, 670. MR661144 (84a:35091)
[152] G. La Nave, and G. Tian, *Soliton-type metrics and Kähler–Ricci flow on symplectic quotients*, preprint, arxiv:0903.2413. To appear in J. Crelle.
[153] R. Lazarsfeld, *Positivity in algebraic geometry*, Vol. I, II Springer, 2004.
[154] J. M. Lee and R. Melrose, *Boundary behaviour of the complex Monge-Ampère equation*, Acta Math. **148** (1982), 159–192, DOI 10.1007/BF02392727. MR666109 (84e:58078)
[155] C. Li, *Remarks on logarithmic K-stability*, preprint, arxiv:1104.0428.
[156] C. Li, *Kähler-Einstein metrics and K-stability*, ProQuest LLC, Ann Arbor, MI, 2012. Thesis (Ph.D.)–Princeton University. MR3078441
[157] C. Li and S. Sun, *Conical Kähler–Einstein metric revisited*, preprint, arxiv:1207.5011.

[158] C. Li and C.-Y. Xu, *Special test configuration and K-stability of Fano varieties*, preprint, arxiv:1111.5398.

[159] H. Li, *On the lower bound of the K-energy and F-functional*, Osaka J. Math. **45** (2008), no. 1, 253–264. MR2416659 (2009a:32036)

[160] A. Lichnerowicz, *Sur les transformations analytiques des variétés kählériennes compactes* (French), C. R. Acad. Sci. Paris **244** (1957), 3011–3013. MR0094479 (20 #996)

[161] J.-W. Liu and X. Zhang, *The conical Kähler–Ricci flow on Fano manifolds*, preprint, arxiv:1402.1832.

[162] Y.-C. Lu, *Holomorphic mappings of complex manifolds*, J. Differential Geometry **2** (1968), 299–312. MR0250243 (40 #3482)

[163] Z. Lu, *On the Futaki invariants of complete intersections*, Duke Math. J. **100** (1999), no. 2, 359–372, DOI 10.1215/S0012-7094-99-10013-5. MR1722959 (2000h:32033)

[164] A. Lunardi, *Analytic semigroups and optimal regularity in parabolic problems*, Progress in Nonlinear Differential Equations and their Applications, 16, Birkhäuser Verlag, Basel, 1995. MR1329547 (96e:47039)

[165] F. Luo and G. Tian, *Liouville equation and spherical convex polytopes*, Proc. Amer. Math. Soc. **116** (1992), no. 4, 1119–1129, DOI 10.2307/2159498. MR1137227 (93b:53034)

[166] H. Maeda, *Classification of logarithmic Fano threefolds*, Compositio Math. **57** (1986), no. 1, 81–125. MR817298 (87h:14035)

[167] T. Mabuchi, *K-energy maps integrating Futaki invariants*, Tohoku Math. J. (2) **38** (1986), no. 4, 575–593, DOI 10.2748/tmj/1178228410. MR867064 (88b:53060)

[168] T. Mabuchi, *Some symplectic geometry on compact Kähler manifolds. I*, Osaka J. Math. **24** (1987), no. 2, 227–252. MR909015 (88m:53126)

[169] T. Mabuchi, *An algebraic character associated with the Poisson brackets*, Recent topics in differential and analytic geometry, Adv. Stud. Pure Math., vol. 18, Academic Press, Boston, MA, 1990, pp. 339–358. MR1145264 (93a:32047)

[170] J. Martínez-García, Dynamic alpha-invariants of del Pezzo surfaces with boundary, Ph.D. thesis, University of Edinburgh, 2013.

[171] Y. Matsushima, *Sur la structure du groupe d'homéomorphismes analytiques d'une certaine variété kählérienne* (French), Nagoya Math. J. **11** (1957), 145–150. MR0094478 (20 #995)

[172] R. Mazzeo, *Elliptic theory of differential edge operators. I*, Comm. Partial Differential Equations **16** (1991), no. 10, 1615–1664, DOI 10.1080/03605309108820815. MR1133743 (93d:58152)

[173] R. Mazzeo, *Regularity for the singular Yamabe problem*, Indiana Univ. Math. J. **40** (1991), no. 4, 1277–1299, DOI 10.1512/iumj.1991.40.40057. MR1142715 (92k:53071)

[174] R. Mazzeo, *Kähler-Einstein metrics singular along a smooth divisor*, Journées "Équations aux Dérivées Partielles" (Saint-Jean-de-Monts, 1999), Univ. Nantes, Nantes, 1999, pp. Exp. No. VI, 10. MR1718970 (2001b:32044)

[175] R. Mazzeo and Y. A. Rubinstein, *The Ricci continuity method for the complex Monge-Ampère equation, with applications to Kähler-Einstein edge metrics* (English, with English and French summaries), C. R. Math. Acad. Sci. Paris **350** (2012), no. 13-14, 693–697, DOI 10.1016/j.crma.2012.07.001. MR2971382

[176] R. Mazzeo and Y. A. Rubinstein, *Kähler–Einstein metrics with crossing edge singularities*, preprint in preparation.

[177] R. Mazzeo, Y. A. Rubinstein, and N. Sešum, *Ricci flow on surfaces with conic singularities*, preprint, arxiv:1306.6688. To appear in Analysis and PDE.

[178] D. McFeron, *The Mabuchi metric and the Kähler-Ricci flow*, Proc. Amer. Math. Soc. **142** (2014), no. 3, 1005–1012, DOI 10.1090/S0002-9939-2013-11856-3. MR3148534

[179] R. C. McOwen, *Point singularities and conformal metrics on Riemann surfaces*, Proc. Amer. Math. Soc. **103** (1988), no. 1, 222–224, DOI 10.2307/2047555. MR938672 (89m:30089)

[180] R. C. McOwen, *Prescribed curvature and singularities of conformal metrics on Riemann surfaces*, J. Math. Anal. Appl. **177** (1993), no. 1, 287–298, DOI 10.1006/jmaa.1993.1258. MR1224820 (94f:53076)

[181] R. B. Melrose, *The Atiyah-Patodi-Singer index theorem*, Research Notes in Mathematics, vol. 4, A K Peters, Ltd., Wellesley, MA, 1993. MR1348401 (96g:58180)

[182] M. Micallef and J. Wolfson, *Area minimizers in a K3 surface and holomorphicity*, Geom. Funct. Anal. **16** (2006), no. 2, 437–452, DOI 10.1007/s00039-006-0555-x. MR2231469 (2007b:53125)

[183] E. A. Mooers, *Heat kernel asymptotics on manifolds with conic singularities*, J. Anal. Math. **78** (1999), 1–36, DOI 10.1007/BF02791127. MR1714065 (2000g:58039)

[184] S. Mori and S. Mukai, *Erratum: "Classification of Fano 3-folds with $B_2 \geq 2$" [Manuscripta Math.* **36** *(1981/82), no. 2, 147–162; MR0641971 (83f:14032)]*, Manuscripta Math. **110** (2003), no. 3, 407, DOI 10.1007/s00229-002-0336-2. MR1969009

[185] J. Moser, *A sharp form of an inequality by N. Trudinger*, Indiana Univ. Math. J. **20** (1970/71), 1077–1092. MR0301504 (46 #662)

[186] J. Moser, *On a nonlinear problem in differential geometry*, Dynamical systems (Proc. Sympos., Univ. Bahia, Salvador, 1971), Academic Press, New York, 1973, pp. 273–280. MR0339258 (49 #4018)

[187] S. Mukai and H. Umemura, *Minimal rational threefolds*, Algebraic geometry (Tokyo/Kyoto, 1982), Lecture Notes in Math., vol. 1016, Springer, Berlin, 1983, pp. 490–518, DOI 10.1007/BFb0099976. MR726439 (85c:14027)

[188] S. B. Myers, *Connections between differential geometry and topology. I. Simply connected surfaces*, Duke Math. J. **1** (1935), no. 3, 376–391, DOI 10.1215/S0012-7094-35-00126-0. MR1545884

[189] A. M. Nadel, *Multiplier ideal sheaves and Kähler-Einstein metrics of positive scalar curvature*, Ann. of Math. (2) **132** (1990), no. 3, 549–596, DOI 10.2307/1971429. MR1078269 (92d:32038)

[190] A. M. Nadel, *Multiplier ideal sheaves and Futaki's invariant*, Geometric theory of singular phenomena in partial differential equations (Cortona, 1995), Sympos. Math., XXXVIII, Cambridge Univ. Press, Cambridge, 1998, pp. 7–16. MR1702085 (2001a:32033)

[191] Y. Nakagawa, *Combinatorial formulae for Futaki characters and generalized Killing forms of toric Fano orbifolds*, The Third Pacific Rim Geometry Conference (Seoul, 1996), Monogr. Geom. Topology, vol. 25, Int. Press, Cambridge, MA, 1998, pp. 223–260. MR1751075 (2001m:32051)

[192] Y. Nakai, *Non-degenerate divisors on an algebraic surface*, J. Sci. Hiroshima Univ. Ser. A **24** (1960), 1–6. MR0136610 (25 #78)

[193] Y. Odaka, C. Spotti, and S. Sun, *Compact moduli spaces of Del Pezzo surfaces and Kähler–Einstein metrics*, preprint, arxiv:1210.0858.

[194] S. T. Paul, *Hyperdiscriminant polytopes, Chow polytopes, and Mabuchi energy asymptotics*, Ann. of Math. (2) **175** (2012), no. 1, 255–296, DOI 10.4007/annals.2012.175.1.7. MR2874643 (2012m:32029)

[195] S. T. Paul, *Stable pairs and coercive estimates for the Mabuchi functional*, preprint, arxiv:1308.4377.

[196] M. Păun, *Regularity properties of the degenerate Monge-Ampère equations on compact Kähler manifolds*, Chin. Ann. Math. Ser. B **29** (2008), no. 6, 623–630, DOI 10.1007/s11401-007-0457-8. MR2470619 (2010m:35157)

[197] P. Petersen, *Riemannian geometry*, 2nd ed., Graduate Texts in Mathematics, vol. 171, Springer, New York, 2006. MR2243772 (2007a:53001)

[198] D. H. Phong, J. Ross, and J. Sturm, *Deligne pairings and the Knudsen-Mumford expansion*, J. Differential Geom. **78** (2008), no. 3, 475–496. MR2396251 (2008k:32072)

[199] D. H. Phong, J. Song, and J. Sturm, *Complex Monge-Ampère equations*, Surveys in differential geometry. Vol. XVII, Surv. Differ. Geom., vol. 17, Int. Press, Boston, MA, 2012, pp. 327–410, DOI 10.4310/SDG.2012.v17.n1.a8. MR3076065

[200] D. H. Phong and J. Sturm, *Lectures on stability and constant scalar curvature*, Current developments in mathematics, 2007, Int. Press, Somerville, MA, 2009, pp. 101–176. MR2532997 (2010j:32042)

[201] D. H. Phong and J. Sturm, *Test configurations for K-stability and geodesic rays*, J. Symplectic Geom. **5** (2007), no. 2, 221–247. MR2377252 (2009a:32033)

[202] D. H. Phong, J. Song, J. Sturm, and B. Weinkove, *The Moser-Trudinger inequality on Kähler-Einstein manifolds*, Amer. J. Math. **130** (2008), no. 4, 1067–1085, DOI 10.1353/ajm.0.0013. MR2427008 (2009e:32027)

[203] D. H. Phong, J. Song, J. Sturm, and B. Weinkove, *The Kähler-Ricci flow and the $\overline{\partial}$ operator on vector fields*, J. Differential Geom. **81** (2009), no. 3, 631–647. MR2487603 (2010b:32035)

[204] D. H. Phong, N. Sesum, and J. Sturm, *Multiplier ideal sheaves and the Kähler-Ricci flow*, Comm. Anal. Geom. **15** (2007), no. 3, 613–632. MR2379807 (2009a:32037)

[205] E. Picard, *De l'équation $\Delta u = ke^u$ sur une surface de Riemann fermée*, J. Math. Pures Appl. (4) 9 (1893), 273–292.

[206] E. Picard, *De l'intégration de l'équation $\Delta u = e^u$ sur une surface de Riemann fermée*, J. Crelle 130 (1905), 243–258.

[207] Yu. G. Prokhorov, *Automorphism groups of Fano 3-folds* (Russian), Uspekhi Mat. Nauk **45** (1990), no. 3 (273), 195–196, DOI 10.1070/RM1990v045n03ABEH002363; English transl., Russian Math. Surveys **45** (1990), no. 3, 222–223. MR1071944 (91j:14034)

[208] D. Ramos, *Gradient Ricci solitons on surfaces*, preprint, arxiv:1304.6391.

[209] F. Rochon and Z. Zhang, *Asymptotics of complete Kähler metrics of finite volume on quasiprojective manifolds*, Adv. Math. **231** (2012), no. 5, 2892–2952, DOI 10.1016/j.aim.2012.08.005. MR2970469

[210] J. Ross and R. Thomas, *An obstruction to the existence of constant scalar curvature Kähler metrics*, J. Differential Geom. **72** (2006), no. 3, 429–466. MR2219940 (2007c:32028)

[211] J. Ross and R. Thomas, *Weighted projective embeddings, stability of orbifolds, and constant scalar curvature Kähler metrics*, J. Differential Geom. **88** (2011), no. 1, 109–159. MR2819757

[212] J. Ross and D. Witt Nyström, *Analytic test configurations and geodesic rays*, J. Symplectic Geom. **12** (2014), 125–169.

[213] Y. A. Rubinstein, *On energy functionals, Kähler-Einstein metrics, and the Moser-Trudinger-Onofri neighborhood*, J. Funct. Anal. **255** (2008), no. 9, 2641–2660, DOI 10.1016/j.jfa.2007.10.009. MR2473271 (2009i:32030)

[214] Y. A. Rubinstein, *Some discretizations of geometric evolution equations and the Ricci iteration on the space of Kähler metrics*, Adv. Math. **218** (2008), no. 5, 1526–1565, DOI 10.1016/j.aim.2008.03.017. MR2419932 (2009m:32045)

[215] Y. A. Rubinstein, *On the construction of Nadel multiplier ideal sheaves and the limiting behavior of the Ricci flow*, Trans. Amer. Math. Soc. **361** (2009), no. 11, 5839–5850, DOI 10.1090/S0002-9947-09-04675-3. MR2529916 (2010i:32021)

[216] Y. A. Rubinstein, *Geometric quantization and dynamical constructions on the space of Kahler metrics*, ProQuest LLC, Ann Arbor, MI, 2008. Thesis (Ph.D.)–Massachusetts Institute of Technology. MR2717532

[217] Y. A. Rubinstein and S. Zelditch, *The Cauchy problem for the homogeneous Monge-Ampère equation, I. Toeplitz quantization*, J. Differential Geom. **90** (2012), no. 2, 303–327. MR2899878

[218] Y. A. Rubinstein and S. Zelditch, *The Cauchy problem for the homogeneous Monge-Ampère equation, II. Legendre transform*, Adv. Math. **228** (2011), no. 6, 2989–3025, DOI 10.1016/j.aim.2011.08.001. MR2844938 (2012m:32042)

[219] Y. A. Rubinstein and S. Zelditch, *The Cauchy problem for the homogeneous Monge-Ampère equation*. III. Lifespan, preprint, arxiv: 1205.4793. To appear in Crelle's J.

[220] J. A. Schouten and D. van Dantzig, *Über unitäre Geometrie* (German), Math. Ann. **103** (1930), no. 1, 319–346, DOI 10.1007/BF01455698. MR1512625

[221] S. Semmes, *Complex Monge-Ampère and symplectic manifolds*, Amer. J. Math. **114** (1992), no. 3, 495–550, DOI 10.2307/2374768. MR1165352 (94h:32022)

[222] N. Sesum and G. Tian, *Bounding scalar curvature and diameter along the Kähler Ricci flow (after Perelman)*, J. Inst. Math. Jussieu **7** (2008), no. 3, 575–587, DOI 10.1017/S1474748008000133. MR2427424 (2009c:53092)

[223] Y. Shi, *On the α-invariants of cubic surfaces with Eckardt points*, Adv. Math. **225** (2010), no. 3, 1285–1307, DOI 10.1016/j.aim.2010.03.024. MR2673731 (2012b:32033)

[224] V. V. Shokurov, *Smoothness of the general anticanonical divisor on Fano 3-fold*, Math USSR, Izv. 14 (1980), 395–405.

[225] V. V. Shokurov, *The existence of a straight line on Fano 3-folds*, Math USSR, Izv. 15 (1980), 173–209.

[226] L. Simon, *Lectures on geometric measure theory*, Proceedings of the Centre for Mathematical Analysis, Australian National University, vol. 3, Australian National University, Centre for Mathematical Analysis, Canberra, 1983. MR756417 (87a:49001)

[227] Y. T. Siu, *Lectures on Hermitian-Einstein metrics for stable bundles and Kähler-Einstein metrics*, DMV Seminar, vol. 8, Birkhäuser Verlag, Basel, 1987. MR904673 (89d:32020)

[228] J. Song and B. Weinkove, *Energy functionals and canonical Kähler metrics*, Duke Math. J. **137** (2007), no. 1, 159–184, DOI 10.1215/S0012-7094-07-13715-3. MR2309146 (2008d:32022)

[229] J. Song and S. Zelditch, *Test configurations, large deviations and geodesic rays on toric varieties*, Adv. Math. **229** (2012), no. 4, 2338–2378, DOI 10.1016/j.aim.2011.12.025. MR2880224

[230] J. Song and X.-W. Wang, *The greatest Ricci lower bound, conical Einstein metrics and the Chern number inequality*, preprint, arxiv:1207.4839.

[231] J. Streets, *Long time existence of minimizing movement solutions of Calabi flow*, Adv. Math. **259** (2014), 688–729, DOI 10.1016/j.aim.2014.03.027. MR3197669

[232] M. Struwe, *Variational methods. Applications to nonlinear partial differential equations and Hamiltonian systems*, 4th ed., Ergebnisse der Mathematik und ihrer Grenzgebiete. 3. Folge. A Series of Modern Surveys in Mathematics [Results in Mathematics and Related Areas. 3rd Series. A Series of Modern Surveys in Mathematics], vol. 34, Springer-Verlag, Berlin, 2008. MR2431434 (2009g:49002)

[233] G. Székelyhidi, *Remarks on conical Kähler–Einstein metrics*, preprint, arxiv:1211.2725.

[234] G. Székelyhidi, *The partial C^0-estimate along the continuity method*, preprint, arxiv:1310.8471.

[235] G. Székelyhidi, *Introduction to extremal metrics*, manuscript, 2013.

[236] M. E. Taylor, *Partial differential equations. Basic theory*, Texts in Applied Mathematics, vol. 23, Springer-Verlag, New York, 1996. MR1395147 (98b:35002a)

[237] R. P. Thomas, *Notes on GIT and symplectic reduction for bundles and varieties*, Surveys in differential geometry. Vol. X, Surv. Differ. Geom., vol. 10, Int. Press, Somerville, MA, 2006, pp. 221–273, DOI 10.4310/SDG.2005.v10.n1.a7. MR2408226 (2010b:14092)

[238] G. Tian, *On Kähler-Einstein metrics on certain Kähler manifolds with $C_1(M) > 0$*, Invent. Math. **89** (1987), no. 2, 225–246, DOI 10.1007/BF01389077. MR894378 (88e:53069)

[239] G. Tian, *On the existence of solutions of a class of Monge-Ampère equations*, Acta Math. Sinica (N.S.) **4** (1988), no. 3, 250–265, DOI 10.1007/BF02560581. A Chinese summary appears in Acta Math. Sinica **32** (1989), no. 4, 576. MR965573 (89k:35095)

[240] G. Tian, *On Calabi's conjecture for complex surfaces with positive first Chern class*, Invent. Math. **101** (1990), no. 1, 101–172, DOI 10.1007/BF01231499. MR1055713 (91d:32042)

[241] G. Tian, *On stability of the tangent bundles of Fano varieties*, Internat. J. Math. **3** (1992), no. 3, 401–413, DOI 10.1142/S0129167X92000175. MR1163733 (93d:32044)

[242] G. Tian, *The K-energy on hypersurfaces and stability*, Comm. Anal. Geom. **2** (1994), no. 2, 239–265. MR1312688 (95m:32030)

[243] G. Tian, *Kähler-Einstein metrics on algebraic manifolds*, Transcendental methods in algebraic geometry (Cetraro, 1994), Lecture Notes in Math., vol. 1646, Springer, Berlin, 1996, pp. 143–185, DOI 10.1007/BFb0094304. MR1603624 (98j:32035)

[244] G. Tian, *Kähler-Einstein metrics with positive scalar curvature*, Invent. Math. **130** (1997), no. 1, 1–37, DOI 10.1007/s002220050176. MR1471884 (99e:53065)

[245] G. Tian, *Bott-Chern forms and geometric stability*, Discrete Contin. Dynam. Systems **6** (2000), no. 1, 211–220, DOI 10.3934/dcds.2000.6.211. MR1739924 (2001a:32034)

[246] G. Tian, *Canonical metrics in Kähler geometry*, Lectures in Mathematics ETH Zürich, Birkhäuser Verlag, Basel, 2000. Notes taken by Meike Akveld. MR1787650 (2001j:32024)

[247] G. Tian, *Extremal metrics and geometric stability*, Houston J. Math. **28** (2002), no. 2, 411–432. Special issue for S. S. Chern. MR1898198 (2003i:53062)

[248] G. Tian, *Existence of Einstein Metrics on Fano Manifolds*, in: Metric and differential geometry (X.-Z. Dai et al., Eds), Springer, 2012, pp. 119–159.

[249] G. Tian, *K-stability and Kähler–Einstein metrics*, preprint, arxiv:1211.4669.

[250] G. Tian, A 3rd derivative estimate for conic Kähler metrics, unpublished notes.

[251] G. Tian, *Stability of pairs*, preprint, arxiv:1310.5544.

[252] G. Tian and S.-T. Yau, *Existence of Kähler-Einstein metrics on complete Kähler manifolds and their applications to algebraic geometry*, Mathematical aspects of string theory (San Diego, Calif., 1986), Adv. Ser. Math. Phys., vol. 1, World Sci. Publishing, Singapore, 1987, pp. 574–628. MR915840

[253] G. Tian and S.-T. Yau, *Complete Kähler manifolds with zero Ricci curvature. I*, J. Amer. Math. Soc. **3** (1990), no. 3, 579–609, DOI 10.2307/1990928. MR1040196 (91a:53096)

[254] G. Tian and Z.-L. Zhang, *Regularity of Kähler–Ricci flows on Fano manifolds*, preprint, arxiv:1310.5897.

[255] G. Tian and X. Zhu, *A nonlinear inequality of Moser-Trudinger type*, Calc. Var. Partial Differential Equations **10** (2000), no. 4, 349–354, DOI 10.1007/s005260010349. MR1767718 (2001f:32044)

[256] G. Tian and X. Zhu, *Convergence of Kähler-Ricci flow*, J. Amer. Math. Soc. **20** (2007), no. 3, 675–699, DOI 10.1090/S0894-0347-06-00552-2. MR2291916 (2007k:53107)

[257] V. Tosatti, *Kähler-Einstein metrics on Fano surfaces*, Expo. Math. **30** (2012), no. 1, 11–31, DOI 10.1016/j.exmath.2011.07.003. MR2899654

[258] V. Tosatti and B. Weinkove, *The Calabi flow with small initial energy*, Math. Res. Lett. **14** (2007), no. 6, 1033–1039, DOI 10.4310/MRL.2007.v14.n6.a11. MR2357473 (2008m:53165)

[259] M. Troyanov, Les surfaces riemanniennes à singularités coniques, Thèse, Université de Genève, 1987.

[260] M. Troyanov, *Prescribing curvature on compact surfaces with conical singularities*, Trans. Amer. Math. Soc. **324** (1991), no. 2, 793–821, DOI 10.2307/2001742. MR1005085 (91h:53059)

[261] H. Tsuji, *Stability of tangent bundles of minimal algebraic varieties*, Topology **27** (1988), no. 4, 429–442, DOI 10.1016/0040-9383(88)90022-5. MR976585 (90d:14021)

[262] A. N. Wang, *A note to Futaki's invariant*, Chinese J. Math. **19** (1991), no. 3, 239–242. MR1126579 (92i:53067)

[263] R. H. Wang and J. Jiang, *Another approach to the Dirichlet problem for equations of Monge-Ampère type*, Northeast. Math. J. **1** (1985), no. 1, 27–40. MR828347 (87f:35106)

[264] Y.-Q. Wang, Smooth approximations of the conical Kähler–Ricci flows, preprint, arxiv:1401.5040.

[265] D. Wu, *Kähler-Einstein metrics of negative Ricci curvature on general quasi-projective manifolds*, Comm. Anal. Geom. **16** (2008), no. 2, 395–435. MR2425471 (2009g:32049)

[266] C.-J. Yao, *Existence of weak conical Kähler–Einstein metrics along smooth hypersurfaces*, preprint, arxiv:1308.4307.

[267] S. T. Yau, *On the curvature of compact Hermitian manifolds*, Invent. Math. **25** (1974), 213–239. MR0382706 (52 #3588)

[268] S. T. Yau, *On the Ricci curvature of a compact Kähler manifold and the complex Monge-Ampère equation. I*, Comm. Pure Appl. Math. **31** (1978), no. 3, 339–411, DOI 10.1002/cpa.3160310304. MR480350 (81d:53045)

[269] H. Yin, *Ricci flow on surfaces with conical singularities*, J. Geom. Anal. **20** (2010), no. 4, 970–995, DOI 10.1007/s12220-010-9136-1. MR2683772 (2012b:53146)

[270] H. Yin, *Ricci flow on surfaces with conical singularities*, part II, preprint, arxiv:1305.4355.

Department of Mathematics, University of Maryland
E-mail address: yanir@umd.edu

Contemporary Mathematics
Volume **630**, 2014
http://dx.doi.org/10.1090/conm/630/12666

Complex b-Manifolds

Gerardo A. Mendoza

ABSTRACT. A complex b-structure on a manifold $\mathcal{M}$ with boundary is an involutive subbundle ${}^b T^{0,1}\mathcal{M}$ of the complexification of ${}^b T\mathcal{M}$ with the property that $\mathbb{C}{}^b T\mathcal{M} = {}^b T^{0,1}\mathcal{M} + \overline{{}^b T^{0,1}\mathcal{M}}$ as a direct sum; the interior of $\mathcal{M}$ is a complex manifold. The complex b-structure determines an elliptic complex of b-operators and induces a rich structure on the boundary of $\mathcal{M}$. We study the cohomology of the indicial complex of the b-Dolbeault complex.

1. Introduction

A complex b-manifold is a smooth manifold with boundary together with a complex b-structure. The latter is a smooth involutive subbundle ${}^b T^{0,1}\mathcal{M}$ of the complexification $\mathbb{C}{}^b T\mathcal{M}$ of Melrose's b-tangent bundle [**5**, **6**] with the property that

$$\mathbb{C}{}^b T\mathcal{M} = {}^b T^{0,1}\mathcal{M} + \overline{{}^b T^{0,1}\mathcal{M}}$$

as a direct sum. Manifolds with complex b-structures generalize the situation that arises as a result of spherical and certain anisotropic (not complex) blowups of complex manifolds at a discrete set of points or along a complex submanifold, cf. [**7**, Section 2], [**9**], as well as (real) blow-ups of complex analytic varieties with only point singularities.

The interior of $\mathcal{M}$ is a complex manifold. Its $\overline{\partial}$-complex determines a b-elliptic complex, the ${}^b\overline{\partial}$-complex, on sections of the exterior powers of the dual of ${}^b T^{0,1}\mathcal{M}$, see Section 2. The indicial families $\overline{\mathcal{D}}(\sigma)$ of the ${}^b\overline{\partial}$-operators at a connected component $\mathcal{N}$ of $\partial\mathcal{M}$ give, for each σ, an elliptic complex, see Section 6. Their cohomology at the various values of σ determine the asymptotics at $\mathcal{N}$ of tempered representatives of cohomology classes of the ${}^b\overline{\partial}$-complex, in particular of tempered holomorphic functions.

Each boundary component $\mathcal{N}$ of $\mathcal{M}$ inherits from ${}^b T^{0,1}\mathcal{M}$ the following objects in the C^∞ category:

(1) an involutive vector subbundle $\overline{\mathcal{V}} \subset \mathbb{C}T\mathcal{N}$ such that $\mathcal{V} + \overline{\mathcal{V}} = \mathbb{C}T\mathcal{N}$;
(2) a real nowhere vanishing vector field $\mathcal{T}$ such that $\mathcal{V} \cap \overline{\mathcal{V}} = \operatorname{span}_{\mathbb{C}} \mathcal{T}$;
(3) a class $\boldsymbol{\beta}$ of sections of $\overline{\mathcal{V}}^*$,

where the elements of $\boldsymbol{\beta}$ have additional properties, described in (4) below. The vector bundle $\overline{\mathcal{V}}$, being involutive, determines a complex of first order differential

2010 *Mathematics Subject Classification.* Primary 32Q99; Secondary 58J10, 32V05.
Key words and phrases. Complex manifolds, b-tangent bundle, cohomology.

operators $\overline{\mathbb{D}}$ on sections of the exterior powers of $\overline{\mathcal{V}}^*$, elliptic because of the second property in (1) above. To that list add

(4) If $\beta \in \boldsymbol{\beta}$ then $\overline{\mathbb{D}}\beta = 0$ and $\Im\langle\beta, \mathcal{T}\rangle = -1$, and if β, $\beta' \in \boldsymbol{\beta}$, then $\beta' - \beta = \overline{\mathbb{D}}u$ with u real-valued.

These properties, together with the existence of a Hermitian metric on $\overline{\mathcal{V}}$ invariant under $\mathcal{T}$ make $\mathcal{N}$ behave in many ways as the circle bundle of a holomorphic line bundle over a compact complex manifold. These analogies are investigated in [**10**–**13**]. The last of these papers contains a detailed account of circle bundles from the perspective of these boundary structures. The paper [**8**], a predecessor of the present one, contains some facts studied here in more detail.

The paper is organized as follows. Section 2 deals with the definition of complex b-structure and Section 3 with holomorphic vector bundles over complex b-manifolds (the latter term just means that the b-tangent bundle takes on a primary role over that of the usual tangent bundle). The associated Dolbeault complexes are defined in these sections accordingly.

Section 4 is a careful account of the structure inherited by the boundary.

In Section 5 we show that complex b-structures have no formal local invariants at boundary points. The issue here is that we do not have a Newlander-Nirenberg theorem that is valid in a neighborhoods of a point of the boundary, so no explicit local model for b-manifolds.

Section 6 is devoted to general aspects of b-elliptic first order complexes A. We introduce here the set $\operatorname{spec}^q_{b,\mathcal{N}}(A)$, the boundary spectrum of the complex in degree q at the component $\mathcal{N}$ of $\mathcal{M}$, and prove basic properties of the boundary spectrum (assuming that the boundary component $\mathcal{N}$ is compact), including some aspects concerning Mellin transforms of A-closed forms. Some of these ideas are illustrated using the b-de Rham complex.

Section 7 is a systematic study of the $\overline{\partial}_b$-complex of CR structures on $\mathcal{N}$ associated with elements of the class $\boldsymbol{\beta}$. Each $\beta \in \boldsymbol{\beta}$ defines a CR structure, $\overline{\mathcal{K}}_\beta = \ker \beta$. Assuming that $\overline{\mathcal{V}}$ admits a $\mathcal{T}$-invariant Hermitian metric, we show that there is $\beta \in \boldsymbol{\beta}$ such that the CR structure $\overline{\mathcal{K}}_\beta$ is $\mathcal{T}$-invariant.

In Section 8 we assume that $\overline{\mathcal{V}}$ is $\mathcal{T}$-invariant and show that for $\mathcal{T}$-invariant CR structures, a theorem proved in [**13**] gives that the cohomology spaces of the associated $\overline{\partial}_b$-complex, viewed as the kernel of the Kohn Laplacian at the various degrees, split into eigenspaces of $-i\mathcal{L}_\mathcal{T}$. The eigenvalues of the latter operator are related to the indicial spectrum of the ${}^b\overline{\partial}$-complex.

In Section 9 we prove a precise theorem on the indicial cohomology and spectrum for the ${}^b\overline{\partial}$-complex under the assumption that $\overline{\mathcal{V}}$ admits a $\mathcal{T}$-invariant Hermitian metric.

Finally, we have included a very short appendix listing a number of basic definitions in connection with b-operators.

2. Complex b-structures

Let $\mathcal{M}$ be a smooth manifold with smooth boundary. An almost CR b-structure on $\mathcal{M}$ is a subbundle $\overline{\mathcal{W}}$ of the complexification, $\mathbb{C}{}^bT\mathcal{M} \to \mathcal{M}$ of the b-tangent bundle of $\mathcal{M}$ (Melrose [**5**, **6**]) such that

$$\mathcal{W} \cap \overline{\mathcal{W}} = 0 \tag{2.1}$$

with $\mathcal{W} = \overline{\overline{\mathcal{W}}}$. If in addition

$$\mathcal{W} + \overline{\mathcal{W}} = \mathbb{C}{}^b T\mathcal{M} \tag{2.2}$$

then we say that $\overline{\mathcal{W}}$ is an almost complex b-structure and write ${}^bT^{0,1}\mathcal{M}$ instead of $\overline{\mathcal{W}}$ and ${}^bT^{1,0}\mathcal{M}$ for its conjugate. As is customary, the adverb "almost" is dropped if $\mathcal{W}$ is involutive. Note that since $C^\infty(\mathcal{M}; {}^bT\mathcal{M})$ is a Lie algebra, it makes sense to speak of involutive subbundles of ${}^bT\mathcal{M}$ (or its complexification).

DEFINITION 2.3. A complex b-manifold is a manifold together with a complex b-structure.

By the Newlander-Nirenberg Theorem [**14**], the interior of complex b-manifold is a complex manifold. However, its boundary is not a CR manifold; rather, as we shall see, it naturally carries a family of CR structures parametrized by the defining functions of $\partial\mathcal{M}$ in $\mathcal{M}$ which are positive in $\overset{\circ}{\mathcal{M}}$.

That $C^\infty(\mathcal{M}; {}^bT\mathcal{M})$ is a Lie algebra is an immediate consequence of the definition of the b-tangent bundle, which indeed can be characterized as being a vector bundle ${}^bT\mathcal{M} \to \mathcal{M}$ together with a vector bundle homomorphism

$$\mathrm{ev} : {}^bT\mathcal{M} \to T\mathcal{M}$$

covering the identity such that the induced map

$$\mathrm{ev}_* : C^\infty(\mathcal{M}; {}^bT\mathcal{M}) \to C^\infty(\mathcal{M}; T\mathcal{M})$$

is a $C^\infty(\mathcal{M}; \mathbb{R})$-module isomorphism onto the submodule $C^\infty_{\tan}(\mathcal{M}; T\mathcal{M})$ of smooth vector fields on $\mathcal{M}$ which are tangential to the boundary of $\mathcal{M}$. Since $C^\infty_{\tan}(\mathcal{M}, T\mathcal{M})$ is closed under Lie brackets, there is an induced Lie bracket on $C^\infty(\mathcal{M}; {}^bT\mathcal{M})$. The homomorphism ev is an isomorphism over the interior of $\mathcal{M}$, and its restriction to the boundary,

$$\mathrm{ev}_{\partial\mathcal{M}} : {}^bT_{\partial\mathcal{M}}\mathcal{M} \to T\partial\mathcal{M} \tag{2.4}$$

is surjective. Its kernel, a fortiori a rank-one bundle, is spanned by a canonical section denoted $\mathfrak{r}\partial_{\mathfrak{r}}$. Here and elsewhere, $\mathfrak{r}$ refers to any smooth defining function for $\partial\mathcal{M}$ in $\mathcal{M}$, by convention positive in the interior of $\mathcal{M}$.

Associated with a complex b-structure on $\mathcal{M}$ there is a Dolbeault complex. Let ${}^b\bigwedge^{0,q}\mathcal{M}$ denote the q-th exterior power of the dual of ${}^bT^{0,1}M$. Then the operator

$$\cdots \to C^\infty(\mathcal{M}; {}^b\textstyle\bigwedge^{0,q}\mathcal{M}) \xrightarrow{{}^b\overline{\partial}} C^\infty(\mathcal{M}; {}^b\textstyle\bigwedge^{0,q+1}\mathcal{M}) \to \cdots$$

is defined by

$$\begin{aligned}(q+1)\,{}^b\overline{\partial}\phi(V_0,\dots,V_q) &= \sum_{j=0}^{q} V_j\phi(V_0,\dots,\hat{V}_j,\dots,V_q) \\ &\quad + \sum_{j<k}(-1)^{j+k}\phi([V_j,V_k],V_0,\dots,\hat{V}_j,\dots,\hat{V}_k,\dots,V_q)\end{aligned} \tag{2.5}$$

as with the standard de Rham differential (see Helgason [**3**, p. 21]) whenever ϕ is a smooth section of ${}^b\bigwedge^q\mathcal{M}$ and $V_0,\dots,V_q \in C^\infty(\mathcal{M}; {}^bT^{0,1}\mathcal{M})$. In this formula V_j acts on functions via the vector field ev_*V_j. The involutivity of ${}^bT^{0,1}\mathcal{M}$ is used in the terms involving brackets, of course. The same proof that $d \circ d = 0$ works here to give that ${}^b\overline{\partial}^2 = 0$. The formula

$${}^b\overline{\partial}(f\phi) = f\,{}^b\overline{\partial}\phi + {}^b\overline{\partial}f \wedge \phi \text{ for } \phi \in C^\infty(\mathcal{M}; {}^b\textstyle\bigwedge^q\mathcal{M}) \text{ and } f \in C^\infty(\mathcal{M}), \tag{2.6}$$

implies that ${}^b\overline{\partial}$ is a first order operator.

Since we do not have at our disposal holomorphic frames (near the boundary) for the bundles of forms of type (p,q) for $p>0$, we define ${}^b\overline{\partial}$ on forms of type (p,q) with $p>0$ with the aid of the b-de Rham complex, exactly as in Folland and Kohn [**2**] for standard complex structures and de Rham complex. The b-de Rham complex, we recall from Melrose [**6**], is the complex associated with the dual, $\mathbb{C}{}^bT^*\mathcal{M}$, of $\mathbb{C}{}^bT\mathcal{M}$,

$$\cdots \to C^\infty(\mathcal{M};{}^b\textstyle\bigwedge^r\mathcal{M}) \xrightarrow{{}^bd} C^\infty(\mathcal{M};{}^b\textstyle\bigwedge^{r+1}\mathcal{M}) \to \cdots$$

where ${}^b\bigwedge^q\mathcal{M}$ denotes the r-th exterior power of $\mathbb{C}{}^bT^*\mathcal{M}$. The operators ${}^bd$ are defined by the same formula as (2.5), now however with the $V_j \in C^\infty(\mathcal{M};\mathbb{C}{}^bT\mathcal{M})$. On functions f we have

$${}^bdf = \mathrm{ev}^* df.$$

More generally,

$$\mathrm{ev}^* \circ d = {}^bd \circ \mathrm{ev}^*$$

in any degree. Also,

$$(2.7)\qquad {}^bd(f\phi) = f\,{}^bd\phi + {}^bdf\wedge\phi \text{ for } \phi\in C^\infty(\mathcal{M};{}^b\textstyle\bigwedge^r\mathcal{M}) \text{ and } f\in C^\infty(\mathcal{M}).$$

It is convenient to note here that for $f\in C^\infty(\mathcal{M})$,

$$(2.8)\qquad {}^bdf \text{ vanishes on } \partial\mathcal{M} \text{ if } f \text{ does.}$$

Now, with the obvious definition,

$$(2.9)\qquad {}^b\textstyle\bigwedge^r\mathcal{M} = \displaystyle\bigoplus_{p+q=r} {}^b\textstyle\bigwedge^{p,q}\mathcal{M}.$$

Using the special cases

$$\begin{aligned} {}^bd &: C^\infty(\mathcal{M};{}^b\textstyle\bigwedge^{0,1}) \to C^\infty(\mathcal{M};{}^b\textstyle\bigwedge^{1,1}) + C^\infty(\mathcal{M};{}^b\textstyle\bigwedge^{0,2}),\\ {}^bd &: C^\infty(\mathcal{M};{}^b\textstyle\bigwedge^{1,0}) \to C^\infty(\mathcal{M};{}^b\textstyle\bigwedge^{2,0}) + C^\infty(\mathcal{M};{}^b\textstyle\bigwedge^{1,1}),\end{aligned}$$

consequences of the involutivity of ${}^bT^{0,1}\mathcal{M}$ and its conjugate, one gets

$${}^bd\phi \in C^\infty(\mathcal{M};{}^b\textstyle\bigwedge^{p+1,q}\mathcal{M}) \oplus C^\infty(\mathcal{M};{}^b\textstyle\bigwedge^{p,q+1}\mathcal{M}) \quad \text{if } \phi\in C^\infty(\mathcal{M};{}^b\textstyle\bigwedge^{p,q}\mathcal{M})$$

for general (p,q). Let $\pi_{p,q} : {}^b\bigwedge^k\mathcal{M} \to {}^b\bigwedge^{p,q}\mathcal{M}$ be the projection according to the decomposition (2.9), and define

$${}^b\partial = \pi_{p+1,q}\,{}^bd, \quad {}^b\overline{\partial} = \pi_{q,p+1}\,{}^bd,$$

so ${}^bd = {}^b\partial + {}^b\overline{\partial}$. The operators ${}^b\overline{\partial}$ are identical to the $\overline{\partial}$-operators over the interior of $\mathcal{M}$ and with the previously defined ${}^b\overline{\partial}$ operators on $(0,q)$-forms, and give a complex

$$(2.10)\qquad \cdots \to C^\infty(\mathcal{M};{}^b\textstyle\bigwedge^{p,q}\mathcal{M}) \xrightarrow{{}^b\overline{\partial}} C^\infty(\mathcal{M};{}^b\textstyle\bigwedge^{p,q+1}\mathcal{M}) \to \cdots$$

for each p. On functions $f:\mathcal{M}\to\mathbb{C}$,

$$(2.11)\qquad {}^b\overline{\partial}f = \pi_{0,1}\,{}^bdf.$$

The formula

$$(2.7')\qquad {}^b\overline{\partial}f\phi = {}^b\overline{\partial}f\wedge\phi + f\,{}^b\overline{\partial}\phi, \quad f\in C^\infty(\mathcal{M}),\ \phi\in C^\infty(\mathcal{M};{}^b\textstyle\bigwedge^{p,q}\mathcal{M}),$$

a consequence of (2.7), implies that ${}^b\overline{\partial}$ is a first order operator. As a consequence of (2.8),

$$(2.8') \qquad {}^b\overline{\partial} f \text{ vanishes on } \partial\mathcal{M} \text{ if } f \text{ does.}$$

The operators of the b-de Rham complex are first order operators because of (2.7), and (2.8) implies that these are b-operators, see (A.1). Likewise, (2.7′) and (2.8′) imply that in any bidegree, the operator $\phi \mapsto \mathfrak{r}^{-1}\,{}^b\overline{\partial}\,\mathfrak{r}\phi$ has coefficients smooth up to the boundary, so

$$(2.12) \qquad {}^b\overline{\partial} \in \operatorname{Diff}^1_b(\mathcal{M}; {}^b\Lambda^{p,q}\mathcal{M}, {}^b\Lambda^{p,q+1}\mathcal{M}),$$

see (A.1). We also get from these formulas that the b-symbol of ${}^b\overline{\partial}$ is

$$(2.13) \qquad {}^b\boldsymbol{\sigma}({}^b\overline{\partial})(\xi)(\phi) = i\pi_{0,1}(\xi) \wedge \phi, \quad x \in \mathcal{M},\ \xi \in {}^bT^*_x\mathcal{M},\ \phi \in {}^b\Lambda^{p,q}_x\mathcal{M},$$

see (A.2). Since $\pi_{0,1}$ is injective on the real b-cotangent bundle (this follows from (2.2)), the complex (2.10) is b-elliptic.

3. Holomorphic vector bundles

The notion of holomorphic vector bundle in the b-category is a translation of the standard one using connections. Let $\rho : F \to \mathcal{M}$ be a complex vector bundle. Recall from [**6**] that a b-connection on F is a linear operator

$${}^b\nabla : C^\infty(\mathcal{M}; F) \to C^\infty(\mathcal{M}; {}^b\Lambda^1\mathcal{M} \otimes F)$$

such that

$$(3.1) \qquad {}^b\nabla f\phi = f\,{}^b\nabla\phi + {}^bdf \otimes \phi$$

for each $\phi \in C^\infty(\mathcal{M}; F)$ and $f \in C^\infty(\mathcal{M})$. This property automatically makes ${}^b\nabla$ a b-operator.

A standard connection $\nabla : C^\infty(\mathcal{M}; F) \to C^\infty(\mathcal{M}; \Lambda^1\mathcal{M} \otimes F)$ determines a b-connection by composition with

$$\mathrm{ev}^* \otimes I : \Lambda^1\mathcal{M} \otimes F \to {}^b\Lambda^1\mathcal{M} \otimes F,$$

but b-connections are more general than standard connections. Indeed, the difference between the latter and the former can be any smooth section of the bundle $\operatorname{Hom}(F, {}^b\Lambda^1\mathcal{M} \otimes F)$. A b-connection ${}^b\nabla$ on F arises from a standard connection if and only if ${}^b\nabla_{\mathfrak{r}\partial_\mathfrak{r}} = 0$ along $\partial\mathcal{M}$.

As in the standard situation, the b-connection ${}^b\nabla$ determines operators

$$(3.2) \qquad {}^b\nabla : C^\infty(\mathcal{M}; {}^b\Lambda^k\mathcal{M} \otimes F) \to C^\infty(\mathcal{M}; {}^b\Lambda^{k+1}\mathcal{M} \otimes F)$$

by way of the usual formula translated to the b setting:

$$(3.3) \qquad {}^b\nabla(\alpha \otimes \phi) = (-1)^k\alpha \wedge {}^b\nabla\phi + {}^bd\alpha \wedge \phi, \quad \phi \in C^\infty(\mathcal{M}; F),\ \alpha \in {}^b\Lambda^k\mathcal{M}.$$

Since

$${}^b\nabla\mathfrak{r}\alpha \otimes \phi = \mathfrak{r}\,{}^b\nabla(\alpha \otimes \phi) + {}^bd\mathfrak{r} \wedge \alpha \otimes \phi$$

is smooth and vanishes on $\partial\mathcal{M}$, also

$${}^b\nabla \in \operatorname{Diff}^1_b(\mathcal{M}; {}^b\Lambda^k\mathcal{M} \otimes F, {}^b\Lambda^{k+1}\mathcal{M} \otimes F).$$

The principal b-symbol of (3.2), easily computed using (3.3) and

$${}^b\boldsymbol{\sigma}({}^b\nabla)({}^bdf)(\phi) = \lim_{\tau\to\infty} \frac{e^{-i\tau f}}{\tau}\,{}^b\nabla e^{i\tau f}\phi$$

for $f \in C^\infty(\mathcal{M};\mathbb{R})$ and $\phi \in C^\infty(\mathcal{M};{}^b\!\bigwedge^k\mathcal{M}\otimes F)$, is

$$ {}^b\boldsymbol{\sigma}({}^b\nabla)(\xi)(\phi) = i\xi\wedge\phi, \quad \xi\in {}^bT^*_x\mathcal{M},\ \phi\in {}^b\!\bigwedge^k_x\mathcal{M}\otimes F_x,\ x\in\mathcal{M}. $$

As expected, the connection is called holomorphic if the component in ${}^b\!\bigwedge^{0,2}\mathcal{M}\otimes F$ of the curvature operator

$$ \Omega = {}^b\nabla^2 : C^\infty(\mathcal{M};F)\to C^\infty(\mathcal{M};{}^b\!\bigwedge^2\mathcal{M}\otimes F), $$

vanishes. Such a connection gives F the structure of a complex b-manifold. Its complex b-structure can be described locally as in the standard situation, as follows. Fix a frame η_μ for F and let the ω^ν_μ be the local sections of ${}^b\!\bigwedge^{0,1}\mathcal{M}$ such that

$$ {}^b\overline{\partial}\eta_\mu = \sum_\nu \omega^\nu_\mu\otimes\eta_\nu. $$

Denote by ζ^μ the fiber coordinates determined by the frame η_μ. Let $V_1,\dots,V_{n+1}$ be a frame of ${}^bT^{0,1}\mathcal{M}$ over U, denote by $\tilde V_j$ the sections of $\mathbb{C}{}^bTF$ over $\rho^{-1}(U)$ which project on the V_j and satisfy $\tilde V_j\zeta^\mu = \tilde V_j\overline{\zeta}^\mu = 0$ for all μ, and by ∂_{ζ^μ} the vertical vector fields such that $\partial_{\zeta^\mu}\zeta^\nu = \delta^\nu_\mu$ and $\partial_{\zeta^\mu}\overline{\zeta}^\nu = 0$. Then the sections

$$ \tilde V_j - \sum_{\mu,\nu}\zeta^\mu\langle\omega^\nu_\mu, V_j\rangle\partial_{\zeta^\nu},\ j=1,\dots,n+1, \quad \partial_{\overline{\zeta}^\nu},\ \nu=1,\dots,k \tag{3.4} $$

of $\mathbb{C}{}^bTF$ over $\rho^{-1}(U)$ form a frame of ${}^bT^{0,1}F$. As in the standard situation, the involutivity of this subbundle of $\mathbb{C}{}^bTF$ is equivalent to the condition on the vanishing of the $(0,2)$ component of the curvature of ${}^b\nabla$. A vector bundle $F\to\mathcal{M}$ together with the complex b-structure determined by a choice of holomorphic b-connection (if one exists at all) is a holomorphic vector bundle.

The $\overline{\partial}$ operator of a holomorphic vector bundle is

$$ {}^b\overline{\partial} = (\pi_{0,q+1}\otimes I)\circ{}^b\nabla : C^\infty(\mathcal{M};{}^b\!\bigwedge^{0,q}\mathcal{M}\otimes F)\to C^\infty(\mathcal{M};{}^b\!\bigwedge^{0,q+1}\mathcal{M}\otimes F). $$

As is the case for standard complex structures, the condition on the curvature of ${}^b\nabla$ implies that these operators form a complex, b-elliptic since

$$ {}^b\boldsymbol{\sigma}({}^b\overline{\partial})(\xi)(\phi) = i\pi_{0,1}(\xi)\wedge\phi, \quad \xi\in{}^bT^*_x\mathcal{M},\ \phi\in{}^b\!\bigwedge^k_x\mathcal{M}\otimes F_x,\ x\in\mathcal{M} $$

and $\pi_{0,1}(\xi)=0$ for $\xi\in{}^bT^*\mathcal{M}$ if and only if $\xi=0$.

Also as usual, a b-connection ${}^b\nabla$ on a Hermitian vector bundle $F\to\mathcal{M}$ with Hermitian form h is Hermitian if

$$ {}^bdh(\phi,\psi) = h({}^b\nabla\phi,\psi)+h(\phi,{}^b\nabla\psi) $$

for every pair of smooth sections ϕ, ψ of F. In view of the definition of ${}^bd$ this means that for every $v\in\mathbb{C}{}^bT\mathcal{M}$ and sections as above,

$$ \operatorname{ev}(v)h(\phi,\psi) = h({}^b\nabla_v\phi,\psi)+h(\phi,{}^b\nabla_{\overline{v}}\psi) $$

On a complex b-manifold $\mathcal{M}$, if an arbitrary connection ${}^b\nabla'$ and the Hermitian form h are given for a vector bundle F, holomorphic or not, then there is a unique *Hermitian* b-connection ${}^b\nabla$ such that $\pi_{0,1}{}^b\nabla = \pi_{0,1}{}^b\nabla'$. Namely, let η_μ be a local orthonormal frame of F, let

$$ (\pi_{0,1}\otimes I)\circ{}^b\nabla'\eta_\mu = \sum_\nu\omega^\nu_\mu\otimes\eta_\nu, $$

and let ${}^b\nabla$ be the connection defined in the domain of the frame by

$$ {}^b\nabla\eta_\mu = (\omega^\nu_\mu - \overline{\omega}^\mu_\nu)\otimes\eta_\nu. \tag{3.5}$$

If the matrix of functions $Q=[q^\mu_\lambda]$ is unitary and $\tilde\eta_\lambda = \sum_\mu q^\mu_\lambda\eta_\mu$, then

$$(\pi_{0,1}\otimes I)\circ {}^b\nabla'\tilde\eta_\lambda = \sum_\nu \tilde\omega^\sigma_\lambda\otimes\tilde\eta_\sigma$$

with

$$\tilde\omega^\sigma_\lambda = \sum_\mu \overline{q}^\mu_\sigma\, {}^b\overline{\partial} q^\mu_\lambda + \sum_{\mu,\nu}\overline{q}^\mu_\sigma q^\nu_\lambda\omega^\mu_\nu,$$

using (3.1), that $Q^{-1}=[\overline{q}^\mu_\lambda]$, and that $\pi_{0,1}{}^b df = {}^b\overline{\partial} f$. Thus

$$\begin{aligned}\tilde\omega^\sigma_\lambda - \overline{\tilde\omega}^\lambda_\sigma &= \sum_\mu(\overline{q}^\mu_\sigma\, {}^b\overline{\partial}q^\mu_\lambda - q^\mu_\lambda\, {}^b\partial\overline{q}^\mu_\sigma) + \sum_{\mu,\nu}(\overline{q}^\mu_\sigma q^\nu_\lambda\omega^\mu_\nu - q^\mu_\lambda\overline{q}^\nu_\sigma\overline{\omega}^\mu_\nu)\\ &= \sum_\mu({}^b\overline{\partial}q^\mu_\lambda + {}^b\partial q^\mu_\lambda)\overline{q}^\mu_\sigma + \sum_{\mu,\nu} q^\nu_\lambda(\omega^\mu_\nu - \overline{\omega}^\nu_\mu)\overline{q}^\mu_\sigma\\ &= \sum_\mu {}^b dq^\mu_\lambda + \overline{q}^\mu_\sigma + \sum_{\mu,\nu} q^\nu_\lambda(\omega^\mu_\nu - \overline{\omega}^\nu_\mu)\overline{q}^\mu_\sigma\end{aligned}$$

using that $\overline{{}^b\overline{\partial} f} = {}^b\partial\overline{f}$ and that $\sum_\mu q^\mu_\lambda\, {}^b\partial\overline{q}^\mu_\sigma = -\sum_\mu {}^b\partial q^\mu_\lambda\, \overline{q}^\mu_\sigma$ because $\sum_\mu q^\mu_\lambda\overline{q}^\mu_\sigma$ is constant, and that ${}^b\overline{\partial}q^\mu_\lambda + {}^b\partial q^\mu_\lambda = {}^b dq^\mu_\lambda$. Thus there is a globally defined Hermitian connection locally given by (3.5). We leave to the reader to verify that this connection is Hermitian. Clearly ${}^b\nabla$ is the unique Hermitian connection such that $\pi_{0,1}{}^b\nabla = \pi_{0,1}{}^b\nabla'$. When ${}^b\nabla'$ is a holomorphic connection, ${}^b\nabla$ is the unique Hermitian holomorphic connection.

LEMMA 3.6. *The vector bundles* ${}^b\bigwedge^{p,0}\mathcal{M}$ *are holomorphic.*

We prove this by exhibiting a holomorphic b-connection. Fix an auxiliary Hermitian metric on ${}^b\bigwedge^{p,0}\mathcal{M}$ and pick an orthonormal frame (η_μ) of ${}^b\bigwedge^{p,0}\mathcal{M}$ over some open set U. Let ω^ν_μ be the unique sections of ${}^b\bigwedge^{0,1}\mathcal{M}$ such that

$${}^b\overline{\partial}\eta_\mu = \sum_\nu \omega^\nu_\mu\wedge\eta_\nu,$$

and let ${}^b\nabla$ be the b-connection defined on U by the formula (3.5). As in the previous paragraph, this gives a globally defined b-connection. That it is holomorphic follows from

$${}^b\overline{\partial}\omega^\nu_\mu + \sum_\lambda \omega^\nu_\lambda\wedge\omega^\lambda_\mu = 0,$$

a consequence of ${}^b\overline{\partial}^2 = 0$. Evidently, with the identifications ${}^b\bigwedge^{0,q}\mathcal{M}\otimes{}^b\bigwedge^{p,0}\mathcal{M} = {}^b\bigwedge^{p,q}\mathcal{M}$, $\pi_{p,q+1}{}^b\nabla$ is the ${}^b\overline{\partial}$ operator in (2.12).

4. The boundary of a complex b-manifold

Suppose that $\mathcal{M}$ is a complex b-manifold and $\mathcal{N}$ is a component of its boundary. We shall assume $\mathcal{N}$ compact, although for the most part this is not necessary.

The homomorphism

$$\mathrm{ev} : \mathbb{C}{}^bT\mathcal{M}\to\mathbb{C}T\mathcal{M}$$

is an isomorphism over the interior of $\mathcal{M}$, and its restriction to $\mathcal{N}$ maps onto $\mathbb{C}T\mathcal{N}$ with kernel spanned by $\mathfrak{r}\partial_\mathfrak{r}$. Write

$$\mathrm{ev}_\mathcal{N} : \mathbb{C}{}^bT_\mathcal{N}\mathcal{M} \to \mathbb{C}T\mathcal{N}$$

for this restriction and

$$\Phi : {}^bT^{0,1}_\mathcal{N}\mathcal{M} \to \overline{\mathcal{V}} \tag{4.1}$$

for of the restriction of $\mathrm{ev}_\mathcal{N}$ to ${}^bT^{0,1}_\mathcal{N}\mathcal{M}$. From (2.1) and the fact that the kernel of $\mathrm{ev}_\mathcal{N}$ is spanned by the real section $\mathfrak{r}\partial_\mathfrak{r}$ one obtains that Φ is injective, so its image,

$$\overline{\mathcal{V}} = \Phi({}^bT^{0,1}_\mathcal{N}\mathcal{M})$$

is a subbundle of $\mathbb{C}T\mathcal{N}$.

Since ${}^bT^{0,1}\mathcal{M}$ is involutive, so is $\overline{\mathcal{V}}$, see [**7**, Proposition 3.12]. From (2.2) and the fact that $\mathrm{ev}_\mathcal{N}$ maps onto $\mathbb{C}T\mathcal{N}$, one obtains that

$$\mathcal{V} + \overline{\mathcal{V}} = \mathbb{C}T\mathcal{N}, \tag{4.2}$$

see [**7**, Lemma 3.13]. Thus

LEMMA 4.3. *$\overline{\mathcal{V}}$ is an elliptic structure.*

This just means what we just said: $\overline{\mathcal{V}}$ is involutive and (4.2) holds, see Trèves [**16**, **17**]; the sum need not be direct. All elliptic structures are locally of the same kind, depending only on the dimension of $\mathcal{V}\cap\overline{\mathcal{V}}$. This is a result of Nirenberg [**15**] (see also Hörmander [**4**]) extending the Newlander-Nirenberg theorem. In the case at hand, $\overline{\mathcal{V}}\cap\mathcal{V}$ has rank 1 because of the relation

$$\mathrm{rank}_\mathbb{C}(\mathcal{V}\cap\overline{\mathcal{V}}) = 2\,\mathrm{rank}_\mathbb{C}\,\overline{\mathcal{V}} - \dim\mathcal{N}$$

which holds whenever (4.2) holds. Nirenberg's in the present situation is:

> Every $p_0 \in \mathcal{N}$ has a neighborhood in which there coordinates $x^1,\dots,x^{2n},t$ such that with $z^j = x^j + ix^{j+n}$, the vector fields
>
> $$\frac{\partial}{\partial\overline{z}^1},\dots,\frac{\partial}{\partial\overline{z}^n},\ \frac{\partial}{\partial t} \tag{4.4}$$
>
> span $\overline{\mathcal{V}}$ near p_0. The function $(z^1,\dots,z^n,t)$ is called a hypoanalytic chart (Baouendi, Chang, and Trèves [**1**], Trèves [**17**]).

The intersection $\overline{\mathcal{V}}\cap\mathcal{V}$ is, in the case we are discussing, spanned by a canonical globally defined real vector field. Namely, let $\mathfrak{r}\partial_\mathfrak{r}$ be the canonical section of ${}^bT\mathcal{M}$ along $\mathcal{N}$. There is a unique section $J\mathfrak{r}\partial_\mathfrak{r}$ of ${}^bT\mathcal{M}$ along $\mathcal{N}$ such that $\mathfrak{r}\partial_\mathfrak{r} + iJ\mathfrak{r}\partial_\mathfrak{r}$ is a section of ${}^bT^{0,1}\mathcal{M}$ along $\mathcal{N}$. Then

$$\mathcal{T} = \mathrm{ev}_\mathcal{N}(J\mathfrak{r}\partial_\mathfrak{r})$$

is a nonvanishing real vector field in $\mathcal{V}\cap\overline{\mathcal{V}}$, (see [**8**, Lemma 2.1]). Using the isomorphism (4.1) we have

$$\mathcal{T} = \Phi(J(\mathfrak{r}\partial_\mathfrak{r}) - i\mathfrak{r}\partial_\mathfrak{r}).$$

Because $\overline{\mathcal{V}}$ is involutive, there is yet another complex, this time associated with the exterior powers of the dual of $\overline{\mathcal{V}}$:

$$\cdots \to C^\infty(\mathcal{N};\textstyle\bigwedge^q\overline{\mathcal{V}}^*) \xrightarrow{\overline{\mathbb{D}}} C^\infty(\mathcal{N};\textstyle\bigwedge^{q+1}\overline{\mathcal{V}}^*) \to \cdots, \tag{4.5}$$

where $\overline{\mathbb{D}}$ is defined by the formula (2.5) where now the V_j are sections of $\overline{\mathcal{V}}$. The complex (4.5) is elliptic because of (4.2). For a function f we have $\overline{\mathbb{D}}f = \iota^*df$, where $\iota^* : \mathbb{C}T^*\mathcal{N} \to \overline{\mathcal{V}}^*$ is the dual of the inclusion homomorphism $\iota : \overline{\mathcal{V}} \to \mathbb{C}T\mathcal{N}$.

For later use we show:

LEMMA 4.6. *Suppose that $\mathcal{N}$ is compact and connected. If $\zeta : \mathcal{N} \to \mathbb{C}$ solves $\overline{\mathbb{D}}\zeta = 0$, then ζ is constant.*

PROOF. Let p_0 be an extremal point of $|\zeta|$. Fix a hypoanalytic chart (z, t) for $\overline{\mathcal{V}}$ centered at p_0. Since $\overline{\mathbb{D}}\zeta = 0$, $\zeta(z,t)$ is independent of t and $\partial_{\overline{z}^\nu}\zeta = 0$. So there is a holomorphic function Z defined in a neighborhood of 0 in $\mathbb{C}^n$ such that $\zeta = Z \circ z$. Then $|Z|$ has a maximum at 0, so Z is constant near 0. Therefore ζ is constant, say $\zeta(p) = c$, near p_0. Let $C = \{p : \zeta(p) = c\}$, a closed set. Let $p_1 \in C$. Since p_1 is also an extremal point of ζ, the above argument gives that ζ is constant near p_1, therefore equal to c. Thus C is open, and consequently ζ is constant on $\mathcal{N}$. □

Since the operators ${}^b\overline{\partial} : C^\infty(\mathcal{M}, {}^b{\textstyle\bigwedge}^{0,q}\mathcal{M}) \to C^\infty(\mathcal{M}, {}^b{\textstyle\bigwedge}^{0,q+1}\mathcal{M})$ are totally characteristic, they induce operators

$$ {}^b\overline{\partial}_b : C^\infty(\mathcal{N}, {}^b{\textstyle\bigwedge}^{0,q}_{\mathcal{N}}\mathcal{M}) \to C^\infty(\mathcal{M}, {}^b{\textstyle\bigwedge}^{0,q+1}_{\mathcal{N}}\mathcal{M}), $$

see (A.3); we use ${}^b{\textstyle\bigwedge}^{0,q}_{\mathcal{N}}\mathcal{M}$ as notation for the part of ${}^b{\textstyle\bigwedge}^{0,q}\mathcal{M}$ over $\mathcal{N}$. These boundary operators define a complex because of (A.4). By way of the dual

$$ \Phi^* : \overline{\mathcal{V}}^* \to {}^b{\textstyle\bigwedge}^{0,1}_{\mathcal{N}}\mathcal{M} \tag{4.7} $$

of the isomorphism (4.1) the operators ${}^b\overline{\partial}_b$ become identical to the operators of the $\overline{\mathbb{D}}$-complex (4.5): The diagram

$$ \begin{array}{ccccccc} \cdots \longrightarrow & C^\infty(\mathcal{N}; {\textstyle\bigwedge}^q\overline{\mathcal{V}}^*) & \xrightarrow{\overline{\mathbb{D}}} & C^\infty(\mathcal{N}; {\textstyle\bigwedge}^{q+1}\overline{\mathcal{V}}^*) & \longrightarrow \cdots \\ & \Phi^*\big\downarrow & & \big\downarrow\Phi^* & \\ \cdots \longrightarrow & C^\infty(\mathcal{N}, {}^b{\textstyle\bigwedge}^{0,q}_{\mathcal{N}}\mathcal{M}) & \xrightarrow{{}^b\overline{\partial}_b} & C^\infty(\mathcal{M}, {}^b{\textstyle\bigwedge}^{0,q+1}_{\mathcal{N}}\mathcal{M}) & \longrightarrow \cdots \end{array} $$

is commutative and the vertical arrows are isomorphisms. This can be proved by writing the ${}^b\overline{\partial}$ operators using Cartan's formula (2.5) for ${}^b\overline{\partial}$ and $\overline{\mathbb{D}}$ and comparing the resulting expressions.

Let $\mathfrak{r} : \mathcal{M} \to \mathbb{R}$ be a smooth defining function for $\partial\mathcal{M}$, $\mathfrak{r} > 0$ in the interior of $\mathcal{M}$. Then ${}^b\overline{\partial}\mathfrak{r}$ is smooth and vanishes on $\partial\mathcal{M}$, so $\frac{{}^b\overline{\partial}\mathfrak{r}}{\mathfrak{r}}$ is also a smooth ${}^b\overline{\partial}$-closed section of ${}^b{\textstyle\bigwedge}^{0,1}\mathcal{M}$. Thus we get a $\overline{\mathbb{D}}$-closed element

$$ \beta_{\mathfrak{r}} = [\Phi^*]^{-1}\frac{{}^b\overline{\partial}\mathfrak{r}}{\mathfrak{r}} \in C^\infty(\partial\mathcal{M}; \overline{\mathcal{V}}^*). \tag{4.8} $$

By definition,

$$ \langle \beta_{\mathfrak{r}}, \mathcal{T}\rangle = \langle \frac{{}^b\overline{\partial}\mathfrak{r}}{\mathfrak{r}}, J(\mathfrak{r}\partial_{\mathfrak{r}}) - i\mathfrak{r}\partial_{\mathfrak{r}}\rangle. $$

Extend the section $\mathfrak{r}\partial_{\mathfrak{r}}$ to a section of ${}^bT\mathcal{M}$ over a neighborhood U of $\mathcal{N}$ in $\mathcal{M}$ with the property that $\mathfrak{r}\partial_{\mathfrak{r}}\mathfrak{r} = \mathfrak{r}$. In U we have

$$ \langle {}^b\overline{\partial}\mathfrak{r}, J(\mathfrak{r}\partial_{\mathfrak{r}}) - i\mathfrak{r}\partial_{\mathfrak{r}}\rangle = (J(\mathfrak{r}\partial_{\mathfrak{r}}) - i\mathfrak{r}\partial_{\mathfrak{r}})\mathfrak{r} = J(\mathfrak{r}\partial_{\mathfrak{r}})\mathfrak{r} - i\mathfrak{r}. $$

The function $J(\mathfrak{r}\partial_{\mathfrak{r}})\mathfrak{r}$ is smooth, real-valued, and vanishes along the boundary. So $\mathfrak{r}^{-1}J(\mathfrak{r}\partial_{\mathfrak{r}})\mathfrak{r}$ is smooth, real-valued. Thus

$$ \langle \beta_{\mathfrak{r}}, \mathcal{T}\rangle = a_{\mathfrak{r}} - i $$

on $\mathcal{N}$ for some smooth function $a_{\mathfrak{r}} : \mathcal{N} \to \mathbb{R}$, see [**8**, Lemma 2.5].

If $\mathfrak{r}'$ is another defining function for $\partial\mathcal{M}$, then $\mathfrak{r}' = \mathfrak{r}e^u$ for some smooth function $u : \mathcal{M} \to \mathbb{R}$. Then

$$
{}^b\overline{\partial}\mathfrak{r}' = e^u\, {}^b\overline{\partial}\mathfrak{r} + e^u \mathfrak{r}\, {}^b\overline{\partial}u
$$

and it follows that

$$
\beta_{\mathfrak{r}'} = \beta_{\mathfrak{r}} + \overline{\mathbb{D}}u.
$$

In particular,

$$
a_{\mathfrak{r}'} = a_{\mathfrak{r}} + \mathcal{T}u.
$$

Let $\mathfrak{a}_t$ denote the one-parameter group of diffeomorphisms generated by $\mathcal{T}$.

PROPOSITION 4.9. *The functions $a_{\mathrm{av}}^{\sup}$, $a_{\mathrm{av}}^{\inf} : \mathcal{N} \to \mathbb{R}$ defined by*

$$
a_{\mathrm{av}}^{\sup}(p) = \limsup_{t\to\infty} \frac{1}{2t}\int_{-t}^{t} a_{\mathfrak{r}}(\mathfrak{a}_s(p))\,ds, \quad a_{\mathrm{av}}^{\inf}(p) = \liminf_{t\to\infty} \frac{1}{2t}\int_{-t}^{t} a_{\mathfrak{r}}(\mathfrak{a}_s(p))\,ds
$$

are invariants of the complex b-structure, that is, they are independent of the defining function $\mathfrak{r}$. The equality $a_{\mathrm{av}}^{\sup} = a_{\mathrm{av}}^{\inf}$ holds for some $\mathfrak{r}$ if and only if it holds for all $\mathfrak{r}$.

Indeed,

$$
\lim_{t\to\infty}\Big(\frac{1}{2t}\int_{-t}^{t} a_{\mathfrak{r}'}(\mathfrak{a}_s(p))\,ds - \frac{1}{2t}\int_{-t}^{t} a_{\mathfrak{r}}(\mathfrak{a}_s(p))\,ds\Big) = \lim_{t\to\infty}\frac{1}{2t}\int_{-t}^{t}\frac{d}{ds}u(\mathfrak{a}_s(p))\,ds = 0
$$

because u is bounded (since $\mathcal{N}$ is compact).

The functions $a_{\mathrm{av}}^{\sup}$, $a_{\mathrm{av}}^{\inf}$ are constant on orbits of $\mathcal{T}$, but they may not be smooth.

EXAMPLE 4.10. Let $\mathcal{N}$ be the unit sphere in $\mathbb{C}^{n+1}$ centered at the origin. Write $(z^1, \dots, z^{n+1})$ for the standard coordinates in $\mathbb{C}^{n+1}$. Fix $\tau_1, \dots, \tau_{n+1} \in \mathbb{R}\backslash 0$, all of the same sign, and let

$$
\mathcal{T} = i\sum_{j=1}^{n+1} \tau_j (z^j \partial_{z^j} - \overline{z}^j \partial_{\overline{z}^j}).
$$

This vector field is real and tangent to $\mathcal{N}$. Let $\overline{\mathcal{K}}$ be the standard CR structure of $\mathcal{N}$ as a submanifold of $\mathbb{C}^{n+1}$ (the part of $T^{0,1}\mathbb{C}^{n+1}$ tangential to $\mathcal{N}$). The condition that the τ_j are different from 0 and have the same sign ensures that $\mathcal{T}$ is never in $\mathcal{K} \oplus \overline{\mathcal{K}}$. Indeed, the latter subbundle of $\mathbb{C}T\mathcal{N}$ is the annihilator of the pullback to $\mathcal{N}$ of $i\overline{\partial}\sum_{\ell=1}^{n+1}|z^\ell|^2$. The pairing of this form with $\mathcal{T}$ is

$$
\langle i\sum_{\ell=1}^{n+1} z^\ell d\overline{z}^\ell, i\sum_{j=1}^{n+1}\tau_j(z^j\partial_{z^j} - \overline{z}^j\partial_{\overline{z}^j})\rangle = \sum_{j=1}^{n+1}\tau_j|z^j|^2,
$$

a function that vanishes nowhere if and only if all τ_j are different from zero and have the same sign. Thus $\overline{\mathcal{V}} = \overline{\mathcal{K}} \oplus \operatorname{span}_{\mathbb{C}}\mathcal{T}$ is a subbundle of $\mathbb{C}T\mathcal{N}$ of rank $n+1$ with the property that $\mathcal{V} + \overline{\mathcal{V}} = \mathbb{C}T\mathcal{N}$. To show that $\overline{\mathcal{V}}$ is involutive we first note that $\overline{\mathcal{K}}$ is the annihilator of the pullback to $\mathcal{N}$ of the span of the differentials $dz^1, \dots, dz^{n+1}$. Let $\mathcal{L}_{\mathcal{T}}$ denote the Lie derivative with respect to $\mathcal{T}$. Then $\mathcal{L}_{\mathcal{T}}dz^j = i\tau_j dz^j$, so if L is a CR vector field, then so is $[L, \mathcal{T}]$. Since in addition $\overline{\mathcal{K}}$ and $\operatorname{span}_{\mathbb{C}}\mathcal{T}$ are themselves involutive, $\overline{\mathcal{V}}$ is involutive. Thus $\overline{\mathcal{V}}$ is an elliptic structure with $\mathcal{V}\cap\overline{\mathcal{V}} = \operatorname{span}_{\mathbb{C}}\mathcal{T}$. Let β be the section of $\overline{\mathcal{V}}^*$ which vanishes on $\overline{\mathcal{K}}$ and satisfies $\langle\beta, \mathcal{T}\rangle = -i$. Let $\overline{\mathbb{D}}$ denote the operators of the associated differential complex. Then $\overline{\mathbb{D}}\beta = 0$, since β vanishes on commutators of sections of $\overline{\mathcal{K}}$ (since $\overline{\mathcal{K}}$ is involutive) and on commutators of $\mathcal{T}$ with sections of $\overline{\mathcal{K}}$ (since such commutators are in $\overline{\mathcal{K}}$).

If the τ_j are positive (negative), this example may be viewed as the boundary of a blowup (compactification) of $\mathbb{C}^{n+1}$, see [**9**].

Let now $\rho : F \to \mathcal{M}$ be a holomorphic vector bundle. Its ${}^b\overline{\partial}$-complex also determines a complex along $\mathcal{N}$,

$$(4.11) \qquad \cdots \to C^\infty(\mathcal{N};\textstyle\bigwedge^q\overline{\mathcal{V}}^*\otimes F_{\mathcal{N}}) \xrightarrow{\overline{\mathbb{D}}} C^\infty(\mathcal{N};\textstyle\bigwedge^{q+1}\overline{\mathcal{V}}^*\otimes F_{\mathcal{N}}) \to \cdots,$$

where $\overline{\mathbb{D}}$ is defined using the boundary operators ${}^b\overline{\partial}_b$ and the isomorphism (4.7):

$$(4.12) \qquad \overline{\mathbb{D}}(\phi\otimes\eta) = (\Phi^*)^{-1}\,{}^b\overline{\partial}_b[\Phi^*(\phi\otimes\eta)]$$

where Φ^* means $\Phi^*\otimes I$. These operators can be expressed locally in terms of the operators of the complex (4.5). Fix a smooth frame η_μ, $\mu = 1,\dots,k$, of F in a neighborhood $U\subset\mathcal{M}$ of $p_0\in\mathcal{N}$, and suppose

$${}^b\overline{\partial}\eta_\mu = \sum_\nu \omega_\mu^\nu\otimes\eta_\nu.$$

The ω_μ^ν are local sections of ${}^b\!\bigwedge^{0,1}\mathcal{M}$, and if $\sum_\mu \phi^\mu\otimes\eta_\mu$ is a section of ${}^b\!\bigwedge^{0,q}\mathcal{M}\otimes F$ over U, then

$${}^b\overline{\partial}\sum\phi^\mu\otimes\eta_\mu = \sum_\nu({}^b\overline{\partial}\phi^\nu + \sum_\mu\omega_\mu^\nu\wedge\phi^\mu)\otimes\eta_\nu.$$

Therefore, using the identification (4.7), the boundary operator ${}^b\overline{\partial}_b$ is the operator given locally by

$$(4.13) \qquad \overline{\mathbb{D}}\sum\phi^\mu\otimes\eta_\mu = \sum_\nu(\overline{\mathbb{D}}\phi^\nu + \sum_\mu\omega_\mu^\nu\wedge\phi^\mu)\otimes\eta_\nu$$

where now the ϕ^μ are sections of ${}^b\!\bigwedge^q\overline{\mathcal{V}}^*$, the ω_μ^ν are the sections of $\overline{\mathcal{V}}^*$ corresponding to the original ω_μ^ν via Φ^*, and $\overline{\mathbb{D}}$ on the right hand side of the formula is the operator associated with $\overline{\mathcal{V}}$.

The structure bundle ${}^b\mathcal{T}^{0,1}F$ is locally given as the span of the sections (3.4). Applying the evaluation homomorphism $\mathbb{C}{}^bT_{\partial F}F \to \mathbb{C}T\partial F$ (over $\mathcal{N}$) to these sections gives vector fields on $F_{\mathcal{N}}$ forming a frame for the elliptic structure $\overline{\mathcal{V}}_F$ inherited by $F_{\mathcal{N}}$. Writing $V_j^0 = \mathrm{ev}V_j$, this frame is just

$$(4.14) \qquad \tilde V_j^0 - \sum_{\mu,\nu}\zeta^\mu\langle\omega_\mu^\nu, V_j^0\rangle\partial_{\zeta^\nu},\ j=1,\dots,n+1, \quad \partial_{\overline{\zeta}^\nu},\ \nu=1,\dots,k,$$

where now the ω_μ^ν are the forms associated to the $\overline{\mathbb{D}}$ operator of $F_{\mathcal{N}}$. Alternatively, one may take the $\overline{\mathbb{D}}$ operators of $F_{\mathcal{N}}$ and use the formula (4.13) to define a subbundle of $\mathbb{C}TF$ locally as the span of the vector fields (4.14), a fortiori an elliptic structure on $F_{\mathcal{N}}$, involutive because

$$\overline{\mathbb{D}}\omega^\nu + \sum_\lambda\omega_\lambda^\nu\wedge\omega_\mu^\lambda = 0.$$

To obtain a formula for the canonical real vector field $\mathcal{T}_F$ in $\overline{\mathcal{V}}_F$, let J_F be the almost complex b-structure of ${}^bTF$ and consider again the sections (3.4); they are defined in an open set $\rho^{-1}(U)$, U a neighborhood in $\mathcal{M}$ of a point of $\mathcal{N}$. Since the elements $\partial_{\overline{\zeta}^\nu}$ are sections of ${}^b\mathcal{T}^{0,1}F$,

$$J_F\Re\partial_{\overline{\zeta}^\nu} = \Im\partial_{\overline{\zeta}^\nu}. \tag{4.15}$$

Pick a defining function $\mathfrak{r}$ for $\mathcal{N}$. Then $\tilde{\mathfrak{r}} = \rho^*\mathfrak{r}$ is a defining function for $F_\mathcal{N}$. We may take $V_{n+1} = \mathfrak{r}\partial_\mathfrak{r} + iJ\mathfrak{r}\partial_\mathfrak{r}$ along $U\cap\mathcal{N}$. Then $\tilde{V}_{n+1} = \tilde{\mathfrak{r}}\partial_{\tilde{\mathfrak{r}}} + i\widetilde{J\mathfrak{r}\partial_\mathfrak{r}}$ along $\rho^{-1}(U)\cap F_\mathcal{N}$ and so

$$J_F\Re\big(\tilde{\mathfrak{r}}\partial_{\tilde{\mathfrak{r}}} + i\widetilde{J\mathfrak{r}\partial_\mathfrak{r}} - \sum_{\mu,\nu}\zeta^\mu\langle\omega_\mu^\nu, \mathfrak{r}\partial_\mathfrak{r} + iJ\mathfrak{r}\partial_\mathfrak{r}\rangle\partial_{\zeta^\nu}\big) = \\ \Im\big(\tilde{\mathfrak{r}}\partial_{\tilde{\mathfrak{r}}} + i\widetilde{J\mathfrak{r}\partial_\mathfrak{r}} - \sum_{\mu,\nu}\zeta^\mu\langle\omega_\mu^\nu, \mathfrak{r}\partial_\mathfrak{r} + iJ\mathfrak{r}\partial_\mathfrak{r}\rangle\partial_{\zeta^\nu}\big)$$

along $\rho^{-1}(U)\cap F_\mathcal{N}$. Using (4.15) this gives

$$J_F\tilde{\mathfrak{r}}\partial_{\tilde{\mathfrak{r}}} = \widetilde{J\mathfrak{r}\partial_\mathfrak{r}} - 2\Im\sum_{\mu,\nu}\zeta^\mu\langle\omega_\mu^\nu, \mathfrak{r}\partial_\mathfrak{r} + iJ\mathfrak{r}\partial_\mathfrak{r}\rangle\partial_{\zeta^\nu}.$$

Applying the evaluation homomorphism gives

$$\mathcal{T}_F = \tilde{\mathcal{T}} - 2\Im\sum_{\mu,\nu}\zeta^\mu\langle\omega_\mu^\nu, \mathfrak{r}\partial_\mathfrak{r} + iJ\mathfrak{r}\partial_\mathfrak{r}\rangle\partial_{\zeta^\nu} \tag{4.16}$$

where $\tilde{\mathcal{T}}$ is the real vector field on $\rho^{-1}(U\cap\mathcal{N}) = \rho^{-1}(U)\cap F_\mathcal{N}$ which projects on $\mathcal{T}$ and satisfies $\tilde{\mathcal{T}}\zeta^\mu = 0$ for all μ.

Let h be a Hermitian metric on F, and suppose that the frame η_μ is orthonormal. Applying $\mathcal{T}_F$ as given in (4.16) to the function $|\zeta|^2 = \sum|\zeta^\mu|^2$ we get that $\mathcal{T}_F$ is tangent to the unit sphere bundle of F if and only if

$$\langle\omega_\mu^\nu, \mathfrak{r}\partial_\mathfrak{r} + iJ\mathfrak{r}\partial_\mathfrak{r}\rangle - \overline{\langle\omega_\nu^\mu, \mathfrak{r}\partial_\mathfrak{r} + iJ\mathfrak{r}\partial_\mathfrak{r}\rangle} = 0$$

for all μ,ν. Equivalently, in terms of the isomorphism (4.7),

$$\langle(\Phi^*)^{-1}\omega_\mu^\nu, \mathcal{T}\rangle + \overline{\langle(\Phi^*)^{-1}\omega_\nu^\mu, \mathcal{T}\rangle} = 0 \quad \text{for all } \mu,\nu. \tag{4.17}$$

DEFINITION 4.18. The Hermitian metric h will be called exact if (4.17) holds.

The terminology in Definition 4.18 is taken from the notion of exact Riemannian b-metric of Melrose [**6**, pg. 31]. For such metrics, the Levi-Civita b-connection has the property that ${}^b\nabla_{\mathfrak{r}\partial_\mathfrak{r}} = 0$ [op. cit., pg. 58]. We proceed to show that the Hermitian holomorphic connection of an exact Hermitian metric on F also has this property. Namely, suppose that h is an exact Hermitian metric, and let η_μ be an orthonormal frame of F. Then for the Hermitian holomorphic connection we have

$$\langle\omega_\mu^\nu - \overline{\omega}_\nu^\mu, \mathfrak{r}\partial_\mathfrak{r}\rangle = \langle\omega_\mu^\nu, \mathfrak{r}\partial_\mathfrak{r}\rangle - \overline{\langle\omega_\nu^\mu, \mathfrak{r}\partial_\mathfrak{r}\rangle} = \frac{1}{2}\big(\langle\omega_\mu^\nu, \mathfrak{r}\partial_\mathfrak{r} + iJ\mathfrak{r}\partial_\mathfrak{r}\rangle - \overline{\langle\omega_\nu^\mu, \mathfrak{r}\partial_\mathfrak{r} + iJ\mathfrak{r}\partial_\mathfrak{r}\rangle}\big)$$

using that the ω_μ^ν are of type $(0,1)$. Thus ${}^b\nabla_{\mathfrak{r}\partial_\mathfrak{r}} = 0$.

5. Local invariants

Complex structures have no local invariants: every point of a complex n-manifold has a neighborhood biholomorphic to a ball in $\mathbb{C}^n$ It is natural to ask the same question about complex b-structures, namely,

> is there a local model depending only on dimension for every complex b-stucture?

In lieu of a Newlander-Nirenberg theorem, we show that complex b-structures have no local formal invariants at the boundary. More precisely:

PROPOSITION 5.1. *Every $p_0 \in \mathcal{N}$ has a neighborhood V in $\mathcal{M}$ on which there are smooth coordinates x^j, $j = 1, \dots, 2n+2$ centered at p_0 with x^{n+1} vanishing on $V \cap \mathcal{N}$ such that with*

$$\overline{L}_j^0 = \frac{1}{2}(\partial_{x^j} + i\partial_{x^{j+n+1}}),\ j \le n, \quad \overline{L}_{n+1}^0 = \frac{1}{2}(x^{n+1}\partial_{x^{n+1}} + i\partial_{x^{2n+2}}) \tag{5.2}$$

there are smooth functions γ_k^j vanishing to infinite order on $V \cap \mathcal{N}$ such that

$$\overline{L}_j = \overline{L}_j^0 + \sum_{k=1}^{n+1} \gamma_j^k L_k^0$$

is a frame for ${}^bT^{0,1}\mathcal{M}$ over V.

The proof will require some preparation. Let $\mathfrak{r} : \mathcal{M} \to \mathbb{R}$ be a defining function for $\partial\mathcal{M}$. Let $p_0 \in \mathcal{N}$, pick a hypoanalytic chart (z, t) (cf. (4.4)) centered at p_0 with $\mathcal{T}t = 1$. Let $U \subset \mathcal{N}$ be a neighborhood of p_0 contained in the domain of the chart, mapped by it to $B \times (-\delta, \delta) \subset \mathbb{C}^n \times \mathbb{R}$, where B is a ball with center 0 and δ is some small positive number. For reference purposes we state

LEMMA 5.3. *On such U, the problem*

$$\overline{\mathbb{D}}\phi = \psi, \quad \psi \in C^\infty(U; \textstyle\bigwedge^{q+1}\overline{\mathcal{V}}^*|_U) \text{ and } \overline{\mathbb{D}}\psi = 0$$

has a solution in $C^\infty(U; \bigwedge^q\overline{\mathcal{V}}^*|_U)$.

Extend the functions z^j and t to a neighborhood of p_0 in $\mathcal{M}$. Shrinking U if necessary, we may assume that in some neighborhood V of p_0 in $\mathcal{M}$ with $V \cap \partial\mathcal{M} = U$, $(z, t, \mathfrak{r})$ maps V diffeomorphically onto $B \times (-\delta, \delta) \times [0, \varepsilon)$ for some δ, $\varepsilon > 0$. Since the form $\beta_\mathfrak{r}$ defined in (4.8) is $\overline{\mathbb{D}}$-closed, there is $\alpha \in C^\infty(U)$ such that

$$-i\overline{\mathbb{D}}\alpha = \beta_\mathfrak{r}.$$

Extend α to V as a smooth function. The section

$${}^b\overline{\partial}(\log \mathfrak{r} + i\alpha) = \frac{{}^b\overline{\partial}\mathfrak{r}}{\mathfrak{r}} + i{}^b\overline{\partial}\alpha \tag{5.4}$$

of ${}^b\bigwedge^{0,1}\mathcal{M}$ over V vanishes on U, since $\beta_\mathfrak{r} + i\overline{\mathbb{D}}\alpha = 0$. So there is a smooth section ϕ of ${}^b\bigwedge^{0,1}\mathcal{M}$ over V such that

$${}^b\overline{\partial}(\log \mathfrak{r} + i\alpha) = \mathfrak{r}e^{i\alpha}\phi.$$

Suppose $\zeta : U \to \mathbb{C}$ is a solution of $\overline{\mathbb{D}}\zeta = 0$ on U, and extend it to V. Then ${}^b\overline{\partial}\zeta$ vanishes on U, so again we have

$${}^b\overline{\partial}\zeta = \mathfrak{r}e^{i\alpha}\psi.$$

for some smooth section ψ of ${}^b\bigwedge^{0,1}\mathcal{M}$ over V. The following lemma will be applied for f_0 equal to $\log \mathfrak{r} + i\alpha$ or each of the functions z^j.

LEMMA 5.5. *Let f_0 be smooth in $V \backslash U$ and suppose that ${}^b\overline{\partial} f_0 = \mathfrak{r}e^{i\alpha}\psi_1$ with ψ_1 smooth on V. Then there is $f : V \to \mathbb{C}$ smooth vanishing at U such that ${}^b\overline{\partial}(f_0 + f)$ vanishes to infinite order on U.*

PROOF. Suppose that $f_1, \dots, f_{N-1}$ are defined on V and that

$$\overline{{}^b\partial} \sum_{k=0}^{N-1} (\mathfrak{r}e^{i\alpha})^k f_k = (\mathfrak{r}e^{i\alpha})^N \psi_N \tag{5.6}$$

holds with ψ_N smooth in V; by the hypothesis, (5.6) holds when $N = 1$. Using (5.4) we get that $\overline{{}^b\partial}(\mathfrak{r}e^{i\alpha}) = (\mathfrak{r}e^{i\alpha})^2\phi$, therefore

$$0 = \overline{{}^b\partial}\big((\mathfrak{r}e^{i\alpha})^N\psi_N\big) = (\mathfrak{r}e^{i\alpha})^N\big[\overline{{}^b\partial}\psi_N + N\mathfrak{r}e^{i\alpha}\phi\wedge\psi_N\big],$$

which implies that $\overline{{}^b\partial}\psi_N = 0$ on U. With arbitrary f_N we have

$$\overline{{}^b\partial}\sum_{k=0}^{N}(\mathfrak{r}e^{i\alpha})^k f_k = (\mathfrak{r}e^{i\alpha})^N(\psi_N + \overline{{}^b\partial} f_N + N\mathfrak{r}e^{i\alpha} f_N\phi).$$

Since $\overline{\mathbb{D}}\psi_N = 0$ and $H^1_{\overline{\mathbb{D}}}(U) = 0$ by Lemma 5.3, there is a smooth function f_N defined in U such that $\overline{\mathbb{D}} f_N = -\psi_N$ in U. So there is χ_N such that $\psi_N + \overline{{}^b\partial} f_N = \mathfrak{r}e^{i\alpha}\chi_N$. With such f_N, (5.6) holds with $N+1$ in place of N and some ψ_{N+1}. Thus there is a sequence $\{f_j\}_{j=1}^\infty$ such that (5.6) holds for each N. Borel's lemma then gives f smooth with

$$f \sim \sum_{k=1}^{\infty}(\mathfrak{r}e^{i\alpha})^k f_k \quad \text{on } U$$

such that $\overline{\mathbb{D}}(f_0 + f)$ vanishes to infinite order on U. □

PROOF OF PROPOSITION 5.1. Apply the lemma with $f_0 = \log\mathfrak{r} + i\alpha$ to get a function f such that $\overline{{}^b\partial}(f_0 + f)$ vanishes to infinite order at U. Let

$$x^{n+1} = \mathfrak{r}e^{-\Im\alpha + \Re f}, \quad x^{2n+2} = \Re\alpha + \Im f.$$

These functions are smooth up to U.

Applying the lemma to each of the functions $f_0 = z^j$, $j = 1, \dots, n$ gives smooth functions ζ^j such that $\zeta^j = z^j$ on U and $\overline{{}^b\partial}\zeta^j = 0$ to infinite order at U. Define

$$x^j = \Re\zeta^j, \quad x^{j+n+1} = \Im\zeta^j, \quad j = 1, \dots, n.$$

The functions x^j, $j = 1\dots, 2n+2$ are independent, and the forms

$$\eta^j = {}^b d\zeta^j, j = 1\dots, n, \quad \eta^{n+1} = \frac{1}{x^{n+1}e^{ix^{2n+2}}}\,{}^b d[x^{n+1}e^{ix^{2n+2}}]$$

together with their conjugates form a frame for $\mathbb{C}{}^bT\mathcal{M}$ near p_0. Let $\eta^j_{1,0}$ and $\eta^j_{0,1}$ be the $(1,0)$ and $(0,1)$ components of η^j according to the complex b-structure of $\mathcal{M}$. Then

$$\eta^j_{0,1} = \sum_k p^j_k\eta^k + q^j_k\overline{\eta}^k.$$

Since $\eta^j_{0,1} = \overline{{}^b\partial}\zeta^j$ vanishes to infinite order at U, the coefficients p^j_k and q^j_k vanish to infinite order at U. Replacing this formula for $\eta^j_{0,1}$ in $\eta^j = \eta^j_{1,0} + \eta^j_{0,1}$ get

$$\sum_k(\delta^j_k - p^j_k)\eta^k - \sum_k q^j_k\overline{\eta}^k = \eta^j_{1,0}.$$

The matrix $I - [p^j_k]$ is invertible with inverse of the form $I + [P^j_k]$ with P^j_k vanishing to infinite order at U. So

$$\eta^j - \sum_k \gamma^j_k\overline{\eta}^k = \sum_k(\delta^j_k + P^j_k)\eta^k_{1,0} \tag{5.7}$$

with suitable γ_k^j vanishing to infinite order on U. Define the vector fields $\overline{L}_j^0$ as in (5.2). The vector fields

$$\overline{L}_j = \overline{L}_j^0 + \sum_k \gamma_j^k L_k^0, \quad j = 1, \dots, n+1$$

are independent and since $\langle \overline{L}_j^0, \eta^k \rangle = 0$ and $\langle L_j^0, \eta^k \rangle = \delta_j^k$, they annihilate each of the forms on the left hand side of (5.7). So they annihilate the forms $\eta_{1,0}^k$, which proves that the $\overline{L}_j$ form a frame of ${}^b\mathcal{T}^{0,1}\mathcal{M}$. □

6. Indicial complexes

Throughout this section we assume that $\mathcal{N}$ is a connected component of the boundary of a compact manifold $\mathcal{M}$. Let

$$\cdots \to C^\infty(\mathcal{M}; E^q) \xrightarrow{A_q} C^\infty(\mathcal{M}; E^{q+1}) \to \cdots \tag{6.1}$$

be a b-elliptic complex of operators $A_q \in \operatorname{Diff}_b^1(\mathcal{M}; E^q, E^{q+1})$; the E^q, $q = 0, \dots, r$, are vector bundles over $\mathcal{M}$.

Note that since A_q is a first order operator,

$$A_q(f\phi) = f A_q \phi - i\, {}^b\boldsymbol{\sigma}(A_q)({}^b df)(\phi). \tag{6.2}$$

This formula follows from the analogous formula for the standard principal symbol and the definition of principal b-symbol. It follows from (6.2) and (2.8) that A_q defines an operator

$$A_{b,q} : \operatorname{Diff}^1(\mathcal{N}; E_{\mathcal{N}}^q, E_{\mathcal{N}}^{q+1}).$$

Fix a smooth defining function $\mathfrak{r} : \mathcal{M} \to \mathbb{R}$ for $\partial\mathcal{M}$, $\mathfrak{r} > 0$ in the interior of $\mathcal{M}$, let

$$\mathcal{A}_q(\sigma) : \operatorname{Diff}_b^1(\mathcal{N}; E_{\mathcal{N}}^q, E_{\mathcal{N}}^{q+1}), \quad \sigma \in \mathbb{C}$$

denote the indicial family of A_q with respect to $\mathfrak{r}$, see (A.5). Using (6.2) and defining

$$\Lambda_{\mathfrak{r},q} = {}^b\boldsymbol{\sigma}(A_q)\Big(\frac{{}^b d\mathfrak{r}}{\mathfrak{r}}\Big),$$

the indicial family of A_q with respect to $\mathfrak{r}$ is

$$\mathcal{A}_q(\sigma) = A_{b,q} + \sigma \Lambda_{\mathfrak{r},q} : C^\infty(\mathcal{N}; E_{\mathcal{N}}^q) \to C^\infty(\mathcal{N}; E_{\mathcal{N}}^{q+1}). \tag{6.3}$$

Because of (A.4), these operators form an elliptic complex

$$\cdots \to C^\infty(\mathcal{N}; E_{\mathcal{N}}^q) \xrightarrow{\mathcal{A}_q(\sigma)} C^\infty(\mathcal{N}; E_{\mathcal{N}}^{q+1}) \to \cdots \tag{6.4}$$

for each σ and each connected component $\mathcal{N}$ of $\partial\mathcal{M}$. The operators depend on $\mathfrak{r}$, but the cohomology groups at a given σ for different defining functions $\mathfrak{r}$ are isomorphic. Indeed, if $\mathfrak{r}'$ is another defining function for $\partial\mathcal{M}$, then $\mathfrak{r}' = e^u \mathfrak{r}$ for some smooth real-valued function u, and a simple calculation gives

$$(A_{b,q} + \sigma \Lambda_{\mathfrak{r},q})(e^{i\sigma u}\phi) = e^{i\sigma u}(A_{b,q} + \sigma \Lambda_{\mathfrak{r}',q})\phi.$$

In analogy with the definition of boundary spectrum of an elliptic operator $A \in \operatorname{Diff}_b^m(\mathcal{M}; E, F)$, we have

DEFINITION 6.5. Let $\mathcal{N}$ be a connected component of $\partial\mathcal{M}$. The family of complexes (6.4), $\sigma \in \mathbb{C}$, is the indicial complex of (6.1) at $\mathcal{N}$. For each $\sigma \in \mathbb{C}$ let $H^q_{\mathcal{A}(\sigma)}(\mathcal{N})$ denote the q-th cohomology group of (6.4) on $\mathcal{N}$. The q-th boundary spectrum of the complex (6.1) at $\mathcal{N}$ is the set

$$\operatorname{spec}^q_{b,\mathcal{N}}(A) = \{\sigma \in \mathbb{C} : H^q_{\mathcal{A}(\sigma)}(\mathcal{N}) \neq 0\}.$$

The q-th boundary spectrum of A is $\operatorname{spec}^q_b(A) = \bigcup_{\mathcal{N}} \operatorname{spec}^q_{b,\mathcal{N}}(A)$.

The spaces $H^q_{\mathcal{A}(\sigma)}(\mathcal{N})$ are finite-dimensional because (6.4) is an elliptic complex and $\mathcal{N}$ is compact. It is convenient to isolate the behavior of the indicial complex according to the components of the boundary, since the sets $\operatorname{spec}^q_{b,\mathcal{N}}(A)$ can vary drastically from component to component.

Suppose that $\mathcal{M}$ is a complex b-manifold. Recall that since

$$\overline{{}^b\partial} \in \operatorname{Diff}^1_b(\mathcal{M}; {}^b\!\Lambda^{0,q}\mathcal{M}, {}^b\!\Lambda^{0,q+1}\mathcal{M}),$$

there are induced boundary operators

$$\overline{{}^b\partial}_b \in \operatorname{Diff}^1(\mathcal{N}; {}^b\!\Lambda^{0,q}_{\mathcal{N}}\mathcal{M}, {}^b\!\Lambda^{0,q+1}_{\mathcal{N}}\mathcal{M})$$

which via the isomorphism (4.1) become the operators of the $\overline{\mathbb{D}}$-complex (4.5). Combining (2.11) and (2.13) we get

$${}^b\boldsymbol{\sigma}(\overline{{}^b\partial})(\frac{{}^b d\mathfrak{r}}{\mathfrak{r}})(\phi) = i\frac{\overline{{}^b\partial}\mathfrak{r}}{\mathfrak{r}} \wedge \phi$$

and using (4.8) we may identify $\widehat{\overline{{}^b\partial}_b}(\sigma)$, given by (6.3), with the operator

$$\overline{\mathcal{D}}(\sigma)\phi = \overline{\mathbb{D}}\phi + i\sigma\beta_{\mathfrak{r}} \wedge \phi. \tag{6.6}$$

If $E \to \mathcal{M}$ is a holomorphic vector bundle, then the indicial family of

$$\overline{{}^b\partial} \in \operatorname{Diff}^1_b(\mathcal{M}; {}^b\!\Lambda^{0,q}\mathcal{M} \otimes E, {}^b\!\Lambda^{0,q+1}\mathcal{M} \otimes E)$$

is again given by (6.6), but using the operator $\overline{\mathbb{D}}$ of the complex (4.11).

Returning to the general complex (6.1), fix a smooth positive b-density $\mathfrak{m}$ on $\mathcal{M}$ and a Hermitian metric on each E^q. Let $\mathcal{A}^\star_q(\sigma)$ be the indicial operator of the formal adjoint, $A^\star_q$, of A_q. The Laplacian $\Box_q$ of the complex (6.1) in degree q belongs to $\operatorname{Diff}^2_b(\mathcal{M}; E^q)$, is b-elliptic, and its indicial operator is

$$\widehat{\Box}_q(\sigma) = \mathcal{A}^\star_q(\sigma)\mathcal{A}_q(\sigma) + \mathcal{A}_{q-1}(\sigma)\mathcal{A}^\star_{q-1}(\sigma).$$

The b-spectrum of $\Box_q$ at $\mathcal{N}$, see Melrose [**6**], is the set

$$\operatorname{spec}_{b,\mathcal{N}}(\Box_q) = \{\sigma \in \mathbb{C} : \widehat{\Box}_q(\sigma) : C^\infty(\mathcal{N}; E^q_{\mathcal{N}}) \to C^\infty(\mathcal{N}; E^q_{\mathcal{N}}) \text{ is not invertible}\}.$$

Note that unless σ is real, $\widehat{\Box}_q(\sigma)$ is not the Laplacian of the complex (6.4).

PROPOSITION 6.7. *For each q, $\operatorname{spec}^q_{b,\mathcal{N}}(A) \subset \operatorname{spec}_{b,\mathcal{N}}(\Box_q)$.*

Note that the set $\operatorname{spec}_{b,\mathcal{N}}(\Box_q)$ depends on the choice of Hermitian metrics and b-density used to construct the Laplacian, but that the subset $\operatorname{spec}^q_{b,\mathcal{N}}(A)$ is independent of such choices. For a general b-elliptic complex (6.1) it may occur that $\operatorname{spec}^q_{b,\mathcal{N}}(A) \neq \operatorname{spec}_{b,\mathcal{N}}(\Box_q)$. In Example 6.13 we show that $\operatorname{spec}^q_{b,\mathcal{N}}({}^b d) \subset \{0\}$. As is well known, $\operatorname{spec}_{b,\mathcal{N}}(\Delta_q)$ is an infinite set if $\dim \mathcal{M} > 1$. At the end of this section we will give an example where $\operatorname{spec}^0_{b,\mathcal{N}}(\overline{{}^b\partial})$ is an infinite set. A full discussion of $\operatorname{spec}^q_{b,\mathcal{N}}({}^b\overline{\partial})$ for any q and other aspects of the indicial complex of complex b-structures is given in Section 9.

PROOF OF PROPOSITION 6.7. Since $\Box_q$ is b-elliptic, the set $\operatorname{spec}_{b,\mathcal{N}}(\Box_q)$ is closed and discrete. Let $H^2(\mathcal{N}; E^q_{\mathcal{N}})$ be the L^2-based Sobolev space of order 2. For $\sigma \notin \operatorname{spec}_{b,\mathcal{N}}(\Box_q)$ let

$$\mathcal{G}_q(\sigma) : L^2(\mathcal{N}; E^q_{\mathcal{N}}) \to H^2(\mathcal{N}; E^q_{\mathcal{N}})$$

be the inverse of $\widehat{\Box}_q(\sigma)$. The map $\sigma \mapsto \mathcal{G}_q(\sigma)$ is meromorphic with poles in $\operatorname{spec}_b(\Box_q)$. Since

$$\mathcal{A}^\star_q(\sigma) = [\mathcal{A}_q(\overline{\sigma})]^\star$$

the operators $\widehat{\Box}_q(\sigma)$ are the Laplacians of the complex (6.4) when σ is real. Thus for $\sigma \in \mathbb{R}\backslash(\operatorname{spec}_{b,\mathcal{N}}(\Box_q) \cup \operatorname{spec}_{b,\mathcal{N}}(\Box_{q+1}))$ we have

$$\mathcal{A}_q(\sigma)\mathcal{G}_q(\sigma) = \mathcal{G}_{q+1}(\sigma)\mathcal{A}_q(\sigma), \quad \mathcal{A}_q(\sigma)^\star\mathcal{G}_{q+1}(\sigma) = \mathcal{G}_q(\sigma)\mathcal{A}^\star_q(\sigma)$$

by standard Hodge theory. Since all operators depend holomorphically on σ, the same equalities hold for $\sigma \in \mathfrak{R} = \mathbb{C}\backslash(\operatorname{spec}_{b,\mathcal{N}}(\Box_q) \cup \operatorname{spec}_{b,\mathcal{N}}(\Box_{q+1}))$. It follows that

$$\mathcal{A}^\star_q(\sigma)\mathcal{A}_q(\sigma)\mathcal{G}_q(\sigma) = \mathcal{G}_q(\sigma)\mathcal{A}^\star_q(\sigma)\mathcal{A}_q(\sigma)$$

in $\mathfrak{R}$. By analytic continuation the equality holds on all of $\mathbb{C}\backslash \operatorname{spec}_{b,\mathcal{N}}(\Box_q)$. Thus if $\sigma_0 \notin \operatorname{spec}_{b,\mathcal{N}}(\Box_q)$ and ϕ is a $\mathcal{A}_q(\sigma_0)$-closed section, $\mathcal{A}_q(\sigma_0)\phi = 0$, then the formula

$$\phi = [\mathcal{A}^\star_q(\sigma_0)\mathcal{A}_q(\sigma_0) + \mathcal{A}_{q-1}(\sigma_0)\mathcal{A}^\star_{q-1}(\sigma_0)]\mathcal{G}_q(\sigma_0)\phi$$

leads to

$$\phi = \mathcal{A}_{q-1}(\sigma_0)[\mathcal{A}^\star_{q-1}(\sigma_0)\mathcal{G}_q(\sigma_0)\phi].$$

Therefore $\sigma_0 \notin \operatorname{spec}^q_{b,\mathcal{N}}(A)$. □

Since $\Box_q$ is b-elliptic, the set $\operatorname{spec}_{b,\mathcal{N}}(\Box_q)$ is discrete and intersects each horizontal strip $a \le \Im\sigma \le b$ in a finite set (Melrose [**6**]). Consequently:

COROLLARY 6.8. *The sets* $\operatorname{spec}^q_{b,\mathcal{N}}(A)$, $q = 0, 1\dots,$ *are closed, discrete, and intersect each horizontal strip* $a \le \Im\sigma \le b$ *in a finite set.*

We note in passing that the Euler characteristic of the complex (6.4) vanishes for each σ. Indeed, let $\sigma_0 \in \mathbb{C}$. The Euler characteristic of the $\mathcal{A}(\sigma_0)$-complex is the index of

$$\mathcal{A}(\sigma_0) + \mathcal{A}(\sigma_0)^\star : \bigoplus_{q \text{ even}} C^\infty(\mathcal{N}; E^q) \to \bigoplus_{q \text{ odd}} C^\infty(\mathcal{N}; E^q).$$

The operator $\mathcal{A}_q(\sigma)$ is equal to $A_{b,q} + \sigma\Lambda_{\mathfrak{r},q}$, see (6.3). Thus $\mathcal{A}_q(\sigma)^\star = A^\star_{b,q} + \overline{\sigma}\Lambda^\star_{\mathfrak{r},q}$, and it follows that for any σ,

$$\mathcal{A}(\sigma) + \mathcal{A}(\sigma)^\star = \mathcal{A}(\sigma_0) + \mathcal{A}(\sigma_0)^\star + (\sigma - \sigma_0)\Lambda_{\mathfrak{r}} + (\overline{\sigma} - \overline{\sigma}_0)\Lambda^\star_{\mathfrak{r}}$$

is a compact perturbation of $\mathcal{A}(\sigma_0) + \mathcal{A}(\sigma_0)^\star$. Therefore, since the index is invariant under compact perturbations, the index of $\mathcal{A}(\sigma) + \mathcal{A}(\sigma)^\star$ is independent of σ. Then it vanishes, since it vanishes when $\sigma \notin \bigcup_q \operatorname{spec}^q_{b,\mathcal{N}}(A)$.

Let $\mathfrak{Mero}^q(\mathcal{N})$ be the sheaf of germs of $C^\infty(\mathcal{N}; E^q)$-valued meromorphic functions on $\mathbb{C}$ and let $\mathfrak{Hol}^q(\mathcal{N})$ be the subsheaf of germs of holomorphic functions. Let $\mathfrak{S}^q(\mathcal{N}) = \mathfrak{Mero}^q(\mathcal{N})/\mathfrak{Hol}^q(\mathcal{N})$. The holomorphic family $\sigma \mapsto \mathcal{A}_q(\sigma)$ gives a sheaf homomorphism $\mathcal{A}_q : \mathfrak{Mero}^q(\mathcal{N}) \to \mathfrak{Mero}^{q+1}(\mathcal{N})$ such that $\mathcal{A}_q(\mathfrak{Hol}^q(\mathcal{N})) \subset \mathfrak{Hol}^{q+1}(\mathcal{N})$ and $\mathcal{A}_{q+1} \circ \mathcal{A}_q = 0$, so we have a complex

$$\cdots \to \mathfrak{S}^q(\mathcal{N}) \xrightarrow{\mathcal{A}_q} \mathfrak{S}^{q+1}(\mathcal{N}) \to \cdots. \tag{6.9}$$

The cohomology sheafs $\mathfrak{H}^q_A(\mathcal{N})$ of this complex contain more refined information about the cohomology of the complex A.

PROPOSITION 6.10. *The sheaf* $\mathfrak{H}^q_A(\mathcal{N})$ *is supported on* $\operatorname{spec}^q_{b,\mathcal{N}}(A)$.

PROOF. Let $\sigma_0 \in \mathbb{C}$ be such that $H^q_{\mathcal{A}(\sigma_0)}(\mathcal{N}) = 0$ and let

$$\phi(\sigma) = \sum_{k=1}^{\mu} \frac{\phi_k}{(\sigma - \sigma_0)^k}, \tag{6.11}$$

$\mu > 0$, $\phi_k \in C^\infty(\mathcal{N}; \bigwedge^q \overline{\mathcal{V}}^*)$, represent the $\mathcal{A}$-closed element $[\phi]$ of the stalk of $\mathfrak{S}^q(\mathcal{N})$ over σ_0. The condition that $\mathcal{A}_q[\phi] = 0$ means that $\mathcal{A}_q(\sigma)\phi(\sigma)$ is holomorphic, that is,

$$\frac{\mathcal{A}_q(\sigma_0)\phi_\mu}{(\sigma - \sigma_0)^\mu} + \sum_{k=1}^{\mu-1} \frac{\mathcal{A}_q(\sigma_0)\phi_k + \Lambda_{\mathfrak{r},q}\phi_{k+1}}{(\sigma - \sigma_0)^k} = 0.$$

In particular $\mathcal{A}_q(\sigma_0)\phi_\mu = 0$. Since $H^q_{\mathcal{A}(\sigma_0)}(\mathcal{N}) = 0$, there is $\psi_\mu \in C^\infty(\mathcal{N}; E^{q-1})$ such that $\mathcal{A}_{q-1}(\sigma_0)\psi_\mu = \phi_\mu$. This shows that if $\mu = 1$, then $[\phi]$ is exact, and that if $\mu > 1$, then letting $\phi'(\sigma) = \phi(\sigma) - \mathcal{A}_{q-1}(\sigma)\psi_\mu/(\sigma - \sigma_0)^\mu$, that ϕ is cohomologous to an element $[\phi']$ represented by a sum as in (6.11) with $\mu - 1$ instead of μ. By induction, $[\phi]$ is exact. □

DEFINITION 6.12. The cohomology sheafs $\mathfrak{H}^q_A(\mathcal{N})$ of the complex (6.9) will be referred to as the indicial cohomology sheafs of the complex A. If $[\phi] \in \mathfrak{h}^q_A(\mathcal{N})$ is a nonzero element of the stalk over σ_0, the smallest μ such that there is a meromorphic function (6.11) representing $[\phi]$ will be called the order of the pole of $[\phi]$.

The relevancy of this notion of pole lies in that it predicts, for any given cohomology class of the complex A, the existence of a representative with the most regular leading term (the smallest power of log that must appear in the expansion at the boundary). We will see later (Proposition 9.5) that for the b-Dolbeault complex, under a certain geometric assumption, the order of the pole of $[\phi] \in \mathfrak{H}^q_{\overline{b\partial}}(\mathcal{N})\backslash 0$ is 1.

EXAMPLE 6.13. For the b-de Rham complex one has $\operatorname{spec}^q_{b,\mathcal{N}}({}^b d) \subset \{0\}$ and

$$H^q_{\mathcal{D}(0)}(\mathcal{N}) = H^q_{\mathrm{dR}}(\mathcal{N}) \oplus H^{q-1}_{\mathrm{dR}}(\mathcal{N})$$

for each component $\mathcal{N}$ of $\partial\mathcal{M}$, and that every element of the stalk of $\mathfrak{H}^q_{bd}(\mathcal{N})$ over 0 has a representative with a simple pole. By way of the residue we get an isomorphism from the stalk over 0 onto $H^q_{\mathrm{dR}}(\mathcal{N})$.

Since the map (2.4) is surjective with kernel spanned by $\mathfrak{r}\partial_{\mathfrak{r}}$, the dual map

$$\mathrm{ev}^*_{\mathcal{N}} : T^*\mathcal{N} \to {}^bT^*_{\mathcal{N}}\mathcal{M} \tag{6.14}$$

is injective with image the annihilator, $\mathcal{H}$, of $\mathfrak{r}\partial\mathfrak{r}$. Let $\mathbf{i}_{\mathfrak{r}\partial_{\mathfrak{r}}} : {}^b\bigwedge^q_{\mathcal{N}}\mathcal{M} \to {}^b\bigwedge^{q-1}_{\mathcal{N}}\mathcal{M}$ denote interior multiplication by $\mathfrak{r}\partial_{\mathfrak{r}}$ Then $\bigwedge^q\mathcal{H} = \ker(\mathbf{i}_{\mathfrak{r}\partial_{\mathfrak{r}}} : {}^b\bigwedge^q_{\mathcal{N}}\mathcal{M} \to {}^b\bigwedge^{q-1}_{\mathcal{N}}\mathcal{M})$. The isomorphism (6.14) gives isomorphisms

$$\mathrm{ev}^*_{\mathcal{N}} : \bigwedge^q\mathcal{N} \to \mathcal{H}^q$$

for each q. Fix a defining function $\mathfrak{r}$ for $\mathcal{N}$ and let $\Pi : {}^b\!\bigwedge^q_{\mathcal{N}}\mathcal{M} \to {}^b\!\bigwedge^q_{\mathcal{N}}\mathcal{M}$ be the projection on $\mathcal{H}^q$ according to the decomposition

$$ {}^b\!\bigwedge^q_{\mathcal{N}}\mathcal{M} = \mathcal{H}^q \oplus \frac{{}^b d\mathfrak{r}}{\mathfrak{r}} \wedge \mathcal{H}^{q-1}, $$

that is,

$$ \Pi\phi = \phi - \frac{{}^b d\mathfrak{r}}{\mathfrak{r}} \wedge \mathbf{i}_{\mathfrak{r}\partial_\mathfrak{r}}\phi. $$

If $\phi^0 \in C^\infty(\mathcal{N}, \mathcal{H}^q)$ and $\phi^1 \in C^\infty(\mathcal{N}, \mathcal{H}^{q-1})$, then

$$ {}^b d(\phi^0 + \frac{{}^b d\mathfrak{r}}{\mathfrak{r}} \wedge \phi^1) = \Pi\, {}^b d\phi^0 + \frac{{}^b d\mathfrak{r}}{\mathfrak{r}} \wedge (-\Pi\, {}^b d\phi^1). $$

Since

$$ \mathfrak{r}^{-i\sigma}\, {}^b d\mathfrak{r}^{i\sigma}\phi = {}^b d\phi + i\sigma \frac{{}^b d\mathfrak{r}}{\mathfrak{r}} \wedge \phi, $$

the indicial operator $\mathcal{D}(\sigma)$ of ${}^b d$ is

$$ \mathcal{D}(\sigma)(\phi_0 + \frac{{}^b d\mathfrak{r}}{\mathfrak{r}} \wedge \phi^1) = \Pi\, {}^b d\phi^0 + \frac{{}^b d\mathfrak{r}}{\mathfrak{r}} \wedge (i\sigma\phi^0 - \Pi\, {}^b d\phi^1). $$

If $\mathcal{D}(\sigma)(\phi_0 + \frac{{}^b d\mathfrak{r}}{\mathfrak{r}} \wedge \phi^1) = 0$, then of course $\Pi^b d\phi^0 = 0$ and $i\sigma\phi^0 = \Pi^b d\phi^1$, and it follows that if $\sigma \neq 0$, then

$$ (\phi_0 + \frac{{}^b d\mathfrak{r}}{\mathfrak{r}} \wedge \phi^1) = \mathcal{D}(\sigma)\frac{1}{i\sigma}\phi^1. $$

Thus all cohomology groups of the complex $\mathcal{D}(\sigma)$ vanish if $\sigma \neq 0$, i.e., $\operatorname{spec}^q_{b,\mathcal{N}}({}^b d) \subset \{0\}$.

It is not hard to verify that

$$ \Pi^b d \operatorname{ev}^*_{\mathcal{N}} = \operatorname{ev}^*_{\mathcal{N}} d. $$

Since

$$ \mathfrak{r}^{-i\sigma}\, {}^b d\mathfrak{r}^{i\sigma}\phi = {}^b d\phi + i\sigma \frac{{}^b d\mathfrak{r}}{\mathfrak{r}} \wedge \phi, $$

the indicial operator of ${}^b d$ at $\sigma = 0$ can be viewed as the operator

$$ \begin{bmatrix} d & 0 \\ 0 & -d \end{bmatrix} : \begin{matrix} \bigwedge^q \mathcal{N} \\ \oplus \\ \bigwedge^{q-1}\mathcal{N} \end{matrix} \to \begin{matrix} \bigwedge^q \mathcal{N} \\ \oplus \\ \bigwedge^{q-1}\mathcal{N} \end{matrix}. $$

From this we get the cohomology groups of $\mathcal{D}(0)$ in terms of the de Rham cohomology of $\mathcal{N}$:

$$ H^q_{\mathcal{D}(0)}(\mathcal{N}) = H^q_{\mathrm{dR}}(\mathcal{N}) \oplus H^{q-1}_{\mathrm{dR}}(\mathcal{N}). $$

Thus the groups $H^q_{\mathcal{D}(0)}(\mathcal{N})$ do not vanish for $q = 0$, 1, $\dim\mathcal{M} - 1$, $\dim\mathcal{M}$ but may vanish for other values of q.

We now show that every element of the stalk of $\mathfrak{H}^q_{{}^b d}(\mathcal{N})$ over 0 has a representative with a simple pole at 0. Suppose that

$$ \phi(\sigma) = \sum_{k=1}^{\mu} \frac{1}{\sigma^k}\left(\phi^0_k + \frac{{}^b d\mathfrak{r}}{\mathfrak{r}} \wedge \phi^1_k\right) \tag{6.15} $$

is such that $\mathcal{D}(\sigma)\phi(\sigma)$ is holomorphic. Then

$$\sum_{k=1}^{\mu} \frac{1}{\sigma^k}\left(d\phi_k^0 - \frac{{}^b d\mathfrak{r}}{\mathfrak{r}} \wedge d\phi_k^1\right) + \frac{{}^b d\mathfrak{r}}{\mathfrak{r}} \wedge \left(\sum_{k=1}^{\mu-1} \frac{i}{\sigma^k}\phi_{k+1}^0\right) = 0,$$

hence $d\phi_1^0 = 0$, $d\phi_\mu^1 = 0$ and $\phi_k^0 = -id\phi_{k-1}^1$, $k = 2, \dots, \mu$. Let

$$\psi(\sigma) = -i\sum_{k=2}^{\mu+1} \frac{1}{\sigma^k}\phi_{k-1}^1.$$

Then

$$\begin{aligned}\mathcal{D}(\sigma)\psi(\sigma) &= -i\sum_{k=2}^{\mu+1} \frac{1}{\sigma^k} d\phi_{k-1}^1 + \frac{{}^b d\mathfrak{r}}{\mathfrak{r}} \wedge \sum_{k=2}^{\mu+1} \frac{1}{\sigma^{k-1}}\phi_{k-1}^1 \\ &= \sum_{k=2}^{\mu} \frac{1}{\sigma^k}\phi_k^0 + \frac{{}^b d\mathfrak{r}}{\mathfrak{r}} \wedge \sum_{k=1}^{\mu} \frac{1}{\sigma^k}\phi_k^1\end{aligned}$$

so

$$\phi(\sigma) - \mathcal{D}(\sigma)\psi(\sigma) = \frac{1}{\sigma}\phi_1^0.$$

The map that sends the class of the $\mathcal{D}(\sigma)$-closed element (6.15) to the class of ϕ_1^0 in $H^q_{\mathrm{dR}}(\mathcal{N})$ is an isomorphism.

EXAMPLE 6.16. As we just saw, the boundary spectrum of the ${}^b d$ complex in degree 0 is just $\{0\}$. In contrast, $\mathrm{spec}^0_{b,\mathcal{N}}({}^b\overline{\partial})$ may be an infinite set. We illustrate this in the context of Example 4.10. The functions

$$z^\alpha = (z^1)^{\alpha_1} \cdots (z^{n+1})^{\alpha_{n+1}},$$

where the α_j are nonnegative integers, are CR functions that satisfy

$$\mathcal{T} z^\alpha = i(\sum \tau_j \alpha_j) z^\alpha.$$

This implies that

$$\overline{\mathbb{D}} z^\alpha + i(-i\sum \tau_j\alpha_j)\beta z^\alpha = 0$$

with β as in Example 4.10, so the numbers $\sigma_\alpha = (-i\sum \tau_j\alpha_j)$ belong to $\mathrm{spec}^0_{b,\mathcal{N}}({}^b\overline{\partial})$.

For the sake of completeness we also show that if $\sigma \in \mathrm{spec}^0_{b,\mathcal{N}}({}^b\overline{\partial})$, then $\sigma = \sigma_\alpha$ for some α as above. To see this, suppose that $\zeta : S^{2n+1} \to \mathbb{C}$ is not identically zero and satisfies

$$\overline{\mathbb{D}}\zeta + i\sigma\zeta\beta = 0$$

for some $\sigma \neq 0$. Then ζ is smooth, because the principal symbol of $\overline{\mathbb{D}}$ on functions is injective. Since $\langle \beta, \mathcal{T}\rangle = -i$,

$$T\zeta + \sigma\zeta = 0.$$

Thus $\zeta(\mathfrak{a}_t(p)) = e^{-\sigma t}\zeta(p)$ for any p. Since $|\zeta(\mathfrak{a}_t(p))|$ is bounded as a function of t and ζ is not identically 0, σ must be purely imaginary. Since ζ is a CR function, it extends uniquely to a holomorphic function $\tilde{\zeta}$ on $B = \{z \in \mathbb{C}^{n+1} : \|z\| < 1\}$, necessarily smooth up to the boundary. Let $\zeta_t = \zeta \circ \mathfrak{a}_t$. This is also a smooth CR function, so it has a unique holomorphic extension $\tilde{\zeta}_t$ to B. The integral curve through $z_0 = (z_0^1, \dots, z_0^{n+1})$ of the vector field $\mathcal{T}$ is

$$t \mapsto \mathfrak{a}_t(z_0) = (e^{i\tau_1 t} z_0^1, \dots, e^{i\tau_{n+1} t} z_0^{n+1})$$

Extending the definition of $\mathfrak{a}_t$ to allow arbitrary $z \in \mathbb{C}^{n+1}$ as argument we then have that $\tilde{\zeta}_t = \tilde{\zeta} \circ \mathfrak{a}_t$. Then

$$\partial_t \tilde{\zeta}_t + \sigma \tilde{\zeta}_t = 0$$

gives

$$\tilde{\zeta}(z) = \sum_{\{\alpha : \boldsymbol{\tau} \cdot \alpha = i\sigma\}} c_\alpha z^\alpha$$

for $|z| < 1$, where $\boldsymbol{\tau} = (\tau_1, \ldots, \tau_{n+1})$. Thus $\sigma = -i \sum \tau_j \alpha_j$ as claimed. Note that $\Im\sigma$ is negative (positive) if the τ_j are positive (negative) and $\alpha \neq 0$.

7. Underlying CR complexes

Again let $\mathfrak{a} : \mathbb{R} \times \mathcal{N} \to \mathcal{N}$ be the flow of $\mathcal{T}$. Let $\mathcal{L}_\mathcal{T}$ denote the Lie derivative with respect to $\mathcal{T}$ on de Rham q-forms or vector fields and let $\mathbf{i}_\mathcal{T}$ denote interior multiplication by $\mathcal{T}$ of de Rham q-forms or of elements of $\bigwedge^q \overline{\mathcal{V}}^*$.

The proofs of the following two lemmas are elementary.

LEMMA 7.1. *If α is a smooth section of the annihilator of $\overline{\mathcal{V}}$ in $\mathbb{C}T^*\mathcal{N}$, then $(\mathcal{L}_\mathcal{T}\alpha)|_{\overline{\mathcal{V}}} = 0$. Consequently, for each $p \in \mathcal{N}$ and $t \in \mathbb{R}$, $d\mathfrak{a}_t : \mathbb{C}T_p\mathcal{N} \to \mathbb{C}T_{\mathfrak{a}_t(p)}\mathcal{N}$ maps $\overline{\mathcal{V}}_p$ onto $\overline{\mathcal{V}}_{\mathfrak{a}_t(p)}$.*

It follows that there is a well defined smooth bundle homomorphism $\mathfrak{a}_t^* : \bigwedge^q \overline{\mathcal{V}}^* \to \bigwedge^q \overline{\mathcal{V}}^*$ covering $\mathfrak{a}_{-t}$. In particular, one can define the Lie derivative $\mathcal{L}_\mathcal{T}\phi$ with respect to $\mathcal{T}$ of an element in $\phi \in C^\infty(\mathcal{N}; \bigwedge^q \overline{\mathcal{V}}^*)$. The usual formula holds:

LEMMA 7.2. *If $\phi \in C^\infty(\mathcal{N}; \bigwedge^q \overline{\mathcal{V}}^*)$, then $\mathcal{L}_\mathcal{T}\phi = \mathbf{i}_\mathcal{T}\overline{\mathbb{D}}\phi + \overline{\mathbb{D}}\mathbf{i}_\mathcal{T}\phi$. Consequently, for each t and $\phi \in C^\infty(\mathcal{N}; \bigwedge^q \overline{\mathcal{V}}^*)$, $\overline{\mathbb{D}}\mathfrak{a}_t^*\phi = \mathfrak{a}_t^*\overline{\mathbb{D}}\phi$.*

For any defining function $\mathfrak{r}$ of $\mathcal{N}$ in $\mathcal{M}$, $\overline{\mathcal{K}}_\mathfrak{r} = \ker \beta_\mathfrak{r}$ is a CR structure of CR codimension 1: indeed, $\mathcal{K}_\mathfrak{r} \cap \overline{\mathcal{K}}_\mathfrak{r} \subset \operatorname{span}_\mathbb{C} \mathcal{T}$ but since $\langle \beta_\mathfrak{r}, \mathcal{T} \rangle$ vanishes nowhere, we must have $\overline{\mathcal{K}} \cap \mathcal{K} = 0$. Since $\mathcal{K} \oplus \overline{\mathcal{K}} \oplus \operatorname{span}_\mathbb{C} \mathcal{T} = \mathbb{C}T\mathcal{N}$, the CR codimension is 1. Finally, if $V, W \in C^\infty(\mathcal{N}; \overline{\mathcal{K}}_\mathfrak{r})$, then

$$\langle \beta_\mathfrak{r}, [V, W] \rangle = V \langle \beta_\mathfrak{r}, W \rangle - W \langle \beta_\mathfrak{r}, V \rangle - 2\overline{\mathbb{D}}\beta(V, W),$$

Since the right hand side vanishes, $[V, W]$ is again a section of $\overline{\mathcal{K}}_\mathfrak{r}$.

Since $\overline{\mathcal{V}} = \overline{\mathcal{K}}_\mathfrak{r} \oplus \operatorname{span}_\mathbb{C} \mathcal{T}$, the dual of $\overline{\mathcal{K}}_\mathfrak{r}$ is canonically isomorphic to the kernel of $\mathbf{i}_\mathcal{T} : \overline{\mathcal{V}}^* \to \mathbb{C}$. We will write $\overline{\mathcal{K}}^*$ for this kernel. More generally, $\bigwedge^q \overline{\mathcal{K}}_\mathfrak{r}^*$ and the kernel, $\bigwedge^q \overline{\mathcal{K}}^*$, of $\mathbf{i}_\mathcal{T} : \bigwedge^q \overline{\mathcal{V}}^* \to \bigwedge^{q-1} \overline{\mathcal{V}}^*$ are canonically isomorphic. The vector bundles $\bigwedge^q \overline{\mathcal{K}}^*$ are independent of the defining function $\mathfrak{r}$. We regard the $\overline{\partial}_b$-operators of the CR structure as operators

$$C^\infty(\mathcal{N}; \textstyle\bigwedge^q \overline{\mathcal{K}}^*) \to C^\infty(\mathcal{N}; \textstyle\bigwedge^{q+1} \overline{\mathcal{K}}^*).$$

They do depend on $\mathfrak{r}$ but we will not indicate this in the notation.

To get a formula for $\overline{\partial}_b$, let

$$\tilde{\beta}_\mathfrak{r} = \frac{i}{i - a_\mathfrak{r}} \beta_\mathfrak{r}$$

(so that $\langle i\tilde{\beta}_\mathfrak{r}, \mathcal{T} \rangle = 1$). The projection $\Pi_\mathfrak{r} : \bigwedge^q \overline{\mathcal{V}}^* \to \bigwedge^q \overline{\mathcal{V}}^*$ on $\bigwedge^q \overline{\mathcal{K}}^*$ according to the decomposition

$$\textstyle\bigwedge^q \overline{\mathcal{V}}^* = \bigwedge^q \overline{\mathcal{K}}^* \oplus i\tilde{\beta}_\mathfrak{r} \wedge \bigwedge^{q-1} \overline{\mathcal{K}}^* \tag{7.3}$$

is

$$\Pi_{\mathfrak{r}}\phi = \phi - i\tilde{\beta}_{\mathfrak{r}} \wedge \mathbf{i}_{\mathcal{T}}\phi. \tag{7.4}$$

LEMMA 7.5. *With the identification of $\bigwedge^q \overline{\mathcal{K}}_{\mathfrak{r}}^*$ with $\bigwedge^q \overline{\mathcal{K}}^*$ described above, the $\overline{\partial}_b$-operators of the CR structure $\overline{\mathcal{K}}_{\mathfrak{r}}$ are given by*

$$\overline{\partial}_b\phi = \Pi_{\mathfrak{r}}\overline{\mathbb{D}}\phi \quad \text{if } \phi \in C^\infty(\mathcal{N}, \textstyle\bigwedge^q \overline{\mathcal{K}}^*), \tag{7.6}$$

PROOF. Suppose that (z,t) is a hypoanalytic chart for $\overline{\mathcal{V}}$ on some open set U, with $\mathcal{T}t = 1$. So $\partial_{\overline{z}^\mu}$, $\mu = 1\dots, n$, $\mathcal{T} = \partial_t$ is a frame for $\overline{\mathcal{V}}$ over U with dual frame $\overline{\mathbb{D}}\overline{z}^\mu$, $\overline{\mathbb{D}}t$. If

$$\beta_{\mathfrak{r}} = \sum_{\mu=1}^{n} \beta_\mu \overline{\mathbb{D}}\overline{z}^\mu + \beta_0\overline{\mathbb{D}}t.$$

then

$$\overline{L}_\mu = \partial_{\overline{z}^\mu} - \frac{\beta_\mu}{\beta_0}\partial_t, \quad \mu = 1, \dots, n$$

is a frame for $\overline{\mathcal{K}}_{\mathfrak{r}}$ over U. Let $\overline{\eta}^\mu$ denote the dual frame (for $\overline{\mathcal{K}}_{\mathfrak{r}}^*$). Since the $\overline{L}_\mu$ commute, $\overline{\partial}_b\overline{\eta}^\mu = 0$, so if $\phi = \sum'_{|I|=q} \phi_I\,\overline{\eta}^I$, then[1]

$$\overline{\partial}_b\phi = \sum_{|J|=q+1}' \sum_{|I|=q}' \sum_\mu \epsilon_J^{\mu I}\overline{L}_\mu\phi_I\,\overline{\eta}^J.$$

On the other hand, the frame of $\overline{\mathcal{V}}^*$ dual to the frame $\overline{L}_\mu$, $\mu = 1, \dots, n$, $\mathcal{T}$ of $\overline{\mathcal{V}}$ is $\overline{\mathbb{D}}\overline{z}^\mu$, $i\tilde{\beta}_{\mathfrak{r}}$, and the identification of $\overline{\mathcal{K}}_{\mathfrak{r}}^*$ with $\overline{\mathcal{K}}^*$ maps the η^μ to the $\overline{\mathbb{D}}\overline{z}^\mu$. So, as a section of $\bigwedge^q\overline{\mathcal{V}}^*$,

$$\phi = \sum_{|I|=q}' \phi_I\,\overline{\mathbb{D}}\overline{z}^I$$

and

$$\overline{\mathbb{D}}\phi = \sum_{|J|=q+1}' \sum_{|I|=q}' \epsilon_J^{\mu I}\overline{L}_\mu\phi_I\,\overline{\mathbb{D}}\overline{z}^J + i\tilde{\beta}_{\mathfrak{r}} \wedge \sum_{|I|=q}' \mathcal{T}\phi_I\,\overline{\mathbb{D}}\overline{z}^I.$$

Thus $\Pi_{\mathfrak{r}}\overline{\mathbb{D}}\phi$ is the section of $\bigwedge^{q+1}\overline{\mathcal{K}}^*$ associated with $\overline{\partial}_b\phi$ by the identifying map. □

Using (7.4) in (7.6) and the fact that $\mathbf{i}_{\mathcal{T}}\overline{\mathbb{D}}\phi = \mathcal{L}_{\mathcal{T}}\phi$ if $\phi \in C^\infty(\mathcal{N}; \bigwedge^q\overline{\mathcal{K}}^*)$ we get

$$\overline{\partial}_b\phi = \overline{\mathbb{D}}\phi - i\tilde{\beta}_{\mathfrak{r}} \wedge \mathcal{L}_{\mathcal{T}}\phi \quad \text{if } \phi \in C^\infty(\mathcal{N}, \textstyle\bigwedge^q\overline{\mathcal{K}}^*). \tag{7.7}$$

The $\overline{\mathbb{D}}$ operators can be expressed in terms of the $\overline{\partial}_b$ operators. Suppose $\phi \in C^\infty(\mathcal{N}; \bigwedge^q\overline{\mathcal{V}}^*)$. Then $\phi = \phi^0 + i\tilde{\beta}_{\mathfrak{r}} \wedge \phi^1$ with unique $\phi^0 \in C^\infty(\mathcal{N}; \bigwedge^q\overline{\mathcal{K}}^*)$ and $\phi^1 \in C^\infty(\mathcal{N}; \bigwedge^{q-1}\overline{\mathcal{K}}^*)$, and

$$\overline{\mathbb{D}}\phi^0 = \overline{\partial}_b\phi^0 + i\tilde{\beta}_{\mathfrak{r}} \wedge \mathcal{L}_{\mathcal{T}}\phi^0,$$

see (7.7). Using

$$\overline{\mathbb{D}}\tilde{\beta}_{\mathfrak{r}} = \frac{\overline{\mathbb{D}}a_{\mathfrak{r}}}{i - a_{\mathfrak{r}}} \wedge \tilde{\beta}_{\mathfrak{r}}$$

[1]The notation is as in, for instance, Folland and Kohn [**2**]: $I = (i_1, \dots, i_q)$ is a tuple of length $|I| = q$ of integers between 1 and n, $\overline{\eta}^I$ means $\overline{\eta}_1^{i_1} \wedge \cdots \wedge \overline{\eta}^{i_q}$, the prime in the summation sign indicates that the sum is taken over strictly increasing indices, μI is the index $(\mu, i_1, \dots, i_q)$, and $\epsilon_K^{\mu I}$ is the sign of the permutation taking μI to K if as sets these indices are equal and have q elements (i.e., no repetition), otherwise it is 0.

and (7.7) again we get

$$\overline{\mathbb{D}}(i\tilde{\beta}_{\mathfrak{r}} \wedge \phi^1) = i\tilde{\beta}_{\mathfrak{r}} \wedge \big(- \frac{\overline{\mathbb{D}} a_{\mathfrak{r}}}{i - a_{\mathfrak{r}}} \wedge \phi^1 - \overline{\mathbb{D}}\phi^1 \big) = i\tilde{\beta}_{\mathfrak{r}} \wedge \big(- \frac{\overline{\partial}_b a_{\mathfrak{r}}}{i - a_{\mathfrak{r}}} \wedge \phi^1 - \overline{\partial}_b \phi^1 \big).$$

This gives

$$(7.8) \qquad \overline{\mathbb{D}} = \begin{bmatrix} \overline{\partial}_b & 0 \\ \mathcal{L}_{\mathcal{T}} & -\overline{\partial}_b - \dfrac{\overline{\partial}_b a_{\mathfrak{r}}}{i - a_{\mathfrak{r}}} \end{bmatrix} : \begin{matrix} C^\infty(\mathcal{N}; \bigwedge^q \overline{\mathcal{K}}^*) \\ \oplus \\ C^\infty(\mathcal{N}; \bigwedge^{q-1} \overline{\mathcal{K}}^*) \end{matrix} \to \begin{matrix} C^\infty(\mathcal{N}; \bigwedge^{q+1} \overline{\mathcal{K}}^*) \\ \oplus \\ C^\infty(\mathcal{N}; \bigwedge^q \overline{\mathcal{K}}^*) \end{matrix}.$$

Since $\mathcal{T}$ itself is $\mathcal{T}$-invariant, $\mathbf{i}_{\mathcal{T}} \mathfrak{a}_t^* = \mathfrak{a}_t^* \mathbf{i}_{\mathcal{T}}$: the subbundle $\overline{\mathcal{K}}^*$ of $\overline{\mathcal{V}}^*$ is invariant under $\mathfrak{a}_t^*$ for each t. This need not be true of $\overline{\mathcal{K}}_{\mathfrak{r}}$, i.e., the statement that for all t, $d\mathfrak{a}_t(\overline{\mathcal{K}}_{\mathfrak{r}}) \subset \overline{\mathcal{K}}_{\mathfrak{r}}$, equivalently,

$$L \in C^\infty(\mathcal{M}; \overline{\mathcal{K}}_{\mathfrak{r}}) \implies [\mathcal{T}, L] \in C^\infty(\mathcal{M}; \overline{\mathcal{K}}_{\mathfrak{r}}),$$

may fail to hold. Since $\overline{\mathbb{D}}\beta_{\mathfrak{r}} = 0$, the formula

$$0 = \mathcal{T}\langle \beta_{\mathfrak{r}}, L\rangle - L\langle \beta_{\mathfrak{r}}, \mathcal{T}\rangle - \langle \beta_{\mathfrak{r}}, [\mathcal{T}, L]\rangle$$

with $L \in C^\infty(\mathcal{N}; \overline{\mathcal{K}}_{\mathfrak{r}})$ gives that $\overline{\mathcal{K}}_{\mathfrak{r}}$ is invariant under $d\mathfrak{a}_t$ if and only if $La_{\mathfrak{r}} = 0$ for each CR vector field, that is, if and only if $a_{\mathfrak{r}}$ is a CR function. This proves the equivalence between the first and last statements in the following lemma. The third statement is the most useful.

LEMMA 7.9. *Let $\mathfrak{r}$ be a defining function for $\mathcal{N}$ in $\mathcal{M}$ and let $\overline{\partial}_b$ denote the operators of the associated CR complex. The following are equivalent:*

(1) The function $a_{\mathfrak{r}}$ is CR;
(2) $\mathcal{L}_{\mathcal{T}} \tilde{\beta}_{\mathfrak{r}} = 0$;
(3) $\mathcal{L}_{\mathcal{T}} \overline{\partial}_b - \overline{\partial}_b \mathcal{L}_{\mathcal{T}} = 0$;
(4) $\overline{\mathcal{K}}_{\mathfrak{r}}$ is $\mathcal{T}$-invariant.

PROOF. From $\beta_{\mathfrak{r}} = (a_{\mathfrak{r}} - i)i\tilde{\beta}_{\mathfrak{r}}$ and $\mathcal{L}_{\mathcal{T}}\beta_{\mathfrak{r}} = \overline{\mathbb{D}} a_{\mathfrak{r}}$ we obtain

$$\overline{\mathbb{D}} a_{\mathfrak{r}} = (\mathcal{L}_{\mathcal{T}} a_{\mathfrak{r}}) i\tilde{\beta}_{\mathfrak{r}} + (a_{\mathfrak{r}} - i) i \mathcal{L}_{\mathcal{T}} \tilde{\beta}_{\mathfrak{r}},$$

so

$$\overline{\partial}_b a_{\mathfrak{r}} = \overline{\mathbb{D}} a_{\mathfrak{r}} - (\mathcal{L}_{\mathcal{T}} a_{\mathfrak{r}}) i\tilde{\beta}_{\mathfrak{r}} = (a_{\mathfrak{r}} - i) i \mathcal{L}_{\mathcal{T}} \tilde{\beta}_{\mathfrak{r}}.$$

Thus $a_{\mathfrak{r}}$ is CR if and only if $\mathcal{L}_{\mathcal{T}} \tilde{\beta}_{\mathfrak{r}} = 0$.

Using $\mathcal{L}_{\mathcal{T}} \overline{\mathbb{D}} = \overline{\mathbb{D}} \mathcal{L}_{\mathcal{T}}$ and the definition of $\overline{\partial}_b$ we get

$$\mathcal{L}_{\mathcal{T}} \overline{\partial}_b \phi = \mathcal{L}_{\mathcal{T}}(\overline{\mathbb{D}}\phi - i\tilde{\beta}_{\mathfrak{r}} \wedge \mathcal{L}_{\mathcal{T}}\phi) = \overline{\partial}_b \mathcal{L}_{\mathcal{T}}\phi - i(\mathcal{L}_{\mathcal{T}} \tilde{\beta}_{\mathfrak{r}}) \wedge \mathcal{L}_{\mathcal{T}}\phi$$

for $\phi \in C^\infty(\mathcal{N}; \bigwedge^q \overline{\mathcal{K}}^*)$. Thus $\mathcal{L}_{\mathcal{T}} \overline{\partial}_b - \overline{\partial}_b \mathcal{L}_{\mathcal{T}} = 0$ if and only if $\mathcal{L}_{\mathcal{T}} \tilde{\beta}_{\mathfrak{r}} = 0$. □

LEMMA 7.10. *Suppose that $\overline{\mathcal{V}}$ admits a $\mathcal{T}$-invariant metric. Then there is a defining function $\mathfrak{r}$ for $\mathcal{N}$ in $\mathcal{M}$ such that $a_{\mathfrak{r}}$ is constant. If $\mathfrak{r}$ and $\mathfrak{r}'$ are defining functions such that $a_{\mathfrak{r}}$ and $a_{\mathfrak{r}'}$ are constant, then $a_{\mathfrak{r}} = a_{\mathfrak{r}'}$. This constant will be denoted $\mathfrak{a}_{\mathrm{av}}$.*

PROOF. Let h be a metric as stated. Let $\mathcal{H}^{0,1}$ be the subbundle of $\overline{\mathcal{V}}$ orthogonal to $\mathcal{T}$. This is $\mathcal{T}$-invariant, and since the metric is $\mathcal{T}$-invariant, $\mathcal{H}^{0,1}$ has a $\mathcal{T}$-invariant metric. This metric gives canonically a metric on $\mathcal{H}^{1,0} = \overline{\mathcal{H}^{0,1}}$. Using the decomposition $\mathbb{C}T\mathcal{N} = \mathcal{H}^{1,0} \oplus \mathcal{H}^{0,1} \oplus \operatorname{span}_{\mathbb{C}} \mathcal{T}$ we get a $\mathcal{T}$-invariant metric on $\mathbb{C}T\mathcal{N}$ for which the decomposition is orthogonal. This metric is induced by a Riemannian metric g. Let $\mathfrak{m}_0$ be the corresponding Riemannian density, which is $\mathcal{T}$-invariant because g is. Since $\overline{\mathbb{D}}$, h, and $\mathfrak{m}_0$ are $\mathcal{T}$-invariant, so are the formal adjoint $\overline{\mathbb{D}}^\star$ of $\overline{\mathbb{D}}$ and the Laplacians of the $\overline{\mathbb{D}}$-complex, and if G denotes the Green's operators for these Laplacians, then G is also $\mathcal{T}$-invariant, as is the orthogonal projection Π on the space of $\overline{\mathbb{D}}$-harmonic forms. Arbitrarily pick a defining function $\mathfrak{r}$ for $\mathcal{N}$ in $\mathcal{M}$. Then

$$a_{\mathfrak{r}} - G\overline{\mathbb{D}}^\star\overline{\mathbb{D}}a_{\mathfrak{r}} = \Pi a_{\mathfrak{r}}$$

where $\Pi a_{\mathfrak{r}}$ is a constant function by Lemma 4.6. Since $\beta_{\mathfrak{r}}$ is $\overline{\mathbb{D}}$-closed, $\overline{\mathbb{D}}a_{\mathfrak{r}} = \mathcal{L}_{\mathcal{T}}\beta_{\mathfrak{r}}$. Thus $G\overline{\mathbb{D}}^\star\overline{\mathbb{D}}a_{\mathfrak{r}} = \mathcal{T}G\overline{\mathbb{D}}^\star\beta_{\mathfrak{r}}$, and since $a_{\mathfrak{r}}$ is real valued and $\mathcal{T}$ is a real vector field,

$$a_{\mathfrak{r}} - \mathcal{T}\Re G\overline{\mathbb{D}}^\star\beta_{\mathfrak{r}} = \Re\Pi a_{\mathfrak{r}}.$$

Extend the function $u = \Re G\overline{\mathbb{D}}^\star\beta_{\mathfrak{r}}$ to $\mathcal{M}$ as a smooth real-valued function. Then $\mathfrak{r}' = e^{-u}\mathfrak{r}$ has the required property.

Suppose that $\mathfrak{r}$, $\mathfrak{r}'$ are defining functions for $\mathcal{N}$ in $\mathcal{M}$ such that $a_{\mathfrak{r}}$ and $a_{\mathfrak{r}'}$ are constant. Then these functions are equal by Proposition 4.9. □

Note that if for some $\mathfrak{r}$, the subbundle $\overline{\mathcal{K}}_{\mathfrak{r}}$ is $\mathcal{T}$-invariant and admits a $\mathcal{T}$ invariant Hermitian metric, then there is a $\mathcal{T}$-invariant metric on $\overline{\mathcal{V}}$.

Suppose now that $\rho : F \to \mathcal{M}$ is a holomorphic vector bundle over $\mathcal{M}$. Using the operators

$$\overline{\mathbb{D}} : C^\infty(\mathcal{N}; {\textstyle\bigwedge}^q\overline{\mathcal{V}}^* \otimes F_{\mathcal{N}}) \to C^\infty(\mathcal{N}; {\textstyle\bigwedge}^{q+1}\overline{\mathcal{V}}^* \otimes F_{\mathcal{N}}),$$

see (4.12), define operators

$$\cdots \to C^\infty(\mathcal{N}; {\textstyle\bigwedge}^q\overline{\mathcal{K}}^* \otimes F_{\mathcal{N}}) \xrightarrow{\overline{\partial}_b} C^\infty(\mathcal{N}; {\textstyle\bigwedge}^{q+1}\overline{\mathcal{K}}^* \otimes F_{\mathcal{N}}) \to \cdots \tag{7.11}$$

by

$$\overline{\partial}_b\phi = \Pi_{\mathfrak{r}}\overline{\mathbb{D}}\phi, \quad \phi \in C^\infty(\mathcal{N}; {\textstyle\bigwedge}^q\overline{\mathcal{K}}^* \otimes F_{\mathcal{N}})$$

where $\Pi_{\mathfrak{r}}$ means $\Pi_{\mathfrak{r}} \otimes I$ with $\Pi_{\mathfrak{r}}$ defined by (7.4). The operators (7.11) form a complex. Define also

$$\mathcal{L}_{\mathcal{T}} = \mathbf{i}_{\mathcal{T}}\overline{\mathbb{D}} + \overline{\mathbb{D}}\mathbf{i}_{\mathcal{T}}$$

where $\mathbf{i}_{\mathcal{T}}$ stands for $\mathbf{i}_{\mathcal{T}} \otimes I$. Then

$$\mathbf{i}_{\mathcal{T}}\mathcal{L}_{\mathcal{T}} = \mathcal{L}_{\mathcal{T}}\mathbf{i}_{\mathcal{T}}, \quad \mathcal{L}_{\mathcal{T}}\overline{\mathbb{D}} = \overline{\mathbb{D}}\mathcal{L}_{\mathcal{T}}.$$

The first of these identities implies that the image of $C^\infty(\mathcal{N}; {\textstyle\bigwedge}^q\overline{\mathcal{K}}^* \otimes F_{\mathcal{N}})$ by $\mathcal{L}_{\mathcal{T}}$ is contained in $C^\infty(\mathcal{N}; {\textstyle\bigwedge}^q\overline{\mathcal{K}}^* \otimes F_{\mathcal{N}})$. With these definitions, $\overline{\mathbb{D}}$ as an operator

$$\overline{\mathbb{D}} : \begin{array}{c} C^\infty(\mathcal{N}; {\textstyle\bigwedge}^q\overline{\mathcal{K}}^* \otimes F_{\mathcal{N}}) \\ \oplus \\ C^\infty(\mathcal{N}; {\textstyle\bigwedge}^{q-1}\overline{\mathcal{K}}^* \otimes F_{\mathcal{N}}) \end{array} \to \begin{array}{c} C^\infty(\mathcal{N}; {\textstyle\bigwedge}^{q+1}\overline{\mathcal{K}}^* \otimes F_{\mathcal{N}}) \\ \oplus \\ C^\infty(\mathcal{N}; {\textstyle\bigwedge}^q\overline{\mathcal{K}}^* \otimes F_{\mathcal{N}}) \end{array}.$$

is given by the matrix in (7.8) with the new meanings for $\overline{\partial}_b$ and $\mathcal{L}_{\mathcal{T}}$.

Assume that there is a $\mathcal{T}$-invariant Riemannian metric on $\mathcal{N}$, that $\mathfrak{r}$ has be chosen so that $a_{\mathfrak{r}}$ is constant, that $\overline{\mathcal{K}}_{\mathfrak{r}}$ is orthogonal to $\mathcal{T}$, and that $\mathcal{T}$ has unit length. Then the term involving $\overline{\partial}_b a_{\mathfrak{r}}$ in the matrix (7.8) is absent, and since $\overline{\mathbb{D}}^2 = 0$,

$$\mathcal{L}_{\mathcal{T}}\overline{\partial}_b = \overline{\partial}_b\mathcal{L}_{\mathcal{T}}.$$

Write $h_{\overline{\mathcal{V}}^*}$ for the metric induced on the bundles ${\textstyle\bigwedge}^q\overline{\mathcal{V}}^*$ or ${\textstyle\bigwedge}^q\overline{\mathcal{K}}^*$.

If η_μ, $\mu = 1, \dots, k$ is a local frame of $F_{\mathcal{N}}$ over an open set $U \subset \mathcal{N}$ and ϕ is a local section of $\bigwedge^q \overline{\mathcal{V}}^* \otimes F_{\mathcal{N}}$ over U, then for some smooth sections ϕ^μ of $\bigwedge^q \overline{\mathcal{V}}^*$ and ω_μ^ν of $\overline{\mathcal{V}}^*$ over U,

$$\phi = \sum_\mu \phi^\mu \otimes \eta_\mu, \quad \overline{\mathbb{D}} \sum_\mu \phi^\mu \otimes \eta_\mu = \sum_\nu (\overline{\mathbb{D}}\phi^\nu + \sum_\mu \omega_\mu^\nu \wedge \phi^\mu) \otimes \eta_\nu.$$

This gives

$$\overline{\partial}_b \sum_\mu \phi^\mu \otimes \eta_\mu = \sum_\nu (\overline{\partial}_b \phi^\nu + \sum_\mu \Pi_{\mathfrak{r}} \omega_\mu^\nu \wedge \phi^\mu) \otimes \eta_\nu$$

and

$$\mathcal{L}_{\mathcal{T}} \sum_\mu \phi^\mu \otimes \eta_\mu = \sum_\nu (\mathcal{L}_{\mathcal{T}} \phi^\nu + \sum_\mu \langle \omega_\mu^\nu, \mathcal{T} \rangle \phi^\mu) \otimes \eta_\nu.$$

Suppose now that h_F is a Hermitian metric on F. With this metric and the metric $h_{\overline{\mathcal{V}}^*}$ we get Hermitian metrics h on each of the bundles $\bigwedge^q \overline{\mathcal{V}}^* \otimes F_{\mathcal{N}}$. If η_μ is an orthonormal frame of $F_{\mathcal{N}}$ and $\phi = \sum \phi^\mu \otimes \eta_\mu$, $\psi = \sum \psi^\mu \otimes \eta_\mu$ are sections of $\bigwedge^q \overline{\mathcal{V}}^* \otimes F_{\mathcal{N}}$, then

$$h(\phi, \psi) = \sum_\nu h_{\overline{\mathcal{V}}^*}(\phi^\mu, \psi^\mu).$$

Therefore

$$\begin{aligned} h(\mathcal{L}_{\mathcal{T}}\phi, \psi) &+ h(\phi, \mathcal{L}_{\mathcal{T}}\psi) \\ &= \sum_\nu h_{\overline{\mathcal{V}}^*}(\mathcal{L}_{\mathcal{T}}\phi^\nu + \sum_\mu \langle \omega_\mu^\nu, \mathcal{T}\rangle \phi^\mu, \psi^\nu) + \sum_\mu h_{\overline{\mathcal{V}}^*}(\phi^\mu, \mathcal{L}_{\mathcal{T}}\psi^\mu + \langle \omega_\nu^\mu, \mathcal{T}\rangle \psi^\nu) \\ &= \sum_\nu \mathcal{T} h_{\overline{\mathcal{V}}^*}(\phi^\nu, \psi^\nu) + \sum_{\mu,\nu} (\langle \omega_\mu^\nu, \mathcal{T}\rangle + \overline{\langle \omega_\nu^\mu, \mathcal{T}\rangle}) h_{\overline{\mathcal{V}}^*}(\phi^\mu, \psi^\nu) \\ &= \mathcal{T} h(\phi, \psi) + \sum_{\mu,\nu} (\langle \omega_\mu^\nu, \mathcal{T}\rangle + \overline{\langle \omega_\nu^\mu, \mathcal{T}\rangle}) h_{\overline{\mathcal{V}}^*}(\phi^\mu, \psi^\nu). \end{aligned}$$

Thus $\mathcal{T} h(\phi, \psi) = h(\mathcal{L}_{\mathcal{T}}\phi, \psi) + h(\phi, \mathcal{L}_{\mathcal{T}}\psi)$ if and only if

$$\langle \omega_\mu^\nu, \mathcal{T}\rangle + \overline{\langle \omega_\nu^\mu, \mathcal{T}\rangle} = 0 \text{ for all } \mu, \nu. \tag{7.12}$$

This condition is (4.17); just note that by the definition of $\overline{\mathbb{D}}$, the forms $(\Phi^*)^{-1}\omega_\mu^\nu$ in (4.17) are the forms that we are denoting ω_μ^ν here. Thus (7.12) holds if and only if h_F is an exact Hermitian metric, see Definition (4.18).

Consequently,

LEMMA 7.13. *The statement*

$$\mathcal{T} h(\phi, \psi) = h(\mathcal{L}_{\mathcal{T}}\phi, \psi) + h(\phi, \mathcal{L}_{\mathcal{T}}\psi) \quad \forall \phi, \psi \in C^\infty(\mathcal{N}; \textstyle\bigwedge^q \overline{\mathcal{V}}^* \otimes F_{\mathcal{N}}) \tag{7.14}$$

holds if and only the Hermitian metric h_F is exact.

8. Spectrum

Suppose that $\overline{\mathcal{V}}$ admits an invariant Hermitian metric. Let $\mathfrak{r}$ be a defining function for $\mathcal{N}$ in $\mathcal{M}$ such that $a_{\mathfrak{r}}$ is constant. By Lemma (7.9) $\overline{\mathcal{K}}_{\mathfrak{r}}$ is $\mathcal{T}$-invariant, so the restriction of the metric to this subbundle gives a $\mathcal{T}$-invariant metric; we use the induced metric on the bundles $\bigwedge^q \overline{\mathcal{K}}^*$ in the following. As in the proof of Lemma 7.10, there is a $\mathcal{T}$-invariant density $\mathfrak{m}_0$ on $\mathcal{N}$.

Let $\rho : F \to \mathcal{M}$ be a Hermitian holomorphic vector bundle, assume that the Hermitian metric of F is exact, so with the induced metric h on the vector bundles $\bigwedge^q \overline{\mathcal{V}}^* \otimes F_{\mathcal{N}}$, (7.14) holds. We will write F in place of $F_{\mathcal{N}}$.

Let $\overline{\partial}_b^\star$ be the formal adjoint of the ${}^b\overline{\partial}$ operator (7.11) with respect to the inner on the bundles $\bigwedge^q \overline{\mathcal{K}}^* \otimes F$ and the density $\mathfrak{m}_0$, and let $\square_{b,q} = \overline{\partial}_b \overline{\partial}_b^\star + \overline{\partial}_b^\star \overline{\partial}_b$ be the formal $\overline{\partial}_b$-Laplacian. Since $-i\mathcal{L}_\mathcal{T}$ is formally selfadjoint and commutes with $\overline{\partial}_b$, $\mathcal{L}_\mathcal{T}$ commutes with $\square_{b,q}$. Let

$$\mathscr{H}^q_{\overline{\partial}_b}(\mathcal{N}; F) = \ker \square_{b,q} = \{\phi \in L^2(\mathcal{N}; \textstyle\bigwedge^q \overline{\mathcal{K}}^* \otimes F) : \square_{b,q}\phi = 0\}$$

and let

$$\operatorname{Dom}_q(\mathcal{L}_\mathcal{T}) = \{\phi \in \mathscr{H}^q_{\overline{\partial}_b}(\mathcal{N}; F) \text{ and } \mathcal{L}_\mathcal{T}\phi \in \mathscr{H}^q_{\overline{\partial}_b}(\mathcal{N}; F)\}.$$

The spaces $\mathscr{H}^q_{\overline{\partial}_b}(\mathcal{N}; F)$ may be of infinite dimension, but in any case they are closed subspaces of $L^2(\mathcal{N}; \bigwedge^q \overline{\mathcal{K}}^* \otimes F)$, so they may be regarded as Hilbert spaces on their own right. If $\phi \in \mathscr{H}^q_{\overline{\partial}_b}(\mathcal{N}; F)$, the condition $\mathcal{L}_\mathcal{T}\phi \in \mathscr{H}^q_{\overline{\partial}_b}(\mathcal{N}; F)$ is equivalent to the condition

$$\mathcal{L}_\mathcal{T}\phi \in L^2(\mathcal{N}; \textstyle\bigwedge^q \overline{\mathcal{K}}^* \otimes F).$$

So we have a closed operator

$$-i\mathcal{L}_\mathcal{T} : \operatorname{Dom}_q(\mathcal{L}_\mathcal{T}) \subset \mathscr{H}^q_{\overline{\partial}_b}(\mathcal{N}; F) \to \mathscr{H}^q_{\overline{\partial}_b}(\mathcal{N}; F). \tag{8.1}$$

The fact that $\square_{b,q} - \mathcal{L}_\mathcal{T}^2$ is elliptic, symmetric, and commutes with $\mathcal{L}_\mathcal{T}$ implies that (8.1) is a selfadjoint Fredholm operator with discrete spectrum (see [**13**, Theorem 2.5]).

Definition 8.2. Let $\operatorname{spec}_0^q(-i\mathcal{L}_\mathcal{T})$ be the spectrum of the operator (8.1), and let $\mathscr{H}^q_{\overline{\partial}_b,\tau}(\mathcal{N}; F)$ be the eigenspace of $-i\mathcal{L}_\mathcal{T}$ in $\mathscr{H}^q_{\overline{\partial}_b}(\mathcal{N}; F)$ corresponding to the eigenvalue τ.

Let $\boldsymbol{\tau}$ denote the principal symbol of $-i\mathcal{T}$. Then the principal symbol of $\mathcal{L}_\mathcal{T}$ acting on sections of $\bigwedge^q \overline{\mathcal{K}}^*$ is $\boldsymbol{\tau} I$. Because $\square_{b,q} - \mathcal{L}_\mathcal{T}^2$ is elliptic, $\operatorname{Char}(\square_{b,q})$, the characteristic variety of $\square_{b,q}$, lies in $\boldsymbol{\tau} \neq 0$. Let

$$\operatorname{Char}^\pm(\square_{b,q}) = \{\nu \in \operatorname{Char}(\square_{b,q}) : \boldsymbol{\tau}(\nu) \gtrless 0\}.$$

By [**13**, Theorem 4.1], if $\square_{b,q}$ is microlocally hypoelliptic on $\operatorname{Char}^\pm(\square_{b,q})$, then

$$\{\tau \in \operatorname{spec}_0^q(-i\mathcal{L}_\mathcal{T}) : \tau \gtrless 0\}$$

is finite. We should perhaps point out that $\operatorname{Char}(\square_{b,q})$ is equal to the characteristic variety, $\operatorname{Char}(\overline{\mathcal{K}}_r)$, of the CR structure.

As a special case consider the situation where F is the trivial line bundle. Let $\theta_\mathfrak{r}$ be the real 1-form on $\mathcal{N}$ which vanishes on $\overline{\mathcal{K}}_\mathfrak{r}$ and satisfies $\langle \theta_\mathfrak{r}, \mathcal{T}\rangle = 1$; thus $\theta_\mathfrak{r}$ is smooth, spans $\operatorname{Char}(\overline{\mathcal{K}}_\mathfrak{r})$, and has values in $\operatorname{Char}^+(\overline{\mathcal{K}}_\mathfrak{r})$. The Levi form of the structure is

$$\operatorname{Levi}_{\theta_\mathfrak{r}}(v, w) = -id\theta_\mathfrak{r}(v, \overline{w}), \quad v,\ w \in \mathcal{K}_{\mathfrak{r},p},\ p \in \mathcal{N}.$$

Suppose that $\operatorname{Levi}_{\theta_\mathfrak{r}}$ is nondegenerate, with k positive and $n-k$ negative eigenvalues. It is well known that then $\square_{b,q}$ is microlocally hypoelliptic at $\nu \in \operatorname{Char}\mathcal{K}_\mathfrak{r}$ for all q except if $q = k$ and $\boldsymbol{\tau}(\nu) < 0$ or if $q = n - k$ and $\boldsymbol{\tau}(\nu) > 0$. Then the already mentioned Theorem 4.1 of [**13**] gives Theorem 6.1 of that paper which reads:

THEOREM 8.3. *Suppose that $\overline{\mathcal{V}}$ admits a Hermitian metric and that for some defining function $\mathfrak{r}$ such that $\mathfrak{a}_\mathfrak{r}$ is constant,* $\mathrm{Levi}_{\theta_\mathfrak{r}}$ *is nondegenerate with k positive and $n-k$ negative eigenvalues. Then*

(1) $\mathrm{spec}_0^q(-i\mathcal{L}_\mathcal{T})$ *is finite if $q \neq k,\ n-k$;*
(2) $\mathrm{spec}_0^k(-i\mathcal{L}_\mathcal{T})$ *contains only finitely many positive elements, and*
(3) $\mathrm{spec}_0^{n-k}(-i\mathcal{L}_\mathcal{T})$ *contains only finitely many negative elements.*

9. Indicial cohomology

Suppose that there is a $\mathcal{T}$-invariant Hermitian metric $\tilde{h}$ on $\overline{\mathcal{V}}$. By Lemma 7.10 there is a defining function $\mathfrak{r}$ such that $\langle \beta_\mathfrak{r}, \mathcal{T}\rangle$ is constant, equal to $a_{\mathrm{av}} - i$. Therefore $\overline{\mathcal{K}}_\mathfrak{r}$ is $\mathcal{T}$-invariant. Let h be the metric on $\overline{\mathcal{V}}$ which coincides with $\tilde{h}$ on $\overline{\mathcal{K}}_\mathfrak{r}$, makes the decomposition $\overline{\mathcal{V}} = \overline{\mathcal{K}}_\mathfrak{r} \oplus \mathrm{span}_\mathbb{C}\, \mathcal{T}$ orthogonal, and for which $\mathcal{T}$ has unit length. The metric h is $\mathcal{T}$-invariant. We fix $\mathfrak{r}$ and such a metric, and let $\mathfrak{m}_0$ be the Riemannian measure associated with h. The decomposition (7.3) of $\bigwedge^q \overline{\mathcal{V}}^*$ is an orthogonal decomposition.

Recall that $\overline{\mathcal{D}}(\sigma)\phi = \overline{\mathbb{D}}\phi + i\sigma\beta_\mathfrak{r} \wedge \phi$. Since $a_\mathfrak{r} = a_{\mathrm{av}}$ is constant (in particular CR),

$$\overline{\mathcal{D}}(\sigma)(\phi^0 + i\tilde{\beta}_\mathfrak{r} \wedge \phi^1) = \overline{\partial}_b\phi_0 + i\tilde{\beta}_\mathfrak{r} \wedge \left[(\mathcal{L}_\mathcal{T} + (1 + ia_{\mathrm{av}})\sigma)\phi^0 - \overline{\partial}_b\phi^1\right]$$

if $\phi^0 \in C^\infty(\mathcal{N}; \bigwedge^q \overline{\mathcal{K}}_\mathfrak{r}^*)$ and $\phi^1 \in C^\infty(\mathcal{N}; \bigwedge^{q-1} \overline{\mathcal{K}}_\mathfrak{r}^*)$. So $\overline{\mathcal{D}}(\sigma)$ can be regarded as the operator

$$\overline{\mathcal{D}}(\sigma) = \begin{bmatrix} \overline{\partial}_b & 0 \\ \mathcal{L}_\mathcal{T} + (1 + ia_{\mathrm{av}})\sigma & -\overline{\partial}_b \end{bmatrix} : \begin{matrix} C^\infty(\mathcal{N}; \bigwedge^q \overline{\mathcal{K}}_\mathfrak{r}^*) \\ \oplus \\ C^\infty(\mathcal{N}; \bigwedge^{q-1} \overline{\mathcal{K}}_\mathfrak{r}^*) \end{matrix} \to \begin{matrix} C^\infty(\mathcal{N}; \bigwedge^{q+1} \overline{\mathcal{K}}_\mathfrak{r}^*) \\ \oplus \\ C^\infty(\mathcal{N}; \bigwedge^q \overline{\mathcal{K}}_\mathfrak{r}^*). \end{matrix} \tag{9.1}$$

Since the subbundles $\bigwedge^q \overline{\mathcal{K}}_\mathfrak{r}$ and $\tilde{\beta} \wedge \bigwedge^{q-1} \overline{\mathcal{K}}_\mathfrak{r}$ are orthogonal with respect to the metric induced by h on $\bigwedge^q \overline{\mathcal{V}}$, the formal adjoint of $\overline{\mathcal{D}}(\sigma)$ with respect to this metric and the density $\mathfrak{m}_0$ is

$$\overline{\mathcal{D}}(\sigma)^\star = \begin{bmatrix} \overline{\partial}_b^\star & -\mathcal{L}_\mathcal{T} + (1 - ia_{\mathrm{av}})\overline{\sigma} \\ 0 & -\overline{\partial}_b^\star \end{bmatrix} : \begin{matrix} C^\infty(\mathcal{N}; \bigwedge^{q+1} \overline{\mathcal{K}}_\mathfrak{r}^*) \\ \oplus \\ C^\infty(\mathcal{N}; \bigwedge^q \overline{\mathcal{K}}_\mathfrak{r}^*) \end{matrix} \to \begin{matrix} C^\infty(\mathcal{N}; \bigwedge^q \overline{\mathcal{K}}_\mathfrak{r}^*) \\ \oplus \\ C^\infty(\mathcal{N}; \bigwedge^{q-1} \overline{\mathcal{K}}_\mathfrak{r}^*) \end{matrix}$$

where $\overline{\partial}_b^\star$ is the formal adjoint of $\overline{\partial}_b$. So the Laplacian, $\Box_{\overline{\mathcal{D}}(\sigma),q}$, of the $\overline{\mathcal{D}}(\sigma)$-complex is the diagonal operator with diagonal entries $P_q(\sigma)$, $P_{q-1}(\sigma)$ where

$$P_q(\sigma) = \Box_{b,q} + (\mathcal{L}_\mathcal{T} + (1 + ia_{\mathrm{av}})\sigma)(-\mathcal{L}_\mathcal{T} + (1 - ia_{\mathrm{av}})\overline{\sigma})$$

acting on $C^\infty(\mathcal{N}; \bigwedge^q \overline{\mathcal{K}}_\mathfrak{r}^*)$ and $P_{q-1}(\sigma)$ is the "same" operator, acting on sections of $\bigwedge^{q-1} \overline{\mathcal{K}}_\mathfrak{r}^*$; recall that $\mathcal{L}_\mathcal{T}$ commutes with $\overline{\partial}_b$ and since $\mathcal{L}_\mathcal{T}^\star = -\mathcal{L}_\mathcal{T}$, also with $\overline{\partial}_b^\star$, and that a_{av} is constant. Note that $P_q(\sigma)$ is an elliptic operator.

Suppose that $\phi \in C^\infty(\mathcal{N}; \bigwedge^q \overline{\mathcal{K}}_\mathfrak{r}^*)$ is a nonzero element of $\ker P_q(\sigma)$; the complex number σ is fixed. Since $P_q(\sigma)$ is elliptic, $\ker P_q(\sigma)$ is a finite dimensional space, invariant under $-i\mathcal{L}_\mathcal{T}$ since the latter operator commutes with $P_q(\sigma)$. As an operator on $\ker P_q(\sigma)$, $-i\mathcal{L}_\mathcal{T}$ is selfadjoint, so there is a decomposition of $\ker P_q(\sigma)$ into eigenspaces of $-i\mathcal{L}_\mathcal{T}$. Thus

$$\phi = \sum_{j=1}^N \phi_j, \quad -i\mathcal{L}_\mathcal{T}\phi_j = \tau_j\phi_j$$

where the τ_j are distinct real numbers and $\phi_j \in \ker P_q(\sigma)$, $\phi_j \neq 0$. In particular,

$$\Box_{b,q}\phi_j + (\mathcal{L}_\mathcal{T} + (1 + ia_{\mathrm{av}})\sigma)(-\mathcal{L}_\mathcal{T} + (1 - ia_{\mathrm{av}})\overline{\sigma})\phi_j = 0,$$

for each j, that is,

$$\Box_{b,q}\phi_j + |i\tau_j + (1 + ia_{\mathrm{av}})\sigma|^2\phi_j = 0.$$

Since $\Box_{b,q}$ is a nonnegative operator and $\phi_j \neq 0$, $i\tau_j + (1 + ia_{\mathrm{av}})\sigma = 0$ and $\phi_j \in \ker \Box_{b,q}$. Since σ is fixed, all τ_j are equal, which means that $N = 1$. Conversely, if $\phi \in C^\infty(\mathcal{N}; \bigwedge^q \overline{\mathcal{K}}_\mathfrak{r}^*)$ belongs to $\ker \Box_{b,q}$ and $-i\mathcal{L}_\mathcal{T}\phi = \tau\phi$, then $P_q(\sigma)\phi = 0$ with σ such that $\tau = (i - a_{\mathrm{av}})\sigma$.

Let $\mathscr{H}^q_{\overline{\mathcal{D}}(\sigma)}(\mathcal{N})$ be the kernel of $\Box_{\overline{\mathcal{D}}(\sigma),q}$.

THEOREM 9.2. *Suppose that $\overline{\mathcal{V}}$ admits a $\mathcal{T}$-invariant metric and let $\mathfrak{r}$ be a defining function for $\mathcal{N}$ in $\mathcal{M}$ such that $\langle\beta_\mathfrak{r}, \mathcal{T}\rangle = a_{\mathrm{av}} - i$ is constant. Then*

$$\operatorname{spec}^q_{b,\mathcal{N}}({}^b\overline{\partial}) = (i - a_{\mathrm{av}})^{-1}\operatorname{spec}^q_0(-i\mathcal{L}_\mathcal{T}) \cup (i - a_{\mathrm{av}})^{-1}\operatorname{spec}^{q-1}_0(-i\mathcal{L}_\mathcal{T}),$$

and if $\sigma \in \operatorname{spec}^q_{b,\mathcal{N}}({}^b\overline{\partial})$, then, with the notation in Definition 8.2

$$\mathscr{H}^q_{\overline{\mathcal{D}}(\sigma)}(\mathcal{N}) = \mathscr{H}^q_{\overline{\partial}_b,\tau(\sigma)}(\mathcal{N}) \oplus \mathscr{H}^{q-1}_{\overline{\partial}_b,\tau(\sigma)}(\mathcal{N})$$

with $\tau(\sigma) = (i - a_{\mathrm{av}})\sigma$.

If the CR structure $\overline{\mathcal{K}}_\mathfrak{r}$ is nondegenerate, Proposition 8.3 gives more specific information on $\operatorname{spec}^q_{b,\mathcal{N}}({}^b\overline{\partial})$. In particular,

PROPOSITION 9.3. *With the hypotheses of Theorem 9.2, suppose that $\mathrm{Levi}_{\theta_\mathfrak{r}}$ is nondegenerate with k positive and $n - k$ negative eigenvalues. If $k > 0$, then $\operatorname{spec}^0_{b,\mathcal{N}} \subset \{\sigma \in \mathbb{C} : \Im\sigma \leq 0\}$, and if $n-k > 0$, then $\operatorname{spec}^0_{b,\mathcal{N}}({}^b\overline{\partial}) \subset \{\sigma \in \mathbb{C} : \Im\sigma \geq 0\}$.*

REMARK 9.4. The b-spectrum of the Laplacian of the ${}^b\overline{\partial}$-complex in any degree can be described explicitly in terms of the joint spectra $\operatorname{spec}(-i\mathcal{L}_\mathcal{T}, \Box_{b,q})$. We briefly indicate how. With the metric h and defining function $\mathfrak{r}$ as in the first paragraph of this section, suppose that h is extended to a metric on ${}^bT^{0,1}\mathcal{M}$. This gives a Riemannian b-metric on $\mathcal{M}$ that in turn gives a b-density $\mathfrak{m}$ on $\mathcal{M}$. With these we get formal adjoints ${}^b\overline{\partial}^\star$ whose indicial families $\overline{\mathcal{D}}^\star(\sigma)$ are related to those of ${}^b\overline{\partial}$ by

$$\overline{\mathcal{D}}^\star(\sigma) = \widehat{{}^b\overline{\partial}^\star}(\sigma) = [\widehat{{}^b\overline{\partial}}(\overline{\sigma})]^\star = \overline{\mathcal{D}}(\overline{\sigma})^\star.$$

By (9.1),

$$\overline{\mathcal{D}}^\star(\sigma) = \begin{bmatrix} \overline{\partial}_b^\star & -\mathcal{L}_\mathcal{T} + (1 - ia_{\mathrm{av}})\sigma \\ 0 & -\overline{\partial}_b^\star \end{bmatrix}.$$

Using this one obtains that the indicial family of the Laplacian $\Box_q$ of the ${}^b\overline{\partial}$-complex in degree q is a diagonal operator with diagonal entries $P'_q(\sigma)$, $P'_{q-1}(\sigma)$ with

$$P'_q(\sigma) = \Box_{b,q} + (\mathcal{L}_\mathcal{T} + (1 + ia_{\mathrm{av}})\sigma)(-\mathcal{L}_\mathcal{T} + (1 - ia_{\mathrm{av}})\sigma)$$

and the analogous operator in degree $q - 1$. The set $\operatorname{spec}_b(\Box_q)$ is the set of values of σ for which either $P'_q(\sigma)$ or $P'_{q-1}(\sigma)$ is not injective. These points can written in terms of the points $\operatorname{spec}(-i\mathcal{L}_\mathcal{T}, \Box_b)$ as asserted. In particular one gets

$$\operatorname{spec}_b(\Box_q) \subset \{\sigma : |\Re\sigma| \leq |a_{\mathrm{av}}||\Im\sigma|\}$$

with $\operatorname{spec}^q_{b,\mathcal{N}}({}^b\overline{\partial})$ being a subset of the boundary of the set on the right.

We now discuss the indicial cohomology sheaf of ${}^b\overline{\partial}$, see Definition 6.12. We will show:

PROPOSITION 9.5. *Let* $\sigma_0 \in \operatorname{spec}^q_{b,\mathcal{N}}({}^b\overline{\partial})$. *Every element of the stalk of* $\mathfrak{H}^q_{{}^b\overline{\partial}}(\mathcal{N})$ *at* σ_0 *has a representative of the form*

$$\frac{1}{\sigma-\sigma_0}\begin{bmatrix}\phi^0\\0\end{bmatrix}$$

where $\phi^0 \in \mathscr{H}^q_{\overline{\partial}_b,\tau_0}(\mathcal{N})$, $\tau_0 = (i - a_{\mathrm{av}})\sigma_0$.

PROOF. Let

$$\phi(\sigma) = \sum_{k=1}^{\mu}\frac{1}{(\sigma-\sigma_0)^k}\begin{bmatrix}\phi^0_k\\ \phi^1_k\end{bmatrix} \tag{9.6}$$

represent an element in the stalk at σ_0 of the sheaf of germs of $C^\infty(\mathcal{N};\bigwedge^q\overline{\mathcal{V}}^*\otimes F)$-valued meromorphic functions on $\mathbb{C}$ modulo the subsheaf of holomorphic elements. Letting $\alpha = 1 + ia_{\mathrm{av}}$ we have

$$\overline{\mathcal{D}}(\sigma)\phi(\sigma) = \sum_{k=1}^{\mu}\frac{1}{(\sigma-\sigma_0)^k}\begin{bmatrix}\overline{\partial}_b\phi^0_k\\ (\mathcal{L}_\mathcal{T}+\alpha\sigma_0)\phi^0_k - \overline{\partial}_b\phi^1_k\end{bmatrix} + \sum_{k=0}^{\mu-1}\frac{\alpha}{(\sigma-\sigma_0)^k}\begin{bmatrix}0\\ \phi^0_{k+1}\end{bmatrix},$$

so the condition that $\overline{\mathcal{D}}(\sigma)\phi(\sigma)$ is holomorphic is equivalent to

$$\overline{\partial}_b\phi^0_k = 0,\ k = 1,\dots,\mu \tag{9.7}$$

and

$$\begin{aligned}(\mathcal{L}_\mathcal{T}+\alpha\sigma_0)\phi^0_\mu - \overline{\partial}_b\phi^1_\mu &= 0,\\ (\mathcal{L}_\mathcal{T}+\alpha\sigma_0)\phi^0_k - \overline{\partial}_b\phi^1_k + \alpha\phi^0_{k+1} &= 0,\ k = 1,\dots,\mu-1.\end{aligned} \tag{9.8}$$

Let $P_{q'} = \square_{b,q'} - \mathcal{L}^2_\mathcal{T}$ in any degree q'. For any $(\tau,\lambda)\in\mathbb{R}^2$ and q' let

$$\mathcal{E}^{q'}_{\tau,\lambda} = \{\psi \in C^\infty(\mathcal{N};\textstyle\bigwedge^{q'}\overline{\mathcal{V}}^*\otimes F) : P_{q'}\psi = \lambda\psi,\ -i\mathcal{L}_\mathcal{T}\psi = \tau\psi\}.$$

This space is zero if (τ,λ) is not in the joint spectrum $\Sigma^{q'} = \operatorname{spec}^{q'}(-i\mathcal{L}_\mathcal{T}, P_{q'})$. Each ϕ^i_k decomposes as a sum of elements in the spaces $\mathcal{E}^{q-i}_{\tau,\lambda}$, $(\tau,\lambda)\in\Sigma^{q-i}$. Suppose that already $\phi^i_k \in \mathcal{E}^{q-i}_{\tau,\lambda}$:

$$P_{q-i}\phi^i_k = \lambda\phi^i_k,\quad -i\mathcal{L}_\mathcal{T}\phi^i_k = \tau\phi^i_k,\quad i = 0,1,\ k = 1,\dots,\mu.$$

Then (9.8) becomes

$$\begin{aligned}(i\tau+\alpha\sigma_0)\phi^0_\mu - \overline{\partial}_b\phi^1_\mu &= 0,\\ (i\tau+\alpha\sigma_0)\phi^0_k - \overline{\partial}_b\phi^1_k + \alpha\phi^0_{k+1} &= 0,\ k = 1,\dots,\mu-1.\end{aligned} \tag{9.9}$$

If $\tau \neq \tau_0$, then $i\tau + \alpha\sigma_0 \neq 0$, and we get $\phi^0_k = \overline{\partial}_b\psi^0_k$ for all k with

$$\psi^0_k = \sum_{j=0}^{\mu-k}\frac{(-\alpha)^j}{(i\tau+\alpha\sigma_0)^{j+1}}\phi^1_{k+j}.$$

Trivially

$$(\mathcal{L}_\mathcal{T}+\alpha\sigma_0)\psi^0_\mu = \phi^1_\mu$$

and also

$$(\mathcal{L}_\mathcal{T}+\alpha\sigma_0)\psi^0_k + \alpha\psi^0_{k+1} = \phi^1_k,\quad k = 1,\dots,\mu-1,$$

so

$$\phi(\sigma)-\overline{\mathcal{D}}(\sigma)\sum_{k=1}^{\mu}\frac{1}{(\sigma-\sigma_0)^k}\begin{bmatrix}\psi_k^0\\0\end{bmatrix}=0$$

modulo an entire element.

Suppose now that the ϕ_k^i are arbitrary and satisfy (9.7)-(9.8). The sum

$$\phi_k^i=\sum_{(\tau,\lambda)\in\Sigma^{q-i}}\phi_{k,\tau,\lambda}^i,\quad \phi_{k,\tau,\lambda}^i\in\mathcal{E}_{\tau,\lambda}^{q-i}\tag{9.10}$$

converges in C^∞, indeed for each N there is $C_{i,k,N}$ such that

$$\sup_{p\in\mathcal{N}}\|\phi_{k,\tau,\lambda}^i(p)\|\le C_{i,k,N}(1+\lambda)^{-N}\quad\text{for all }\tau,\lambda.\tag{9.11}$$

Since $\overline{\mathcal{D}}(\sigma)$ preserves the spaces $\mathcal{E}_{\tau,\lambda}^q\oplus\mathcal{E}_{\tau,\lambda}^{q-1}$, the relations (9.9) hold for the $\phi_{k,\tau,\lambda}^i$ for each (τ,λ). Therefore, with

$$\psi_k^0=\sum_{\substack{(\tau,\lambda)\in\Sigma^{q-1}\\ \tau\neq\tau_0}}\sum_{j=0}^{\mu-k}\frac{(-\alpha)^j}{(i\tau+\alpha\sigma_0)^{j+1}}\phi_{k+j,\tau,\lambda}^1\tag{9.12}$$

we have formally that

$$\phi(\sigma)-\overline{\mathcal{D}}(\sigma)\sum_{k=1}^{\mu}\frac{1}{(\sigma-\sigma_0)^k}\begin{bmatrix}\psi_k\\0\end{bmatrix}=\sum_{k=1}^{\mu}\frac{1}{(\sigma-\sigma_0)^k}\begin{bmatrix}\tilde\phi_k^0\\\tilde\phi_k^1\end{bmatrix}$$

with

$$\tilde\phi_k^i=\sum_{\substack{(\tau,\lambda)\in\Sigma^{q-1}\\ \tau=\tau_0}}\phi_{k,\tau,\lambda}^i,\quad \phi_{k,\tau,\lambda}^i\in\mathcal{E}_{\tau,\lambda}^{q-i}.\tag{9.13}$$

However, the convergence in C^∞ of the series (9.12) is questionable since there may be a sequence $\{(\tau_\ell,\lambda_\ell)\}_{\ell=1}^\infty\subset\operatorname{spec}(-i\mathcal{L}_\mathcal{T},P_{q-1})$ of distinct points such that $\tau_\ell\to\tau_0$ as $\ell\to\infty$, so that the denominators $i\tau_\ell+\alpha\sigma_0$ in the formula for ψ_k^0 tend to zero so fast that for some nonnegative N, $\lambda_\ell^{-N}/(i\tau_\ell+\alpha\sigma_0)$ is unbounded. To resolve this difficulty we will first show that $\phi(\sigma)$ is $\overline{\mathcal{D}}(\sigma)$-cohomologous (modulo holomorphic terms) to an element of the same form as $\phi(\sigma)$ for which in the series (9.10) the terms $\phi_{k,\tau,\lambda}^i$ vanish if $\lambda-\tau^2>\varepsilon$; the number $\varepsilon>0$ is chosen so that

$$(\tau_0,\lambda)\in\Sigma^q\cup\Sigma^{q-1}\implies\lambda=\tau_0^2\text{ or }\lambda\ge\tau_0^2+\varepsilon.\tag{9.14}$$

Recall that $\operatorname{spec}^{q'}(-i\mathcal{L}_\mathcal{T},P_{q'})\subset\{(\tau,\lambda):\lambda\ge\tau^2\}$.

For any $V\subset\bigcup_{q'}\Sigma^{q'}$ let

$$\Pi_V^{q'}:L^2(\mathcal{N};\textstyle\bigwedge^{q'}\overline{\mathcal{V}}^*\otimes F)\to L^2(\mathcal{N};\bigwedge^{q'}\overline{\mathcal{V}}^*\otimes F)$$

be the orthogonal projection on $\bigoplus_{(\tau,\lambda)\in V}\mathcal{E}_{\tau,\lambda}^{q'}$. If $\psi\in C^\infty(\mathcal{N};\bigwedge^{q'}\overline{\mathcal{V}}^*\otimes F)$, then the series

$$\Pi_V^{q'}\psi=\sum_{(\tau,\lambda)\in V}\psi_{\tau,\lambda},\quad\psi_{\tau,\lambda}\in\mathcal{E}_{\tau,\lambda}^{q'}$$

converges in C^∞. It follows that $\Box_{b,q'}$ and $\mathcal{L}_\mathcal{T}$ commute with $\Pi_V^{q'}$ and that $\overline{\partial}_b\Pi_V^{q'}=\Pi_V^{q'+1}\overline{\partial}_b$. Since the $\Pi_V^{q'}$ are selfadjoint, also $\overline{\partial}_b^\star\Pi_V^{q'+1}=\Pi_V^{q'}\overline{\partial}_b^\star$.

Let

$$U=\{(\tau,\lambda)\in\Sigma^q\cup\Sigma^{q-1}:\lambda<\tau^2+\varepsilon\},\quad U^c=\Sigma^q\cup\Sigma^{q-1}\backslash U.$$

Then, for any sequence $\{(\tau_\ell,\lambda_\ell)\}\subset U$ of distinct points we have $|\tau_\ell|\to\infty$ as $\ell\to\infty$. Define

$$G^{q'}_{U^c}\psi=\sum_{(\tau,\lambda)\in U^c}\frac{1}{\lambda-\tau^2}\psi_{\tau,\lambda}$$

In this definition the denominators $\lambda-\tau^2$ are bounded from below by ε, so $G^{q'}_{U^c}$ is a bounded operator in L^2 and maps smooth sections to smooth sections because the components of such sections satisfy estimates as in (9.11). The operators are analogous to Green operators: we have

$$\Box_{b,q'}G^{q'}_{U^c}=G^{q'}_{U^c}\Box_{b,q'}=I-\Pi^{q'}_U \tag{9.15}$$

so if $\overline{\partial}_b\psi=0$, then

$$\Box_{b,q'}G^{q'}_{U^c}\psi=\overline{\partial}_b\overline{\partial}_b^\star G^{q'}_{U^c}\psi \tag{9.16}$$

since $\overline{\partial}_bG^{q'}_{U^c}=G^{q'+1}_{U^c}\overline{\partial}_b$.

Write $\phi(\sigma)$ in (9.6) as

$$\phi(\sigma)=\Pi_{U^c}\phi(\sigma)+\Pi_U\phi(\sigma)$$

where

$$\Pi_{U^c}\phi(\sigma)=\sum_{k=1}^{\mu}\frac{1}{(\sigma-\sigma_0)^k}\begin{bmatrix}\Pi^q_{U^c}\phi^0_k\\ \Pi^{q-1}_{U^c}\phi^1_k\end{bmatrix},\quad \Pi_U\phi(\sigma)=\sum_{k=1}^{\mu}\frac{1}{(\sigma-\sigma_0)^k}\begin{bmatrix}\Pi^q_U\phi^0_k\\ \Pi^{q-1}_U\phi^1_k\end{bmatrix}.$$

Since $\overline{\mathcal{D}}(\sigma)\phi(\sigma)$ is holomorphic, so are $\overline{\mathcal{D}}(\sigma)\Pi_{U^c}\phi(\sigma)$ and $\overline{\mathcal{D}}(\sigma)\Pi_U\phi(\sigma)$.

We show that $\Pi_{U^c}\phi(\sigma)$ is exact modulo holomorphic functions. Using (9.7), (9.15), and (9.16), $\Pi^q_{U^c}\phi^0_k=\overline{\partial}_b^\star\overline{\partial}_b\Pi^q_{U^c}\phi^0_k$. Then

$$\Pi_{U^c}\phi(\sigma)-\overline{\mathcal{D}}(\sigma)\sum_{k=1}^{\mu}\frac{1}{(\sigma-\sigma_0)^k}\begin{bmatrix}\overline{\partial}_b^\star G^q_{U^c}\Pi^q_U\phi^0_k\\ 0\end{bmatrix}=\sum_{k=1}^{\mu}\frac{1}{(\sigma-\sigma_0)^k}\begin{bmatrix}0\\ \hat{\phi}^1_k\end{bmatrix}$$

modulo a holomorphic term for some $\hat{\phi}^1_k$ with $\Pi^{q-1}_{U^c}\hat{\phi}^1_k=\hat{\phi}^1_k$. The element on the right is $\overline{\mathcal{D}}(\sigma)$-closed modulo a holomorphic function, so its components satisfy (9.7), (9.8), which give that the $\tilde{\phi}^1_k$ are $\overline{\partial}_b$-closed. Using again (9.15) and (9.16) we see that $\Pi_{U^c}\phi(\sigma)$ represent an exact element.

We may thus assume that $\Pi^q_{U^c}\phi(\sigma)=0$. If this is the case, then the series (9.12) converges in C^∞, so $\phi(\sigma)$ is cohomologous to the element

$$\tilde{\phi}(\sigma)=\sum_{k=1}^{\mu}\frac{1}{(\sigma-\sigma_0)^k}\begin{bmatrix}\tilde{\phi}^0_k\\ \tilde{\phi}^1_k\end{bmatrix}$$

where the $\tilde{\phi}^i_k$ are given by (9.13) and satisfy $\Pi^{q-i}_{U^c}\tilde{\phi}^i_k=0$. By (9.14), $\tilde{\phi}^i_k\in\mathcal{E}^{q-i}_{\tau_0,\tau_0^2}$. In particular, $\Box_{b,q-i}\phi^i_k=0$.

Assuming now that already $\phi^i_k\in\mathcal{E}^{q-i}_{\tau_0,\tau_0^2}$, the formulas (9.9) give (since $\tau=\tau_0$ and $i\tau_0+\alpha\sigma_0=0$)

$$\overline{\partial}_b\phi^1_\mu=0,\quad \phi^0_k=\overline{\partial}_b\frac{1}{\alpha}\phi^1_{k-1},\ k=2,\dots,\mu.$$

Then

$$\phi(\sigma) - \frac{1}{\alpha}\overline{\mathcal{D}}(\sigma)\sum_{k=2}^{\mu+1}\frac{1}{(\sigma-\sigma_0)^k}\begin{bmatrix}\phi^1_{k-1}\\ 0\end{bmatrix} = \frac{1}{\sigma-\sigma_0}\begin{bmatrix}\phi^0_1\\ 0\end{bmatrix}$$

with $\square_{b,q}\phi^0_1 = 0$. □

Appendix A. Totally characteristic differential operators

We review here some basic definitions and notation concerning totally characteristic differential operators.

Let E, $F \to \mathcal{M}$ be vector bundles and let $\operatorname{Diff}^m(\mathcal{M}; E, F)$ be the space of differential operators $C^\infty(\mathcal{M}; E) \to C^\infty(\mathcal{M}; F)$ of order m. Then

(A.1) $\operatorname{Diff}^m_b(\mathcal{M}; E, F)$, the space of totally characteristic differential operators of order m, consists of those elements $P \in \operatorname{Diff}^m(\mathcal{M}; E, F)$ with the property
$$\mathfrak{r}^{-\nu}P\mathfrak{r}^{\nu} \in \operatorname{Diff}^m(\mathcal{M}; E, F), \quad \nu = 1, \dots, m$$
i.e., $\mathfrak{r}^{-\nu}P\mathfrak{r}^{\nu}$ has coefficients smooth up to the boundary.

Let $\pi : T^*\mathcal{M} \to \mathcal{M}$ and ${}^b\pi : {}^bT^*\mathcal{M} \to \mathcal{M}$ be the canonical projections. Suppose $P \in \operatorname{Diff}^m_b(\mathcal{M}; E, F)$. Since P is in particular a differential operator, it has a principal symbol

$$\boldsymbol{\sigma}(P) \in C^\infty(T^*\mathcal{M}; \operatorname{Hom}(\pi^*E, \pi^*F)).$$

The fact that P is totally characteristic implies that $\boldsymbol{\sigma}(P)$ lifts to a section

$${}^b\boldsymbol{\sigma}(P) \in C^\infty({}^bT^*\mathcal{M}; \operatorname{Hom}({}^b\pi^*E, {}^b\pi^*F)),$$

the principal b-symbol of P, characterized by

$$ {}^b\boldsymbol{\sigma}(P)(\operatorname{ev}^*\xi) = \boldsymbol{\sigma}(P)(\xi). \tag{A.2}$$

If $P \in \operatorname{Diff}^m_b(\mathcal{M}; E, F)$, then P induces a differential operator

$$P_b \in \operatorname{Diff}^m_b(\partial\mathcal{M}; E_{\partial\mathcal{M}}, F_{\partial\mathcal{M}}), \tag{A.3}$$

as follows. If $\phi \in C^\infty(\partial\mathcal{M}; E_{\partial\mathcal{M}})$, let $\tilde{\phi} \in C^\infty(\mathcal{M}; E)$ be an extension of ϕ and let

$$P_b\phi = (P\tilde{\phi})|_{\partial\mathcal{M}}.$$

The condition (A.1) ensures that $P\tilde{\phi}|_{\partial\mathcal{M}}$ is independent of the extension of ϕ used. Clearly if P and Q are totally characteristic differential operators, then so is PQ, and

$$(PQ)_b = P_bQ_b. \tag{A.4}$$

The indicial family of $P \in \operatorname{Diff}^m_b(\mathcal{M}; E, F)$ is defined as follows. Fix a defining function $\mathfrak{r}$ for $\partial\mathcal{M}$. Then for any $\sigma \in \mathbb{C}$,

$$P(\sigma) = \mathfrak{r}^{-i\sigma}P\mathfrak{r}^{i\sigma} \in \operatorname{Diff}^m_b(\mathcal{M}; E, F).$$

Let

$$\widehat{P}(\sigma) = P(\sigma)_b. \tag{A.5}$$

References

[1] M. S. Baouendi, C. H. Chang, and F. Trèves, *Microlocal hypo-analyticity and extension of CR functions*, J. Differential Geom. **18** (1983), no. 3, 331–391. MR723811 (85h:32030)

[2] G. B. Folland and J. J. Kohn, *The Neumann problem for the Cauchy-Riemann complex*, Princeton University Press, Princeton, N.J.; University of Tokyo Press, Tokyo, 1972. Annals of Mathematics Studies, No. 75. MR0461588 (57 #1573)

[3] Sigurdur Helgason, *Differential geometry, Lie groups, and symmetric spaces*, Pure and Applied Mathematics, vol. 80, Academic Press, Inc. [Harcourt Brace Jovanovich, Publishers], New York-London, 1978. MR514561 (80k:53081)

[4] Lars Hörmander, *The Frobenius-Nirenberg theorem*, Ark. Mat. **5** (1965), 425–432 (1965). MR0178222 (31 #2480)

[5] Richard B. Melrose, *Transformation of boundary problems*, Acta Math. **147** (1981), no. 3-4, 149–236, DOI 10.1007/BF02392873. MR639039 (83f:58073)

[6] Richard B. Melrose, *The Atiyah-Patodi-Singer index theorem*, Research Notes in Mathematics, vol. 4, A K Peters, Ltd., Wellesley, MA, 1993. MR1348401 (96g:58180)

[7] Gerardo A. Mendoza, *Strictly pseudoconvex b-CR manifolds*, Comm. Partial Differential Equations **29** (2004), no. 9-10, 1437–1503, DOI 10.1081/PDE-200037762. MR2103843 (2006c:35049)

[8] Gerardo A. Mendoza, *Boundary structure and cohomology of b-complex manifolds*, Partial Differential Equations and Inverse Problems, Contemp. Math., vol. 362, Amer. Math. Soc., Providence, RI, 2004, pp. 303–320, DOI 10.1090/conm/362/06620. MR2091666 (2005h:58033)

[9] Gerardo A. Mendoza, *Anisotropic blowup and compactification*, Recent progress on some problems in several complex variables and partial differential equations, Contemp. Math., vol. 400, Amer. Math. Soc., Providence, RI, 2006, pp. 173–187, DOI 10.1090/conm/400/07539. MR2222474 (2007b:58033)

[10] Gerardo A. Mendoza, *Characteristic classes of the boundary of a complex b-manifold*, Complex analysis, Trends Math., Birkhäuser/Springer Basel AG, Basel, 2010, pp. 245–262, DOI 10.1007/978-3-0346-0009-5_15. MR2885120

[11] Gerardo A. Mendoza, *A Gysin sequence for Manifolds with $\mathbb{R}$-Action*, Geometric Analysis of Several Complex Variables and Related Topics, Contemp. Math., vol. 550, Amer. Math. Soc., Providence, RI, 2011, pp. 139–154, DOI 10.1090/conm/550/10870. MR2868558

[12] Gerardo A. Mendoza, *Two embedding theorems*, From Fourier analysis and number theory to radon transforms and geometry, Dev. Math., vol. 28, Springer, New York, 2013, pp. 399–429, DOI 10.1007/978-1-4614-4075-8_20. MR2986969

[13] Gerardo A. Mendoza, *Hypoellipticity and vanishing theorems*, Bull. Inst. Math. Acad. Sin. (N.S.) **8** (2013), no. 2, 231–258. MR3098538

[14] A. Newlander and L. Nirenberg, *Complex analytic coordinates in almost complex manifolds*, Ann. of Math. (2) **65** (1957), 391–404. MR0088770 (19,577a)

[15] L. Nirenberg, *A complex Frobenius theorem*, Seminar on Analytic Functions I, Princeton, N.J., 1957, 172–189.

[16] François Trèves, *Approximation and representation of functions and distributions annihilated by a system of complex vector fields*, École Polytechnique, Centre de Mathématiques, Palaiseau, 1981. MR716137 (84k:58008)

[17] François Trèves, *Hypo-analytic structures. Local theory*, Princeton Mathematical Series, vol. 40, Princeton University Press, Princeton, NJ, 1992. MR1200459 (94e:35014)

E-mail address: gmendoza@temple.edu

DEPARTMENT OF MATHEMATICS, TEMPLE UNIVERSITY, PHILADELPHIA, PENNSYLVANIA 19122

Contemporary Mathematics
Volume **630**, 2014
http://dx.doi.org/10.1090/conm/630/12667

Iterative Structures on Singular Manifolds

Bert-Wolfgang Schulze

ABSTRACT. We outline elements of an iterative approach to operators on categories of stratified spaces, referred to as manifolds with singularities. The sources of our iterative methods come from boundary value problems. In this case we have a two-component principal symbolic hierarchy, consisting of the interior and the boundary symbol. The boundary symbol is operator-valued and encodes ellipticity of boundary conditions. If the underlying space M is a manifold with singularities of order $k \in \mathbb{N}$, characterised by a stratification $s(M) := \{s_0(M), \ldots, s_k(M)\}$, then the principal symbolic hierarchy of an operator A on M is a sequence $\sigma(A) := \{\sigma_0(A), \ldots, \sigma_k(A)\}$ which determines ellipticity and solvability under additional edge conditions, analogues of boundary conditions. The operators are corner-degenerate in stretched representation, and they are studied in connection with suitable quantisations (operator conventions) to turn them to elements of algebras of pseudo-differential operators controlled by principal and also complete symbolic hierarchies.

Introduction

The analysis of partial differential equations on manifolds with singularities has a long history. It is motivated by diverse applications, e.g., in mechanics, heat conduction, particle physics, numerical analysis and structure mathematics, in particular, geometry, topology, or index theory. In this exposition we focus on operators on manifolds with higher singularities in connection with principal symbolic hierarchies, quantisations, actions in weighted distribution spaces, and ellipticity and parametrices.

The approach that we point out in the present paper has its origins in the theory of elliptic boundary value problems. An account on this development is given in [**25**, Chapter 10]. The first steps in this direction came from Agranovich and Dynin [**1**], Kondratiev [**30**], Vishik and Eskin [**74**], [**75**], Agranovich and Vishik [**2**], Sternin [**71**], [**72**], Eskin [**16**], Boutet de Monvel [**9**], and joint work with Rempel [**43**], [**45**]. Also other schools contributed to the analysis on singular manifolds, see the work of Melrose [**36**], [**38**], Melrose and Mendoza [**37**], Mazzeo [**35**], or Plamenevskij [**40**], [**41**]. During the past decades the development has been promoted by numerous activities in different directions, e.g., on geometric operators on corner manifolds

2010 *Mathematics Subject Classification.* Primary 35S35; Secondary 35J70.
Key words and phrases. Stratified spaces, conical, edge, and corner singularities, degenerate operators, principal symbolic hierarchies, weighted Sobolev spaces, meromorphic operator-valued symbols.

and index theory, but also on boundary value problems, see Schulze [**59**], Schulze and Seiler [**66**], Melrose and Rochon [**39**], Albin and Melrose [**3**].

Acknowledgement: This exposition has been written along the lines of a talk during the Workshop "Spectral Invariants on Singular and Non-compact Spaces", organised by Pierre Albin and Frédéric Rochon at the CENTE DE RECHERCHE MATHÉMATIQUE in Montreal, July 23 - 27, 2012. The author cordially thanks for the hospitality and the support by the CRM.

1. Manifolds with singularities of higher order

1.1. Motivation and preliminary observations. Let us first recall some generalities around pseudo-differential operators. Classical pseudo-differential operators on an open smooth manifold M appear as parametrices of elliptic operators $A \in \mathrm{Diff}^\mu(M)$. Here $\mathrm{Diff}^\mu(M)$ means the space of all differential operators of order μ with smooth coefficients in local representations

$$A = \sum_{|\alpha|\le\mu} a_\alpha(x) D_x^\alpha \tag{1.1}$$

associated with a chart $\chi : U \to \Omega$ on M, $\Omega \subseteq \mathbb{R}^n$ open. We do not insist utmost generality on the kind of smooth manifolds. For convenience we confine ourselves to the category $\mathfrak{M}_0$ of smooth manifolds which can be written as a countable union of compact sets. We consider $\mathrm{Diff}^\mu(M)$ in its natural Fréchet topology. Ellipticity of an $A \in \mathrm{Diff}^\mu(M)$ means

$$\sigma_0(A)(x,\xi) \ne 0 \quad \text{for all} \quad (x,\xi) \in T^*M \setminus 0$$

where T^*M is the cotangent bundle of M and 0 the zero section, where

$$\sigma_0(A)(x,\xi) = \sum_{|\alpha|=\mu} a_\alpha(x)\xi^\alpha$$

is the invariantly defined homogeneous principal symbol of A. Subscript 0 indicates the relation to $\mathfrak{M}_0$. The symbol here is scalar. Later on we also consider operators between distributional sections of (smooth complex) vector bundles E, F on M. In this case $\sigma_0(A) : \pi_M^* E \to \pi_M^* F$, $\pi_M^* : T^*M \setminus 0 \to M$, is a bundle morphism. The symbol $\sigma_0(A)$ is (positively) homogeneous in $\xi \ne 0$ of order μ, i.e.,

$$\sigma_0(A)(x,\delta\xi) = \delta^\mu \sigma_0(A)(x,\xi) \tag{1.2}$$

for all $\delta \in \mathbb{R}_+$.

On M we fix a Riemannian metric and identify smoothing operators on M with $C^\infty(M \times M)$ by $u \mapsto \int c(x,x')u(x')dx'$ for $c(x,x') \in C^\infty(M \times M)$. The space $L^\mu(M)$ of pseudo-differential operators on M of order $\mu \in \mathbb{R}$ is defined as the set of those $A : C_0^\infty(M) \to C^\infty(M)$ that are modulo smoothing operators locally in $\Omega \subseteq \mathbb{R}^n$, defined by

$$Au(x) = \mathrm{Op}(a)u(x) := \iint e^{i(x-x')\xi} a(x,\xi)u(x')dx d\!\!\,\bar{}\,\xi, \tag{1.3}$$

$d\!\!\,\bar{}\,\xi := (2\pi)^{-n} d\xi$. Here $a(x,\xi)$ belongs to Hörmander's symbol space $S^\mu(\Omega \times \mathbb{R}^n)$, defined as the set of all $a(x,\xi) \in C^\infty(\Omega \times \mathbb{R}^n)$ such that

$$|D_x^\alpha D_\xi^\beta a(x,\xi)| \le c\,\langle\xi\rangle^{\mu-|\beta|}$$

for all $(x,\xi) \in K \times \mathbb{R}^n$, $K \subset\subset \Omega$, $\alpha, \beta \in \mathbb{N}^n$, for constants $c = c(\alpha,\beta,K) > 0$. If a function $a_{(\mu)}(x,\xi) \in C^\infty(\Omega \times (\mathbb{R}^n \setminus \{0\}))$ is homogeneous of order μ in the sense

$$a_{(\mu)}(x,\delta\xi) = \delta^\mu a_{(\mu)}(x,\xi), \quad \delta \in \mathbb{R}_+,$$

then for any excision function χ (i.e., $\chi \in C^\infty(\mathbb{R}^n)$, $\chi(\xi) = 0$ for $|\xi| < c_0$, $\chi(\xi) = 1$ for $|\xi| > c_1$ for some $0 < c_0 < c_1$) we have $\chi(\xi)a_{(\mu)}(x,\xi) \in S^\mu(\Omega \times \mathbb{R}^n)$. A symbol $a(x,\xi) \in S^\mu(\Omega \times \mathbb{R}^n)$ is called classical, if there are components $a_{(\mu-j)}(x,\xi)$ of homogeneity order $\mu - j$, $j \in \mathbb{N} (= \{0,1,2,\dots\})$ such that

$$a(x,\xi) - \chi(\xi)\sum_{j=0}^{N} a_{(\mu-j)}(x,\xi) \in S^{\mu-(N+1)}(\Omega \times \mathbb{R}^n)$$

for every $N \in \mathbb{N}$ and any excision function χ. Let $S^\mu_{\mathrm{cl}}(\Omega \times \mathbb{R}^n)$ denote the subspace of $S^\mu(\Omega \times \mathbb{R}^n)$ of all classical symbols. Moreover, let $L^\mu_{\mathrm{cl}}(M)$ be the space of all $A \in L^\mu(M)$ such that the local symbols in the representation (1.3) are classical. Let $L^{-\infty}(M) := \bigcap_{\mu\in\mathbb{R}} L^\mu(M)$ be identified with $C^\infty(M \times M)$. If a consideration is valid both in the classical and general case we write subscript (cl). The definitions of $S^\mu_{(\mathrm{cl})}(\Omega \times \mathbb{R}^n)$ and $L^\mu_{(\mathrm{cl})}(M)$ extend to the parameter-dependent case, with parameters $\lambda \in \mathbb{R}^l$. The dimension of covariables ξ in symbols may be independent of n. Thus we have symbol spaces $S^\mu_{(\mathrm{cl})}(\Omega \times \mathbb{R}^{n+l}_{\xi,\lambda})$. Moreover, we form $L^{-\infty}(M;\mathbb{R}^l) := \mathcal{S}(\mathbb{R}^l, L^{-\infty}(M))$, and we define $L^\mu_{(\mathrm{cl})}(M;\mathbb{R}^l)$ to be the space of all operator families $A(\lambda)$ that are modulo $L^{-\infty}(M;\mathbb{R}^l)$ of the form (1.3) for symbols $a(x,\xi,\lambda) \in S^\mu_{(\mathrm{cl})}(\Omega \times \mathbb{R}^{n+l}_{\xi,\lambda})$.

An $A(\lambda) \in L^\mu_{(\mathrm{cl})}(M;\mathbb{R}^l)$ is called parameter-dependent elliptic if for its local representations $\mathrm{Op}(a)(\lambda)$ for $a(x,\xi,\lambda) \in S^\mu_{(\mathrm{cl})}(\Omega \times \mathbb{R}^{n+l}_{\xi,\lambda})$ modulo smoothing elements there is a $p(x,\xi,\lambda) \in S^{-\mu}_{(\mathrm{cl})}(\Omega \times \mathbb{R}^{n+l}_{\xi,\lambda})$ such that $(ap)(x,\xi,\lambda) = 1$ modulo $S^{-1}_{(\mathrm{cl})}(\Omega \times \mathbb{R}^{n+l}_{\xi,\lambda})$. In the classical case that means that the parameter-dependent homogeneous principal symbol $\sigma_0(A)(x,\xi,\lambda)$ does not vanish for all x and all $(\xi,\lambda) \neq 0$. As is well-known a parameter-dependent elliptic $A(\lambda) \in L^\mu_{(\mathrm{cl})}(M;\mathbb{R}^l)$ has a (properly supported) parameter-dependent elliptic parametrix $P(\lambda) \in L^{-\mu}_{(\mathrm{cl})}(M;\mathbb{R}^l)$ such that $A(\lambda)P(\lambda) - 1$, $P(\lambda)A(\lambda) - 1 \in L^{-\infty}(M;\mathbb{R}^l)$.

Let us now consider some examples on how an embedded stratum $Y \hookrightarrow M$, i.e., a smooth submanifold Y of any codimension ≥ 1 in a smooth manifold M, gives rise to a symbolic response of an operator $A \in \mathrm{Diff}^\mu(M)$ and to a corresponding new notion of ellipticity. The case of pseudo-differential operators is not so straightforward though in principle analogous. However, we then have to establish specific quantisations; those will be postponed for the moment. First let Y be the boundary of a submanifold M_+ with boundary Y, $\dim M = \dim M_+$. In local coordinates $\Omega := \mathbb{R}^n$ we assume that M_+ corresponds to the half-space $\overline{\mathbb{R}}^n_+ = \{x = (x_1,\dots,x_n) \in \mathbb{R}^n : x_n \geq 0\}$ and Y to $x_n = 0$. For convenience we assume that the coefficients of (1.1) are independent of x for large $|x|$. Then besides $\sigma_0(A)(x,\xi)$ the operator A over M_+ acquires the principal boundary symbol, namely,

$$\sigma_1(A)(y,\eta) := \sigma_0(A)(y,0,\eta,D_{x_n}), \tag{1.4}$$

$y := (y_1,\dots,x_{n-1})$, $\eta := (\xi_1,\dots,\xi_{n-1})$, regarded as a family of operators

$$\sigma_1(A)(y,\eta) : H^s(\mathbb{R}_+) \to H^{s-\mu}(\mathbb{R}_+), \tag{1.5}$$

$\eta \neq 0$, $H^s(\mathbb{R}_+) := H^s(\mathbb{R})|_{\mathbb{R}_+}$, $s \in \mathbb{R}$. As a substitute of the homogeneity (1.2) we have the relation

$$\sigma_1(A)(y, \delta\eta) = \delta^\mu \kappa_\delta \sigma_1(A)(y, \eta) \kappa_\delta^{-1}, \tag{1.6}$$

for all $\delta \in \mathbb{R}_+$. Here

$$\kappa_\delta u(x_n) := \delta^{1/2} u(\delta x_n), \tag{1.7}$$

defines a group $\kappa = \{\kappa_\delta\}_{\delta \in \mathbb{R}_+}$ of isomorphisms $\kappa_\delta : H^s(\mathbb{R}_+) \to H^s(\mathbb{R}_+)$, strongly continuous in δ. The ellipticity of A with respect to $\sigma_0(\cdot)$ entails the Fredholm property of (1.5) where we assume s not to be too small, in this case $s > \mu - 1/2$. Because of (1.6) the Fredholm index of (1.5) is the same as that of $\sigma_1(A)(y, \eta/|\eta|)$. The boundary symbol is invariantly defined as an operator function parametrised by $T^*Y \setminus 0$. If Y is compact the index $\operatorname{ind} \sigma_1(A)(y, \eta)$ for elliptic A only depends on $\eta/|\eta|$. Thus the Fredholm operators $\sigma_1(A)(y, \eta/|\eta|)$ continuously depend on the points of the compact unit cosphere bundle S^*Y of the boundary. There is then a K-theoretic index

$$\operatorname{ind}_{S^*Y} \sigma_1(A) \in K(S^*Y)$$

(independent of the choice of s). If $\pi : S^*Y \to Y$, $(y, \eta) \mapsto y$, is the canonical projection, it is interesting to know whether or not we have

$$\operatorname{ind}_{S^*Y} \sigma_1(A) \in \pi^* K(Y), \tag{1.8}$$

where $\pi^* : K(Y) \to K(S^*Y)$ is induced by the bundle pull back $\pi^* : \operatorname{Vect}(Y) \to \operatorname{Vect}(S^*Y)$. Here $\operatorname{Vect}(\cdot)$ means the set of smooth complex (finite-dimensional) vector bundles over the space in parentheses. The condition (1.8) can be violated, e.g., for important geometric operators A, cf. [**4**], [**6**]. In any case, if A is an elliptic differential operator, then (1.5) is always surjective. Thus $\ker_{S^*Y} \sigma_1(A) := \{\ker \sigma_1(A)(y, \eta) : (y, \eta) \in S^*Y)\} \in \operatorname{Vect}(S^*Y)$, and $\operatorname{ind}_{S^*Y} \sigma_1(A)$ can be represented by the pair $(\ker_{S^*Y} \sigma_1(A), 0)$; here N indicates the bundle of fibre-dimension $N \in \mathbb{N}$. The condition (1.8) is then equivalent to the existence of a $J \in \operatorname{Vect}(Y)$ such that $\ker \sigma_1(A) \oplus \mathbb{C}^N = \pi^* J$ for a suitable $N \in \mathbb{N}$.

Let us now assume that a smooth manifold Y is embedded in a smooth manifold M, $\dim Y < \dim M = n$, and show how an $A \in \operatorname{Diff}^\mu(M)$ acquires from Y an additional principal symbolic component $\sigma_1(A)$. If $Y =: \{c\}$ is a single point, then we interpret c as an embedded conical singularity. Locally close to c we identify M with $\mathbb{R}^n$ and c with the origin. Then, writing the operator (1.1) in polar coordinates in $\mathbb{R}^n \setminus \{0\}$ for $r := |x|$ we obtain A in the form

$$A = r^{-\mu} \sum_{j=0}^{\mu} a_j(r)(-r\partial_r)^j \tag{1.9}$$

for coefficients $a_j \in C^\infty(\overline{\mathbb{R}}_+, \operatorname{Diff}^{\mu-j}(S^{n-1}))$, with S^{n-1} being the unit sphere in $\mathbb{R}^n$. The minus sign at the r-derivative is motivated by the Mellin transform

$$Mu(z) := \int_0^\infty r^z u(r) dr/r, \tag{1.10}$$

first for $u \in C_0^\infty(\mathbb{R}_+)$, with the covariable $z \in \mathbb{C}$, and the inverse $M^{-1}g(r) := \int_{\Gamma_{1/2}} r^{-z} u(z) đz$, $đz := (2\pi)^{-1} dz$,

$$\Gamma_\beta := \{z \in \mathbb{C} : \operatorname{Re} z = \beta\}. \tag{1.11}$$

The Mellin transform extends by continuity to an isomorphism $M : L^2(\mathbb{R}_+) \to L^2(\Gamma_{1/2})$. On the subspaces $\mathcal{H}^{s,0}(\mathbb{R}_+)$, $s \in \mathbb{N}$, defined by $\{u \in L^2(\mathbb{R}_+) : (r\partial_r)^j u \in L^2(\mathbb{R}_+),\ j = 0, \ldots, s\}$, for $s \geq 1$ we have

$$-r\partial_r = M^{-1} z M. \tag{1.12}$$

Now besides the standard homogenoeous principal symbol $\sigma_0(A)$ there is another symbol, namely,

$$\sigma_1(A)(z) := \sum_{j=0}^{\mu} a_j(0) z^j, \tag{1.13}$$

called the conormal symbol of A. The notation comes from its role in boundary value problems without the transmission property, as a function on the conormal bundle of the boundary, see [**43**]. For convenience we fix the chart $\chi : U \to \mathbb{R}^n$ of the neighbourhood $U \subset M$ containing c. We observe (1.13) for all z on a weight line $\Gamma_{n/2-\gamma}$ for a weight $\gamma \in \mathbb{R}$ which is an extra information to be feeded in. We interpret (1.13) as a family of continuous operators

$$\sigma_1(A)(z) : H^s(S^{n-1}) \to H^{s-\mu}(S^{n-1}), \tag{1.14}$$

$s \in \mathbb{R}$. For another $B \in \mathrm{Diff}^{\nu}(M)$ written in an analogous form as (1.9) we have

$$\sigma_1(AB)(z) = \sigma_1(A)(z + \nu)\sigma_1(B)(z).$$

In the case $Y \hookrightarrow M$ for $\dim Y = q > 0$ we identify Y in local coordinates in $\mathbb{R}^n$ with the hyperplane $\{\tilde{x} = 0\}$ for $\tilde{x} := (x_{q+1}, \ldots, x_n)$, and write $y := (x_1, \ldots, x_q)$, $\eta := (\xi_1, \ldots, \xi_q)$. In polar coordinates in $\mathbb{R}^d_{\tilde{x}} \setminus \{0\}$, $d = n - q$, $r := |\tilde{x}|$, our operator takes the form

$$A := r^{-\mu} \sum_{j+|\alpha| \leq \mu} a_{j\alpha}(r, y)(-r\partial_r)^j (rD_y)^{\alpha}, \tag{1.15}$$

for coefficients $a_{j\alpha}(r, y) \in C^{\infty}(\overline{\mathbb{R}}_+ \times \mathbb{R}^q, \mathrm{Diff}^{\mu-(j+|\alpha|)}(S^{d-1}))$. The symbolic component $\sigma_1(A)$ contributed by the stratum Y is

$$\sigma_1(A)(y, \eta) := r^{-\mu} \sum_{j+|\alpha| \leq \mu} a_{j\alpha}(0, y)(-r\partial_r)^j (r\eta)^{\alpha}, \tag{1.16}$$

$(y, \eta) \in T^*Y \setminus 0$.

In the following we first discuss symbols of a similar structure on stratified spaces. After that in Section 2 we develop some tools on pseudo-differential operators on a manifold with conical singularities. More information can be found in [**55**] or [**58**]. In the present exposition we need it for the edge pseudo-differential calculus in Section 3. It was first developed in this form in [**52**]. Both the cone and the edge calculus are algebras, containing the parametrices of elliptic elements (algebraic operations make sense when, e.g., in compositions, the range of the first factor fits to the domain of the one, etc.). It is not the intention here to give complete descriptions. In Section 5 we illustrate other iterative constructions that are suggested by our approach. Another aspect of the structures is that the ellipticity is connected with extra elliptic edge conditions of trace and potential type, analogues of boundary conditions. Section 4 is devoted to Shapiro-Lopatinskij ellipticity of edge conditions, Section 6 to the case when an analogue of the Atiyah-Bott obstruction of [**4**] does not vanish. While boundary value problems in this context for pseudo-differential operators with the transmission property at the boundary have been introduced first in [**59**] as an algebra, closed under the construction of parametrices of elliptic elements, a similar algebra of edge problems with global projection conditions has been established in a joint work with Seiler [**64**], also unifying the concepts of Shapiro-Lopatinskij ellipticity with that of global projection conditions in this case. Another paper [**66**] of a similar orientation studied the case of boundary value problems without the transmission property at the boundary. Those are treated as edge problems, but they have additional interesting properties. A very remarkable aspect is that a truncated operator $\mathrm{r}^+ A \mathrm{e}^+$ for an $A \in L^{\mu}_{\mathrm{cl}}(M)$ on a smooth manifold M with respect to an embedded smooth manifold M_+ with boundary Y (where e^+ is the extension by 0 from $\mathrm{int}\, M_+$ to M and r^+ the restriction from M to $\mathrm{int}\, M_+$) belongs to the edge calculus over M_+, regarded as a manifold with edge, with all the structure details of the edge calculus, see Section 3. This phenomenon has been established in [**65**]. A simpler special case of truncated operators of order zero on the half axis has been studied by Eskin in [**16**]. Later on, in joint work with Rempel [**44**], truncated operators on the half-axis have been characterised in the higher order case in terms of full Mellin expansions with meromorphic symbols. In [**57**] it was proved that the truncated operators are elements of the cone algebra on the half-axis. All these things belong to the background of the cone and edge calculus, including the invention of edge spaces, connected with strongly continuous group actions on Hilbert spaces, appearing as Kegel-spaces and subspaces with asymptotics, see also [**47**] and [**52**].

1.2. The categories $\mathfrak{M}_k$. We define a sequence of categories of stratified spaces where PDEs can be formulated in an iterative manner. General references on stratified spaces are Fulton and MacPherson [**19**], or Weinberger [**76**].

DEFINITION 1.1. A topological space M (under some natural conditions on the topology in general) is said to be a manifold with singularities of order $k \in \mathbb{N}, k \geq 1$, if

(i) M contains a subspace $Y \in \mathfrak{M}_0$ such that $M \setminus Y \in \mathfrak{M}_{k-1}$;

(ii) Y has a neighbourhood $V \subseteq M$ which is a (locally trivial) cone bundle over Y with fibre X^Δ for some $X \in \mathfrak{M}_{k-1}$.

Transition maps $X^\Delta \rightarrow X^\Delta$ are induced by restrictions of $\mathfrak{M}_{k-1}$-isomorphisms $\mathbb{R} \times X \rightarrow \mathbb{R} \times X$ to $\overline{\mathbb{R}}_+ \times X$. This gives rise to corresponding transition maps $X^\Delta \times \Omega \rightarrow X^\Delta \times \tilde{\Omega}$ for the respective X^Δ -bundles over Y.

REMARK 1.2. $\mathfrak{M}_k$ is a category with a natural notion of morphisms and isomorphisms.

The notation "manifold" for objects in $\mathfrak{M}_k$ is only used here for convenience. In general those are no topological manifolds (for instance, X^Δ for a torus $X = S^1 \times S^1$); it is also common to speak about pseudo-manifolds.

REMARK 1.3. $Y =: Y^k$ is called the minimal stratum of M. The space $M \backslash Y^k \in \mathfrak{M}_{k-1}$ contains a space $Y^{k-1} \in \mathfrak{M}_0$, such that $(M \setminus Y^k) \setminus Y^{k-1} \in \mathfrak{M}_{k-2}$, etc. This yields a representation

$$M = Y^0 \cup Y^1 \cup Y^2 \cup ... \cup Y^k$$

as a disjoint union of strata $Y^j \in \mathfrak{M}_0$. We set $\operatorname{int} M := Y^0$, called the maximal stratum of M and $\dim M := \dim(\operatorname{int} M)$. Moreover, M is locally near Y^j modelled on an X^Δ_{j-1}-bundle over Y^j, for $X^\Delta_{j-1} \in \mathfrak{M}_{j-1}$ with trivialisations $X^\Delta_{j-1} \times U^j, U^j \subseteq \mathbb{R}^{\dim Y^j}$ open.

REMARK 1.4. The same topological space M can be stratified in different ways. For instance, we have $M = \mathbb{R}^n \in \mathfrak{M}_0$ but also $M \in \mathfrak{M}_1$ when we set $Y^1 = \{0\}, Y^0 = \mathbb{R}^n \setminus \{0\}$.

REMARK 1.5. $M \in \mathfrak{M}_k$, $L \in \mathfrak{M}_l$ implies $M \times L \in \mathfrak{M}_{k+l}$.

Let us now consider another example. Assume $X_0 \in \mathfrak{M}_0$, and form the space

$$M := (\cdots((X_0^\Delta \times \Omega_1)^\Delta \times \Omega_2)^\Delta \times \cdots)^\Delta \times \Omega_{k-1})^\Delta \times \Omega_k \tag{1.17}$$

for open manifolds $\Omega_j \in \mathfrak{M}_0$ of dimension $q_j, j = 1, \dots, k$. Then we have

$$\begin{aligned} Y^k &= \Omega_k, \\ Y^{k-1} &= (\mathbb{R}_+ \times \Omega_{k-1}) \times \Omega_k, \\ Y^{k-2} &= (\mathbb{R}_+ \times \Omega_{k-2}) \times (\mathbb{R}_+ \times \Omega_{k-1}) \times \Omega_k, \dots, \\ Y^1 &= (\mathbb{R}_+ \times \Omega_1) \times \cdots \times (\mathbb{R}_+ \times \Omega_{k-1}) \times \Omega_k, \\ Y^0 &= (\mathbb{R}_+ \times X_0) \times (\mathbb{R}_+ \times \Omega_1) \times \cdots \times (\mathbb{R}_+ \times \Omega_{k-1}) \times \Omega_k. \end{aligned} \tag{1.18}$$

The space (1.17) has a typical corner singularity of order k and can be used for local descriptions of any $M \in \mathfrak{M}_k$.
For $M \in \mathfrak{M}_k$ in general we often write

$$s_j(M) := Y^j, j = 1, \dots, k, \tag{1.19}$$

and

$$M' := M \setminus s_0(M), M'' := M' \setminus s_0(M'), \dots, \tag{1.20}$$

using $M' \in \mathfrak{M}_{k-1}$, such that the maximal stratum $s_0(M')$ of M' again makes sense, etc., or also $M^{(j)} := M'^{\dots\prime}$ (with j primes), and $M^{(0)} := M$. Then $s_{k-1}(M') = s_k(M), \dots, s_0(M') = s_1(M)$, etc. The spaces M^j form a chain

$$M = M^{(0)} \supset M^{(1)} \supset \cdots \supset M^{(k)} = s_k(M) \tag{1.21}$$

where $M^j \in \mathfrak{M}_{k-j}$ for all j,

$$\dim M^{(j)} > \dim M^{(l)},$$

and

$$\dim s_l(M) < \dim s_j(M) \quad \text{for } 0 \le j < l \le k.$$

1.3. Corner-degenerate operators and their symbolic hierarchies. Let $M \in \mathfrak{M}_k, k \ge 1$, and $Y := Y^k$; then $\operatorname{Diff}^\mu_{\deg}(M)$ denotes the set of all $A \in \operatorname{Diff}^\mu_{\deg}(M \setminus Y)$, $(\operatorname{Diff}^\mu_{\deg}(\cdot) = \operatorname{Diff}^\mu(\cdot)$ in the smooth case) such that close to Y in the local variables

$$(r, x) \in \mathbb{R}_+ \times X \quad \text{for} \quad \dim Y = 0, \quad \text{and} \quad (r, x, y) \in \mathbb{R}_+ \times X \times \Omega \quad \text{for} \quad \dim Y > 0,$$

we have

$$A = r^{-\mu} \sum_{j=0}^{\mu} a_j(r)(-r\partial_r)^j \quad \text{for} \quad \dim Y = 0,$$

for coefficients $a_j \in C^\infty(\overline{\mathbb{R}}_+, \mathrm{Diff}^{\mu-j}_{\mathrm{deg}}(X))$, and

$$A = r^{-\mu} \sum_{j+|\alpha|\le\mu} a_{j\alpha}(r,y)(-r\partial_r)^j (rD_y)^\alpha \quad \text{for} \quad \dim Y > 0,$$

for coefficients $a_{j\alpha} \in C^\infty(\overline{\mathbb{R}}_+ \times \Omega, \mathrm{Diff}^{\mu-(j+|\alpha|)}_{\mathrm{deg}}(X))$, respectively. The principal symbolic hierarchies are iteratively defined by

$$\sigma(A) := (\sigma(A|_{M\setminus Y}), \sigma_k(A)) \tag{1.22}$$

where $\sigma(A|_{M\setminus Y})$ is known by the steps before, while

$$\sigma_k(A)(z) = \sum_{j=0}^{\mu} a_j(0) z^j \quad \text{for} \quad \dim Y = 0,\ z \in \mathbb{C}, \tag{1.23}$$

and

(1.24)

$$\sigma_k(A)(y,\eta) = r^{-\mu} \sum_{j+|\alpha|\le\mu} a_{j\alpha}(0,y)(-r\partial_r)^j (r\eta)^\alpha \text{ for } \dim Y > 0, (y,\eta) \in T^*\Omega \setminus 0.$$

The component $\sigma_k(A)(z)$ for $\dim Y = 0$ takes values in $\mathrm{Diff}^\mu_{\mathrm{deg}}(X)$ and $\sigma_k(A)(y,\eta)$ for $\dim Y > 0$ in $\mathrm{Diff}^\mu_{\mathrm{deg}}(X^\wedge)$, $X^\wedge = \mathbb{R}_+ \times X$.

REMARK 1.6. If we dissolve the information in (1.22) with respect to the stratification

$$s(M) := (s_0(M), s_1(M), \dots, s_k(M)) \tag{1.25}$$

of M, we obtain $k+1$ components of $\sigma(A)$, namely,

$$\sigma(A) = (\sigma_0(A), \sigma_1(A), \dots, \sigma_k(A)), \tag{1.26}$$

with $\sigma_0(A) \in C^\infty(T^*(\mathrm{int}\, M)\setminus 0)$ being the standard homogeneous principal symbol of A on the maximal stratum $\mathrm{int}\, M = s_0(M)$ of M, while the other components $\sigma_j(A)$ are operator-valued and associated with $Y^j = s_j(M)$, where $\sigma_j(A)$ is of analogous form as (1.24) for $\dim s_j(M) > 0$ (in the latter case parametrised by $T^*s_j(M)\setminus 0$). The symbol $\sigma_k(A)$ for $\dim s_k(M) = 0$ acts between Sobolev spaces over X_{k-1} with a $(k-1)$-tuple of weights, where M is locally near $s_k(M)$ modelled on X^Δ_{k-1}. Moreover, $\sigma_j(A)$ for $0 < j < k$ where $\dim s_j(M) > 0$ is analogous to (1.24), depending on points in $T^*s_j(M)\setminus 0$, act between spaces over $X^\wedge_{j-1}$ with a $(j-1)$-tuple of weights, and M in this case is modelled on $X^\wedge_{j-1} \times U^j$. For the moment we do not discuss the nature of the weighted spaces.

REMARK 1.7. Let $A \in \mathrm{Diff}^\mu_{\mathrm{deg}}(M), \tilde{A} \in \mathrm{Diff}^{\tilde{\mu}}_{\mathrm{deg}}(M)$; then $A\tilde{A} \in \mathrm{Diff}^{\mu+\tilde{\mu}}_{\mathrm{deg}}(M)$, and we have $\sigma(A\tilde{A}) = \sigma(A)\sigma(\tilde{A})$ (with componentwise composition and the rule $\sigma_k(A\tilde{A}) = (T^{\tilde{\mu}}\sigma(A))\sigma(\tilde{A})$ when $\dim s_k(M) = 0$ where $(T^\beta f)(z) = f(z+\beta)$ for any real β).

In fact, for $j = 0$ the assertion is clear anyway. For $k > 0$ we first consider the case $q_k = \dim s_k(M) > 0$. We interpret the operators A and $\tilde{A}$ as operators in y with operator-valued coefficients, i.e.,

$$A = \sum_{|\beta|\le\mu} b_\beta(y) D_y^\beta, \tilde{A} = \sum_{|\delta|\le\tilde{\mu}} \tilde{b}_\delta(y) D_y^\delta; \tag{1.27}$$

then $A\tilde{A} = \sum_{|\vartheta|\le\mu+\tilde{\mu}} c_\vartheta(y)D_y^\vartheta$. Setting

$$(1.28)\qquad b(y,\eta) = \sum_{|\beta|\le\mu} b_\beta(y)\eta^\beta, \tilde{b}(y,\eta) = \sum_{|\delta|\le\tilde{\mu}} \tilde{b}_\delta(y)\eta^\delta, c(y,\eta) = \sum_{|\vartheta|\le\mu+\tilde{\mu}} c_\vartheta(y)\eta^\vartheta,$$

it follows that $c(y,\eta) = \sum_{\alpha\in\mathbb{N}^q} 1/\alpha!(\partial_\eta^\alpha b)(y,\eta)D_y^\alpha\tilde{b}(y,\eta)$ which is a finite sum. In the computation of σ_k only the term for $|\alpha| = 0$ survives, since the definition of the principal symbol contains coefficients frozen at $r = 0$, and η-differentiations produce powers of r such that the corresponding contribution to the edge symbol disappears. In a similar manner, rewriting the coefficients in A or $\tilde{A}$ as functions of r in the form $a(r,\cdot) = a(0,\cdot) + ra_1(r,\cdot)$, and $\tilde{a}(r,\cdot) = \tilde{a}(0,\cdot) + r\tilde{a}_1(r,\cdot)$, respectively, for functions $a_1(r,\cdot), \tilde{a}_1(r,\cdot)$ that are smooth up to $r = 0$ we see again that the contributions with higher r-powers vanish. The composition rule for symbols σ_j for $j < k$ can be verified in a similar manner.

In the case $\dim s_k(M) = 0$ we express our operators in terms of the Mellin transform (1.10) which has the property (1.12). For functions with compact support in r we obtain an entire function in z. This allows us to restrict $Mu(z)$ to any parallel $\Gamma_\beta := \{z \in \mathbb{C} : \operatorname{Re} z = \beta\}$ to the imaginary axis. For $\beta = 1/2 - \gamma$ we obtain the weighted Mellin transform $M_\gamma u := Mu|_{\Gamma_{1/2-\gamma}}$. Setting

$$(1.29)\qquad \operatorname{op}_M^\gamma(h)u(z) := M_\gamma^{-1}h(r,z)(M_\gamma u)(z)$$

for a corresponding (in general operator-valued) Mellin amplitude function $h(r,z)$ we can write

$$(1.30)\qquad A = r^{-\mu}\operatorname{op}_M^{\gamma-\tilde{\mu}}(h),\ \tilde{A} = r^{-\tilde{\mu}}\operatorname{op}_M^\gamma(\tilde{h})$$

for $h(r,z) = \sum_{j=0}^{\mu} a_j(r)z^j, \tilde{h}(r,z) = \sum_{l=0}^{\tilde{\mu}} \tilde{a}_l(r)z^l$. The computation of σ_k refers again to coefficients frozen at $r = 0$. For convenience we assume $h = h(0,z), \tilde{h} = \tilde{h}(0,z)$ where $\sigma_k(A)(z) = h(0,z), \sigma_k(\tilde{A})(z) = \tilde{h}(0,z)$. Now for the composition it follows that

$$(1.31)\qquad A\tilde{A} = r^{-\mu}\operatorname{op}_M^{\gamma-\tilde{\mu}}(h)r^{-\tilde{\mu}}\operatorname{op}_M^\gamma(\tilde{h}) = r^{-(\mu+\tilde{\mu})}\operatorname{op}_M^\gamma((T^{-\tilde{\mu}}h)\tilde{h})$$

which gives us $\sigma_k(A\tilde{A})(z) = h(z+\tilde{\mu})\tilde{h}(z)$.

By Definition 1.1 a space $M \in \mathfrak{M}_k, k \ge 1$, is locally near points of $s_k(M)$ modelled on $X_{k-1}^\Delta \times \Omega_k$, for an open set $\Omega_k \subseteq \mathbb{R}^{q_k}, q_k = \dim s_k(M)$, and by stretched variables we understand a corresponding splitting over $s_0(X_{k-1})^\wedge \times \Omega_k$ into variables (r_k, x_{k-1}, y_k). Denoting by $(\rho_k, \xi_{k-1}, \eta_k)$ the corresponding covariables we obtain $\sigma_0(A)$ as a function

$$(1.32)\qquad \sigma_0(A) = \sigma_0(A)(r_k, x_{k-1}, y_k, \rho_k, \xi_{k-1}, \eta_k),$$

where

$$(1.33)\qquad s_k\sigma_0(A)(r_k, x_{k-1}, y_k, \rho_k, \xi_{k-1}, \eta_k) := r_k^\mu\sigma_0(A)(r_k, x_{k-1}, y_k, r_k^{-1}\rho_k, \xi_{k-1}, r_k^{-1}\eta_k),$$

is smooth up to $r_k = 0$. The space $X_{k-1} \in \mathfrak{M}_{k-1}$ is locally near $s_{k-1}(X_{k-1})$ modelled on $X_{k-2}^\Delta \times \Omega_{k-1}$ for some $X_{k-2} \in \mathfrak{M}_{k-2}$ and an open set $\Omega_{k-1} \subseteq \mathbb{R}^{q_{k-1}}, q_{k-1} = \dim s_{k-1}(X_{k-1})$, and we can decompose (x_{k-1}, ξ_{k-1}) into

$$(1.34)\qquad (x_{k-1}, \xi_{k-1}) = (r_{k-1}, x_{k-2}, y_{k-1}, \rho_{k-1}, \xi_{k-2}, \eta_{k-1})$$

for $(r_{k-1}, x_{k-2}, y_{k-1}) \in \mathbb{R}_+ \times s_0(X_{k-2}) \times \Omega_{k-1}$. This gives us $\sigma_0(A)$ as a function

$$(1.35)\qquad \sigma_0(A) = \sigma_0(A)(r_{k-1}, r_k, x_{k-2}, y_{k-1}, y_k, \rho_{k-1}, \rho_k, \xi_{k-2}, \eta_{k-1}, \eta_k),$$

and

$$\begin{aligned}&s_{k-1}s_k\sigma_0(A)(r_{k-1},r_k,x_{k-2},y_{k-1},y_k,\rho_{k-1},\rho_k,\xi_{k-2},\eta_{k-1},\eta_k)\\&:=(r_{k-1}r_k)^\mu\sigma_0(A)(r_k,r_{k-1},x_{k-2},y_{k-1},y_k,(r_{k-1}r_k)^{-1}\rho_k,r_{k-1}^{-1}\rho_{k-1},\\&\xi_{k-2},r_{k-1}^{-1}\eta_{k-1},(r_{k-1}r_k)^{-1}\eta_k)\end{aligned}\tag{1.36}$$

is smooth up to $r_{k-1}=0, r_k=0$. This process can be continued, the next step concerns a neighbourhood of $s_0(X_{k-2})$ in variables and covariables

$$(x_{k-2},\xi_{k-2})=(r_{k-2},x_{k-3},y_{k-2},\rho_{k-2},\xi_{k-3},\eta_{k-2}),\tag{1.37}$$

and so on. At the end we obtain

$$\begin{aligned}\sigma_0(A)=&\sigma_0(A)(r_1,\dots,r_{k-1},r_k,x_0,y_1,\dots,y_{k-1},y_k,\\&\rho_1,\dots,\rho_{k-1},\rho_k,\xi_0,\eta_1,\dots,\eta_{k-1},\eta_k),\end{aligned}\tag{1.38}$$

and

(1.39)

$$\begin{aligned}&s_1\cdots s_{k-1}s_k\sigma_0(A)(r_1,\dots,r_{k-1},r_k,x_0,y_1,\dots,y_{k-1},y_k,\rho_1,\dots,\rho_{k-1},\rho_k,\xi_0,\\&\eta_1,\dots,\eta_{k-1},\eta_k):=(r_1\dots r_{k-1}r_k)^\mu\sigma_0(A)(r_1,\dots,r_{k-1},r_k,x_0,y_1,\dots,y_{k-1},y_k,\\&r_1^{-1}\rho_1,\dots,(r_1\dots r_{k-1})^{-1}\rho_{k-1},(r_1\dots r_{k-1}r_k)^{-1}\rho_k,\xi_0,r_1^{-1}\eta_1,\dots,\\&(r_1\dots r_{k-1})^{-1}\eta_{k-1},(r_1\dots r_{k-1}r_k)^{-1}\eta_k)\end{aligned}$$

is smooth up to $r_1=0,\dots,r_k=0$. The functions

$$\{s_k\sigma_0(A),\dots,s_1\cdots s_{k-1}s_k\sigma_0(A)\}\tag{1.40}$$

are also referred to as the reduced interior symbols. Also $\sigma_1(A),\dots,\sigma_{k-1}(A)$ admit reduced variants, namely,

$$\begin{aligned}&\{s_k\sigma_1(A),\dots,s_2\cdots s_k\sigma_1(A)\},\\&\{s_k\sigma_2(A),\dots,s_3\cdots s_k\sigma_2(A)\},\\&\dots,\\&\{s_k\sigma_{k-2}(A),s_{k-1}s_k\sigma_{k-2}(A)\},\\&s_k\sigma_{k-1}(A),\end{aligned}\tag{1.41}$$

while $\sigma_k(A)=\sigma_k(A)(y_k,\eta_k)$ remains as it is, namely, an operator function

$$C_0^\infty(\operatorname{int}X_{k-1}^\wedge)\to C_0^\infty(\operatorname{int}X_{k-1}^\wedge)$$

with $\operatorname{int}X_{k-1}^\wedge=(\operatorname{int}X_{k-1})^\wedge$ being described by variables (r_k,x_{k-1}). Let us now illustrate the definition in the case $\sigma_{k-1}(A)$, which is an operator function depending on points in $T^*s_{k-1}(M\setminus s_k(M))\setminus 0$. The reduced symbol refers to a neighbourhood close to $s_k(M)$ in the representation of $M\setminus s_k(M)$ as $G_{k-1}\times X_{k-2}^\wedge$ with G_{k-1} corresponding to a chart on $s_{k-1}(M)$ in the variables (r_k,y_{k-1},y_k) (cf. the notation in (1.18) where $G_{k-1}=(\mathbb{R}_+\times\Omega_{k-1})\times\Omega_k)$. In these variables we have

$$\sigma_{k-1}(A)(r_k,y_{k-1},y_k,\rho_k,\eta_{k-1},\eta_k):C_0^\infty(\operatorname{int}X_{k-2}^\wedge)\to C_0^\infty(\operatorname{int}X_{k-2}^\wedge)\tag{1.42}$$

where $\operatorname{int}X_{k-2}^\wedge$ is considered in the variables (r_{k-1},x_{k-2}). The degenerate behaviour of the operator A has the consequence that

(1.43)

$$s_k\sigma_{k-1}(A)(r_k,y_{k-1},y_k,\rho_k,\eta_{k-1},\eta_k):=r_k^\mu\sigma_{k-1}(A)(r_k,y_{k-1},y_k,r_k^{-1}\rho_k,\eta_{k-1},r_k^{-1}\eta_k)$$

is smooth up to $r_k = 0$.
Let us continue this process one step more. Then it should be clear how it works in general. First the symbol $\sigma_{k-2}(A)$ is an operator function $C_0^\infty(\operatorname{int} X_{k-3}^\wedge) \to C_0^\infty(\operatorname{int} X_{k-3}^\wedge)$ depending on variables and covariables in $T^* s_{k-2}(M \setminus (s_k(M) \cup s_{k-1}(M))) \setminus 0$. Close to $s_k(M)$ the manifold $s_{k-2}(M \setminus (s_k(M) \cup s_{k-1}(M)))$ admits local variables of the form (r_k, v_{k-1}, y_k) with covariables $(\rho_k, \theta_{k-1}, \eta_k)$. Similarly as before we obtain
(1.44)
$$s_k\sigma_{k-2}(A)(r_k, v_{k-1}, y_k, \rho_k, \theta_{k-1}, \eta_k) := r_k^\mu \sigma_{k-2}(A)(r_k, v_{k-1}, y_k, r_k^{-1}\rho_k, \theta_{k-1}, r_k^{-1}\eta_k)$$
as a smooth function up to $r_k = 0$. However, close to $s_{k-1}(M)$ the variables v_{k-1} can be splitted up once again into $v_{k-1} = (r_{k-1}, y_{k-1})$ with covariables $\theta_{k-1} = (\rho_{k-1}, \eta_{k-1})$ which allows us to form

$$\begin{aligned} &s_{k-1}s_k\sigma_{k-2}(A)(r_{k-1}, r_k, y_{k-1}, y_k, \rho_{k-1}, \rho_k, \eta_{k-1}, \eta_k) \\ &:= (r_{k-1}r_k)^\mu \sigma_{k-2}(A)(r_{k-1}, r_k, y_{k-1}, y_k, r_{k-1}^{-1}\rho_{k-1}, (r_{k-1}r_k)^{-1}\rho_k, \\ &\quad r_{k-1}^{-1}\eta_{k-1}, (r_{k-1}r_k)^{-1}\eta_k) \end{aligned} \tag{1.45}$$

as a smooth function up to $r_{k-1} = r_k = 0$. It is now clear how to define (1.41) in general.

REMARK 1.8. A role of the principal symbols (1.26) including the reduced components (1.40) and (1.41) is to define an adequate notion of ellipticity of the operator A. Let us call the operator σ_0-elliptic if $\sigma_0(A)$ never vanishes on $T^*s_0(M) \setminus 0$, and if in addition $s_j \cdots s_k\sigma_0(A)$ does not vanish up to $r_j = \cdots = r_k = 0$ for every $j = 1, \ldots, k$. The right notion of ellipticity for the other components of (1.26) is an element of a more complete theory of the operators that we only sketch in this article. However, there is a non-trivial special case where we simply require the operator functions $\sigma_j(A)$ to be bijective between the above-mentioned weighted spaces for all points in $T^*s_j(M) \setminus 0$ for $j = 1, \ldots, k$ (when $\dim s_k(M) > 0$, otherwise for all z on a suitable parallel to the imaginary axis), together with a similar bijectivity condition for all $s_j \cdots s_k\sigma_l(A), j = l+1, \ldots, k$, and $1 \le l \le k-1$, up to $r_j = \cdots = r_k = 0$. In general when A is σ_0-elliptic we cannot expect that the other symbolic components satisfy such bijectivity conditions. Under a suitable topological condition on $\sigma_0(A)$, for instance, in the case $k = 1$ and $\dim s_k(M) > 0$, we can generate extra edge conditions, a substitute of (Shapiro-Lopatinskij) elliptic boundary conditions, that fill up the Fredholm family $\sigma_k(A)$ to a family of isomorphisms, provided that we (are able to) avoid some exceptional weights. The complete story is long and not the focus of the present paper. For $\dim s_k(M) = 0$ things are a little easier; then we have bijectivity up to some discrete set of exceptional weights.

Ellipticity under such aspects have been studied for the case $k = 1$ in many variants. For instance, in boundary value problems in smooth domains the role of elliptic boundary conditions is to have a bijective boundary symbol, cf. Boutet de Monvel [**9**]. However, the criterion on σ_0 that makes this possible may be violated, cf. Atiyah, Bott [**4**] for a K-theoretic formulation of the respective topological obstruction in the case of differential operators, Boutet de Monvel [**9**] for operators with the transmission property at the boundary, and Schulze [**55**] for edge problems. A new algebra of boundary value problems that extends the one of Boutet de Monvel and admits arbitrary σ_0-elliptic operators with the transmission property (including those with non-vanishing topological obstruction) has been established by the author in [**59**], see also [**61**]. The approach of [**59**] unifies the concept of Shapiro-Lopatinskij elliptic conditions and global projection conditions, containing as special cases conditions of Atiyah-Patodi-Singer type [**6**]. A generalisation to edge problems for smooth edges was given in the author's joint paper [**64**] with Seiler, see the considerations in Section 6 below and also [**66**] which concerns boundary value problems without the transmission property. The discussion of all this material is worth to be continued in the future.

1.4. Edge spaces and operator-valued symbols. In this subsection we develop some material which belongs to the iterative concept of operators on corner manifolds. Hilbert spaces in the following general constructions will assumed to be separable.

DEFINITION 1.9. (i) A Hilbert space H is said to be endowed with a group action $\kappa = \{\kappa_\lambda\}_{\lambda \in \mathbb{R}_+}$, if $\kappa_\lambda : H \to H$, $\lambda \in \mathbb{R}_+$, is a group of isomorphisms, $\kappa_\lambda \kappa_\nu = \kappa_{\lambda\nu}$ for all $\lambda, \nu \in \mathbb{R}_+$, and if for every fixed $u \in H$ the map $\lambda \to \kappa_\lambda u$ defines a continuous function $\mathbb{R}_+ \to H$ (i.e., $\{\kappa_\lambda\}_{\lambda \in \mathbb{R}_+}$ is strongly continuous).

(ii) Let E be a Fréchet space, written as a projective limit of Hilbert spaces E^j, $j \in \mathbb{N}$, with continuous embeddings $E^j \hookrightarrow E^0$ for all j. Then E is said to be endowed with a group action $\kappa = \{\kappa_\lambda\}_{\lambda \in \mathbb{R}_+}$, if κ is a group action on E^0 which restricts to a group action on E^j for every $j \in \mathbb{N}$.

EXAMPLE 1.10. (i) In the case $H := \mathcal{K}^{s,\gamma}(X^\wedge)$, cf. Subsection 2.1 below, we often consider the group action

$$(\kappa_\lambda u)(r, x) := \lambda^{(n+1)/2} u(\lambda r, x),\ \lambda \in \mathbb{R}_+, \ \text{for } n := \dim X. \tag{1.46}$$

(ii) For $H := \mathbb{C}^k$, $k \in \mathbb{N}$, in most cases we set $\kappa_\lambda := \mathrm{id}$ for all $\lambda \in \mathbb{R}_+$.

(iii) Let us represent the Schwartz space as a projective limit

$$\mathcal{S}(\mathbb{R}^n) = \varprojlim_{j \in \mathbb{N}} H^{j,j}(\mathbb{R}^n)$$

for $H^{j,j}(\mathbb{R}^n) := \langle x \rangle^{-j} H^j(\mathbb{R}^n)$. Then

$$(\kappa_\lambda u)(x) := \lambda^{n/2} u(\lambda x),\ \lambda \in \mathbb{R}_+,$$

determines a group action in $\mathcal{S}(\mathbb{R}^n)$ in the sense of Definition 1.9 (ii).

REMARK 1.11. The group action of Definition 1.9 (i) in the space $\mathcal{K}^{0,0}(X^\wedge) = r^{-n/2} L^2(\mathbb{R}_+ \times X)$ is unitary.

REMARK 1.12. Let H be a Hilbert space with group action $\kappa = \{\kappa_\lambda\}_{\lambda \in \mathbb{R}_+}$. Then there are constants $c, M > 0$ such that

$$\left\| \kappa_\lambda \right\|_{\mathcal{L}(H)} \le c(\max\{\lambda, \lambda^{-1}\})^M. \tag{1.47}$$

A proof can be found in [**26**].

Let us now define operator-valued symbols referring to group actions in the involved spaces. Let H and $\widetilde{H}$ be Hilbert spaces with group action $\kappa = \{\kappa_\lambda\}_{\lambda \in \mathbb{R}_+}$ and $\tilde{\kappa} = \{\tilde{\kappa}_\lambda\}_{\lambda \in \mathbb{R}_+}$, respectively. A function $f \in C^\infty(\mathbb{R}^q \setminus \{0\}, \mathcal{L}(H, \widetilde{H}))$ is said to be (twisted) homogeneous of order $\mu \in \mathbb{R}$, if

$$f(\lambda \eta) = \lambda^\mu \tilde{\kappa}_\lambda f(\eta) \kappa_\lambda^{-1} \tag{1.48}$$

for all $\lambda \in \mathbb{R}_+$, $\eta \in \mathbb{R}^q \setminus \{0\}$. Let $S^{(\mu)}(\mathbb{R}^q \setminus \{0\}; H, \widetilde{H})$ denote the space of all those functions f. The space $S^{(\mu)}(\mathbb{R}^q \setminus \{0\}; H, \widetilde{H})$ is Fréchet in a natural semi-norm system that immediately follows from the definition. We then set

$$S^{(\mu)}(U \times (\mathbb{R}^q \setminus \{0\}); H, \widetilde{H}) := C^\infty(U, S^{(\mu)}(\mathbb{R}^q \setminus \{0\}; H, \widetilde{H})) \tag{1.49}$$

for any open $U \subseteq \mathbb{R}^p$. Observe that when $f(y,\eta)$ is an element of the latter space, we have $D_y^\alpha D_\eta^\beta f(y,\eta) \in S^{(\mu-|\beta|)}(U \times (\mathbb{R}^q \setminus \{0\}); H, \widetilde{H})$ for arbitrary multi-indices $\alpha \in \mathbb{N}^p, \beta \in \mathbb{N}^q$. To verify the latter relation for y-independent f it suffices to differentiate (1.48) on both sides with respect to η; the y-dependent case is easy as well.

REMARK 1.13. For every $f(y,\eta) \in S^{(\mu)}(U \times (\mathbb{R}^q \setminus \{0\}); H, \widetilde{H})$ we have

$$f(y,\eta) = |\eta|^\mu \kappa_{|\eta|} f(y, \eta/|\eta|)\kappa_{|\eta|}^{-1} \text{ for all } (y,\eta) \in U \times (\mathbb{R}^q \setminus \{0\}). \tag{1.50}$$

In other words, an $f \in S^{(\mu)}(U \times (\mathbb{R}^q \setminus \{0\}); H, \widetilde{H})$ is uniquely determined by its values $f_1(y,\eta) := f(y,\eta/|\eta|) \in C^\infty(U \times S^{q-1}, \mathcal{L}(H,\widetilde{H}))$. In general, not every element $f_1(y,\eta) \in C^\infty(U \times S^{q-1}, \mathcal{L}(H,\widetilde{H}))$ can be extended by homogeneity to an $f \in S^{(\mu)}(U \times (\mathbb{R}^q \setminus \{0\}); H, \widetilde{H})$. For instance, an $f_1 \in \mathcal{L}(L^2(\mathbb{R}), L^2(\mathbb{R}))$ which is independent of (y,η), belongs at the same time to $C^\infty(U \times S^{q-1}, \mathcal{L}(L^2(\mathbb{R}), L^2(\mathbb{R}))$. Assume that $(f_1 u)(r) := m(r) \int_{\mathbb{R}} l(r')u(r')dr'$ for some $l, m \in L^2(\mathbb{R})$. Define $(\kappa_\lambda u)(r) := u(\lambda r)$, $\lambda \in \mathbb{R}_+$; then the formula (1.50) gives us

$$f(\eta)u(r) : |\eta|^{\mu+1} m(r|\eta|) \int_{\mathbb{R}} l(r'|\eta|)u(r')dr', \tag{1.51}$$

and we see immediately that we lose smoothness in $\eta \neq 0$. However, if we assume $l, m \in \mathcal{S}(\mathbb{R})$ then it follows that (1.51) belongs to $S^{(\mu)}(\mathbb{R}^q \setminus \{0\}; L^2(\mathbb{R}), L^2(\mathbb{R}))$.

DEFINITION 1.14. The space

$$S^\mu(U \times \mathbb{R}^q; H, \widetilde{H}) \tag{1.52}$$

for $\mu \in \mathbb{R}$, $U \subseteq \mathbb{R}^p$ open, is defined to be the set of all $a(y,\eta) \in C^\infty(U \times \mathbb{R}^q, \mathcal{L}(H,\widetilde{H}))$ such that the expressions

$$\sup_{y\in K, \eta \in \mathbb{R}^q} \langle\eta\rangle^{-\mu-|\beta|} \|\tilde{\kappa}_{\langle\eta\rangle}^{-1}\{D_y^\alpha D_\eta^\beta a(y,\eta)\}\kappa_{\langle\eta\rangle}\|_{\mathcal{L}(H,\widetilde{H})} \tag{1.53}$$

are finite for every $K \subset\subset U$, $\alpha \in \mathbb{N}^p$, $\beta \in \mathbb{N}^q$, $\langle\eta\rangle := (1+|\eta|^2)^{1/2}$. Moreover, by $S^\mu_{\mathrm{cl}}(U \times \mathbb{R}^q; H, \widetilde{H})$ the space of classical symbols we denote the set of all $a(y,\eta)$ in (1.52) such that there are $a_{(\mu-j)}(y,\eta) \in S^{(\mu-j)}(U \times (\mathbb{R}^q \setminus \{0\}); H, \widetilde{H})$, $j \in \mathbb{N}$, with the property

$$r_N(y,\eta) := a(y,\eta) - \chi(\eta) \sum_{j=0}^N a_{(\mu-j)}(y,\eta) \in S^{\mu-(N+1)}(U \times \mathbb{R}^q; H, \widetilde{H}) \tag{1.54}$$

for every $N \in \mathbb{N}$ (and any fixed excision function χ). For a classical symbol $a(y,\eta)$ we often set

$$\sigma_\wedge(a)(y,\eta) := a_{(\mu)}(y,\eta).$$

The elements of (1.52) are called operator-valued symbols (with twisted symbolic estimates), and classical if they belong to $S^\mu_{\mathrm{cl}}(U \times \mathbb{R}^q; H, \widetilde{H})$. The functions

$a_{(\mu-j)}(y,\eta)$ are called the (twisted) homogeneous components of the classical symbol $a(y,\eta)$. If a consideration concerns both (1.52) and the classical case we often write $S^{\mu}_{(\mathrm{cl})}(U\times\mathbb{R}^q;H,\widetilde{H})$.

If necessary we indicate the dependence on the group actions κ and $\tilde{\kappa}$ by writing

$$S^{\mu}_{(\mathrm{cl})}(U\times\mathbb{R}^q;H,\widetilde{H})_{\kappa,\tilde{\kappa}}. \tag{1.55}$$

The subspaces of y-independent elements (with 'constant' coefficients) will be denoted by

$$S^{\mu}_{(\mathrm{cl})}(\mathbb{R}^q;H,\widetilde{H}) \quad (\text{or } S^{\mu}_{(\mathrm{cl})}(\mathbb{R}^q;H,\widetilde{H})_{\kappa,\tilde{\kappa}}).$$

We write

$$S^{\mu}_{(\mathrm{cl})}(U\times\mathbb{R}^q;H,\widetilde{H})_{\mathrm{id},\mathrm{id}}$$

for $\kappa_\lambda=\mathrm{id}_H$ and $\tilde{\kappa}_\lambda=\mathrm{id}_{\widetilde{H}}$, $\lambda\in\mathbb{R}_+$.

DEFINITION 1.15. Given a Hilbert space H with group action $\kappa=\{\kappa_\lambda\}_{\lambda\in\mathbb{R}_+}$ the space $\mathcal{W}^s(\mathbb{R}^q,H)$ for $s\in\mathbb{R}$ is defined to be the completion of $\mathcal{S}(\mathbb{R}^q,H)$ (the Schwartz space of H-valued functions in $\mathbb{R}^q$) with respect to the norm

$$\|u\|_{\mathcal{W}^s(\mathbb{R}^q,H)}=\Big\{\int\langle\eta\rangle^{2s}\|\kappa^{-1}_{\langle\eta\rangle}\hat{u}(\eta)\|^2_H đ\eta\Big\}^{1/2}. \tag{1.56}$$

Here $\hat{u}(\eta)=(F_{y\to\eta}u)(\eta)$ is the Fourier transform in $\mathbb{R}^q$. We also write

$$\mathcal{W}^s(\mathbb{R}^q,H)_\kappa \tag{1.57}$$

to indicate the dependence on the group action κ. We call $\mathcal{W}^s(\mathbb{R}^q,H)$ the abstract edge space (with the edge $\mathbb{R}^q$) of smoothness s. In the case $\kappa_\lambda=\mathrm{id}_H$, $\lambda\in\mathbb{R}_+$, the corresponding edge space is also denoted by $H^s(\mathbb{R}^q,H)$, or, alternatively, $\mathcal{W}^s(\mathbb{R}^q,H)_{\mathrm{id}}$.

REMARK 1.16. For $H=\mathbb{C}$ and $\kappa_\lambda=\mathrm{id}_H$ for $\lambda\in\mathbb{R}_+$ we have

$$\mathcal{W}^s(\mathbb{R}^q,\mathbb{C})=H^s(\mathbb{R}^q).$$

Observe that for the constant $M>0$ of Proposition 1.12 we have continuous embeddings

$$H^{s+M}(\mathbb{R}^q,H)\hookrightarrow\mathcal{W}^s(\mathbb{R}^q,H)\hookrightarrow H^{s-M}(\mathbb{R}^q,H) \tag{1.58}$$

for every $s\in\mathbb{R}$.

We set $\mathcal{W}^\infty(\mathbb{R}^q,H):=\bigcap_{s\in\mathbb{R}}\mathcal{W}^s(\mathbb{R}^q,H)$, and in a similar manner we define

$$H^\infty(\mathbb{R}^q,H):=\bigcap_{s\in\mathbb{R}}H^s(\mathbb{R}^q,H).$$

From (1.58) it follows that

$$\mathcal{W}^\infty(\mathbb{R}^q,H)=H^\infty(\mathbb{R}^q,H). \tag{1.59}$$

REMARK 1.17. (i) The space $\mathcal{W}^s(\mathbb{R}^q,H)$, $s\in\mathbb{R}$, of Definition 1.15 is a Hilbert space with the scalar product

$$(u,v)_{\mathcal{W}^s(\mathbb{R}^q,H)}=\int\langle\eta\rangle^{2s}\big(\kappa^{-1}_{\langle\eta\rangle}\hat{u}(\eta),\kappa^{-1}_{\langle\eta\rangle}\hat{v}(\eta)\big)_H đ\eta$$

where $(\cdot,\cdot)_H$ is the scalar product in H.

(ii) If $\{\kappa_\lambda\}_{\lambda\in\mathbb{R}_+}$ is a unitary group in H we have $\mathcal{W}^s(\mathbb{R}^q,H)=H^s(\mathbb{R}^q,H)$ for all $s\in\mathbb{R}$, and, in particular,

$$\mathcal{W}^0(\mathbb{R}^q,H)=L^2(\mathbb{R}^q,H). \tag{1.60}$$

Observe that for every $s\in\mathbb{N}$ we have $\mathcal{W}^s(\mathbb{R}^q,H)=\{u\in\mathcal{W}^0(\mathbb{R}^q,H):D_y^\alpha u\in\mathcal{W}^0(\mathbb{R}^q,H),|\alpha|\le s\}$ and an equivalence of norms

$$\|u\|_{\mathcal{W}^s(\mathbb{R}^q,H)}\sim\{\|u\|^2_{\mathcal{W}^0(\mathbb{R}^q,H)}+\sum_{|\alpha|\le s}\|D_y^\alpha u\|^2_{\mathcal{W}^0(\mathbb{R}^q,H)}\}^{1/2}.$$

The latter in turn is equivalent to

$$\{\|u\|^2_{\mathcal{W}^0(\mathbb{R}^q,H)}+\sum_{|\alpha|=s}\|D_y^\alpha u\|^2_{\mathcal{W}^0(\mathbb{R}^q,H)}\}^{1/2}.$$

REMARK 1.18. Definition 1.15 also makes sense for a Banach space H with group action. Also some conclusions on edge spaces remain valid in the Banach space case. However, here we focus on Hilbert spaces H.

REMARK 1.19. For $H=H^s(\mathbb{R}^m)$, $(\kappa_\lambda u)(x):=\lambda^{\frac{m}{2}}u(\lambda x)$, $\lambda\in\mathbb{R}_+$, we have

$$\mathcal{W}^s(\mathbb{R}^q,H^s(\mathbb{R}^m))=H^s(\mathbb{R}^m\times\mathbb{R}^q). \tag{1.61}$$

It suffices to show the equivalence of the corresponding norms for $u(x,y)\in\mathcal{S}(\mathbb{R}^m\times\mathbb{R}^q)$ $(=\mathcal{S}(\mathbb{R}^q_y,\mathcal{S}(\mathbb{R}^m_x))$. Let us set $f(x,\eta):=(F_{y\to\eta}u)(x,\eta)$ and $g(\xi,\eta):=(F_{x\to\xi}f)(\xi,\eta)$. Then (up to equivalence of norms)we have

$$\begin{aligned}\|u\|^2_{\mathcal{W}^s(\mathbb{R}^q,H^s(\mathbb{R}^m))}&=\int\langle\eta\rangle^{2s}\|\kappa^{-1}_{\langle\eta\rangle}f(\cdot,\eta)\|^2_{H^s(\mathbb{R}^m)}d\eta\\&=\int\langle\eta\rangle^{2s}\Big\{\int\langle\xi\rangle^{2s}\langle\eta\rangle^m|g(\langle\eta\rangle\xi,\eta)|^2d\xi\Big\}d\eta.\end{aligned}\tag{1.62}$$

Here we employed the relation

$$\begin{aligned}\|\kappa^{-1}_{\langle\eta\rangle}f(\cdot,\eta)\|^2_{H^s(\mathbb{R}^m)}&=\int\langle\xi\rangle^{2s}|(F_{x\to\xi}\kappa^{-1}_{\langle\eta\rangle}f)(\xi,\eta)|^2d\xi\\&=\int\langle\xi\rangle^{2s}|\kappa_{\langle\eta\rangle}(F_{x\to\xi}f)(\xi,\eta)|^2d\xi,\end{aligned}$$

coming from the identity $F\kappa_\lambda^{-1}=\kappa_\lambda F$. Using $\langle\eta\rangle^2\langle\frac{\xi}{\langle\eta\rangle}\rangle^2=\langle\eta\rangle^2+|\xi|^2=\langle\xi,\eta\rangle^2$, for the right hand side of (1.62) we obtain

$$\int\langle\eta\rangle^{2s}\Big\{\int\langle\frac{\xi}{\langle\eta\rangle}\rangle^{2s}|g(\xi,\eta)|^2d\xi\Big\}d\eta=\int\langle\xi,\eta\rangle^{2s}|g(\xi,\eta)|^2d\xi d\eta=\|u\|^2_{H^s(\mathbb{R}^{m+q})}.$$

THEOREM 1.20. *Let H be a Hilbert space with group action $\kappa=\{\kappa_\lambda\}_{\lambda\in\mathbb{R}_+}$, and set $(\chi_\lambda u)(y):=\lambda^{\frac{q}{2}}\kappa_\lambda u(\lambda y)$, $\lambda\in\mathbb{R}_+$, for $u(y)\in\mathcal{W}^s(\mathbb{R}^q,H)$. Then $\chi=(\chi_\lambda)_{\lambda\in\mathbb{R}_+}$ is a group action on $\mathcal{W}^s(\mathbb{R}^q,H)$, and we have*

$$\mathcal{W}^s(\mathbb{R}^p,\mathcal{W}^s(\mathbb{R}^q,H)_\kappa)_\chi=\mathcal{W}^s(\mathbb{R}^{p+q},H)_\kappa$$

for every $s\in\mathbb{R}$.

The proof is a slight generalisation of that of Proposition 1.19.

REMARK 1.21. We have

$$\mathcal{W}^\infty(\mathbb{R}^q,H)=\{u\in L^2(\mathbb{R}^q,H):D_y^\alpha u\in L^2(\mathbb{R}^q,H)\text{ for all }\alpha\in\mathbb{N}^q\}. \tag{1.63}$$

In fact, we have $H^s(\mathbb{R}^q, H) = \{u \in \mathcal{S}'(\mathbb{R}^q, H) : \int \langle\eta\rangle^{2s} \|\hat{u}(\eta)\|_H^2 d\eta < \infty\}$. For every $s \in \mathbb{R}$ there is an $s_1 \in \mathbb{N}$ such that $H^{s_1}(\mathbb{R}^q, H) \hookrightarrow \mathcal{W}^s(\mathbb{R}^q, H)$ is a continuous embedding. Moreover, for every $s_1 \in \mathbb{N}$ there is an $\tilde{s} \in \mathbb{N}$ with a continuous embedding $\mathcal{W}^{\tilde{s}}(\mathbb{R}^q, H) \hookrightarrow H^{s_1}(\mathbb{R}^q, H)$. Thus $\mathcal{W}^\infty(\mathbb{R}^q, H) = \bigcap_{s\in\mathbb{R}} \mathcal{W}^s(\mathbb{R}^q, H) = \bigcap_{s\in\mathbb{N}} H^s(\mathbb{R}^q, H)$. By virtue of

$$H^s(\mathbb{R}^q, H) = \{u \in L^2(\mathbb{R}^q, H) : D_y^\alpha u \in L^2(\mathbb{R}^q, H) \text{ for all } \alpha \in \mathbb{N}^q,\ |\alpha| \le s\}$$

for every $s \in \mathbb{N}$ we then obtain the relation (1.63).

REMARK 1.22. Let $\eta \mapsto [\eta]$ be a strictly positive smooth function in $\mathbb{R}^q$ such that $[\eta] = |\eta|$ for $|\eta| > C$ for some $C > 0$. We have an isomorphism

$$K := F^{-1} \kappa_{[\eta]} F : H^s(\mathbb{R}^q, H) \to \mathcal{W}^s(\mathbb{R}^q, H) \tag{1.64}$$

(see the notation in Definition 1.15). For convenience, in (1.64) we refer to $[\eta]$ rather than $\langle\eta\rangle$), see the formula (1.64). In other words, we have

$$\kappa_{[\eta]}^{-1} F_{y\to\eta} u(\eta) = F_{y\to\eta} \kappa_{[\eta]}^{-1} u(\eta) \in F_{y\to\eta} H^s(\mathbb{R}^q, H) \tag{1.65}$$

for every $u \in \mathcal{W}^s(\mathbb{R}^q, H)$. In particular,

$$F_{\eta\to y}^{-1} \{\kappa_{[\eta]}^{-1} \hat{u}(\eta) : u \in \mathcal{W}^0(\mathbb{R}^q, H)\} = L^2(\mathbb{R}^q, H). \tag{1.66}$$

Set $\mathcal{S}'(\mathbb{R}^q, H) := \mathcal{L}(\mathcal{S}(\mathbb{R}^q), H)$, endowed with the topology of uniform convergence on bounded sets. Moreover, let $\hat{\mathcal{F}}(\mathbb{R}^q, H)$ denote the space of all measurable functions $f : \mathbb{R}^q \to H$ such that for every $N \in \mathbb{N}$ there is a constant $c_N > 0$ such that

$$\|f(\eta)\|_H \le c_N \langle\eta\rangle^{-N} \text{ for all } N \in \mathbb{N}, \tag{1.67}$$

and let $\mathcal{F}(\mathbb{R}^q, H)$ defined to be the set of all $u \in \mathcal{S}'(\mathbb{R}^q, H)$ such that $\hat{u} \in \hat{\mathcal{F}}(\mathbb{R}^q, H)$. The space $\hat{\mathcal{F}}(\mathbb{R}^q, H)$ is Fréchet with the best constants c_N in (1.67) as the semi-norms. The isomorphism $F : \mathcal{F}(\mathbb{R}^q, H) \to \hat{\mathcal{F}}(\mathbb{R}^q, H)$ gives rise to a Fréchet topology also in the space $\mathcal{F}(\mathbb{R}^q, H)$.

THEOREM 1.23. *Let H be a Hilbert space with group action $\kappa = \{\kappa_\lambda\}_{\lambda\in\mathbb{R}_+}$. Then $\mathcal{W}^s(\mathbb{R}^q, H)$, $s \in \mathbb{R}$, coincides with the space of all $u \in \mathcal{S}'(\mathbb{R}^q, H)$ such that $Fu = \hat{u}$ is a measurable function $\mathbb{R}^q \to H$ and*

$$\Big\{\int_{\mathbb{R}^q} \langle\eta\rangle^{2s} \|\kappa_{\langle\eta\rangle}^{-1} \hat{u}(\eta)\|_H^2 đ\eta\Big\}^{\frac{1}{2}} < \infty. \tag{1.68}$$

Concerning a proof see [**26**, Proposition 2.4.(c), combined with Definition 2.3].

2. The cone algebra

2.1. Weighted spaces on a manifold with conical singularities. Let us first recall some notation in connection with weighted Mellin Sobolev spaces. By $\mathcal{A}(U, E)$ for an open $U \subseteq \mathbb{C}$ and a Fréchet space E we denote the space of all holomorphic functions in U with values in E. Consider the Mellin transform (1.10) first for $u \in C_0^\infty(\mathbb{R}_+)$. It is well-known that M induces a continuous operator $M : C_0^\infty(\mathbb{R}_+) \to \mathcal{A}(\mathbb{C})$ where $Mu|_{\Gamma_\beta} \in \mathcal{S}(\Gamma_\beta)$ for every β, uniformly in compact β-intervals, cf. the notation (1.11). The operator

$$M : C_0^\infty(\mathbb{R}_+) \to \mathcal{S}(\Gamma_{1/2-\gamma}),\ \gamma \in \mathbb{R},$$

extends by continuity to an isomorphism

$$M_\gamma : r^\gamma L^2(\mathbb{R}_+) \to L^2(\Gamma_{1/2-\gamma}).$$

Let $\mathcal{H}^{s,\gamma}(\mathbb{R}_+)$ defined to be the completion of $C_0^\infty(\mathbb{R}_+)$ with respect to the norm

$$\|u\|_{\mathcal{H}^{s,\gamma}(\mathbb{R}_+)} := \Big\{ \int_{\Gamma_{1/2-\gamma}} (1+|z|^2)^s |Mu(z)|^2 đz \Big\}^{1/2},$$

$đz := (2\pi i)^{-1}dz$. Set

$$S_\gamma u(x) := e^{-(1/2-\gamma)x}u(e^{-x}),\ x \in \mathbb{R}.$$

The transformation S_γ induces an isomorphism

$$S_\gamma : C_0^\infty(\mathbb{R}_+) \to C_0^\infty(\mathbb{R}) \tag{2.1}$$

with inverse

$$S_\gamma^{-1}v(r) = r^{\gamma-1/2}v(-\log r).$$

We have

$$(M_\gamma u)(\frac{1}{2} - \gamma + i\rho) = (FS_\gamma u)(\rho), \tag{2.2}$$

first for $u \in C_0^\infty(\mathbb{R}_+)$. Here $(Fv)(\rho) = \int_{-\infty}^{\infty} e^{-i\rho x}v(x)dx$ is the standard Fourier transform with its inverse $(F^{-1}g)(x) = (2\pi)^{-1}\int e^{ix\rho}g(\rho)d\rho$.

REMARK 2.1. The transformation (2.1) extends to an isomorphism

$$S_\gamma : \mathcal{H}^{s,\gamma}(\mathbb{R}_+) \to H^s(\mathbb{R})$$

for every $s, \gamma \in \mathbb{R}$. In particular, we get an isomorphism $S_\gamma : r^\gamma L^2(\mathbb{R}_+) \to L^2(\mathbb{R})$.

In fact, using (2.2) we can write $\big(M_{\gamma,r\to z}u\big)(1/2-\gamma+i\rho) = \big(F_{x\to\rho}S_\gamma u\big)(\rho)$ and hence the norm

$$\|u\|_{\mathcal{H}^{s,\gamma}(\mathbb{R}_+)} = \Big\{ \iint \langle 1/2-\gamma+i\rho\rangle^{2s} \big|\big(F_{x\to\rho}S_\gamma u\big)(\rho)\big|^2 đ\rho \Big\}^{1/2},$$

$đ\rho = (2\pi)^{-1}d\rho$, is equivalent to $\|S_\gamma u\|_{H^s(\mathbb{R}_x)}$.

Since $H^s_{\rm loc}$-spaces are invariant under diffeomorphisms we have $\mathcal{H}^{s,\gamma}(\mathbb{R}_+) \subset H^s_{\rm loc}(\mathbb{R}_+)$. Thus the Kegel-spaces

$$\mathcal{K}^{s,\gamma}(\mathbb{R}_+) := \{u = \omega u_0 + (1-\omega)u_\infty : u_0 \in \mathcal{H}^{s,\gamma}(\mathbb{R}_+), u_\infty \in H^s(\mathbb{R})\} \tag{2.3}$$

for any cut-off function $\omega(t)$ are independent of the choice of ω. Moreover, define Kegel-spaces with weight $e \in \mathbb{R}$ at infinity, namely,

$$\mathcal{K}^{s,\gamma;e}(\mathbb{R}_+) := [r]^{-e}\mathcal{K}^{s,\gamma}(\mathbb{R}_+),$$

where $r \mapsto [r]$ is any strictly positive smooth function that is equal to r for sufficiently large r and set

$$\mathcal{S}^\gamma(\mathbb{R}_+) := \varprojlim_{l\in\mathbb{N}} \mathcal{K}^{l,\gamma;l}(\mathbb{R}_+).$$

First let N be a smooth closed manifold, and form the cone $N^\Delta := (\overline{\mathbb{R}}_+ \times N)/(\{0\}\times N)$ where $\{0\}\times N$ represents the vertex v_0. Moreover, set $N^\wedge := \mathbb{R}_+ \times N$. A manifold M with conical singularity v is a topological space such that $M \setminus \{v\}$ is a smooth manifold, and v has a neighbourhood V in M such that there is a homeomorphism

$$\chi^\Delta : V \to N^\Delta \tag{2.4}$$

with $\chi^\Delta(v) = v_0$, and (2.4) restricts to a diffeomorphism

$$\chi^\wedge : V \setminus \{v\} \to N^\wedge. \tag{2.5}$$

The choice of (2.5) determines a splitting of (local) variables (r, x) on $V \setminus \{v\}$ near v. For practical reasons we keep it fixed. Another choice of

$$\tilde{\chi}^\Delta : V \to N^\Delta,\ \tilde{\chi}^\wedge : V \setminus \{v\} \to N^\wedge$$

is said to define an equivalent conical structure of M near v if

$$\tilde{\chi}^\wedge \circ (\chi^\wedge)^{-1} : \mathbb{R}_+ \times N \to \mathbb{R}_+ \times N$$

extends to a diffeomorphism $\overline{\mathbb{R}}_+ \times N \to \overline{\mathbb{R}}_+ \times N$ between the respective manifolds with boundary. Usually we do not employ other splitting of variables. However, this notion of equivalence allows us to distinguish the chosen conical structure from diverse cuspidal configurations relative to the given conical one.

For simplicity we consider here the case of a single conical singularity. This automatically covers the case of finitely many conical singularities since N may have finitely many connected components.

From M we can pass to the stretched manifold $\mathbb{M}$ which is a smooth manifold with boundary $\partial\mathbb{M} \cong N$, obtained by invariantly attaching a copy of N at $M \setminus \{v\}$. Then $\partial\mathbb{M}$ has a collar neighbourhood $\mathbb{V}$ in $\mathbb{M}$ such that there is a diffeomorphism

$$\chi^{\|} : \mathbb{V} \to \overline{\mathbb{R}}_+ \times N \tag{2.6}$$

between the corresponding manifolds with smooth boundary.

A simple example is $M := N^\Delta$. Then $\mathbb{M} = \overline{\mathbb{R}}_+ \times N$.

Note that the quotient map $\overline{\mathbb{R}}_+ \times N \to N^\Delta$ allows us to define a continuous map

$$\pi : \mathbb{M} \to M$$

such that $\pi|_{\mathbb{M}\setminus\partial\mathbb{M}} : \mathbb{M} \setminus \partial\mathbb{M} \to M \setminus \{v\}$ is a diffeomorphism and $\pi|_{\partial\mathbb{M}} : \partial\mathbb{M} \to v$ collapses $\partial\mathbb{M}$ to the point v.

It is often convenient to interpret M as a stratified space, here as a disjoint union

$$M = s_0(M) \cup s_1(M)$$

for the strata $s_0(M) := M \setminus \{v\}$, $s_1(M) := \{v\}$.

Let $\mathrm{Diff}^\mu(\Omega)$ for an open manifold Ω denote the space of all differential operators of order μ with smooth coefficients in local coordinates.

Let $\mathcal{H}^{s,\gamma}(N^\wedge)$ first for $s \in \mathbb{N}$, $\gamma \in \mathbb{R}$, defined to be the set of all $u(r, x) \in r^{\gamma-n/2}L^2(N^\wedge)$ (with $L^2(N^\wedge)$ referring to the measure $drdx$), $n = \dim N$, such that

$$(r\partial_r)^j D_x^\alpha u(r, x) \in r^{\gamma-n/2}L^2(N^\wedge)$$

for all $D_x^\alpha \in \mathrm{Diff}^{|\alpha|}(N)$, $j \in \mathbb{N}$, $j + |\alpha| \leq s$. By duality with respect to the scalar product of $\mathcal{H}^{0,0}(N^\wedge) = r^{-n/2}L^2(N^\wedge)$ we can define $\mathcal{H}^{s,\gamma}(N^\wedge)$ for $-s \in \mathbb{N}$, $\gamma \in \mathbb{R}$, and complex interpolation gives us the spaces

$$\mathcal{H}^{s,\gamma}(N^\wedge) \quad \text{for all} \quad s, \gamma \in \mathbb{R}.$$

It is often useful to employ other equivalent definitions. Let $\mathcal{H}^{s,\gamma}(\mathbb{R}^{n+1}_+)$ be the completion of $C_0^\infty(\mathbb{R}^{n+1}_+)$ with respect to the norm

$$\Big\{ \int_{\Gamma_{\frac{n+1}{2}-\gamma}} \int_{\mathbb{R}^n} \langle z, \xi\rangle^{2s} |(M_{r\to z} F_{x\to\xi} u)(z, \xi)|^2 đz đ\xi \Big\}^{1/2}.$$

Moreover, let $\chi : U \to \mathbb{R}^n$ be a chart on N and form

$$1 \times \chi : \mathbb{R}_+ \times U \to \mathbb{R}^{n+1}_+, \ \ (1 \times \chi)(r, x) := (r, \chi(x)).$$

Then $u(r, x) \in \mathcal{H}^{s,\gamma}(N^\wedge)$ is equivalent to

$$(\varphi u) \circ (1 \times \chi)^{-1} \in \mathcal{H}^{s,\gamma}(\mathbb{R}^{n+1}_+)$$

for any such chart χ and every $\varphi \in C_0^\infty(U)$. Finally, let $R^s(\lambda) \in L^s_{\mathrm{cl}}(N; \mathbb{R}_\lambda)$ be a parameter-dependent elliptic family of operators of order $s \in \mathbb{R}$ that induces isomorphisms

$$R^s(\lambda) : H^m(N) \to H^{m-s}(N)$$

for all $m \in \mathbb{R}$, $\lambda \in \mathbb{R}$. Then $\mathcal{H}^{s,\gamma}(N^\wedge)$ can be equivalently defined as the completion of $C_0^\infty(N^\wedge)$ with respect to the norm

$$\Big\{ \int_{\Gamma_{\frac{n+1}{2}-\gamma}} \| R^s(\operatorname{Im} z)(M_{r\to z} u)(z) \|^2_{L^2(N)} đz \Big\}^{1/2}.$$

Clearly there always exists such an $R^s(\lambda)$.

Definition 2.2. Let M be a compact manifold with conical singularity v, locally near v modelled on N^Δ as described at the beginning. Then

$$H^{s,\gamma}(M) \ \text{ for } s, \gamma \in \mathbb{R} \tag{2.7}$$

denotes the subspace of all $u \in H^s_{\mathrm{loc}}(M \setminus \{v\})$ such that for (2.5) we have

$$\omega(u|_{V\setminus\{v\}} \circ (\chi^\wedge)^{-1}) \in \mathcal{H}^{s,\gamma}(N^\wedge) \tag{2.8}$$

for any cut-off function ω.

Here and in the sequel a cut-off function ω on the half-axis is any real-valued $\omega \in C_0^\infty(\overline{\mathbb{R}}_+)$ such that $\omega \equiv 1$ close to 0.

A cut-off function ω_1 on M means any $\omega_1 \in C(M)$ with compact support in V, $\omega_1|_{V\setminus\{v\}} \in C^\infty(V \setminus \{v\})$, such that $\omega_1 \equiv 1$ close to v.

The spaces $\mathcal{H}^{s,\gamma}(N^\wedge)$ encode some control with respect to the dependence of distributions on r not only for $r \to 0$ but also for $r \to \infty$. Therefore, we distinguish between the notation $\mathcal{H}^{s,\gamma}$ and $H^{s,\gamma}$.

Let us now turn to other spaces globally on the infinite stretched cone $N^\wedge$. Let $\chi : U \to \mathbb{R}^n$ be a chart on N and form

$$\chi_{\mathrm{cone}} : \mathbb{R}_+ \times U \to \mathbb{R}^{n+1}_+, \ \ \chi_{\mathrm{cone}}(r, x) := (r, r\chi(x)).$$

Then $H^s_{\mathrm{cone}}(N^\wedge)$ is defined to be the set of all

$$u \in H^s_{\mathrm{loc}}(\mathbb{R} \times N)|_{N^\wedge} \ \text{ such that } ((1-\omega)\varphi u) \circ \chi^{-1}_{\mathrm{cone}} \in H^s(\mathbb{R}^{n+1}) \tag{2.9}$$

for every such chart χ_{cone}, every $\varphi \in C_0^\infty(U)$, and some cut-off function ω. Observe that for $N = S^n$ (the unit sphere in $\mathbb{R}^{n+1}$) the condition (2.9) is equivalent to

$$u \in H^s_{\mathrm{loc}}(\mathbb{R}^{n+1} \setminus \{0\}) \ \text{ such that } (1-\omega)u \in H^s(\mathbb{R}^{n+1}).$$

In the latter formula the cut-off function is used in the meaning $\omega = \omega(|\tilde{x}|)$, $\tilde{x} \in \mathbb{R}^{n+1}$. Incidentally we also employ spaces with weight $e \in \mathbb{R}$ at infinity:

$$H^{s;e}_{\mathrm{cone}}(N^\wedge) := [r]^{-e} H^s_{\mathrm{cone}}(N^\wedge).$$

Now we set

$$\mathcal{K}^{s,\gamma}(N^\wedge) := \{\omega u_0 + (1-\omega)u_\infty : u_0 \in \mathcal{H}^{s,\gamma}(N^\wedge), u_\infty \in H^s_{\mathrm{cone}}(N^\wedge)\},$$

s, $\gamma \in \mathbb{R}$. The notation $\mathcal{K}^{s,\gamma}$ is chosen to remind of "Kegel".

2.2. Mellin pseudo-differential operators. The cone calculus can be developed in terms of Mellin pseudo-differential operators in $r \in \mathbb{R}_+$. In the case of scalar amplitude functions $f(r, r', z) \in S^{\mu}_{(\mathrm{cl})}(\mathbb{R}_+ \times \mathbb{R}_+ \times \Gamma_{1/2-\gamma})$, $z = 1/2 - \gamma + i\rho$, we set

$$\mathrm{op}^{\gamma}_{M}(f)u(r) := \int_{\mathbb{R}} \int_{\mathbb{R}_+} \left(\frac{r}{r'}\right)^{1/2-\gamma+i\rho} f(r, r', 1/2-\gamma+i\rho)u(r')\frac{dr'}{r'} đ\rho,$$

$đ\rho = (2\pi)^{-1} d\rho$, first for $u \in C^{\infty}_{0}(\mathbb{R}_+)$ and then extended to larger distribution spaces. In the case of an r'-independent Mellin symbol f we have

$$\mathrm{op}^{\gamma}_{M}(f) = M^{-1}_{\gamma} f M_{\gamma}$$

for the weighted Mellin transform M_{γ}. More generally we admit operator-valued amplitude functions

$$f(r, r', z) \in C^{\infty}(\mathbb{R}_+ \times \mathbb{R}_+, L^{\mu}_{(\mathrm{cl})}(N; \Gamma_{1/2-\gamma})).$$

Most of the generalities are valid both for the classical and non-classical case. However, in the cone calculus there is a specific point where we need classical symbols. Therefore, from now on we concentrate on the classical case.

Definition 2.3. The space $M^{\mu}_{\mathcal{O}}(N; \mathbb{R}^l)$, $\mu \in \mathbb{R}$, is defined to be the set of all $h \in \mathcal{A}(\mathbb{C}, L^{\mu}_{\mathrm{cl}}(N; \mathbb{R}^l))$ such that

$$h|_{\Gamma_\beta} \in L^{\mu}_{\mathrm{cl}}(N; \Gamma_\beta \times \mathbb{R}^l)$$

for every real β, uniformly in compact intervals.

The space $M^{\mu}_{\mathcal{O}}(N; \mathbb{R}^l)$ is (nuclear) Fréchet with a semi-norm system that follows directly from the definition. For $l = 0$ we write $M^{\mu}_{\mathcal{O}}(N)$.

2.3. Spaces and operators with asymptotics. In this subsection we discuss discrete asymptotics of weighted distributions on a manifold with conical singularity and operators that are associated with such asymptotics.

A sequence

$$\mathcal{P} := \{(p_j, m_j)\}_{j=0,\ldots,J} \subset \mathbb{C} \times \mathbb{N},$$

$J = J(\mathcal{P}) \in \mathbb{N} \cup \{\infty\}$, is called a (discrete) asymptotic type, associated with the weight data (γ, Θ), $\Theta = (\vartheta, 0]$, $-\infty \le \vartheta < 0$, $\gamma \in \mathbb{R}$, if the following holds:

$$\pi_{\mathbb{C}}\mathcal{P} := \{p_j\}_{j=0,\ldots,J} \subset \Big\{\frac{n+1}{2} - \gamma + \vartheta < \mathrm{Re}\, z < \frac{n+1}{2} - \gamma\Big\}.$$

If Θ is finite then $\pi_{\mathbb{C}}\mathcal{P}$ is finite, otherwise $\mathrm{Re}\, p_j \to -\infty$ as $j \to \infty$.

Singular functions with asymptotics of type $\mathcal{P}$ will be of the form

$$\omega(r)c(x)r^{-p_j}\log^k r, \ \ 0 \le k \le m_j, \ \ c \in C^{\infty}(N),$$

for $(p_j, m_j) \in \mathcal{P}$ and a cut-off function ω. For finite Θ we set

$$\mathcal{E}_{\mathcal{P}}(N^{\wedge}) := \Big\{\sum_{j=0}^{J}\sum_{k=0}^{m_j} \omega(r)c_{jk}(x)r^{-p_j}\log^k r : 0 \le k \le m_j, c_{jk} \in C^{\infty}(N)\Big\}.$$

The space $\mathcal{E}_{\mathcal{P}}(N^{\wedge})$ is Fréchet as a direct sum of finitely many copies of $C^{\infty}(N)$. Observe that $\pi_{\mathbb{C}}\mathcal{P} \subset \{\mathrm{Re}\, z < \frac{n+1}{2} - \gamma\}$ entails $\mathcal{E}_{\mathcal{P}}(N^{\wedge}) \subset \mathcal{K}^{\infty,\gamma}(N^{\wedge})$. Moreover, let us consider spaces of functions of flatness Θ with respect to the weight γ, namely,

$$\mathcal{K}^{s,\gamma}_{\Theta}(N^{\wedge}) := \varprojlim_{i \in \mathbb{N}} \mathcal{K}^{s,\gamma-\vartheta-(1+i)^{-1}}(N^{\wedge}).$$

For $\Theta = (-\infty, 0]$ we may set $\mathcal{K}^{s,\gamma}_{\Theta}(N^\wedge) := \mathcal{K}^{s,\infty}(N^\wedge) = \bigcap_{\delta\in\mathbb{R}} \mathcal{K}^{s,\delta}(N^\wedge)$. Set

$$\mathcal{K}^{s,\gamma}_{\mathcal{P}}(N^\wedge) := \mathcal{K}^{s,\gamma}_{\Theta}(N^\wedge) + \mathcal{E}_{\mathcal{P}}(N^\wedge)$$

as a direct sum of Fréchet spaces. In the case $\Theta = (-\infty, 0]$ we first form $\Theta_l := (-(l+1), 0]$, $l \in \mathbb{N}$, $\mathcal{P}_l := \{(p, m) \in \mathcal{P} : \operatorname{Re} p > \frac{n+1}{2} - \gamma - (l+1)\}$, and set

$$\mathcal{K}^{s,\gamma}_{\mathcal{P}}(N^\wedge) := \varprojlim_{l\in\mathbb{N}} \mathcal{K}^{s,\gamma}_{\mathcal{P}_l}(N^\wedge).$$

Moreover, let

$$\mathcal{K}^{s,\gamma;e}(N^\wedge) := [r]^{-e}\mathcal{K}^{s,\gamma;e}(N^\wedge), \quad \mathcal{K}^{s,\gamma;e}_{\mathcal{P}}(N^\wedge) := [r]^{-e}\mathcal{K}^{s,\gamma;e}_{\mathcal{P}}(N^\wedge),$$

$e \in \mathbb{R}$. For a manifold M with conical singularities we set

$$H^{s,\gamma}_{\mathcal{P}}(M) := \{u \in H^{s,\gamma}(M) : \omega(u|_{V\setminus\{v\}} \circ (\chi^\wedge)^{-1}) \in \mathcal{K}^{s,\gamma}_{\mathcal{P}}(N^\wedge)\},$$

for any cut-off function ω, cf. analogously the Definition 2.2.

In the future we will often identify $V\setminus\{v\}$ with $N^\wedge$ and for abbreviation drop the pull backs under $\chi^\wedge$ when we talk about functions or operators in a neighbourhood of the conical singularity v. Then we simply write

$$H^{s,\gamma}_{\mathcal{P}}(M) = \{u \in H^{s,\gamma}(M) : \omega u \in \mathcal{K}^{s,\gamma}_{\mathcal{P}}(N^\wedge)\}.$$

DEFINITION 2.4. (i) A $G \in \bigcap_{s\in\mathbb{R}} \mathcal{L}(\mathcal{K}^{s,\gamma}(N^\wedge), \mathcal{K}^{\infty,\gamma-\mu}(N^\wedge))$ is called a Green operator on $N^\wedge$ for a closed smooth manifold N if there are (G-dependent) asymptotic types $\mathcal{P}$ and $\mathcal{Q}$ such that

$$G \in \bigcap_{s,e\in\mathbb{R}} \mathcal{L}(\mathcal{K}^{s,\gamma;e}(N^\wedge), \mathcal{K}^{\infty,\gamma-\mu;\infty}_{\mathcal{P}}(N^\wedge)),$$

$$G^* \in \bigcap_{s,e\in\mathbb{R}} \mathcal{L}(\mathcal{K}^{s,-\gamma+\mu;e}(N^\wedge), \mathcal{K}^{\infty,-\gamma;\infty}_{\mathcal{Q}}(N^\wedge)).$$

(ii) A $G \in \bigcap_{s\in\mathbb{R}} \mathcal{L}(H^{s,\gamma}(M), H^{\infty,\gamma-\mu}(M))$ is called a Green operator on the (compact) manifold M with conical singularity if there are asymptotic types $\mathcal{P}$ and $\mathcal{Q}$ such that

$$G \in \bigcap_{s\in\mathbb{R}} \mathcal{L}(H^{s,\gamma}(M), H^{\infty,\gamma-\mu}_{\mathcal{P}}(M)), \quad G^* \in \bigcap_{s\in\mathbb{R}} \mathcal{L}(H^{s,-\gamma+\mu}(M), H^{\infty,-\gamma}_{\mathcal{Q}}(M)).$$

By $L_G(\cdot, \boldsymbol{g})$ for $\boldsymbol{g} := (\gamma, \gamma-\mu, \Theta)$ we denote the space of Green operators on $N^\wedge$ and M, respectively. The subspaces with fixed $\mathcal{P}$, $\mathcal{Q}$ will be denoted by $L_G(\cdot, \boldsymbol{g})_{\mathcal{P},\mathcal{Q}}$.

Green operators play the role of smoothing operators in the cone calculus. Those are compact as operators

$$G : \mathcal{K}^{s,\gamma}(N^\wedge) \to \mathcal{K}^{s-\mu,\gamma-\mu}(N^\wedge)$$

and

$$G : H^{s,\gamma}(M) \to H^{s-\mu,\gamma-\mu}(M),$$

respectively. Another ingredient are the smoothing Mellin operators which are, in general, not compact.

A sequence

$$\mathcal{R} := \{(r_j, l_j)\}_{j\in\mathbb{J}} \subset \mathbb{C} \times \mathbb{N},$$

$\mathbb{J} = \mathbb{J}(\mathcal{R}) \subseteq \mathbb{Z}$, is called a (discrete) Mellin asymptotic type if $\pi_{\mathbb{C}}\mathcal{R} := \{r_j\}_{j\in\mathbb{J}}$ intersects every strip $\{c < \operatorname{Re} z < c'\}$, $c < c'$, in a finite set.

Let $M^{-\infty}_{\mathcal{R}}(N)$ denote the subspace of all

$f \in \mathcal{A}(\mathbb{C} \setminus \pi_{\mathbb{C}}\mathcal{R}, L^{-\infty}(N))$ that are meromorphic with poles at the points r_j of multiplicity $l_j + 1$, where the Laurent coefficients at $(z - r_j)^{-(k+1)}$, $0 \le k \le m_j$, are operators in $L^{-\infty}(N)$ of finite rank, and for any $\pi_{\mathbb{C}}\mathcal{R}$-excision function χ we have $\chi f|_{\Gamma_\beta} \in \mathcal{S}(\Gamma_\beta, L^{-\infty}(N))$ for every $\beta \in \mathbb{R}$, uniformly in finite β-intervals.

We formulate smoothing Mellin operators first for finite $\Theta := (-(k+1), 0]$, $k \in \mathbb{N}$. Starting with a sequence

$$f_j \in M^{-\infty}_{\mathcal{R}_j}(N),\ j = 0, \ldots, k,$$

of Mellin symbols for Mellin asymptotic types $\mathcal{R}_j$ we form operators

$$M := r^{-\mu} \sum_{j=0}^{k} \omega r^j \mathrm{op}_M^{\gamma_j - n/2}(f_j)\omega' \tag{2.10}$$

for weights $\gamma_j \in \mathbb{R}$ such that

$$\gamma - j \le \gamma_j \le \gamma,\ \ \Gamma_{\frac{n+1}{2} - \gamma_j} \cap \pi_{\mathbb{C}}\mathcal{R}_j = \emptyset,\ j = 0, \ldots, k, \tag{2.11}$$

for cut-off functions ω, ω'.

REMARK 2.5. The Mellin symbols f_j, $j = 0, \ldots, k$, are uniquely determined by the operator (2.10) as a map $C_0^\infty(N^\wedge) \to C^\infty(N^\wedge)$. More precisely, if $\tilde{M}$ is another operator with the same Mellin symbols but other weights $\tilde{\gamma}_j$ satisfying analogous conditions as (2.11), and other cut-off functions $\tilde{\omega}$, $\tilde{\omega}'$ rather than ω, ω', then we have

$$M - \tilde{M} \in L_G(N^\wedge, \boldsymbol{g})$$

for $\boldsymbol{g} = (\gamma, \gamma - \mu, \Theta)$.

The way of recovering the sequence of f_j can be found in [**55**, Subsection 1.3.1, Theorem 4].

We define $L_{M+G}(\cdot, \boldsymbol{g})$, $\boldsymbol{g} = (\gamma, \gamma - \mu, \Theta)$, $\Theta := (-(k+1), 0]$, as the set of all operators $M + G$ for arbitrary $G \in L_G(\cdot, \boldsymbol{g})$ and M of the form (2.10); here $\cdot$ stands for $N^\wedge$ or M.

For $\Theta = (-\infty, 0]$ we form $\Theta_k := (-(k+1), 0]$, $k \in \mathbb{N}$, and then define

$$L_{M+G}(\cdot, \boldsymbol{g}) := \bigcap_{k \in \mathbb{N}} L_{M+G}(\cdot, \boldsymbol{g}_k)$$

for $\boldsymbol{g}_k := (\gamma, \gamma - \mu, \Theta_k)$.

The operators in $L_{M+G}(\cdot, \boldsymbol{g})$ are called smoothing Mellin plus Green operators of the cone calculus.

The pseudo-differential calculus on a manifold M with conical singularities is a substructure of $L^\mu_{\mathrm{cl}}(M \setminus \{v\})$, $\mu \in \mathbb{R}$, consisting of pseudo-differential operators with the typical Fuchs type degenerate behaviour near v, locally in the splitting of variables $(r, x) \in N^\wedge$. A first case of a manifold with conical singularity is the infinite cone N^Δ where we talk about the calculus over the open stretched cone $N^\wedge$ containing a specific control of operators for $r \to \infty$. Moreover, we have the case of a compact M with conical singularities. In the following we write $\varphi \prec \varphi'$ for functions φ, φ' if the function φ' is equal to 1 on $\mathrm{supp}\,\varphi$.

DEFINITION 2.6. The space $L^\mu(N^\wedge, \boldsymbol{g})$ for $\boldsymbol{g} = (\gamma, \gamma - \mu, \Theta)$, $\Theta = (-(k+1), 0]$, $k \in \mathbb{N} \cup \{\infty\}$, is defined to be the set of all operators

$$A = \omega r^{-\mu} \mathrm{op}_M^{\gamma - n/2}(h)\omega' + (1 - \omega)A_\infty(1 - \omega'') + M + G \tag{2.12}$$

where $\omega'' \prec \omega \prec \omega'$ are cut-off functions and $h(r,z) \in C^\infty(\overline{\mathbb{R}}_+, M^\mu_{\mathcal{O}}(N))$, cf. Definition 2.3, $M+G \in L_{M+G}(N^\wedge, \boldsymbol{g})$, and $A_\infty \in L^{\mu;0}_{\mathrm{cl}}(N^\asymp)|_{N^\wedge}$, in the sense of [**58**, page 168] where $N^\asymp \cong \mathbb{R} \times N$ is treated as a manifold with conical exits $r \to \pm\infty$.

For a compact manifold M with conical singularity v the space $L^\mu(M, \boldsymbol{g})$ is defined to be the set of all $A \in L^\mu_{\mathrm{cl}}(M \setminus \{v\})$ that are modulo $L_G(M, \boldsymbol{g})$ locally near v of the form (2.12).

The symbols of operators $A \in L^\mu(N^\wedge, \boldsymbol{g})$ which determine ellipticity as well as lower order terms form a triple

$$\sigma(A) = (\sigma_0(A), \sigma_1(A), \sigma_{\mathrm{e}}(A)) \tag{2.13}$$

where e indicates the triple of principal exit symbolic components, explained in [**58**]. Moreover, $\sigma_0(A)$ is simply the homogeneous principal symbol of A of order μ as an operator in $L^\mu_{\mathrm{cl}}(N^\wedge)$. Clearly it is degenerate in the sense that locally near $r=0$ in the splitting of variables $(r,x) \in \mathbb{R}_+ \times X$ and covariables (ρ, ξ)

$$\sigma_0(A)(r,x,\rho,\xi) = r^{-\mu}\tilde{\sigma}_0(A)(r,x,r\rho,\xi)$$

for a function $\tilde{\sigma}_0(A)(r,x,\tilde{\rho},\xi)$, $(\tilde{\rho},\xi) \neq 0$, smooth up to $r=0$. Moreover,

$$\sigma_1(A)(z) := h(0,z) + f_0(z) : H^s(N) \to H^{s-\mu}(N) \tag{2.14}$$

is an operator function between Sobolev spaces on N, depending on $z \in \Gamma_{(n+1)/2-\gamma}$.

In the case $A \in L^\mu(M, \boldsymbol{g})$ we set

$$\sigma(A) = (\sigma_0(A), \sigma_1(A)).$$

The calculus of operators $A \in L^\mu(\cdot, \boldsymbol{g})$ defines continuous operators

$$A : \mathcal{K}^{s,\gamma}(N^\wedge) \to \mathcal{K}^{s-\mu,\gamma-\mu}(N^\wedge) \quad \text{and} \quad H^{s,\gamma}(M) \to H^{s-\mu,\gamma-\mu}(M), \tag{2.15}$$

respectively, and

$$A : \mathcal{K}^{s,\gamma}_{\mathcal{P}}(N^\wedge) \to \mathcal{K}^{s-\mu,\gamma-\mu}_{\mathcal{Q}}(N^\wedge) \quad \text{and} \quad H^{s,\gamma}_{\mathcal{P}}(M) \to H^{s-\mu,\gamma-\mu}_{\mathcal{Q}}(M),$$

for all s and asymptotic types $\mathcal{P}$ and some resulting $\mathcal{Q}$. Moreover, we have compositions and formal adjoints within the calculus.

The ellipticity with respect to $\sigma(A)$ is equivalent to the Fredholm property of the operators (2.15), there is a parametrix $P \in L^{-\mu}(\cdot, \boldsymbol{g}^{-1})$ for $\boldsymbol{g}^{-1} := (\gamma-\mu, \gamma, \Theta)$. If A is elliptic, then

$$Au = f \in \mathcal{K}^{s-\mu,\gamma-\mu}(N^\wedge) \ (H^{s-\mu,\gamma-\mu}(M))$$

and $u \in \mathcal{K}^{-\infty,\gamma}(N^\wedge)$ $(H^{-\infty,\gamma-\mu}(M))$ entails

$$u \in \mathcal{K}^{s,\gamma}(N^\wedge) \ (H^{s,\gamma-\mu}(M)).$$

In addition, if the right hand side f belongs to a corresponding subspace with asymptotics of type $\mathcal{Q}$, then the solution u belongs to a corresponding subspace with asymptotics of a resulting type $\mathcal{P}$. All this may be found, for instance, in [**55**] or [**58**]. In the present exposition we mainly refer to the cone calculus on $N^\wedge$ as the background of the edge calculus.

2.4. Continuous asymptotics. The edge calulus below contains cone pseudo-differential operators smoothly depending on variables $y \in \Omega$ for an open set $\Omega \subseteq \mathbb{R}^q$. Smooth dependence of an operator (2.12) on y can be defined by

$$h(r, y, z) \in C^\infty(\overline{\mathbb{R}}_+ \times \Omega, M^\mu_{\mathcal{O}}(X)),$$

$$A_\infty(y) \in C^\infty(\Omega, L^{\mu;0}_{\mathrm{cl}}(N^\asymp)|_{N^\wedge}),$$

together with smooth dependence of M via $f_j \in C^\infty(\Omega, M^{-\infty}_{\mathcal{R}_j}(N))$ in (2.10) and $G(y) \in \bigcup_{\mathcal{P},\mathcal{Q}} C^\infty(\Omega, L_G(\cdot, \boldsymbol{g})_{\mathcal{P},\mathcal{Q}})$. However, in such a definition the involved asymptotic types $\mathcal{R}_j$ and $\mathcal{P}$, $\mathcal{Q}$ are independent of y.

The ellipticity of $A(y)$ with respect to σ_1 requires the bijectivity of (2.14) for $z \in \Gamma_{(n+1)/2-\gamma}$ for all $y \in \Omega$, but $\sigma_1(A)^{-1}(y, z)$ is a family of meromorphic operator functions with poles in the complex plane that are variable in y and, in general, with variable and branching multiplicities. A tool for understanding asymptotics of solutions to families of elliptic equations $A(y)u = f$ under varying $y \in \Omega$ is the scenario of continuous asymptotics, first introduced and studied in [**45**], [**47**]. Later on it has been systematically developed in [**51**], [**55**], see also [**28**, Subsection 2.3], or [**58**]. We do not present here the details of continuous and branching asymptotics. Let us only say that the singular functions of continuous asymptotics are superpositions of singular functions with discrete asymptotics with "densities" that are analytic functionals in the complex Mellin plane.

For a first impression on the nature of our algebras of singular pseudo-differential operators it suffices to imagine discrete asymptotics. Nevertheless, for parametrices of elliptic operators and elliptic regularity of solutions with asymptotics the notion of asymptotics has to be generalised. However, under a suitable extra condition on the poles of $\sigma_1(A)^{-1}(y, z)$ also parametrices belong to the spaces with discrete asymptotics. In any case, in order to have simpler formulations, from now in this exposition we admit continuous asymptotic types $\mathcal{P}$ for weighted distributions, associated with weight data (γ, Θ). Those belong to corresponding Fréchet spaces $\mathcal{K}^{s,\gamma}_{\mathcal{P}}(N^\wedge)$, $H^{s,\gamma}_{\mathcal{P}}(M)$, etc. We also have spaces of Mellin symbols $M^\mu_{\mathcal{R}}(N)$ with continuous asymptotic types $\mathcal{R}$.

There are very interesting substructures with variable discrete asymptotic types $\mathcal{P}(y)$ and $\mathcal{R}(y)$, respectively. We do not deepen here the information, but refer to several articles on variable discrete asymptotics, see, for instance, [**69**], [**68**], and the bibliographies there. The main reason to employ here continuous asymptotics is that the edge calculus can be described in a more concise way in that framework, without elaborate explanations on the kind of y-dependent poles of conormal symbols of elliptic operators.

3. The edge algebra

3.1. Weighted spaces on a manifold with edge. If N is a smooth closed manifold then the wedge $W := N^\Delta \times \Omega$ for any open $\Omega \subseteq \mathbb{R}^q$ is an example of a manifold with edge, here Ω. The space W is at the same time a trivial N^Δ-bundle over Ω where the fibre is just the cone $N^\Delta = (\overline{\mathbb{R}}_+ \times N)/(\{0\} \times N)$. Let us also illustrate some other notation in this example, e.g., the stretched manifold $\mathbb{W} := \overline{\mathbb{R}}_+ \times N \times \Omega$ associated with W which is a smooth manifold with boundary $\partial\mathbb{W} = N \times \Omega$ (identified with $\{0\} \times N \times \Omega$) which is in turn a trivial N-bundle over Ω. There is then a canonical projection

$$\pi : \mathbb{W} \to W$$

which restricts to the bundle projection $\pi|_{\partial\mathbb{W}} : \partial\mathbb{W} \to \Omega$ and a diffeomorphism $\pi|_{\operatorname{int}\mathbb{W}} : \operatorname{int}\mathbb{W} \to W \setminus \Omega$. We may interpret W also as a stratified space with the strata

$$s_0(W) := W \setminus \Omega = N^\wedge \times \Omega, \;\; s_1(W) := \Omega;$$

both are smooth manifolds, and we have

$$W = s_0(W) \cup s_1(W)$$

as a disjoint union.

More generally, a topological space M is a manifold with edge $Y := s_1(M)$ if both $s_1(M)$ and $s_0(M) := M \setminus s_1(M)$ are smooth manifolds, and Y has a neighbourhood V in M which has the structure of a (locally trivial) N^Δ-bundle for a smooth closed manifold N. Denoting any trivialisation by

$$\chi_G^\Delta : V|_G \to N^\Delta \times \Omega$$

which restricts to a chart $\chi_G : G \to \Omega$ on Y, for some coordinate neighbourhood G on Y, we have a restriction

$$\chi_G^\wedge : (V \setminus Y)|_G \to N^\wedge \times \Omega. \tag{3.1}$$

Let $(r, x, y) \in N^\wedge \times \Omega$ denote the corresponding local splitting of variables. Another choice of a trivialisation

$$\tilde{\chi}_G^\Delta : V|_G \to N^\Delta \times \tilde{\Omega}$$

is said to belong to an equivalent wedge structure if

$$\tilde{\chi}_G^\wedge \circ (\chi_G^\wedge)^{-1} : \mathbb{R}_+ \times N \times \Omega \to \mathbb{R}_+ \times N \times \tilde{\Omega},$$

$(r, x, y) \to (\tilde{r}, \tilde{x}, \tilde{y})$, is the restriction of a diffeomorphism $\overline{\mathbb{R}}_+ \times N \times \Omega \to \overline{\mathbb{R}}_+ \times N \times \tilde{\Omega}$ to $\mathbb{R}_+ \times N \times \Omega$. This gives us a cocycle of transition maps $N \times \Omega \to N \times \tilde{\Omega}$ of a corresponding N-bundle over Y. Similarly as in the case of conical singularities we fix a system of trivialisations (not necessarily a maximal one) and assume for simplicity that the transition maps $\tilde{r} = \tilde{r}(r, x, y)$, $\tilde{x} = \tilde{x}(r, x, y)$, $\tilde{y} = \tilde{y}(r, x, y)$ are independent of r for r in a small neighbourhood of 0 (at $r = 0$ we have $\tilde{r}(0, x, y) = 0$ and $\tilde{y}(0, x, y)$ is independent of x).

By invariantly attaching the above-mentioned N-bundle over Y to $M \setminus Y$ we obtain the stretched manifold $\mathbb{M}$ associated with M. It is a smooth manifold with boundary $\partial\mathbb{M}$ (just our N-bundle), and we have again a canonical projection

$$\pi : \mathbb{M} \to M \tag{3.2}$$

which restricts to the bundle projection

$$\pi_{\partial\mathbb{M}} : \partial\mathbb{M} \to Y$$

and to a diffeomorphism $\pi_{\operatorname{int}\mathbb{M}} : \operatorname{int}\mathbb{M} \to W \setminus Y$.

Similarly as in BVPs it makes sense to study operators between distributional sections of vector bundles over the manifold M with edge Y. It will be convenient to admit bundles $E, F \in \operatorname{Vect}(\mathbb{M})$, not only on M itself. The case of bundles over M is a special case; we can pass to pull backs to $\mathbb{M}$ under the projection (3.2).

For any $E \in \operatorname{Vect}(\mathbb{M})$ we have the restriction $E' := E|_{\partial\mathbb{M}} \in \operatorname{Vect}(\partial\mathbb{M})$ and restrictions to the fibres N_y of $\partial\mathbb{M}$ over $y \in Y$, denoted by $E''_y := E'|_{N_y}$. If $\partial\mathbb{M}$ is locally described by $N \times \Omega$ then we simply write E'' rather than E''_y.

For brevity we denote the restriction of an $E \in \operatorname{Vect}(\mathbb{M})$ to $M \setminus \partial\mathbb{M} \cong M \setminus Y$ again by E.

Let $\mathrm{Diff}^\mu(\cdot\,; J, G)$ denote the space of differential operators of order μ between the C^∞ sections of bundles J, G over the respective smooth manifold, indicated by "$\cdot$".

If M is a manifold with edge Y, and $E, F \in \mathrm{Vect}(\mathbb{M})$, an operator $A \in \mathrm{Diff}^\mu(M\setminus Y; E, F)$ is said to be edge-degenerate if close to Y in the above-mentioned splitting of variables $(r, x, y) \in N^\wedge \times \Omega$ it is of the form

$$A = r^{-\mu} \sum_{j+|\alpha|\le\mu} a_{j\alpha}(r, y)(-r\partial_r)^j (rD_y)^\alpha \tag{3.3}$$

for coefficients $a_{j\alpha} \in C^\infty(\overline{\mathbb{R}}_+ \times \Omega, \mathrm{Diff}^{\mu-(j+|\alpha|)}(N; E'', F''))$. The general idea of the edge calculus is to establish a subalgebra of $\bigcup_{\mu\in\mathbb{R}} L^\mu_{\mathrm{cl}}(M\setminus Y; E, F)$ that contains all edge-degenerate differential operators together with the parametrices of elliptic elements.

The ellipticity in this situation is again determined by a principal symbolic hierarchy, here denoted by

$$\sigma(A) = (\sigma_0(A), \sigma_1(A)),$$

where the components are contributed by the strata $M\setminus Y$ and Y, respectively. The first component $\sigma_0(A) : \pi^*_{M\setminus Y}E \to \pi^*_{M\setminus Y}F$ is the homogeneous principal symbol of A over $M \setminus Y$. The principal edge symbol σ_1 in the pseudo-differential case will be defined below. The situation is similar to boundary value problems; the boundary symbol is an analogue of σ_1. In the case of differential operators (3.3) it is defined as

$$\sigma_1(A)(y, \eta) := r^{-\mu} \sum_{j+|\alpha|\le\mu} a_{j\alpha}(0, y)(-r\partial_r)^j (r\eta)^\alpha,$$

as an operator function

$$\sigma_1(A)(y, \eta) : \mathcal{K}^{s,\gamma}(N^\wedge, E^\wedge) \to \mathcal{K}^{s-\mu,\gamma-\mu}(N^\wedge, F^\wedge), \tag{3.4}$$

$(y, \eta) \in \Omega \times (\mathbb{R}^q \setminus \{0\})$, where in this case $E^\wedge$, $F^\wedge$ are interpreted as pull backs from N to $N^\wedge$ of the former bundles E'', F'' under the canonical projection $N^\wedge \to N$.

The smoothness s is arbitrary, while the weight γ will be fixed in the discussion of ellipticity. From BVPs we already know that ellipticity requires additional conditions, here edge conditions in place of boundary conditions. This will be the topic in Section 4. In the present subsection we introduce weighted Sobolev spaces that play a major role for the edge algebra.

Recall that in boundary value problems the presence of a boundary contributes some anisotropic aspects to the calculus. More precisely, we distinguish between the tangential variables and a variable transversal to the boundary. Nevertheless in operators with the transmission property we could argue in terms of the negative counterpart of the configuration with boundary, and the Sobolev spaces over a neighbouring manifold do not "feel" the presence of the boundary. However, in edge-degenerate operators and, as we shall see below, in associated distribution spaces we have a more relevant difference between variables on the edge and the transversal cone.

REMARK 3.1. The space $H := \mathcal{K}^{s,\gamma;e}(N^\wedge)$ for any fixed s, γ, $e \in \mathbb{R}$ is a Hilbert space with group action $\kappa = \{\kappa_\lambda\}_{\lambda\in\mathbb{R}_+}$,

$$(\kappa_\lambda u)(r, x) = \lambda^{(n+1)/2} u(\lambda r, x), \lambda \in \mathbb{R}_+, n = \dim N.$$

It induces a group action on the Fréchet subspace $\mathcal{K}_{\mathcal{P}}^{s,\gamma;e}(N^\wedge)$ for any asymptotic type $\mathcal{P}$ associated with the weight data (γ,Θ).

For $e=0$ we obtain weighted edge spaces

$$\mathcal{W}^s(\mathbb{R}^q,\mathcal{K}^{s,\gamma}(N^\wedge)) \tag{3.5}$$

of smoothness $s\in\mathbb{R}$ and weight $\gamma\in\mathbb{R}$, and also subspaces $\mathcal{W}^s(\mathbb{R}^q,\mathcal{K}_{\mathcal{P}}^{s,\gamma}(N^\wedge))$ with edge asymptotics of type $\mathcal{P}$.

Note that Proposition 1.19 has the consequence that

$$H^s_{\text{comp}}(N^\wedge\times\mathbb{R}^q)\subset\mathcal{W}^s(\mathbb{R}^q,\mathcal{K}^{s,\gamma}(N^\wedge))\subset H^s_{\text{loc}}(N^\wedge\times\mathbb{R}^q)$$

for every s, $\gamma\in\mathbb{R}$, cf. [**58**]. This allows us to define weighted spaces modelled on (3.5) on a manifold M with edge Y. For convenience we first assume that M is compact.

Definition 3.2. Let M be a compact manifold with edge Y, locally near Y modelled on $N^\Delta\times\mathbb{R}^q$ as described at the beginning. Then

$$H^{s,\gamma}(M) \text{ for } s,\ \gamma\in\mathbb{R} \tag{3.6}$$

denotes the subspace of all $u\in H^s_{\text{loc}}(M\setminus Y)$ such that for all "charts" of the kind (3.1) for $\Omega:=\mathbb{R}^q$ and a finite open covering of Y by such coordinate neighbourhoods G we have

$$\omega(u|_{V|_G\setminus G}\circ(\chi_G^\wedge)^{-1})\in\mathcal{W}^s(\mathbb{R}^q,\mathcal{K}^{s,\gamma}(N^\wedge)) \tag{3.7}$$

for any cut-off function ω.

Recall that cut-off functions on the half-axis have been defined after Definition 2.2.

By a cut-off function θ on M we understand any $\theta\in C(M)$ with compact support in V such that $\theta\equiv1$ close to Y and $\theta|_{V\setminus Y}\in C^\infty(V\setminus Y)$.

The notation (3.6) is motivated in a similar manner as (2.7). In fact, although the spaces (3.5) are locally close to Y modelled on (3.6) the notation $\mathcal{W}^s(\mathbb{R}^q,\mathcal{K}^{s,\gamma}(N^\wedge))$ indicates some extra control for $|y|\to\infty$ and $r\to\infty$ which is cut off in the meaning of (3.6).

For convenience we often drop the substitution of variables coming from $\chi_G^\wedge$; instead of (3.7) we simply write

$$\omega u\in\mathcal{W}^s(\mathbb{R}^q,\mathcal{K}^{s,\gamma}(N^\wedge)) \tag{3.8}$$

for any cut-off function ω.

There is a straightforward extension of the definition of our spaces to distributional sections of vector bundles $E\in\text{Vect}(\mathbb{M})$. Remark 3.1 is valid in analogous form for the spaces $\mathcal{K}^{s,\gamma;e}(N^\wedge,E^\wedge)$, cf. also the notation in formula (3.4), which gives us the spaces

$$\mathcal{W}^s(\mathbb{R}^q,\mathcal{K}^{s,\gamma}(N^\wedge,E^\wedge)). \tag{3.9}$$

For a compact manifold M with edge Y and $E\in\text{Vect}(\mathbb{M})$ we define (in abuse of notation)

$$H^{s,\gamma}(M,E) \text{ for } s,\ \gamma\in\mathbb{R} \tag{3.10}$$

as the subspace of all $u\in H^s_{\text{loc}}(M\setminus Y,E)$ such that an analogue of (3.7) holds, now for the space (3.9) on the right and for a generalisation of the meaning of $(\chi_G^\wedge)^{-1}$ which involves also the corresponding trivialisations of E over $V|_G\setminus G$.

3.2. Edge amplitude functions. Edge amplitude functions will be specific operator-valued symbols in the sense of Definition 1.14. The values of the symbols will belong to the cone algebra $L^\mu(N^\wedge, \boldsymbol{g}; E^\wedge, F^\wedge)$ for $\boldsymbol{g} = (\gamma, \gamma - \mu, \Theta)$, cf. Definition 2.6, and we take into account also the Green and smoothing Mellin plus Green subclasses. What concerns the Green symbols we admit from the very beginning 2×2 block matrix symbols of a similar meaning as the Green symbols in Definition 2.4, which also contain entries of trace and potential type. The following definition takes into account Remark 3.1.

DEFINITION 3.3. By $\mathcal{R}_G^\nu(\Omega \times \mathbb{R}^q, \boldsymbol{g}; \boldsymbol{w})$ for an open set $\Omega \subseteq \mathbb{R}^q$, weight data $\boldsymbol{g} := (\gamma, \gamma - \mu, \Theta)$ of similar meaning as in Definition 2.4, μ, ν, $\gamma \in \mathbb{R}$ and $\boldsymbol{w} := (E'', F''; j_1, j_2)$, we denote the space of all operator-valued symbols

$$g(y, \eta) \in \bigcap_{s,e \in \mathbb{R}} S^\mu_{\mathrm{cl}}(\Omega \times \mathbb{R}^q; \mathcal{K}^{s,\gamma;e}(N^\wedge, E^\wedge) \oplus \mathbb{C}^{j_1}, \mathcal{K}^{\infty,\gamma-\mu;\infty}(N^\wedge, F^\wedge) \oplus \mathbb{C}^{j_2}))$$

such that

$$g(y, \eta) \in \bigcap_{s,e \in \mathbb{R}} S^\nu_{\mathrm{cl}}(\Omega \times \mathbb{R}^q; \mathcal{K}^{s,\gamma;e}(N^\wedge, E^\wedge) \oplus \mathbb{C}^{j_1}, \mathcal{K}_{\mathcal{P}}^{\infty,\gamma-\mu;\infty}(N^\wedge, F^\wedge) \oplus \mathbb{C}^{j_2})),$$

$$g^*(y, \eta) \in \bigcap_{s,e \in \mathbb{R}} S^\nu_{\mathrm{cl}}(\Omega \times \mathbb{R}^q; \mathcal{K}^{s,-\gamma+\mu;e}(N^\wedge, F^\wedge) \oplus \mathbb{C}^{j_2}, \mathcal{K}_{\mathcal{Q}}^{\infty,-\gamma;\infty}(N^\wedge, E^\wedge) \oplus \mathbb{C}^{j_1})).$$

for g-dependent asymptotic types $\mathcal{P}$ and $\mathcal{Q}$. The pointwise formal adjoint refers to the sesquilinear pairings of the kind

$$\mathcal{K}^{s,\gamma;e}(N^\wedge, E^\wedge) \times \mathcal{K}^{-s,-\gamma;-e}(N^\wedge, E^\wedge) \to \mathbb{C}$$

induced by the $\mathcal{K}^{0,0;0}(N^\wedge, E^\wedge)(= r^{-n/2}L^2(\mathbb{R}_+ \times N, E^\wedge))$-scalar product. The elements of

$$\mathcal{R}_G^\nu(\Omega \times \mathbb{R}^q, \boldsymbol{g}; \boldsymbol{w})$$

are called Green symbols of the edge calculus. We write $\mathcal{R}_G^\nu(\Omega \times \mathbb{R}^q, \boldsymbol{g}; \boldsymbol{w})_{\mathcal{P},\mathcal{Q}}$ for the subspace of Green symbols with fixed asymptotic types $\mathcal{P}$, $\mathcal{Q}$.

In the case of $\boldsymbol{w} := (E'', F''; 0, 0)$, i.e., Green symbols in upper left corners, we also use the notation

$$\mathcal{R}_G^\nu(\Omega \times \mathbb{R}^q, \boldsymbol{g}; E'', F'') \text{ and } \mathcal{R}_G^\nu(\Omega \times \mathbb{R}^q, \boldsymbol{g}; E'', F'')_{\mathcal{P},\mathcal{Q}},$$

respectively. If $\mathcal{P} = \mathcal{O}$ and $\mathcal{Q} = \mathcal{O}$ are the trivial asymptotic types (i.e., $\pi_{\mathbb{C}}\mathcal{P} = \emptyset$, $\pi_{\mathbb{C}}\mathcal{Q} = \emptyset$) and $\Theta = (-\infty, 0]$, then $\mathcal{R}_G^\nu(\Omega \times \mathbb{R}^q, \boldsymbol{g}; E'', F'')_{\mathcal{O},\mathcal{O}}$ is independent of $\boldsymbol{g}$, and we write

$$\mathcal{R}_G^\nu(\Omega \times \mathbb{R}^q; E'', F'')_{\mathcal{O},\mathcal{O}} \tag{3.11}$$

for the corresponding class.

For a Green symbol we form the (twisted) homogeneous principal symbol of order μ

$$\sigma_1(g)(y, \eta) : \begin{matrix} \mathcal{K}^{s,\gamma}(N^\wedge, E^\wedge) \\ \oplus \\ \mathbb{C}^{j_1} \end{matrix} \longrightarrow \begin{matrix} \mathcal{K}^{\infty,\gamma-\mu}(N^\wedge, F^\wedge) \\ \oplus \\ \mathbb{C}^{j_2} \end{matrix}, \tag{3.12}$$

according to the homogeneous component of order μ of the classical symbol g.

Other operator-valued symbols are obtained in terms of smoothing Mellin operators of the cone calculus. Consider first a finite weight interval $\Theta = (-(k+1), 0]$, $k \in \mathbb{N}$, and start with a sequence

$$f_{j\alpha} \in C^\infty(\Omega, M^{-\infty}_{\mathcal{R}_{j\alpha}}(N; E'', F'')),\ \alpha \in \mathbb{N}^q,\ |\alpha| \le j,\ j = 0, \ldots, k. \tag{3.13}$$

For any cut-off function $\omega(r)$ we write $\omega_\eta(r) := \omega(r[\eta])$; here $\eta \to [\eta]$ is any strictly positive function in $C^\infty(\mathbb{R}^q)$ such that $[\eta] = |\eta|$ for $|\eta| \ge c$ for some $c > 0$. Let us form

$$m(y,\eta) := r^{-\mu} \sum_{j=0}^{k} r^j \sum_{|\alpha| \le j} \omega_\eta \mathrm{op}_M^{\gamma_{j\alpha}-n/2}(f_{j\alpha}) \eta^\alpha \omega'_\eta \tag{3.14}$$

for weights $\gamma_{j\alpha}$ such that

$$\gamma - j \le \gamma_{j\alpha} \le \gamma,\ \Gamma_{\frac{n+1}{2}-\gamma_{j\alpha}} \cap \pi_{\mathbb{C}}\mathcal{R}_{j\alpha} = \emptyset \text{ for all } j, \alpha, \tag{3.15}$$

and cut-off functions ω, ω'.

REMARK 3.4. We have

$$m(y,\eta) \in \bigcap_{s,e\in\mathbb{R}} S^\mu_{\mathrm{cl}}(\Omega \times \mathbb{R}^q; \mathcal{K}^{s,\gamma;e}(N^\wedge, E^\wedge), \mathcal{K}^{\infty,\gamma-\mu;\infty}(N^\wedge, F^\wedge))$$

and

$$m(y,\eta) \in \bigcap_{s,e\in\mathbb{R}} S^\mu_{\mathrm{cl}}(\Omega \times \mathbb{R}^q; \mathcal{K}^{s,\gamma;e}_{\mathcal{P}}(N^\wedge, E^\wedge), \mathcal{K}^{\infty,\gamma-\mu;\infty}_{\mathcal{Q}}(N^\wedge, F^\wedge))$$

for every asymptotic type $\mathcal{P}$ with some resulting $\mathcal{Q}$.

REMARK 3.5. Let $\tilde{m}(y,\eta)$ be of analogous form as (3.14) for the same Mellin symbols (3.13), however, with other cut-off functions $\tilde{\omega}$, $\tilde{\omega}'$, other $\eta \to \widetilde{[\eta]}$, and weights $\tilde{\gamma}_{j\alpha}$ satisfying conditions like (3.15). Then we have

$$m(y,\eta) - \tilde{m}(y,\eta) \in \mathcal{R}^\mu_G(\Omega \times \mathbb{R}^q, \boldsymbol{g}; E'', F'')$$

(the notation on the right means the space of Green symbols of the form of upper left corners).

DEFINITION 3.6. By $\mathcal{R}^\mu_{M+G}(\Omega \times \mathbb{R}^q, \boldsymbol{g}; \boldsymbol{w})$, $\boldsymbol{g} = (\gamma, \gamma-\mu, \Theta)$, $\Theta = (-(k+1), 0]$, we denote the space of all operator functions of the form

$$\begin{pmatrix} m(y,\eta) & 0 \\ 0 & 0 \end{pmatrix} + g(y,\eta)$$

for arbitrary $m(y,\eta)$ of the form (3.14) and $g(y,\eta) \in \mathcal{R}^\mu_G(\Omega \times \mathbb{R}^q, \boldsymbol{g}; \boldsymbol{w})$.

We set

$$\sigma_1(m)(y,\eta) := r^{-\mu} \sum_{j=0}^{k} r^j \sum_{|\alpha| = j} \omega_{|\eta|} \mathrm{op}_M^{\gamma_{j\alpha}-n/2}(f_{j\alpha})(y) \eta^\alpha \omega'_{|\eta|}, \tag{3.16}$$

$\omega_{|\eta|}(r) := \omega(r|\eta|)$, etc. For describing the structure of full edge amplitude functions we need some auxiliary material on the Mellin quantisation of edge-degenerate operator functions.

Let

$$\tilde{p}(r, y, \tilde{\rho}, \tilde{\eta}) \in C^\infty(\overline{\mathbb{R}}_+ \times \Omega, L^\mu_{\mathrm{cl}}(N; E'', F''; \mathbb{R}^{1+q}_{\tilde{\rho},\tilde{\eta}})), \tag{3.17}$$

$\Omega \subseteq \mathbb{R}^q$ open, and form

$$p(r, y, \rho, \eta) := \tilde{p}(r, y, r\rho, r\eta). \tag{3.18}$$

DEFINITION 3.7. The space $M_{\mathcal{O}}^{\mu}(N; E'', F''; \mathbb{R}^q_{\eta})$, $\mu \in \mathbb{R}$, is defined to be the set of all $h(z, \eta) \in \mathcal{A}(\mathbb{C}, L^{\mu}_{\mathrm{cl}}(N; E'', F''; \mathbb{R}^q_{\eta}))$ such that

$$h(z, \eta)|_{\Gamma_\beta} \in L^{\mu}_{\mathrm{cl}}(N; E'', F''; \Gamma_\beta \times \mathbb{R}^q)$$

for every $\beta \in \mathbb{R}$, uniformly in compact β-intervals.

Similarly as $M_{\mathcal{O}}^{\mu}(X)$, cf. Definition 2.3, the space $M_{\mathcal{O}}^{\mu}(N; E'', F''; \mathbb{R}^q)$ is a (nuclear) Fréchet space.

We frequently employ the following Mellin quantisation result.

THEOREM 3.8. *For every $p(r, y, \rho, \eta)$ of the form* (3.18) *with* (3.17) *there is an*

$$\tilde{h}(r, y, z, \tilde{\eta}) \in C^{\infty}(\overline{\mathbb{R}}_+ \times \Omega, M_{\mathcal{O}}^{\mu}(N; E'', F''; \mathbb{R}^q_{\tilde{\eta}})) \tag{3.19}$$

such that for

$$h(r, y, z, \eta) := \tilde{h}(r, y, z, r\eta) \tag{3.20}$$

we have

$$\mathrm{Op}_r(p)(y, \eta) = \mathrm{op}_M^{\beta}(h)(y, \eta)$$

mod $C^{\infty}(\Omega, L^{-\infty}(N^\wedge; E^\wedge, F^\wedge; \mathbb{R}^q_{\eta}))$, *for every $\beta \in \mathbb{R}$.*

If p and h are related to each other as in Theorem 3.8 we call h a Mellin quantisation of p.

Observe that when we set

$$p_0(r, y, \rho, \eta) := \tilde{p}(0, y, r\rho, r\eta), \quad h_0(r, y, z, \eta) := \tilde{h}(0, y, z, r\eta), \tag{3.21}$$

also h_0 is a Mellin quantisation of p_0.

Let us fix cut-off functions $\omega'' \prec \omega \prec \omega'$, set $\chi := 1 - \omega$, $\chi' := 1 - \omega''$, and write

$$r^{-\mu}\mathrm{Op}_r(p)(y, \eta) = r^{-\mu}\omega_\eta \mathrm{Op}_r(p)(y, \eta)\omega'_\eta + r^{-\mu}\chi_\eta \mathrm{Op}_r(p)(y, \eta)\chi'_\eta + c(y, \eta) \tag{3.22}$$

where $c(y, \eta) \in C^{\infty}(\Omega, L^{-\infty}(N^\wedge; E^\wedge, F^\wedge; \mathbb{R}^q_{\eta}))$. Theorem 3.8 allows us to write

$$r^{-\mu}\mathrm{Op}_r(p)(y, \eta) = r^{-\mu}\omega_\eta \mathrm{op}_M^{\gamma - n/2}(h)(y, \eta)\omega'_\eta + r^{-\mu}\chi_\eta \mathrm{Op}_r(p)(y, \eta)\chi'_\eta \tag{3.23}$$

mod $C^{\infty}(\Omega, L^{-\infty}(N^\wedge; E^\wedge, F^\wedge; \mathbb{R}^q))$.

Now let us choose cut-off functions

$$\epsilon'' \prec \epsilon \prec \epsilon'.$$

Similarly as (3.22) we can decompose once again the left hand side of (3.23) as

$$r^{-\mu}\mathrm{Op}_r(p)(y, \eta) = r^{-\mu}\epsilon \mathrm{Op}_r(p)(y, \eta)\epsilon' + r^{-\mu}(1 - \epsilon)\mathrm{Op}_r(p)(y, \eta)(1 - \epsilon'')$$

mod $C^{\infty}(\Omega, L^{-\infty}(N^\wedge; E^\wedge, F^\wedge; \mathbb{R}^q))$. Combined with (3.23) we obtain

$$r^{-\mu}\mathrm{Op}_r(p)(y, \eta) = w(y, \eta) + b(y, \eta) + c(y, \eta) \tag{3.24}$$

for

$$\begin{aligned} w(y, \eta) &= r^{-\mu}\epsilon\{\omega_\eta \mathrm{op}_M^{\gamma - n/2}(h)(y, \eta)\omega'_\eta + \chi_\eta \mathrm{Op}_r(p)(y, \eta)\chi'_\eta\}\epsilon', \\ b(y, \eta) &= r^{-\mu}(1 - \epsilon)\mathrm{Op}_r(p)(y, \eta)(1 - \epsilon'') \end{aligned} \tag{3.25}$$

for another $c(y, \eta) \in C^{\infty}(\Omega, L^{-\infty}(N^\wedge; E^\wedge, F^\wedge; \mathbb{R}^q))$.

REMARK 3.9. The operator function $b(y,\eta)$ can be written in the form

$$b(y,\eta) = \mathrm{Op}_r(p_{\mathrm{int}})(y,\eta) \tag{3.26}$$

mod $C^\infty(\Omega, L^{-\infty}(N^\wedge; E^\wedge, F^\wedge; \mathbb{R}^q))$ for some

$$p_{\mathrm{int}}(r,y,\rho,\eta) \in C^\infty(\mathbb{R}_+ \times \Omega, L^\mu_{\mathrm{cl}}(N^\wedge; E^\wedge, F^\wedge; \mathbb{R}^{1+q}_{\rho,\eta})).$$

REMARK 3.10. We have

$$w(y,\eta) \in S^\mu(\Omega \times \mathbb{R}^q; \mathcal{K}^{s,\gamma}(N^\wedge, E^\wedge), \mathcal{K}^{s-\mu,\gamma-\mu}(N^\wedge, F^\wedge)) \tag{3.27}$$

as well as

$$\varphi b(y,\eta)\varphi' \in S^\mu(\Omega \times \mathbb{R}^q; \mathcal{K}^{s,\gamma}(N^\wedge, E^\wedge), \mathcal{K}^{s-\mu,\gamma-\mu}(N^\wedge, F^\wedge)) \tag{3.28}$$

for every φ, $\varphi' \in C_0^\infty(\mathbb{R}_+)$ and $b(y,\eta)$ as in (3.26).

The way to construct from (3.18) an operator function $w(y,\eta)$ of the form (3.25) with the property (3.27) is interpreted as a quantisation of the edge-degenerate operator family (3.18). The summand $b(y,\eta)$ in (3.24) is localised off $r=0$ and only contributes to the interior part of the operators.

REMARK 3.11. An alternative to the edge quantisation $r^{-\mu}\mathrm{Op}_r(p)(y,\eta) \to w(y,\eta)$ is to set

$$v(y,\eta) := r^{-\mu}\epsilon \mathrm{op}_M^{\gamma-n/2}(h)(y,\eta)\epsilon'.$$

It can be proved, cf. [**20**], that

$$w(y,\eta) - v(y,\eta) \in \mathcal{R}^\mu_G(\Omega \times \mathbb{R}^q; E'', F'')_{\mathcal{O},\mathcal{O}},$$

cf. (3.11).

DEFINITION 3.12. Let $\mathcal{R}^\mu(\Omega \times \mathbb{R}^q, \boldsymbol{g}; \boldsymbol{w})$, $\boldsymbol{g} = (\gamma, \gamma-\mu, \Theta)$, $\Theta = (-(k+1), 0]$, $\boldsymbol{w} = (E'', F''; j_1, j_2)$, denote the set of all operator functions

$$a(y,\eta) = \begin{pmatrix} w(y,\eta) + m(y,\eta) & 0 \\ 0 & 0 \end{pmatrix} + g(y,\eta) \tag{3.29}$$

for arbitrary $w(y,\eta)$ like (3.25), $m(y,\eta)$ of the form (3.14), and $g(y,\eta) \in \mathcal{R}^\mu_G(\Omega \times \mathbb{R}^q, \boldsymbol{g}; \boldsymbol{w})$.

The space $\mathbb{R}^q, \boldsymbol{g}; \boldsymbol{w})$ for $\Theta = (-\infty, 0]$ is defined as intersection of the corresponding spaces for $\boldsymbol{g}_k = (\gamma, \gamma-\mu, (-(k+1), 0])$, $k \in \mathbb{N}$. The final results make sense also for this case, however, it is more convenient first to consider the case of finite Θ.

For a function $a(y,\eta) \in \mathcal{R}^\mu(\Omega \times \mathbb{R}^q, \boldsymbol{g}; \boldsymbol{w})$ written in the form (3.29) we set

$$\sigma_1(a)(y,\eta) = \begin{pmatrix} \sigma_1(w+m)(y,\eta) & 0 \\ 0 & 0 \end{pmatrix} + \sigma_1(g)(y,\eta) \tag{3.30}$$

where $\sigma_1(g)(y,\eta)$ is given by (3.12), $\sigma_1(m)(y,\eta)$ by (3.16) for $\boldsymbol{g} = (\gamma, \gamma-\mu, \Theta)$, and

$$\sigma_1(w)(y,\eta) := r^{-\mu}\omega_{|\eta|}\mathrm{op}_M^{\gamma-n/2}(h_0)(y,\eta)\omega'_{|\eta|} + r^{-\mu}\chi_{|\eta|}\mathrm{Op}_r(p_0)(y,\eta)\chi'_{|\eta|}. \tag{3.31}$$

3.3. The edge calculus. Let M be a compact manifold with edge Y, let $E,\ F \in \mathrm{Vect}(\mathbb{M})$, $J_1,\ J_2 \in \mathrm{Vect}(Y)$, and fix weight data $\boldsymbol{g} = (\gamma, \gamma - \mu, \Theta)$.

Before we give a definition of the edge algebra we have a look at general pseudo-differential operators globally on $M \setminus Y$, acting between distributional sections of $E,\ F$. Let $\{G_j\}_{j=1,\dots,N}$ be an open covering of Y by coordinate neighbourhoods, $\{\varphi_j\}_{j=1,\dots,N}$ a subordinate partition of unity, and $\{\varphi'_j\}_{j=1,\dots,N}$ a system of functions $\varphi'_j \in C_0^\infty(G_j)$, $\varphi'_j \prec \varphi_j$ for all j. Moreover, let $\theta \prec \theta'$ be cut-off functions, and choose $\psi \prec \psi'$ in $C_0^\infty(M \setminus Y)$ such that $\operatorname{dist}(\operatorname{supp}\psi, Y)$ is small enough. Then every $A \in L^\mu_{\mathrm{cl}}(M \setminus Y; E, F)$ can be written in the form

$$A = \sum_{j=1}^{N} \theta\varphi_j B_j \theta' \varphi'_j + \psi A_{\mathrm{int}} \psi' + C$$

with the following ingredients: $C \in L^{-\infty}(M \setminus Y; E, F)$, $A_{\mathrm{int}} \in L^\mu_{\mathrm{cl}}(M \setminus Y; E, F)$, and $B_j = ((\chi^\wedge_{G_j})^{-1})_* \mathrm{Op}_{r,y}(b_j)$, cf. the notation (3.1),

$$b_j(r, y, \rho, \eta) \in C^\infty(\mathbb{R}_+ \times \mathbb{R}^q, L^\mu_{\mathrm{cl}}(N; E'', F''; \mathbb{R}^{1+q}_{\rho,\eta})),$$

where the operator push forwards also take into account the transition maps of the bundles $E,\ F$ close to Y.

Let $L^{-\infty}(M, \boldsymbol{g}; \boldsymbol{v})$ for $\boldsymbol{v} := (E, F; J_1, J_2)$ and $\boldsymbol{g} := (\gamma, \gamma - \mu, \Theta)$ defined to be the set of all continuous operators

$$\mathcal{C} : \begin{matrix} H^{s,\gamma}(M, E) \\ \oplus \\ H^s(Y, J_1) \end{matrix} \longrightarrow \begin{matrix} H^{\infty,\gamma-\mu}_{\mathcal{P}}(M, F) \\ \oplus \\ H^\infty(Y, J_2) \end{matrix}$$

such that

$$\mathcal{C}^* : \begin{matrix} H^{s,-\gamma+\mu}(M, F) \\ \oplus \\ H^s(Y, J_2) \end{matrix} \longrightarrow \begin{matrix} H^{\infty,-\gamma}_{\mathcal{Q}}(M, E) \\ \oplus \\ H^\infty(Y, J_1) \end{matrix}$$

for all $s \in \mathbb{R}$, for certain asymptotic types $\mathcal{P}$ and $\mathcal{Q}$, associated with the weight data $(\gamma - \mu, \Theta)$ and $(-\gamma, \Theta)$, respectively.

Now let $g_i(y, \eta) \in \mathcal{R}^\mu_G(\mathbb{R}^q \times \mathbb{R}^q, \boldsymbol{g}; \boldsymbol{w})$ for $\boldsymbol{w} = (E'', F''; j_1, j_2)$ be a Green symbol. Assume for simplicity that $g_i(y, \eta)$ is independent of y for large $|y|$. Then we obtain a continuous operator

$$\mathrm{Op}_y(g_i) : \begin{matrix} \mathcal{W}^s(\mathbb{R}^q, \mathcal{K}^{s,\gamma}(N^\wedge, E^\wedge)) \\ \oplus \\ H^s(\mathbb{R}^q, \mathbb{C}^{j_1}) \end{matrix} \longrightarrow \begin{matrix} \mathcal{W}^s(\mathbb{R}^q, \mathcal{K}^{\infty,\gamma-\mu}(N^\wedge, F^\wedge)) \\ \oplus \\ H^{s-\mu}(\mathbb{R}^q, \mathbb{C}^{j_2}) \end{matrix}$$

for every $s \in \mathbb{R}$. For charts $\chi_{G_i} : G_i \to \mathbb{R}^q$ on the edge Y and $\chi^\wedge_{G_i} : (V \setminus Y)|_{G_i} \to N^\wedge \times \mathbb{R}^q$, cf. (3.1), and using corresponding bundle pull backs we form the operators

$$\mathcal{G}_i := \begin{pmatrix} \theta\varphi_i & 0 \\ 0 & \varphi_i \end{pmatrix} \begin{pmatrix} (\chi^\wedge_{G_i})^{-1}_* & 0 \\ 0 & (\chi^\wedge_{G_i})^{-1}_* \end{pmatrix} \mathrm{Op}_y(g_i) \begin{pmatrix} \theta'\varphi'_i & 0 \\ 0 & \varphi'_i \end{pmatrix} :$$
$$\begin{matrix} H^{s,\gamma}(M, E) \\ \oplus \\ H^s(Y, J_1) \end{matrix} \longrightarrow \begin{matrix} H^{s-\mu,\gamma-\mu}(M, F) \\ \oplus \\ H^{s-\mu}(Y, J_2) \end{matrix}, i = 1, \dots, N,$$

$g_i \in \mathcal{R}^{\mu}_{G}(\mathbb{R}^q \times \mathbb{R}^q, \boldsymbol{g}; \boldsymbol{w})$. Let $L^{\mu}_{G}(M, \boldsymbol{g}; \boldsymbol{v})$, $\boldsymbol{v} := (E, F; J_1, J_2)$, defined to be the set of all operators of the form

$$\mathcal{G} := \sum_{j=1}^{N} \mathcal{G}_i + \mathcal{C}$$

for arbitrary $g_i \in \mathcal{R}^{\mu}_{G}(\mathbb{R}^q \times \mathbb{R}^q, \boldsymbol{g}; \boldsymbol{w})$ and $\mathcal{C} \in L^{-\infty}(M, \boldsymbol{g})$.

The elements of $L^{\mu}_{G}(M, \boldsymbol{g}; \boldsymbol{v})$ are called Green operators of order μ of the edge calculus on M. From the local (twisted) homogeneous principal symbols of Green amplitude functions we get an invariantly defined homogeneous principal symbol

$$\sigma_1(\mathcal{G})(y,\eta) : \begin{matrix} \mathcal{K}^{s,\gamma}(N^{\wedge}_y, E^{\wedge}_y) \\ \oplus \\ J_{1,y} \end{matrix} \longrightarrow \begin{matrix} \mathcal{K}^{\infty,\gamma-\mu}(N^{\wedge}_y, F^{\wedge}_y) \\ \oplus \\ J_{2,y} \end{matrix},$$

$(y, \eta) \in T^*Y \setminus 0$, for a Green operator $\mathcal{G} \in L^{\mu}_{G}(M, \boldsymbol{g}; \boldsymbol{v})$. Homogeneity means

$$\sigma_1(\mathcal{G})(y, \lambda\eta) = \lambda^{\mu} \begin{pmatrix} \kappa_\lambda & 0 \\ 0 & 1 \end{pmatrix} \sigma_1(\mathcal{G})(y,\eta) \begin{pmatrix} \kappa_\lambda^{-1} & 0 \\ 0 & 1 \end{pmatrix}$$

for every $(y, \eta) \in T^*Y \setminus 0$, $\lambda \in \mathbb{R}_+$.

Definition 3.13. The space $L^{\mu}(M, \boldsymbol{g}; \boldsymbol{v})$ for $\boldsymbol{v} := (E, F; J_1, J_2)$ is defined to be the set of all operators

$$\mathcal{A} = \begin{pmatrix} A & 0 \\ 0 & 0 \end{pmatrix} + \mathcal{G}$$

for $\mathcal{G} \in L^{\mu}_{G}(M, \boldsymbol{g}; \boldsymbol{v})$ and $A = \sum_{j=1}^{N} \theta\varphi_j \mathrm{Op}_y(a_j)\theta'\varphi'_j + \psi A_{\mathrm{int}}\psi'$, for arbitrary

$A_{\mathrm{int}} \in L^{\mu}_{\mathrm{cl}}(M \setminus Y; E, F)$, ψ, $\psi' \in C^{\infty}_{0}(M \setminus Y)$, $a_j \in \mathcal{R}^{\mu}(\Omega \times \mathbb{R}^q, \boldsymbol{g}; E'', F'')$.

Theorem 3.14. *An* $\mathcal{A} \in L^{\mu}(M, \boldsymbol{g}; \boldsymbol{v})$ *for* $\boldsymbol{g} := (\gamma, \gamma - \mu, \Theta)$, $\boldsymbol{v} := (E, F; J_1, J_2)$, *induces continuous operators*

$$\mathcal{A} : \begin{matrix} H^{s,\gamma}(M, E) \\ \oplus \\ H^s(Y, J_1) \end{matrix} \longrightarrow \begin{matrix} H^{s-\mu,\gamma-\mu}(M, F) \\ \oplus \\ H^{s-\mu}(Y, J_2) \end{matrix} \tag{3.32}$$

and

$$\mathcal{A} : \begin{matrix} H^{s,\gamma}_{\mathcal{P}}(M, E) \\ \ominus \\ H^s(Y, J_1) \end{matrix} \longrightarrow \begin{matrix} H^{s-\mu,\gamma-\mu}_{\mathcal{Q}}(M, F) \\ \oplus \\ H^{s-\mu}(Y, J_2) \end{matrix} \tag{3.33}$$

for $s \in \mathbb{R}$ *and every asymptotic type* $\mathcal{P}$ *for some resulting* $\mathcal{Q}$.

Every $\mathcal{A} \in L^{\mu}(M, \boldsymbol{g}; \boldsymbol{v})$ has a pair of principal symbols, namely,

$$\sigma(\mathcal{A}) = (\sigma_0(\mathcal{A}), \sigma_1(\mathcal{A}))$$

with $\sigma_0(\mathcal{A}) := \sigma_0(A)$ being the standard homogeneous principal symbol of $A \in L^{\mu}_{\mathrm{cl}}(M \setminus Y; E, F)$ of order μ,

$$\sigma_0(\mathcal{A}) : \pi^*_{M \setminus Y} E \to \pi^*_{M \setminus Y} F,$$

$\pi_{M \setminus Y} : T^*(M \setminus Y) \setminus 0 \to M \setminus Y$. The homogeneous principal edge symbol represents a family of continuous operators

$$\sigma_1(\mathcal{A})(y,\eta) : \begin{matrix} \mathcal{K}^{s,\gamma}(N^{\wedge}_y, E^{\wedge}_y) \\ \oplus \\ J_{1,y} \end{matrix} \longrightarrow \begin{matrix} \mathcal{K}^{s-\mu,\gamma-\mu}(N^{\wedge}_y, F^{\wedge}_y) \\ \oplus \\ J_{2,y} \end{matrix},$$

$(y,\eta) \in T^*Y \setminus 0$. This follows from the definition of the various amplitude functions involved in $\sigma_1(\cdot)$.

Let us set for the moment $\sigma(\mathcal{A}) =: \sigma^\mu(\mathcal{A}) = (\sigma_0^\mu(\mathcal{A}), \sigma_1^\mu(\mathcal{A}))$, and form

$$L^{\mu-1}(M, \boldsymbol{g}; \boldsymbol{v}) := \{\mathcal{A} \in L^\mu(M, \boldsymbol{g}; \boldsymbol{v}) : \sigma^\mu(\mathcal{A}) = 0\}. \tag{3.34}$$

Every $\mathcal{A} \in L^{\mu-1}(M, \boldsymbol{g}; \boldsymbol{v})$ has again a pair of principal symbols, now denoted by $\sigma^{\mu-1}(\mathcal{A}) = (\sigma_0^{\mu-1}(\mathcal{A}), \sigma_1^{\mu-1}(\mathcal{A}))$ of order $\mu - 1$. This gives us a subspace

$$L^{\mu-2}(M, \boldsymbol{g}; \boldsymbol{v}) := \{\mathcal{A} \in L^{\mu-1}(M, \boldsymbol{g}; \boldsymbol{v}) : \sigma^{\mu-1}(\mathcal{A}) = 0\}, \tag{3.35}$$

and then, successively,

$$L^{\mu-j}(M, \boldsymbol{g}; \boldsymbol{v}) := \{\mathcal{A} \in L^{\mu-(j-1)}(M, \boldsymbol{g}; \boldsymbol{v}) : \sigma^{\mu-(j-1)}(\mathcal{A}) = 0\}, \tag{3.36}$$

for every $j \in \mathbb{N}, j \geq 1$. Set

$$L^{-\infty}(M, \boldsymbol{g}; \boldsymbol{v}) := \bigcap_{j\in\mathbb{N}} L^{\mu-j}(M, \boldsymbol{g}; \boldsymbol{v}).$$

REMARK 3.15. Every $\mathcal{A} \in L^{\mu-1}(M, \boldsymbol{g}; \boldsymbol{v})$ is compact as an operator (3.32) for every $s \in \mathbb{R}$.

REMARK 3.16. Let $\mathcal{A}_j \in L^{\mu-j}(M, \boldsymbol{g}; \boldsymbol{v})$, $j \in \mathbb{N}$, be an arbitrary sequence, $\boldsymbol{g} := (\gamma, \gamma-\mu, (-(k+1), 0])$ for a finite k, and assume that the asymptotic types involved in the Green operators are independent of j. Then there exists an $\mathcal{A} \in L^\mu(M, \boldsymbol{g}; \boldsymbol{v})$ such that

$$\mathcal{A} - \sum_{j=0}^{N} \mathcal{A}_j \in L^{\mu-(N+1)}(M, \boldsymbol{g}; \boldsymbol{v})$$

for every $N \in \mathbb{N}$, and $\mathcal{A}$ is unique mod $L^{-\infty}(M, \boldsymbol{g}; \boldsymbol{v})$.

THEOREM 3.17. (i) $\mathcal{A} \in L^\mu(M, \boldsymbol{g}_0; \boldsymbol{v}_0)$, $\mathcal{B} \in L^\rho(M, \boldsymbol{g}_1; \boldsymbol{v}_1)$ *for* $\boldsymbol{g}_0 = (\gamma - \rho, \gamma-(\mu+\rho), \Theta)$, $\boldsymbol{g}_1 = (\gamma, \gamma-\rho, \Theta)$, *implies* $\mathcal{AB} \in L^{\mu+\rho}(M, \boldsymbol{g}_0 \circ \boldsymbol{g}_1; \boldsymbol{v}_0 \circ \boldsymbol{v}_1)$ *(when the bundle data in the middle fit together such that* $\boldsymbol{v}_0 \circ \boldsymbol{v}_1$ *makes sense), and we have* $\sigma(\mathcal{AB}) = \sigma(\mathcal{A})\sigma(\mathcal{B})$ *with componentwise multiplication.*

(ii) $\mathcal{A} \in L^\mu(M, \boldsymbol{g}; \boldsymbol{v})$ *for* $\boldsymbol{g} = (\gamma, \gamma - \mu, \Theta)$, *and* $\boldsymbol{v} = (V_1, V_2; J_1, J_2)$ *implies* $\mathcal{A}^* \in L^\mu(M, \boldsymbol{g}^*; \boldsymbol{v}^*)$ *for* $\boldsymbol{v}^* = (V_2, V_1; J_2, J_1)$, *where* $\mathcal{A}^*$ *is the formal adjoint in the sense*

$$(u, \mathcal{A}^* v)_{H^{0,0}(M,V_1)\oplus H^0(Y,J_1)} = (\mathcal{A}u, v)_{H^{0,0}(M,V_2)\oplus H^0(Y,J_2)}$$

for all $u \in H^{\infty,\infty}(M, V_1) \oplus H^\infty(Y, J_1)$, $v \in H^{\infty,\infty}(M, V_2) \oplus H^\infty(Y, J_2)$, *and we have* $\sigma(\mathcal{A}^*) = \sigma(\mathcal{A})^*$ *with componentwise formal adjoint.*

4. Shapiro-Lopatinskij edge-ellipticity

4.1. Interior ellipticity and an analogue of the Atiyah-Bott condition.

We saw in the preceding section that operators $\mathcal{A}$ in the edge calculus have a pair

$$\sigma(\mathcal{A}) = (\sigma_0(\mathcal{A}), \sigma_1(\mathcal{A})) \tag{4.1}$$

of principal symbols. Both components will take part in the notion of ellipticity. The ellipticity with respect to σ_0, is a condition to the upper left corner A of the 2×2 block matrix operator $\mathcal{A}$.

DEFINITION 4.1. Let M be a manifold with edge Y. An operator

$$A \in L^{\mu}(M, \boldsymbol{g}; E, F), \quad E,\, F \in \mathrm{Vect}(\mathbb{M}),$$

is called σ_0-elliptic if

(i) the standard homogeneous principal symbol

$$\sigma_0(A) : \pi^*_{M\setminus Y} E \to \pi^*_{M\setminus Y} F,$$

$\pi_{M\setminus Y} : T^*(M \setminus Y) \setminus 0 \to M \setminus Y$, is an isomorphism (here E and F mean the restrictions $E|_{M\setminus Y}$ and $F|_{M\setminus Y}$, respectively, $M \setminus Y \cong \mathbb{M} \setminus \partial\mathbb{M}$);

(ii) locally near $\partial\mathbb{M}$ in the splitting of variables $(r, x, y) \in \overline{\mathbb{R}}_+ \times \Sigma \times \Omega$ with the covariables (ρ, ξ, η) the reduced symbol

$$\tilde{\sigma}_0(A)(r, x, y, \rho, \xi, \eta) := r^{\mu}\sigma_0(A)(r, x, y, r^{-1}\rho, \xi, r^{-1}\eta)$$

induces isomorphisms

$$\tilde{\sigma}_0(A)(r, x, y, \rho, \xi, \eta) : E_{(r,x,y)} \to F_{(r,x,y)}$$

for all $(r, x, y, \rho, \xi, \eta) \in \overline{\mathbb{R}}_+ \times \Sigma \times (\mathbb{R}^{1+n+q}_{\rho,\xi,\eta} \setminus \{0\})$ (here $\Sigma \subseteq \mathbb{R}^n$ and $\Omega \subseteq \mathbb{R}^q$ correspond to charts on N and Y, respectively, and $E_{(r,x,y)}, \dots$ mean the fibres of $E, \dots$ over the corresponding points in local coordinates).

Recall that $\sigma_1(A)(y, \eta)$ for $A = P + M + G$ with P being the non-smoothing pseudo-differential part and $M + G$ the smoothing Mellin plus Green part defines an operator family

$$\sigma_1(A)(y, \eta) : \mathcal{K}^{s,\gamma}(N^{\wedge}, E^{\wedge}) \to \mathcal{K}^{s-\mu,\gamma-\mu}(N^{\wedge}, F^{\wedge}) \tag{4.2}$$

of the form $\sigma_1(A) = \sigma_1(P) + \sigma_1(M) + \sigma_1(G)$ for

$$\sigma_1(P)(y, \eta) = r^{-\mu}\{\omega_{|\eta|}\mathrm{op}_M^{\gamma-n/2}(h_0)(y, \eta)\omega'_{|\eta|} + \chi_{|\eta|}\mathrm{Op}_r(p_0)(y, \eta)\chi'_{|\eta|}\},$$

moreover, $\sigma_1(M)(y, \eta)$ defined by (3.16) and the homegeneous principal part

$$\sigma_1(G)(y, \eta) = g_{(\mu)}(y, \eta)$$

of the Green symbol $g(y, \eta)$ of the operator G, $(y, \eta) \in T^*Y \setminus 0$. We have

$$\sigma_1(A)(y, \eta) \in L^{\mu}(N^{\wedge}, \boldsymbol{g}; E^{\wedge}, F^{\wedge}),$$

cf. Definition 2.6. The cone calculus also has a symbolic structure, in the case of an infinite stretched cone a triple

$$\sigma = (\sigma_0, \sigma_1, \sigma_{\mathrm{e}}),$$

cf. (2.13).

REMARK 4.2. Let $A \in L^{\mu}(M, \boldsymbol{g}; E, F)$ be σ_0-elliptic in the sense of Definition 4.1. Then $\sigma_1(A)(y, \eta)$ for every $(y, \eta) \in T^*Y \setminus 0$ is σ_0- and σ_{e}-elliptic with respect to the corresponding subordinate symbolic components of the cone calculus over $N^{\wedge}$, defined in (2.13). This allows us to consider $(\sigma_0\sigma_1(A))(y, z)$ and $(\sigma_1\sigma_1(A))(y, z)$. For every $y \in Y$ there is a discrete set $D_A(y) \subset \mathbb{C}$ such that

$$(\sigma_1\sigma_1(A))(y, z) : H^s(N, E'') \to H^{s-\mu}(N, F'')$$

is an isomorphism if and only if $z \notin D_A(y)$, for all $s \in \mathbb{R}$.

This observation entails the following result.

THEOREM 4.3. *Let $A \in L^{\mu}(M, \boldsymbol{g}; E, F)$ be σ_0-elliptic. Then (4.2) is a Fredholm operator for $\eta \neq 0$, $s \in \mathbb{R}$, and every $\gamma \in \mathbb{R}$, $y \in Y$, such that*

$$\Gamma_{\frac{n+1}{2}-\gamma} \cap D_A(y) = \emptyset. \tag{4.3}$$

In the following we assume that our operator $A \in L^{\mu}(M, \boldsymbol{g}; E, F)$ satisfies the condition (4.3) for every $y \in Y$. Because of the homogeneity

$$\sigma_1(y, \lambda\eta) = \lambda^{\mu}\kappa_{\lambda}\sigma_1(A)(y,\eta)\kappa_{\lambda}^{-1}$$

we have

$$\operatorname{ind}\sigma_1(A)(y,\eta) = \sigma_1(A)(y, \eta/|\eta|).$$

More precisely, the dimensions of $\ker\sigma_1(A)(y,\eta)$ and $\operatorname{coker}\sigma_1(A)(y,\eta)$ only depend on $\eta/|\eta|$. Therefore, (4.2) may be regarded as a family of Fredholm operators depending on the parameters $(y,\eta) \in S^*Y$, the unit cosphere bundle of Y which is a compact topological space. Analogously as in boundary value problems this gives rise to an index element $\operatorname{ind}_{S^*Y}\sigma_1(A) \in K(S^*Y)$. The property

$$\operatorname{ind}_{S^*Y}\sigma_1(A) \in \pi^*K(Y), \tag{4.4}$$

$\pi : S^*Y \to Y$, is an analogue of a corresponding condition known from boundary value problems, cf. [**4**], [**9**], and it is of analogous meaning for the edge calculus. If A satisfies the relation (4.4) we say that the Atiyah-Bott obstruction vanishes.

THEOREM 4.4. *Let $A \in L^{\mu}(M, \boldsymbol{g}; E, F)$ be σ_0-elliptic, and (4.3) be satisfied for some $\gamma \in \mathbb{R}$ and all $y \in Y$, and denote the family of Fredholm operators (4.2) for the moment by*

$$\sigma_1(A)^{\gamma}(y,\eta) : \mathcal{K}^{s,\gamma}(N^{\wedge}, E^{\wedge}) \to \mathcal{K}^{s-\mu,\gamma-\mu}(N^{\wedge}, F^{\wedge}),$$

$(y,\eta) \in S^*Y$. Then if $\tilde{\gamma} \in \mathbb{R}$ is another weight satisfying $\Gamma_{\frac{n+1}{2}-\tilde{\gamma}} \cap D_A(y) = \emptyset$ for all $y \in Y$, we have

$$\operatorname{ind}_{S^*Y}\sigma_1(A)^{\gamma} \in \pi^*K(Y) \Leftrightarrow \operatorname{ind}_{S^*Y}\sigma_1(A)^{\tilde{\gamma}} \in \pi^*K(Y).$$

In addition if $\tilde{A} \in L^{\mu}(M, \boldsymbol{g}; E, F)$ satisfies $\tilde{A} = A \bmod L^{\mu}_{M+G}(M, \boldsymbol{g}; E, F)$ then

$$\Gamma_{\frac{n+1}{2}-\gamma} \cap D_{\tilde{A}}(y) = \emptyset, \quad \Gamma_{\frac{n+1}{2}-\gamma} \cap D_A(y) = \emptyset$$

for some $\gamma \in \mathbb{R}$ and all $y \in Y$ has the consequence

$$\operatorname{ind}_{S^*Y}\sigma_1(\tilde{A})^{\gamma} \in \pi^*K(Y) \Leftrightarrow \operatorname{ind}_{S^*Y}\sigma_1(A)^{\gamma} \in \pi^*K(Y).$$

The idea of the proof is the same as for [**33**, Theorem 1.11].

4.2. Elliptic edge conditions, parametrices, and the Fredholm property. Let us now study ellipticity with respect to both components of (4.1) from the point of view of an analogue of the Shapiro-Lopatinskij ellipticity in the edge calculus, also referred to as SL-ellipticity. References of the theorems in this subsection are [**58**, Subsection 3.5.2], [**28**, Subsection 4.5.3], [**25**, Section 7.2] and Dorschfeldt [**14**, Subsection 3.3] (concerning the necessity of ellipticity for the Fredholm property).

DEFINITION 4.5. An operator $\mathcal{A} \in L^{\mu}(M, \boldsymbol{g}; \boldsymbol{v})$ for $\boldsymbol{v} := (E, F; J_1, J_2)$ is called elliptic if

(i) its upper left corner $A \in L^{\mu}(M, \boldsymbol{g}; E, F)$ is σ_0-elliptic in the sense of Definition 4.1;

(ii) $\mathcal{A}$ is σ_1-elliptic, i.e.,

$$\sigma_1(\mathcal{A})(y,\eta) : \begin{matrix} \mathcal{K}^{s,\gamma}(N_y^\wedge, E_y^\wedge) \\ \oplus \\ J_{1,y} \end{matrix} \longrightarrow \begin{matrix} \mathcal{K}^{s-\mu,\gamma-\mu}(N_y^\wedge, F_y^\wedge) \\ \oplus \\ J_{2,y} \end{matrix} \tag{4.5}$$

is a family of isomorphisms for all $(y,\eta) \in T^*Y \setminus 0$ and some $s \in \mathbb{R}$.

REMARK 4.6. (i) If (4.5) is a family of isomorphisms for some $s \in \mathbb{R}$ then this is the case for all $s \in \mathbb{R}$, and $\ker \sigma_1(\mathcal{A})$ as well as $\dim \operatorname{coker} \sigma_1(\mathcal{A})$ are independent of s.

(ii) The condition of isomorphisms (4.5) for all $(y,\eta) \in T^*Y \setminus 0$ is equivalent to the one for all $(y,\eta) \in S^*Y$.

(iii) The condition of isomorphisms (4.5) for all $(y,\eta) \in S^*Y$ has the consequence that

$$\sigma_1(A)(y,\eta) : \mathcal{K}^{s,\gamma}(N_y^\wedge, E^\wedge) \to \mathcal{K}^{s-\mu,\gamma-\mu}(N_y^\wedge, F^\wedge)$$

is a family of Fredholm operators, and we have

$$\operatorname{ind}_{S^*Y}\sigma_1(A) = [J_2] - [J_1] \in \pi^*K(Y).$$

THEOREM 4.7. *Let $A \in L^\mu(M, \boldsymbol{g}; E, F)$ be σ_0-elliptic. Then the property*

$$\operatorname{ind}_{S^*Y}\sigma_1(A) \in \pi^*K(Y)$$

entails the existence of an SL*-elliptic operator*

$$\mathcal{A} \in L^\mu(M, \boldsymbol{g}; \boldsymbol{v}), \quad \boldsymbol{v} := (E, F; J_1, J_2),$$

for suitable J_1, $J_2 \in$ Vect(Y) *which contains A as the upper left corner.*

Let M be a compact manifold with edge.

DEFINITION 4.8. Let $\mathcal{A} \in L^\mu(M, \boldsymbol{g}; \boldsymbol{v})$, $\boldsymbol{v} := (E, F; J_1, J_2)$. A

$$\mathcal{P} \in L^{-\mu}(M, \boldsymbol{g}^{-1}; \boldsymbol{v}^{-1}), \quad \boldsymbol{v}^{-1} := (F, E; J_2, J_1),$$

is called a parametrix of $\mathcal{A}$ if

$$\mathcal{C}_{\mathrm{L}} := \mathcal{I} - \mathcal{P}\mathcal{A} \in L^{-\infty}(M, \boldsymbol{g}_{\mathrm{L}}; \boldsymbol{v}_{\mathrm{L}}), \; \mathcal{C}_{\mathrm{R}} := \mathcal{I} - \mathcal{A}\mathcal{P} \in L^{-\infty}(M, \boldsymbol{g}_{\mathrm{R}}; \boldsymbol{v}_{\mathrm{R}}) \tag{4.6}$$

for $\boldsymbol{v}_{\mathrm{L}} := (E, E; J_1, J_1)$, $\boldsymbol{v}_{\mathrm{R}} := (F, F; J_2, J_2)$, with $\mathcal{I}$ denoting the respective identity operators.

THEOREM 4.9. *Let $\mathcal{A} \in L^\mu(M, \boldsymbol{g}; \boldsymbol{v})$, $\boldsymbol{v} := (E, F; J_1, J_2)$, be elliptic. Then there is a parametrix $\mathcal{B} \in L^{-\mu}(M, \boldsymbol{g}^{-1}; \boldsymbol{v}^{-1})$, $\boldsymbol{v}^{-1} := (F, E; J_2, J_1)$, such that the operators $\mathcal{C}_{\mathrm{L}}$ and $\mathcal{C}_{\mathrm{R}}$ in the relation* (4.6) *are projections*

$$\mathcal{C}_{\mathrm{L}} : H^{s,\gamma}(M, E) \oplus H^s(Y, J_1) \to \mathcal{V}, \; \mathcal{C}_{\mathrm{R}} : H^{s-\mu,\gamma-\mu}(M, F) \oplus H^{s-\mu}(Y, J_2) \to \mathcal{W}$$

for all $s \in \mathbb{R}$.

THEOREM 4.10. *For an operator $\mathcal{A} \in L^\mu(M, \boldsymbol{g}; \boldsymbol{v})$, $\boldsymbol{v} := (E, F; J_1, J_2)$, the following conditions are equivalent:*

(i) *$\mathcal{A}$ is elliptic in the sense of Definition* 4.5.

(ii) *The operator*

$$\mathcal{A} : H^{s,\gamma}(M, E) \oplus H^s(Y, J_1) \to H^{s-\mu,\gamma-\mu}(M, F) \oplus H^{s-\mu}(Y, J_2) \tag{4.7}$$

is Fredholm for some $s = s_0 \in \mathbb{R}$.

The assertion (i) $\Rightarrow$ (ii) follows from Theorem 4.9. A proof of (ii) $\Rightarrow$ (i) is given in Dorschfeldt [**14**].

REMARK 4.11. Let $\mathcal{A} \in L^{\mu}(M, \boldsymbol{g}; \boldsymbol{v})$, $\boldsymbol{v} := (E, F; J_1, J_2)$, be elliptic.

(i) The operator (4.7) is Fredholm for every $s \in \mathbb{R}$.

(ii) $\mathcal{V} := \ker_s \mathcal{A} = \{u \in H^{s,\gamma}(M, E) \oplus H^s(Y, J_1) : \mathcal{A}u = 0\}$ is a finite dimensional subspace of $H^{\infty,\gamma}(M, E) \oplus H^{\infty}(Y, J_1)$ independent of s, and there is a finite-dimensional $\mathcal{W} \subset H^{\infty,\gamma-\mu}(M, F) \oplus H^{\infty}(Y, J_2)$ independent of s such that

$$\operatorname{im}_s \mathcal{A} + \mathcal{W} = H^{s-\mu,\gamma-\mu}(M, F) \oplus H^{s-\mu}(Y, J_2)$$

for every s; here $\operatorname{im}_s \mathcal{A} = \{\mathcal{A}u : u \in H^{s,\gamma}(M, E) \oplus H^s(Y, J_1)\}$.

REMARK 4.12. There is a variant of the edge pseudo-differential calculus containing a parameter $\lambda \in \mathbb{R}^l$ in edge-degenerate form analogously as η, cf. Dorschfeldt [**14**]. Most of the constructions and assertions from the case $l = 0$ have a natural analogue in the case $l > 0$.

There are many examples of elliptic edge problems. Theorem 4.7 is a way to characterise all of them. The proof employs the idea of filling up a family of Fredholm operators, parametrised by a compact topological space, by operators of finite rank to a 2×2 block matrix family of isomorphisms, see [**25**, Subsection 3.3.4]. Other concrete examples are contained in [**12**]. More examples follow from parameter-dependent σ_0-elliptic families by taking the parameter large enough, see [**48**].

5. Iterated Mellin quantisation

5.1. Mellin quantisation for second order singularities. We now formulate an iterative Mellin quantisation result for corner-degenerate symbols, proved in a joint paper with N. Habal [**24**]. The general scheme of Mellin quantisation is the one of [**58**, Theorem 3.2.7]. Later on there have been established other ways to Mellin quantise degenerate operator families, see, e.g., Krainer [**31**]. However, because of some technical difficulties in the higher corner case we avoided that.

Degenerate symbols for second order corners have the form

$$p(r, x, y, \rho, \xi, \eta) = \tilde{p}(r, x, y, r_1\rho_1, r_1r_2\rho_2, \xi, r_1\eta^1, r_1r_2\eta^2) \tag{5.1}$$

for $r := (r_1, r_2)$, $y := (y^1, y^2) \in \mathbb{R}^q$ for $q := q_1 + q_2$, and we intend to show a Mellin quantisation result analogously as the above theorem for first order singularities.

Let us give a brief motivation of symbols of that kind. If we consider, for instance, a Riemannian metric on $\mathbb{R}_+ \times \mathbb{R}_+ \times \mathbb{R}^n_x \times \mathbb{R}^{q_1} \times \mathbb{R}^{q_1} \ni (r_1, r_2, x, y^1, y^2)$ of the form

$$dr_2^2 + r_2^2\big(dr_1^2 + r_1^2dx^2 + (dy^1)^2\big) + (dy^2)^2,$$

then the Laplace-Beltrami operator has a symbol like $r_1^{-2}r_2^{-2}p(r, x, y, \rho, \xi, \eta)$ with p being of the form (5.1). In this case $\tilde{p}(r, x, y, \tilde{\rho}, \xi, \tilde{\eta})$ is elliptic of order 2 with respect to the covariables $(\tilde{\rho}, \xi, \tilde{\eta})$, up to $r = 0$.

Another motivation of such symbols comes from the analysis of elliptic (non-degenerate) differential operators of order μ in a corner configuration, say, embedded in an Euclidean space. Then symbols in degenerate form appear by iteratively introducing polar coordinates into the operator, here twice, and together with a weight factor $r_1^{-\mu} r_2^{-\mu}$. In a similar manner there occur higher order degenerate symbols as mentioned at the beginning when we have corners of order k.

From the spaces $M^\mu_{\mathcal{O}_{z_1}}(X;\mathbb{R}^q)$, cf. Definition 2.3, we can pass to certain derived spaces, for instance $C^\infty(\overline{\mathbb{R}}_+ \times \overline{\mathbb{R}}_+ \times \mathbb{R}^q_y, M^\mu_{\mathcal{O}_{z_1}}(X;\Gamma_\delta \times \mathbb{R}^q_\eta))$ for Γ_δ in the variable $z_2 \in \mathbb{C}$, $\delta \in \mathbb{R}$.

DEFINITION 5.1. Let $M^\mu_{\mathcal{O}_{z_1},\mathcal{O}_{z_2}}(X;\mathbb{R}^q_\eta)$ defined to be the space of all

$$h(z_1, z_2, \eta) \in \mathcal{A}(\mathbb{C}_{z_2}, M^\mu_{\mathcal{O}_{z_1}}(X;\mathbb{R}^q_\eta))$$

such that

$$h(z_1, z_2, \eta)|_{\Gamma_\delta} \in M^\mu_{\mathcal{O}_{z_1}}(X;\Gamma_\delta \times \mathbb{R}^q_\eta),$$

for every $\delta \in \mathbb{R}$, uniformly in compact δ-intervals.

This space is Fréchet in a natural way. In the above definition we may interchange the role of z_1 and z_2 and get the same space, in other words

$$M^\mu_{\mathcal{O}_{z_1},\mathcal{O}_{z_2}}(X;\mathbb{R}^q_\eta) \cong M^\mu_{\mathcal{O}_{z_2},\mathcal{O}_{z_1}}(X;\mathbb{R}^q_\eta).$$

The following result tells us that the introduced spaces are very rich for every $\mu \in \mathbb{R}$.

THEOREM 5.2. *For every*

$$m(r_1, r_2, y, z_1, z_2, \eta) \in C^\infty(\overline{\mathbb{R}}_+ \times \overline{\mathbb{R}}_+ \times \mathbb{R}^q_y, L^\mu(X;\Gamma_\beta \times \Gamma_\delta \times \mathbb{R}^q_\eta))$$

for fixed $\beta, \delta \in \mathbb{R}$ *there exists an*

$$h(r_1, r_2, y, z_1, z_2, \eta) \in C^\infty(\overline{\mathbb{R}}_+ \times \overline{\mathbb{R}}_+ \times \mathbb{R}^q_y, M^\mu_{\mathcal{O}_{z_1},\mathcal{O}_{z_2}}(X;\mathbb{R}^q_\eta))$$

such that

$$h|_{\Gamma_\beta \times \Gamma_\delta} = m \bmod C^\infty(\mathbb{R}_+ \times \mathbb{R}_+ \times \mathbb{R}^q, L^{-\infty}(X;\Gamma_\beta \times \Gamma_\delta \times \mathbb{R}^q)),$$

and h *is unique modulo* $C^\infty(\mathbb{R}_+ \times \mathbb{R}_+ \times \mathbb{R}^q, M^{-\infty}_{\mathcal{O}_{z_1},\mathcal{O}_{z_2}}(X;\mathbb{R}^q_\eta))$.

Consider corner-degenerate operator functions along X, i.e.,

$$p(r_1, r_2, y, \rho_1, \rho_2, \eta^1, \eta^2) = \tilde{p}(r_1, r_2, y, r_1\rho_1, r_1 r_2 \rho_2, r_1\eta^1, r_1 r_2 \eta^2) \tag{5.2}$$

for

$$\tilde{p}(r, y, \tilde{\rho}_1, \tilde{\tilde{\rho}}_2, \tilde{\eta}^1, \tilde{\tilde{\eta}}^2) \in C^\infty(\overline{\mathbb{R}}_+ \times \overline{\mathbb{R}}_+ \times \mathbb{R}^q_y, L^\mu_{\mathrm{cl}}(X;\mathbb{R}^{2+q}_{\tilde{\rho}_1,\tilde{\tilde{\rho}}_2,\tilde{\eta}^1,\tilde{\tilde{\eta}}^2})).$$

THEOREM 5.3. *For every* p *of the form* (5.2) *there exists an*

$$\tilde{h}(r, y, z_1, \tilde{z}_2, \tilde{\eta}^1, \tilde{\tilde{\eta}}^2) \in C^\infty(\overline{\mathbb{R}}_+ \times \overline{\mathbb{R}}_+ \times \mathbb{R}^q_y, M^\mu_{\mathcal{O}_{z_1},\mathcal{O}_{\tilde{z}_2}}(X;\mathbb{R}^q_{\tilde{\eta}^1,\tilde{\tilde{\eta}}^2}))$$

such that for

$$h(r, y, z_1, \tilde{z}_2, \eta^1, \eta^2) := \tilde{h}(r, y, z_1, r_1 z_2, r_1\eta^1, r_1 r_2 \eta^2)$$

we have

$$\mathrm{Op}_{r_2,r_1}(p)(y,\eta) = \mathrm{op}^{\beta_2}_{M_{r_2}} \mathrm{op}^{\beta_1}_{M_{r_1}}(h)(y,\eta)$$

modulo $C^\infty(\mathbb{R}^q_y, L^{-\infty}(\mathbb{R}_+ \times \mathbb{R}_+ \times X;\mathbb{R}^q_\eta))$ *for any reals* β_1, β_2.

5.2. Mellin quantisation for higher order singularities. The method of proving the Mellin quantisation result is iterative. As before let X be a closed C^∞ manifold.

DEFINITION 5.4. The space $M^\mu_{\mathcal{O}_z}(X;\mathbb{R}^q_\eta)$ for $z=(z_1,\dots,z_k)$, $\eta\in\mathbb{R}^q$, $k\geq 2$, $\mu\in\mathbb{R}$, is defined to be the set of all $h(z',z_k,\eta)$, $z'=(z_1,\dots,z_{k-1})$, such that $h(z',z_k,\eta)\in\mathcal{A}(\mathbb{C}_{z_k},M^\mu_{\mathcal{O}_{z'}}(X;\mathbb{R}^q_\eta))$ with

$$h(z',z_k,\eta)|_{\Gamma_\delta}\in M^\mu_{\mathcal{O}_{z'}}(X;\Gamma_\delta\times\mathbb{R}^q_\eta)$$

for every $\delta\in\mathbb{R}$, uniformly in compact δ-intervals. We also write $M^\mu_{\mathcal{O}}(X;\mathbb{R}^q)$ rather than $M^\mu_{\mathcal{O}_z}(X;\mathbb{R}^q)$.

Consider an operator function

$$\begin{aligned}&p(r,y,\rho,\eta)\\&=\tilde{p}(r,y,r_1\rho_1,r_1r_2\rho_2,\dots,r_1\dots r_k\rho_k,r_1\eta^1,r_1r_2\eta^2,\dots,r_1\dots r_k\eta^k)\end{aligned}$$

for

$$\tilde{p}(r,y,\tilde{\rho},\tilde{\eta})\in C^\infty((\overline{\mathbb{R}}_+)^k\times\Omega,L^\mu(X;\mathbb{R}^{k+q}_{\tilde{\rho},\tilde{\eta}})),$$

$\Omega\subseteq\mathbb{R}^q$ open, $(\overline{\mathbb{R}}_+)^k=\overline{\mathbb{R}}_+\times\dots\times\overline{\mathbb{R}}_+$ (k factors).

THEOREM 5.5. *For every p of the above-mentioned kind there exists an*

$$\tilde{h}(r,y,\tilde{z},\tilde{\eta})\in C^\infty((\overline{\mathbb{R}}_+)^k\times\Omega,M^\mu_{\mathcal{O}_{\tilde{z}}}(X;\mathbb{R}^q_{\tilde{\eta}})),$$

$\tilde{z}=(\tilde{z}_1,\dots,\tilde{z}_k)$, *such that for*

$$\begin{aligned}h(r,y,\tilde{z},\eta):=\tilde{h}(r,y,z_1,r_1z_2,r_1r_2z_3,\dots,r_1r_2\dots r_{k-1}z_k,&\\ r_1\eta^1,r_1r_2\eta^2,\dots,r_1r_2\dots r_k\eta^k)&\end{aligned}$$

we have

$$\mathrm{Op}_{r_k,\dots,r_1}(p)(y,\eta)=\mathrm{op}^{\beta_k}_{M_{r_k}}\dots\mathrm{op}^{\beta_1}_{M_{r_1}}(h)(y,\eta)$$

mod $C^\infty(\Omega,L^{-\infty}(\mathbb{R}^k_+\times X;\mathbb{R}^q))$ *for any reals* $\beta_1,\dots,\beta_k$.

REMARK 5.6. Our theorem refers to an equivalence of (y,η)-depending families of operators $C^\infty_0(\mathbb{R}^k_+\times X)\to C^\infty(\mathbb{R}^k_+\times X)$ modulo smoothing operators. The role of the Mellin quantisation is to modify the original operator by removing some smoothing (possibly singular) error and to obtain someone that admits a continuous extension between suitable weighted Sobolev spaces. This aspect cannot be discussed in detail here; it requires more voluminous considerations.

Let us finally observe that the spaces $M^\mu_{\mathcal{O}}(X;\mathbb{R}^q)$ admit a concept of ellipticity and parametrices that is also a part of the elliptic theory in higher corner operators. First note that

$$h_j(z,\eta)\in M^{\mu_j}_{\mathcal{O}}(X;\mathbb{R}^q),\ j=1,2,$$

implies $h_1(z,\eta)h_2(z,\eta)\in M^{\mu_1+\mu_2}_{\mathcal{O}}(X;\mathbb{R}^q)$ for arbitrary $\mu_1,\mu_2\in\mathbb{R}$.

An element $h(z,\eta)\in M^\mu_{\mathcal{O}}(X;\mathbb{R}^q)$ is called elliptic if there is a tuple $\beta=(\beta_1,\dots,\beta_k)\in\mathbb{R}^k$ such that

$$h(z,\eta)|_{\Gamma_\beta}\in L^\mu(X;\Gamma_\beta\times\mathbb{R}^q)\ \text{for}\ \Gamma_\beta:=\Gamma_{\beta_1}\times\ldots\times\Gamma_{\beta_k}$$

is parameter-dependent elliptic of order μ with the parameters $(z_1,\dots,z_k,\eta)\in\Gamma_\beta\times\mathbb{R}^q$. This definition is independent of the choice of β, i.e., the condition is equivalent to the one with respect to $\Gamma_{\beta'}$ for any other $\beta'\in\mathbb{R}^k$.

THEOREM 5.7. *Let $h(z,\eta) \in M^{\mu}_{\mathcal{O}}(X;\mathbb{R}^q)$ be elliptic; then there exists an*

$$h^{(-1)}(z,\eta) \in M^{-\mu}_{\mathcal{O}}(X;\mathbb{R}^q)$$

such that

$$h(z,\eta)h^{(-1)}(z,\eta) = 1,\ h^{(-1)}(z,\eta)h(z,\eta) = 1 \bmod M^{-\infty}_{\mathcal{O}}(X;\mathbb{R}^q).$$

6. Toeplitz edge problems

6.1. Projection data. In this subsection we recall some material on pseudo-differential projections from [**59**] which is necessary to studying a Toeplitz analogue of the edge calculus, cf. also the article [**66**]. Let M be an oriented closed compact C^∞ manifold.

THEOREM 6.1. *Let $J \in \mathrm{Vect}(M)$, and let $p : \pi_M^* J \to \pi_M^* J$ for $\pi_M : T^*M \setminus 0 \to M$ be a smooth bundle morphism of homogeneity 0 in the covariable $\xi \neq 0$, i.e., $p(x,\lambda\xi) = p(x,\xi)$ for all $(x,\xi) \in T^*M\setminus 0$, $\lambda \in \mathbb{R}_+$, and $p^2 = p$. Then there exists a $P \in L^0_{\mathrm{cl}}(M;J,J)$ such that $\sigma_0(P) = p$ and $P^2 = P$. Moreover, if p satisfies the condition $p^* = p$, the operator P can be chosen in such a way that $P^* = P$.*

Concerning more material and other observations on pseudo-differential projections, see also Birman, Solomjak [**8**].

Let $J \in \mathrm{Vect}(M)$ and

$$p : \pi_M^* J \to \pi_M^* J \tag{6.1}$$

be a projection as in Theorem 6.1. Then we have

$$L := \operatorname{im} p \in \mathrm{Vect}(T^*M\setminus 0) \tag{6.2}$$

which is a subbundle of $\pi_M^* J$. Conversely for every $L \in \mathrm{Vect}(T^*M\setminus 0)$ there exists a $J \in \mathrm{Vect}(M)$ such that L is a subbundle of $\pi_M^* J$. In fact, there exists an N and an $L^\perp \in \mathrm{Vect}(T^*M\setminus 0)$ such that $L \oplus L^\perp = (T^*M\setminus 0) \times \mathbb{C}^N$.

DEFINITION 6.2. A triple $\mathbb{L} := (P,J,L)$ will be called projection data on M when $P \in L^0_{\mathrm{cl}}(M;J,J)$ is a projection as in Theorem 6.1, and L defined by (6.2). Let $\mathbb{P}(M)$ denote the set of all such projection data.

THEOREM 6.3. (i) *For every $J \in \mathrm{Vect}(M)$ we have $(\mathrm{id}, J, \pi_M^* J) \in \mathbb{P}(M)$.*

(ii) *For every $\mathbb{L} = (P,J,L)$, $\widetilde{\mathbb{L}} = (\tilde{P},\tilde{J},\tilde{L}) \in \mathbb{P}(M)$ we have*

$$\mathbb{L} \oplus \widetilde{\mathbb{L}} := (P \oplus \tilde{P}, J \oplus \tilde{J}, L \oplus \tilde{L}) \in \mathbb{P}(M).$$

(iii) *Every $\mathbb{L} = (P,J,L) \in \mathbb{P}(M)$ admits complementary projection data $\mathbb{L}^\perp \in \mathbb{P}(M)$ in the sense that $\mathbb{L} \oplus \mathbb{L}^\perp = (\mathrm{id}, F, \pi_M^* F)$ for some $F \in \mathrm{Vect}(M)$.*

(iv) *Every $\mathbb{L} = (P,J,L) \in \mathbb{P}(M)$ has an adjoint $\mathbb{L}^* = (P^*, J, L^*) \in \mathbb{P}(M)$ where P^* is the adjoint of P in $L^2(M,J)$ and $L^* \in \mathrm{Vect}(T^*M\setminus 0)$ given by $\operatorname{im} p^*$ for $p^* = \sigma_0(P^*)$. Note that $(\mathbb{L}^*)^* = \mathbb{L}$.*

(v) *For every subbundle L of $\pi_M^* J$, $J \in \mathrm{Vect}(M)$, there exist projection data $\mathbb{L} = (P,J,L) \in \mathbb{P}(M)$.*

The proof is straightforward.

REMARK 6.4. For every $\mathbb{L} = (P,J,L) \in \mathbb{P}(M)$ we find complementary projection data $\mathbb{L}^\perp = (1-P, J, L^\perp)$ by setting $L^\perp = \operatorname{im} \sigma_0(1-P)$ where we have $L \oplus L^\perp = \pi_M^* J$.

REMARK 6.5. For every $\mathbb{L} = (P, J, L)$ we can form continuous projections also in Sobolev spaces $H^s(M, J)$ of distributional sections in the bundle J,

$$P : H^s(M, J) \to H^s(M, J),$$

$s \in \mathbb{R}$. Let us set

$$H^s(M, \mathbb{L}) := PH^s(M, J). \tag{6.3}$$

This is a closed subspace of $H^s(M, J)$, in fact, a Hilbert space with the scalar product induced by $H^s(M, J)$.

REMARK 6.6. Let $\mathbb{L} = (P, J, L)$ and $\tilde{\mathbb{L}} = (\tilde{P}, \tilde{J}, L) \in \mathbb{P}(M)$ where J is a subbundle of $\tilde{J}$ (such that L is a subbundle also of $\pi^*_M \tilde{J}$) and $P : H^s(M, J) \to H^s(M, \mathbb{L})$ the restriction of $\tilde{P} : H^s(M, \tilde{J}) \to H^s(M, \tilde{\mathbb{L}})$ to $H^s(M, J)$. Then $H^s(M, \mathbb{L}) = H^s(M, \tilde{\mathbb{L}})$.

6.2. Edge operators with global projection conditions. We now outline some material on edge operators where the Atiyah-Bott obstruction for the existence of Shapiro-Lopatinskij elliptic conditions does not vanish. We refer here to results of [**65**], see also [**64**], [**59**]. Recall that one of the aspects in these papers is the algebra structure, closed under constructing parametrices. Moreover, the case of Shapiro-Loptinskij ellipticity is a special case. Note that the consideration of cases where the the Atiyah-Bott obstruction does not vanish has been studied by numerous authors in different situations, see, in particular, Grubb and Seeley, [**22**], Savin and Sternin [**49**], Melrose and Rochon [**39**], Albin and Melrose [**3**].

Let M be a compact manifold with edge Y.

DEFINITION 6.7. Let $\mathbb{L}_i := (P_i, J_i, L_i) \in \mathbb{P}(Y)$ be projection data, $V_i \in \mathrm{Vect}(\mathbb{M})$, $i = 1, 2$, and set

$$\boldsymbol{v} := (V_1, V_2; J_1, J_2), \ \ \boldsymbol{l} := (V_1, V_2; \mathbb{L}_1, \mathbb{L}_2).$$

Then $\mathcal{T}^\mu(M, \boldsymbol{g}; \boldsymbol{l})$, $\mu \in \mathbb{R}$, for $\boldsymbol{g} = (\gamma, \gamma - \mu, \Theta)$ is defined to be the set of all operators

$$\mathcal{A} := \mathcal{P}_2 \tilde{\mathcal{A}} \mathcal{E}_1$$

for $\mathcal{P}_2 := \mathrm{diag}(1, P_2)$, $\mathcal{E}_1 := \mathrm{diag}(1, E_1)$, for the maps $E_1 : H^s(Y, \mathbb{L}_1) \to H^s(Y, J_1)$, $P_2 : H^{s-\mu}(Y, J_2) \to H^{s-\mu}(Y, \mathbb{L}_2)$, and arbitrary $\tilde{\mathcal{A}} \in L^\mu(M, \boldsymbol{g}; \boldsymbol{v})$. The elements of $\mathcal{T}^\mu(M, \boldsymbol{g}; \boldsymbol{l})$ will be called edge problems of order μ with global projection conditions. Moreover, set

$$\mathcal{T}^{-\infty}(M, \boldsymbol{g}; \boldsymbol{l}) := \{\mathcal{P}_2 \tilde{\mathcal{C}} \mathcal{E}_1 : \tilde{\mathcal{C}} \in L^{-\infty}(M, \boldsymbol{g}; \boldsymbol{l})\}. \tag{6.4}$$

Observe that the space (6.4) can be equivalently characterised as the set of all $\mathcal{A} \in \mathcal{T}^\mu(M, \boldsymbol{g}; \boldsymbol{l})$, $\mathcal{A} := \mathcal{P}_2 \tilde{\mathcal{A}} \mathcal{E}_1$ for some $\tilde{\mathcal{A}} \in L^\mu(M, \boldsymbol{g}; \boldsymbol{v})$, such that $\mathcal{P}_2 \tilde{\mathcal{A}} \mathcal{P}_1 \in L^{-\infty}(M, \boldsymbol{g}; \boldsymbol{v})$; then $\mathcal{A} = \mathcal{P}_2(\mathcal{P}_2 \tilde{\mathcal{A}} \mathcal{P}_1)\mathcal{E}_1$. Moreover,

$$\mathcal{P}_2(\mathcal{P}_2 \tilde{\mathcal{A}} \mathcal{P}_1)\mathcal{E}_1 \in \mathcal{T}^{-\infty}(M, \boldsymbol{g}; \boldsymbol{l}) \Rightarrow \mathcal{P}_2 \tilde{\mathcal{A}} \mathcal{P}_1 \in L^{-\infty}(M, \boldsymbol{g}; \boldsymbol{v}).$$

THEOREM 6.8. *Every $\mathcal{A} \in \mathcal{T}^\mu(M, \boldsymbol{g}; \boldsymbol{l})$ induces continuous operators*

$$\mathcal{A} : \begin{array}{c} H^{s,\gamma}(M, V_1) \\ \oplus \\ H^s(Y, \mathbb{L}_1) \end{array} \longrightarrow \begin{array}{c} H^{s-\mu,\gamma-\mu}(M, V_2) \\ \oplus \\ H^{s-\mu}(Y, \mathbb{L}_2) \end{array} \tag{6.5}$$

for every $s \in \mathbb{R}$.

The proof is evident after Theorem 3.14.

REMARK 6.9. Given $V_i \in \mathrm{Vect}(\mathbb{M})$, $i = 1, 2$, there is a canonical isomorphism

$$\mathcal{T}^\mu(M, \boldsymbol{g}; \boldsymbol{l}) \to \{\mathcal{P}_2 \tilde{\mathcal{A}} \mathcal{P}_1 : \tilde{\mathcal{A}} \in L^\mu(M, \boldsymbol{g}; \boldsymbol{v})\}.$$

In other words we have an identification

$$\mathcal{T}^\mu(M, \boldsymbol{g}; \boldsymbol{l}) = L^\mu(M, \boldsymbol{g}; \boldsymbol{v})/\sim \tag{6.6}$$

with the equivalence relation

$$\tilde{\mathcal{A}} \sim \tilde{\mathcal{B}} \Leftrightarrow \mathcal{P}_2 \tilde{\mathcal{A}} \mathcal{P}_1 = \mathcal{P}_2 \tilde{\mathcal{B}} \mathcal{P}_1. \tag{6.7}$$

The space $\mathcal{T}^\mu(M, \boldsymbol{g}; \boldsymbol{l})$ is equipped with the principal symbolic structure

$$\sigma(\mathcal{A}) = (\sigma_0(\mathcal{A}), \sigma_1(\mathcal{A}))$$

with the interior and the edge symbolic component. Writing $\mathcal{A} = (A_{ij})_{i,j=1,2}$ we first set $\sigma_0(\mathcal{A}) := \sigma_0(A_{11})$, i.e.,

$$\sigma_0(\mathcal{A}) : \pi^*_{M \setminus Y} V_1 \to \pi^*_{M \setminus Y} V_2$$

for $\pi_{M \setminus Y} : T^*(M \setminus Y) \setminus 0 \to M \setminus Y$.

The edge symbol of $\mathcal{A}$, represented as $\mathcal{A} = \mathcal{P}_2 \tilde{\mathcal{A}} \mathcal{E}_1$, is defined as

$$\sigma_1(\mathcal{A}) = \mathrm{diag}(1, p_2) \sigma_1(\tilde{\mathcal{A}}) \mathrm{diag}(1, e_1),$$

$$\sigma_1(\mathcal{A})(y, \eta) : \begin{matrix} \mathcal{K}^{s,\gamma}(N_y^\wedge, V_{1,y}) \\ \oplus \\ L_{1,(y,\eta)} \end{matrix} \longrightarrow \begin{matrix} \mathcal{K}^{s-\mu,\gamma-\mu}(N_y^\wedge, V_{2,y}) \\ \oplus \\ L_{2,(y,\eta)} \end{matrix} \tag{6.8}$$

where $p_2(y, \eta)$ is the homogeneous principal symbol of order zero of the projection $P_2 \in L^0_{\mathrm{cl}}(Y; J_2, J_2)$, and $e_1 : L_{1,(y,\eta)} \to (\pi_Y^* J_1)_{(y,\eta)}$ is the canonical embedding.

REMARK 6.10. Identifying an operator $\mathcal{A} = \mathcal{P}_2 \tilde{\mathcal{A}} \mathcal{E}_1 \in \mathcal{T}^\mu(M, \boldsymbol{g}; \boldsymbol{l})$ with $\tilde{\tilde{\mathcal{A}}} := \mathcal{P}_2 \tilde{\mathcal{A}} \mathcal{P}_1 \in L^\mu(M, \boldsymbol{g}; \boldsymbol{v})$, cf. Remark 6.9, then $\sigma(\tilde{\tilde{\mathcal{A}}}) = 0$ in the sense of $L^\mu(M, \boldsymbol{g}; \boldsymbol{v})$ is equivalent to $\sigma(\mathcal{A}) = 0$ in the sense of $T^\mu(M, \boldsymbol{g}; \boldsymbol{l})$.

Using (3.36) we define $T^\nu(M, \boldsymbol{g}; \boldsymbol{l})$ for $\boldsymbol{g} = (\gamma, \gamma - \mu, \Theta)$, $\mu - \nu \in \mathbb{N}$,

$$T^\nu(M, \boldsymbol{g}; \boldsymbol{l}) := \{\mathcal{P}_2 \tilde{\mathcal{A}} \mathcal{E}_1 : \tilde{\mathcal{A}} \in L^\nu(M, \boldsymbol{g}; \boldsymbol{v})\}.$$

REMARK 6.11. $\mathcal{A} \in \mathcal{T}^\mu(M, \boldsymbol{g}; \boldsymbol{l})$ and $\sigma(\mathcal{A}) = 0$ imply $\mathcal{A} \in \mathcal{T}^{\mu-1}(M, \boldsymbol{g}; \boldsymbol{l})$, and the operator (6.5) is compact for every $s \in \mathbb{R}$.

REMARK 6.12. Let $\mathcal{A}_j \in \mathcal{T}^{\mu-j}(M, \boldsymbol{g}; \boldsymbol{l})$, $j \in \mathbb{N}$, be an arbitrary sequence, $\boldsymbol{g} := (\gamma, \gamma-\mu, (-(k+1), 0])$ for a finite k, and assume that the asymptotic types involved in the Green operators are independent of j. Then there exists an $\mathcal{A} \in \mathcal{T}^\mu(M, \boldsymbol{g}; \boldsymbol{l})$ such that

$$\mathcal{A} - \sum_{j=0}^{N} \mathcal{A}_j \in \mathcal{T}^{\mu-(N+1)}(M, \boldsymbol{g}; \boldsymbol{l})$$

for every $N \in \mathbb{N}$, and $\mathcal{A}$ is unique mod $\mathcal{T}^{-\infty}(M, \boldsymbol{g}; \boldsymbol{l})$.

THEOREM 6.13. (i) *$\mathcal{A} \in \mathcal{T}^\mu(M, \boldsymbol{g}_0; \boldsymbol{l}_0)$, $\mathcal{B} \in \mathcal{T}^\rho(M, \boldsymbol{g}_1; \boldsymbol{l}_1)$ for $\boldsymbol{g}_0 = (\gamma - \rho, \gamma - (\mu + \rho), \Theta)$, $\boldsymbol{g}_1 = (\gamma, \gamma - \rho, \Theta)$, implies $\mathcal{AB} \in \mathcal{T}^{\mu+\rho}(M, \boldsymbol{g}_0 \circ \boldsymbol{g}_1; \boldsymbol{l}_0 \circ \boldsymbol{l}_1)$ (when the projection data in the middle fit together such that $\boldsymbol{l}_0 \circ \boldsymbol{l}_1$ makes sense), and we have $\sigma(\mathcal{AB}) = \sigma(\mathcal{A})\sigma(\mathcal{B})$ with componentwise multiplication.*

(ii) *$\mathcal{A} \in \mathcal{T}^\mu(M, \boldsymbol{g}; \boldsymbol{l})$ for $\boldsymbol{g} = (\gamma, \gamma-\mu, \Theta)$, and $\boldsymbol{l} = (V_1, V_2; \mathbb{L}_1, \mathbb{L}_2)$ implies $\mathcal{A}^* \in \mathcal{T}^\mu(M, \boldsymbol{g}^*; \boldsymbol{l}^*)$ for $\boldsymbol{l}^* = (V_2, V_1; \mathbb{L}_2^*, \mathbb{L}_1^*)$, where $\mathcal{A}^*$ is the formal adjoint in the sense*

$$(u, \mathcal{A}^* v)_{H^{0,0}(M,V_1) \oplus H^0(Y,\mathbb{L}_1)} = (\mathcal{A}u, v)_{H^{0,0}(M,V_2) \oplus H^0(Y,\mathbb{L}_2)}$$

for all $u \in H^{\infty,\infty}(M, V_1) \oplus H^\infty(Y, \mathbb{L}_1)$, $v \in H^{\infty,\infty}(M, V_2) \oplus H^\infty(Y, \mathbb{L}_2)$, and we have $\sigma(\mathcal{A}^*) = \sigma(\mathcal{A})^*$ with componentwise formal adjoint.

Theorem 6.13 is a consequence of Theorem 3.17. We have a natural notion of direct sum

$$\mathcal{T}^\mu(M, \boldsymbol{g}; \boldsymbol{l}) \oplus \mathcal{T}^\mu(M, \boldsymbol{g}; \boldsymbol{m}) = \mathcal{T}^\mu(M, \boldsymbol{g}; \boldsymbol{l} \oplus \boldsymbol{m})$$

where $\sigma(\mathcal{A} \oplus \mathcal{B}) = \sigma(\mathcal{A}) \oplus \sigma(\mathcal{B})$ with the componentwise direct sum of symbols.

6.3. Ellipticity, parametrices, and the Fredholm property. We now turn to ellipticity in the Toeplitz calculus of edge problems.

DEFINITION 6.14. Let

$$\mathcal{A} \in \mathcal{T}^\mu(M, \boldsymbol{g}; \boldsymbol{l}),\ \boldsymbol{g} := (\gamma, \gamma - \mu, \Theta),\ \boldsymbol{l} := (V_1, V_2; \mathbb{L}_1, \mathbb{L}_2),$$

$V_i \in \mathrm{Vect}(\mathrm{M})$, $\mathbb{L}_i = (P_i, J_i, L_i) \in \mathbb{P}(Y)$, $i = 1, 2$. The operator $\mathcal{A}$ is called elliptic if the upper left corner $A \in L^\mu(M, \boldsymbol{g}; V_1, V_2)$ is σ_1-elliptic in the sense of Definition 4.1 and if the edge symbol (6.8) is an isomorphism for every $(y, \eta) \in T^*Y \setminus 0$ and some $s = s_0 \in \mathbb{R}$.

REMARK 6.15. The bijectivity of (6.8) for some $s = s_0 \in \mathbb{R}$ is equivalent to the bijectivity of (6.8) for all $s \in \mathbb{R}$ or of

$$\sigma_1(\mathcal{A})(y, \eta) : \begin{array}{c} \mathcal{K}^{\infty,\gamma;\infty}(N_y^\wedge, V_{1,y}) \\ \oplus \\ L_{1,(y,\eta)} \end{array} \longrightarrow \begin{array}{c} \mathcal{K}^{\infty,\gamma-\mu;\infty}(N_y^\wedge, V_{2,y}) \\ \oplus \\ L_{2,(y,\eta)} \end{array}. \tag{6.9}$$

THEOREM 6.16. *For every σ_0-elliptic $A \in L^\mu(M, \boldsymbol{g}; V_1, V_2)$, where*

$$\sigma_1(A)(y, \eta) : \mathcal{K}^{\infty,\gamma;\infty}(N_y^\wedge, V_{1,y}) \to \mathcal{K}^{\infty,\gamma-\mu;\infty}(N_y^\wedge, V_{2,y}) \tag{6.10}$$

is a family of Fredholm operators, there exist projection data $\mathbb{L}_1, \mathbb{L}_2 \in \mathbb{P}(Y)$, a Green operator $G \in L_G^\mu(M, \boldsymbol{g}; V_1, V_2)$, and an element $\mathcal{A} \in \mathcal{T}^\mu(M, \boldsymbol{g}; \boldsymbol{l})$, $\boldsymbol{l} := (V_1, V_2; \mathbb{L}_1, \mathbb{L}_2)$, with $A + G$ in the upper left corner, such that $\mathcal{A}$ is elliptic in the sense of Definition 6.14.

This result is contained in [**64**, Remark 3.10].

In boundary value problems it is justified to interpret Green operators as boundary conditions. This is also the case in edge problems, in particular, in the calculus with projection data. In other words, Theorem 6.16 essentially says that for every σ_0-elliptic $A \in L^\mu(M, \boldsymbol{g}; V_1, V_2)$ such that (6.10) is a family of Fredholm operators, there exists an elliptic edge problem $\mathcal{A}$ with global projection edge conditions.

DEFINITION 6.17. Let $\mathcal{A} \in \mathcal{T}^\mu(M, \boldsymbol{g}; \boldsymbol{l})$ be as in Definition 6.14. An operator

$$\mathcal{P} \in \mathcal{T}^{-\mu}(M, \boldsymbol{g}^{-1}; \boldsymbol{l}^{-1})$$

for $\boldsymbol{g}^{-1} = (\gamma - \mu, \gamma, \Theta)$ and $\boldsymbol{l}^{-1} = (V_2, V_1; \mathbb{L}_2, \mathbb{L}_1)$, is called a parametrix of $\mathcal{A}$, if

$$\mathcal{C}_{\mathrm{L}} := \mathcal{I} - \mathcal{P}\mathcal{A} \in \mathcal{T}^{-\infty}(M, \boldsymbol{g}_{\mathrm{L}}; \boldsymbol{l}_{\mathrm{L}}),\ \mathcal{C}_{\mathrm{R}} := \mathcal{I} - \mathcal{A}\mathcal{P} \in \mathcal{T}^{-\infty}(M, \boldsymbol{g}_{\mathrm{R}}; \boldsymbol{l}_{\mathrm{R}})$$

for $\boldsymbol{g}_{\mathrm{L}} = (\gamma, \gamma, \Theta)$ or $\boldsymbol{g}_{\mathrm{R}} = (\gamma - \mu, \gamma - \mu, \Theta)$, with $\mathcal{I}$ being the respective identity operators, and $\boldsymbol{l}_{\mathrm{L}} := (V_1, V_1; \mathbb{L}_1, \mathbb{L}_1)$, $\boldsymbol{l}_{\mathrm{R}} := (V_2, V_2; \mathbb{L}_2, \mathbb{L}_2)$.

THEOREM 6.18. *Let* $\mathcal{A} \in \mathcal{T}^\mu(M, \boldsymbol{g}; \boldsymbol{l})$, $\mu \in \mathbb{R}$, $\boldsymbol{l} := (V_1, V_2; \mathbb{L}_1, \mathbb{L}_2)$ *for* $V_1, V_2 \in \mathrm{Vect}(\mathbb{M})$, $\mathbb{L}_1, \mathbb{L}_2 \in \mathbb{P}(Y)$.

(i) *The operator* $\mathcal{A}$ *is elliptic if and only if*

$$\mathcal{A} : \begin{array}{c} H^{s,\gamma}(M, V_1) \\ \oplus \\ H^s(Y, \mathbb{L}_1) \end{array} \longrightarrow \begin{array}{c} H^{s-\mu,\gamma-\mu}(X, V_2) \\ \oplus \\ H^{s-\mu}(Y, \mathbb{L}_2) \end{array} \tag{6.11}$$

is a Fredholm operator for an $s = s_0 \in \mathbb{R}$.

(ii) *If* $\mathcal{A}$ *is elliptic,* (6.11) *is Fredholm for all* $s \in \mathbb{R}$, *and* $\dim\ker\mathcal{A}$ *and* $\dim\operatorname{coker}\mathcal{A}$ *are independent of* s.

(iii) *An elliptic* $\mathcal{A} \in \mathcal{T}^\mu(M, \boldsymbol{g}; \boldsymbol{l})$ *has a parametrix* $\mathcal{P} \in \mathcal{T}^{-\mu}(M, \boldsymbol{g}^{-1}; \boldsymbol{l}^{-1})$ *in the sense of* Definition 6.17.

This corresponds to [**64**, Theorem 3.13], except for the necessity of ellipticity for the Fredholm property. The latter can be proved along the lines of [**66**, Theorem 38].

As a corollary we obtain the following result. For every $\mu, \gamma \in \mathbb{R}$, $\boldsymbol{g} := (\gamma, \gamma - \mu, \Theta)$, $V \in \mathrm{Vect}(\mathbb{M})$ and $\mathbb{L} \in \mathbb{P}(Y)$ there exists an elliptic element $\mathcal{R}^\mu_{V,\mathbb{L}} \in \mathcal{T}^\mu(M, \boldsymbol{g}; \boldsymbol{l})$ for $\boldsymbol{l} := (V, V; \mathbb{L}, \mathbb{L})$ which induces a Fredholm operator

$$\mathcal{R}^\mu_{V,\mathbb{L}} : \begin{array}{c} H^{s,\gamma}(X, V) \\ \oplus \\ H^s(Y, \mathbb{L}) \end{array} \rightarrow \begin{array}{c} H^{s-\mu,\gamma-\mu}(X, V) \\ \oplus \\ H^{s-\mu}(Y, \mathbb{L}) \end{array}$$

for every $s \in \mathbb{R}$.

References

[1] M. S. Agranovič and A. S. Dynin, *General boundary-value problems for elliptic systems in higher-dimensional regions* (Russian), Dokl. Akad. Nauk SSSR **146** (1962), 511–514. MR0140820 (25 #4234)

[2] M. S. Agranovič and M. I. Višik, *Elliptic problems with a parameter and parabolic problems of general type* (Russian), Uspehi Mat. Nauk **19** (1964), no. 3 (117), 53–161. MR0192188 (33 #415)

[3] P. Albin and R. Melrose, *Fredholm realizations of elliptic symbols on manifolds with boundary*, J. Reine Angew. Math. **627** (2009), 155–181, DOI 10.1515/CRELLE.2009.014. MR2494931 (2011e:58046)

[4] M. F. Atiyah and R. Bott, *The index problem for manifolds with boundary*, Differential Analysis, Bombay Colloq., 1964, Oxford Univ. Press, London, 1964, pp. 175–186. MR0185606 (32 #3069)

[5] M. F. Atiyah and I. M. Singer, *The index of elliptic operators I, II, III*, Ann. of Math. (2), **87** (1968), 483-530, 531-545, 546-604.

[6] M. F. Atiyah, V. Patodi, and I.M. Singer, *Spectral asymmetry and Riemannian geometry I, II, III* Math. Proc. Cambridge Philos. Soc., **77, 78, 79** (1975, 1976, 1976), 43-69, 405-432, 315-330.

[7] K. Bekka and D. Trotman, *On metric properties of stratified sets*, Manuscripta Math. **111** (2003), no. 1, 71–95, DOI 10.1007/s00229-003-0354-8. MR1981597 (2004c:58013)

[8] M. S. Birman and M. Z. Solomjak, *On the subspaces admitting a pseudodifferential projection*, Vestnik LGU **1** (1982), 18-25.

[9] L. Boutet de Monvel, *Boundary problems for pseudo-differential operators*, Acta Math. **126** (1971), no. 1-2, 11–51. MR0407904 (53 #11674)

[10] D. Calvo, C.-I. Martin, and B.-W. Schulze, *Symbolic structures on corner manifolds*, RIMS Conf. dedicated to L. Boutet de Monvel on "Microlocal Analysis and Asymptotic Analysis", Kyoto, August 2004, Keio University, Tokyo, 2005, pp. 22-35.

[11] D. Calvo and B.-W. Schulze, *Edge symbolic structures of second generation*, Math. Nachr. **282** (2009), no. 3, 348–367, DOI 10.1002/mana.200710741. MR2503157 (2011d:58064)

[12] D.-C. Chang, N. Habal, and B.-W. Schulze, *The edge algebra structure of the Zaremba problem*, J. Pseudo-Differ. Oper. Appl. **5** (2014), no. 1, 69–155, DOI 10.1007/s11868-013-0088-7. MR3170601

[13] N. Dines, *Elliptic operators on corner manifolds*, Ph.D. thesis, University of Potsdam, 2006.

[14] Ch. Dorschfeldt, *Algebras of pseudo-differential operators near edge and corner singularities*, Math. Res. **102**, Akademie Verlag, Berlin, 1998.

[15] Y. V. Egorov and B.-W. Schulze, *Pseudo-differential operators, singularities, applications*, Operator Theory: Advances and Applications, vol. 93, Birkhäuser Verlag, Basel, 1997. MR1443430 (98e:35181)

[16] G. I. Eskin, *Boundary value problems for elliptic pseudodifferential equations*, Transl. of Nauka, Moskva, 1973, Math. Monographs, Amer. Math. Soc. **52**, Providence, Rhode Island 1980.

[17] H.-J. Flad, G. Harutyunyan, R. Schneider, and B.-W. Schulze, *Explicit Green operators for quantum mechanical Hamiltonians. I. The hydrogen atom*, Manuscripta Math. **135** (2011), no. 3-4, 497–519, DOI 10.1007/s00229-011-0429-x. MR2813446 (2012e:81059)

[18] H.-J. Flad and G. Harutyunyan, *Ellipticity of quantum mechanical Hamiltonians in the edge algebra*, Discrete Contin. Dyn. Syst. **Suppl. Vol. I** (2011), 420–429. MR2987424

[19] W. Fulton and R. MacPherson, *Categorical framework for the study of singular spaces*, Mem. Amer. Math. Soc. **31** (1981), no. 243, vi+165, DOI 10.1090/memo/0243. MR609831 (83a:55015)

[20] J. B. Gil, B. W. Schulze, and J. Seiler, *Cone pseudodifferential operators in the edge symbolic calculus*, Osaka J. Math. **37** (2000), no. 1, 221–260. MR1750278 (2001a:58038)

[21] I. C. Gohberg and E. I. Sigal, *An operator generalization of the logarithmic residue theorem and the theorem of Rouché*, Math. USSR Sbornik **13**, 4 (1971), 603-625.

[22] G. Grubb and R. T. Seeley, *Weakly parametric pseudodifferential operators and Atiyah-Patodi-Singer boundary problems*, Invent. Math. **121** (1995), no. 3, 481–529, DOI 10.1007/BF01884310. MR1353307 (96k:58216)

[23] N. Habal and B.-W. Schulze, *Holomorphic corner symbols*, J. Pseudo-Differ. Oper. Appl. **2** (2011), no. 4, 419–465, DOI 10.1007/s11868-011-0040-7. MR2854840

[24] N. Habal and B.-W. Schulze, *Mellin quantisation in corner operators*, Operator theory, pseudo-differential equations, and mathematical physics, Oper. Theory Adv. Appl., vol. 228, Birkhäuser/Springer Basel AG, Basel, 2013, pp. 151–172, DOI 10.1007/978-3-0348-0537-7_8. MR3025494

[25] G. Harutyunyan and B.-W. Schulze, *Elliptic mixed, transmission and singular crack problems*, EMS Tracts in Mathematics, vol. 4, European Mathematical Society (EMS), Zürich, 2008. MR2376895 (2009k:35002)

[26] T. Hirschmann, *Functional analysis in cone and edge Sobolev spaces*, Ann. Global Anal. Geom. **8** (1990), no. 2, 167–192, DOI 10.1007/BF00128002. MR1088510 (92d:58203)

[27] D. Kapanadze, B.-W. Schulze, and I. Witt, *Coordinate invariance of the cone algebra with asymptotics*, Parabolicity, Volterra calculus, and conical singularities, Oper. Theory Adv. Appl., vol. 138, Birkhäuser, Basel, 2002, pp. 307–358. MR1966208 (2004e:58046)

[28] D. Kapanadze and B.-W. Schulze, *Crack theory and edge singularities*, Mathematics and its Applications, vol. 561, Kluwer Academic Publishers Group, Dordrecht, 2003. MR2023308 (2004j:58026)

[29] H. C. King and D. Trotman, *Poincare-Hopf theorems on singular spaces*, manuscript (2007).

[30] V. A. Kondrat′ev, *Boundary value problems for elliptic equations in domains with conical or angular points* (Russian), Trudy Moskov. Mat. Obšč. **16** (1967), 209–292. MR0226187 (37 #1777)

[31] T. Krainer, *The calculus of Volterra Mellin pseudodifferential operators with operator-valued symbols*, Parabolicity, Volterra calculus, and conical singularities, Oper. Theory Adv. Appl., vol. 138, Birkhäuser, Basel, 2002, pp. 47–91. MR1966205 (2004d:35262)

[32] H. Kumano-go, *Pseudo-differential operators*, The MIT Press, Cambridge, Massachusetts and London, England, 1981.

[33] X. Liu and B.-W. Schulze, *Boundary value problems in edge representation*, Math. Nachr. **280** (2007), no. 5-6, 581–621, DOI 10.1002/mana.200610504. MR2308486 (2009f:47074)

[34] L. Maniccia and B.-W. Schulze, *An algebra of meromorphic corner symbols* (English, with English and French summaries), Bull. Sci. Math. **127** (2003), no. 1, 55–99, DOI 10.1016/S0007-4497(02)00007-6. MR1957798 (2003k:58040)

[35] R. Mazzeo, *Elliptic theory of differential edge operators. I*, Comm. Partial Differential Equations **16** (1991), no. 10, 1615–1664, DOI 10.1080/03605309108820815. MR1133743 (93d:58152)

[36] R. Melrose, *Transformation of boundary problems*, Acta Math. **147** (1981), no. 3-4, 149–236, DOI 10.1007/BF02392873. MR639039 (83f:58073)

[37] R. Melrose and G. A. Mendoza, *Elliptic operators of totally characteristic type*, Preprint, MSRI 1983.

[38] R. Melrose, *The Atiyah-Patodi-Singer index theorem*, Research Notes in Mathematics, vol. 4, A. K. Peters, Ltd., Wellesley, MA, 1993. MR1348401 (96g:58180)

[39] R. Melrose and F. Rochon, *Index in K-theory for families of fibred cusp operators*, K-Theory **37** (2006), no. 1-2, 25–104, DOI 10.1007/s10977-006-0003-6. MR2274670 (2008c:58013)

[40] B. A. Plamenevskiĭ, *The boundedness of singular integrals in spaces with weight* (Russian), Mat. Sb. (N.S.) **76(118)** (1968), 573–592. MR0511933 (58 #23547)

[41] B. A. Plamenevskij, *Algebras of pseudo-differential operators*, Nauka, Moscow, 1986.

[42] S. Rempel and B.-W. Schulze, *Index theory of elliptic boundary problems*, Akademie-Verlag, Berlin, 1982. MR690065 (85b:58126)

[43] S. Rempel and B.-W. Schulze, *Parametrices and boundary symbolic calculus for elliptic boundary problems without the transmission property*, Math. Nachr. **105** (1982), 45–149, DOI 10.1002/mana.19821050105. MR670511 (84a:58083)

[44] S. Rempel and B.-W. Schulze, *Complete Mellin symbols and the conormal asymptotics in boundary value problems*, Proc. Journées "Equ. aux Dériv. Part.", St.-Jean de Monts, Conf. No. V , 1984.

[45] S. Rempel and B.-W. Schulze, *Branching of asymptotics for elliptic operators on manifolds with edges*, Partial differential equations (Warsaw, 1984), Banach Center Publ., vol. 19, PWN, Warsaw, 1987, pp. 231–236. MR1055175

[46] S. Rempel and B.-W. Schulze, *Complete Mellin and Green symbolic calculus in spaces with conormal asymptotics*, Ann. Global Anal. Geom. **4** (1986), no. 2, 137–223, DOI 10.1007/BF00129908. MR910551 (89f:58132)

[47] S. Rempel and B.-W. Schulze, *Asymptotics for elliptic mixed boundary problems. Pseudo-differential and Mellin operators in spaces with conormal singularity*, Mathematical Research, vol. 50, Akademie-Verlag, Berlin, 1989. MR1002573 (90h:35083)

[48] W. Rungrottheera and B.-W. Schulze, *Holomorphic operator families on a manifold with edge*, J. Pseudo-Differ. Oper. Appl. **4** (2013), no. 3, 297–315, DOI 10.1007/s11868-013-0064-2. MR3089831

[49] A. Savin and B. Sternin, *Index defects in the theory of spectral boundary value problems*, Aspects of boundary problems in analysis and geometry, Oper. Theory Adv. Appl., vol. 151, Birkhäuser, Basel, 2004, pp. 170–238. MR2072500 (2005e:58039)

[50] B.-W. Schulze, *Regularity with continuous and branching asymptotics for elliptic operators on manifolds with edges*, Integral Equations Operator Theory **11** (1988), no. 4, 557–602, DOI 10.1007/BF01199307. MR950517 (89k:58283)

[51] B.-W. Schulze, *Ellipticity and continuous conormal asymptotics on manifolds with conical singularities*, Math. Nachr. **136** (1988), 7–57, DOI 10.1002/mana.19881360102. MR952461 (89j:58137)

[52] B.-W. Schulze, *Pseudo-differential operators on manifolds with edges*, Symposium "Partial Differential Equations" (Holzhau, 1988), Teubner-Texte Math., vol. 112, Teubner, Leipzig, 1989, pp. 259–288. MR1105817

[53] B.-W. Schulze, *Corner Mellin operators and reduction of orders with parameters*, Ann. Scuola Norm. Sup. Pisa Cl. Sci. (4) **16** (1989), no. 1, 1–81. MR1056128 (91k:58127)

[54] B.-W. Schulze, *Mellin representations of pseudo-differential operators on manifolds with corners*, Ann. Global Anal. Geom. **8** (1990), no. 3, 261–297, DOI 10.1007/BF00127939. MR1089238 (92b:58226)

[55] B.-W. Schulze, *Pseudo-differential operators on manifolds with singularities*, Studies in Mathematics and its Applications, vol. 24, North-Holland Publishing Co., Amsterdam, 1991. MR1142574 (93b:47109)

[56] B.-W. Schulze, *The Mellin pseudo-differential calculus on manifolds with corners*, Symposium "Analysis on Manifolds with Singularities" (Breitenbrunn, 1990), Teubner-Texte Math., vol. 131, Teubner, Stuttgart, 1992, pp. 208–289, DOI 10.1007/978-3-663-11577-9_22. MR1217731 (94d:47053)

[57] B.-W. Schulze, *Pseudo-differential boundary value problems, conical singularities, and asymptotics*, Mathematical Topics, vol. 4, Akademie Verlag, Berlin, 1994. MR1282496 (95e:58172)

[58] B.-W. Schulze, *Boundary value problems and singular pseudo-differential operators*, Pure and Applied Mathematics (New York), John Wiley & Sons, Ltd., Chichester, 1998. MR1631763 (99m:35281)

[59] B.-W. Schulze, *An algebra of boundary value problems not requiring Shapiro-Lopatinskij conditions*, J. Funct. Anal. **179** (2001), no. 2, 374–408, DOI 10.1006/jfan.2000.3684. MR1809115 (2002c:58038)

[60] B.-W. Schulze, *Operators with symbol hierarchies and iterated asymptotics*, Publ. Res. Inst. Math. Sci. **38** (2002), no. 4, 735–802. MR1917163 (2003k:58041)

[61] B.-W. Schulze, *Toeplitz operators, and ellipticity of boundary value problems with global projection conditions*, Aspects of boundary problems in analysis and geometry, Oper. Theory Adv. Appl., vol. 151, Birkhäuser, Basel, 2004, pp. 342–429. MR2072486 (2005g:58047)

[62] B.-W. Schulze, *Corner Mellin operators and reduction of orders with parameters*, Report MATH, vol. 88, Akademie der Wissenschaften der DDR, Karl-Weierstrass-Institut für Mathematik, Berlin, 1988. With a German summary. MR942221 (89h:58186)

[63] B.-W. Schulze, *The iterative structure of the corner calculus*, Pseudo-differential operators: analysis, applications and computations, Oper. Theory Adv. Appl., vol. 213, Birkhäuser/Springer Basel AG, Basel, 2011, pp. 79–103, DOI 10.1007/978-3-0348-0049-5_5. MR2867418

[64] B.-W. Schulze and J. Seiler, *Edge operators with conditions of Toeplitz type*, J. Inst. Math. Jussieu **5** (2006), no. 1, 101–123, DOI 10.1017/S1474748005000198. MR2195947 (2006m:58033)

[65] B. W. Schulze and J. Seiler, *The edge algebra structure of boundary value problems*, Ann. Global Anal. Geom. **22** (2002), no. 3, 197–265, DOI 10.1023/A:1019939316595. MR1925589 (2003h:35292)

[66] B.-W. Schulze and J. Seiler, *Boundary value problems with global projection conditions*, J. Funct. Anal. **206** (2004), no. 2, 449–498, DOI 10.1016/S0022-1236(03)00140-X. MR2021855 (2005a:58036)

[67] J. Seiler, *The cone algebra and a kernel characterization of Green operators*, Approaches to singular analysis (Berlin, 1999), Oper. Theory Adv. Appl., vol. 125, Birkhäuser, Basel, 2001, pp. 1–29. MR1827169 (2002d:47072)

[68] B.-W. Schulze and L. Tepoyan, *The singular functions of branching edge asymptotics*, Pseudo-differential operators, generalized functions and asymptotics, Oper. Theory Adv. Appl., vol. 231, Birkhäuser/Springer Basel AG, Basel, 2013, pp. 27–53, DOI 10.1007/978-3-0348-0585-8_2. MR3075931

[69] B.-W. Schulze and A. Volpato, *Branching asymptotics on manifolds with edge*, J. Pseudo-Differ. Oper. Appl. **1** (2010), no. 4, 433–493, DOI 10.1007/s11868-010-0019-9. MR2747905 (2012c:35497)

[70] B.-W. Schulze and Y. Wei, *The Mellin-edge quantisation for corner operators*, Complex Anal. Oper. Theory **8** (2014), no. 4, 803–841, DOI 10.1007/s11785-013-0289-3. MR3178034

[71] B. Ju. Sternin, *Topological aspects of a problem of S.L. Sobolev*, (Russ) Moscow Institute of Electric Engineering, Moscow 1971.

[72] B. Ju. Sternin, *A relative elliptic theory, and S. L. Sobolev's problem* (Russian), Dokl. Akad. Nauk SSSR **230** (1976), no. 2, 287–290. MR0426067 (54 #14013)

[73] B.-W. Schulze and M. W. Wong, *Mellin and Green operators of the corner calculus*, J. Pseudo-Differ. Oper. Appl. **2** (2011), no. 4, 467–507, DOI 10.1007/s11868-011-0036-3. MR2854841

[74] M. I. Višik and G. I. Èskin, *Convolution equations in a bounded region* (Russian), Uspehi Mat. Nauk **20** (1965), no. 3 (123), 89–152. MR0185273 (32 #2741)

[75] M. I. Višik and G. I. Èskin, *Convolution equations in a bounded region in spaces with weight norms* (Russian), Mat. Sb. (N.S.) **69 (111)** (1966), 65–110. MR0218851 (36 #1935)
[76] S. Weinberger, *The topological classification of stratified spaces*, Chicago Lectures in Mathematics, University of Chicago Press, Chicago, IL, 1994. MR1308714 (96b:57024)

INSTITUTE OF MATHEMATICS, UNIVERSITY OF POTSDAM, AM NEUEN PALAIS 10, POTSDAM D-14469, GERMANY
E-mail address: schulze@math.uni-potsdam.de

Contemporary Mathematics
Volume **630**, 2014
http://dx.doi.org/10.1090/conm/630/12668

The Fundamental Gap and One-Dimensional Collapse

Zhiqin Lu and Julie Rowlett

Abstract. The *fundamental gap* of a bounded, connected domain in $\mathbb{R}^n$ is the difference between the first two (positive) eigenvalues of the Euclidean Laplacian with Dirichlet boundary condition. The *one dimensional collapse* considered here is the degeneration of a family of convex, bounded domains in $\mathbb{R}^n$ to a domain in $\mathbb{R}^{n-1}$. The boundary of the domains need not be smooth, merely Lipschitz continuous. Our results show that the fundamental gap detects the geometry of one-dimensional collapse, and that depending upon the geometry of the collapse, the gap can either diverge or remain bounded.

Contents

1. Motivation and results

Our sign convention and boundary conditions for the Laplace equation on a domain $\Omega \subset \mathbb{R}^n$ shall be

$$\Delta u = \sum_{k=1}^{n} \frac{\partial^2 u}{\partial x_k^2} = -\lambda u, \quad \text{Dirichlet boundary condition: } u = 0 \text{ on } \partial\Omega.$$

This equation arises from physics by separating the time and space variables in the wave equation, and the eigenvalues correspond to the frequencies of standing waves on Ω. The set of all eigenvalues is known as *the spectrum.* Based on the physical interpretation Marc Kac posed the now well-known question, "Can one hear the shape of a drum?" [**21**]. The mathematical formulation is: if two Euclidean domains have the same spectrum, do they have the same shape? For arbitrary bounded, connected domains with piecewise smooth boundary Gordon, Webb, and Wolpert proved that the answer is "no" [**14, 15**]. For convex domains, however,

2010 *Mathematics Subject Classification.* Primary 58C40.

Key words and phrases. Polygonal domain, triangle, sector, eigenvalues, Dirichlet Laplacian, collapsing domain, degenerating domain, fundamental gap.

Kac's question is an open problem. In some cases one *can* "hear the shape of a drum" by restricting to "drums" with specific geometric properties. Durso used the heat and wave traces to prove that isospectral triangular domains are congruent [**8**]. Grieser and Maronna gave a proof using only the heat trace [**20**].

For both proofs one must know that the entire spectra of two triangles coincide to prove that they are congruent. Since three independent parameters determine a triangle (up to congruence), it is natural to conjecture that the first three eigenvalues suffice to determine if two triangles are congruent. Antunes and Freitas demonstrated strong numerical evidence in support of this conjecture in [**3**].

In some cases a finite set of eigenvalues can indeed detect geometric features such as symmetry. Pólya and Szegő proved that the first eigenvalue detects the regular n-gon among all convex n-gons with fixed area for $n = 3$ and 4 [**29**]. For $n \geq 5$, the analogous result has not been proven and is known as the Pólya-Szegő Conjecture.

The next natural object to study after the first eigenvalue is the difference between it and the rest of the spectrum, known as the *fundamental gap.* If one scales a bounded domain $\Omega \subset \mathbb{R}^n$ by a constant factor c, then the eigenvalues change according to:

$$\lambda_k(c\Omega) = c^{-2}\lambda_k(\Omega).$$

This motivates the definition of the scale-invariant *gap function*

$$\xi(\Omega) := d^2(\Omega)\left(\lambda_2(\Omega) - \lambda_1(\Omega)\right),$$

where d is the diameter of the domain.

In [**26**] we demonstrated that the gap function detects the equilateral triangle among all Euclidean triangles.

THEOREM 1 ([**26**]). *Let T be a Euclidean triangle. Then*

$$\xi(T) \geq \frac{64\pi^2}{9}$$

with equality if and only if T is equilateral.

A fundamental result concerning the gap function is due to Andrews and Clutterbuck [**2**] who proved that for any convex domain $\Omega \subset \mathbb{R}^n$,

$$\xi(\Omega) \geq 3\pi^2.$$

It is a straightforward exercise to compute that the right side of the above inequality is equal to the gap function on a one dimensional bounded domain (a segment). Andrews's and Clutterbuck's Theorem shows that the gap function in n dimensions is bounded below by the gap function in 1 dimension. This motivated us to investigate the behavior of the gap function on families of collapsing domains and to determine if and when it converges to the gap function on a lower dimensional domain. Further motivation to study eigenvalues of thin and collapsing domains comes from physics; see for example [**16**], [**17**], and [**18**].

1.1. Examples. Although the following examples may be well-known, we include them for the sake of completeness and to give intuition for that which follows.

1.1.1. *Rectangles.* There are very few domains for which one may explicitly compute the spectrum. The most elementary example of a planar domain with computable spectrum is a rectangle. For a rectangle $R \cong [0, L] \times [0, W]$ the set of eigenvalues and eigenfunctions may be computed by separation of variables. The eigenvalues are thus:

$$\lambda_{j,k} = \frac{k^2\pi^2}{L^2} + \frac{j^2\pi^2}{W^2}, \quad j, k \in \mathbb{N}.$$

Without loss of generality we assume $W \leq L$. Then, the gap function

$$\xi(R) = (L^2 + W^2)\left(\frac{4\pi^2}{L^2} + \frac{\pi^2}{W^2} - \frac{\pi^2}{L^2} - \frac{\pi^2}{W^2}\right) = \frac{3\pi^2(L^2 + W^2)}{L^2}.$$

The gap function detects the shape of a square since for any rectangle R

$$\xi(R) \leq 6\pi^2,$$

with equality if and only if R is a square. For a family of rectangles undergoing one-dimensional collapse by letting $W \to 0$ the gap function converges to $3\pi^2$ which is the gap function for a segment.

1.1.2. *Circular sectors.* Circular sectors may be the most natural example to consider after rectangular domains because in polar coordinates a circular sector looks like a rectangle.

The Laplacian in polar coordinates (r, θ) is

$$\Delta = \partial_r^2 + r^{-1}\partial_r + r^{-2}\partial_\theta^2.$$

Separating variables in the Laplace equation leads to the equations

$$g''(\theta) = -\mu g(\theta), \quad r^2 f''(r) + rf'(r) + \lambda r^2 f(r) = \mu f(r), \qquad \mu > 0. \tag{1.1}$$

For a circular sector of opening angle $\alpha\pi$ and radius 1, the boundary conditions are

$$g(0) = g(\alpha\pi) = 0, \quad f(0) = f(1) = 0.$$

The solutions to the first equation in (1.1) are

$$g(\theta) = \sin(k\theta/\alpha), \quad k \in \mathbb{N}, \quad \mu = \mu_k = \frac{k^2}{\alpha^2}.$$

The second equation in (1.1) can be re-arranged to

$$r^2 f''(r) + rf'(r) + (\lambda r^2 - \mu_k)f = 0, \quad f(0) = f(1) = 0,$$

which one recognizes as a Bessel equation. The solutions are

$$J_{k/\alpha}(\sqrt{\lambda}r), \quad J_{-k/\alpha}(\sqrt{\lambda}r),$$

where J_x denotes the Bessel function of order x. To satisfy the boundary conditions the second solution is not allowed since it does not vanish at $r = 0$. Consequently,

$$\sqrt{\lambda} = \sqrt{\lambda_{j,k}} = Z_{j,k},$$

is the jth zero of the Bessel function of order k/α. The eigenvalues of the sector are therefore

$$\lambda_{j,k} = Z_{j,k}^2.$$

The first and second zeros of the Bessel function of order ν are

$$Z_{\nu,i} = \nu - \frac{a_i}{2^{1/3}}\nu^{1/3} + O(\nu^{-1/3}), \quad i = 1, 2, \tag{1.2}$$

where a_i is the ith zero of the Airy function of the first kind. These formulae can be found in [**23**, **25**], and [**30**] for Bessel functions of real order and were previously demonstrated in [**34**] and [**28**] for Bessel functions of integer order. Since

$$a_1 \approx -2.34, \text{ and } a_2 \approx -4.09, \tag{1.3}$$

the first two Dirichlet eigenvalues of the circular sector of opening angle $\alpha\pi$ and radius one are approximately

$$\lambda_i(S_\alpha) = \frac{1}{\alpha^2} + \frac{c_i}{\alpha^{4/3}} + O(\alpha^{-1}), \quad i = 1, 2, \tag{1.4}$$

where

$$c_1 = -a_1 2^{2/3} \approx 3.71 \text{ and } c_2 = -a_2 2^{2/3} \approx 6.49. \tag{1.5}$$

Consequently the gap function of a circular sector S_α of opening angle $\alpha\pi$ is asymptotic to

$$\xi(S_\alpha) = \frac{c_2 - c_1}{\alpha^{4/3}} + O(\alpha^{-1}), \quad \alpha \to 0. \tag{1.6}$$

In this case the gap function is unbounded as the sector collapses to the segment.

These examples show that the gap function is sensitive to the geometry of convex planar domains which collapse to a segment. Motivated by the strikingly different asymptotic behavior of the gap function on rectangles and circular sectors, in [**26**] we investigated the behavior of the gap function on simplicial domains and proved the following.

THEOREM 2 ([**26**]). *Let Y be an $n-1$ simplex for some $n \geq 2$. Let $\{X_j\}_{j\in\mathbb{N}}$ be a sequence of n-simplices each of which is a graph over Y. Assume the height of X_j over Y vanishes as $j \to \infty$. Then $\xi(X_j) \to \infty$ as $j \to \infty$. More precisely, there is a constant $C > 0$ depending only on n and Y such that $\xi(X_j) \geq Ch(X_j)^{-4/3}$, where $h(X_j)$ is the height of X_j.*

Any triangle with unit diameter is a 2-simplex with base

$$Y = \{(x, 0) : 0 \leq x \leq 1\}.$$

By Theorem 2, the gap function on a family of triangles which collapses to a segment diverges with at least the same asymptotic rate as a collapsing sector (1.6). In §2, we explore different ways in which triangles may collapse to the segment and present an elementary proof that the gap function on any sequence of collapsing triangles diverges. In §3, we prove that the gap function is sensitive to the geometry of convex planar domains which collapse to a segment. The proof of our main result comprises §4.

To state this result we shall identify $\mathbb{R}^{n-1}$ with the subset

$$\{(x_1, \cdots, x_{n-1}, 0) \mid (x_1, \cdots, x_{n-1}) \in \mathbb{R}^{n-1}\} \subset \mathbb{R}^n.$$

Let $E \subset \mathbb{R}^{n-1}$ be a convex domain, and let h_ε^- and h_ε^+ be respectively one-parameter families of convex, non-positive and concave non-negative Lipschitz continuous functions. Assume that $h_\varepsilon^\pm$ are uniformly bounded from above and below, and that their moduli of continuity are uniformly bounded from above for all ε. Define

$$h_\varepsilon(x) := |h_\varepsilon^+(x) - h_\varepsilon^-(x)|, \quad \sigma_\varepsilon := \max\{h_\varepsilon(x) \mid x \in E\},$$

and

$$\Omega_\varepsilon := \{(x, y) \mid x \in E, \quad \varepsilon h_\varepsilon^-(x) \leq y \leq \varepsilon h_\varepsilon^+(x)\}.$$

THEOREM 3. *Let E, $h_\varepsilon^\pm$, σ_ε, and Ω_ε be as above, and let*

$$U_\varepsilon(\delta) := \{x \in E \mid h_\varepsilon(x) \geq \sigma_\varepsilon - \delta\}.$$

If

$$\operatorname{diam} U_\varepsilon(\delta) \to 0, \quad as\ \delta \to 0, \quad uniformly\ in\ \varepsilon,$$

then

$$\lambda_2(\Omega_\varepsilon) - \lambda_1(\Omega_\varepsilon) \to \infty, \qquad as\ \varepsilon \to 0.$$

A more refined theorem was proven by Borisov and Freitas in a slightly different setting [**4**]. They determined the asymptotics of both the eigenvalues and eigenfunctions for domains in $\mathbb{R}^d$ with smooth boundary which collapse to a domain in $\mathbb{R}^{d-1}$. Since simplices and polygonal domains do not satisfy the hypotheses of [ibid] due to the presence of corners, those results cannot be used to prove Theorem 3. We note however that in two dimensions Theorem 2 can be deduced from Friedlander and Solomyak's results [**13**]. Both [**4**] and [**13**] require sophisticated functional analysis, whereas our proofs are rather elementary.

2. Collapsing triangular domains

2.1. Preliminaries. The variational principle, also known as the mini-max principle, states that the eigenvalues are given by the infima of the Rayleigh-Ritz quotient,

$$\lambda_{k+1} = \inf\left\{ \frac{\int_\Omega |\nabla f|^2}{\int_\Omega f^2} \,\middle|\, 0 = \int_\Omega f f_j, j = 0, \ldots, k \right\} \tag{2.1}$$

where $f_0 \equiv 0$, and f_j is an eigenfunction for λ_j for $j \geq 1$. We refer to [**6**] for the proof. The test functions above and in the equivalent "maxi-min" principle (2.2) are $\mathcal{C}^2$ functions which vanish on the boundary of Ω and are not identically zero,

$$\lambda_k = \inf_{\dim(L)=k} \left\{ \sup_{f \in L} \frac{\int_\Omega |\nabla f|^2}{\int_\Omega f^2} \right\}. \tag{2.2}$$

The maxi-min principle can be used to prove domain monotonicity (see [**7**])

$$\Omega \subset \Omega' \implies \lambda_k(\Omega) \geq \lambda_k(\Omega'), \quad \forall k \in \mathbb{N}.$$

We shall use the standard "big-O" and "little-o" from asymptotic analysis. We shall also use the following notations for functions $f, g : \mathbb{R} \to \mathbb{R}$,

$$f \lesssim g \text{ as } t \to t_0 \iff \exists C, \varepsilon > 0 \text{ such that } f(t) \leq Cg(t) \quad \forall t \in [t_0 - \varepsilon, t_0 + \varepsilon],$$
$$f \gtrsim g \quad \text{as } t \to t_0 \iff -f \lesssim -g \text{ as } t \to t_0,$$
$$f \sim g \quad \text{as } t \to L \iff f \lesssim g \text{ and } g \lesssim f \text{ as } t \to t_0,$$
$$f \approx g \iff \lim_{t \to t_0} \frac{f}{g} = 1.$$

In some arguments we will use the following constants[1]

$$c_1' = -a_1' 2^{2/3} \approx 1.62, \quad c_2' = -a_2' 2^{2/3} \approx 5.16. \tag{2.3}$$

Above a_i' is the ith negative zero of the derivative of the Airy function of the first kind.

[1]This constant arises from the asymptotic formula for the first two zeros of the derivative of the Bessel function which is related to the first two Dirichlet eigenvalue of an obtuse isosceles triangle; see [**11**].

2.2. The moduli space of triangles. Since the gap function is invariant under scaling, we restrict to triangles with diameter one. Such a triangle has angles

$$0 < \alpha\pi \leq \beta\pi \leq \pi - \alpha\pi - \beta\pi.$$

The moduli space of triangles, which we shall call P, can thus be identified with a triangle in the $\alpha \times \beta$ plane; see Figure 1. We are interested in the behavior of ξ approaching the boundary of P which is the dashed vertical segment in Figure 1 and corresponds to triangles which degenerate to a segment.

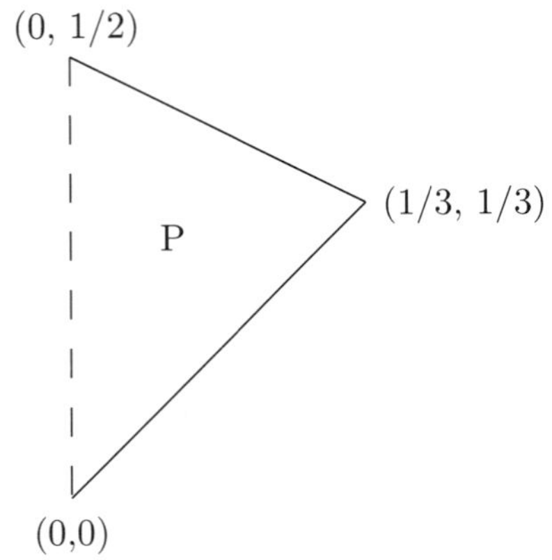

FIGURE 1. Moduli space of triangles.

Consider the triangle T with angles $0 < \alpha\pi \leq \beta\pi \leq \pi - \alpha\pi - \beta\pi$, and assume for some fixed $\varepsilon > 0$, $\beta \geq \varepsilon$. Let the side opposite $\alpha\pi$ have length A, the side opposite $\beta\pi$ have length B, and the third side have length one. The Law of Sines states that

$$\frac{\sin(\alpha\pi)}{A} = \frac{\sin(\beta\pi)}{B} = \frac{\sin(\pi - \alpha\pi - \beta\pi)}{1}.$$

Then

$$B = \frac{\sin(\beta\pi)}{\sin(\alpha\pi + \beta\pi)} \implies |1 - B| = O(\alpha), \text{ as } \alpha \to 0. \tag{2.4}$$

We approximate $\xi(T)$ using domain monotonicity with two sectors both of opening angle $\alpha\pi$. The larger sector S has radius 1 and contains T, whereas the smaller sector σ has radius B and is contained in T. By domain monotonicity,

$$\lambda_2(T) \geq \lambda_2(S), \quad \lambda_1(T) \leq \lambda_1(\sigma).$$

Then,

$$\lambda_2(S) - \lambda_1(\sigma) \leq \xi(T) \leq \lambda_2(\sigma) - \lambda_1(S).$$

Since

$$\lambda_k(\sigma) = B^{-2}\lambda_k(S), \quad \forall k \in \mathbb{N},$$

we have the estimate

$$\lambda_2(S) - B^{-2}\lambda_1(S) \leq \xi(T) \leq B^{-2}\lambda_2(S) - \lambda_1(S). \tag{2.5}$$

By (2.4) and (1.4), there are constants c, $c' > 0$ such that

$$\xi(S) - \frac{c}{\alpha} = \frac{c_2 - c_1}{\alpha^{4/3}} - \frac{c}{\alpha} \leq \xi(T) \leq \xi(S) + \frac{c'}{\alpha} = \frac{c_2 - c_1}{\alpha^{4/3}} + \frac{c'}{\alpha}.$$

Consequently,

$$\lim_{\alpha\to 0}\frac{\xi(T)}{\xi(S)}=1. \tag{2.6}$$

More generally, we have the following.

PROPOSITION 1. *Let $\{T_n\}$ be a sequence of triangles with diameter one and angles $\alpha_n\pi$, $\beta_n\pi$ and $\pi(1-\alpha_n-\beta_n)$. Let S_n be the sector with opening angle $\alpha_n\pi$ and radius 1. Assume that $\alpha_n=o(\beta_n^3)$ and $\beta_n\to 0$ as $n\to\infty$. Then*

$$\xi(T_n)\approx\xi(S_n)=\frac{c_2-c_1}{\alpha_n^{4/3}}\quad \text{as } \beta_n\to 0.$$

PROOF. For simplicity, we shall abuse notation and drop the subscript n. Estimating with two sectors S and σ as above with opening angle $\alpha\pi$ and radii 1 and B, respectively, we again have (2.5). In this case since both α and β tend to 0, by (2.4)

$$|1-B|=O\left(\frac{\alpha}{\beta}\right).$$

Therefore (2.5) becomes

$$\xi(S)-\frac{c\alpha}{\beta}\lambda_1(S)\le\xi(T)\le\xi(S)+\frac{c'\alpha}{\beta}\lambda_2(S),$$

for some positive constants c and c'. By (1.4) for some positive constants C and C' we have

$$\xi(S)-\frac{C}{\alpha\beta}=\frac{c_2-c_1}{\alpha^{4/3}}-\frac{C}{\alpha\beta}\le\xi(T)\le\xi(S)+\frac{C'}{\alpha\beta}=\frac{c_2-c_1}{\alpha^{4/3}}+\frac{C'}{\alpha\beta}.$$

Since $\alpha=o(\beta^3)$, (2.6) follows. □

We next consider obtuse isosceles triangles with diameter one. By [**11**],

$$\lambda_i(T)=\frac{4}{\alpha^2}+\frac{4c_i'}{\alpha^{4/3}}+O(\alpha^{-2/3}),\quad i=1,2.$$

Then,

$$\xi(T)\approx\frac{4(c_2'-c_1')}{\alpha^{4/3}}+O(\alpha^{-2/3})\quad \text{as } \alpha\to 0.$$

More generally, we have the following.

PROPOSITION 2. *Let $\{T_n\}$ be a sequence of triangles with diameter one and angles $\alpha_n\pi$, $\beta_n\pi$ and $\pi(1-\alpha_n-\beta_n)$. Assume that $\alpha_n=\beta_n+o(\beta_n^3)$ and $\beta_n\to 0$ as $n\to\infty$. Then*

$$\xi(T_n)\approx\frac{4(c_2'-c_1')}{\alpha_n^{4/3}}\quad \text{as } n\to\infty.$$

PROOF. By abuse of notation we shall again drop the subscript n. We estimate the gap function on the triangle PRT in Figure 2 using a smaller isosceles triangle QRT and a larger isosceles triangle PST.

By domain monotonicity

$$\lambda_k(PST)\le\lambda_k(PRT)\le\lambda_k(QRT).$$

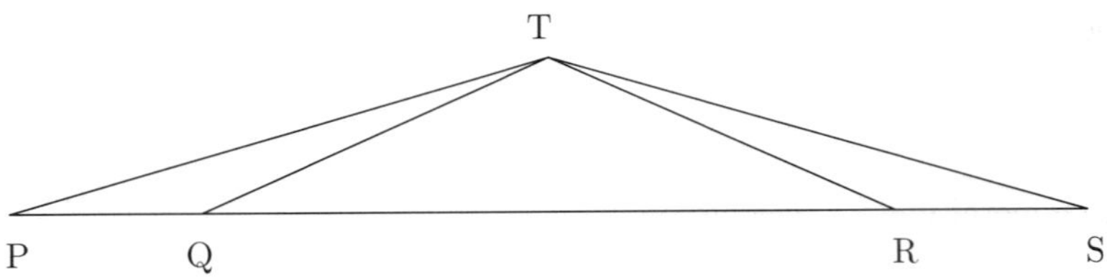

FIGURE 2. Triangle with two angles collapsing and approximating isosceles triangles.

The triangle QRT has angles $\beta\pi$, $\beta\pi$, and $\pi - 2\beta\pi$, whereas the triangle PST has angles $\alpha\pi$, $\alpha\pi$, and $\pi - 2\alpha\pi$. Let B denote the length of QT, A the length of PT, Γ the length of QR, and γ the length of PQ. By **[11]**,

$$\lambda_k(QRT) = \frac{4}{\Gamma^2}\left(\frac{1}{\beta^2} + \frac{c_k'}{\beta^{4/3}}\right) + O(\beta^{-2/3}), \quad k = 1, 2, \tag{2.7}$$

and

$$\lambda_k(PST) = \frac{4}{(\Gamma + 2\gamma)^2}\left(\frac{1}{\alpha^2} + \frac{c_k'}{\alpha^{4/3}}\right) + O(\alpha^{-2/3}), \quad k = 1, 2. \tag{2.8}$$

By the Law of Cosines

$$\Gamma = 2B\cos(\beta\pi),$$

and by the Law of Sines

$$B = \frac{A\sin(\alpha\pi)}{\sin(\beta\pi)} \implies |A - B| = o(\beta^2).$$

By the Law of Sines

$$\gamma = \frac{B\sin(\beta\pi - \alpha\pi)}{\sin(\alpha\pi)} = o(\beta^2).$$

Since PRT has unit diameter,

$$\Gamma + \gamma = 1, \quad |\Gamma - 1| = o(\beta^2).$$

The estimate for $\xi(PRT)$ is

$$\lambda_2(PST) - \lambda_1(QRT) \le \xi(PRT) \le \lambda_2(QRT) - \lambda_1(PST). \tag{2.9}$$

By (2.5) and (1.4)

$$\xi(PRT) \approx 4(c_2' - c_1')\beta^{-4/3}, \quad \text{as } \beta \to 0.$$

□

The above propositions imply the following.

COROLLARY 1. *The gap function is not polyhomogeneous on P.*

PROOF. Polyhomogeneity would imply that the gap function $\lambda_2 - \lambda_1$ on triangles with unit diameter admits an expansion of the form

$$\xi(\alpha, \beta) \sim \sum_{k=1}^{\infty} \alpha^{z_k} \log(\alpha)^{p_k} b_k(\beta - \alpha), \quad \alpha \to 0. \tag{2.10}$$

Above, the coefficient functions b_k are smooth in $\beta - \alpha$, the real parts of the (possibly complex) powers z_k tend toward infinity as $k \to \infty$, and the powers p_k

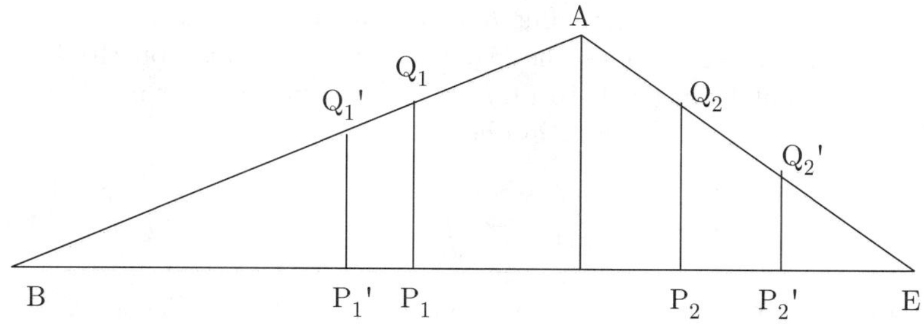

FIGURE 3. Arbitrary collapsing triangle.

are non-negative integers. For all points away from $(\alpha, \beta) = (0, 0)$, ξ is asymptotic to

$$\xi(\alpha, \beta) \approx \frac{c_2 - c_1}{\alpha^{4/3}}, \qquad \text{as } \alpha \to 0.$$

This corresponds to an expansion (2.10) with $z_1 = -4/3$, $p_1 = 0$, and $b_1(x)$ a smooth function with $b_1(x) > 0$ for $x > 0$, and $b_1(0) = c_2 - c_1$. However, along trajectories approaching $(\alpha, \beta) = (0, 0)$ with $\beta - \alpha = o(\beta^3)$, by Proposition 2 the gap is asymptotic to $4(c_2' - c_1')\alpha^{-4/3}$. This would require $b_1(0) = 4(c_2' - c_1')$. Since

$$4(c_2' - c_1') \approx 14, \quad c_2 - c_1 \approx 3,$$

there can be no smooth function b_1 which satisfies these conditions. □

REMARK 1. In forthcoming work by Grieser and Melrose the eigenvalues are shown to be polyhomogeneous on the moduli space of triangles with an appropriate blow-up at the point $(\alpha, \beta) = (0, 0)$ [**19**].

2.3. A general collapsing triangle. For the general case consider the triangle $T = ABE$ in Figure 3. We make the following assumptions: the smallest angle of the triangle is at the vertex B and measures $\alpha\pi$, $|DE| \leq |BD|$, and $|BE| = 1$. Fix

$$0 < \varepsilon < \frac{2}{9}.$$

Let P_1 be between B and D so that $|P_1D| = \alpha^\varepsilon$, and P_1' be between B and P_1 so that $|P_1'D| = 2\alpha^\varepsilon$. If $|DE| > 2\alpha^\varepsilon$, let P_2 be between D and E so that $|P_2D| = \alpha^\varepsilon$, and P_2' be between P_2 and E so that $|P_2'D| = 2\alpha^\varepsilon$. If $|DE| \leq 2\alpha^\varepsilon$, we do not define or use the points P_2, P_2'. If $|DE| > 2\alpha^\varepsilon$ let U be the trapezoid $AQ_1P_1P_2Q_2$ and similarly let $U' = AQ_1'P_1'P_2'Q_2'$. If $|DE| \leq 2\alpha^\varepsilon$, we let $U = AQ_1P_1E$ and $U' = AQ_1'P_1'E$. Let $V = ABE \setminus U$ and $V' = ABE \setminus U'$. In the estimates to follow, we show that we may estimate $\lambda_2(ABE) - \lambda_1(ABE)$ using $\lambda_2(U') - \lambda_1(U')$.

Let f_i be the eigenfunction for $\lambda_i = \lambda_i(ABE)$, for $i = 1, 2$. Assume that

$$\int_{ABE} f_i^2 dxdy = 1, \quad i = 1, 2,$$

and define

$$\eta_i := \int_V f_i^2 dxdy, \quad i = 1, 2. \tag{2.11}$$

In the following arguments the notation "$dxdy$" shall be suppressed whenever it is clear from context. We identify the base of the triangle BE with the segment from $(0,0) \in \mathbb{R}^2$ to $(1,0) \in \mathbb{R}^2$. Let $h(x)$ denote the height of the triangle over a point $(x,0)$ in the base. Since the eigenfunctions f_i vanish on the boundary of the vertical segment from $(x,0)$ to $(x,h(x))$, the variational principle for the first eigenvalue of this vertical segment implies

$$\frac{\int_0^{h(x)} |\nabla f_i|^2}{\int_0^{h(x)} f_i^2} dy \geq \frac{\pi^2}{h(x)^2} \implies \int_0^{h(x)} |\nabla f_i|^2 dy \geq \frac{\pi^2}{h(x)^2} \int_0^{h(x)} f_i^2 dy. \tag{2.12}$$

This is known as the one-dimensional Poincaré inequality. The heights of V, U, and U' satisfy

$$h(V) \lesssim (1-\alpha^\varepsilon)\alpha, \quad h(U) \lesssim \alpha, \quad h(U') \lesssim \alpha.$$

Integrating the inequality (2.12) with respect to x over the points $(x,0)$ in the base of each of V, U, and U', we have the estimates

$$\frac{\int_V |\nabla f_i|^2}{\int_V f_i^2} \gtrsim \frac{1}{(1-\alpha^\varepsilon)^2\alpha^2}, \quad \frac{\int_U |\nabla f_i|^2}{\int_U f_i^2} \gtrsim \frac{1}{\alpha^2}, \text{ and } \frac{\int_{U'} |\nabla f_i|^2}{\int_{U'} f_i^2} \gtrsim \frac{1}{\alpha^2}.$$

By the variational principle (2.1),

$$\frac{\eta_i}{(1-\alpha^\varepsilon)^2\alpha^2} + \frac{1-\eta_i}{\alpha^2} \lesssim \int_V |\nabla f_i|^2 + \int_U |\nabla f_i|^2 = \lambda_i.$$

Since the triangle ABD is a right triangle, by [**11**] and domain monotonicity

$$\lambda_i \leq \lambda_2(ABD) \lesssim \frac{1}{\alpha^2} + \frac{c_2}{\alpha^{4/3}} + O(\alpha^{-1}).$$

Therefore,

$$\eta_i \lesssim \frac{\alpha^{2/3} c_2 (1-\alpha^\varepsilon)^2}{\alpha^\varepsilon(2-\alpha^\varepsilon)} \lesssim \alpha^{2/3-\varepsilon}, \quad i = 1,2. \tag{2.13}$$

2.3.1. *Estimate for* λ_1. To estimate $\lambda_2 - \lambda_1 = \xi(T)$ from below, by domain monotonicity

$$\lambda_1 \leq \lambda_1(U') \implies \xi(T) \geq \lambda_2 - \lambda_1(U').$$

In the following arguments, we will show that we can estimate $\lambda_2(U')$ from above in terms of λ_1 and λ_2.

2.3.2. *Estimate for* λ_2. Let ρ be a smooth compactly supported function so that

$$\rho|_U \equiv 1, \quad \rho|_{V'} \equiv 0. \tag{2.14}$$

We may choose ρ so that

$$|\nabla \rho| \leq \frac{1}{\alpha^\varepsilon}, \quad \text{and} \quad |\Delta\rho|, |\Delta(\rho^2)| \leq \frac{1}{\alpha^{2\varepsilon}}. \tag{2.15}$$

Note that

$$-(\rho f_i)\Delta(\rho f_i) = \lambda_i \rho^2 f_i^2 - f_i^2 \rho\Delta\rho - 2f_i\rho(\nabla\rho)(\nabla f_i). \tag{2.16}$$

We would like to use ρf_2 in the variational principle (2.1) for $\lambda_2(U')$, but since ρf_2 is not à priori orthogonal to the first eigenfunction for U', we must modify it a bit. Since ρf_1 is not orthogonal to the first eigenfunction for U' because both are positive, there is some $a \in \mathbb{R}$ such that $\rho f_2 + a\rho f_1$ is orthogonal to the first eigenfunction for U'. We may then use $\rho f_2 + a\rho f_1$ as a test function for the variational principle (2.1) for $\lambda_2(U')$,

$$\lambda_2(U') \leq \frac{\int_{U'} |\nabla(\rho f_2 + a\rho f_1)|^2}{\int_{U'} (\rho f_2 + a\rho f_1)^2}. \tag{2.17}$$

Since

$$\lambda_2 = \lambda_2 \frac{\int_{U'} (\rho f_2 + a\rho f_1)^2}{\int_{U'} (\rho f_2 + a\rho f_1)^2} = \frac{\lambda_2 \int_{U'} \rho^2 f_2^2 + a^2 \lambda_2 \int_{U'} \rho^2 f_1^2 + \lambda_2 \int_{U'} 2a\rho^2 f_1 f_2}{\int_{U'} (\rho f_2 + a\rho f_1)^2},$$

expanding the numerator and denominator in (2.17) we have

$$\lambda_2(U') \leq \lambda_2 + \frac{I + II + III}{\int_{U'} (\rho f_2 + a\rho f_1)^2},$$

where

$$I := \int_{U'} |\nabla(\rho f_2)|^2 - \lambda_2 \int_{U'} \rho^2 f_2^2, \quad II := a^2 \left(\int_{U'} |\nabla(\rho f_1)|^2 - \lambda_2 \int_{U'} \rho^2 f_1^2 \right),$$

and

$$III := \int_{U'} 2a\nabla(\rho f_1) \cdot \nabla(\rho f_2) - \lambda_2 \int_{U'} 2a\rho^2 f_1 f_2.$$

In the arguments below we estimate I, II, and III from above. Using integration by parts

$$\int_{U'} |\nabla \rho f_i|^2 = -\int_{U'} \rho f_i \Delta(\rho f_i) = \lambda_i \int_{U'} \rho^2 f_i^2 - \frac{1}{2} \int_{U'} \nabla \rho^2 \nabla f_i^2 - \int_{U'} f_i^2 \rho \Delta \rho. \tag{2.18}$$

We estimate using integration by parts, the fact that both $\Delta\rho$ and $\nabla\rho$ vanish identically on U, and (2.11), (2.13),

$$\begin{aligned} \left| \int_{U'} |\nabla(\rho f_i)|^2 - \lambda_i \int_{U'} \rho^2 f_i^2 \right| &\leq \frac{1}{2} \left| \int_{U'} f_i^2 \Delta \rho^2 \right| + \int_{U'} |\rho \Delta \rho| f_i^2, \\ &\lesssim \alpha^{-2\varepsilon} \int_{U' \setminus U} f_i^2 \lesssim \alpha^{-2\varepsilon} \int_V f_i^2 \lesssim \alpha^{2/3 - 3\varepsilon}. \end{aligned} \tag{2.19}$$

We compute using integration by parts: $\int_{U'} \nabla(\rho f_1) \nabla(\rho f_2) =$

$$-\int_{U'} \rho f_1 \Delta(\rho f_2) = \lambda_2 \int_{U'} \rho^2 f_1 f_2 - 2 \int_{U'} \rho f_1 \nabla \rho \cdot \nabla f_2 - \int_{U'} \rho f_1 f_2 \Delta \rho, \tag{2.20}$$

and $\int_{U'} \nabla(\rho f_1) \nabla(\rho f_2) =$

$$-\int_{U'} \rho f_2 \Delta(\rho f_1) = \lambda_1 \int_{U'} \rho^2 f_1 f_2 - 2 \int_{U'} \rho f_2 \nabla \rho \cdot \nabla f_1 - \int_{U'} \rho f_2 f_1 \Delta \rho. \tag{2.21}$$

This gives the inequality

$$\begin{aligned} &\left| 2 \int_{U'} \nabla(\rho f_1) \nabla(\rho f_2) - (\lambda_1 + \lambda_2) \int_{U'} \rho^2 f_1 f_2 \right| \\ &\lesssim \left| \int_{U' \setminus U} \frac{f_1 f_2}{\alpha^{2\varepsilon}} \right| + 2 \left| \int_{U' \setminus U} \rho \nabla \rho \nabla (f_1 f_2) \right| \\ &\lesssim \alpha^{-2\varepsilon} \left| \int_V f_1 f_2 \right| \lesssim \alpha^{2/3 - 3\varepsilon}, \end{aligned} \tag{2.22}$$

which follows from integration by parts, the Schwarz inequality, and (2.11), (2.13). By (2.19),

$$|I| \lesssim \alpha^{2/3-3\varepsilon}. \tag{2.23}$$

To estimate II, we note that $\lambda_1 \leq \lambda_2$ so that

$$II \leq a^2 \left(\int_{U'} |\nabla \rho f_1|^2 - \lambda_1 \int_{U'} \rho^2 f_1^2 \right) \lesssim a^2 \alpha^{2/3-3\varepsilon}, \tag{2.24}$$

which follows from (2.19).

Note that by the orthogonality of f_1 and f_2, the Schwarz inequality and (2.13),

$$\left| \int_{U'} \rho^2 f_1 f_2 \right| = \left| \int_{ABE} (1-\rho^2) f_1 f_2 \right| \leq \left| \int_V f_1 f_2 \right| \lesssim \alpha^{2/3-\varepsilon}. \tag{2.25}$$

By (2.22), (2.25), and adding and subtracting $\lambda_1 \int_{U'} a\rho^2 f_1 f_2$, we estimate III,

$$III \leq |a| \left| 2 \int_{U'} \nabla(\rho f_1) \cdot \nabla(\rho f_2) - (\lambda_1 + \lambda_2) \int_{U'} \rho^2 f_1 f_2 \right| + |a|(\lambda_2 - \lambda_1) \left| \int_{U'} \rho^2 f_1 f_2 \right|$$

$$\lesssim |a|(\alpha^{2/3-3\varepsilon} + (\lambda_2 - \lambda_1)\alpha^{2/3-\varepsilon}). \tag{2.26}$$

Next we estimate the denominator in (2.17). By (2.25), and (2.13) which implies $\int_{U'} f_i^2 \gtrsim 1 - \alpha^{2/3-\varepsilon}$,

$$\int_{U'} (\rho f_2 + a\rho f_1)^2 \gtrsim (1+a^2)(1-\alpha^{2/3-\varepsilon}) - 2|a|\alpha^{2/3-\varepsilon}. \tag{2.27}$$

This inequality together with our estimates (2.23), (2.24), (2.26), and (2.27) in the variational principle for $\lambda_2(U')$ (2.17) show that

$$\lambda_2(U') \lesssim \lambda_2 + \frac{\alpha^{2/3-3\varepsilon} + a^2\alpha^{2/3-3\varepsilon} + |a|\alpha^{2/3-3\varepsilon} + |a|(\lambda_2 - \lambda_1)\alpha^{2/3-\varepsilon}}{(1+a^2)(1-\alpha^{2/3-\varepsilon}) - 2|a|\alpha^{2/3-\varepsilon}}.$$

This simplifies a bit to

$$\lambda_2(U') \lesssim \lambda_2 + \alpha^{2/3-3\varepsilon} \left(\frac{1 + a^2 + |a| + |a|(\lambda_2 - \lambda_1)\alpha^{2\varepsilon}}{1 + a^2 - (1+a^2)\alpha^{2/3-\varepsilon} - 2|a|\alpha^{2/3-\varepsilon}} \right).$$

We therefore have the bound

$$\lambda_2(U') \lesssim \lambda_2 + \alpha^{2/3-3\varepsilon} + (\lambda_2 - \lambda_1)\alpha^{2/3-\varepsilon}.$$

2.3.3. *Gap estimate.* Using our estimates for λ_1 and λ_2, and the choice of $\varepsilon \in (0, 2/9)$ which implies $\alpha^{2/3-3\varepsilon} \to 0$ as $\alpha \to 0$, we have

$$\lambda_2 - \lambda_1 \gtrsim \lambda_2(U') - \lambda_1(U') - (\lambda_2 - \lambda_1)\alpha^{2/3-\varepsilon}.$$

We then have

$$(\lambda_2 - \lambda_1) \gtrsim \frac{\lambda_2(U') - \lambda_1(U')}{1 + \alpha^{2/3-\varepsilon}}.$$

By the main theorem of [**31**], since the diameter of U' is at most $4\alpha^\varepsilon$,

$$\lambda_2(U') - \lambda_1(U') \geq \frac{\pi^2}{64\alpha^{2\varepsilon}},$$

which shows that

$$\lambda_2 - \lambda_1 \gtrsim C\alpha^{-2\varepsilon}$$

and is therefore unbounded as $\alpha \to 0$. We have shown that for any triangle with one or two small angles, $\xi(T)$ becomes unbounded as the triangle collapses to a segment. □

3. Collapsing polygonal domains

The geometric feature which seems to determine whether the gap remains bounded or diverges as a polygonal domain collapses to a segment is the rate at which boundary points approach the segment. In the familiar example of rectangles this rate is perfectly uniform, and the gap remains bounded. In the case of triangles, points near B collapse "more quickly" than points near A (see Figure 3), and the gap diverges. The theorem below generalizes this observation to convex domains.

To state the theorem we shall require the following notions. Let $\{Q_n\}_{n\in\mathbb{N}}$ be a family of convex domains in $\mathbb{R}^2$. The arguments in both this section and the following can be generalized to convex domains with diameter 1 which collapse to the diameter, but for the sake of simplicity we shall assume Q_n is the convex hull of the graph of a Lipschitz continuous function over the unit interval and identify the base of Q_n with $S_n = \{(x,0) : 0 \le x \le 1\}$. Without loss of generality we assume that the domains are contained in the upper half plane.

DEFINITION 3.1. *The* height h_n *of* Q_n *is defined by*

$$h_n := \max\{y \mid (x,y) \in Q_n\}. \tag{3.2}$$

We say that the domains Q_n collapse *as* $n \to \infty$ *if*

$$\lim_{n\to\infty} h_n = 0.$$

The following definition describes collapse in which all boundary points are the same distance from the base up to $o(h^2)$, where h denotes the height.

DEFINITION 3.3. *We say that* Q_n *are* asymptotically rectangular *as* $n \to \infty$ *if for each* n*, there exist rectangles* $R_n = [C_n, C_n + A_n] \times [0, h_n] \supset Q \supset r_n = [c_n, c_n + a_n] \times [0, b_n]$ *and a constant* $\gamma > 0$ *such that* $a_n \ge \gamma$ *for all* $n \in \mathbb{N}$ *and*

$$\frac{h_n}{b_n} = 1 + o(h_n^2) \quad \text{as } n \to \infty. \tag{3.4}$$

Triangular domains do not collapse asymptotically rectangularly. The following notion generalizes the "non-uniform" collapse of triangular domains.

DEFINITION 3.5. *We say that* Q_n *collapse* non-uniformly *as* $n \to \infty$ *if there exist convex inscribed polygons* $U_n \subset Q_n$ *such that the following hold.*

1. *The diameter of* $U_n \to 0$ *as* $n \to \infty$.
2. *The longest side* Σ_n *of* U_n *is contained in* S_n.
3. *The height of* $U_n = h_n$.
4. *The height of* $V_n := Q_n \setminus U_n$*, satisfies* $h(V_n) \sim (1 - h_n^{\varepsilon})h_n$ *for some* $\varepsilon \in (0, 2/9)$.

These two types of collapse determine whether the gap remains bounded or diverges as the domains collapse.

THEOREM 4. *Let* $\{Q_n\}_{n\in\mathbb{N}}$ *be a family of convex domains as above, and assume that* Q_n *collapse as* $n \to \infty$. *If* Q_n *are asymptotically rectangular as* $n \to \infty$*, then* *the gap function* $\xi(Q_n)$ *is bounded as* $n \to \infty$. *If* Q_n *collapse non-uniformly as* $n \to \infty$, *then the gap function* $\xi(Q_n) \to \infty$ *as* $n \to \infty$.

PROOF. Assume first that the collapsing domains are asymptotically rectangular. By domain monotonicity

$$\lambda_2(Q_n) - \lambda_1(Q_n) \leq \lambda_2(r_n) - \lambda_1(R_n).$$

For simplicity we shall drop the subscript and denote $\lambda_i = \lambda_i(Q)$ for $i = 1, 2$. Since $b \leq h \to 0$, we may assume that $a \geq b$. Then

$$\lambda_2(r) - \lambda_1(R) = \pi^2 \left(\frac{h^2 - b^2}{h^2 b^2} + \frac{4}{a^2} - \frac{1}{A^2} \right).$$

Since $Q \subset R$, $A \geq 1$, and by assumption a is also bounded below as $n \to \infty$. Thus

$$\xi(Q) \lesssim \pi^2 \left(\frac{(h/b)^2 - 1}{h^2} + \frac{4}{a^2} - \frac{1}{A^2} \right).$$

By the definition of asymptotically rectangular (3.4), it follows that $\xi(Q)$ is bounded as $n \to \infty$.

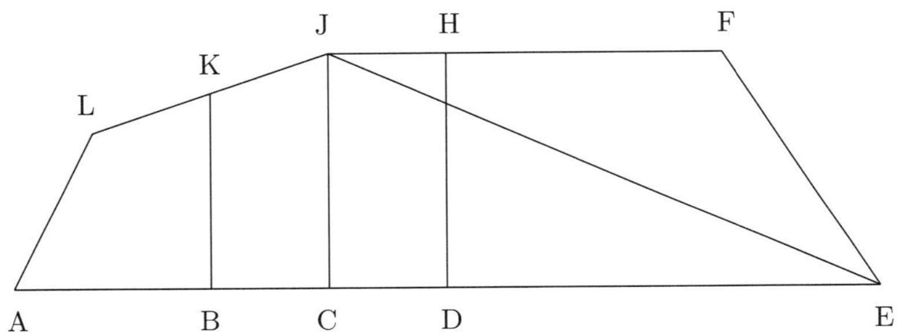

FIGURE 4. Collapsing polygon with unbounded gap.

For the case of non-uniform collapse we shall generalize the estimates used in the preceding section for triangles. Let $U' \supset U$ be a convex inscribed polygon so that one side Γ of U' satisfies $S \supset \Gamma \supset \Sigma$ and $|\Gamma - \Sigma| \sim h^{\varepsilon}$, where $\varepsilon \in (0, 2/9)$ is defined by Definition (3.4). We can define such an inscribed polygon since the diameter of $U \to 0$, and S has length 1. With this definition the diameter of $U' \to 0$ as $h \to 0$.

Let f_i be the eigenfunction for Q with eigenvalue λ_i, for $i = 1, 2$. By convexity, Q contains an inscribed right triangle T of height $h = h(U)$ and base at least $1/2$. For example in Figure 4, $Q = AEFJL$, $T = JCE$, $h = |JC|$, and $U = JKBDH$. Assume f_i are normalized so that

$$\int_Q f_i^2 = 1, \text{ and let } \eta_i := \int_V f_i^2, \quad i = 1, 2.$$

By the one dimensional Poincaré inequality,

$$\frac{\int_V |\nabla f_i|^2}{\int_V f_i^2} \gtrsim \frac{\pi^2}{(h(V))^2},$$

and

$$\frac{\int_U |\nabla f_i|^2}{\int_U f_i^2} \gtrsim \frac{\pi^2}{(h(U))^2}, \quad \frac{\int_{U'} |\nabla f_i|^2}{\int_{U'} f_i^2} \gtrsim \frac{\pi^2}{(h(U))^2},$$

since U and U' have the same height. For $h \approx 0$, the measure of the smallest angle of $T \sim h = h(U)$. By domain monotonicity

$$\lambda_i(Q) \le \lambda_2(T) \le \lambda_2(\sigma) \lesssim \frac{\pi^2}{(h(U))^2} + \frac{c_2}{(h(U))^{4/3}} + O(h(U)^{-1}),$$

where σ is a sector of the same opening angle as T which is contained in T and has radius bounded below by a fixed constant.

Estimating η_i as we did for triangles,

$$\frac{\eta_i \pi^2}{(h(V))^2} + \frac{\pi^2(1-\eta_i)}{(h(U))^2} \lesssim \int_V |\nabla f_i|^2 + \int_U |\nabla f_i|^2 = \lambda_i \le \lambda_2(T)$$

$$\lesssim \frac{\pi^2}{(h(U))^2} + \frac{c_1}{(h(U))^{4/3}} + O(h(U)^{-1}).$$

This gives the following estimate for η_i,

$$\eta_i \lesssim \frac{h(U)^{2/3} h(V)^2}{h(U)^2 - h(V)^2}, \quad i = 1, 2. \tag{3.6}$$

By assumption $h = h(U)$ and $h(V) \sim h(1 + h^\varepsilon)$ for some $\varepsilon \in (0, 2/9)$. It follows that

$$\eta := \max\{\eta_1, \eta_2\} \lesssim h^{2/3-\varepsilon}. \tag{3.7}$$

Let

$$\delta := \frac{2}{3} - \varepsilon. \tag{3.8}$$

3.0.4. *Estimates for* λ_1. By domain monotonicity

$$\lambda_1 \le \lambda_1(U') \implies \lambda_2 - \lambda_1 \ge \lambda_2 - \lambda_1(U').$$

3.0.5. *Estimates for* λ_2. We shall again define a cut-off function ρ as in (2.14) so that,

$$|\nabla \rho| \le h^{-\varepsilon}, \quad |\Delta \rho| \le h^{-2\varepsilon}.$$

Since the function ρf_2 is not à priori an admissible test function for the variational principle for $\lambda_2(U')$ we again modify it to make it orthogonal to the first eigenfunction on U'. Since the first eigenfunction of U' and the first eigenfunction of Q are both positive on the interior, we shall consider the test function

$$\rho f_2 + a\rho f_1,$$

where a is chosen so that $\rho f_2 + a\rho f_1$ is $\mathcal{L}^2$ orthogonal to the first eigenfunction of U'.

The variational principle for $\lambda_2(U')$ using $\rho f_2 + a\rho f_1$ as a test function

$$\lambda_2(U') \le \frac{\int_{U'} |\nabla(\rho f_2 + a\rho f_1)|^2}{\int_{U'} (\rho f_2 + a\rho f_1)^2}. \tag{3.9}$$

Since

$$\lambda_2 = \lambda_2 \frac{\int_{U'} (\rho f_2 + a\rho f_1)^2}{\int_{U'} (\rho f_2 + a\rho f_1)^2},$$

expanding the numerator and denominator in (3.9) we have

$$\lambda_2(U') \le \lambda_2 + \frac{I + II + III}{\int_{U'} (\rho f_2 + a\rho f_1)^2},$$

where

$$I := \int_{U'} |\nabla(\rho f_2)|^2 - \lambda_2 \int_{U'} \rho^2 f_2^2, \quad II := a^2 \left(\int_{U'} |\nabla(\rho f_1)|^2 - \lambda_2 \int_{U'} \rho^2 f_1^2 \right)$$

and

$$III := \int_{U'} 2a\nabla(\rho f_1) \cdot \nabla(\rho f_2) - \lambda_2 \int_{U'} 2a\rho^2 f_1 f_2.$$

Using (2.18), we estimate

$$\left| \int_{U'} |\nabla(\rho f_i)|^2 - \lambda_i \int_{U'} \rho^2 f_i^2 \right| \leq \frac{1}{2} \left| \int_{U'} \Delta \rho^2 f_i^2 \right| + \int_{U'} |\rho \Delta \rho| f_i^2,$$

$$\lesssim h^{-2\varepsilon} \int_{U'-U} f_i^2 \lesssim h^{-2\varepsilon} \int_V f_i^2 \lesssim h^{\delta - 2\varepsilon}, \tag{3.10}$$

since $\Delta\rho$ and $\nabla\rho$ vanish identically on U. By (2.20) and (2.21),

$$\left| 2 \int_{U'} \nabla(\rho f_1) \nabla(\rho f_2) - (\lambda_1 + \lambda_2) \int_{U'} \rho^2 f_1 f_2 \right|$$
$$\lesssim \int_{U'-U} \frac{|f_1 f_2|}{h^{2\varepsilon}} + 2 \left| \int_{U'-U} \rho \nabla \rho \nabla (f_1 f_2) \right|$$
$$\lesssim h^{-2\varepsilon} \left(\int_V |f_1 f_2| + 2 \int_V |f_1 f_2| \right) \lesssim h^{\delta - 2\varepsilon}, \tag{3.11}$$

which follows from integration by parts and the Schwarz inequality.

By our calculations for triangles, our estimate for η, and the estimates (3.10) and (3.11),

$$I \lesssim h^{\delta - 2\varepsilon}, \tag{3.12}$$

$$II \lesssim a^2 h^{\delta - 2\varepsilon}, \text{ and} \tag{3.13}$$

$$III \lesssim |a|(h^{\delta - 2\varepsilon} + (\lambda_2 - \lambda_1) h^\delta). \tag{3.14}$$

Moreover,

$$\int_{U'} (\rho f_2 + a\rho f_1)^2 \gtrsim (1 + a^2)(1 - h^\delta) - 2|a|h^\delta. \tag{3.15}$$

Using these estimates, we estimate the Rayleigh-Ritz quotient for $\rho f_2 + a\rho f_1$,

$$\lambda_2(U') \lesssim \lambda_2 + \frac{h^{\delta-2\varepsilon} + a^2 h^{\delta - 2\varepsilon} + |a| h^{\delta - 2\varepsilon} + |a|(\lambda_2 - \lambda_1) h^\delta}{(1 + a^2)(1 - h^\delta) - 2|a|h^\delta}. \tag{3.16}$$

This simplifies to

$$\lambda_2(U') \lesssim \lambda_2 + h^{\delta - 2\varepsilon} \left(\frac{1 + a^2 + |a| + |a|(\lambda_2 - \lambda_1) h^{2\varepsilon}}{1 + a^2 - (1 + a^2) h^\delta - 2|a| h^\delta} \right).$$

We therefore have the bound as $h \to 0$,

$$\lambda_2(U') \lesssim \lambda_2 + h^{\delta - 2\varepsilon} + h^\delta (\lambda_2 - \lambda_1).$$

3.0.6. *Gap estimate.* Using our estimates for $\lambda_i(U')$

$$\lambda_2 - \lambda_1 \gtrsim \lambda_2 - \lambda_1(U') \gtrsim \lambda_2(U') - \lambda_1(U') - h^{\delta-2\varepsilon} - (\lambda_2 - \lambda_1)h^{\delta},$$

Since

$$\delta = 2/3 - \varepsilon, \quad \varepsilon < 2/9,$$

it follows that

$$\delta - 2\varepsilon = 2/3 - 3\varepsilon > 0.$$

Consequently,

$$\lambda_2 - \lambda_1 \gtrsim \frac{\lambda_2(U') - \lambda_1(U')}{1 + h^{\delta}}. \tag{3.17}$$

By hypothesis on U and definition of U', the diameter of U' vanishes as $h \to 0$, so (3.17) and the main theorem of [**31**] imply

$$\xi \sim \lambda_2 - \lambda_1 \to \infty, \qquad \text{as } h \to 0.$$

□

The theorem shows that the gap function is sensitive to the rate at which boundary points collapse to the base. We shall demonstrate in the following section that the same phenomenon is true for domains in $\mathbb{R}^n$ under one dimensional collapse.

4. The general case

To prove Theorem 3 we shall require the following proposition.

PROPOSITION 3. *Let $\Omega \subset \mathbb{R}^n$ be a convex bounded domain with Lipschitz boundary, and let ϕ be the first eigenfunction for the Euclidean Laplacian on Ω with the Dirichlet boundary condition. Let U be a convex subset of Ω. Define*

$$\mu := \inf\left\{ \frac{\int_U |\nabla\varphi|^2\phi^2}{\int_U \varphi^2\phi^2} \,\middle|\, \varphi \in C^2(U), \quad \int_U \varphi\phi^2 = 0 \right\}.$$

Then

$$\mu \geq \frac{\pi^2}{4d(U)^2}, \quad d(U) = \text{ diameter of } U.$$

PROOF. By the variational principle (see [**27**] §2), μ is the first positive eigenvalue for the Bakry-Émery Laplace operator

$$\mathcal{L} := \Delta + 2\nabla \log\phi\nabla,$$

with the Neumann boundary condition on ∂U. Analogous to the classical variational principle for the Laplacian, the infimum is achieved by a function φ which satisfies

$$\Delta\varphi + 2\nabla\log\phi\nabla\varphi = -\mu\varphi, \qquad \left.\frac{\partial\varphi}{\partial n}\right|_{\partial U} = 0. \tag{4.1}$$

We define for $\varepsilon > 0$

$$g := \frac{1}{2}\left(|\nabla\varphi|^2 + (\mu+\varepsilon)\varphi^2\right).$$

CLAIM 1. *Let x_0 be the point at which g achieves its maximum. Then* $\nabla\varphi(x_0) = 0$.

PROOF. To prove the claim we first note that if g attains its maximum at $x_0 \in U \setminus \partial U$, then $\nabla\varphi(x_0) = 0$, and $\Delta g(x_0) \le 0$. On the other hand if g achieves its maximum at $x_0 \in \partial U$, then we have

$$0 \le \nabla g \cdot \vec{n} = \left((\mu+\varepsilon)\varphi\nabla\varphi + \left(\sum_{i=1}^n \varphi_i\varphi_{ij}\right)_{j=1}^n \right) \cdot \vec{n},$$

where $\vec{n}$ is the outward normal vector at x_0. Since φ satisfies the Neumann boundary condition,

$$0 \le \nabla g \cdot \vec{n} = \sum_{i.j=1}^n \varphi_i\varphi_{ij} \cdot n_j = -\sum_{i,j=1}^n h_{ij}\varphi_i\varphi_j,$$

where h_{ij} is the second fundamental form. By convexity of U, h_{ij} is non-negative definite so we have

$$0 \ge \nabla g \cdot \vec{n} = \sum_{i.j=1}^n \varphi_i\varphi_{ij} \cdot n_j = -\sum_{i,j=1}^n h_{ij}\varphi_i\varphi_j \le 0.$$

This implies that $\nabla g(x_0) = 0$. To show that $\Delta g(x_0) \le 0$, note that

$$\Delta g(x_0) = \Delta' g(x_0) + g_{nn}(x_0) + 2H\nabla g(x_0) \cdot \vec{n},$$

where H is the mean curvature, g_{nn} is the second derivative of g in the normal direction, and Δ' is the Laplace operator for ∂U. At the maximum point x_0,

$$\Delta' g(x_0) \le 0.$$

We have shown above that $\nabla g(x_0) \cdot \vec{n} = 0$, and since x_0 is the maximum point, $g_{nn}(x_0) \le 0$. It follows that

$$\Delta g(x_0) \le 0.$$

Consequently, in any case we have $\nabla g(x_0) =$

$$(\mu+\varepsilon)\varphi\nabla\varphi + \left(\sum_{i=1}^n \varphi_i\varphi_{ij}\right)_{j=1}^n = 0 \implies (\mu+\varepsilon)\varphi\varphi_j = -\sum_{i=1}^n \varphi_i\varphi_{ij}, \tag{4.2}$$

for all $j = 1, \ldots, n$. Note that the above equalities and the calculations below are all for the maximum point x_0, so for the sake of simplicity we shall not write x_0. We also have at the maximum point

$$\Delta g = (\mu+\varepsilon)|\nabla\varphi|^2 + (\mu+\varepsilon)\varphi\Delta\varphi + \sum_{i,j=1}^n (\varphi_{ij}^2 + \varphi_i\varphi_{ijj}) \le 0. \tag{4.3}$$

For each $j = 1, \ldots, n$ in (4.2), multiplying by φ_j and summing over j, we have

$$\left(\sum_{i,j=1}^n \varphi_i\varphi_{ij}\varphi_j\right)^2 = \left(\sum_{j=1}^n -(\mu+\varepsilon)\varphi\varphi_j^2\right)^2 = (\mu+\varepsilon)^2\varphi^2|\nabla\varphi|^4. \tag{4.4}$$

To show that $\nabla\varphi(x_0) = 0$ we will assume for the sake of contradiction that $\nabla\varphi(x_0) \ne 0$. Then by the Cauchy Inequality in (4.4)

$$\sum_{i,j=1}^n (\varphi_{ij})^2 = \sum_{i,j=1}^n (\varphi_{ij})^2 \frac{|\nabla\varphi|^4}{|\nabla\varphi|^4} \ge \frac{\left(\sum_{i,j=1}^n \varphi_{ij}\varphi_i\varphi_j\right)^2}{|\nabla\varphi|^4} = (\mu+\varepsilon)^2\varphi^2.$$

Next we have by (4.1)

$$\sum_{i,j=1}^{n} \varphi_i \varphi_{ijj} = \sum_{i=1}^{n} \varphi_i (\Delta\varphi)_i = \sum_{i=1}^{n} \varphi_i(-\mu\varphi - 2\nabla \log \phi \cdot \nabla\varphi)_i.$$

By [**5**] the logarithm of ϕ is concave and therefore the Hessian of $\log \phi$ is negative definite, and we have

$$\sum_{i,j=1}^{n} -2\varphi_i\varphi_j \frac{\phi_{ij}}{\phi} \geq 0.$$

Consequently,

$$\sum_{i=1}^{n} \varphi_i(-\mu\varphi - 2\nabla \log \phi \cdot \nabla\varphi)_i \geq -\mu|\nabla\varphi|^2 - 2\sum_{i,j=1}^{n} \varphi_i \frac{\phi_j}{\phi}\varphi_{ij}.$$

By (4.2),

$$\sum_{i,j=1}^{n} \varphi_i \frac{\phi_j}{\phi}\varphi_{ij} = -\sum_{j=1}^{n} (\mu+\varepsilon)\varphi\varphi_j\frac{\phi_j}{\phi} = -(\mu+\varepsilon)\varphi\nabla\log\phi\nabla\varphi,$$

so we have

$$\sum_{i=1}^{n} \varphi_i(-\mu\varphi - 2\nabla \log \phi \cdot \nabla\varphi)_i \geq -\mu|\nabla\varphi|^2 + 2(\mu+\varepsilon)\varphi\nabla\log\phi\nabla\varphi.$$

Putting it all together in (4.3) we have

$$0 \geq (\mu+\varepsilon)|\nabla\varphi|^2 + (\mu+\varepsilon)\varphi\Delta\varphi + \sum_{i,j=1}^{n}(\varphi_{ij}^2 + \varphi_i\varphi_{ijj}) \geq$$

$$(\mu+\varepsilon)|\nabla\varphi|^2 + (\mu+\varepsilon)\varphi(-\mu\varphi - 2\nabla\log\phi\nabla\varphi) + (\mu+\varepsilon)^2\varphi^2 - \mu|\nabla\varphi|^2 + 2(\mu+\varepsilon)\varphi\nabla\log\phi\nabla\varphi$$
$$= \varepsilon^2\varphi^2 + +\varepsilon|\nabla\varphi|^2 + \mu\varepsilon\varphi^2.$$

Since $\varepsilon, \mu > 0$, we clearly must have $\nabla\varphi = 0$ which is a contradiction. □

It follows that at the maximum point x_0 for g, $\nabla\varphi$ vanishes and hence

$$g(x) \leq g(x_0) = \frac{1}{2}(\mu+\varepsilon)\varphi^2(x_0) \leq \frac{1}{2}(\mu+\varepsilon)\max(\varphi^2).$$

Assume we have normalized φ such that the maximum of φ^2 is 1. Then,

$$\frac{1}{2}\left(|\nabla\varphi|^2 + (\mu+\varepsilon)\varphi^2\right) = g \leq \frac{1}{2}(\mu+\varepsilon)$$

and so

$$|\nabla\varphi|^2 \leq (\mu+\varepsilon)(1-\varphi^2).$$

Letting $\varepsilon \to 0$ we have

$$|\nabla\varphi|^2 \leq \mu(1-\varphi^2) \implies \frac{|\nabla\varphi|}{\sqrt{1-\varphi^2}} \leq \sqrt{\mu}. \tag{4.5}$$

Since

$$\int_U \varphi\phi^2 = 0,$$

and $\phi > 0$ on the interior of U, there exist points $y_0, y_1 \in U$ such that $\varphi(y_0) = 0$ and $\varphi(y_1)^2 = 1$. Without loss of generality we may assume that $\varphi(y_1) = 1$. Let $\gamma(t)$ be a unit-speed geodesic in U such that $\gamma(0) = y_0$ and $\gamma(1) = y_1$. Then

$$\left|\int_\gamma \frac{\nabla\varphi}{\sqrt{1-\varphi^2}}d\sigma\right| = \left|\int_0^1 \frac{\nabla\varphi(\gamma(t))\dot\gamma(t)}{\sqrt{1-\varphi(\gamma(t))^2}}dt\right|$$
$$= |\arcsin(\varphi(\gamma(1))) - \arcsin(\varphi(\gamma(0)))| = \frac{\pi}{2}.$$

By (4.5),

$$\left|\int_\gamma \frac{\nabla\varphi}{\sqrt{1-\varphi^2}}d\sigma\right| \le l(\gamma)\sqrt{\mu} \le d(U)\sqrt{\mu}.$$

It follows that

$$\frac{\pi}{2} \le d(U)\sqrt{\mu} \implies \frac{\pi^2}{4d(U)^2} \le \mu.$$

□

We now have the necessary ingredients to prove our main result.

PROOF OF THEOREM 3. In what follows, the constant C is independent of δ and ε but may be different from line to line and within the same line. Let u_1 and u_2 be the eigenfunctions of Ω_ε for the eigenvalues $\lambda_1(\Omega_\varepsilon), \lambda_2(\Omega_\varepsilon)$, respectively, and assume

$$\int_{\Omega_\varepsilon} u_i^2 = 1, \quad i = 1, 2.$$

Define

$$E(\delta) := \{x \in E \mid h_\varepsilon(x) > \sigma_\varepsilon - \delta\};$$
$$U(\delta) := (E(\delta) \times \mathbb{R}) \cap \Omega_\varepsilon;$$
$$V(\delta) := \Omega_\varepsilon \setminus U(\delta)$$

We shall again estimate the $\mathcal{L}^2$ norm of the eigenfunctions on $V(\delta)$ using the one dimensional Poincaré inequality which implies

$$\lambda_i(\Omega_\varepsilon) = \int_{\Omega_\varepsilon} |\nabla u_i|^2 \gtrsim \frac{\pi^2}{(\sigma_\varepsilon - \delta)^2\varepsilon^2}\int_{V(\delta)} u_i^2 + \frac{\pi^2}{\sigma_\varepsilon^2\varepsilon^2}\left(1 - \int_{V(\delta)} u_i^2\right), \tag{4.6}$$

for $i = 1, 2$. By the Lipschitz continuity of $h_\varepsilon^\pm$ and the uniform boundedness of the modulus of continuity, it follows that Ω_ε contains a generalized cylinder of height $\sim \varepsilon(\sigma_\varepsilon - \delta^2)$ over $E(\delta^2)$. The eigenvalues of such a cylinder are bounded above by $\lambda_i(E(\delta^2)) + \frac{\pi^2}{(\sigma_\varepsilon - \delta^2)^2\varepsilon^2}$. By domain monotonicity,

$$\lambda_i(\Omega_\varepsilon) \le \lambda_i(U(\delta^2)) \lesssim \lambda_i(E(\delta^2)) + \frac{\pi^2}{(\sigma_\varepsilon - \delta^2)^2\varepsilon^2}. \tag{4.7}$$

Let x_0 be the maximum point of $h_\varepsilon(x)$ so that $h_\varepsilon(x_0) = \sigma_\varepsilon$. Then by the Lipschitz continuity of $h_\varepsilon^\pm$ and the uniform boundedness of the modulus of continuity, there is a constant C, such that

$$E(\delta^2) \supset B_{x_0}(C\delta^2).$$

Therefore, we have by domain monotonicity

$$\lambda_i(E(\delta^2)) \lesssim \delta^{-4}.$$

Combining the above inequality with with (4.6) and (4.7), we have

$$\frac{\pi^2}{(\sigma_\varepsilon - \delta)^2\varepsilon^2}\int_{V(\delta)} u_i^2 + \frac{\pi^2}{\sigma_\varepsilon^2\varepsilon^2}\left(1 - \int_{V(\delta)} u_i^2\right) \lesssim \frac{\pi^2}{\delta^4} + \frac{\pi^2}{(\sigma_\varepsilon - \delta^2)^2\varepsilon^2},$$

and hence we have

$$\int_{V(\delta)} u_i^2 \lesssim \delta + \frac{\varepsilon^2}{\delta^5}.$$

We choose $\delta = \varepsilon^{1/3}$. Then we have

$$\int_{V(\delta)} u_i^2 \lesssim \varepsilon^{1/3}. \tag{4.8}$$

Since u_1 is orthogonal to u_2, we have

$$\left|\int_{U(\delta)} u_1 u_2\right| = \left|\int_{V(\delta)} u_1 u_2\right| \lesssim \varepsilon^{1/3}, \tag{4.9}$$

where the last inequality follows from (4.8) and the Schwarz inequality.

Let

$$\psi := \frac{u_2}{u_1}$$

and let

$$\alpha := \frac{\int_{U(\delta)} \psi u_1^2}{\int_{U(\delta)} u_1^2}, \quad \tilde{\psi} := \psi - \alpha.$$

By the estimate (4.9) and definition of ψ,

$$\alpha = \frac{\int_{U(\delta)} u_1 u_2}{\int_{U(\delta)} u_1^2} \lesssim \frac{\varepsilon^{1/3}}{1 - \varepsilon^{1/3}}.$$

Let $\mu = \mu(\delta)$ be defined as in Proposition 3, where in the statement of the proposition $\phi = u_1$, and $U = U(\delta)$. Note that by the definition of $h_\varepsilon^{\pm}$ as concave, non-positive and convex non-negative functions and the convexity of E it follows that U is a convex subset of $\mathbb{R}^n$. We compute that

$$\int_U \tilde{\psi} u_1^2 = 0.$$

Since μ is the infimum and $\nabla\tilde{\psi} = \nabla\psi$,

$$\mu \leq \frac{\int_U |\nabla\psi|^2 u_1^2}{\int_U \tilde{\psi}^2 u_1^2} \leq \frac{\int_{\Omega_\varepsilon} |\nabla\psi|^2 u_1^2}{\int_U \tilde{\psi}^2 u_1^2} = \frac{\lambda_2(\Omega_\varepsilon) - \lambda_1(\Omega_\varepsilon)}{\int_U \tilde{\psi}^2 u_1^2}.$$

The final equality in the numerator follows from §2 of [**27**]. We compute

$$\begin{aligned}\int_U \tilde{\psi}^2 u_1^2 &= \int_U u_2^2 + \alpha^2 \int_U u_1^2 - 2\int_U \alpha u_2 u_1\\ &= \int_U u_2^2 - \alpha \int_U u_1 u_2 \gtrsim 1 - \varepsilon^{1/3}.\end{aligned}$$

Consequently we have the estimate

$$\mu \lesssim \frac{(\lambda_2(\Omega_\varepsilon) - \lambda_1(\Omega_\varepsilon))}{1 - \varepsilon^{1/3}}.$$

By Proposition 3

$$\mu \geq \frac{\pi^2}{4d(U)^2}.$$

By the assumption that $d(U) \to 0$ uniformly as $\varepsilon \to 0$, it follows that $\lambda_2(\Omega_\varepsilon) - \lambda_1(\Omega_\varepsilon) \to \infty$ as $\varepsilon \to 0$. Finally we note that the diameter of Ω_ε is uniformly bounded from above and from below (by the diameter of E, for example), and so it follows that

$$\xi(\Omega_\varepsilon) \sim \lambda_2(\Omega_\varepsilon) - \lambda_1(\Omega_\varepsilon) \to \infty, \qquad \text{as } \varepsilon \to 0.$$

□

REMARK 2. The assumption that the diameter of $U_\varepsilon(\delta)$ tends to zero is necessary. For example, assume there exists a fixed constant $c > 0$ such that

$$\operatorname{diam}\{x \in E \mid h_\varepsilon(x) \equiv \sigma_\varepsilon\} \geq c, \quad \forall \varepsilon.$$

By domain monotonicity estimates using two generalized cylinders analogous to the two rectangles used in the proof of the bounded-gap case of Theorem 4, it is straightforward to show that in this case the gap remains bounded under one dimensional collapse. The details are left to the reader.

REMARK 3. In [**4**] and [**13**] much more precise results for the behavior of the eigenvalues and the eigenfunctions under one-dimensional collapse were obtained. We note that those results require that the collapsing domains are defined by

$$\Omega_\varepsilon = \{(x,y) \mid x \in E, \quad 0 \leq y \leq \varepsilon h(x)\} \to E \qquad \text{as } \varepsilon \to 0. \tag{4.10}$$

Since the collapse is described by scaling a *fixed* non-negative function h by the parameter ε, those results do not immediately apply to the cases considered here. It would be interesting to investigate whether the results of [**4**] and [**13**] may be extended to more general collapsing geometry.

Acknowledgements

The first author is supported by NSF grant DMS-12-06748, and the second author acknowledges the support of the Max Planck Institut für Mathematik in Bonn and the Universität Göttingen. Both authors are grateful to the anonymous referee for comments and suggestions which improved the quality of the paper.

References

[1] Milton Abramowitz and Irene A. Stegun, *Handbook of mathematical functions with formulas, graphs, and mathematical tables*, National Bureau of Standards Applied Mathematics Series, vol. 55, For sale by the Superintendent of Documents, U.S. Government Printing Office, Washington, D.C., 1964. MR0167642 (29 #4914)

[2] Ben Andrews and Julie Clutterbuck, *Proof of the fundamental gap conjecture*, J. Amer. Math. Soc. **24** (2011), no. 3, 899–916, DOI 10.1090/S0894-0347-2011-00699-1. MR2784332 (2012d:35051)

[3] Pedro R. S. Antunes and Pedro Freitas, *On the inverse spectral problem for Euclidean triangles*, Proc. R. Soc. Lond. Ser. A Math. Phys. Eng. Sci. **467** (2011), no. 2130, 1546–1562, DOI 10.1098/rspa.2010.0540. MR2795790 (2012e:65242)

[4] Denis Borisov and Pedro Freitas, *Asymptotics of Dirichlet eigenvalues and eigenfunctions of the Laplacian on thin domains in* $\mathbb{R}^d$, J. Funct. Anal. **258** (2010), no. 3, 893–912, DOI 10.1016/j.jfa.2009.07.014. MR2558181 (2011a:35362)

[5] Herm Jan Brascamp and Elliott H. Lieb, *On extensions of the Brunn-Minkowski and Prékopa-Leindler theorems, including inequalities for log concave functions, and with an application to the diffusion equation*, J. Functional Analysis **22** (1976), no. 4, 366–389. MR0450480 (56 #8774)

[6] Isaac Chavel, *Eigenvalues in Riemannian geometry*, Pure and Applied Mathematics, vol. 115, Academic Press, Inc., Orlando, FL, 1984. Including a chapter by Burton Randol; With an appendix by Jozef Dodziuk. MR768584 (86g:58140)

[7] R. Courant and D. Hilbert, *Methods of Mathematical Physics, Volume 1,* Interscience Publishers, (1937).

[8] C. Durso, *Solution of the inverse spectral problem for triangles,* PhD Thesis, Massachusetts Institute of Technology, (1990).

[9] S. Finch, *Airy Function Zeroes,* supplementary material for *Mathematical Constants,* Cambridge University Press, (2003).

[10] S. Finch, *Bessel Function Zeroes,* supplementary material for *Mathematical Constants,* Cambridge University Press, (2003).

[11] Pedro Freitas, *Precise bounds and asymptotics for the first Dirichlet eigenvalue of triangles and rhombi*, J. Funct. Anal. **251** (2007), no. 1, 376–398, DOI 10.1016/j.jfa.2007.04.012. MR2353712 (2008k:35339)

[12] Pedro Freitas and David Krejčiřík, *A sharp upper bound for the first Dirichlet eigenvalue and the growth of the isoperimetric constant of convex domains*, Proc. Amer. Math. Soc. **136** (2008), no. 8, 2997–3006, DOI 10.1090/S0002-9939-08-09399-4. MR2399068 (2009b:35309)

[13] Leonid Friedlander and Michael Solomyak, *On the spectrum of the Dirichlet Laplacian in a narrow strip*, Israel J. Math. **170** (2009), 337–354, DOI 10.1007/s11856-009-0032-y. MR2506330 (2010f:35271)

[14] Carolyn Gordon, David L. Webb, and Scott Wolpert, *One cannot hear the shape of a drum*, Bull. Amer. Math. Soc. (N.S.) **27** (1992), no. 1, 134–138, DOI 10.1090/S0273-0979-1992-00289-6. MR1136137 (92j:58111)

[15] C. Gordon, D. Webb, and S. Wolpert, *Isospectral plane domains and surfaces via Riemannian orbifolds*, Invent. Math. **110** (1992), no. 1, 1–22, DOI 10.1007/BF01231320. MR1181812 (93h:58172)

[16] Daniel Grieser, *Spectra of graph neighborhoods and scattering*, Proc. Lond. Math. Soc. (3) **97** (2008), no. 3, 718–752, DOI 10.1112/plms/pdn020. MR2448245 (2011c:58059)

[17] Daniel Grieser, *Thin tubes in mathematical physics, global analysis and spectral geometry*, Analysis on graphs and its applications, Proc. Sympos. Pure Math., vol. 77, Amer. Math. Soc., Providence, RI, 2008, pp. 565–593, DOI 10.1090/pspum/077/2459891. MR2459891 (2010g:58025)

[18] Daniel Grieser and David Jerison, *Asymptotics of eigenfunctions on plane domains*, Pacific J. Math. **240** (2009), no. 1, 109–133, DOI 10.2140/pjm.2009.240.109. MR2485476 (2010b:35061)

[19] D. Grieser and R. Melrose, *Eigenvalues of collapsing triangles,* in preparation.

[20] Daniel Grieser and Svenja Maronna, *Hearing the shape of a triangle*, Notices Amer. Math. Soc. **60** (2013), no. 11, 1440–1447, DOI 10.1090/noti1063. MR3154630

[21] Mark Kac, *Can one hear the shape of a drum?*, Amer. Math. Monthly **73** (1966), no. 4, 1–23. MR0201237 (34 #1121)

[22] J. R. Kuttler and V. G. Sigillito, *Eigenvalues of the Laplacian in two dimensions*, SIAM Rev. **26** (1984), no. 2, 163–193, DOI 10.1137/1026033. MR738929 (85k:65086)

[23] T. Lang and R. Wong, *"Best possible" upper bounds for the first two positive zeros of the Bessel function $J_\nu(x)$: the infinite case*, J. Comput. Appl. Math. **71** (1996), no. 2, 311–329, DOI 10.1016/0377-0427(95)00220-0. MR1399899 (97h:33016)

[24] Peter Li and Shing Tung Yau, *Estimates of eigenvalues of a compact Riemannian manifold*, Geometry of the Laplace operator (Proc. Sympos. Pure Math., Univ. Hawaii, Honolulu, Hawaii, 1979), Proc. Sympos. Pure Math., XXXVI, Amer. Math. Soc., Providence, R.I., 1980, pp. 205–239. MR573435 (81i:58050)

[25] Lee Lorch and Riccardo Uberti, *"Best possible" upper bounds for the first positive zeros of Bessel functions—the finite part*, J. Comput. Appl. Math. **75** (1996), no. 2, 249–258, DOI 10.1016/S0377-0427(96)00065-9. MR1426265 (98b:33010)

[26] Zhiqin Lu and Julie Rowlett, *The fundamental gap of simplices*, Comm. Math. Phys. **319** (2013), no. 1, 111–145, DOI 10.1007/s00220-013-1670-9. MR3034027

[27] Zhiqin Lu and Julie Rowlett, *Eigenvalues of collapsing domains and drift Laplacians*, Math. Res. Lett. **19** (2012), no. 3, 627–648, DOI 10.4310/MRL.2012.v19.n3.a10. MR2998145

[28] F. W. J. Olver, *Bessel Functions Part II, Zeros and Associated Values,* Royal Society Math Tables Vol. 7, Cambridge University Press, (1960).

[29] G. Pólya and G. Szegö, *Isoperimetric Inequalities in Mathematical Physics*, Annals of Mathematics Studies, no. 27, Princeton University Press, Princeton, N. J., 1951. MR0043486 (13,270d)

[30] C. K. Qu and R. Wong, *"Best possible" upper and lower bounds for the zeros of the Bessel function $J_\nu(x)$*, Trans. Amer. Math. Soc. **351** (1999), no. 7, 2833–2859, DOI 10.1090/S0002-9947-99-02165-0. MR1466955 (99j:33006)

[31] I. M. Singer, Bun Wong, Shing-Tung Yau, and Stephen S.-T. Yau, *An estimate of the gap of the first two eigenvalues in the Schrödinger operator*, Ann. Scuola Norm. Sup. Pisa Cl. Sci. (4) **12** (1985), no. 2, 319–333. MR829055 (87j:35280)

[32] Bartłomiej Siudeja, *Sharp bounds for eigenvalues of triangles*, Michigan Math. J. **55** (2007), no. 2, 243–254, DOI 10.1307/mmj/1187646992. MR2369934 (2008k:35341)

[33] M. van den Berg, *On condensation in the free-boson gas and the spectrum of the Laplacian*, J. Statist. Phys. **31** (1983), no. 3, 623–637, DOI 10.1007/BF01019501. MR711491 (85i:82022)

[34] G. N. Watson, *A treatise on the theory of Bessel functions*, Cambridge Mathematical Library, Cambridge University Press, Cambridge, 1995. Reprint of the second (1944) edition. MR1349110 (96i:33010)

[35] Shing-Tung Yau, *An estimate of the gap of the first two eigenvalues in the Schrödinger operator*, Lectures on partial differential equations, New Stud. Adv. Math., vol. 2, Int. Press, Somerville, MA, 2003, pp. 223–235. MR2055851 (2005c:35219)

[36] Shing-Tung Yau, *Nonlinear analysis in geometry*, Enseign. Math. (2) **33** (1987), no. 1-2, 109–158. MR896385 (88g:58003)

[37] Qi Huang Yu and Jia Qing Zhong, *Lower bounds of the gap between the first and second eigenvalues of the Schrödinger operator*, Trans. Amer. Math. Soc. **294** (1986), no. 1, 341–349, DOI 10.2307/2000135. MR819952 (87f:35179)

Department of Mathematics, 410D Rowland Hall, University of California, Irvine, California 92697-3875

E-mail address: `zlu@uci.edu`

Max Planck Institut für Mathematik, Vivatgasse 7, 53111 Bonn, Germany

E-mail address: `rowlett@mpim-bonn.mpg.de`

Contemporary Mathematics
Volume **630**, 2014
http://dx.doi.org/10.1090/conm/630/12669

The Logarithmic Singularities of the Green Functions of the Conformal Powers of the Laplacian

Raphaël Ponge

ABSTRACT. Green functions play an important role in conformal geometry. In this paper, we explain how to compute explicitly the logarithmic singularities of the Green functions of the conformal powers of the Laplacian. These operators include the Yamabe and Paneitz operators, as well as the conformal fractional powers of the Laplacian arising from scattering theory for Poincaré-Einstein metrics. The results are formulated in terms of Weyl conformal invariants arising from the ambient metric of Fefferman-Graham. As applications we obtain characterizations in terms of Green functions of locally conformally flat manifolds and a spectral theoretic characterization of the conformal class of the round sphere.

Introduction

Motivated by the analysis of the singularity of the Bergman kernel of a strictly pseudoconvex domain $\Omega \subset \mathbb{C}^n$, Fefferman [**Fe3**] launched the program of determining *all* local invariants of a strictly pseudoconvex CR structure. This program was subsequently extended to deal with local invariants of other parabolic geometries, including conformal geometry [**FG1**, **BEG**]. It has since found connections with various areas of mathematics and mathematical physics such as geometric PDEs, geometric scattering theory and conformal field theory. For instance, the Poincaré-Einstein metric of Fefferman-Graham [**FG1**, **FG3**] was a main impetus for the AdS/CFT correspondence.

Green functions of conformally invariant operators plays a fundamental role in conformal geometry. Parker-Rosenberg [**PR**] computed the logarithmic singularity of Yamabe operator in low dimension. In [**Po1**, **Po2**] it was shown that the logarithmic singularities of Green functions of conformally invariant ΨDOs are linear combinations of Weyl conformal invariants. Those invariants are obtained from complete metric constructions of the covariant derivatives of the curvature tensor of ambient metric of Fefferman-Graham [**FG1**, **FG3**]. The approach of [**PR**] was based on results of Gilkey [**Gi3**] on heat invariants for Laplace-type operators. It is

2010 *Mathematics Subject Classification.* Primary 53A55; Secondary 53A30.
Key words and phrases. conformal invariants, Fefferman's program.
The research of this paper was partially supported by NSERC discovery grant 341328-07 (Canada), JSPS grant in aid 30549291(Japan), research resettlement and foreign faculty research grants from Seoul National University, and NRF grant for basic research 2013R1A1A2008802 (South Korea).

not clear how to extend this approach to higher order GJMS operators, leave aside conformal fractional powers of the Laplacian.

Exploiting the invariant theory for conformal structures, the main result of this paper is an explicit, and surprinsingly simple, formula for the logarithmic singularities of the Green functions of the conformal powers of the Laplacian in terms of Weyl conformal invariants obtained from the heat invariants of the Laplace operator (Theorem 7.3). Here by conformal powers we mean the operators of Graham-Jenne-Mason-Sparling [**GJMS**] and, more generally, the conformal fractional powers of the Laplacian [**GZ**]. These operators include the Yamabe and Paneitz operators.

Granted this result, it becomes straightforward to compute the logarithmic singularities of the Green functions of the conformal powers of the Laplacian from the sole knowledge of the heat invariants of the Laplace operators (see Theorem 7.8 and Theorem 7.9). These results have several important consequences.

In dimension $n \geq 5$ the logarithmic singularity of Green function of the k-th conformal power of the Laplacian with $k = \frac{n}{2} - 2$ is of special interest since this is a scalar multiple of the norm-square of the Weyl tensor. As a result we see that the vanishing of this logarithmic singularity is equivalent to local conformal-flatness (Theorem 7.11). This can be seen as version in conformal geometry of the conjecture of Radamanov [**Ra**] on the vanishing of the logarithmic singularity of the Bergman kernel. This also enables us to obtain a spectral theoretic characterization of the conformal class of the round sphere amount compact simply connected manifolds of dimension ≥ 5 (see Theorem 7.13).

The idea behind the proof of Theorem 7.3 is the following. In the Riemannian case, there is a simple relationship between the logarithmic singularities of the Green functions of the powers of the Laplace operator and its heat invariant (see Eq. (6.7)). We may expect that by some analytic continuation of the signature this relationship still pertains in some way in non-Riemannian signature. The GJMS operators are obtained from the powers of the Laplacian associated to the ambient metric (which has Lorentzian signature), so there ought to be some relations between the logarithmic singularities of the Green functions of the conformal powers of the Laplace operator and its heat invariants.

One way to test this conjecture is to look at the special case of Ricci-flat metrics. In this case the computation follows easily from the Riemannian case. The bulk of the proof then is to show that the Ricci-flat case implies the general case. Thus, the analytic extension of the signature is replaced by the new principle that in order to prove an equality between conformal invariants it is enough to prove it for Ricci-flat metrics.

This "Ricci-flat principle" is actually fairly general, and so the results should also for other type of conformally invariant operators that are conformal analogues of elliptic covariant operators. In particular, it should hold for conformal powers of the Hodge Laplacian on forms [**BG, Va2**] and the conformal powers of the square of the Dirac operator [**GMP1, GMP2**].

In addition, this approach can be extended to the setting of CR geometry. In particular, in [**Po1, Po2**] it was shown that the logarithmic singularities of the Green functions of CR invariant hypoelliptic Ψ_HDOs are linear combinations of Weyl CR invariants. We may also expect relate the logarithmic singularities of the Green functions of the CR invariant powers of the sub-Laplacian [**GG**] to the heat invariants of the sub-Laplacian [**BGS**]. However, the heat invariants for the sub-Laplacian are much less known than that for the Laplace operator. In particular, in order to recapture the Chern tensor requires computing the 3rd coefficient in the heat kernel asymptotics for the sub-Laplacian. In fact, for our purpose it would be enough to carry out the computation in the special case of circle bundles over Ricci-flat Kähler manifolds.

This paper is organized as follows. In Section 1, we recall the main facts on the heat kernel asymptotics for the Laplace operator. In Section 2, we recall the geometric description of the singularity of the Bergman kernel of a strictly pseudoconvex complex domain and the construction of local CR invariants by means of Fefferman's ambient Kähler-Lorentz metric. In Section 3, we recall the construction of the Fefferman-Graham's ambient metric and GJMS operators. In Section 4, we recall the construction of local conformal invariants by means of the ambient metric. In Section 5, we explain the construction of conformal fractional powers of the Laplacian and its connection with scattering theory and the Poincaré-Einstein metric. In Section 6, we gather basic facts about Green functions and their relationship with heat kernels. In Section 7, we state the main result and derive various consequences. In Section 8, we give an outline of the proof of the main result.

Acknowledgements. It is a pleasure to thank Pierre Albin, Charles Fefferman, Rod Gover, Robin Graham, Colin Guillarmou, Kengo Hirachi, Dmitry Jakobson, Andreas Juhl, Andras Vasy, and Maciej Zworski for various discussions related to the subject matter of this paper. Part of the research of this paper was carried out during visits to Kyoto University, McGill University, MSRI, UC Berkeley, and the University of Tokyo. The author wish to thank these institutions for their hospitality.

1. The Heat Kernel of the Laplace Operator

The most important differential operator attached to a closed Riemannian manifold (M^n, g) is the Laplace operator,

$$\Delta_g u = \frac{-1}{\sqrt{\det g(x)}} \sum_{i,j} \partial_i \left(g^{ij}(x) \sqrt{\det g(x)}\, \partial_j u \right).$$

This operator lies at the interplay between Riemannian geometry and elliptic theory. On the one hand, a huge amount of geometric information can be extracted from the analysis of the Laplace operator. For instance, if $0 = \lambda_0(\Delta_g) < \lambda_1(\Delta_g) \leq \lambda_2(\Delta_g) \leq \cdots$ are the eigenvalues of Δ_g counted with multiplicity, then Weyl's Law asserts that, as $k \to \infty$,

$$\lambda_k(\Delta_g) \sim (ck)^{\frac{2}{n}}, \qquad c := (4\pi)^{-\frac{n}{2}} \Gamma\left(\frac{n}{2}+1\right)^{-1} \operatorname{Vol}_g(M), \tag{1.1}$$

where $\operatorname{Vol}_g(M)$ is the Riemannian volume of (M, g). On the other hand, Riemannian geometry is used to describe singularities or asymptotic behavior of solutions of PDEs associated to the Laplace operator. This aspect is well illustrated by the asymptotics of the heat kernel of Δ_g.

Denote by $e^{-t\Delta_g}$, $t > 0$, the semigroup generated by Δ_g. That is, $u(x,t) = (e^{-t}\Delta_g u_0)(x)$ is solution of the heat equation,

$$(\partial_t + \Delta_g)u(x,t) = 0, \qquad u(x,t) = u_0(x) \quad \text{at } t = 0.$$

The heat kernel $K_{\Delta_g}(x,y;t)$ is the kernel function of the heat semigroup,

$$e^{-t\Delta_g}u(x)=\int_M K_{\Delta_g}(x,y;t)u(y)v_g(y),$$

where $v_g(y) = \sqrt{g(x)}|dx|$ is the Riemannian volume density. Equivalently, $K_{\Delta_g}(x,y;t)$ provides us with a fundamental solution for the heat equation (1). Furthermore,

$$\sum e^{-t\lambda_j(\Delta_g)}=\operatorname{Tr}e^{-t\Delta_g}=\int_M K_{\Delta_g}(x,x;t)u(x)v_g(x). \tag{1.2}$$

THEOREM 1.1 ([**ABP**, **Gi1**, **Mi**]). *Let (M^n,g) be a closed manifold. Then*

(1) *As $t\to 0^+$,*

$$K_{\Delta_g}(x,x;t)\sim(4\pi t)^{-\frac{n}{2}}\sum_{j\geq 0}t^j a_{2j}(\Delta_g;x), \tag{1.3}$$

where the asymptotics holds with respect to the Fréchet-space topology of $C^\infty(M)$.

(2) *Each coefficient $a_{2j}(\Delta_g;x)$ is a local Riemannian invariant of weight $2j$.*

REMARK 1.2. There is an asymptotics similar to (1.3) for the heat kernel of any selfadjoint differential operators with positive-definite principal symbol (see [**Gi3**, **Gr**]).

REMARK 1.3. A *local Riemannian invariant* is a (smooth) function $I_g(x)$ of $x\in M$ and the metric tensor g which, in any local coordinates, has an universal expression of the form,

$$I_g(x)=\sum a_{\alpha\beta}\left(g(x)\right)\left(\partial^\beta g(x)\right)^\alpha, \tag{1.4}$$

where the sum is finite and the $a_{\alpha\beta}$ are smooth functions on $\mathrm{Gl}_n(\mathbb{R})$ that are *independent* of the choice of the local coordinates. We further say that $I_g(x)$ has weight w when

$$I_{\lambda^2 g}(x)=\lambda^{-w}I_g(x)\quad\forall\lambda>0.$$

These definitions continue to make sense for pseudo-Riemannian structures of non-positive signature. Note also that with our convention for the weight, the weight is always an even nonnegative integer.

Examples of Riemannian invariants are provided by complete metric contractions of the tensor products of covariant derivatives of the curvature tensors $R_{ijkl}=\langle R(\partial_i,\partial_j)\partial_k,\partial_l\rangle$. Namely,

$$\operatorname{Contr}_g(\nabla^{k_1}R\otimes\cdots\otimes\nabla^{k_l}R).$$

Such an invariant has weight $w=(k_1+\cdots+k_l)+2l$ and is called a *Weyl Riemannian invariant*. Up to a sign factor, the only Weyl Riemannian invariants of weight $w=0$ and $w=2$ are the constant function 1 and the scalar curvature $\kappa := R^{ij}{}_{ji}$ respectively. In weight $w=4$ we obtain the following four invariants,

$$\kappa^2,\qquad |\operatorname{Ric}|^2:=\operatorname{Ric}^{ij}\operatorname{Ric}_{ij},\qquad |R|^2:=R^{ijkl}R_{ijkl},\qquad \Delta_g\kappa,$$

where $\operatorname{Ric}_{ij}:=R^k{}_{ijk}$ is the Ricci tensor. In weight $w=6$ they are 17 such invariants; the only invariants that do not involve the Ricci tensor are

$$|\nabla R|^2:=\nabla^m R^{ijkl}\nabla_m R_{ijkl},\qquad R_{ij}{}^{kl}R^{ij}{}_{pq}R^{pq}{}_{kl},\qquad R_{ijkl}R^i{}_p{}^k{}_qR^{pjql}.$$

THEOREM 1.4 (Atiyah-Bott-Patodi [**ABP**]). *Any local Riemannian invariant of weight w, $w \in 2\mathbb{N}_0$, is a universal linear combination of Weyl Riemannian invariants of same weight.*

REMARK 1.5. By using normal coordinates the proof is reduced to determining all invariant polynomial of the orthogonal group $O(n)$ (or $O(p,q)$ in the pseudo-Riemannian case). They are determined thanks to Weyl's invariant theory for semisimple groups.

Combining Theorem 1.1 and Theorem 1.4 we obtain the following structure theorem for the heat invariants.

THEOREM 1.6 (Atiyah-Bott-Patodi [**ABP**]). *Each heat invariant $a_{2j}(\Delta_g;x)$ in (1.3) is a universal linear combination of Weyl Riemannian invariants of weight $2j$.*

REMARK 1.7. There is a similar structure result for the heat invariants of any selfadjoint elliptic covariant differential operator with positive principal symbol.

In addition, the first few of the heat invariants are computed.

THEOREM 1.8 ([**BGM, MS, Gi2**]). *The first four heat invariants $a_{2j}(\Delta_g;x)$, $j=0,\ldots,3$ are given by the following formulas,*

$$a_0(\Delta_g;x) = 1, \quad a_2(\Delta_g;x) = -\frac{1}{6}\kappa,$$

$$a_4(\Delta_g;x) = \frac{1}{180}|R|^2 - \frac{1}{180}|\operatorname{Ric}|^2 + \frac{1}{72}\kappa^2 - \frac{1}{30}\Delta_g\kappa,$$

$$a_6(\Delta_g;x) = \frac{1}{9\cdot 7!}\left(81|\nabla R|^2 + 64R_{ij}{}^{kl}R^{ij}{}_{pq}R^{pq}{}_{kl} + 352R_{ijkl}R^{i}{}_{p}{}^{k}{}_{q}R^{pjql}\right) \tag{1.5}$$
$$+ \text{Weyl Riemannian invariants involving the Ricci tensor.}$$

REMARK 1.9. We refer to Gilkey's monograph [**Gi3**] for the full formula for $a_6(\Delta_g;x)$ and formulas for the heat invariants of various Laplace type operators, including the Hodge Laplacian on forms. We mention that Gilkey's formulas involve a constant multiple of $-\langle R, \Delta_g R\rangle = g^{pq}R_{ijkl}R_{ijkl;pq}$, but this Weyl Riemannian invariant is a linear combination of other Weyl Riemannian invariants. In particular, using the Bianchi identities we find that, modulo Weyl Riemannian involving the Ricci tensor,

$$-\langle R, \Delta_g R\rangle = R_{ij}{}^{kl}R^{ij}{}_{pq}R^{pq}{}_{kl} + 4R_{ijkl}R^{i}{}_{p}{}^{k}{}_{q}R^{pjql}.$$

This relation is incorporated into (1.5).

REMARK 1.10. Polterovich [**Po**] established formulas for *all* the heat invariants $a_j(\Delta_g;x)$ in terms of the Riemannian distance function.

REMARK 1.11. Combining the equality $a_0(\Delta_g;x) = 1$ with (1.2) and (1.3) shows that, as $t \to 0^+$,

$$\sum e^{-t\lambda_j(\Delta_g)} \sim (4\pi t)^{-\frac{n}{2}} \int_M v_g(x) = (4\pi t)^{-\frac{n}{2}} \operatorname{Vol}_g(M).$$

We then can apply Karamata's Tauberian theorem to recover Weyl's Law (1.1).

REMARK 1.12. A fundamental application of the Riemannian invariant theory of the heat kernel asymptotics is the proof of the local index theorem by Atiyah-Bott-Patodi [**ABP**] (see also [**Gi3**]).

2. The Bergman Kernel of a Strictly Pseudoconvex Domain

A fundamental problem in several complex variables is to find local computable biholomorphic invariants of a strictly pseudoconvex domain $\Omega \subset \mathbb{C}^n$. One approach to this issue is to look at the boundary singularity of the Bergman metric of Ω or equivalently its Bergman kernel. Recall that the Bergman kernel is the kernel function of the orthogonal projection,

$$B_\Omega : L^2(\Omega) \longrightarrow L^2(\Omega) \cap \mathrm{Hol}(\Omega),$$
$$B_\Omega u(z) = \int K_\Omega(z,w)u(w)dw.$$

For instance, in the case of the unit ball $\mathbb{B}^{2n} = \{|z| < 1\} \subset \mathbb{C}^n$ we have

$$B_{\mathbb{B}^{2n}}(z,w) = \frac{n!}{\pi^n}(1 - z\overline{w})^{-(n+1)}.$$

The Bergman kernel lies at the interplay of complex analysis and differential geometry. On the one hand, it provides us with the reproducing kernel of the domain Ω and it plays a fundamental role in the analysis of the $\overline{\partial}$-Neuman problem on Ω. On the other hand, it provides us with a biholomorphic invariant Kähler metric, namely, the Bergman metric,

$$ds^2 = \sum \frac{\partial^2}{\partial z^j \partial z^{\overline{k}}} \log B(z,z) dz^j dz^{\overline{k}}.$$

In what follows we let ρ be a defining function of Ω so that $\Omega = \{\rho < 0\}$ and $i\partial\overline{\partial}\rho > 0$.

THEOREM 2.1 (Fefferman, Boutet de Monvel-Sjöstrand). *Near the boundary* $\partial\Omega = \{\rho = 0\}$,

$$K_\Omega(z,z) = \varphi(z)\rho(z)^{-(n+1)} - \psi(z)\log\rho(z),$$

where $\varphi(z)$ *and* $\psi(z)$ *are smooth up to* $\partial\Omega$.

Motivated by the analogy with the heat asympototics (1.3), where the role of the time variable t is played by the defining function $\rho(z)$, Fefferman [**Fe3**] launched the program of giving a geometric description of the singularity of the Bergman kernel similar to the description provided by Theorem 1.1 and Theorem 1.6 for the heat kernel asymptotics.

A first issue at stake concerns the choice of the defining function. We would like to make a biholomorphically invariant choice of defining function. This issue is intimately related to the complex Monge-Ampère equation on Ω:

$$J(u) := (-1)^n \det \begin{pmatrix} u & \partial_{z^{\overline{k}}} u \\ \partial_{z^j} u & \partial_{z^j}\partial_{z^{\overline{k}}} u \end{pmatrix} = 1, \qquad u_{|\partial\Omega} = 0. \tag{2.1}$$

A solution of the Monge-Ampère equation is unique and biholomorphically invariant in the sense that, given any bilohomorphism $\Phi : \Omega \to \Omega$, we have

$$u(\Phi(z)) = |\det \Phi'(z)|^{\frac{2}{n+1}} u(z).$$

We mention the following important results concerning the Monge-Ampère equation.

THEOREM 2.2 (Cheng-Yau [**CY**]). *Let* $\Omega \subset \mathbb{C}^n$ *be a strictly pseudoconvex domain.*

(1) *There is a unique exact solution* $u_0(z)$ *of the complex Monge-Ampère equation* (2.1).
(2) *The solution* $u_0(z)$ *is* C^∞ *on* Ω *and belongs to* $C^{n+\frac{3}{2}-\epsilon}\left(\overline{\Omega}\right)$ *for all* $\epsilon > 0$.
(3) *The metric* $ds^2 = \sum \frac{\partial^2}{\partial z^j \partial \overline{z^k}} \log\left(u_0(z)^{-(n+1)}\right) dz^j d\overline{z^k}$ *is a Kähler-Einstein metric on* Ω.

THEOREM 2.3 (Lee-Melrose [**LM**]). *Let* $\Omega \subset \mathbb{C}^n$ *be a strictly pseudoconvex domain with defining function* $\rho(z)$. *Then, near the boundary* $\partial\Omega = \{\rho = 0\}$, *the Cheng-Yau solution to the Monge-Ampère equation has a behavior of the form,*

$$u_0(z) \sim \rho(z) \sum_{k \geq 0} \eta_k(z) \left(\rho(z)^{n+1} \log \rho(z)\right)^k, \tag{2.2}$$

where the functions $\eta_k(z)$ *are smooth up to the boundary.*

The Cheng-Yau solution is not smooth up to the boundary, but if we only seek for asymptotic solutions then we do get smooth solutions.

THEOREM 2.4 (Fefferman [**Fe1**]). *Let* $\Omega \subset \mathbb{C}^n$ *be a strictly pseudoconvex domain with defining function* $\rho(z)$. *Then there are functions in* $C^\infty\left(\overline{\Omega}\right)$ *that are solutions to the asymptotic Monge-Ampère equation,*

$$J(u) = 1 + O(\rho^{n+1}) \text{ near } \partial\Omega, \qquad u_{|\partial\Omega} = 0. \tag{2.3}$$

REMARK 2.5. As it can be seen from (2.2), the error term $O(\rho^{n+1})$ cannot be improved in general if we seek for a smooth approximate solution.

Any smooth solution $u(z)$ of (2.3) is a defining function for Ω and it is asymptotically biholomorphically invariant in the sense that, for any biholomorphism $\Phi : \Omega \to \Omega$,

$$u\left(\Phi(z)\right) = \left|\det \Phi'(z)\right|^{\frac{2}{n+1}} u(z) + O(\rho(z)^{n+1})$$

The complex structure of Ω induces on the boundary $\partial\Omega$ a strictly pseudoconvex CR structure. As a consequence of a well-known result of Fefferman [**Fe2**] there is one-to-one correspondence between biholomorphisms of Ω and CR diffeomorphisms of $\partial\Omega$. Therefore, boundary values of biholomorphic invariants of Ω gives rise to CR invariants of $\partial\Omega$. We refer to [**Fe3**] for the precise definition of a local CR invariant. Let us just mention that a local CR invariant $I(z)$ has weight ω when, for any biholomorphism $\Phi : \Omega \to \Omega$, we have

$$I\left(\Phi(z)\right) = \left|\det \Phi'(z)\right|^{-\frac{w}{n+1}} I(z) \qquad \text{on } \partial\Omega.$$

PROPOSITION 2.6 (Fefferman [**Fe3**]). *Let* $\Omega \subset \mathbb{C}^n$ *be a strictly pseudoconvex domain. Then*

(1) *Near the boundary* $\partial\Omega$,

$$K_\Omega(z, z) = u(z)^{-(n+1)} \sum_{j=0}^{n+1} I_u^{(j)}(z) u(z)^j + O\left(\log u(z)\right),$$

where $u(z)$ *is a smooth solution of* (2.3).
(2) *For each* j, *the boundary value of* $I_u^{(j)}(z)$ *is a local CR invariant of weight* $2j$.

This leads us to the geometric part of program of Fefferman: determining all CR invariants of strictly pseudoconvex domain so as to have an analogue of Theorem 1.4 for CR invariants. However, the CR invariant theory is much more involved than the Riemannian invariant theory. The geometric problem reduces to the invariant theory for a parabolic subgroup $P \subset \mathrm{SU}(n+1,1)$. However, as P is not semisimple, Weyl's invariant theory does not apply anymore. The corresponding invariant theory was developed by Fefferman [**Fe3**] and Bailey-Eastwood-Graham [**BEG**].

The Weyl CR invariants are constructed as follows. Let $u(z)$ be a smooth approximate solution to the Monge-Ampère equation in the sense of (2.3). On the ambient space $\mathbb{C}^* \times \Omega$ consider the potential,

$$U(z_0, z) = |z_0|^2 u(z), \qquad (z_0, z) \in \mathbb{C}^* \times \Omega.$$

Using this potential we define the Kähler-Lorentz metric,

$$\tilde{g} = \sum_{0 \leq j,k \leq n} \frac{\partial^2 U}{\partial z^j \partial z^{\overline{k}}} dz^j dz^{\overline{k}}.$$

A biholomorphism $\Phi : \Omega \to \Omega$ acts on the ambient space by

$$\tilde{\Phi}(z_0, z) = \left(|\det \Phi'(z)|^{-\frac{1}{n+1}}, \Phi(z) \right), \qquad (z_0, z) \in \mathbb{C}^* \times \Omega.$$

This is the transformation law under biholomorphisms for $\mathcal{K}_\Omega^{\frac{1}{n+1}}$, where $\mathcal{K}_\Omega$ is the canonical line bundle of Ω.

THEOREM 2.7 (Fefferman [**Fe1**]).

(1) *The Kähler-Lorentz metric $\tilde{g}$ is asymptotically Ricci flat and invariant under biholomorphisms in the sense that*

$$\tilde{\Phi}^* \tilde{g} = \tilde{g} + \mathrm{O}\left(\rho(z)^{n+1}\right) \quad \text{and} \quad \mathrm{Ric}(\tilde{g}) = \mathrm{O}\left(\rho(z)^n\right) \qquad \text{near } \mathbb{C}^* \times \partial\Omega,$$

where Φ is any biholomorphism of Ω.

(2) *The restriction of $\tilde{g}$ to $S^1 \times \partial\Omega$ is a Lorentz metric g whose conformal class is invariant under biholomorphisms.*

REMARK 2.8. The above construction of the ambient metric is a special case of the ambient metric construction associated to a conformal structure due to Fefferman-Graham [**FG1**, **FG3**] (see also Section 3).

REMARK 2.9. We refer to [**Le**] for an intrinsic construction of the Fefferman circle bundle $S^1 \times \partial\Omega$ and its Lorentz metric g for an arbitrary nondegenerate CR manifold.

It follows from Theorem 2.7 that, given a Weyl Riemannian invariant $I_{\tilde{g}}(z_0, z)$ constructed out of the covariant derivatives of the curvature tensor of $\tilde{g}$, the boundary value of $I_{\tilde{g}}(z_0, z)$ is a local CR invariant provided that it does not involved covariant derivatives of too high order. This condition is fullfilled if $I_{\tilde{g}}(z_0, z)$ has weight $w \leq 2n$, in which case its boundary value is a CR invariant of same weight. Such an invariant is called a *Weyl CR invariant.* We observe that the Ricci-flatness of the ambient metric kills all Weyl invariants involving the Ricci tensor of $\tilde{g}$. Therefore, there are much fewer Weyl CR invariants than Riemannian Weyl invariants.

The following results are the analogues of Theorem 1.4 and Theorem 1.6 for the Bergman kernel.

THEOREM 2.10 (Fefferman [**Fe3**], Bayley-Eastwood-Graham [**BEG**]). *Any local CR invariant of weight $w \leq 2n$ is a linear combination of Weyl CR invariants of same weight.*

THEOREM 2.11 (Fefferman [**Fe3**], Bayley-Eastwood-Graham [**BEG**]). *Let $\Omega \subset \mathbb{C}^n$ be a strictly pseudoconvex domain. Then, near the boundary $\partial\Omega$,*

$$K_\Omega(z,z) = u(z)^{-(n+1)} \sum_{j=0}^{n+1} I_u^{(j)}(z) u(z)^j + \mathrm{O}\left(\log u(z)\right),$$

where $u(z)$ is a smooth solution of (2.3) and, for $j = 0, \ldots, n+1$,

$$I_u^{(j)}(z) = J_{\tilde{g}}^{(j)}(z_0, z)_{|z_0=1},$$

where $J_{\tilde{g}}^{(j)}(z_0,z)$ is a linear combination of complete metric contractions of weight w of the covariant derivatives of the Kähler-Lorentz metric $\tilde{g}$.

REMARK 2.12. We refer to [**Hi1**] for an invariant description of the logarithmic singularity $\psi(z)$ of the Bergman kernel.

3. Ambient Metric and Conformal Powers of the Laplacian

Let (M^n, g) be a Riemannian manifold. In dimension $n = 2$ the Laplace operator is conformally invariant in the sense that, if we make a conformal change of metric $\hat{g} = e^{2\Upsilon} g$, $\Upsilon \in C^\infty(M, \mathbb{R})$, then

$$\Delta_{\hat{g}} = e^{-2\Upsilon} \Delta_g.$$

This conformal invariance breaks down in dimension $n \geq 3$. A conformal version of the Laplacian is provided by the Yamabe operator,

$$P_{1,g} := \Delta_g + \frac{n-2}{4(n-1)}\kappa,$$
$$P_{1,e^{2\Upsilon}g} = e^{-(\frac{n}{2}+1)\Upsilon} P_{1,g} e^{(\frac{n}{2}-1)\Upsilon} \qquad \forall \Upsilon \in C^\infty(M, \mathbb{R}).$$

The operator of Paneitz [**Pa**] is a conformally invariant operator with principal part Δ_g^2. Namely,

$$P_{2,g} := \Delta_g^2 + \delta V d + \frac{n-4}{2}\left\{\frac{1}{2(n-1)}\Delta_g R_g + \frac{n}{8(n-1)^2} R_g^2 - 2|P|^2\right\},$$
$$P_{2,e^{2\Upsilon}g} = e^{-(\frac{n}{2}+2)\Upsilon} P_{2,g} e^{(\frac{n}{2}-2)\Upsilon} \qquad \forall \Upsilon \in C^\infty(M, \mathbb{R}).$$

Here $P_{ij} = \frac{1}{n-2}(\mathrm{Ric}_{g_{ij}} - \frac{R_g}{2(n-1)} g_{ij})$ is the Schouten-Weyl tensor and V is the tensor $V_{ij} = \frac{n-2}{2(n-1)} R_g g_{ij} - 4P_{ij}$ acting on 1-forms (i.e., $V(\omega_i dx^i) = (V_i{}^j \omega_j) dx^i$).

More generally, Graham-Jenne-Mason-Sparling [**GJMS**] constucted higher order conformal powers of the Laplacian by using the ambient metric of Fefferman-Graham [**FG1**, **FG3**]. This metric extends Fefferman's Kähler-Lorentz metric to any ambient space associated to a conformal structure. It is constructed as follows.

Consider the ray-bundle,

$$G := \bigsqcup_{x \in M} \{t^2 g(x);\ t > 0\} \subset S^2 T^* M,$$

where S^2T^*M is the bundle of symmetric $(0,2)$-tensors. We note that G depends only the conformal class $[g]$. Moreover, on G there is a natural family of dilations δ_s, $s>0$, given by

$$\delta_s(\hat{g}) = s\hat{g} \qquad \forall \hat{g} \in G. \tag{3.1}$$

Let $\pi : G \to M$ be the canonical submersion of G onto M (i.e., the restriction to G of the canonical submersion of the bundle S^2T^*M). Let $d\pi^t : S^2T^*M \to S^2T^*G$ be the differential of π on symmetric $(0,2)$-tensors. Then G carries a canonical symmetric $(0,2)$-tensor g_0 defined by

$$g_0(x,\hat{g}) = d\pi(x)^t\hat{g} \qquad \forall (x,\hat{g}) \in G. \tag{3.2}$$

The datum of the representative metric g defines a fiber coordinate t on G such that $\hat{g} = t^2 g(x)$ for all $(x,\hat{g}) \in G$. If $\{x^j\}$ are local coordinates on M, then $\{x^j, t\}$ are local coordinates on G and in these coordinates the tensor g_0 is given by

$$g_0 = t^2 g_{ij}(x) dx^i dx^j,$$

where g_{ij} are the coefficients of the metric tensor g in the local coordinates $\{x^j\}$.

The *ambient space* is the $(n+2)$-dimensional manifold

$$\tilde{G} := G \times (-1,1)_\rho,$$

where we denote by ρ the variable in $(-1,1)$. We identify G with the hypersurface $G_0 := \{\rho = 0\}$. We also note that the dilations δ_s in (3.1) lifts to dilations on $\tilde{G}$.

THEOREM 3.1 (Fefferman-Graham [**FG1**,**FG3**]). *Near $\rho = 0$ there is a Lorentzian metric $\tilde{g}$, called the ambient metric, which is defined up to infinite order in ρ when n is odd, and up to order $\frac{n}{2}$ when n is even, such that*

(i) *In local coordinates $\{t, x^j, \rho\}$,*

$$\tilde{g} = 2\rho(dt)^2 + t^2 g_{ij}(x,\rho) + 2tdtd\rho,$$

where $g_{ij}(x,\rho)$ is a family of symmetric $(0,2)$-tensors such that $g(x,\rho)|_{\rho=0} = g_{ij}(x)$.

(ii) *The ambient metric is asymptotically Ricci-flat, in the sense that*

$$\operatorname{Ric}(\tilde{g}) = \begin{cases} \mathrm{O}(\rho^\infty) & \text{if } n \text{ is odd,} \\ \mathrm{O}(\rho^{\frac{n}{2}}) & \text{if } n \text{ is even.} \end{cases} \tag{3.3}$$

REMARK 3.2. We note that (i) implies that

$$\tilde{g}_{|TG_0} = g_0 \qquad \text{and} \qquad \delta_s^*\tilde{g} = s^2\tilde{g} \quad \forall s > 0, \tag{3.4}$$

where g_0 is the symmetric tensor (3.2).

REMARK 3.3. The ambient metric is unique up to its order of definition near $\rho = 0$. Solving the equation (3.3) leads us to a system of nonlinear PDEs for $g_{ij}(x,\rho)$. When n is even, there is an obstruction to solve this system at infinite order which given by the a conformally invariant tensor. This tensor is the Bach tensor in dimension 4. Moreover, it vanishes on conformally Einstein metrics.

EXAMPLE 3.4. Suppose that (M,g) is Einstein with $\operatorname{Ric}(g) = 2\lambda(n-1)g$. Then

$$\tilde{g} = 2\rho(dt)^2 + t^2(1+\lambda\rho)^2 g_{ij}(x) + 2tdtd\rho.$$

Conformal powers of the Laplacian are obtained from the powers of the ambient metric Laplacian $\tilde{\Delta}_{\tilde{g}}$ on $\tilde{G}$ as follows.

PROPOSITION 3.5 (Graham-Jenne-Mason-Sparling [**GJMS**]). *Let $k \in \mathbb{N}_0$ and further assume $k \leq \frac{n}{2}$ when n is even. Then*

(1) *We define a operator $P_{k,g} : C^\infty(M) \to C^\infty(M)$ by*

$$P_{k,g}u(x) := t^{-\left(\frac{n}{2}+k\right)} \tilde{\Delta}_{\tilde{g}}^{k} \left(t^{\frac{n}{2}-k}\tilde{u}(x,\rho)\right)\Big|_{\rho=0}, \qquad u \in C^\infty(M), \tag{3.5}$$

where $\tilde{u}$ is any smooth extension of u to $M \times \mathbb{R}$ (i.e., the r.h.s. above is independent of the choice of $\tilde{u}$).

(2) *$P_{k,g}$ is a differential operator of order $2k$ with same leading part as Δ_g^k.*

(3) *$P_{k,g}$ is a conformally covariant differential operator such that*

$$P_{k,e^{2\Upsilon}g} = e^{-\left(\frac{1}{2}n+k\right)\Upsilon}(P_{k,g})e^{\left(\frac{1}{2}n-k\right)\Upsilon} \qquad \forall \Upsilon \in C^\infty(M,\mathbb{R}). \tag{3.6}$$

REMARK 3.6. The homogeneity of the ambient metric $\tilde{g}$ implies that the r.h.s. is idenpendent of t and obeys (3.6).

REMARK 3.7. The operator $P_{k,g}$ is called the k-th GJMS operator. For $k = 1$ and $k = 2$ we recover the Yamabe operator and Paneitz operator respectively.

REMARK 3.8. The GJMS operators are selfadjoint [**FG2**, **GZ**].

REMARK 3.9. The GJMS operator $P_{k,g}$ can also be obtained as the obstruction to extending a function $u \in C^\infty(M)$ into a homogeneous harmonic function on the ambient space. In addition, we refer to [**FG4**, **GP**, **Ju1**, **Ju2**] for various features and properties of the GJMS operarors.

REMARK 3.10. When the dimension n is even, the ambient metric construction is obstructed at finite order, and so the GJMS construction breaks down for $k > \frac{n}{2}$. As proved by Graham [**Gr1**] in dimension 4 for $k = 3$ and by Gover-Hirachi [**GH**] in general, there do not exist conformally invariant operators with same leading part as Δ_g^k for $k > \frac{n}{2}$ when n is even.

REMARK 3.11. The operator $P_{\frac{n}{2},g}$ is sometimes called the *critical GJMS operator.* Note that for $P_{\frac{n}{2}}$ the transformation law (3.6) becomes

$$P_{\frac{n}{2},e^{2\Upsilon}g} = e^{-n\Upsilon}P_{\frac{n}{2},g} \qquad \forall \Upsilon \in C^\infty(M,\mathbb{R}).$$

It is tempting to use the critical GJMS operator as a candidate for a subsitute to the Laplacian to prove a version of the Knizhnik-Polyakov-Zamolodchikov (KPZ) formula in dimension ≥ 3. We refer to [**CJ**] for results in this direction.

EXAMPLE 3.12 (Graham [**Gr2**, **FG3**], Gover [**Go**]). Suppose that (M^n, g) is a Einstein manifold with $\mathrm{Ric}_g = \lambda(n-1)g$, $\lambda \in \mathbb{R}$. Then, for all $k \in \mathbb{N}_0$, we have

$$P_{k,g} = \prod_{1\leq j\leq k} \left(\Delta_g - \frac{1}{4}\lambda(n+2j-2)(n-2j)\right). \tag{3.7}$$

4. Local Conformal Invariants

In this section, we describe local scalar invariants of a conformal structure and explain how to construct them by means of the ambient metric of Fefferman-Graham [**FG1**, **FG3**].

DEFINITION 4.1. A local conformal invariant of weight w, $w \in 2\mathbb{N}_0$, is a local Riemannian invariant $I_g(x)$ such that

$$I_{e^{2\Upsilon}g}(x) = e^{-w\Upsilon(x)}I_g(x) \qquad \forall \Upsilon \in C^\infty(M,\mathbb{R}).$$

The most important conformally invariant tensor is the *Weyl tensor*,

$$W_{ijkl} = R_{ijkl} - (P_{jk}g_{il} + P_{il}g_{jk} - P_{jl}g_{ik} - P_{ik}g_{jl}), \tag{4.1}$$

where $P_{jk} = \frac{1}{n-2}(\mathrm{Ric}_{jk} - \frac{\kappa_g}{2(n-1)}g_{jk})$ is the Schouten-Weyl tensor. In particular, in dimension $n \geq 4$ the Weyl tensor vanishes if and only if (M, g) is locally conformally flat. Moreover, as the Weyl tensor is conformally invariant of weight 2, we get scalar conformal invariants by taking complete metric contractions of tensor products of the Weyl tensor.

Let $I_g(x)$ be a local Riemannian invariant of weight w. By using the ambient metric, Fefferman-Graham [**FG1**, **FG3**] produced a recipe for constructing local conformal invariant from $I_g(x)$ as follows.

Step 1: Thanks to Theorem 1.4 we know that $I_g(x)$ is a linear combination of complete metric contractions of covariant derivatives of the curvature tensor. Substituting into this complete metric contractions the ambient metric $\tilde{g}$ and the covariant derivatives of its curvature tensor we obtain a local Riemannian invariant $I_{\tilde{g}}(t, x, \rho)$ on the ambient metric space $(\tilde{G}, \tilde{g})$. For instance,

$$\mathrm{Contr}_g(\nabla^{k_1} R \otimes \cdots \otimes \nabla^{k_l} R) \longrightarrow \mathrm{Contr}_{\tilde{g}}(\tilde{\nabla}^{k_1} \tilde{R} \otimes \cdots \otimes \tilde{\nabla}^{k_l} \tilde{R})$$

where $\tilde{\nabla}$ is the ambient Levi-Civita connection and $\tilde{R}$ is the ambient curvature tensor.

Step 2: We define a function on M by

$$\tilde{I}_g(x) := t^{-w} I_{\tilde{g}}(t, x, \rho)\big|_{\rho=0} \qquad \forall x \in M. \tag{4.2}$$

If n is odd this is always well defined. If n is even this is well defined provided only derivatives of $\tilde{g}$ of not too high order are involved; this is the case if $w \leq n$. Note also that thanks to the homogeneity of the ambient metric the r.h.s. of (4.2) is always independent of the variable t.

PROPOSITION 4.2 ([**FG1**, **FG3**]). *The function $\tilde{I}_g(x)$ is a local conformal invariant of weight w.*

REMARK 4.3. The fact that $\tilde{I}_g(x)$ transforms conformally under conformal change of metrics is a consequence of the homogeneity of the ambient metric in (3.4).

We shall refer to the rule $I_g(x) \rightarrow \tilde{I}_g(x)$ as *Fefferman-Graham's rule.* When $I_g(x)$ is a Weyl Riemannian invariant we shall call $\tilde{I}_g(x)$ a *Weyl conformal invariant.*

EXAMPLE 4.4. Let us give a few examples of using Fefferman-Graham's rule (when n is even we further assume that the invariants $I_g(x)$ below have weight $\leq n$).

(a) If $I_g(x) = \mathrm{Contr}_g(\nabla^{k_1} \mathrm{Ric} \otimes \nabla^{k_2} R \otimes \cdots \otimes \nabla^{k_l} R)$, then the Ricci flatness of $\hat{g}$ implies that

$$\tilde{I}_g(x) = 0.$$

(b) (See [**FG1**, **FG3**].) If $I_g(x) = \mathrm{Contr}_g(R \otimes \cdots \otimes R)$ (i.e., no covariant derivatives are involved), then $\tilde{I}_g(x)$ is obtained by substituting the Weyl tensor for the curvature tensor, that is,

$$\tilde{I}_g(x) = \mathrm{Contr}_g(W \otimes \cdots \otimes W).$$

(c) (See [**FG1**, **FG3**].) For $I_g(x) = |\nabla R|^2$, we have

$$\tilde{I}_g(x) = \Phi_g(x) := |V|^2 + 16\langle W, U\rangle + 16|C|^2,$$

where $C_{jkl} = \nabla_l P_{jk} - \nabla_k P_{jl}$ is the Cotton tensor and the tensors U and V are defined by

$$V_{mijkl} := \nabla_s W_{ijkl} - g_{im} C_{jkl} + g_{jm} C_{ikl} - g_{km} C_{lij} + g_{lm} C_{kij},$$
$$U_{mjkl} := \nabla_m C_{jkl} + g^{rs} P_{mr} W_{sjkl}.$$

THEOREM 4.5 (Bailey-Eastwood-Graham [**BEG**], Fefferman-Graham [**FG3**]). *Let $w \in 2\mathbb{N}_0$, and further assume $w \leq n$ when n is even. Then every local conformal invariant of weight w is a linear combination of Weyl conformal invariants of same weight.*

REMARK 4.6. As in the Riemannian case, the strategy for the proof of Theorem 4.5 has two main steps. The first step is a reduction of the geometric problem to the invariant theory for a parabolic subgroup $P \subset \mathrm{O}(n+1,1)$. This reduction is a consequence of the jet isomorphism theorem in conformal geometry, which is proved in [**FG3**]. The relevant invariant theory for the parabolic subgroup P is developped in [**BEG**].

REMARK 4.7. We refer to [**GrH**] for a description of the local conformal invariants in even dimension beyond the critical weight $w = n$.

5. Scattering Theory and Conformal Fractional Powers of the Laplacian

The ambient metric construction realizes a conformal structure as a hypersuface in the ambient space. An equivalent approach is to realize a conformal structure as the conformal infinity of an asymptotically hyperbolic Einstein (AHE) manifold. This enables us to use scattering theory to construct conformal fractional powers of the Laplacian.

Let (M^n, g) be a compact Riemannian manifold. We assume that M is the boundary of some manifold $\overline{X}$ with interior X. Note we always can realize the double $M \sqcup M$ as the boundary of $[0,1] \times M$. We assume that X carries a AHE metric g^+. This means that

(1) There is a definining function ρ for $M = \partial X$ such that, near M, the metric g^+ is of the form,

$$g^+ = \rho^{-2} g(\rho, x) + \rho^{-2}(d\rho)^2,$$

where $g(\rho, x)$ is smooth up ∂X and agrees with $g(x)$ at $\rho = 0$.

(2) The metric g^+ is Einstein, i.e.,

$$\mathrm{Ric}(g^+) = -ng^+.$$

In particular (X, g^+) has conformal boundary $(M, [g])$, where $[g]$ is the conformal class of g.

The scattering matrix for an AHE metric g^+ is constructed as follows. We denote by Δ_{g^+} the Laplacian on X for the metric g^+. The continuous spectrum of Δ_{g^+} agrees with $[\frac{1}{4}n^2, \infty)$ and its pure point spectrum $\sigma_{\mathrm{pp}}(\Delta_{g^+})$ is a finite subset of $\left(0, \frac{1}{4}n^2\right)$ (see [**Ma1**, **Ma2**]). Define

$$\Sigma = \left\{ s \in \mathbb{C};\ \Re s \geq \frac{n}{2},\ s \notin \frac{n}{2} + \mathbb{N}_0,\ s(n-s) \notin \sigma_{\mathrm{pp}}(\Delta_{g^+}) \right\}.$$

For $s \in \Sigma$, consider the eigenvalue problem,

$$\Delta_{g^+} - s(n-s) = 0.$$

For any $f \in C^\infty(M)$, the above eigenvalue problem as a unique solution of the form,

$$u = F\rho^{n-s} + G\rho^s, \quad F, G \in C^\infty(\overline{X}), \quad F_{|\partial X} = f.$$

The scattering matrix is the operator $S_g(s) : C^\infty(M) \to C^\infty(M)$ given by

$$S_g(s)f := G_{|M} \qquad \forall f \in C^\infty(M).$$

Thus $S_g(s)$ can be seen as a generalized Dirichlet-to-Neuman operator, where $F_{|\partial X}$ represents the "Dirichlet data" and $G_{|\partial X}$ represents the "Neuman data".

THEOREM 5.1 (Graham-Zworski [**GZ**]). *For $z \in -\frac{n}{2} + \Sigma$ define*

$$P_{z,g} := 2^{2z}\frac{\Gamma(z)}{\Gamma(-z)}S_g\left(\frac{1}{2}n + z\right).$$

The family $(P_{z,g})$ uniquely extends to a holomorphic family of pseudodifferential operators $(P_{z,g})_{\Re z>0}$ in such way that

(i) *The operator $P_{z,g}$ is a ΨDO of order $2z$ with same principal symbol as Δ_g^z.*

(ii) *The operator $P_{z,g}$ is conformally invariant in the sense that*

$$P_{z,e^{2\Upsilon}g} = e^{-(\frac{n}{2}+z)\Upsilon}(P_{z,g})e^{(\frac{n}{2}-z)\Upsilon} \qquad \forall \Upsilon \in C^\infty(M,\mathbb{R}). \tag{5.1}$$

(iii) *The family $(P_{z,g})_{z\in\Sigma}$ is a holomorphic family of ΨDOs and has a meromorphic extension to the half-space $\Re z > 0$ with at worst finite rank simple pole singularities.*

(iv) *Let $k \in \mathbb{N}$ be such that $-k^2 + \frac{1}{4}n^2 \notin \sigma_{\text{pp}}(\Delta_{g^+})$. Then*

$$\lim_{z\to k} P_{z,g} = P_{k,g},$$

where $P_{k,g}$ is the k-th GJMS operator defined by (3.5).

REMARK 5.2. Results of Joshi-Sá Barreto [**JS**] show that each operator $P_{z,g}$ is a Riemannian invariant ΨDO in the sense that all the homogeneous components are given by universal expressions in terms of the partial derivatives of the components of the metric g (see [**Po2**] for the precise definition).

REMARK 5.3. If $k \in \mathbb{N}$ is such that $\lambda_k := -k^2 + \frac{1}{4}n^2 \in \sigma_{\text{pp}}(\Delta_{g^+})$, then (5.1) holds modulo a finite rank smoothing operator obtained as the restriction to M of the orthogonal projection onto the eigenspace $\ker\left(\Delta_{g^+} - \lambda_k\right)$.

REMARK 5.4. The analysis of the scattering matrix $S_g(s)$ by Graham-Zworski [**GZ**] relies on the analysis of the resolvent $\Delta_{g^+} - s(n-s)$ by Mazzeo-Melrose [**MM**]. Guillarmou [**Gu1**] established the meromorphic continuation of the resolvent to the whole complex plane $\mathbb{C}$. There are only finite rank poles when the AHE metric is even. This gives the meromorphic continuation of the scattering matrix $S_g(z)$ and the operators $P_{z,g}$ to the whole complex plane. We refer to [**Va1, Va2**] for alternative approaches to these questions.

REMARK 5.5. We refer to [**CM**] for an interpretation of the operators $P_{z,g}$ as fractional Laplacians in the sense of Caffarelli-Sylvestre [**CS**].

EXAMPLE 5.6 (Branson [**Br**]). Consider the round sphere $\mathbb{S}^n$ seen as the boundary of the unit ball $\mathbb{B}^{n+1} \subset \mathbb{R}^{n+1}$ equipped with its standard hyperbolic metric. Then

$$P_{z,g} = \frac{\Gamma\left(\sqrt{\Delta_g + \frac{1}{4}(n-1)^2} + 1 + z\right)}{\Gamma\left(\sqrt{\Delta_g + \frac{1}{4}(n-1)^2} + 1 - z\right)}, \qquad z \in \mathbb{C}.$$

We refer to [**GN**] for an extension of this formula to manifolds with constant sectional curvature.

A general conformal structure cannot always be realized as the conformal boundary of an AHE manifold. However, as showed by Fefferman-Graham [**FG1**, **FG3**] it always can be realized as a conformal boundary formally.

Let (M^n, g) be a Riemannian manifold. Define $X = M \times (0, \infty)$ and let $\overline{X} = M \times [0, \infty)$ be the closure of X. We shall denote by r the variable in $[0, \infty)$ and we identify M with the boundary $r = 0$.

THEOREM 5.7 (Fefferman-Graham [**FG1**, **FG3**]). *Assume M has odd dimension. Then, near the boundary $r = 0$, there is a Riemannian metric g^+, called Poincaré-Einstein metric, which is defined up to infinite order in r and satisfies the following properties:*

(i) *In local coordinates $\{r, x^j\}$,*
$$g^+ = r^{-2}(dr)^2 + r^{-2} g_{ij}(x, \rho),$$
where $g_{ij}(x, r)$ is a family of symmetric $(0, 2)$-tensors such that $g(x, r)|_{r=0} = g_{ij}(x)$.

(ii) *It is asymptotically Einstein in the sense that, near the boundary $r = 0$,*

$$\mathrm{Ric}(g^+) = -ng^+ + \mathrm{O}(r^\infty). \tag{5.2}$$

REMARK 5.8. When the dimension n is even there is also a Poincaré-Einstein metric which is defined up to order $n-2$ in r and satisfies the Einstein equation (5.2) up to an $\mathrm{O}(r^{n-2})$ error.

REMARK 5.9. The constructions of the Poincaré-Einstein metric and ambient metric are equivalent. We can obtain either metric from the other. For instance, the map $\phi : (r, x) \to \left(r^{-1}, x, -\frac{1}{2}r^2\right)$ is a smooth embedding of X into the hypersurface,

$$\mathcal{H}^+ = \{\tilde{g}(T, T) = -1\} = \left\{(t, x\rho);\ 2\rho t^2 = -1\right\} \subset \tilde{G},$$

where $T = t\partial_t$ is the infinitesimal generator for the dilations (3.1). Then g^+ is obtained as the pullback to X of $\tilde{g}_{|T\mathcal{H}}$. In particular, the metric g^+ is even in the sense that the tensor $g(x, r)$ has a Taylor expansion near $r = 0$ involving *even* powers of r only.

EXAMPLE 5.10. Assume that (M, g) is Einstein, with $\mathrm{Ric}(g) = 2\lambda(n-1)g$. Then

$$g^+ = r^{-2}(dr)^2 + r^{-2}\left(1 - \frac{1}{2}\lambda r^2\right) g_{ij}(x) dx^i dx^j.$$

PROPOSITION 5.11. *Assume n is odd. Then there is an entire family $(P_{g,z})_{z \in \mathbb{C}}$ of ΨDOs on M uniquely defined up to smoothing operators such that*

(i) *Each operator $P_{g,z}$, $z \in \mathbb{C}$, is a ΨDO of order $2z$ and has same principal symbol as Δ_g^z.*

(ii) *Each operator $P_{g,z}$ is a conformally invariant ΨDO in the sense of* [**Po2**] *in such a way that, for all $\Upsilon \in C^\infty(M, \mathbb{R})$,*

$$P_{z,e^{2\Upsilon}g} = e^{-(\frac{n}{2}+z)\Upsilon}(P_{z,g})e^{(\frac{n}{2}-z)\Upsilon} \qquad \text{mod } \Psi^{-\infty}(M), \tag{5.3}$$

where $\Psi^{-\infty}(M)$ is the space of smoothing operators on M.

(iii) *For all $k \in \mathbb{N}$, the operator $P_{z,g}$ agrees at $z = k$ with the k-th GJMS operator $P_{k,g}$.*

(iv) *For all $z \in \mathbb{C}$,*

$$P_{z,g}P_{-z,g} = 1 \qquad \text{mod } \Psi^{-\infty}(M). \tag{5.4}$$

REMARK 5.12. The equation (5.4) is an immediate consequence of the functional equation for the scattering matrix,

$$S_g(s)S_g(n-s) = 1.$$

This enables us to extend the family $(P_{z,g})$ beyond the halfspace $\Re z > 0$.

REMARK 5.13. When n is even, we also can construct operators $P_{z,g}$ as above except that each operator $P_{z,g}$ is unique and satisfies (5.3) modulo ΨDOs of order $z - 2n$.

REMARK 5.14. By using a polynomial continuation of the homogeneous components of the symbols of the GJMS operators, Petterson [**Pe**] also constructed a holomorphic family of ΨDOs satisfying the properties (i)–(iv) above.

EXAMPLE 5.15 (Joshi-Sá Barreto [**JS**]). Suppose that (M, g) is Ricci flat and compact. Then, for all $z \in \mathbb{C}$,

$$P_{g,z} = \Delta_g^z \qquad \text{mod } \Psi^{-\infty}(M).$$

As mentioned above, the constructions of the ambient metric and Poincaré-Einstein metric are equivalent. Thus, it should be possible to define the GJMS operators directly in terms of the Poincaré-Einstein metric, so as to interpret the GJMS operators as boundary operators. This interpretation is actually implicit in [**GZ**]. Indeed, if we combine the definition (3.5) of the GJMS operators with the formula on the 2nd line from the top on page 113 of [**GZ**], then we arrive at the following statement.

PROPOSITION 5.16 ([**GJMS**], [**GZ**]). *Let $k \in \mathbb{N}$ and further assume $k \leq \frac{n}{2}$ when n is even. Then, for all $u \in C^\infty(M)$,*

$$P_{k,g}u = r^{-\frac{1}{2}n-k}\prod_{j=0}^{k-1}\left(\Delta_{g^+} + \left(-\frac{1}{2}n + k - 2j\right)\left(\frac{1}{2}n + k - 2j\right)\right)(r^{\frac{1}{2}n-k}u^+)\Big|_{r=0},$$

where u^+ is any function in $C^\infty(\overline{X})$ which agrees with u on the boundary $r = 0$.

REMARK 5.17. Guillarmou [**Gu2**] obtained this formula via scattering theory. We also refer to [**GW**] for a similar formula in terms of the tractor bundle. In addition, using a similar approach as above, Hirachi [**Hi2**] derived an analogous formula for the CR GJMS operators of Gover-Graham [**GG**].

6. Green Functions of Elliptic Operators

In this section, we describe the singularities of the Green function of an elliptic operator and its closed relationship with the heat kernel asymptotics.

Let (M^n, g) be a Riemannian manifold. Let $P : C_c^\infty(M) \to C^\infty(M)$ be an elliptic pseudodifferential operator (ΨDO) of (integer) order $m > 0$. Its *Green function* $G_P(x,y)$ (when it exists) is the fundamental solution of P, that is,

$$P_x G_P(x,y) = \delta(x-y). \tag{6.1}$$

Equivalently, $G_P(x,y)$ is the inverse kernel of P, i.e.,

$$P_x\left(\int_M G_P(x,y)u(y)v_g(y)\right) = u(x).$$

In general, P may have a nontrivial kernel (e.g., the kernel of the Laplace operator Δ_g consists of constant functions). Therefore, by Green function we shall actually mean a solution of (6.1) modulo a C^∞-error, i.e.,

$$P_x G_P(x,y) = \delta(x-y) \qquad \mod C^\infty(M\times M).$$

This means that using $G_P(x,y)$ we always can solve the equation $Pv = u$ modulo a smooth error. In other words, $G_P(x,y)$ is the kernel function of parametrix for P. As such it always exists.

The Green function $G_P(x,y)$ is smooth off the diagonal $y = x$. As this is the kernel function of a pseudodifferential operator of order $-m$ (namely, a parametrix for P), the theory of pseudodifferential operators enables us to describe the form of its singularity near the diagonal.

PROPOSITION 6.1 (See [**Ta**]). *Suppose that P has order $m \leq n$. Then, in local coordinates and near the diagonal $y = x$,*

$$G_P(x,y) = |x-y|^{-n+m}\sum_{0\leq j<n-m} a_j(x,\theta)|x-y|^j - \gamma_P(x)\log|x-y| + \mathrm{O}(1), \tag{6.2}$$

where $\theta = |x-y|^{-1}(x-y)$ and the functions $a_j(x,\theta)$ and $\gamma_P(x)$ are smooth. Moreover,

$$\gamma_P(x) = (2\pi)^{-n}\int_{S^{n-1}} p_{-n}(x,\xi)d^{n-1}\xi, \tag{6.3}$$

where $p_{-n}(x,\xi)$ is the symbol of degree $-n$ of any parametrix for P.

REMARK 6.2. The various terms in (6.2) depends on the various components of the symbol of a parametrix for P, and so, at the exception of the leading term $a_0(x,\theta)$, they transform in some cumbersome way under changes of local coordinates. However, it can be shown, that the coefficient $\gamma_P(x)$ makes sense intrinsically on M.

REMARK 6.3. Suppose that P is a differential operator. Then any parametrix for P is odd class in the sense of [**KV**]. This implies that its symbols are homogeneous with respect to the symmetry $\xi \to -\xi$, e.g., $p_{-n}(x,-\xi) = (-1)^n p_{-n}(x,\xi)$. In particular, when the dimension n is odd, we get

$$\int_{S^{n-1}} p_{-n}(x,\xi)d^{n-1}\xi = \int_{S^{n-1}} p_{-n}(x,-\xi)d^{n-1}\xi = -\int_{S^{n-1}} p_{-n}(x,\xi)d^{n-1}\xi = 0,$$

and hence $\gamma_P(x) = 0$ for all $x \in M$.

The formula (6.3) provides us with a close relationship with the noncommutative residue trace of Guillemin [**Gui1**, **Gui2**] and Wodzicki [**Wo1**, **Wo2**], since the right-hand side of (6.3) gives the noncommutative residue density of a parametrix for P. This noncommutative residue trace is the residual functional induced on integer order ΨDOs by the analytic extension of the ordinary trace to non-integer order ΨDOs. This enables us to obtain the following statement.

PROPOSITION 6.4. *Assume that M is compact. Then the zeta function $\zeta(P;s) = \operatorname{Tr} P^{-s}$, $\Re s > \frac{n}{m}$, has a meromorphic extension to the whole complex plane $\mathbb{C}$ with at worst simple pole singularities. Moreover, at $s=1$, we have*

$$m \operatorname{Res}_{s=1} \operatorname{Tr} P^{-s} = \int_M \gamma_P(x) v_g(x). \tag{6.4}$$

REMARK 6.5. We refer to [**Se**] for the construction of the complex powers P^s, $s \in \mathbb{C}$. The construction depends on the choice of a spectral cutting for both P and its principal symbol, but the residues at integer points of the zeta function $\zeta(P;s) = \operatorname{Tr} P^{-s}$ do not depend on this choice.

REMARK 6.6. Proposition 6.4 has a local version (see [**Gui2**, **KV**, **Wo2**]). If we denote by $K_{P^{-s}}(x,y)$ the kernel of P^{-s} for $\Re s > \frac{n}{m}$, then the map $s \to K_{P^{-s}}(x,x)$ has a meromorphic extension to $\mathbb{C}$ with at worst simple pole singularities in such a way that

$$m \operatorname{Res}_{s=1} K_{P^{-s}}(x,x) = \gamma_P(x). \tag{6.5}$$

We also note that the above equality continues to hold if we replace P^{-s} by any holomorphic ΨDO family $P(s)$ such that $\operatorname{ord} P(s) = -ms$ and $P(1)$ is a parametrix for P. This way Eq. (6.5) continues to hold when M is not compact.

Suppose now that M is compact and P has a positive leading symbol, so that we can form the heat semigroup e^{-tP} and the heat kernel $K_P(x,y;t)$ associated to P. The heat kernel asymptotics for P then takes the form,

$$K_P(x,x;t) \sim (4\pi t)^{-\frac{n}{m}} \sum_{j\geq 0} t^{\frac{j}{m}} a_j(P;x) + \log t \sum_{k\geq 1} t^k b_j(P;x) \qquad \text{as } t \to 0^+. \tag{6.6}$$

We refer to [**Wi**, **GS**] for a derivation of the above heat kernel asymptotics for ΨDOs. The residues of the local zeta function $\zeta(P;s;x)(x) = K_{P^{-s}}(x,x)$ are related to the coefficient in the heat kernel asymptotic for P as follows.

By Mellin's formula, for $\Re s > 0$ we have

$$\Gamma(s) P^{-s} = \int_0^\infty t^{s-1}(1-\Pi_0)e^{-tP}dt, \qquad \Re s > 0,$$

where Π_0 is the orthogonal projection onto the nullspace of P. Together with the heat kernel asymptotics (6.6) this implies that, for $\Re s > \frac{n}{m}$,

$$\begin{aligned}
\Gamma(s) K_{P^{-s}}(x,x) &= \int_0^1 t^{s-1} K_P(x,x;t)dt + h(x;s) \\
&= (4\pi)^{-\frac{n}{m}} \sum_{0\leq j<n} \int_0^1 t^{s+\frac{j-n}{m}} a_j(P;x) t^{-1} dt + h(x;s) \\
&= (4\pi)^{-\frac{n}{m}} \sum_{0\leq j<n} \frac{m}{ms+j-n} a_j(P;x) + h(x;s) \operatorname{Hol}(\Re s > 0),
\end{aligned}$$

where $h(x;s)$ is a general notation for a holomorphic function on the halfspace $\Re s > 0$. Thus, for $j = 0, 1, \ldots, n-1$, we have

$$\Gamma\left(\frac{n-j}{m}\right) \operatorname{Res}_{s=\frac{n-j}{m}} K_{P^{-s}}(x,x) = (4\pi)^{-\frac{n}{m}} a_j(P;x).$$

Combining this with (6.5) we arrive at the following result.

PROPOSITION 6.7 (Compare [**PR**]). *Let $k \in m^{-1}\mathbb{N}$ be such that $mk - n \in \mathbb{N}_0$. Then, under the above assumptions, we have*

$$\gamma_{P^k}(x) = (4\pi)^{-\frac{n}{m}} \frac{m}{\Gamma(k)} a_{n-mk}(P;x). \tag{6.7}$$

The above result provides us with a way to compute the logarithmic singularities of the Green functions of the powers of P from the knowledge of coefficients of the heat kernel asymptotics. For instance, in the case of the Laplace operator combining Theorem 1.8 and Proposition 6.7 gives the following result.

PROPOSITION 6.8. *For $k = \frac{n}{2}, \frac{n}{2}-1, \frac{n}{2}-2$ and provided that $k > 0$, we have*

$$\gamma_{\Delta_g^{\frac{n}{2}}}(x) = (4\pi)^{-\frac{n}{2}} \frac{2}{\Gamma\left(\frac{n}{2}\right)}, \qquad \gamma_{\Delta_g^{\frac{n}{2}-1}}(x) = (4\pi)^{-\frac{n}{2}} \frac{2}{\Gamma\left(\frac{n}{2}-1\right)} \cdot \frac{-1}{6}\kappa,$$

$$\gamma_{\Delta_g^{\frac{n}{2}-2}}(x) = (4\pi)^{-\frac{n}{2}} \frac{2}{\Gamma\left(\frac{n}{2}-2\right)} \left(\frac{1}{180}|R|^2 - \frac{1}{180}|\operatorname{Ric}|^2 + \frac{1}{72}\kappa^2 - \frac{1}{30}\Delta_g\kappa\right).$$

7. Green Functions and Conformal Geometry

Green functions of the Yamabe operator and other conformal powers of the Laplacian play an important role in conformal geometry. This is illustrated by the solution to the Yamabe problem by Schoen [**Sc**] or by the work of Okikoliu [**Ok**] on variation formulas for the zeta-regularized determinant of the Yamabe operator in odd dimension. Furthermore, Parker-Rosenberg [**PR**] computed the logarithmic singularity of the Green function of the Yamabe operator in even low dimension.

THEOREM 7.1 (Parker-Rosenberg [**PR**, Proposition 4.2]). *Let (M^n, g) be a closed Riemannian manifold. Then*

(1) *In dimension $n = 2$ and $n = 4$ we have*

$$\gamma_{P_{1,g}}(x) = 2(4\pi)^{-1} \ (n=2) \qquad \text{and} \qquad \gamma_{P_1}(x) = 0 \ (n=4).$$

(2) *In dimension $n = 6$,*

$$\gamma_{P_{1,g}}(x) = (4\pi)^{-3} \frac{1}{90}|W|^2,$$

where $|W|^2 = W^{ijkl}W_{ijkl}$ is the norm-square of the Weyl tensor.

(3) *In dimension $n = 8$,*

$$\gamma_{P_{1,g}}(x) = (4\pi)^{-4} \frac{2}{9\cdot 7!}\left(81\Phi_g + 64 W_{ij}{}^{kl}W^{ij}{}_{pq}W^{pq}{}_{kl} + 352 W_{ijkl}W^{i}{}_{p}{}^{k}{}_{q}W^{pjql}\right),$$

where Φ_g is given by (4.4).

REMARK 7.2. Parker and Rosenberg actually computed the coefficient $a_{\frac{n}{2}-1}(P_{1,g};x)$ of t^{-1} in the heat kernel asymptotics for the Yamabe operator when M is closed. We obtain the above formulas for $\gamma_{P_{1,g}}(x)$ by using (6.7). We also note that Parker and Rosenberg used a different sign convention for the curvature tensor.

The computation of Parker and Rosenberg in [**PR**] has two main steps. The first step uses the formulas for Gilkey [**Gi2, Gi3**] for the heat invariants of Laplace type operator, since the Yamabe operator is such an operator. This expresses the coefficient $a_{\frac{n}{2}-1}(P_{1,g};x)$ as a linear combination of Weyl Riemannian invariants. For $n=8$ there are 17 such invariants. The 2nd step consists in rewriting these linear combinations in terms of Weyl conformal invariant, so as to obtain the much simpler formulas above.

It is not clear how to extend Parker-Rosenberg's approach for computing the logarithmic singularities $\gamma_{P_{k,g}}(x)$ of the Green functions of other conformal powers of the Laplacian. This would involve computing the coefficients of the heat kernel asymptotics for *all* these operators, including conformal fractional powers of the Laplacian in odd dimension.

As the following result shows, somewhat amazingly, in order to compute the $\gamma_{P_{k,g}}(x)$ the *sole* knowledge of the coefficients heat kernel asymptotics for the Laplace operator is enough.

THEOREM 7.3. *Let (M^n,g) be a Riemannian manifold and let $k\in\frac{1}{2}\mathbb{N}$ be such that $\frac{n}{2}-k\in\mathbb{N}_0$. Then the logarithmic singularity of the Green function of $P_{k,g}$ is given by*

$$\gamma_{P_{k,g}}(x)=\frac{2}{\Gamma(k)}(4\pi)^{-\frac{n}{2}}\tilde{a}_{n-2k}(\Delta_g;x),$$

where $\tilde{a}_{n-2k}(\Delta_g;x)$ is the local conformal invariant obtained by applying Fefferman-Graham's rule to the heat invariant $a_{n-2k}(\Delta_g;x)$ in (1.3).

REMARK 7.4. When n is odd the condition $\frac{n}{2}-k\in\mathbb{N}_0$ imposes k to be an half-integer, and so in this case $P_{k,g}$ is a conformal fractional powers of the Laplacian.

Using Theorem 7.3 and the knowledge of the heat invariants $a_{2j}(\Delta_g;x)$ it becomes straighforward to compute the logarithmic singularities $\gamma_{P_{k,g}}(x)$. In particular, we recover the formulas of Parker-Rosenberg [**PR**] stated in Theorem 7.1.

THEOREM 7.5 ([**Po1, Po2**]). *Let (M^n,g) be a Riemannian manifold of dimension $n\geq 3$. Then*

$$\gamma_{P_{\frac{n}{2},g}}(x)=\frac{2}{\Gamma\left(\frac{n}{2}\right)}(4\pi)^{-\frac{n}{2}}\qquad \text{and}\qquad \gamma_{P_{\frac{n}{2}-1,g}}(x)=0.$$

PROOF. We know from Theorem 1.8 that $a_0(\Delta_g;x)=1$ and $a_2(\Delta_g;x)=-\frac{1}{6}\kappa$. Thus $\tilde{a}_0(\Delta_g;x)=1$. Moreover, as κ is the trace of Ricci tensor, from Example 4.4.(a) we see that $\tilde{a}_2(\Delta_g;x)=0$. Combining this with Theorem 7.3 gives the result. □

REMARK 7.6. The equality $\gamma_{P_{\frac{n}{2}}}(x)=2\Gamma\left(\frac{n}{2}\right)^{-1}(4\pi)^{-\frac{n}{2}}$ can also be obtained by direct computation using (6.3) and the fact that in this case $p_{-n}(x,\xi)$ is the principal symbol of a parametrix for $P_{\frac{n}{2},g}$. As $P_{\frac{n}{2},g}$ has same principal symbol as $\Delta_g^{\frac{n}{2}}$, the principal symbol of any parametrix for $P_{\frac{n}{2},g}$ is equal to $|\xi|_g^{-n}$, where $|\xi|_g^2=g^{ij}(x)\xi_i\xi_j$.

REMARK 7.7. The equality $\gamma_{P_{\frac{n}{2}-1,g}}(x)=0$ is obtained in [**Po1, Po2**] by showing this is a conformal invariant weight 1 and using the fact there is no nonzero conformal invariant of weight 1. Indeed, by Theorem 1.4 any Riemannian invariant of weight 1 is a scalar multiple of the scalar curvature, but the scalar curvature is not a conformal invariant. Therefore, any local conformal of weight 1 must be zero.

THEOREM 7.8. *Let (M^n, g) be a Riemannian manifold of dimension $n \geq 5$. Then*

$$\gamma_{P_{\frac{n}{2}-2}}(x) = \frac{(4\pi)^{-\frac{n}{2}}}{\Gamma\left(\frac{n}{2}-2\right)} \cdot \frac{1}{90}|W|^2. \tag{7.1}$$

PROOF. By Theorem 1.8 we have

$$a_4(\Delta_g; x) = \frac{1}{180}|R|^2 - \frac{1}{180}|\operatorname{Ric}|^2 + \frac{1}{72}\kappa^2 - \frac{1}{30}\Delta_g \kappa.$$

As Example 4.4.(b) shows, the Weyl conformal invariant associated to $|R|^2$ by Fefferman-Graham's rule is $|W|^2$. Moreover, as $|\operatorname{Ric}|^2$, κ^2, and $\Delta_g \kappa$ involve the Ricci tensor, the corresponding Weyl conformal invariants are equal to 0. Thus,

$$\tilde{a}_4(\Delta_g; x) = \frac{1}{180}|W|^2.$$

Applying Theorem 7.3 then completes the proof. □

THEOREM 7.9. *Let (M^n, g) be a Riemannian manifold of dimension $n \geq 7$. Then*

$$\gamma_{P_{\frac{n}{2}-3}}(x) = \frac{(4\pi)^{-\frac{n}{2}}}{\Gamma\left(\frac{n}{2}-3\right)} \cdot \frac{2}{9 \cdot 7!}\left(81\Phi_g + 64 W_{ij}{}^{kl} W^{ij}{}_{pq} W^{pq}{}_{kl} + 352 W_{ijkl} W^{i}{}_{p}{}^{k}{}_{q} W^{pjql}\right),$$

where $\Phi_g(x)$ is given by (4.4).

PROOF. Thanks to Theorem 1.8 we know that

$$a_6(\Delta_g; x) = \frac{1}{9 \cdot 7!}\left(81|\nabla R|^2 + 64 R_{ij}{}^{kl} R^{ij}{}_{pq} R^{pq}{}_{kl} + 352 R_{ijkl} R^{i}{}_{p}{}^{k}{}_{q} R^{pjql}\right) + I^{(3)}_{g,0}(x),$$

where $I^{(3)}_{g,0}(x)$ is a linear combination of Weyl Riemannian invariants involving the Ricci tensor. Thus $\tilde{a}_6(\Delta_g; x) = 0$ by Example 4.4.(a). By Example 4.4.(b), the Weyl conformal invariants corresponding to $R_{ij}{}^{kl} R_{kl}{}^{pq} R_{pq}{}^{ij}$ and $R^{i}{}_{jk}{}^{q} R_{i}{}^{pk}{}_{l} R_{j}{}^{pl}{}_{q}$ are $W_{ij}{}^{kl} W_{kl}{}^{pq} W_{pq}{}^{ij}$ and $W^{i}{}_{jk}{}^{q} W_{i}{}^{pk}{}_{l} W_{j}{}^{pl}{}_{q}$ respectively. Moreover, by Example 4.4.(c) applying Fefferman-Graham's rule to $|\nabla R|^2$ yields the conformal invariant Φ_g given by (4.4). Thus,

$$\tilde{a}_6(\Delta_g; x) = \frac{1}{9 \cdot 7!}\left(81\Phi_g + 64 W_{ij}{}^{kl} W^{ij}{}_{pq} W^{pq}{}_{kl} + 352 W_{ijkl} W^{i}{}_{p}{}^{k}{}_{q} W^{pjql}\right).$$

Combining this with Theorem 7.3 proves the result. □

REMARK 7.10. In the case of Yamabe operator (i.e., $k = 1$) the above results give back the results of Parker-Rosenberg [**PR**] as stated in Theorem 7.1.

Let us present some applications of the above results. Recall that a Riemannian manifold (M^n, g) is said to be locally conformally flat when it is locally conformally equivalent to the flat Euclidean space $\mathbb{R}^n$. As is well known, in dimension $n \geq 4$, local conformal flatness is equivalent to the vanishing of the Weyl tensor. Therefore, as an immediate consequence of Theorem 7.8 we obtain the following result.

THEOREM 7.11. *Let (M^n, g) be a Riemannian manifold of dimension $n \geq 5$. Then the following are equivalent:*

(1) *(M^n, g) is locally conformally flat.*
(2) *The logarithmic singularity $\gamma_{P_{\frac{n}{2}-2}, g}(x)$ vanishes identically on M.*

REMARK 7.12. A well known conjecture of Radamanov [**Ra**] asserts that, for a strictly pseudoconvex domain $\Omega \subset \mathbb{C}^n$, the vanishing of the logarithmic singularity of the Bergman kernel is equivalent to Ω being biholomorphically equivalent to the unit ball $\mathbb{B}^{2n} \subset \mathbb{C}^n$. Therefore, we may see Theorem 7.11 as an analogue of Radamanov conjecture in conformal geometry in terms of Green functions of conformal powers of the Laplacian.

In the compact case, we actually obtain a spectral theoretic characterization of the conformal class of the round sphere as follows.

THEOREM 7.13. *Let (M^n, g) be a compact simply connected Riemannian manifold of dimension $n \geq 5$. Then the following are equivalent:*

(1) *(M^n, g) is conformally equivalent to the round sphere $\mathbb{S}^n$.*
(2) $\int_M \gamma_{P_{\frac{n}{2}-2},g}(x) v_g(x) = 0$.
(3) $\operatorname{Res}_{s=1} \operatorname{Tr}\left[(P_{\frac{n}{2}-2})^{-s}\right] = 0$.

PROOF. The equivalence between (2) and (3) is a consequence of (6.4). By Theorem 7.8, we have

$$\int_M \gamma_{P_{\frac{n}{2}-2},g}(x) v_g(x) = \frac{(4\pi)^{-\frac{n}{2}}}{\Gamma\left(\frac{n}{2}-2\right)} \cdot \frac{1}{90} \int_M |W(x)|^2 v_g(x).$$

As $|W(x)|^2 \geq 0$ we see that $\int_M \gamma_{P_{\frac{n}{2}-2},g}(x) v_g(x) = 0$ if and only if $W(x) = 0$ identically on M. As M is compact and simply connected, a well known result of Kuiper [**Ku**] asserts that the vanishing of Weyl tensor is equivalent to the existence of a conformal diffeomorphism from (M, g) onto the round sphere $\mathbb{S}^n$. Thus (1) and (2) are equivalent. The proof is complete. □

8. Proof of Theorem 7.3

In this section, we outline the proof of Theorem 7.3. The strategy of the proof is divided into 9 main steps.

STEP 1. *Let $k \in \frac{1}{2}\mathbb{N}$ be such that $n - 2k \in \mathbb{N}_0$. Then*

$$\gamma_{\Delta_g^k}(x) = 2\Gamma(k)^{-1}(4\pi)^{-\frac{n}{2}} a_{n-2k}(\Delta_g; x).$$

This follows from Proposition 6.7. This is the Riemannian version of Theorem 7.3 and the main impetus for that result.

STEP 2. *Let $k \in \frac{1}{2}\mathbb{N}$ be such that $n - 2k \in \mathbb{N}_0$. Then $\gamma_{P_{k,g}}(x)$ is a linear combination of Weyl conformal invariants of weight $n - 2k$.*

This step is carried out in [**Po1**, **Po2**]. This is a general result for the logarithmic singularities of the Green functions of conformally invariant ΨDOs. For the Yamabe operator, the transformation of $\gamma_{P_{k,g}}(x)$ under conformal change of metrics was observed by Parker-Rosenberg [**PR**]. Their argument extends to general conformally invariant operators.

STEP 3. *Let $I_g(x)$ be a local Riemannian invariant of any weight $w \in 2\mathbb{N}_0$ if n is odd or of weight $w \leq n$ if n is even. If (M, g) is Ricci-flat, then*

$$I_g(x) = \tilde{I}_g(x) \qquad \text{on } M,$$

where $\tilde{I}_g(x)$ is the local conformal invariant associted to $I_g(x)$ by Fefferman-Graham's rule.

This result seems to be new. It uses the fact that if (M,g) is Ricci flat, then by Example 3.4 the ambient metric is given by

$$\tilde{g} = 2\rho(dt)^2 + t^2 g_{ij}(x) + 2tdtd\rho.$$

STEP 4. *Theorem 7.3 holds when (M^n, g) is Ricci-flat.*

If (M,g) is Ricci-flat, then it follows from Example 3.12 and Example 5.15 that $P_{k,g} = \Delta_g^k$, and hence $\gamma_{P_{k,g}}(x) = \gamma_{\Delta_g^k}(x)$. Combining this with Step 1 and Step 3 then gives

$$\gamma_{P_{k,g}}(x) = \gamma_{\Delta_g^k}(x) = 2\Gamma(k)^{-1}(4\pi)^{-\frac{n}{2}} a_{n-2k}(\Delta_g; x) = 2\Gamma(k)^{-1}(4\pi)^{-\frac{n}{2}} \tilde{a}_{n-2k}(\Delta_g; x).$$

This proves Step 4.

In order to prove Theorem 7.3 for general Riemannian metrics we actually need to establish pointwise versions of Step 4 and Step 3. This is the purpose of the next three steps.

STEP 5. *Let $I_g(x) = \operatorname{Contr}_g\left(\nabla^{k_1} R \otimes \cdots \otimes \nabla^{k_l} R\right)$ be a Weyl Riemannian invariant of weight $< n$ and assume that $\operatorname{Ric}(g) = \mathrm{O}(|x-x_0|^{n-2})$ near a point $x_0 \in M$. Then $\tilde{I}_g(x) = I_g(x)$ at $x = x_0$.*

This step follows the properties of the ambient metric and complete metric contractions of ambient curvature tensors described in [**FG3**].

STEP 6. *Let $k \in \frac{1}{2}\mathbb{N}$ be such that $n - 2k \in 2\mathbb{N}_0$ and assume that $\operatorname{Ric}(g) = \mathrm{O}(|x - x_0|^{n-2})$ near a point $x_0 \in M$. Then $\gamma_{P_{k,g}}(x) = \gamma_{\Delta_g^k}(x)$ at $x = x_0$.*

If P is an elliptic ΨDO, then (6.3) gives an expression in local coordinates for $\gamma_P(x)$ in terms of the symbol of degree $-n$ of a parametrix for P. The construction of the symbol of a parametrix shows that this symbol is a polynomial in terms of inverse of the principal symbol of P and partial derivatives of its other homogeneous symbols. It then follows that if P is Riemannian invariant, then $\gamma_P(x)$ is of the form (1.4), that is, this is a Riemannian invariant (see [**Po1**, **Po2**]).

Moreover, it follows from the ambient metric construction of the GJMS operator [**GJMS**, **FG3**] that, when k is an integer, $P_{k,g}$ and Δ_g^k differs by a differential operator whose coefficients are polynomials in the covariant derivatives of order $< n-2$ of the Ricci tensor. As a result $\gamma_{P_{k,g}}(x)$ and $\gamma_{\Delta_g^k}(x)$ differ by a linear combination of complete metric contraction of tensor powers of covariant derivatives of the Ricci tensor. Thus $\gamma_{P_{k,g}}(x)$ and $\gamma_{\Delta_g^k}(x)$ agree at $x = x_0$ if $\operatorname{Ric}(g)$ vanishes at order $< n-2$ near x_0. This is also true for the conformal fractional powers of the Laplacian thanks to the properties of the scattering matrix for Poincaré-Einstein metric in [**JS**] and [**GZ**].

STEP 7. *Let $I_g(x) = \operatorname{Contr}_g\left(\nabla^{k_1} R \otimes \cdots \otimes \nabla^{k_l} R\right)$ be a Weyl Riemannian invariant of weight $w < n$. Then the following are equivalent:*

(i) *$\tilde{I}_g(x)$ at $x = 0$ for all metrics of signature $(n,0)$ on $\mathbb{R}^n$.*
(ii) *$I_g(x) = 0$ at $x = 0$ for all metrics of signature $(n,0)$ on $\mathbb{R}^n$ such that $\operatorname{Ric}(g) = \mathrm{O}\left(|x|^{n-2}\right)$ near $x = 0$.*
(iii) *$I_g(x) = 0$ at $x = 0$ for all metrics of signature $(n+1,1)$ on $\mathbb{R}^{n+2}$ such that $\operatorname{Ric}(g) = \mathrm{O}\left(|x|^{n-2}\right)$ near $x = 0$.*

The implication (i) $\Rightarrow$ (ii) is an immediate consequence of Step 5. The implication (iii) $\Rightarrow$ (i) follows from the construction of the conformal invariant $\tilde{I}_g(x)$, since the ambient metric is of the form given in (iii) if we use the variable $r = \sqrt{2\rho}$ instead of ρ (cf. [**FG3**, Chapter 4]; see in particular Theorem 4.5 and Proposition 4.7). Therefore, the bulk of Step 7 is proving that (ii) implies (iii).

It can be given sense to a vector space structure on formal linear combinations of complete metric contractions of tensor without any reference to dimension or the signature of the metric. If the weight is less than n, then the formal vanishing is equivalent to the algebraic vanishing by inputing tensors in dimension n. The idea in the proof of the proof of the implication (ii) $\Rightarrow$ (iii) is showing that (ii) and (iii) are equivalent to the same system of linear equations on the space of formal linear combinations of complete metric contractions of Ricci-flat curvature tensors. This involves proving a version of the "2nd main theorem of invariant theory" for Ricci-flat curvature tensors, rather than for collections of trace-free tensors satisfying the Young symmetries of Riemannian curvature tensors. In other words, we need to establish a *nonlinear* version of [**BEG**, Theorem B.3].

STEP 8. *Proof of Theorem* 7.3.

By Step 2 we know that $\gamma_{P_{k,g}}(x)$ is a linear combination of Weyl conformal invariants of weight $2n-k$, so by Theorem 4.5 there is a Riemannian invariant $I_g(x)$ of weight $2n-k$ such that

$$\gamma_{P_{k,g}}(x) = 2\Gamma(k)^{-1}(4\pi)^{-\frac{n}{2}}\tilde{I}_g(x).$$

We need to show that $I_g(x) = a_{n-2k}(\Delta_g;x)$. If (M,g) is Ricci flat, then by Step 3 and Step 4,

$$I_g(x) = \tilde{I}_g(x) = \frac{1}{2}\Gamma(k)(4\pi)^{\frac{n}{2}}\gamma_{P_{k,g}}(x) = \frac{1}{2}\Gamma(k)(4\pi)^{\frac{n}{2}}\gamma_{\Delta_g^k}(x) = a_{n-2k}(\Delta_g;x).$$

By Step 6 this result continues to hold at a point $x_0 \in M$ if $\operatorname{Ric}(g) = \mathrm{O}(|x-x_0|^{n-2})$ near x_0. Therefore, in this case the Riemannian invariant $I_g(x) - a_{n-2k}(\Delta_g;x)$ vanishes at x_0. Using Step 7 we then deduce that the associated conformal invariant vanishes. That is,

$$\gamma_{P_{k,g}}(x) = 2\Gamma(k)^{-1}(4\pi)^{-\frac{n}{2}}\tilde{a}_{n-2k}(\Delta_g;x).$$

This proves Theorem 7.3.

References

[ABP] M. Atiyah, R. Bott, and V. K. Patodi, *On the heat equation and the index theorem*, Invent. Math. **19** (1973), 279–330. MR0650828 (58 #31287)

[BEG] Toby N. Bailey, Michael G. Eastwood, and C. Robin Graham, *Invariant theory for conformal and CR geometry*, Ann. of Math. (2) **139** (1994), no. 3, 491–552, DOI 10.2307/2118571. MR1283869 (95h:53016)

[BGS] Richard Beals, Peter C. Greiner, and Nancy K. Stanton, *The heat equation on a CR manifold*, J. Differential Geom. **20** (1984), no. 2, 343–387. MR788285 (86g:58135)

[BGM] Marcel Berger, Paul Gauduchon, and Edmond Mazet, *Le spectre d'une variété riemannienne* (French), Lecture Notes in Mathematics, Vol. 194, Springer-Verlag, Berlin-New York, 1971. MR0282313 (43 #8025)

[Br] Thomas P. Branson, *Sharp inequalities, the functional determinant, and the complementary series*, Trans. Amer. Math. Soc. **347** (1995), no. 10, 3671–3742, DOI 10.2307/2155203. MR1316845 (96e:58162)

[BG] Thomas Branson and A. Rod Gover, *Conformally invariant operators, differential forms, cohomology and a generalisation of Q-curvature*, Comm. Partial Differential Equations **30** (2005), no. 10-12, 1611–1669, DOI 10.1080/03605300500299943. MR2182307 (2006k:58060)

[CS] Luis Caffarelli and Luis Silvestre, *An extension problem related to the fractional Laplacian*, Comm. Partial Differential Equations **32** (2007), no. 7-9, 1245–1260, DOI 10.1080/03605300600987306. MR2354493 (2009k:35096)

[CM] Sun-Yung Alice Chang and María del Mar González, *Fractional Laplacian in conformal geometry*, Adv. Math. **226** (2011), no. 2, 1410–1432, DOI 10.1016/j.aim.2010.07.016. MR2737789 (2012k:58057)

[CJ] L. Chen and D. Jakobson, *Gaussian free fields and KPZ relation in* $\mathbb{R}^4$. E-print, arXiv, October 2012.

[CY] Shiu Yuen Cheng and Shing Tung Yau, *On the regularity of the Monge-Ampère equation* $\det(\partial^2 u/\partial x_i \partial sx_j) = F(x, u)$, Comm. Pure Appl. Math. **30** (1977), no. 1, 41–68. MR0437805 (55 #10727)

[Fe1] Charles L. Fefferman, *Monge-Ampère equations, the Bergman kernel, and geometry of pseudoconvex domains*, Ann. of Math. (2) **103** (1976), no. 2, 395–416. MR0407320 (53 #11097a)

[Fe2] Charles Fefferman, *The Bergman kernel and biholomorphic mappings of pseudoconvex domains*, Invent. Math. **26** (1974), 1–65. MR0350069 (50 #2562)

[Fe3] Charles Fefferman, *Parabolic invariant theory in complex analysis*, Adv. in Math. **31** (1979), no. 2, 131–262, DOI 10.1016/0001-8708(79)90025-2. MR526424 (80j:32035)

[FG1] Charles Fefferman and C. Robin Graham, *Conformal invariants. The mathematical heritage of Élie Cartan (Lyon, 1984)*, Astérisque **Numero Hors Serie** (1985), 95–116. MR837196 (87g:53060)

[FG2] Charles Fefferman and C. Robin Graham, *Q-curvature and Poincaré metrics*, Math. Res. Lett. **9** (2002), no. 2-3, 139–151, DOI 10.4310/MRL.2002.v9.n2.a2. MR1909634 (2003f:53053)

[FG3] Charles Fefferman and C. Robin Graham, *The ambient metric*, Annals of Mathematics Studies, vol. 178, Princeton University Press, Princeton, NJ, 2012. MR2858236

[FG4] Charles Fefferman and C. Robin Graham, *Juhl's formulae for GJMS operators and Q-curvatures* . To appear in J. Amer. Soc.. DOI: 10.1090/S0894-0347-2013-00765-1.

[Gi1] Peter B. Gilkey, *Curvature and the eigenvalues of the Laplacian for elliptic complexes*, Advances in Math. **10** (1973), 344–382. MR0324731 (48 #3081)

[Gi2] Peter B. Gilkey, *The spectral geometry of a Riemannian manifold*, J. Differential Geometry **10** (1975), no. 4, 601–618. MR0400315 (53 #4150)

[Gi3] Peter B. Gilkey, *Invariance theory, the heat equation, and the Atiyah-Singer index theorem*, 2nd ed., Studies in Advanced Mathematics, CRC Press, Boca Raton, FL, 1995. MR1396308 (98b:58156)

[Go] A. R. Gover, *Laplacian operators and Q-curvature on conformally Einstein manifolds*, Math. Ann. **336** (2006), no. 2, 311–334, DOI 10.1007/s00208-006-0004-z. MR2244375 (2007f:58046)

[GG] A. Rod Gover and C. Robin Graham, *CR invariant powers of the sub-Laplacian*, J. Reine Angew. Math. **583** (2005), 1–27, DOI 10.1515/crll.2005.2005.583.1. MR2146851 (2006f:32049)

[GH] A. Rod Gover and Kengo Hirachi, *Conformally invariant powers of the Laplacian—a complete nonexistence theorem*, J. Amer. Math. Soc. **17** (2004), no. 2, 389–405, DOI 10.1090/S0894-0347-04-00450-3. MR2051616 (2005c:58062)

[GP] A. Rod Gover and Lawrence J. Peterson, *Conformally invariant powers of the Laplacian, Q-curvature, and tractor calculus*, Comm. Math. Phys. **235** (2003), no. 2, 339–378, DOI 10.1007/s00220-002-0790-4. MR1969732 (2004d:58047)

[GW] A. Rod Gover and A. Waldron, *Boundary calculus for conformally compact manifolds.* E-print, arXiv, April 2011.

[Gr1] C. Robin Graham, *Conformally invariant powers of the Laplacian. II. Nonexistence*, J. London Math. Soc. (2) **46** (1992), no. 3, 566–576, DOI 10.1112/jlms/s2-46.3.566. MR1190439 (94c:58227)

[Gr2] C. Robin Graham, Talk at the workshop *Conformal structure in geometry, analysis, and physics.* AIM, Palo Alto, Aug. 12–16, 2003.

[GJMS] C. Robin Graham, Ralph Jenne, Lionel J. Mason, and George A. J. Sparling, *Conformally invariant powers of the Laplacian. I. Existence*, J. London Math. Soc. (2) **46** (1992), no. 3, 557–565, DOI 10.1112/jlms/s2-46.3.557. MR1190438 (94c:58226)

[GrH] C. Robin Graham and Kengo Hirachi, *Inhomogeneous ambient metrics*, Symmetries and overdetermined systems of partial differential equations, IMA Vol. Math. Appl., vol. 144, Springer, New York, 2008, pp. 403–420, DOI 10.1007/978-0-387-73831-4_20. MR2384722 (2009f:53012)

[GZ] C. Robin Graham and Maciej Zworski, *Scattering matrix in conformal geometry*, Invent. Math. **152** (2003), no. 1, 89–118, DOI 10.1007/s00222-002-0268-1. MR1965361 (2004c:58064)

[Gr] Peter Greiner, *An asymptotic expansion for the heat equation*, Arch. Rational Mech. Anal. **41** (1971), 163–218. MR0331441 (48 #9774)

[GS] Gerd Grubb and Robert T. Seeley, *Weakly parametric pseudodifferential operators and Atiyah-Patodi-Singer boundary problems*, Invent. Math. **121** (1995), no. 3, 481–529, DOI 10.1007/BF01884310. MR1353307 (96k:58216)

[Gu1] Colin Guillarmou, *Meromorphic properties of the resolvent on asymptotically hyperbolic manifolds*, Duke Math. J. **129** (2005), no. 1, 1–37, DOI 10.1215/S0012-7094-04-12911-2. MR2153454 (2006k:58051)

[Gu2] Colin Guillarmou, Personal communication.

[GMP1] Colin Guillarmou, Sergiu Moroianu, and Jinsung Park, *Eta invariant and Selberg zeta function of odd type over convex co-compact hyperbolic manifolds*, Adv. Math. **225** (2010), no. 5, 2464–2516, DOI 10.1016/j.aim.2010.05.004. MR2680173 (2011m:58038)

[GMP2] Guillarmou, C.; Moroianu, S.; Park, J.: *Calderon and Bergman projectors on spin manifolds with boundary.* J. Geom. Anal., doi:10.1007/s12220-012-9338-9.

[GN] Colin Guillarmou and Frédéric Naud, *Wave 0-trace and length spectrum on convex co-compact hyperbolic manifolds*, Comm. Anal. Geom. **14** (2006), no. 5, 945–967. MR2287151 (2008f:58032)

[Gui1] Victor Guillemin, *A new proof of Weyl's formula on the asymptotic distribution of eigenvalues*, Adv. in Math. **55** (1985), no. 2, 131–160, DOI 10.1016/0001-8708(85)90018-0. MR772612 (86i:58135)

[Gui2] Victor Guillemin, *Gauged Lagrangian distributions*, Adv. Math. **102** (1993), no. 2, 184–201, DOI 10.1006/aima.1993.1064. MR1252031 (95c:58171)

[Ju1] Andreas Juhl, *Families of conformally covariant differential operators, Q-curvature and holography*, Progress in Mathematics, vol. 275, Birkhäuser Verlag, Basel, 2009. MR2521913 (2010m:58048)

[Ju2] Andreas Juhl, *Explicit formulas for GJMS-operators and Q-curvatures*, Geom. Funct. Anal. **23** (2013), no. 4, 1278–1370, DOI 10.1007/s00039-013-0232-9. MR3077914

[Hi1] Kengo Hirachi, *Construction of boundary invariants and the logarithmic singularity of the Bergman kernel*, Ann. of Math. (2) **151** (2000), no. 1, 151–191, DOI 10.2307/121115. MR1745015 (2001f:32003)

[Hi2] Kengo Hirachi, Personal communication.

[JS] Mark S. Joshi and Antônio Sá Barreto, *Inverse scattering on asymptotically hyperbolic manifolds*, Acta Math. **184** (2000), no. 1, 41–86, DOI 10.1007/BF02392781. MR1756569 (2002g:58052)

[KV] Maxim Kontsevich and Simeon Vishik, *Geometry of determinants of elliptic operators*, Functional analysis on the eve of the 21st century, Vol. 1 (New Brunswick, NJ, 1993), Progr. Math., vol. 131, Birkhäuser Boston, Boston, MA, 1995, pp. 173–197. MR1373003 (96m:58264)

[Ku] N. H. Kuiper, *On conformally-flat spaces in the large*, Ann. of Math. (2) **50** (1949), 916–924. MR0031310 (11,133b)

[MS] H. P. McKean Jr. and I. M. Singer, *Curvature and the eigenvalues of the Laplacian*, J. Differential Geometry **1** (1967), no. 1, 43–69. MR0217739 (36 #828)

[Le] John M. Lee, *The Fefferman metric and pseudo-Hermitian invariants*, Trans. Amer. Math. Soc. **296** (1986), no. 1, 411–429, DOI 10.2307/2000582. MR837820 (87j:32063)

[LM] John M. Lee and Richard Melrose, *Boundary behaviour of the complex Monge-Ampère equation*, Acta Math. **148** (1982), 159–192, DOI 10.1007/BF02392727. MR666109 (84e:58078)

[Ma1] Rafe Mazzeo, *The Hodge cohomology of a conformally compact metric*, J. Differential Geom. **28** (1988), no. 2, 309–339. MR961517 (89i:58005)

[Ma2] Rafe Mazzeo, *Unique continuation at infinity and embedded eigenvalues for asymptotically hyperbolic manifolds*, Amer. J. Math. **113** (1991), no. 1, 25–45, DOI 10.2307/2374820. MR1087800 (92f:58187)

[MM] Rafe R. Mazzeo and Richard B. Melrose, *Meromorphic extension of the resolvent on complete spaces with asymptotically constant negative curvature*, J. Funct. Anal. **75** (1987), no. 2, 260–310, DOI 10.1016/0022-1236(87)90097-8. MR916753 (89c:58133)

[Mi] S. Minakshisundaram, *Eigenfunctions on Riemannian manifolds*, J. Indian Math. Soc. (N.S.) **17** (1953), 159–165 (1954). MR0061750 (15,877d)

[Ok] K. Okikiolu, *Critical metrics for the determinant of the Laplacian in odd dimensions*, Ann. of Math. (2) **153** (2001), no. 2, 471–531, DOI 10.2307/2661347. MR1829756 (2002j:58056)

[Pa] Stephen M. Paneitz, *A quartic conformally covariant differential operator for arbitrary pseudo-Riemannian manifolds (summary)*, SIGMA Symmetry Integrability Geom. Methods Appl. **4** (2008), Paper 036, 3 pp., DOI 10.3842/SIGMA.2008.036. MR2393291 (2009b:58080)

[PR] Thomas Parker and Steven Rosenberg, *Invariants of conformal Laplacians*, J. Differential Geom. **25** (1987), no. 2, 199–222. MR880183 (89e:58118)

[Pe] Lawrence J. Peterson, *Conformally covariant pseudo-differential operators*, Differential Geom. Appl. **13** (2000), no. 2, 197–211, DOI 10.1016/S0926-2245(00)00023-1. MR1783963 (2001k:58050)

[Po] Iosif Polterovich, *Heat invariants of Riemannian manifolds*, Israel J. Math. **119** (2000), 239–252, DOI 10.1007/BF02810670. MR1802656 (2002b:58031)

[Po1] Raphaël Ponge, *Logarithmic singularities of Schwartz kernels and local invariants of conformal and CR structures operators*. E-print, arXiv, October 2007.

[Po2] Raphaël Ponge, *A microlocal approach to Fefferman's program in conformal and CR geometry*, Motives, quantum field theory, and pseudodifferential operators, Clay Math. Proc., vol. 12, Amer. Math. Soc., Providence, RI, 2010, pp. 221–244. MR2762532 (2011m:32062)

[Ra] I. P. Ramadanov, *A characterization of the balls in* $\mathbf{C}^n$ *by means of the Bergman kernel*, C. R. Acad. Bulgare Sci. **34** (1981), no. 7, 927–929. MR639487 (83b:32022)

[Sc] Richard Schoen, *Conformal deformation of a Riemannian metric to constant scalar curvature*, J. Differential Geom. **20** (1984), no. 2, 479–495. MR788292 (86i:58137)

[Se] R. T. Seeley, *Complex powers of an elliptic operator*, Singular Integrals (Proc. Sympos. Pure Math., Chicago, Ill., 1966), Amer. Math. Soc., Providence, R.I., 1967, pp. 288–307. MR0237943 (38 #6220)

[Ta] Michael E. Taylor, *Partial differential equations. II. Qualitative studies of linear equations*, Applied Mathematical Sciences, vol. 116, Springer-Verlag, New York, 1996. MR1395149 (98b:35003)

[Va1] András Vasy, *Microlocal analysis of asymptotically hyperbolic spaces and high-energy resolvent estimates*, Inverse problems and applications: inside out. II, Math. Sci. Res. Inst. Publ., vol. 60, Cambridge Univ. Press, Cambridge, 2013, pp. 487–528. MR3135765

[Va2] András Vasy, *Analytic continuation and high energy estimates for the resolvent of the Laplacian on forms on asymptotically hyperbolic spaces*. E-print, arXiv, June 2012.

[Wi] Harold Widom, *A complete symbolic calculus for pseudodifferential operators* (English, with French summary), Bull. Sci. Math. (2) **104** (1980), no. 1, 19–63. MR560744 (81m:58078)

[Wo1] Mariusz Wodzicki, *Local invariants of spectral asymmetry*, Invent. Math. **75** (1984), no. 1, 143–177, DOI 10.1007/BF01403095. MR728144 (85g:58089)

[Wo2] Mariusz Wodzicki, *Noncommutative residue. I. Fundamentals*, K-theory, arithmetic and geometry (Moscow, 1984), Lecture Notes in Math., vol. 1289, Springer, Berlin, 1987, pp. 320–399, DOI 10.1007/BFb0078372. MR923140 (90a:58175)

Department of Mathematical Sciences, Seoul National University, Seoul, South Korea

E-mail address: ponge.snu@gmail.com

Contemporary Mathematics
Volume **630**, 2014
http://dx.doi.org/10.1090/conm/630/12670

A Symbolic Calculus for Fourier Integral Operators

Yuri Safarov

Abstract. The paper develops a symbolic calculus for Fourier integral operators associated with canonical transformations.

Introduction

The paper deals with Fourier integral operators on a closed n-dimensional C^∞-manifold M associated with homogeneous canonical transformations. We shall abbreviate the words 'Fourier integral operator' and 'pseudodifferential operator' to FIO and ψDO and shall always be assuming that the operators act in the space of half-densities on M. Recall that, for the half-densities, the inner product $\int_M u(x)\,\overline{v(x)}\,\mathrm{d}x$ is well defined, and so are the adjoint operators.

It is well known that FIOs associated with canonical transformations form a *-algebra. However, explicit formulae for the principal symbols of their adjoints and compositions are not obvious, partly due to the fact that there are many possible definitions of the symbol. Most textbooks (in particular, [**D**], [**H2**], [**Tr**]) define the principle symbol of a FIO as a half-density on a Lagrangian manifold with values in the Keller–Maslov bundle. This definition is convenient for theoretical purposes, but is not suitable when we need to know the exact value of the symbol at a given point.

Furthermore, the standard definition makes it impossible for the principal symbol of the composition to be equal to the product of principal symbols, as the product of half-densities is not a half-density. One can identify half-densities with functions by fixing a model positive density on the Lagrangian manifold but, since there are three FIOs involved in the composition formula, the result obtained with this approach may depend on the choice of the model densities. Also, it is not immediately clear how to choose local trivializations of the three Keller–Maslov bundles, in which the product formula would hold.

The aim of this paper is to present a relatively simple approach which allows one to develop an explicit symbolic calculus. One of its main ideas is to avoid considering the most general phase functions associated with a canonical transformation Φ and to use only phase functions from the class $\mathfrak{F}_\Phi$ defined in Subsection 2.2.

To make our point more clear, let us recall that the classical theory of ψDOs deals only with the phase functions of the form $(x-y)\cdot\xi$. Instead, one could

2010 *Mathematics Subject Classification.* Primary 35S30.

treat ψDOs as FIOs associated with the identity transformation, introduce general phase functions associated with this transformation, define principal symbols as half-densities on its graph with values in the Keller–Maslov bundle, and try to prove results in this general setting. However, such a general approach would unlikely have any advantages, and would only make proofs and results less transparent. Likewise, for a general canonical transformation Φ, one does not need to consider all possible phase functions. The phase functions from the class $\mathfrak{F}_\Phi$, which can be thought of as analogues of $(x-y)\cdot\xi$, turn out to be sufficient to define FIOs associated with Φ and to carry out all calculations.

In the first two sections we briefly review some basic notions and results from symplectic geometry and introduce the class of phase function $\mathfrak{F}_\Phi$. Section 3 is devoted to the definition of FIOs and their principal symbols. The main results are stated and proved in Section 4.

Acknowlegements. The paper can be regarded as a continuation of the joint project [**JSS**], in which we needed explicit formulae for the principal symbols of adjoints and compositions of FIOs. The results of Section 4 would have made our task much easier but, unfortunately, at that time they were not available.

I am grateful to S. Eswarathasan and A. Strohmaier for helpful discussions, and to CRM for sponsoring my visits to Montreal.

1. Notation and definitions

Let M be a smooth n-dimensional manifold. We denote points of M by x, y, z, and the covectors from T^*_xM, T^*_yM, T^*_zM by ξ, η, ζ respectively. The same letters are used for local coordinates on M and the corresponding dual coordinates in the fibres of the cotangent bundle T^*M. If $f=(f_1,\dots,f_n)$ is a vector-function of n-dimensional variable $\theta=(\theta_1,\dots,\theta_n)$ then

- f_θ denotes the $n\times n$-matrix function with entries $(f_i)_{\theta_j}$, where j enumerates elements of the ith row.

If C is a real symmetric matrix then

- $\kappa_+(C)$ and $\kappa_-(C)$ are the numbers of its strictly positive and strictly negative eigenvalues.

1.1. Lagrangian subspaces of $T_\vartheta T^*M$. Let us fix a point $\vartheta=(x,\xi)\in T^*M$ and consider the tangent space $T_\vartheta T^*M$ over this point. Denote by Π^V_ϑ the differential of the projection $T^*M\mapsto M$ at the point ϑ. The kernel V_ϑ of the mapping $\Pi^V_\vartheta : T_\vartheta T^*M \mapsto T^*_xM$ is said to be the *vertical subspace* of $T_\vartheta T^*M$. Clearly, V_ϑ is the Lagrangian subspace spanned by vectors of the form $\mathbf{v}\cdot\nabla_\xi$ where $\mathbf{v}\in\mathbb{R}^n$.

Recall that Lagrangian subspaces are said to be transversal if their intersection is zero. It is well known that for every finite collection of Lagrangian subspaces there exists a Lagrangian subspace transversal to all of them.

A Lagrangian subspace of $T_\vartheta T^*M$ is transversal to the vertical subspace V_ϑ if and only if it consists of vectors of the form $\mathbf{x}\cdot\nabla_x + A\mathbf{x}\cdot\nabla_\xi$, where $\mathbf{x}\in\mathbb{R}^n$ and $A\in\mathbb{R}^{n\times n}$ is a fixed symmetric matrix.

Under a change of coordinates $x\to\tilde x$ the components of a vector from $T_\vartheta T^*M$ transform as follows. If $\mathbf{x}\cdot\nabla_x+\mathbf{v}\cdot\nabla_\xi\in T_\vartheta T^*M$ in the local coordinates x with some $\mathbf{x}, \mathbf{v} \in \mathbb{R}^n$ then

$$\mathbf{x} \cdot \nabla_x + \mathbf{v} \cdot \nabla_\xi = J\mathbf{x} \cdot \nabla_{\tilde{x}} + (J^{-1})^T \mathbf{v} \cdot \nabla_{\tilde{\xi}} + CJ\mathbf{x} \cdot \nabla_{\tilde{\xi}} \tag{1.1}$$

where J denotes the Jacobi matrix $\tilde{x}_x$ and C is the symmetric matrix with entries $C_{ij} = \sum_k \xi_k \frac{\partial^2 x_k}{\partial \tilde{x}_j \partial \tilde{x}_j}$. This implies that

$$\{\mathbf{x} \cdot \nabla_x + A\mathbf{x} \cdot \nabla_\xi \mid \mathbf{x} \in \mathbb{R}^n\} = \left\{\mathbf{x} \cdot \nabla_{\tilde{x}} + \tilde{A}\mathbf{x} \cdot \nabla_{\tilde{\xi}} \mid \mathbf{x} \in \mathbb{R}^n\right\}$$

where $\tilde{A} = (J^{-1})^T A J^{-1} + C$. If $\xi \neq 0$ then, for any given symmetric matrix $A \in \mathbb{R}^{n \times n}$, we can choose coordinates $\tilde{x}$ in a neighbourhood of ϑ in such a way that $\tilde{A} = 0$. Thus we have

LEMMA 1.1. *If $\vartheta = (x, \xi)$ with $\xi \neq 0$ then for every Lagrangian subspace $H_\vartheta \subset T_\vartheta T^*M$ transversal to V_ϑ there exist local coordinates $\tilde{x}$ in which H_ϑ is spanned by the vectors $\partial_{\tilde{x}_i}$.*

One can think of H_ϑ as a horizontal subspace of $T_\vartheta T^*M$ associated with the coordinates $\tilde{x}$.

1.2. Kashiwara index. Let L_1 and L_2 be arbitrary Lagrangian subspaces of $T_\vartheta T^*M$. Consider the quadratic form

$$Q_{L_1,V_\vartheta,L_2}[\theta_1, \theta, \theta_2] = \omega(\theta_1, \theta) + \omega(\theta, \theta_2) + \omega(\theta_2, \theta_1)$$

on the vector space $L_1 \oplus V_\vartheta \oplus L_2$, where ω denotes the symplectic form $\mathrm{d}x \wedge \mathrm{d}\xi$, $\theta \in V_\vartheta$ and $\theta_j \in L_j$. Its signature $\operatorname{sgn} Q_{L_1,V_\vartheta,L_2}$ is called the *Kashiwara index* of the triple $\{L_1, V_\vartheta, L_2\}$ (see, for instance, [**RS**, Section 7.8]). Since the vertical subspace is fixed, we shall drop V_ϑ from the notation and denote

$$\kappa(L_1, L_2) = \operatorname{sgn} Q_{L_1,V_\vartheta,L_2}\,. \tag{1.2}$$

Assume that $\vartheta = (x, \xi)$ with $\xi \neq 0$, and let H_ϑ be the horizontal subspace associated with coordinates x. A general Lagrangian subspace $L \in T^*_\vartheta M$ can be written in the form

$$L = \{B\mathbf{w} \cdot \nabla_x + C\mathbf{w} \cdot \nabla_\xi \mid \mathbf{w} \in \mathbb{R}^n\}$$

where B and C are real $n \times n$-matrices such that $\operatorname{rank}(B, C) = n$ and $B^T C = C^T B$. One can easily show that

$$\kappa(L, H_\vartheta) = \kappa_+(B^T C)\,. \tag{1.3}$$

Denote

$$r(L_1, L_2) = n + \dim\left(L_1 \bigcap V_\vartheta\right) - \dim\left(L_2 \bigcap V_\vartheta\right) - \dim\left(L_1 \bigcap L_2\right),$$

and let

$$\varkappa(L_1, L_2) = \frac{1}{2}\left(\kappa(L_1, L_2) + r(L_1, L_2)\right). \tag{1.4}$$

If H_ϑ is a horizontal subspace transversal to the Lagrangian subspaces L_1 and L_2 then $L_j = \{A_j \mathbf{v}_j \cdot \nabla_x + \mathbf{v}_j \cdot \nabla_\xi \mid \mathbf{v}_j \in \mathbb{R}^n\}$ in the associated coordinates x, where $A_j \in \mathbb{R}^{n \times n}$ are symmetric matrices. Obviously, $\operatorname{rank} A_j = n - \dim(L_j \bigcap V_\vartheta)$ and $\operatorname{rank}(A_1 - A_2) = n - \dim(L_1 \bigcap L_2)$, so that

$$r(L_1, L_2) = \operatorname{rank} A_2 - \operatorname{rank} A_1 + \operatorname{rank}(A_1 - A_2)\,. \tag{1.5}$$

Parametrizing L_j by $\mathbf{v}_j \in \mathbb{R}^n$ and V_ϑ by $\mathbf{v} \in \mathbb{R}^n$, we obtain

$$Q_{L_1,V_\vartheta,L_2}[\mathbf{v}_1,\mathbf{v},\mathbf{v}_2] = \frac{1}{2}\langle \mathcal{Q}(\mathbf{v}_1,\mathbf{v},\mathbf{v}_2),(\mathbf{v}_1,\mathbf{v},\mathbf{v}_2)\rangle,$$

where

$$\mathcal{Q} = \begin{pmatrix} 0 & A_1 & A_2 - A_1 \\ A_1 & 0 & -A_2 \\ A_2 - A_1 & -A_2 & 0 \end{pmatrix}$$

If $J = \begin{pmatrix} 0 & 1 & 1 \\ 1 & 0 & 1 \\ 1 & 1 & 0 \end{pmatrix}$ then $J^T \mathcal{Q} J = \begin{pmatrix} -2A_2 & 0 & 0 \\ 0 & 2(A_2 - A_1) & 0 \\ 0 & 0 & 2A_1 \end{pmatrix}$. It follows that

$$\kappa(L_1,L_2) = \operatorname{sgn} A_1 - \operatorname{sgn} A_2 - \operatorname{sgn}(A_1 - A_2)\,. \tag{1.6}$$

and, consequently,

$$\varkappa(L_1,L_2) = \kappa_-(A_2) - \kappa_-(A_1) + \kappa_-(A_1 - A_2)\,. \tag{1.7}$$

In particular, the above equality implies that $\varkappa(L_1,L_2) \in \mathbb{Z}$ for all Lagrangian subspaces $L_1, L_2 \subset T_\vartheta T^*M$

2. Canonical transformations in T^*M

2.1. Definitions. Recall that a mapping

$$\Phi : (y,\eta) \mapsto (x^\star(y,\eta), \xi^\star(y,\eta))$$

from an open subset $\mathcal{D}(\Phi) \subset T^*M$ into T^*M is said to be a canonical transformation if it preserves the symplectic 2-form $\mathrm{d}x \wedge \mathrm{d}\xi$ on TT^*M or, in other words, if

$$(\xi_y^\star)^T x_y^\star - (x_y^\star)^T \xi_y^\star = (\xi_\eta^\star)^T x_\eta^\star - (x_\eta^\star)^T \xi_\eta^\star = 0 \tag{2.1}$$

and

$$(\xi_\eta^\star)^T x_y^\star - (x_\eta^\star)^T \xi_y^\star = I\,. \tag{2.2}$$

A canonical transformation is nondegenerate. It is said to be homogeneous if

$$(x^\star(y,\lambda\eta), \xi^\star(y,\lambda\eta)) = (x^\star(y,\eta), \lambda\xi^\star(y,\eta))\,, \qquad \forall \lambda > 0\,. \tag{2.3}$$

A homogeneous canonical transformation also preserves the symplectic 1-form $\xi \cdot \mathrm{d}x$, that is,

$$(x_\eta^\star)^T \xi^\star = 0 \quad \text{and} \quad (x_y^\star)^T \xi^\star = \eta\,. \tag{2.4}$$

Let us fix $\vartheta = (y,\eta) \in \mathcal{D}(\Phi)$ and local coordinates x in a neighbourhood of $x^\star(y,\eta)$. The differential $\mathrm{d}\Phi$ maps the vertical subspace V_ϑ onto a Lagrangian subspace $\mathrm{d}\Phi(V_\vartheta) \subset T_{\Phi(\vartheta)}T^*M$. In the local coordinates, the restriction $\mathrm{d}\Phi|_{V_\vartheta}$ is given by the matrix $(x_\eta^\star, \xi_\eta^\star) \in \mathbb{R}^{2n\times n}$ with columns $(x_{\eta_k}^\star, \xi_{\eta_k}^\star)$, $k = 1,\dots,n$, and

$$\mathrm{d}\Phi(V_\vartheta) = \left\{ x_\eta^\star \mathbf{w} \cdot \nabla_x + \xi_\eta^\star \mathbf{w} \cdot \nabla_\xi \mid \mathbf{w} \in \mathbb{R}^n \right\}. \tag{2.5}$$

The matrix $x_\eta^\star \in \mathbb{R}^{n\times n}$ is the coordinate representation of the composition $\Pi_{\Phi(\vartheta)}^V \circ \mathrm{d}\Phi : V_\vartheta \mapsto T_{x^\star}^* M$. It is invariantly defined and behaves as a tensor under change of coordinates x and y. Since Φ is nondegenerate, $\ker x_\eta^\star = \mathrm{d}\Phi(V_\vartheta) \bigcap V_{\Phi(\theta)}$.

Similarly, the matrix $\xi_\eta^\star$ is the coordinate representation of the mapping $\Pi_{\Phi(\vartheta)}^H \circ \mathrm{d}\Phi : V_\vartheta \mapsto V_{\Phi(\vartheta)}$, where $\Pi_{\Phi(\vartheta)}^H : T_{\Phi(\vartheta)}T^*M \mapsto V_{\Phi(\vartheta)}$ is the projection onto the vertical subspace $V_{\Phi(\vartheta)}$ along $H_{\Phi(\vartheta)}$. The matrix $\xi_\eta^\star$ also behaves as a tensor under a change of coordinates y. The transformation law for $\xi^\star_\eta$ under a change of coordinates x is given by (1.1). It is nondegenerate if and only if $H_{\Phi(\vartheta)}$ is transversal to the image $\mathrm{d}\Phi(V_\vartheta)$.

2.2. Phase functions associated with homogeneous canonical transformations. Let $\mathfrak{F}_\Phi$ be the set of functions

$$\varphi(x;y,\eta) \in C^\infty\left(M \times \mathcal{D}(\Phi) \setminus \{0\}\right)$$

with $\operatorname{Im}\varphi \geq 0$, which satisfy the following conditions.

(a$_1$) φ is positively homogeneous in η of degree 1;
(a$_2$) $\varphi(x;y,\eta) = (x-x^\star)\cdot\xi^\star + O\left(|x-x^\star|^2\right)$ as $x \to x^\star$;
(a$_3$) $\det\varphi_{x\eta}(x^\star(y,\eta);y,\eta) \neq 0$ for all $(y,\eta) \in \mathcal{D}(\Phi)\setminus\{0\}$.

The first equality (2.4) and **(a$_2$)** imply that

$$\varphi_{\eta_k}(x;y,\eta) \;=\; (x-x^\star)\cdot\left(\xi^\star_{\eta_k} - \varphi_{xx}(x^\star;y,\eta)\,x^\star_{\eta_k}\right) + O\left(|x-x^\star|^2\right) \tag{2.6}$$

as $x \to x^\star$ for all $k = 1,\dots,n$. Therefore

$$\varphi_{x\eta}(x^\star;y,\eta) \;=\; \xi^\star_\eta \;-\; \varphi_{xx}(x^\star;y,\eta)\,x^\star_\eta\,, \tag{2.7}$$

$$\varphi_{\eta\eta}(x^\star;y,\eta) \;=\; -\,(x^\star_\eta)^T\varphi_{x\eta}(x^\star;y,\eta) \tag{2.8}$$

and, in view of **(a$_3$)**,

$$\operatorname{rank}\varphi_{\eta\eta}(x^\star;y,\eta) \;=\; \operatorname{rank}x^\star_\eta(y,\eta)\,. \tag{2.9}$$

Let $\vartheta = (y,\eta) \in \mathcal{D}(\Phi)\setminus\{0\}$. Every function φ satisfying the conditions **(a$_1$)** and **(a$_2$)** defines a horizontal bundle H^φ over the image $\Phi(T^*M)$ formed by the subspaces

$$H^\varphi_{\Phi(\vartheta)} \;=\; \{\mathbf{x}\cdot\nabla_x - A\mathbf{x}\cdot\nabla_\xi \mid \mathbf{x}\in\mathbb{R}^n\} \;\subset\; T_{\Phi(\vartheta)}T^*M\,, \tag{2.10}$$

where $A = (\operatorname{Re}\varphi)_{xx}(x^\star;y,\eta)$. Lemma 1.1 and (2.10) imply

COROLLARY 2.1. *If φ satisfies* **(a$_1$)** *and* **(a$_2$)** *then for each fixed point $(y,\eta) \in \mathcal{D}(\Phi)\setminus\{0\}$ there exist local coordinates x on a neighbourhood of $x^\star(y,\eta)$ such that $(\operatorname{Re}\varphi)_{xx}(x^\star;y,\eta) = 0$.*

The other way around, for every horizontal bundle H over $\Phi(T^*M)\setminus\{0\}$ there exists a phase function satisfying the conditions **(a$_1$)** and **(a$_2$)** such that $H = H^\varphi$.

The matrix $(\operatorname{Re}\varphi)_{x\eta}(x^\star;y,\eta)$ is the coordinate representation of the mapping $\Pi^{H^\varphi}_{\Phi(\vartheta)}\circ\mathrm{d}\Phi|_{V_\vartheta}$ where $\Pi^{H^\varphi}_{\Phi(\vartheta)}$ is the projection onto $V_{\Phi(\vartheta)}$ along the horizontal subspace $H^\varphi_{\Phi(\vartheta)}$. It is nondegenerate if and only if $H^\varphi_{\Phi(\vartheta)}$ is transversal to $\mathrm{d}\Phi(V_\vartheta)$. This shows that the existence of a real phase function φ satisfying **(a$_1$)**–**(a$_3$)** on an open set $\mathcal{O}\subset\mathcal{D}(\Phi)$ is equivalent to the following condition

(C$_1$) there is smooth family of Lagrangian subspaces $H_{\Phi(\vartheta)}$ transversal both to $V_{\Phi(\vartheta)}$ and $\mathrm{d}\Phi(V_\vartheta)$ for all $\vartheta\in\mathcal{O}$.

The condition **(C$_1$)** can always be satisfied locally (that is, by choosing a sufficiently small $\mathcal{O}$). However, it may not be fulfilled globally, even if $\mathcal{D}(\Phi)$ is connected and simply connected.

On the other hand, **(a$_3$)** holds for almost all complex-valued phase functions satisfying **(a$_1$)** and **(a$_2$)**. The following is [**SV**, Corollary 2.4.5] (alternatively, see [**LSV**, Lemma 1.4]).

LEMMA 2.2. *If φ satisfies* $(\mathbf{a}_1)$ *and* $(\mathbf{a}_2)$ *and the symmetric matrix* $(\operatorname{Im}\varphi)_{xx}(x^\star; y, \eta)$ *is positive definite then* $\det \varphi_{x\eta}(x^\star; y, \eta) \neq 0$.

REMARK 2.3. The above result implies, in particular, that the set $\mathfrak{F}_\Phi$ is connected and simply connected. This allows one to prove various results by continuously transforming one phase function into another (see, for instance, [**SV**, Section 2.7.4]).

EXAMPLE 2.4. Let $\mathcal{O} \subset \mathcal{D}(\Phi)$ be an open conic set. If $\mathcal{O}$ is sufficiently small, we can choose local coordinates x on a neighbourhood of the projection of $\Phi(\mathcal{O})$ onto M and define $\varphi(x; y, \eta) = (x - x^\star) \cdot \xi^\star$. Obviously, this function satisfies the conditions $(\mathbf{a}_1)$, $(\mathbf{a}_2)$, $\varphi_{xx} \equiv 0$, $\varphi_{x\eta} = \xi^\star_\eta$ and, in view of (2.1),

$$\varphi_{\eta\eta}(x^\star; y, \eta) \;=\; -(x^\star_\eta)^T \xi^\star_\eta \;=\; -(\xi^\star_\eta)^T x^\star_\eta \,. \tag{2.11}$$

The equality $\varphi_{xx} \equiv 0$ means that the horizontal bundle H^φ over $\Phi(\mathcal{O})$ coincides with the horizontal bundle H associated with the coordinates x. The function φ satisfies $(\mathbf{a}_3)$ on $\mathcal{O}$ if and only if $H_{\Phi(\vartheta)}$ is transversal to $\mathrm{d}\Phi(V_\vartheta)$ for all $\vartheta \in \mathcal{O}$. In this case, by (2.5),

$$\mathrm{d}\Phi(V_\vartheta) \;=\; \left\{ x^\star_\eta\,(\xi^\star_\eta)^{-1}\mathbf{w}\cdot\nabla_x \,+\, \mathbf{w}\cdot\nabla_\xi \mid \mathbf{w}\in\mathbb{R}^n \right\}, \quad \forall\vartheta\in\mathcal{O}\,. \tag{2.12}$$

Example 2.4 will play a crucial role in the proof of Theorem 4.1. We shall first obtain a local result, using the phase function $(x - x^\star) \cdot \xi^\star$, and then show that it holds globally because all the objects involved do not depend on the choice of phase function.

2.3. Maslov index. Let $C = C_1 + iC_2 \in \mathbb{C}^{n\times n}$ be a symmetric matrix with a nonnegative (in the sense of operator theory) real part C_1, and let Π_C be the orthogonal projection on $\ker C$. We shall denote

$$\det\nolimits_+ C \;= \det(C + \Pi_C)\,,$$

assuming that the branch of the argument $\arg\det_+ C$ is chosen in such a way that it is continuous with respect to C on the set of matrices with a fixed kernel and is equal to zero when $C_2 = 0$.

REMARK 2.5. If $\dim\ker C \;=\; m$ then $\det_+ C \;=\; \varepsilon^{-m}\det(C+\varepsilon I)|_{\varepsilon=0}$. This implies that

(i) $\arg\det_+ C = \arg\det_+(J^T C J)$ for all nondegenerate $J \in \mathbb{R}^{n\times n}$,
(ii) the restriction of $\arg\det_+ C$ to the set of matrices C with a fixed $\dim\ker C$ continuously depends on C.

REMARK 2.6. If $C_1 = 0$ then $\arg\det_+ C = \frac{\pi}{2}\operatorname{sgn} C_2$ where $\operatorname{sgn} C_2$ is the signature of C_2 (see [**H2**, Section 3.4]).

Note that the matrices $\varphi_{x\eta}$ and $\varphi_{\eta\eta}$ behave as tensors under change of coordinates x and y. It follows that the functions $\arg(\det^2\varphi_{x\eta})$ and, in view of Remark 2.5(i), $\arg\det_+(\varphi_{\eta\eta}/i)$ do not depend on the choice of local coordinates. Let us denote

$$\Theta^{\mathrm{r}}_\varphi(y,\eta) \;=\; (2\pi)^{-1}\arg\det{}^2\varphi_{x\eta}(x^\star, y, \eta), \tag{2.13}$$

$$\Theta^{\mathrm{s}}_\varphi(y,\eta) \;=\; \pi^{-1}\arg\det\nolimits_+(\varphi_{\eta\eta}(x^\star, y, \eta)/i) - \left(\operatorname{rank} x^\star_\eta\right)/2\,. \tag{2.14}$$

The following is [**LSV**, Proposition 2.3].

PROPOSITION 2.7. *The multi-valued function*

$$\Theta_\Phi \;=\; \Theta^{\mathrm{r}}_\varphi \,-\, \Theta^{\mathrm{s}}_\varphi$$

takes integer values and does not depend on the choice of the phase function $\varphi \in \mathfrak{F}_\Phi$ *and local coordinates. The branches of* Θ_Φ *are continuous along any path on which* $\operatorname{rank} x^\star_\eta$ *is constant.*

By the above, the function Θ_Φ is uniquely determined by the canonical transformation Φ. It is multi-valued only due to the fact that $\arg\det^2\varphi_{x\eta}$ is multi-valued. Note that, in view of (2.8) and Remark 2.5(ii), Θ^{s}_Φ continuously depends on $\varphi \in \mathfrak{F}_\Phi$ and, obviously, so does Θ^{r}_Φ. It follows that the branches of Θ_Φ do not change under a continuous transformation of $\varphi \in \mathfrak{F}$.

EXAMPLE 2.8. If $(\operatorname{Im}\varphi)_{xx}(x^\star; y, \eta) = 0$ on an open set $\Omega \subset \mathcal{D}(\Phi)$ then, in view of (2.9) and Remark 2.6,

$$\Theta^{\mathrm{s}}_\varphi(y,\eta) \;=\; -\,\kappa_+(\varphi_{\eta\eta}(x^\star; y, \eta)) \tag{2.15}$$

and, consequently,

$$\Theta_\Phi(y,\eta) \;=\; m \,+\, \kappa_+(\varphi_{\eta\eta}(x^\star; y, \eta)) \tag{2.16}$$

for all $(y,\eta) \in \mathcal{O}$, where m is an integer depending on the choice of the branch of $\arg\det^2\varphi_{x\eta}$.

DEFINITION 2.9. If γ is a path in $\mathcal{D}(\Phi)$ then $-\int_\gamma \mathrm{d}\Theta_\Phi$ (understood as a Stieltjes integral) is said to the Maslov index of γ.

Since the function $\Theta^{\mathrm{s}}_\varphi$ is single-valued, we have $\int_\gamma \mathrm{d}\Theta_\Phi = \int_\gamma \mathrm{d}\Theta^{\mathrm{r}}_\varphi$ for any closed path $\gamma \subset \mathcal{D}(\Phi)$ and any $\varphi \in \mathfrak{F}_\Phi$. It follows that the de Rham cohomology class of the 1-form $-(2\pi)^{-1}\mathrm{d}\Theta^{\mathrm{r}}_\varphi$ on $\mathcal{D}(\Phi)$ does not depend on the choice of $\varphi \in \mathfrak{F}_\Phi$. It is usually called the *Maslov class*. If the Maslov class is trivial then the Maslov index of any path $\gamma : [0,1] \to \mathcal{D}(\Phi)$ is equal to $\Theta_\Phi(\gamma(0)) - \Theta_\Phi(\gamma(1))$.

REMARK 2.10. Let ΛT^*M be the bundle of Lagrangian Grassmanians over T^*M, and let $\Lambda^{(k)}T^*M$ be the subbundle whose fibre $\Lambda^{(k)}_\vartheta T^*M$ over the point $\vartheta \in T^*M$ consists of subspaces $L \subset \Lambda_\vartheta T^*M$ such that $\dim L \bigcap V_\vartheta \ge k$. The condition (2.3) implies that $\mathrm{d}\Phi(V_\vartheta) \in \Lambda^{(1)}_\vartheta T^*M$ for all $\vartheta \in T^*M$. Instead of the path $\gamma \subset \mathcal{D}(\Phi)$, one can think of the corresponding path $\mathrm{d}\Phi(V_\gamma)$ in ΛT^*M, and then the Maslov index can be interpreted as the index of intersection of $\mathrm{d}\Phi(V_\gamma)$ with the set $\Lambda^{(2)}T^*M$. Alternatively, one can consider the complex structure on TT^*M with real and imaginary subspaces over $\vartheta \in T^*M$ being V_ϑ and H^φ_θ. After that the Maslov class can be defined in the spirit of [**Ar**], as the cohomology class of the 1-form $-\mathrm{d}(\arg\det^2 U_\varphi)$ where the unitary matrix U_φ is the radial part of the nondegenerate matrix $\varphi_{x\eta}(x^\star; y, \eta)$.

REMARK 2.11. Proposition 2.7 together with Remark 2.6 implies the following well known result (see, for instance, [**H1**, Theorem 3.2.1]).

PROPOSITION 2.12. *Let* $\mathcal{O}_j$ *be connected and simply connected open subsets of* $\mathcal{D}(\Phi)$, *and let* φ_j *be real-valued phase functions satisfying the conditions* $(\mathbf{a}_1)$–$(\mathbf{a}_3)$ *on* $\mathcal{O}_j$. *Then there exist integer numbers* m_{jk} *such that*

$$\frac{1}{2}\,\operatorname{sgn}(\varphi_j)_{\eta\eta}(x^\star; y, \eta) \;-\; \frac{1}{2}\,\operatorname{sgn}(\varphi_k)_{\eta\eta}(x^\star; y, \eta) \;=\; m_{jk} \tag{2.17}$$

for all $(y,\eta) \in \mathcal{O}_j \bigcap \mathcal{O}_k$.

Choosing an open cover $\{\mathcal{O}_j\}$ of $\mathcal{D}(\Phi)$ and real-valued phase functions φ_j satisfying $(\mathbf{a}_1)$–$(\mathbf{a}_3)$ on $\mathcal{O}_j$, one obtains an integer cocycle on $\mathcal{D}(\Phi)$ defined by (2.17). The corresponding Čech cohomology class is the image of the Maslov class under the standard isomorphism between de Rham and Čech cohomology groups. Corollary 2.1, (1.3) and (2.5) imply that

$$\operatorname{sgn}\varphi_{\eta\eta}(x^\star;y,\eta) \;=\; -\,\kappa\left(\mathrm{d}\Phi(V_\vartheta),H^{\varphi}_{\Phi(\vartheta)}\right) \tag{2.18}$$

for all real-valued phase functions φ satisfying $(\mathbf{a}_1)$ and $(\mathbf{a}_2)$. The equality (2.18) allows one to define the above Čech cohomology class in an invariant manner, without using the phase functions.

There are many other definitions and generalizations of the concept of Maslov index (see, for instance [**CLM**, **D**, **H2**, **RS**]). We won't elaborate further on this topic, since they are not needed for our purposes.

3. Fourier integral operators

Further on, we are always assuming that the symbols and amplitudes belong to Hörmander's classes S^m_{phg} (see, for instance, [**H2**, Chapter 18]). Recall that the conic support $\operatorname{cone\,supp} p$ of a function $p\in S^m_{\mathrm{phg}}$ is defined as the closure of the union $\bigcup_j \operatorname{supp} p_j$, where p_j are the positively homogeneous functions appearing in the asymptotic expansion $p\sim\sum_j p_j$.

3.1. Definition. Let V be an operator in the space of half-densities on M with Schwartz kernel $\mathcal{V}(x,y)$ (that is, $Vu(x)=\langle\mathcal{V}(x,\cdot),u(\cdot)\rangle$). The operator V is said to be a FIO of order m associated with Φ if $\mathcal{V}(x,y)$ can be represented modulo a smooth half-density by an oscillatory integral of the form

$$(2\pi)^{-n}\int_{T^*_yM} e^{i\varphi(x;y,\eta)}p(y,\eta)\,\left|\det\varphi_{x\eta}(x;y,\eta)\right|^{1/2}\,\varsigma(x;y,\eta)\,d\eta\,, \tag{3.1}$$

where $\varphi\in\mathfrak{F}_\Phi$, $p\in S^m_{\mathrm{phg}}$ is an amplitude with $\operatorname{cone\,supp} p\subset\mathcal{D}(\Phi)$, and ς is an arbitrary cut-off function such that

- $(\mathbf{a}_4)$ ς is positively homogeneous of degree 0 for large η,
- $(\mathbf{a}_5)$ ς is identically equal to 1 in a small neighbourhood of the set $\{x=x^\star(y,\eta)\}$ and vanishes outside another small neighbourhood of the set $\{x=x^\star(y,\eta)\}$,
- $(\mathbf{a}_6)$ on $\operatorname{supp}\varsigma$, the equation $\varphi_\eta(x;y,\eta)=0$ has the only solution $x=x^\star$ and $\det\varphi_{x\eta}(x;y,\eta)\neq 0$.

Note that

- in view of $(\mathbf{a}_3)$, the conditions $(\mathbf{a}_6)$ are fulfilled whenever $\operatorname{supp}\varsigma$ is sufficiently small,
- the right hand side of (3.1) behaves as a half-density with respect to x and y and, consequently, the corresponding operator acts in the space of half-densities.

Remark 3.1. The above definition of a FIO was introduced in [**LSV**] (see also [**SV**, Chapter 2]). It is equivalent to the traditional one, which is given in terms of local real-valued phase functions parametrizing the Lagrangian manifold

$$\{(y,\eta;x,\xi)\in\mathcal{D}(\Phi)\times T^*M\mid(x,\xi)=(x^\star(y,\eta),\xi^\star(y,\eta))\} \tag{3.2}$$

(see, for example, [**H1**] or [**Tr**]).

REMARK 3.2. One can define a FIO using (3.1) with an amplitude $\tilde{p}(x;y,\eta) \in S^m_{\mathrm{phg}}$ depending on $x \in M$ instead of $p(y,\eta)$. These two definitions are equivalent. Indeed, since $(x-x^\star)e^{i\varphi} = B\,\nabla_\eta\, e^{i\varphi}$ with some smooth matrix-function B, one can always remove the dependence on x by expanding $\tilde{p}$ into Taylor's series at the point $x = x^\star$, replacing $(x-x^\star)e^{i\varphi}$ with $B\,\nabla_\eta\, e^{i\varphi}$ and integrating by parts. In particular, this procedure shows that (3.1) with an x-dependent amplitude $\tilde{p}(x;y,\eta)$ defines an infinitely smooth half-density whenever $p \equiv 0$ in a conic neighbourhood of the set $\{x = x^\star\}$.

The following is [**LSV**, Theorem 1.8].

LEMMA 3.3. *The Schwartz kernel of a FIO associated with Φ can be represented modulo a smooth half-density by an integral of the form* (3.1) *with any phase function $\varphi \in \mathfrak{F}_\Phi$ and any cut-off function ς satisfying* $(\mathbf{a}_4)$–$(\mathbf{a}_6)$.

One can find all homogeneous terms in the expansion of the amplitude $p(y,\eta)$ by analysing asymptotic behaviour of the Fourier transforms of localizations of the distribution (3.1). This implies that $p(y,\eta)$ is determined modulo a rapidly decreasing function by the FIO and the phase function φ. It is not difficult to show that the conic support $\operatorname{cone\,supp} p$ does not depend on the choice of φ and is determined only by the FIO V itself (see, for instance, [**SV**, Section 2.7.4]). We shall denote it by $\operatorname{cone\,supp} V$.

3.2. Singular principal symbol. Let $\mathcal{V}(x,y)$ be given by (3.1) with $p \in S^m_{\mathrm{phg}}$, and let p_m be the leading homogeneous term of p. Let us fix a point $(y_0,\eta_0) \in \mathcal{D}(\Phi)$, denote $\Phi(y_0,\eta_0) = (x_0,\xi_0)$ and choose arbitrary local coordinates in a neighbourhood of x_0 such that $\det \xi^\star_\eta(y_0,\eta_0) \neq 0$. If ρ is a C^∞-function supported in a sufficiently small neighbourhood of x_0 then, applying the stationary phase formula, we see that

$$(3.3)\qquad \int e^{-i\lambda x\cdot\xi_0}\rho(x)\,\mathcal{V}(x,y_0)\,dx \;=\; c\,\rho(x_0)\,i^{\,-\Theta^{\mathrm{s}}_\varphi(y_0,\eta_0)}\,p_m(y_0,\eta_0)\,\lambda^m \;+\; O(\lambda^{m-1})$$

as $\lambda \to +\infty$, where $\Theta^{\mathrm{s}}_\varphi$ is the function defined by (2.14) and

$$c \;=\; e^{-i\lambda x_0\cdot\xi_0}\left(i^{\,\kappa_-((x^\star_\eta)^T\cdot\,\xi^\star_\eta)}\;|\det \xi^\star_\eta|^{-1/2}\right)\Big|_{(y,\eta)=(y_0,\eta_0)}$$

(see [**LSV**, Lemma 1.18] for details). Since c depends only on Φ and the choice of local coordinates, it follows that the function

$$(3.4)\qquad s_V(y,\eta) \;=\; i^{\,-\Theta^{\mathrm{s}}_\varphi(y,\eta)}\,p_m(y,\eta)$$

on $\mathcal{D}(\Phi)$ is uniquely defined by the operator V. We shall call it the *singular principal symbol* of the FIO V. In view of Remark 2.5(ii), the function s_V is continuous along any path on which $\operatorname{rank} x^\star_\eta$ is constant, but it may have jumps at the points where $x^\star_\eta$ changes its rank.

REMARK 3.4. The singular symbol s_V determines the FIO V modulo lower terms and is uniquely determined by V. However, in the general case, it is a discontinuous function which makes it inconvenient to deal with in applications.

3.3. Classical principal symbol. Let $\mathcal{G}(\Phi)$ be the smooth principal $\mathbb{Z}$-bundle over $\mathcal{D}(\Phi)$ whose fibre at a point $(y,\eta)\in\mathcal{D}(\Phi)$ is the additive group mapping different branches of $\Theta^{\mathrm{r}}_\varphi(y,\eta)$ into each other. In view of Remark 2.3, $\mathcal{G}(\Phi)$ does not depend on the choice of $\varphi\in\mathfrak{F}$. It is usually called the *Maslov principal bundle* (another definition and further discussions can be found, for instance, in [**RS**, Section 12.6]).

Factoring out $4\mathbb{Z}$, we obtain a principal $\mathbb{Z}_4$-bundle $\mathcal{G}_4(\Phi)$ over $\mathcal{D}(\Phi)$, which can be thought of as a smooth 4-fold covering of $\mathcal{D}(\Phi)$ (the group structure is not important for our purposes). It is trivial (that is, consists of four disconnected components) if and only if

(C$_2$) the Maslov cohomology class of $\mathcal{D}(\Phi)$ is trivial modulo $4\mathbb{Z}$.

Note that **(C$_2$)** holds true whenever $\mathcal{D}(\Phi)$ is simply connected.

It is convenient to consider $\Theta^{\mathrm{r}}_\varphi$ and Θ_Φ as single-valued functions on $\mathcal{G}_4(\Phi)$ and to define the function $q=i^{-\Theta^{\mathrm{r}}_\varphi}p$ on $\mathcal{G}_4(\Phi)$ to be the *full symbol* of the FIO V. Under this definition,

$$q(y,\eta)\,\left(\det{}^2\varphi_{x\eta}(x;y,\eta)\right)^{1/4} \;=\; p(y,\eta)\,\left|\det\varphi_{x\eta}(x;y,\eta)\right|^{1/2}$$

is a single-valued function on $M\times\mathcal{D}(\Phi)$ and (3.1) can be rewritten in the form

$$(3.5)\qquad (2\pi)^{-n}\int_{T_y^*M} e^{i\varphi(x;y,\eta)}q(y,\eta)\,\left(\det{}^2\varphi_{x\eta}(x;y,\eta)\right)^{1/4}\,\varsigma(x;y,\eta)\,d\eta\,.$$

The leading homogeneous term $i^{-\Theta^{\mathrm{r}}_\varphi}p_m$ of the amplitude q is said to be the *classical principal symbol* of the FIO V. We shall denote it by σ_V. Clearly,

$$(3.6)\qquad \sigma_V \;=\; i^{-\Theta_\Phi}\,s_V\,.$$

Proposition 2.7 implies that σ_V is uniquely determined by the operator V as a function on $\mathcal{G}_4(\Phi)$.

REMARK 3.5. The classical principal symbol σ_V can also be interpreted as a multi-valued function on $\mathcal{D}(\Phi)$ or a section of the linear bundle associated with the Maslov cohomology class (see, for instance, [**H1**] and [**Tr**]). However, under any approach, its precise value at a given point is not uniquely defined and depends on the choice of branch of the function Θ_Φ.

Given a point $\vartheta\in\mathcal{D}(\Phi)$, one can enumerate the branches of the function Θ_Φ (or, equivalently, the fibres of $\mathcal{G}_4(\Phi)$) over any connected and simply connected neighbourhood of ϑ by the value of Θ_Φ at the point ϑ. Let us denote by $\Theta_{\Phi,\vartheta}$ the branch of Θ_Φ which is equal to 0 at ϑ and define

$$(3.7)\qquad \sigma_{V,\vartheta}(y,\eta) \;=\; i^{-\Theta_{\Phi,\vartheta}(y,\eta)}\,s_V(y,\eta)\,.$$

In other words, $\sigma_{V,\vartheta}$ is the branch of σ_V such that

$$(3.8)\qquad \sigma_{V,\vartheta}(\vartheta) \;=\; s_V(\vartheta)\,.$$

EXAMPLE 3.6. If Φ is the identity transformation then $\varphi_{x\eta}(x^\star;y,\eta)\equiv I$ for all $\varphi\in\mathfrak{F}_\Phi$ and the corresponding FIO V is a ψDO (see, for example, [**Sh**, Theorem 19.1]). If we put $\Theta^{\mathrm{r}}_\varphi\equiv 0$ then, for all $\vartheta\in T^*M$, $\sigma_{V,\vartheta}$ coincides with the principal symbol σ_V of the ψDO V in the sense of the theory of pseudodifferential operators.

EXAMPLE 3.7. More generally, if Φ is homotopic to the identity transformation then Φ satisfies $(\mathbf{C}_2)$ and one can single out a branch of Θ_Φ by assigning to $\Theta_\Phi(\vartheta)$ the value $m(\vartheta)$ obtained from 0 by a continuous transformation of a phase function corresponding to $\Phi = I$ into φ. This idea works, in particular, if Φ is the shift along geodesics or billiard trajectories (see, for instance, [**SV**, Section 2.6.3] or [**JSS**]). Under this definition, $\sigma_V(y,\eta) = i^{-m(\vartheta)}\sigma_{V,\vartheta}(y,\eta)$.

4. Symbolic calculus for FIOs

4.1. Main theorem and corollaries. The parts (1) of the following statements are well known and can be found in textbooks on FIOs. The only novelty is the explicit formulae for the principal symbols in the parts (2) and (3).

THEOREM 4.1. *Let V_1, V_2 be FIOs of order m_1, m_2 associated with canonical transformations Φ_1, Φ_2. If $\Phi_1(T^*M) \subset \mathcal{D}(\Phi_2^{-1})$ then*

(1) *the composition $V_2^*V_1$ is a FIO of order $m_1 + m_2$ associated with the canonical transformation $\Phi = \Phi_2^{-1} \circ \Phi_1$ such that* $\operatorname{cone\,supp}(V_2^*V_1) \subset \big(\operatorname{cone\,supp} V_1 \bigcap \Phi^{-1}(\operatorname{cone\,supp} V_2)\big)$,

(2) $s_{V_2^*V_1}(\vartheta) = i^{\,k(\vartheta)}\, s_{V_1}(\vartheta)\, \overline{s_{V_2}(\Phi(\vartheta))}$ *for all* $\vartheta \in \mathcal{D}(\Phi)$,

(3) $\sigma_{V_2^*V_1,\vartheta}(y,\eta) = i^{\,k(\vartheta)}\, \sigma_{V_1,\vartheta}(y,\eta)\, \overline{\sigma_{V_2,\Phi(\vartheta)}(\Phi(y,\eta))}$ *for all* $\vartheta \in \mathcal{D}(\Phi)$ *and* $(y,\eta) \in \mathcal{D}(\Phi)$,

where $k(\vartheta) = \varkappa\big(\mathrm{d}\Phi_1(V_\vartheta), \mathrm{d}\Phi_2(V_{\Phi(\vartheta)})\big)$ *is the modified Kashiwara index defined by* (1.4).

The proof of Theorem 4.1 is given in Subsection 4.2.

Taking $V_1 = I$ and noting that $\varkappa\big(V_\vartheta, \mathrm{d}\Phi_2(V_{\Phi(\vartheta)})\big) = 0$, we obtain

COROLLARY 4.2. *Let V be a FIO of order m associated with a canonical transformation Φ. Then the adjoint operator V^* is a FIO of order m associated with the inverse canonical transformation Φ^{-1} such that*

(1) $\operatorname{cone\,supp} V^* \subset \Phi(\operatorname{cone\,supp} V)$,

(2) $s_{V^*}(\vartheta) = \overline{s_V(\Phi^{-1}(\vartheta))}$,

(3) $\sigma_{V^*,\vartheta}(y,\eta) = \overline{\sigma_{V,\Phi^{-1}(\vartheta)}(\Phi^{-1}(y,\eta))}$.

Theorem 4.1 and Corollary 4.2 immediately imply

COROLLARY 4.3. *Let V_1, V_2 be as in Theorem 4.1. If $\Phi_1(T^*M) \subset \mathcal{D}(\Phi_2)$ then*

(1) *the composition V_2V_1 is a FIO of order m_1+m_2 associated with the canonical transformation $\Phi = \Phi_2 \circ \Phi_1$ such that* $\operatorname{cone\,supp}(V_2V_1) \subset \big(\operatorname{cone\,supp} V_1 \bigcap \Phi_1^{-1}(\operatorname{cone\,supp} V_2)\big)$,

(2) $s_{V_2V_1}(\vartheta) = i^{\,k(\vartheta)}\, s_{V_1}(\vartheta)\, s_{V_2}(\Phi_1(\vartheta))$,

(3) $\sigma_{V_2V_1,\vartheta}(y,\eta) = i^{\,k(\vartheta)}\, \sigma_{V_1,\vartheta}(y,\eta)\, \sigma_{V_2,\Phi_1(\vartheta)}(\Phi_1(y,\eta))$,

where $k(\vartheta) = \varkappa\big(\mathrm{d}\Phi_1(V_\vartheta), \mathrm{d}\Phi_2^{-1}(V_{\Phi(\vartheta)})\big)$.

The above results imply the following refined version of Egorov's theorem.

COROLLARY 4.4. *Let V_k be FIOs associated with a canonical transformation Φ. If A is a ψDO with* $\operatorname{cone\,supp} A \subset \Phi_1(T^*M)$ *then the composition $B = V_2^*AV_1$ is a ψDO with the principal symbol*

$$\sigma_B(y,\eta) \;=\; \sigma_{V_1,\vartheta}(y,\eta)\, \sigma_A(\Phi(y,\eta))\, \overline{\sigma_{V_2,\vartheta}(y,\eta)}\,, \qquad \forall \vartheta \in \mathcal{D}(\Phi)\,. \tag{4.1}$$

PROOF. The formula (4.1) for the principal symbol is obtained by applying Corollary 4.3 to the composition AV_1 and then Theorem 4.1 to the composition of V_2^* and AV_1. □

4.2. Proof of Theorem 4.1. Let

$$\Phi_1 : (y,\eta) \mapsto \left(z^{(1)}(y,\eta), \zeta^{(1)}(y,\eta)\right),$$
$$\Phi_2 : (x,\xi) \mapsto \left(z^{(2)}(x,\xi), \zeta^{(2)}(x,\xi)\right),$$
$$\Phi : (y,\eta) \mapsto (x^\star(y,\eta), \xi^\star(y,\eta)),$$

so that

$$z^{(2)}(x^\star,\xi^\star) = z^{(1)}(y,\eta) \quad \text{and} \quad \zeta^{(2)}(x^\star,\xi^\star) = \zeta^{(1)}(y,\eta)\,. \tag{4.2}$$

The proof proceeds in several steps.

4.2.1. *Step 1.* Since we can split V_1 into the sum of FIOs with small conic supports, it is sufficient to prove the theorem assuming that $\operatorname{cone\,supp} V_1$ lies in an arbitrarily small conic neighbourhood $\mathcal{O}_1$ of a fixed point $(y_0,\eta_0) \in \mathcal{D}(\Phi_1)$.

Let us represent the Schwartz kernels $\mathcal{V}_1(z,y)$ and $\mathcal{V}_2(z,x)$ of the FIO V_j by oscillatory integrals

$$\begin{aligned} &(2\pi)^{-n} \int_{T_y^*M} e^{i\varphi^{(1)}(z;y,\eta)} p_1(y,\eta) \left|\det \varphi^{(1)}_{z\eta}\right|^{1/2} \varsigma_1(z;y,\eta)\, d\eta\,, \\ &(2\pi)^{-n} \int_{T_x^*M} e^{i\varphi^{(2)}(z;x,\xi)} p_2(x,\xi) \left|\det \varphi^{(2)}_{x\xi}\right|^{1/2} \varsigma_2(z;x,\xi)\, d\xi \end{aligned} \tag{4.3}$$

of the form (3.1) with phase functions $\varphi_j \in \mathfrak{F}_{\Phi_j}$ and $p_j \in S^{m_j}_{\mathrm{phg}}$. Then the Schwartz kernel of the composition $V_2^* V_1$ coincides with

$$\begin{aligned} &(2\pi)^{-2n} \iiint e^{i\psi(x,\xi;z;y,\eta)} b(x,\xi;z;y,\eta)\, d\eta\, dz\, d\xi \\ &\quad = (2\pi)^{-2n} \iiint |\eta|^{-n} e^{i\psi(x,|\eta|\xi;z;y,\eta)} b(x,|\eta|\xi;z;y,\eta)\, d\eta\, dz\, d\xi\,, \end{aligned} \tag{4.4}$$

where the integrals are taken over $T_x^*M \times M \times T_y^*M$,

$$\psi(x,\xi;z;y,\eta) \;=\; \varphi^{(1)}(z;y,\eta) - \varphi^{(2)}(z;x,\xi) \tag{4.5}$$

and

$$\begin{aligned} &b(x,\xi;z;y,\eta) \\ &\quad = p_1(y,\eta)\, \overline{p_2(x,|\eta|\xi)} \left|\det \varphi^{(1)}_{x\eta}\right|^{1/2} \left|\det \varphi^{(2)}_{x\eta}\right|^{1/2} \varsigma_1(z;y,\eta)\, \varsigma_2(z;x,\xi)\,. \end{aligned} \tag{4.6}$$

Now we are going to apply the stationary phase method to the integral with respect to the variables z and ξ, considering $|\eta|$ as a large parameter. A rigorous justification of the stationary phase formula for non-convergent integrals of this type can be found, for instance, in [**SV**, Appendix C].

In view of $(\mathbf{a}_6)$, the equations $\psi_\xi = 0$ and $\psi_z = 0$ are equivalent to

$$z = z^{(2)}(x,\xi) \quad \text{and} \quad \varphi^{(1)}_z(z;y,\eta) = \varphi^{(2)}_z(z;x,\xi)\,. \tag{4.7}$$

If the cut-off functions ς_j have sufficiently small supports then, in view of $(\mathbf{a}_6)$, the equations (4.7) imply that the points $\left(z^{(1)}(y,\eta), \zeta^{(1)}(y,\eta)\right)$ and $\left(z^{(2)}(x,\xi), \zeta^{(2)}(x,\xi)\right)$

are close to each other. Thus we can assume without loss of generality that $\operatorname{cone\,supp} V_2$ lies in a small conic neighbourhood $\mathcal{O}_2$ of the set $\Phi(\mathcal{O}_1)$.

Let $\Phi_1(y_0,\eta_0) = (z_0,\zeta_0) = \Phi_2(x_0,\xi_0)$, and let $z = (z_1,\dots,z_n)$ be local coordinates in a neighbourhood of z_0 such that the corresponding horizontal subspace $H_{(z_0,\zeta_0)}$ is transversal to the images $\mathrm{d}\Phi_1(V_{(y_0,\eta_0)})$ and $\mathrm{d}\Phi_2(V_{(x_0,\xi_0)})$. Then $\det \zeta^{(1)}_\eta(y_0,\eta_0) \neq 0$ and $\det \zeta^{(1)}_\eta(y_0,\eta_0) \neq 0$ (see Example 2.4) and, consequently, we have $\det \zeta^{(1)}_\eta(y,\eta) \neq 0$ for all $(y,\eta)\in\mathcal{O}_1$ and $\det \zeta^{(2)}_\xi(x,\xi) \neq 0$ for all $(x,\xi)\in\mathcal{O}_2$ provided that the neighbourhoods $\mathcal{O}_j$ are small enough. It follows that the phase functions

$$
\begin{aligned}
\varphi^{(1)}(z;y,\eta) &= (z - z^{(1)}(y,\eta))\cdot\zeta^{(1)}(y,\eta)\,,\\
\varphi^{(2)}(z;x,\xi) &= (z - z^{(2)}(x,\xi))\cdot\zeta^{(2)}(x,\xi)
\end{aligned}
\tag{4.8}
$$

satisfy the conditions $(\mathbf{a}_1)$–$(\mathbf{a}_3)$ for all $(y,\eta)\in\mathcal{O}_1$ and $(x,\xi)\in\mathcal{O}_2$.

4.2.2. *Step 2.* Since the singular and classical principal symbols do not depend on the choice of the phase function, it is sufficient to prove the theorem assuming that $\varphi^{(j)}$ are given by (4.8). In this case

$$
\varphi^{(1)}_{z\eta}(z;y,\eta) = \zeta^{(1)}_\eta(y,\eta)\,,\quad \varphi^{(2)}_{z\xi}(z;x,\xi) = \zeta^{(2)}_\xi(x,\xi)
\tag{4.9}
$$

and the equations (4.7) turn into

$$
z = z^{(2)}(x,\xi) \quad\text{and}\quad \zeta^{(2)}(x,\xi) = \zeta^{(1)}(y,\eta)\,.
\tag{4.10}
$$

Since $\det\zeta^{(2)}_\xi \neq 0$, if the sets $\mathcal{O}_1$ and $\operatorname{supp}\varsigma_j$ are small enough then the second equation (4.10) has a unique ξ-solution $\hat\xi(x;y,\eta)$ such that

$$
\hat\xi(x^\star(y,\eta);y,\eta) = \xi^\star(y,\eta)\,.
\tag{4.11}
$$

Thus the stationary point is $(z,\xi) = (z^{(2)}(x,\hat\xi),\hat\xi)$. It is unique and non-degenerate because, in view of (4.8) and (4.9),

$$
\begin{pmatrix} \psi_{zz} & \psi_{z\xi} \\ \psi_{\xi z} & \psi_{\xi\xi} \end{pmatrix} = \begin{pmatrix} 0 & \zeta^{(2)}_\xi \\ \left(\zeta^{(2)}_\xi\right)^T & \psi_{\xi\xi} \end{pmatrix}.
\tag{4.12}
$$

Now, applying the stationary phase formula, we see that (4.4) coincides modulo a smooth function with

$$
(2\pi)^{-n}\int e^{i\varphi(x;y,\eta)}\left|\det\zeta^{(2)}_\xi(x,\hat\xi)\right|^{-1}\tilde p(x;y,\eta)\,\varsigma(x;y,\eta)\,\mathrm{d}\eta\,,
\tag{4.13}
$$

where

$$
\varphi(x;y,\eta) = \psi(x,\hat\xi;z^{(2)}(x,\hat\xi);y,\eta) = \left(z^{(2)}(x,\hat\xi) - z^{(1)}(y,\eta)\right)\cdot\zeta^{(1)}(y,\eta)\,,
\tag{4.14}
$$

$\tilde p$ is an amplitude of class $S^{m_1+m_2}_{\mathrm{phg}}$ with

$$
\operatorname{cone\,supp}\tilde p \subset \{(x;y,\eta)\mid (y,\eta)\in\operatorname{cone\,supp} p_1,\ (x,\hat\xi)\in\operatorname{cone\,supp} p_2\},
$$

and

$$
\varsigma(x;y,\eta) = \varsigma_1(z^{(2)}(x,\hat\xi);y,\eta)\,\varsigma_2(z^{(2)}(x,\hat\xi);x,\hat\xi)\,.
$$

is a cut-off function satisfying $(\mathbf{a}_4)$ and $(\mathbf{a}_5)$. The leading homogeneous term of the amplitude $\tilde p$ is equal to

$$
\tilde p_m(x;y,\eta) = p_1(y,\eta)\,\overline{p_2(z^{(2)}(x,\hat\xi),\hat\xi)}\left|\det\zeta^{(1)}_\eta(y,\eta)\right|^{1/2}\left|\det\zeta^{(2)}_\xi(x,\hat\xi)\right|^{1/2},
\tag{4.15}
$$

where p_{j,m_j} are the leading homogeneous terms of the amplitudes p_j (here we have used the fact that the signature of the Hessian (4.12) is equal to zero).

4.2.3. *Step 3.* In view of (4.2) and (4.11),

$$\varphi(x^\star; y, \eta) \;=\; 0\,. \tag{4.16}$$

Since

$$\psi_z(x, \hat\xi, z^{(2)}(x, \hat\xi), y, \eta) = \psi_\xi(x, \hat\xi, z^{(2)}(x, \hat\xi), y, \eta) = 0 \tag{4.17}$$

for all x, y, η, we also have

$$\varphi_x(x, y, \eta) = \psi_x(x, \hat\xi, z^{(2)}(x, \hat\xi), y, \eta) = (z_x^{(2)}(x, \hat\xi))^T\, \zeta^{(2)}(x, \hat\xi)\,.$$

This equality and (2.4) imply that $\varphi_x(x, y, \eta) = \hat\xi(x, y, \eta)$. Now, by (4.11),

$$\varphi_x(x^\star; y, \eta) \;=\; \xi^\star\,. \tag{4.18}$$

Similarly, from (4.17) and (2.4) it follows that

$$\begin{aligned} \varphi_\eta(x, y, \eta) &= \psi_\eta(x, \hat\xi, z^{(2)}(x, \hat\xi), y, \eta) \\ &= \nabla_\eta \left((z - z^{(1)}(y, \eta)) \cdot \zeta^{(1)}(y, \eta) \right)\Big|_{z = z^{(2)}(x, \hat\xi)} \\ &= (z^{(2)}(x, \hat\xi) - z^{(1)}(y, \eta)) \cdot \zeta_\eta^{(1)}(y, \eta)\,. \end{aligned} \tag{4.19}$$

Differentiating the identity $\zeta^{(2)}(x, \hat\xi) \equiv \zeta^{(1)}(y, \eta)$, we obtain

$$\begin{aligned} \hat\xi_x &= -\left(\zeta_\xi^{(2)}(x, \hat\xi) \right)^{-1} \zeta_x^{(2)}(x, \hat\xi)\,, \\ \hat\xi_\eta &= \left(\zeta_\xi^{(2)}(x, \hat\xi) \right)^{-1} \zeta_\eta^{(1)}(y, \eta)\,. \end{aligned} \tag{4.20}$$

The first equality (4.20) and (4.19) imply that

$$\begin{aligned} \varphi_{x\eta}(x; y, \eta) &= \left(\nabla_x\, z^{(2)}(x, \hat\xi) \right)^T \zeta_\eta^{(1)}(y, \eta) \\ &= \left(z_x^{(2)}(x, \hat\xi) - z_\xi^{(2)}(x, \hat\xi) \left(\zeta_\xi^{(2)}(x, \hat\xi) \right)^{-1} \zeta_x^{(2)}(x, \hat\xi) \right)^T \zeta_\eta^{(1)}(y, \eta)\,. \end{aligned}$$

In view of (2.1) and (2.2),

$$\begin{aligned} & z_x^{(2)} - z_\xi^{(2)} \left(\zeta_\xi^{(2)} \right)^{-1} \zeta_x^{(2)} \\ &\qquad = \left((\zeta_\xi^{(2)})^T \right)^{-1} \left((\zeta_\xi^{(2)})^T z_x^{(2)} - (z_\xi^{(2)})^T \zeta_x^{(2)} \right) = \left((\zeta_\xi^{(2)})^T \right)^{-1} \end{aligned}$$

Therefore

$$\varphi_{x\eta}(x; y, \eta) \;=\; \left(\zeta_\xi^{(2)}(x, \hat\xi) \right)^{-1} \zeta_\eta^{(1)}(y, \eta)\,. \tag{4.21}$$

The second equality (4.20), (4.19) and (4.11) imply that

$$\begin{aligned} \varphi_{\eta\eta}(x^\star; y, \eta) &= \left(\nabla_\eta z^{(2)}(x, \hat\xi) - z_\eta^{(1)}(y, \eta) \right)^T \zeta_\eta^{(1)}(y, \eta) \Big|_{x = x^\star} \\ &= \left(z_\xi^{(2)}(x^\star, \xi^\star) \left(\zeta_\xi^{(2)}(x^\star, \xi^\star) \right)^{-1} \zeta_\eta^{(1)}(y, \eta) - z_\eta^{(1)}(y, \eta) \right)^T \zeta_\eta^{(1)}(y, \eta)\,. \end{aligned}$$

Since the matrix $\varphi_{\eta\eta}$ is symmetric, the above can be rewritten as

$$(4.22)\quad \varphi_{\eta\eta}(x^\star;y,\eta) = (\zeta_\eta^{(1)})^T \left(z_\xi^{(2)}(x^\star,\xi^\star)\left(\zeta_\xi^{(2)}(x^\star,\xi^\star)\right)^{-1} - z_\eta^{(1)}\left(\zeta_\eta^{(1)}\right)^{-1}\right)\zeta_\eta^{(1)},$$

where $x^\star$, $\xi^\star$, $z_\eta^{(1)}$ and $\zeta_\eta^{(1)}$ are evaluated at (y,η).

4.2.4. *Step 4.* Since $\det \zeta_\eta^{(j)} \neq 0$, the equalities (4.16), (4.18) and (4.21) imply that $\varphi \in \mathfrak{F}_\Phi$ and ς satisfies $(\mathbf{a}_6)$ provided that $\mathcal{O}_1$ is small enough. Thus (4.13) defines the Schwartz kernel of a FIO associated with the canonical transformation Φ. Applying the procedure described in Remark 3.2, we can remove the dependence of $\tilde p$ on x and rewrite (4.13) in the form

$$(4.23)\qquad (2\pi)^{-n}\int e^{i\varphi(x;y,\eta)}\, p(y,\eta)\, \left|\det \varphi_{x\eta}(x;y,\eta)\right|^{1/2}\, \varsigma(x;y,\eta)\, d\eta$$

where $p(y,\eta)$ is an amplitude of class $S^{m_1+m_2}_{\mathrm{phg}}$. From (4.11) it follows that

$$\operatorname{cone\,supp} p \subset \{(y,\eta) \in \operatorname{cone\,supp} p_1 \mid \Phi(y,\eta) \in \operatorname{cone\,supp} p_2\}.$$

This completes the proof of the first statement of the theorem.

4.2.5. *Step 5.* In order to prove the second statement, let us note that, in view of (4.11), (4.15) and (4.21), the leading homogeneous term p_m of the amplitude p in (4.23) is equal to

$$(4.24)\qquad p_m(y,\eta) = p_{1,m_1}(y,\eta)\,\overline{p_{2,m_2}\left(\Phi(y,\eta)\right)},$$

where p_{j,m_j} are the leading homogeneous terms of the amplitudes p_j. Since the phase functions $\varphi^{(j)}$ and φ are real, (2.15) implies that

$$\begin{aligned} p_{1,m_1}(y,\eta) &= i^{-\kappa_+\left(\varphi_{\eta\eta}^{(1)}(y,\eta)\right)}\, s_{V_1}(y,\eta)\,, \\ p_{2,m_2}\left(\Phi(y,\eta)\right) &= i^{-\kappa_+\left(\varphi_{\xi\xi}^{(2)}(\Phi(y,\eta))\right)}\, s_{V_2}\left(\Phi(y,\eta)\right)\,, \\ s_{V_2^*V_1}(y,\eta) &= i^{\kappa_+(\varphi_{\eta\eta}(x^\star;y,\eta))} p_m(y,\eta)\,. \end{aligned}$$

In view of (2.11), we have

$$\kappa_+\left(\varphi_{\eta\eta}^{(1)}(y,\eta)\right) = \kappa_-(A_1) \quad\text{and}\quad \kappa_+\left(\varphi_{\xi\xi}^{(2)}\left(\Phi(y,\eta)\right)\right) = \kappa_-(A_2)\,,$$

where

$$A_1 = z_\eta^{(1)}(y,\eta)\left((\zeta_\eta^{(1)}(y,\eta)\right)^{-1}, \quad A_2 = z_\xi^{(2)}\left(\Phi(y,\eta)\right)\left(\zeta_\xi^{(2)}\left(\Phi(y,\eta)\right)\right)^{-1}.$$

Also, by (4.22), $\kappa_+\left(\varphi_{\eta\eta}(x^\star;y,\eta)\right) = \kappa_-(A_1 - A_2)$. Therefore from (4.24) it follows that

$$s_{V_2^*V_1}(y,\eta) = i^{\,h(y,\eta)}\, s_{V_1}(y,\eta)\,\overline{s_{V_2}\left(\Phi(y,\eta)\right)}\,,$$

with $h(y,\eta) = \kappa_-(A_2) - \kappa_-(A_1) + \kappa_-(A_1 - A_2)$. It remains to notice that, in view of (1.7) and (2.12), $h(y,\eta) = \varkappa\left(\mathrm{d}\Phi_1(V_{(y,\eta)}), \mathrm{d}\Phi_2(V_{\Phi(y,\eta)})\right)$.

4.2.6. *Step 6.* Finally, for each fixed $\vartheta_j \in \mathcal{D}(\Phi_j)$ and $\vartheta \in \mathcal{D}(\Phi)$, the principal symbols σ_{V_j,ϑ_j} and $\sigma_{V_2^*V,\vartheta}$ are smooth single-valued functions on simply connected neighbourhoods of these points. From (3.6), (3.8) and part (2) it follows that

- $\sigma_{V_2^*V_1,\vartheta}(y,\eta) = i^m\,\sigma_{V_1,\vartheta_1}(y,\eta)\,\overline{\sigma_{V_2,\vartheta_2}(\Phi(y,\eta))}$ with an integer m independent of (y,η);
- $\sigma_{V_2^*V_1,\vartheta}(\vartheta) = i^{k(\vartheta)}\,\sigma_{V_1,\vartheta}(\vartheta)\,\overline{\sigma_{V_2,\Phi(\vartheta)}(\Phi(\vartheta))}$.

These two statements imply the required formula for the classical principal symbols. □

References

[Ar] V. I. Arnol′d, *On a characteristic class entering into conditions of quantization* (Russian), Funkcional. Anal. i Priložen. **1** (1967), 1–14. MR0211415 (35 #2296)

[CLM] S. E. Cappell, R. Lee, and E. Y. Miller, *On the Maslov index*, Comm. Pure Appl. Math. **47** (1994), no. 2, 121–186, DOI 10.1002/cpa.3160470202. MR1263126 (95f:57045)

[D] J. J. Duistermaat, *Fourier integral operators*, Progress in Mathematics, vol. 130, Birkhäuser Boston, Inc., Boston, MA, 1996. MR1362544 (96m:58245)

[JSS] D. Jakobson, Y. Safarov and A. Strohmaier, *The Semiclassical theory of discontinuous systems and ray-splitting billiards,* http://arxiv.org/abs/1301.6783

[H1] L. Hörmander, *Fourier integral operators. I*, Acta Math. **127** (1971), no. 1-2, 79–183. MR0388463 (52 #9299)

[H2] L. Hörmander, *The Analysis of Linear Partial Differential Operators.* Springer-Verlag, New York, 1984.

[LSV] A. Laptev, Yu. Safarov, and D. Vassiliev, *On global representation of Lagrangian distributions and solutions of hyperbolic equations*, Comm. Pure Appl. Math. **47** (1994), no. 11, 1411–1456, DOI 10.1002/cpa.3160471102. MR1296784 (95m:58121)

[RS] G. Rudolph and M. Schmidt, *Differential geometry and mathematical physics. Part I. Manifolds, Lie groups and Hamiltonian systems*, Theoretical and Mathematical Physics, Springer, Dordrecht, 2013. MR3012162

[SV] Yu. Safarov and D. Vassiliev, *The asymptotic distribution of eigenvalues of partial differential operators*, Translations of Mathematical Monographs, vol. 155, American Mathematical Society, Providence, RI, 1997. Translated from the Russian manuscript by the authors. MR1414899 (97h:58163)

[Sh] M. A. Shubin, *Pseudodifferential operators and spectral theory*, Springer Series in Soviet Mathematics, Springer-Verlag, Berlin, 1987. Translated from the Russian by Stig I. Andersson. MR883081 (88c:47105)

[Tr] F. Trèves, *Introduction to pseudodifferential and Fourier integral operators. Vol. 2. Fourier integral operators*, Plenum Press, New York-London, 1980. The University Series in Mathematics. MR597145 (82i:58068)

Department of Mathematics, King's College London, Strand, London WC2R 2LS, United Kingdom

E-mail address: `yuri.safarov@kcl.ac.uk`

Contemporary Mathematics
Volume **630**, 2014
http://dx.doi.org/10.1090/conm/630/12671

Strichartz Estimates on Exterior Polygonal Domains

Dean Baskin, Jeremy L. Marzuola, and Jared Wunsch

ABSTRACT. Using a new local smoothing estimate of the first and third authors, we prove local-in-time Strichartz and smoothing estimates without a loss exterior to a large class of polygonal obstacles with arbitrary boundary conditions and global-in-time Strichartz estimates without a loss exterior to a large class of polygonal obstacles with Dirichlet boundary conditions. In addition, we prove a global-in-time local smoothing estimate in exterior wedge domains with Dirichlet boundary conditions and discuss some nonlinear applications.

1. Introduction

In this paper we prove a family of local- and global-in-time Strichartz estimates for solutions to the Schrödinger equation

$$\begin{cases} (D_t + \Delta)\, u(t,x) = 0 \\ \qquad\qquad u(0,x) = f(x), \end{cases} \tag{1}$$

where $D_t = i^{-1}\partial_t$, Δ is the negative definite Laplace-Beltrami operator on domains of the form $X = \mathbb{R}^2 \setminus P$ for P any non-trapping polygonal region such that no three vertices are collinear (as defined in the recent work of the first and last author [**2**]), and where we take either Dirichlet or Neumann boundary conditions for the Laplacian for the local result and only Dirichlet boundary conditions for the global result.[1] These assumptions and the resulting restrictions on allowed obstacles P are discussed in detail as Assumptions 1, 2 and 3 in Section 2 of [**2**], to which we refer the reader for more details. The main tools we require for the proof are the local smoothing estimate on such domains (due to the first and third authors) and the Strichartz estimates on wedge domains (due to Ford [**18**]).

2010 *Mathematics Subject Classification.* 35Q41, 58J50.

The first author acknowledges the support of NSF postdoctoral fellowship DMS-1103436, and the third author was partially supported by NSF grant DMS-1001463.

The second author was partially supported by an IBM Junior Faculty Development Award through the University of North Carolina and would like to thank the University of Chicago for hosting him during the completion of part of this work. The third author is grateful to Fabrice Planchon and John Stalker for a series of discussions of smoothing and Strichartz estimates on cones over a period of several years. In addition, we thank Fabrice Planchon for a careful reading of the result and providing several references for the scattering theory discussion in the final section.

[1]The essential distinction between the Dirichlet and Neumann cases for our purposes is that low-energy resolvent estimates seem to be readily available in the literature in the Dirichlet case (e.g. in [**7**]) but not in the Neumann case.

We note here that to define the Laplacian, we use the standard Friedrichs extension, which is the canonical self-adjoint extension of a non-negative densely defined symmetric operator as defined in for instance [**4**, **16**]. The Neumann Laplacian is taken to be the usual Friedrichs extension of the Laplace operator acting on smooth functions which vanish in a neighborhood of the vertices. The Dirichlet Laplacian is taken to be the typical Friedrichs extension of the Laplace operator acting on smooth functions which have compact support contained in X.

We now briefly discuss the geometric restrictions on P needed to apply the results of [**2**]. In particular, we review the non-trapping assumption on the exterior of P as it is an important restriction in all exterior domain results. Let P be a polygonal domain in $\mathbb{R}^2$, not necessarily connected. A *geometric geodesic* on $\mathbb{R}^2 \backslash P$ that does not pass through the vertices of P is defined as a continuous curve that is a concatenation of maximally extended straight line segments in $\mathbb{R}^2 \backslash P$, such that on ∂P, successive segments make equal angles with the boundary ("specular reflection"). More generally, a geometric geodesic is one that may pass through the vertices of P in such a way that it is locally a uniform limit of geometric geodesics missing the vertices. This means that in general such a geodesic has two possible continuations each time it hits a vertex, corresponding to taking the limit of families approaching the vertex from the left and right sides.

We let B be a closed ball containing P in its interior. We say that P is *non-trapping* if there exists $T > 0$ such that every geometric geodesic starting in B leaves B in time less than T (this condition is of course independent of the choice of B). We assume henceforth that P is non-trapping.

In addition to assuming that P is non-trapping, we also require the assumption that no three vertices of P are collinear along geometric geodesics. We further remark that the third assumption of [**2**], requiring that cone points be pairwise non-conjugate, is automatically satisfied for the Euclidean domains under consideration here.

We recall that admissible Strichartz exponents for the Schrödinger equation in dimension $n = 2$ are given by the following:

$$\frac{2}{p} + \frac{2}{q} = 1, \quad p, q \geq 2, \quad (p, q) \neq (2, \infty). \tag{2}$$

We are now ready to state the main result of this note.

THEOREM 1. *For any (p, q) satisfying equation (2), there is a constant $C_{p,q,T}$ so that on $I = [0, T]$*

$$\left\|e^{-it\Delta} f\right\|_{L^p(I, L^q(X))} \leq C_{p,q,T} \, \|f\|_{L^2(X)}$$

for all $u_0 \in L^2(X)$. If X has Dirichlet boundary conditions, we can take $I = \mathbb{R}$.

REMARK 1. Using a now standard application of the Christ-Kiselev lemma [**12**], we can conclude that for a solution u to the inhomogeneous Schrödinger IBVP

$$\begin{cases} (D_t + \Delta)\, u(t, x) = F(t, x) \\ \qquad\qquad u(0, x) = f(x) \end{cases} \tag{3}$$

satisfying either Dirichlet or Neumann homogeneous boundary conditions, the estimate

$$\|u\|_{L^{p_1}(I; L^{q_1}(X))} \leq C \left(\|f\|_{L^2(X)} + \|F\|_{L^{p_2'}(I; L^{q_2'}(X))} \right) \tag{4}$$

holds for $\frac{2}{p_j} + \frac{2}{q_j} = 1$ for $j = 1, 2$. Here, $(\cdot)'$ denotes the dual exponent, e.g. $\frac{1}{p_1} + \frac{1}{p_1'} = 1$.

2. Global Strichartz Estimates for the Model Problems

The proof of the theorem relies on Strichartz estimates on $\mathbb{R}^2$, as well as Strichartz estimates on a two-dimensional cone, which we recall the result here for completeness. Here and in what follows, we denote by $C(\mathbb{S}^1_\rho)$ the cone over the circle of circumference ρ, equipped with the conic metric $dr^2 + r^2\, d\theta^2$.

THEOREM 2 (Strichartz estimates on $\mathbb{R}^2$, $C(\mathbb{S}^1_\rho)$; see, e.g., Keel–Tao [**22**] and Ford [**18**]). *Suppose that (p, q) and $(\tilde{p}, \tilde{q})$ are admissible Strichartz exponents in the sense of equation* (2). *If u is a solution to the Schrödinger equation*

$$\begin{cases} (D_t + \Delta_Y)\, u(t, x) = F(t, x), \\ \qquad\qquad u(0, x) = f(x), \end{cases}$$

on $Y = \mathbb{R}^2$ or $C(\mathbb{S}^1_\rho)$, then

$$\|u\|_{L^p(\mathbb{R}; L^q(Y))} \leq C(\|f\|_{L^2} + \|F\|_{L^{\tilde{p}'}(\mathbb{R}; L^{\tilde{q}'}(Y))}),$$

where $\tilde{p}'$ and $\tilde{q}'$ are the conjugate exponents to $\tilde{p}$ and $\tilde{q}$, respectively.

REMARK 2. Our results are closely related to the work on smoothing and Strichartz estimates for potentials with inverse-square singularities by Burq–Planchon–Stalker–Tahvildar-Zadeh and Planchon–Stalker–Tahvildar-Zadeh in [**10**] and [**28**]. For work on smoothing estimates for the Schrödinger equation in *smooth* exterior domains, we refer the reader to the early works of Burq [**8**], as well as Burq–Gerard–Tzvetkov [**9**] and Anton [**1**] who constructed parametrices for exterior domain problems that proved Strichartz estimates with errors controlled by local smoothing estimates. Local smoothing results were later extended by Robbiano–Zuily [**31**] to include quadratic potential wells. Scale invariant Strichartz estimates for exterior domains first appeared in Planchon–Vega [**29**] and Blair–Smith–Sogge [**6**], though not for the full range of admissible Strichartz pairs. For Strichartz estimates exterior to a smooth, convex obstacle however, scale invariant estimates have been established in the full range of estimates in Ivanovici [**20**], Ivanovici–Planchon [**21**], and Blair [**3**].

REMARK 3. Strichartz estimates exist for the wave equation on both $\mathbb{R}^2$ and $C(\mathbb{S}^1_\rho)$, but the analog of Theorem 1 for the wave equation on exterior domains can be directly computed from the analysis done by Blair, Ford and the second author in [**5**] due to the finite speed of propagation. Hence, quantifying the effects of diffraction as in [**2**] plays a much larger role in Schrödinger dynamics than in the corresponding wave dynamics on such domains.

3. Local Smoothing Estimates for X and on Euclidean Cones

The proof of Theorem 1 will also rely upon local smoothing estimates for $\mathbb{R}^2$, $C(\mathbb{S}^1_\rho)$ as well as on the space X in order to glue together similar dispersive results on model problems. We begin with the local result that is independent of choice of boundary conditions for X as a consequence of being local-in-time (and thus requiring no low-energy resolvent estimates):

THEOREM 3 ("Local" local smoothing estimate; see B.–W. [**2**]). *If X is a domain exterior to a non-trapping polygon, u is a solution of the Schrödinger equation*

$$\begin{cases} D_t u(t,z) + \Delta u(t,z) = 0, \\ u(0,z) = f(z), \end{cases}$$

with Dirichlet or Neumann boundary conditions and $\chi \in C^\infty(X)$ is a smooth, compactly supported function, then u satisfies a local smoothing estimate

$$\|\chi u\|_{L^2([0,T];\mathcal{D}_{1/2})} \leq C_T \|f\|_{L^2},$$

where $\mathcal{D}_{1/2}$ is the domain of $(-\Delta)^{1/4}$.

In the case of Dirichlet boundary conditions, we can significantly strengthen the above result to gain global control over the local smoothing norm.

THEOREM 4 ("Global" local smoothing estimate). *If X is a domain exterior to a non-trapping polygon, u is a solution of the Schrödinger equation*

$$\begin{cases} D_t u(t,z) + \Delta u(t,z) = 0, \\ u(0,z) = f(z), \end{cases}$$

with Dirichlet boundary conditions and $\chi \in C^\infty(X)$ is a smooth, compactly supported function, then u satisfies a local smoothing estimate

$$\|\chi u\|_{L^2(\mathbb{R};\mathcal{D}_{1/2})} \leq C \|f\|_{L^2},$$

where $\mathcal{D}_{1/2}$ is the domain of $(-\Delta)^{1/4}$.

REMARK 4. Note that Theorems 3, 4 imply the dual estimate

$$\left\| \int_I e^{is\Delta} \chi F(s)\, ds \right\|_{L^2} \leq C \|F\|_{L^2(\mathbb{R};\mathcal{D}_{-1/2})}, \tag{5}$$

for I either $[0,T]$ or $\mathbb{R}$ respectively.

PROOF. We rely on the high-frequency resolvent estimates of the first and third authors [**2**], estimates due to Morawetz [**26**] for intermediate frequencies and Burq [**7**] for small frequencies, then apply a now-standard TT^* argument.

Consider the operator $Tu_0 = \chi e^{-it\Delta} u_0$. We wish to show that T is a bounded operator from $L^2(X)$ to $L^2(\mathbb{R};\mathcal{D}_{1/2})$. It suffices to show that TT^* is bounded from $L^2(\mathbb{R};\mathcal{D}_{-1/2})$ to $L^2(\mathbb{R};\mathcal{D}_{1/2})$. The operator TT^* is given by

$$\begin{aligned} TT^* f &= \chi \int_{\mathbb{R}} e^{-i(t-s)\Delta} \chi f(s)\, ds \\ &= \chi \int_{s<t} e^{-i(t-s)\Delta} \chi f(s)\, ds + \chi \int_{s>t} e^{-i(t-s)\Delta} \chi f(s)\, ds = \chi T_+ f + \chi T_- f. \end{aligned}$$

Observe that $T_\pm f$ are both solutions of the inhomogeneous Schrödinger equation

$$D_t u + \Delta u = \frac{1}{i} \chi f.$$

Suppose for now that f is compactly supported in time, i.e., $f(t,x) = 0$ for $t \notin [-t_0, t_0]$. In this case, $T_+ f$ vanishes for $t < -t_0$ and $T_- f$ vansishes for $t > t_0$.

We wish to show that there is a constant C, independent of t_0, so that

$$\int_{\mathbb{R}} \|\chi T_{\pm} f(t,x)\|^2_{\mathcal{D}_{1/2}} \, dt \leq C \int_{\mathbb{R}} \|f(t,x)\|^2_{\mathcal{D}_{-1/2}} \, dt.$$

By Plancherel's theorem, it suffices to show that

$$\int_{\mathbb{R}} \left\| \chi \widehat{T_{\pm} f}(E,x) \right\|^2_{\mathcal{D}_{1/2}} dE \leq C \int_{\mathbb{R}} \left\| \hat{f}(E,x) \right\|^2_{\mathcal{D}_{-1/2}} dE,$$

where $\hat{f}$ denotes the Fourier transform of f in t.

Observe that $\widehat{T_{\pm} f}(E,x)$ solves

$$(\Delta + E)\widehat{T_{\pm} f} = \frac{1}{i} \chi \hat{f}.$$

Moreover, the condition on the support of f implies that $\widehat{T_+ f}$ is holomorphic in the lower half-plane, while $\widehat{T_- f}$ is holomorphic in the upper half-plane. In particular, if $R(z) = (\Delta + z)^{-1}$ where it is invertible,

$$\widehat{T_{\pm} f}(E,x) = \lim_{\beta \downarrow 0} R(E \mp i\beta) \left(\frac{1}{i} \chi \hat{f}(E,x) \right).$$

We must thus estimate $\chi R(E \mp i0)\chi$ as an operator $\mathcal{D}_{-1/2} \to \mathcal{D}_{1/2}$. The high-frequency estimates from [**2**] imply that there is some E_0 so that for $E \geq E_0$,

$$\|\chi R(E \mp i0)\chi\|_{L^2 \to L^2} \leq \frac{C}{\sqrt{E}}.$$

Using this bound and the identity

$$\Delta \chi R(E \pm i\beta) = \chi - (E \pm i\beta)\chi R(E \pm i\beta) + [\Delta, \chi] R(E \pm i\beta)$$

yields the following high-energy estimate for the resolvent:

$$\|\chi R(E \pm i0)\chi\|_{L^2 \to \mathcal{D}_2} \leq C\sqrt{E}.$$

Interpolating the two estimates shows that $R(E \pm i0)$ is bounded (with uniform bound for $E \geq E_0$) as an operator from L^2 to $\mathcal{D}_1$ and thus from $\mathcal{D}_{-1/2}$ to $\mathcal{D}_{1/2}$.

The argument of Morawetz [**26**, Lemmas 15 and 16], which remains valid in our setting, shows that the same bound holds at intermediate energies as well. For uniform bounds down to $E = 0$, we rely upon an argument of Burq [**7**, Appendix B.2].

We now apply the resolvent estimates, which shows that there is a constant C independent of λ and t_0 so that

$$\left\| \chi \widehat{T_{\pm} f(E,x)} \right\|^2_{\mathcal{D}_{1/2}} \leq C \|f\|^2_{\mathcal{D}_{-1/2}}.$$

Integrating in E then finishes the proof in the compactly supported setting. For the general setting, we simply note that the constant is independent of the support and that compactly supported functions are dense in L^2_t. □

We will need one result that we have not been able to find explicitly in the literature, but whose proof uses standard methods. This result concerns global-in-time local smoothing for the Schrödinger equation on an infinite wedge domain, which of course serves as a local model for our polygon near a vertex and is equivalent to $C(\mathbb{S}^1_\rho)$ (Cf. [**19**], for instance). Let $X_\rho = \{\theta \in [0, \rho/2]\} \subset \mathbb{R}^2$ for $\rho \in [0, 4\pi)$.

LEMMA 1. *A solution to the Schrödinger equation on X_ρ with Dirichlet or Neumann conditions satisfies the local smoothing estimates of Theorem 4. Consequently, the dual estimate* (5) *is satisfied on X_ρ as well.*

PROOF. We first note that solutions to the Schrödinger equation on X_ρ with Dirichlet boundary conditions are equivalent, by extending in an odd manner to the cone over the circle of circumference ρ obtained by "doubling" the wedge X_ρ. That is to say, we may identify solutions to the Schrödinger equation on $\mathbb{R} \times X_\rho$ to solutions on the "edge manifold" $\mathbb{R} \times C(\mathbb{S}^1_\rho)$, where as usual $C(\mathbb{S}^1_\rho)$ has the metric $ds^2 = dr^2 + r^2 d\theta^2$ with $\theta \in \mathbb{S}^1_\rho$. We make this identification by extending the solutions to be odd or even under the involution

$$\mathbb{S}^1_\rho \ni \theta \to \rho - \theta$$

according to the choice of boundary condition. Thus, it will suffice to consider solutions to the Schrödinger equation on the cone,

$$\text{(6)} \qquad \begin{cases} (D_t + \partial_r^2 + r^{-1}\partial_r + r^{-2}\Delta_{\mathbb{S}^1_\rho})u(t,x) = 0, \\ u(0,x) = f(x). \end{cases}$$

We refer the reader to [**4, 5, 18**] as well as [**24**] for a discussion of Sobolev spaces on cones and the nature of the operator $\Delta_{C(\mathbb{S}^1_\rho)}$. In particular, we briefly recall the characterization given in [**24**] of the nature of the domains of powers of the Friedrichs Laplacian on the cone.[2] First we recall the definition of b-vector fields and operators. The space of *b-vector fields*, denoted $\mathcal{V}_b(C(\mathbb{S}^1_\rho))$ is the vector space of vector fields on $[0,\infty) \times \mathbb{S}^1_\rho$ tangent to $0 \times \mathbb{S}^1_\rho$. In local coordinates (r,θ) near ∂M, they are generated over $C^\infty([0,\infty) \times \mathbb{S}^1_\rho)$ by the vector fields $r\partial_r$ and ∂_θ. One easily verifies that $\mathcal{V}_b(C(\mathbb{S}^1_\rho))$ forms a Lie algebra. The set of b-differential operators, $\mathrm{Diff}^*_{\mathrm{b}}(C(\mathbb{S}^1_\rho))$, is the universal enveloping algebra of this Lie algebra: it is the filtered algebra consisting of operators of the form

$$\text{(7)} \qquad A = \sum_{|\alpha|+j \le m} a_{j,\alpha}(r,\theta)(rD_r)^j D_\theta^\alpha \in \mathrm{Diff}^m_{\mathrm{b}}(C(\mathbb{S}^1_\rho)).$$

Now let $L^2_b(C(\mathbb{S}^1_\rho))$ be the space of square-integrable functions with respect to the "b-density" $r^{-1}\,dr\,d\theta$. We define the b-Sobolev spaces $H^m_{\mathrm{b}}(C(\mathbb{S}^1_\rho))$ for $m \in \mathbb{N}$ as

$$\{u : Au \in L^2_b(C(\mathbb{S}^1_\rho)) \text{ for } A \in \mathrm{Diff}^m_{\mathrm{b}}(C(\mathbb{S}^1_\rho))\}.$$

This definition can be extended to a definition of $H^s_{\mathrm{b}}(C(\mathbb{S}^1_\rho))$ for $s \in \mathbb{R}$ by interpolation and duality, or, better yet, by developing the b-pseudodifferential calculus as in [**25**].

Finally, we recall that Lemma 3.2 of [**24**] tells us that if we let $\mathcal{D}_s$ denote the domain of $\Delta^{s/2}_{C(\mathbb{S}^1_\rho)}$ (again, with $\Delta_{C(\mathbb{S}^1_\rho)}$ denoting the Friedrichs Laplacian) then

$$\mathcal{D}_s = r^{-1+s} H^s_{\mathrm{b}}(C(\mathbb{S}^1_\rho)), \quad |s| < 1.$$

This identification does break down at $s = 1$—see §3 of [**25**] for details.

[2] As discussed in Remark 1.2[4] we take the Neumann Laplacian on the original planar domain to be the Friedrichs extension from smooth functions vanishing at the vertex or vertices, satisfying Neumann conditions at edge; thus, upon doubling to a cone, we are working with the Friedrichs extension of the Laplacian on smooth functions compactly supported away from the cone tip.

Finally we are ready to prove a local smoothing estimate. We will prove an estimate of the form

$$\int_0^\infty \left(\left\| \langle r \rangle^{-\frac{3}{2}} \partial_r u \right\|_{L^2}^2 + \left\| r^{-\frac{3}{2}} \sqrt{-\Delta_{\mathbb{S}^1_\rho}} u \right\|_{L^2}^2 \right) dt \le C \left\| f \right\|_{\mathcal{D}_{1/2}}^2 . \tag{8}$$

Taken together with its time-reversed version, this will yield the estimate

$$\|\chi u\|_{L^2(\mathbb{R};\mathcal{D}_1)} \le C\|f\|_{\mathcal{D}_{1/2}}$$

(indeed it is somewhat stronger than the needed estimate, being global in space, with weights, rather than local). Now to obtain the estimate stated in the theorem, we simply shift Sobolev exponents by applying this estimate to the solution $\langle \Delta \rangle^{-1/4} u$. Thus to prove the lemma it will suffice to obtain (8).

To do this, we separate variables and treat the small and large angular frequencies separately.

Let us first introduce the commutants A_0 and B_0, given by

$$A_0 = \partial_r, \quad B_0 = \frac{r}{\langle r \rangle} \partial_r$$

and observe that acting on smooth functions compactly supported away from the cone tip,

$$\begin{aligned} [A_0, -\Delta_{C(\mathbb{S}^1_\rho)}] &= \frac{1}{r^2}\partial_r + \frac{2}{r^3}\partial_\theta^2, \\ [B_0, -\Delta_{C(\mathbb{S}^1_\rho)}] &= 2\left(\frac{1}{\langle r \rangle^3}\partial_r^2 + \frac{1}{r^2\langle r \rangle}\partial_\theta^2 \right) + \frac{1}{r\langle r \rangle}\left(1 + \frac{1}{\langle r \rangle^2} - \frac{3r^2}{\langle r \rangle^4} \right)\partial_r. \end{aligned}$$

Note that the formal adjoints (with respect to the usual volume form) are given by

$$\begin{aligned} A_0^* &= -\partial_r - \frac{1}{r}, \\ B_0^* &= -\frac{r}{\langle r \rangle}\partial_r - \frac{2+r^2}{\langle r \rangle^3}. \end{aligned}$$

Now setting $A = (A_0 - A_0^*)/2$ and $B = (B_0 - B_0^*)/2$, we have

$$\begin{aligned} \left[A, -\Delta_{C(\mathbb{S}^1_\rho)}\right] &= \frac{2}{r^3}\partial_\theta^2 + \frac{1}{2r^3}, \\ \left[B, -\Delta_{C(\mathbb{S}^1_\rho)}\right] &= -2\partial_r^* \langle r \rangle^{-3}\partial_r + 2\langle r \rangle^{-1} r^{-2}\partial_\theta^2 + g, \end{aligned} \tag{9}$$

with

$$g(r) = \frac{r^4 + 8r^2 - 8}{2\langle r \rangle^7}.$$

Now consider a function $u \in \mathcal{C}^\infty(\mathbb{R}; \mathcal{C}_c^\infty((0,\infty) \times \mathbb{S}^1_\rho)$ with

$$(D_t + \Delta_{C(\mathbb{S}^1_\rho)})u = F, \ u(0,x) = f. \tag{10}$$

We will separate $u(t)$ in angular modes, preserving a single high-angular-frequency component and separating out all low-angular-frequency components: given J, we let $e_j(\theta) = e^{2\pi i j\theta/\rho}$ and we Fourier analyze u into

$$u = \sum_j a_j(t,r)e_j(\theta) \equiv \sum u_j(t,r,\theta).$$

We then split

$$u = \tilde{u} + \sum_{|j|<J} u_j(t,r)$$

with

$$\tilde{u} = \sum_{|j|\geq J} a_j(t,r)e_j(\theta)$$

denoting the high angular frequencies. Let $\widetilde{F}$, $\tilde{f}$ etc. denote the corresponding decompositions of F, f.

Now for j sufficiently large,

$$\langle -\partial_\theta^2 u_j, u_j \rangle \gg 1,$$

and hence by (9), there exists $c > 0$ with

$$\left\langle [B, \Delta_{C(\mathbb{S}^1_\rho)}]\tilde{u}, \tilde{u}\right\rangle \geq c\left(\left\| \langle r\rangle^{-\frac{3}{2}}\partial_r u\right\|^2_{L^2} + \left\| r^{-\frac{3}{2}}\sqrt{-\Delta_{\mathbb{S}^1_\rho}}u\right\|^2_{L^2}\right). \tag{11}$$

Thus, we compute

$$i^{-1}\partial_t\langle B\tilde{u},\tilde{u}\rangle = \left\langle [B, -\Delta_{C(\mathbb{S}^1_\rho)}]\tilde{u},\tilde{u}\right\rangle + 2\operatorname{Im}\left\langle \widetilde{F}, B\tilde{u}\right\rangle.$$

Integration then yields

$$\begin{aligned}\int_0^T \langle [B, \Delta_{C(\mathbb{S}^1_\rho)}]\tilde{u},\tilde{u}\rangle dt \leq |\langle B\tilde{u}(T),\tilde{u}(T)\rangle| + |\langle B\tilde{u}(0),\tilde{u}(0)\rangle| \\ + 2\int_0^T \left|\left\langle \widetilde{F}, B\tilde{u}\right\rangle\right| dt.\end{aligned}$$

Since $B \in \operatorname{Diff}^1_{\mathrm{b}}(C(\mathbb{S}^1_\rho))$, we certainly have

$$B : \mathcal{D}_{1/2} \to \mathcal{D}_{-1/2} = \mathcal{D}^*_{1/2},$$

by the identification $\mathcal{D}_{\pm 1/2} = r^{-1\pm 1/2}H_{\mathrm{b}}^{\pm 1/2}(C(\mathbb{S}^1_\rho))$. Thus, for all T,

$$\begin{aligned}\int_0^T \left(\left\| \langle r\rangle^{-\frac{3}{2}}\partial_r u\right\|^2_{L^2} + \left\| r^{-\frac{3}{2}}\sqrt{-\Delta_{\mathbb{S}^1_\rho}}u\right\|^2_{L^2}\right) dt \leq \\ c\int_0^T \|u\|_{\mathcal{D}_{1/2}}\|F\|_{\mathcal{D}_{1/2}}\,dt + \|f\|^2_{\mathcal{D}_{1/2}} + \|u(T)\|^2_{\mathcal{D}_{1/2}}.\end{aligned}$$

An elementary energy estimate for the inhomogeneous equation shows that since $e^{-it\Delta} : \mathcal{D}_s \to \mathcal{D}_s$ for all s, for all $t > 0$ we have

$$\|u(t)\|_{\mathcal{D}_{1/2}} \leq \|f\|_{\mathcal{D}_{1/2}} + \int_0^\infty \|F(s)\|_{\mathcal{D}_{1/2}}\,ds.$$

Thus for all $T > 0$,

$$\int_0^T \left(\left\| \langle r\rangle^{-\frac{3}{2}}\partial_r u\right\|^2_{L^2} + \left\| r^{-\frac{3}{2}}\sqrt{-\Delta_{\mathbb{S}^1_\rho}}u\right\|^2_{L^2}\right) dt \leq c\left(\int_0^\infty \|F\|_{\mathcal{D}_{1/2}}\right)^2 + c\|f\|^2_{\mathcal{D}_{1/2}}, \tag{12}$$

with the constant c independent of T. Thus, the map

$$(f, F) \in \mathcal{D}_{1/2} \oplus L^1(\mathbb{R}; \mathcal{D}_{1/2}) \to u$$

extends by continuity to yield (12) in particular[3] with any $f \in \mathcal{D}_{1/2}$ and $F = 0$. Letting $T \to \infty$, this is the desired estimate for the $\tilde{u}$ term.

We now turn to the u_j terms, for which we will need the commutant A as well. We treat the cases $j \neq 0$ and $j = 0$ separately. For $j \neq 0$ we note that $\langle \partial_\theta^2 e_j, e_j \rangle < -1/4$ owing to our assumption that $\rho < 4\pi$. Thus

$$\Big\langle [A, -\Delta_{C(\mathbb{S}^1_\rho)}] u_j, u_j \Big\rangle \geq c\Big(\Big\| r^{-3/2} \partial_\theta u_j \Big\|^2_{L^2} + \Big\| r^{-3/2} u_j \Big\|^2_{L^2} \Big).$$

Consequently, using the commutant $B + \epsilon^{-1} A$ in the argument above with ϵ sufficiently small formally yields the desired estimate on u_j, for $j \neq 0$. Note that it is essential that $A \in r^{-1}\mathrm{Diff}^1_{\mathrm{b}}(C(\mathbb{S}^1_\rho))$, hence $A : \mathcal{D}_{1/2} \to \mathcal{D}_{-1/2} = \mathcal{D}^*_{1/2}$ so that we may proceed just as above.

Finally, we deal with $j = 0$. Here we employ the commutant $B - \epsilon^{-1} A$, where we remark that on u_0, a crucial sign flips and we have

$$\Big\langle [A, -\Delta_{C(\mathbb{S}^1_\rho)}] u_0, u_0 \Big\rangle \leq -c\Big(\Big\| r^{-3/2} \partial_\theta u_0 \Big\|^2_{L^2} + \Big\| r^{-3/2} u_0 \Big\|^2_{L^2} \Big)$$

(where of course we really have $\partial_\theta u_0 = 0$). □

REMARK 5. The above proof, while appealingly simple, does not extend to the case of a slit obstacle, i.e. $\rho = 4\pi$ or indeed to any product cone in which $-1/4$ is in the spectrum of the Laplacian on the link. However, addition of a third operator,

$$W_0 = f(r)\partial_r,$$

with f to be determined allows these cases to be handled in the same fashion as above. We calculate in general:

$$W_0^* = -W_0 - \frac{f(r)}{r} - f'(r),$$

giving the multiplier

$$W = \frac{W_0 - W_0^*}{2} = f(r)\partial_r + \frac{f(r)}{2r} + \frac{f'(r)}{2}.$$

Thus,

$$\Big\langle [W, -\Delta_{C(\mathbb{S}^1_\rho)}] u, u \Big\rangle = -\langle 2f'(r)\partial_r u, \partial_r u \rangle + \Big\langle \frac{f''(r)}{r} u, u \Big\rangle + \Big\langle \frac{f'''(r)}{2} u, u \Big\rangle - \Big\langle \frac{f'(r)}{2r^2} u, u \Big\rangle.$$

Consequently, provided

- $f \in C^3$ with uniformly bounded derivatives,
- $f' < 0$,
- $f''(r)/r + f'''(r)/2 - f'(r)/(2r^2) > 0$

---

[3]The reader may note that in obtaining these estimates we have not pursued optimality in F : the solution u should be one derivative more regular. We have avoided this issue owing to the breakdown of the identification of the domain $\mathcal{D}_1$ with a weighted b-Sobolev space (principally relevant in our analysis of the zero-angular-mode below). We of course obtain the correct mapping properties of the inhomogeneous Schrödinger equation from $L^2\mathcal{D}_{-1/2}$ to $L^2\mathcal{D}_{1/2}$ ex post facto by duality.

we obtain an estimate. In particular, taking $f(s) = (1+r)^{-\frac{1}{2}}$ gives a positive operator satisfying bounds as in (8) with slightly different weights,

$$\int_0^\infty \left(\left\| (1+r)^{-\frac{3}{4}} \partial_r u \right\|_{L^2}^2 + \left\| r^{-1} \langle r \rangle^{-\frac{3}{4}} \sqrt{-\Delta_{\mathbb{S}^1_\rho}} u \right\|_{L^2}^2 \right) dt \leq C \|f\|_{\mathcal{D}_{1/2}}^2 \tag{13}$$

on this mode. Note, we have made no attempt to optimize weights here.

4. Proof of Theorem 1

We are now ready to prove the main result, which follows from the same arguments whether $I = [0,T]$ or $\mathbb{R}$. Suppose that the polygon is contained in a ball of radius R_0 in $Z_0 = \mathbb{R}^2$ and let $U_0 = \mathbb{R}^2 \setminus \overline{B(0,R_0)}$. For each vertex of the polygon, we let U_j be a neighborhood of that vertex in X so that U_j can be considered as a neighborhood of the cone point in a wedge domain Z_j given by $\{\theta \in [0, \rho/2]\} \subset \mathbb{R}^2$. We may assume that the union of the U_j covers $X \setminus U_0$. Let $\chi_0, \chi_1, \dots, \chi_N$ be a partition of unity subordinate to this cover of X.

Set u to be the solution of the Schrödinger equation with initial data f, i.e.,

$$\begin{cases} (D_t + \Delta)\, u = 0, \\ \qquad u(0,z) = f(z). \end{cases}$$

Consider now $u_j = \chi_j u_j$. Note that u_j solves the following inhomogeneous Schrödinger equation on Z_j:

$$\begin{cases} D_t u_j + \Delta u_j = [\Delta, \chi_j] u, \\ \qquad u_j(0,z) = \chi_j(z) f(z). \end{cases}$$

We write $u_j = u_j' + u_j''$, where u_j' is the solution of the homogeneous equation on Z_j with the same initial data and u_j'' is the solution of the inhomogeneous equation with zero initial data. We know by [**22**] (for Z_0) and by [**18**] (for Z_j, $j > 0$) that u_j satisfies the homogeneous Strichartz estimate.

We now set $v_j(t,z) = [\Delta, \chi_j]u$. Then by Duhamel's Principle,

$$u_j'' = \int_0^t e^{-i(t-s)\Delta_{Z_j}} v_j(s)\, ds.$$

Note that $[\Delta, \chi_j]$ is a compactly supported differential operator of order 1 supported away from the vertices and so the local smoothing estimate for X implies that there is a constant C so that

$$\|v_j\|_{L^2(\mathbb{R}, \mathcal{D}_{-1/2})} \leq C \|f\|_{L^2(X)}\,.$$

We wish to show that u_j'' obeys the Strichartz estimates. As we are assuming that $p > 2$, the Christ–Kiselev lemma [**12**] implies that it is enough to show the estimate for

$$\int_{\mathbb{R}} e^{-i(t-s)\Delta_{Z_j}} v_j(s)\, ds = e^{-it\Delta_{Z_j}} \int_{\mathbb{R}} e^{is\Delta_{Z_j}} v_j(s)\, ds.$$

By the dual local smoothing estimate for Z_j from Lemma 1,

$$\left\| \int_{\mathbb{R}} e^{is\Delta_{Z_j}} v_j(s) \right\|_{L^2} \leq C \|v_j\|_{L^2(\mathbb{R}; \mathcal{D}_{-1/2})} \leq C \|f\|_{L^2}\,.$$

Applying the homogeneous Strichartz estimate for the propagator then finishes the proof.

5. A brief comment on nonlinear applications

Let $X = \mathbb{R}^2 \setminus P$, for P a non-trapping polygonal domain with either Dirichlet or Neumann boundary conditions on each edge. We use our loss-less local in time Strichartz estimates to extend well-posedness results to problems of the form

$$D_t u + \Delta u = \beta(|u|^2)u, \quad u_0 \in H^s(X) \tag{14}$$

for some $s \geq 0$, where β is any polynomial such that $\beta(0) = 0$. For a more general set of assumptions on β, see for instance the treatments in the books of Cazenave [**11**] and Tao [**35**] and the references therein, as well as a succinct exposition in the recent survey article of D'Ancona [**15**]. Largely, what follows will mirror the discussions in $\mathbb{R}^2$ from [**11**]. The arguments throughout implicitly rely upon Sobolev embeddings and Gagliardo-Nirenberg inequalities extending from $\mathbb{R}^2$ to X. Note, the nonlinear Schrödinger equation (14) has conservation of mass

$$M(u) = \|u\|_{L^2(X)} \tag{15}$$
$$= M(u_0) \tag{16}$$

and conservation of energy

$$E(u) = \|\nabla u\|_{L^2(X)} + \int_X B(u^2) d\text{Vol}_X \tag{17}$$
$$= E(u_0),$$

where

$$B(z) = \int_0^z \beta(y)dy.$$

These conservation laws that are quite useful for the study of well-posedness in that they allow one to control the L^2 and H^1 norms of the solution for a variety of nonlinearities β.

5.1. Local Nonlinear Results. The main results are of the following form.

THEOREM 5. *Let $0 \leq \sigma < 1$. Given $\|u_0\|_{H^\sigma} < M$, there exists a $T_{max}(M)$, $T_{min}(M) \leq \infty$ such that (14) has a unique solution $u \in C((-T_{min}, T_{max}), H^s) \cap L^q_{(-T_{min},T_{max})} L^r(X)$ with continuous dependence upon the initial data. In addition, if T_{max} (T_{min}) $< \infty$, then $\|u\|_{H^s} \to \infty$ as $t \to T_{max}$ (T_{min}). In particular, for L^2-subcritical nonlinearities, $T_{\max} = T_{\min} = \infty$.*

PROOF. We discuss the proof in some special cases, citing proper references for further details.

Given the conservation laws, if $\sigma = 1$, the results follow from careful estimates using the density of smooth functions in $H^1(X)$ (see Theorem 3.3.5 in [**11**]). Hence, the primary contribution to well-posedness theory easily derived from the Strichartz estimates is the component of uniqueness of the evolution (see Theorems 4.3.1 and 3.3.9 in [**11**]).

LEMMA 2. *Given the assumptions on β, there exist ρ_1, $\rho_2 \in [2, \infty)$ such that for each $0 < M < \infty$, there exists $C(M) < \infty$ such that*

$$\|\beta(|u|^2)u - \beta(|v|)^2 v\|_{L^{\rho_1'}} \leq C(M)\|u - v\|_{L^{\rho_2}}$$

for all $u, v \in H^1(X)$ such that $\|u\|_{H^1}, \|v\|_{H^1} \leq M$. As a result, if u and v are weak solutions of (14) on a time interval I with initial data $u_0 \in H^1$, then $u = v$.

The lemma follows by applying the Strichartz estimates to the equation for $u - v$ to obtain an estimate

$$\|u - v\|_{L^q L^r} \leq C\|u - v\|_{L^{q'} L^r} \tag{18}$$

for (q, r) an admissible Strichartz pair and the observation using Hölder's inequality that if for $0 \in I$, a finite interval,

$$\|f\|_{L^b(I)} \leq C\|f\|_{L^a(I)}$$

with $b > a$, then $f = 0$ almost everywhere in I.

Let us now consider $\sigma = 0$. Let us take

$$\beta(z^2) = g z^{\rho-1}, \tag{19}$$

for $\rho > 1$, $g \in \mathbb{R}$. Since we are in two dimensions, for $1 < \rho < 3$, using Theorem 1 and (4), we are able to apply standard bootstrapping arguments to such nonlinear problems, which we include here for completeness. We take instead of (14) the corresponding integral equation

$$S(u) = e^{-it\Delta} u_0 + i \int_0^t e^{-i(t-s)\Delta} \beta(|u|^2) u(s) ds,$$

which we show serves as a contraction mapping $S : Y_T \to Y_T$, for

$$Y_T = C([-T, T], L^2(X)) \cap L^p([-T, T], L^q(X))$$

with p, q an allowed Strichartz pair. Indeed, from (4) and Hölder's inequality as the equation is by scaling invariance computed to be sub-L^2-critical there exists p_1, q_1 Strichartz pairs such that

$$\begin{cases} p_1' \rho < p, \ q_1' \rho = q, \\ \frac{2}{p} + \frac{2}{q} = 1, \ \frac{2}{p_1} + \frac{2}{q_1} = 1, \\ p, p_1 \in [2, \infty], \ q, q_1 \in [2, \infty) \end{cases}$$

and

$$\|S(u)\|_{Y_T} \leq C(\|u_0\|_{L^2} + T^\rho \|u\|_{Y_T}),$$

$$\|S(u) - S(v)\|_{Y_T} \leq \tilde{C} T^\lambda \|u - v\|_{Y_T} \| |u| + |v| \|_{Y_T}^{\rho - 1}.$$

See, for instance, the discussion in [**11**], Section 4.6 for a detailed example of how to apply the bootstrapping principle once such bounds of the solution map are proved. Hence, for each $u_0 \in L^2$, there exists a time interval T and an upper bound U such that given $u, v \in B(0, U) \subset Y_T$, we have

$$S : B(0, U) \to B(0, U),$$
$$\|S(u) - S(v)\|_{Y_T} \leq \frac{1}{2} \|u - v\|_{Y_T}.$$

For $0 < \sigma < 1$, in $\mathbb{R}^2$, interpolative results up the H^s critical exponent $\rho = 1 + 2/(1 - \sigma)$ hold in Besov type spaces in which it is simple to take advantage of the Sobolev embeddings in H^s, see for instance Section 4.9 of [**11**]. Such spaces can be defined on $C(\mathbb{S}^1_\rho)$, but doing so goes beyond the scope of this note. □

5.2. Global Nonlinear Results. Let us take β as in (19) for simplicity and assume Dirichlet boundary conditions. A remaining open problem is to determine how much of the above linear analysis can be extended to domains X with Neumann boundary conditions; this would permit us to address global nonlinear questions in the Neumann case as well.

For the L^2 sub-critical case $1 < \rho < 3$, the natural L^2 conservation gives global well-posedness in L^2 for such equations by simply iteration of the argument over this uniform time interval. Up to this point, our analysis has cared very little about the leading order sign in the nonlinearity, which is generally irrelevant to finite time results. For the L^2 critical/supercritical case ($3 \leq \rho < \infty$), however, one must rely upon the natural H^1 conservation laws of such a system; such results are known to hold provided the nonlinearity is defocusing ($g > 0$)—see [**11**], Theorem 6.1.1, taking $\sigma = 1$.

For the defocusing case ($g > 0$), global well-posedness in H^1 holds immediately for any initial data for all powers of the nonlinearity since the positive conserved energy in (17) means that the H^1 norm can bounded for all time at any scale. For the focusing case ($g < 0$), global well-posedness is a subtle phenomenon owing to the potential existence of nonlinear bound states. With sufficient regularity in weighted spaces, a calculation by Weinstein [**36**] showed that in the case $\rho = 3$ on $\mathbb{R}^2$ there is a finite threshold of L^2 mass below which a solution exists for all time. The threshold is related to a nonlinear bound state that gives an optimal constant for the Gagliardo-Nirenberg inequality

$$\|f\|_{L^p(X)} \leq C_{opt}(X)\|u\|^{\alpha(p)}_{\dot{H}^1(X)}\|u\|^{1-\alpha(p)}_{L^2(X)},$$

where $p < \infty$ in two dimensions.[4] Following [**11**], Theorem 6.1.1, if one can use the conservation of energy and mass to provide an a priori bound on the H^1 norm throughout the evolution, then the methods sketched above yield the following theorem about Schrödinger evolution on X, where we employ the notation from Theorem 5:

THEOREM 6. *Assume that there exists $0 < \epsilon < 1$, $M > 0$ and $0 < C(M)$ such that*

$$\left|\int_X B(u^2)d\,Vol_X\right| \leq (1-\epsilon)\|u\|^2_{H^1(X)} + C(M) \tag{20}$$

for $u \in H^1(X)$ such that $\|u\|_{L^2(X)} \leq M$. Then, given $\|u_0\|_{L^2} \leq M$ in (14), we have $T_{min} = T_{max} = \infty$ in Theorem 5.

As we have *global*-in-time Strichartz estimates, it is natural to pursue the question of *scattering* of solutions with general critical or supercritical nonlinearities in two settings:

- In the focusing case, for small enough data in H^1 (possibly with the condition of finite variance or $\|xu\|^2_{L^2(X)} < \infty$),
- In the defocusing case, for general data.

[4]It would be of interest to see how such a calculation translates to the settings of exterior polygonal domains and in particular if the relevant Gagliardo-Nirenberg constant changes at all on product cones or polygonal exterior domains.

Note that a scattering state, say u_+, can be easily seen to depend upon global dispersive results in the sense that global existence implies that in the H^1 norm we construct

$$\begin{aligned} u_+ &= \lim_{t\to\infty} e^{-it\Delta}u(x,t) \\ &= u_0 - i\int_0^\infty e^{-is\Delta}\beta(|u|^2)u(s)ds, \end{aligned}$$

provided the integral is bounded. See for instance Tao [**35**], Chapter 3.6 for a discussion. We consequently propose the following:

CONJECTURE 1. *Given $\epsilon > 0$ sufficiently small and $u_0 \in H^1$, $\|u_0\|_{H^1(X)} < \epsilon$, there exist u_+, $u_- \in H^1$ such that given u a global solution to (14), we have*

$$\|u(t) - e^{it\Delta}u_\pm\|_{H^1} \to 0$$

as $t \to \pm\infty$ with

$$M(u_+) = M(u_-) = M(u_0) \quad \text{and} \quad \|\nabla u_+\|^2_{L^2(X)} = \|\nabla u_-\|^2_{L^2(X)} = E(u_0).$$

For more details,[5] we refer the reader to the treatment of scattering theory in higher dimensions of Strauss [**32**], [**33**] and in particular to the treatment by Nakanishi in [**27**], where the result was obtained in $\mathbb{R}^2$. See also [**14**] for a one-dimensional scattering result and [**13**], [**29**] for two- and three-dimensional scattering results using Interaction Morawetz style estimates, which have potential for applying to exterior domains. In two dimensions with a *smooth* star-shaped obstacle, scattering is proved in [**30**]; as the techniques there rely only upon integration by parts, they should extend to star-shaped polygonal domains (and possibly a broader class of non-trapping polygons) as well. In the case of finite variance on $\mathbb{R}^2$, this result follows from an extension of the pseudoconformal transformation, which is based upon a natural commuting vector field constructed from the Hamilton flow defined by the Schrödinger operator globally on $\mathbb{R}^2$; thus we do not currently have a version of this conservation law for exterior domains X.

Scattering can be shown in the defocussing case for any initial data given that the nonlinearity is critical/supercritical using a variant of the Morawetz estimates, which give local energy decay in the form of an estimate

$$\int_X \frac{|u|^p}{|x|^q}dx \le \left(\frac{p}{2-q}\right)^q \|u\|^{p-q}_{L^p}\|\nabla u\|^q_{L^p} \tag{21}$$

for $u \in W^{1,p}(X)$ where $q \le 2$ and $0 \le q \le p$, $1 \le p < \infty$. In recent works, such questions have been approached on $\mathbb{R}^2$ for less regularity using concentration compactness techniques and interaction Morawetz estimates in the works of [**17**, **23**]; one might hope to generalize these results to product cones and exterior domains.

References

[1] Ramona Anton, *Global existence for defocusing cubic NLS and Gross-Pitaevskii equations in three dimensional exterior domains* (English, with English and French summaries), J. Math. Pures Appl. (9) **89** (2008), no. 4, 335–354, DOI 10.1016/j.matpur.2007.12.006. MR2401142 (2009f:35312)

[2] Dean Baskin and Jared Wunsch, *Resolvent estimates and local decay of waves on conic manifolds*, J. Differential Geom. **95** (2013), no. 2, 183–214. MR3128982

[5]Excellent introductory treatments of scattering are available in [**11**], Sections 7.8 and 7.9 and [**34**], Section 3.3.

[3] M. D. Blair. On refined local smoothing estimates for the Schrödinger equation in exterior domains. 2011.

[4] Matthew D. Blair, G. Austin Ford, Sebastian Herr, and Jeremy L. Marzuola, *Strichartz estimates for the Schrödinger equation on polygonal domains*, J. Geom. Anal. **22** (2012), no. 2, 339–351, DOI 10.1007/s12220-010-9187-3. MR2891729

[5] Matthew D. Blair, G. Austin Ford, and Jeremy L. Marzuola, *Strichartz estimates for the wave equation on flat cones*, Int. Math. Res. Not. IMRN **3** (2013), 562–591. MR3021793

[6] M. D. Blair, H. F. Smith, and C. D. Sogge. *Strichartz estimates and the nonlinear Schrödinger equation in exterior domains. arXiv*, math.AP, Jan 2010. 22 pages.

[7] Nicolas Burq, *Décroissance de l'énergie locale de l'équation des ondes pour le problème extérieur et absence de résonance au voisinage du réel* (French), Acta Math. **180** (1998), no. 1, 1–29, DOI 10.1007/BF02392877. MR1618254 (99j:35119)

[8] N. Burq, *Smoothing effect for Schrödinger boundary value problems* (English, with English and French summaries), Duke Math. J. **123** (2004), no. 2, 403–427, DOI 10.1215/S0012-7094-04-12326-7. MR2066943 (2006e:35026)

[9] N. Burq, P. Gérard, and N. Tzvetkov, *On nonlinear Schrödinger equations in exterior domains* (English, with English and French summaries), Ann. Inst. H. Poincaré Anal. Non Linéaire **21** (2004), no. 3, 295–318, DOI 10.1016/S0294-1449(03)00040-4. MR2068304 (2005g:35264)

[10] Nicolas Burq, Fabrice Planchon, John G. Stalker, and A. Shadi Tahvildar-Zadeh, *Strichartz estimates for the wave and Schrödinger equations with the inverse-square potential*, J. Funct. Anal. **203** (2003), no. 2, 519–549, DOI 10.1016/S0022-1236(03)00238-6. MR2003358 (2004m:35025)

[11] T. Cazenave. *Semilinear Schrödinger equations*, volume 10 of *Courant Lecture Notes in Mathematics*. New York University Courant Institute of Mathematical Sciences, New York, 2003.

[12] Michael Christ and Alexander Kiselev, *Maximal functions associated to filtrations*, J. Funct. Anal. **179** (2001), no. 2, 409–425, DOI 10.1006/jfan.2000.3687. MR1809116 (2001i:47054)

[13] J. Colliander, M. Grillakis, and N. Tzirakis, *Tensor products and correlation estimates with applications to nonlinear Schrödinger equations*, Comm. Pure Appl. Math. **62** (2009), no. 7, 920–968, DOI 10.1002/cpa.20278. MR2527809 (2010c:35175)

[14] James Colliander, Justin Holmer, Monica Visan, and Xiaoyi Zhang, *Global existence and scattering for rough solutions to generalized nonlinear Schrödinger equations on* $\mathbb{R}$, Commun. Pure Appl. Anal. **7** (2008), no. 3, 467–489, DOI 10.3934/cpaa.2008.7.467. MR2379437 (2009c:35433)

[15] P. D'Ancona. *Dispersion phenomena in partial differential equations. Math. Comp. and Dyn. Sys.*, pages 160–174, 2009. Springer, New York.

[16] E. B. Davies, *Heat kernels and spectral theory*, Cambridge Tracts in Mathematics, vol. 92, Cambridge University Press, Cambridge, 1990. MR1103113 (92a:35035)

[17] B. Dodson. *Global well-posedness and scattering for the defocusing, L^2-critical, nonlinear Schrödinger equation when $d = 2$*, 2010.

[18] G. Austin Ford, *The fundamental solution and Strichartz estimates for the Schrödinger equation on flat Euclidean cones*, Comm. Math. Phys. **299** (2010), no. 2, 447–467, DOI 10.1007/s00220-010-1050-7. MR2679817 (2011i:35204)

[19] Andrew Hassell, Luc Hillairet, and Jeremy Marzuola, *Eigenfunction concentration for polygonal billiards*, Comm. Partial Differential Equations **34** (2009), no. 4-6, 475–485, DOI 10.1080/03605300902768909. MR2530705 (2011e:58053)

[20] Oana Ivanovici, *On the Schrödinger equation outside strictly convex obstacles*, Anal. PDE **3** (2010), no. 3, 261–293, DOI 10.2140/apde.2010.3.261. MR2672795 (2011j:58037)

[21] Oana Ivanovici and Fabrice Planchon, *On the energy critical Schrödinger equation in 3D non-trapping domains*, Ann. Inst. H. Poincaré Anal. Non Linéaire **27** (2010), no. 5, 1153–1177, DOI 10.1016/j.anihpc.2010.04.001. MR2683754 (2012a:35297)

[22] Markus Keel and Terence Tao, *Endpoint Strichartz estimates*, Amer. J. Math. **120** (1998), no. 5, 955–980. MR1646048 (2000d:35018)

[23] Rowan Killip, Terence Tao, and Monica Visan, *The cubic nonlinear Schrödinger equation in two dimensions with radial data*, J. Eur. Math. Soc. (JEMS) **11** (2009), no. 6, 1203–1258, DOI 10.4171/JEMS/180. MR2557134 (2010m:35487)

[24] Richard Melrose and Jared Wunsch, *Propagation of singularities for the wave equation on conic manifolds*, Invent. Math. **156** (2004), no. 2, 235–299, DOI 10.1007/s00222-003-0339-y. MR2052609 (2005e:58048)

[25] Richard B. Melrose, *The Atiyah-Patodi-Singer index theorem*, Research Notes in Mathematics, vol. 4, A. K. Peters, Ltd., Wellesley, MA, 1993. MR1348401 (96g:58180)

[26] Cathleen S. Morawetz, *Decay for solutions of the exterior problem for the wave equation*, Comm. Pure Appl. Math. **28** (1975), 229–264. MR0372432 (51 #8641)

[27] Kenji Nakanishi, *Energy scattering for nonlinear Klein-Gordon and Schrödinger equations in spatial dimensions* 1 *and* 2, J. Funct. Anal. **169** (1999), no. 1, 201–225, DOI 10.1006/jfan.1999.3503. MR1726753 (2000m:35141)

[28] Fabrice Planchon, John G. Stalker, and A. Shadi Tahvildar-Zadeh, *L^p estimates for the wave equation with the inverse-square potential*, Discrete Contin. Dyn. Syst. **9** (2003), no. 2, 427–442. MR1952384 (2003j:35195)

[29] Fabrice Planchon and Luis Vega, *Bilinear virial identities and applications* (English, with English and French summaries), Ann. Sci. Éc. Norm. Supér. (4) **42** (2009), no. 2, 261–290. MR2518079 (2010b:35441)

[30] Fabrice Planchon and Luis Vega, *Scattering for solutions of NLS in the exterior of a 2D star-shaped obstacle*, Math. Res. Lett. **19** (2012), no. 4, 887–897, DOI 10.4310/MRL.2012.v19.n4.a12. MR3008422

[31] Luc Robbiano and Claude Zuily, *The Kato smoothing effect for Schrödinger equations with unbounded potentials in exterior domains*, Int. Math. Res. Not. IMRN **9** (2009), 1636–1698, DOI 10.1093/imrn/rnn169. MR2500973 (2010e:35063)

[32] Walter A. Strauss, *Nonlinear scattering theory at low energy*, J. Funct. Anal. **41** (1981), no. 1, 110–133, DOI 10.1016/0022-1236(81)90063-X. MR614228 (83b:47074a)

[33] Walter A. Strauss, *Nonlinear scattering theory at low energy: sequel*, J. Funct. Anal. **43** (1981), no. 3, 281–293, DOI 10.1016/0022-1236(81)90019-7. MR636702 (83b:47074b)

[34] Catherine Sulem and Pierre-Louis Sulem, *The nonlinear Schrödinger equation. Self-focusing and wave collapse*, Applied Mathematical Sciences, vol. 139, Springer-Verlag, New York, 1999. MR1696311 (2000f:35139)

[35] Terence Tao, *Nonlinear dispersive equations. Local and global analysis*, CBMS Regional Conference Series in Mathematics, vol. 106, Published for the Conference Board of the Mathematical Sciences, Washington, DC; by the American Mathematical Society, Providence, RI, 2006. MR2233925 (2008i:35211)

[36] Michael I. Weinstein, *Nonlinear Schrödinger equations and sharp interpolation estimates*, Comm. Math. Phys. **87** (1982/83), no. 4, 567–576. MR691044 (84d:35140)

Mathematics Department, Northwestern University, Evanston, Illinois 60208

Current address: Mathematics Department, Texas A&M University, College Station, TX 77843

E-mail address: dbaskin@math.northwestern.edu

Mathematics Department, University of North Carolina, Phillips Hall, CB#3250, Chapel Hill, North Carolina 27599

E-mail address: marzuola@math.unc.edu

Mathematics Department, Northwestern University, Evanston, Illinois 60208

E-mail address: jwunsch@math.northwestern.edu

Contemporary Mathematics
Volume **630**, 2014
http://dx.doi.org/10.1090/conm/630/12672

Probabilistic Sobolev Embeddings, Applications to Eigenfunctions Estimates

Nicolas Burq and Gilles Lebeau

ABSTRACT. The purpose of this article is to present some probabilistic versions of Sobolev embeddings and an application to the growth rate of the L^p norms of Spherical harmonics on spheres. More precisely, we prove that (for natural probability measures), almost every Hilbert base of $L^2(\mathbb{S}^d)$ made of spherical harmonics has all its elements uniformly bounded in any L^p space ($p < +\infty$). We also show that most of the analysis extends to the case of Riemanian manifolds with groups of isometries acting transitively.

1. Introduction

Consider (M, g) a compact Riemanian manifold of dimension d and $\mathbf{\Delta}$ the Laplace operator on (M, g). Since the works by Hörmander [**10**] and Sogge [**14**], it is known that the eigenfunctions satisfy

THEOREM. *For any $2 \leq p \leq +\infty$, there exists $C > 0$ such that for any eigenfunctions of the Laplace operator u, $-\mathbf{\Delta}_g u = \lambda^2 u$, one has*

$$\|u\|_{L^p(M)} \leq C\lambda^{\delta(p)}\|u\|_{L^2} \tag{1.1}$$

with

$$\delta(p) = \begin{cases} \frac{(d-1)}{2} - \frac{d}{p} \text{ if } p \geq \frac{2(d+1)}{d-1} \\ \frac{(d-1)}{2}\left(\frac{1}{2} - \frac{1}{p}\right) \text{ if } p \leq \frac{2(d+1)}{d-1} \end{cases} \tag{1.2}$$

These estimates are also known to be optimal on the spheres (endowed with their standard metric). In the first regime, the so called zonal spherical harmonics (which concentrate on two opposite points on a diameter of the sphere) are optimizing (1.1) while in the second regime, the highest weight spherical harmonics ($u(x_1, \cdots x_{d+1}) = (x_1 + ix_2)^n$) which concentrate on the equator $(x_3, \cdot, x_{d+1}) = 0$ are saturating (1.1). On the other hand, the case of the tori shows that there exists manifolds on which these estimates can be improved. Indeed, on tori, we have by simple number theory arguments

$$\delta(p) \leq \frac{d-2}{2}\Big(1 - \frac{2}{p}\Big) + \epsilon \tag{1.3}$$

2010 *Mathematics Subject Classification.* Primary 34I20, 35P20, 28AXX.

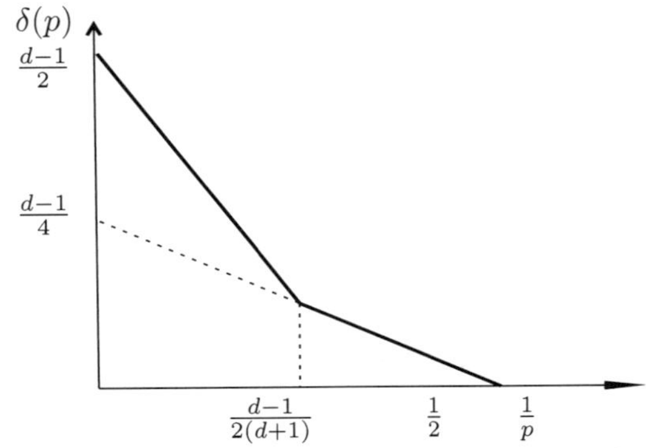

FIGURE 1.1. Eigenfunction estimates

(for $d \geq 5$, the ϵ can be dropped). On the other hand, it is standard and can easily be shown (see e.g. Section 4.1 and the works by Bourgain [**2, 3**] and Bourgain-Rudnick [**4**] for many more results on eigenfunctions on tori) that on tori

$$\delta(p) \geq \frac{d-2}{2} - \frac{d}{p}, \qquad p \geq \frac{2d}{d-2} \tag{1.4}$$

and consequently, as soon as $d \geq 3$, even on the torus, there exists sequences of L^2-normalized eigenfunctions having unbounded L^p norms. On the other hand, of course, the usual eigenbasis of eigenfunctions

$$e_n(x) = e^{in\cdot x}, n \in \mathbb{Z}^d$$

has all its L^p norms bounded and consequently a natural question is to ask whether this phenomenon is unique or there exists other manifolds exhibiting the same kind of behaviour (existence of both sequences having unbounded L^p norms and sequences having bounded L^p norms). Our result is that it is true on all spheres $\mathbb{S}^d$ endowed with their standard metric (for any $p < +\infty$), despite the optimality of (1.1). We actually prove a stronger result. Let us recall that the eigenfunctions of the Laplace operator on the spheres $\mathbb{S}^d$ (with its standard metric) are the spherical harmonics of degree k (restrictions to the sphere $\mathbb{S}^d$ of the harmonic polynomials of degree k) and they satisfy

- For any $k \in \mathbb{N}$, the vector space of spherical harmonics of degree k, E_k, has dimension

$$N_k = \binom{k+d}{d} - \binom{k+d-2}{d} \sim_{k\to+\infty} \frac{2e}{(d-1)!}\left(\frac{k}{e}\right)^{d-1} \tag{1.5}$$

- For any $e \in E_k$, $-\mathbf{\Delta}_{\mathbb{S}^d} e = -k(k+d-1)e$.
- The vector space spanned by the spherical harmonics is dense in $L^2(\mathbb{S}^d)$

As a consequence, any family $B = (B_k)_{k\in\mathbb{N}} = ((b_{k,l})_{l=1,\dots,N_k})_{k\in\mathbb{N}}$ of orthonormal bases of (E_{N_k}) is a Hilbert base of $L^2(\mathbb{S}^d)$. Our result is the following (with $\Gamma(z) = \int_0^{+\infty} t^{z-1}e^{-t}dt$ the classical Gamma function).

THEOREM 1. *For any $d \geq 2$, consider the space of spherical harmonics of degree k, E_k endowed with the $L^2(\mathbb{S}^d)$ norm and $\mathbb{S}_k$ its unit sphere endowed with the uniform probability measure P_k. Consider the space of Hilbert bases of $L^2(\mathbb{S}^d)$ having all elements harmonic polynomials, with real coefficients, endowed with its* *natural probability measure, ν (see Section 3 for a precise definition). Then for any $2 \leq q < +\infty$, we have*

$$(1.6)\quad \mathcal{A}^q_{q,k} = \mathbb{E}(\|u\|^q_{L^q(\mathbb{S}^d)}) = \int_{u \in \mathbb{S}_k} \|u\|^q_{L^q(\mathbb{S}^d)} dP_k = \frac{N_k^{q/2}}{Vol(\mathbb{S}^d)^{\frac{q}{2}-1}} \Big(\frac{\Gamma(\frac{q+1}{2})\Gamma(\frac{N_k}{2})}{\Gamma(\frac{1}{2})\Gamma(\frac{N_k+q}{2})} \Big),$$

$$\mathcal{A}_{q,k} = \frac{e^{-e}}{Vol(\mathbb{S}^d)^{\frac{1}{2}}} \sqrt{q} \Big(1 + \mathcal{O}(\frac{q}{N_k}) + \mathcal{O}(\frac{1}{q}) \Big)$$

and there exists $\mathcal{M}_{q,k} > 0$ such that

$$(1.7)\qquad \Big| \mathcal{M}_{q,k} - \mathcal{A}_{q,k} \Big| \leq \begin{cases} C \frac{\sqrt{q}}{N_k^{\frac{d}{(d-1)q}}} & \text{if } \frac{2(d+1)}{d-1} \leq q < +\infty \\ C \frac{\sqrt{q}}{N_k^{\frac{1}{4}+\frac{1}{2q}}} & \text{if } 2 \leq q \leq \frac{2(d+1)}{d-1} \end{cases}$$

$$(1.8)\quad P_k(u \in S_k; \Big| \|u\|_{L^q} - \mathcal{M}_{q,k} \Big| > \Lambda) \leq \begin{cases} 2e^{-N_k^{\frac{2d}{(d-1)q}} r^2} & \text{if } \frac{2(d+1)}{d-1} \leq q < +\infty \\ 2e^{-N_k^{\frac{1}{2}+\frac{1}{q}} r^2} & \text{if } 2 \leq q \leq \frac{2(d+1)}{d-1} \end{cases}$$

and

$$(1.9)\quad \nu(\{B = (b_{k,l})_{k \in \mathbb{N}, l=1,\cdots N_k} \in \mathcal{B}; \exists k, l; \Big| \|b_{k,l}\|_{L^q(\mathbb{S}^d)} - \mathcal{M}_{q,k} \Big| > r\}) \leq Ce^{-cr^2}$$

Furthermore, there exists $C, c, c_0 > 0$ such that

$$(1.10)\quad \nu(\{B = (b_{k,l})_{k \in \mathbb{N}, l=1,\cdots N_k} \in \mathcal{B}; \exists k, l; \|b_{k,l}\|_{L^\infty(\mathbb{S}^d)} > (c_0 + r)\sqrt{\log(k)}\}) \leq Ce^{-cr^2}$$

REMARK 1.1. *VanderKam* [**15**] *(see also Zelditch* [**16**]*) proves an estimate for the L^∞-norm of spherical harmonics on $\mathbb{S}^2$ which involves a logarithmic loss ($\log^2(k)$), in the same context. Though the results are in spirit very close, the methods of proof appear to differ significantly: VanderKam's approach (which do not appear to be flexible enough to tackle the case $p < +\infty$, or to give lower bounds) is based upon the precise descriptions of some particular spherical harmonics, implied by Ramanujan's conjectures proved by Deligne* [**7**]. *As explained above, our proof is very general (see Section 4) and the only sphere-specific fact we use is the exact spectral clustering*

$$\sigma(-\boldsymbol{\Delta}) = \{k(k+d-1), k \in \mathbb{N}^*\}.$$

REMARK 1.2. *We actually prove a stronger result. Indeed, when proving the first part of Theorem 1 (equations (1.6), (1.7), (1.8)), modulo a small change in the estimates, we only use the fact that the group of isometries on the manifold M acts transitively, whereas for (1.9) and (1.10), we use only the fact that the dimension of the eigenspaces is bounded from below by a positive power of the eigenvalue (see Theorem 3 and Remark 3.2)*

Let us end this introduction by mentioning that most of the work presented here is extracted from more general results on probabilistic Sobolev embeddings and applications to the study of the behaviour of solutions to Partial Differential Equations [**5**].

2. Probabilistic estimates

2.1. The spectral function. In this section, we consider a compact Riemanian manifold (M, g). We assume that the group of isometries of M acts transitively. Denote by $E_\lambda = \mathrm{Ker}(-\mathbf{\Delta}_g - \lambda^2 \mathrm{Id})$, $\lambda \in \sigma(\sqrt{-\mathbf{\Delta}})$, an eigenspace of the Laplace operator of dimension N_λ. Let $(e_{j,\lambda})_{j=1}^{N_\lambda}$ be an orthonormal basis of the space E_λ, and denote by $F_\lambda = (e_{j,\lambda})_{j=1}^{N_\lambda}$, and

$$f_\lambda(x) = \Big(\sum_{j=1}^{N_\lambda} e_{j,\lambda}^2(x)\Big)^{1/2}.$$

We can identify a point $U = (u_j)_{j=1}^{N_\lambda} \in \mathbb{S}^{N_\lambda - 1}$ on the unit sphere in $\mathbb{R}^{N_\lambda}$ with the function

$$u(x) = \sum_{j=1}^{N_\lambda} u_j e_{j,\lambda}(x) = U \cdot F_\lambda(x) \in \mathbb{S}_k, \tag{2.1}$$

the sphere of functions in E_λ having $L^2(M)$-norm equal to 1.

LEMMA 2.1. *Under the assumptions above, we have*

$$f_\lambda^2(x) = \sum_{j=1}^{N_\lambda} e_{j,\lambda}^2(x) = \frac{N_\lambda}{Vol(\mathbb{S}^d)}.$$

PROOF. We notice that

$$K_\lambda(x, y) = \sum_{j=1}^{N_\lambda} e_{j,\lambda}(x) e_{j,\lambda}(y)$$

is the kernel of the orthogonal projector on the space E_λ which is invariant by the action of any isometry R. As a consequence, we deduce $K_\lambda(x, R^{-1}y) = K_\lambda(Rx, y)$, which implies that $f_\lambda^2(x)$ is invariant by the isometries of the manifold M, hence constant on M, with integral N_λ. □

LEMMA 2.2. *Under the same assumptions as in Lemma 2.1, there exists $C > 0$ such that for any $\lambda \in \sigma(-\mathbf{\Delta})$, for any $u \in E_\lambda$ we have*

$$\|u\|_{L^\infty(M)} \leq \Big(\frac{N_\lambda}{Vol(M)}\Big)^{\frac{1}{2}} \|u\|_{L^2(M)} \tag{2.2}$$

PROOF. Let us denote by T the orthogonal projector on E_k. By the usual TT^* argument the norm of T seen as an operator from $L^2(M)$ to $L^\infty(M)$ is equal to the square root of the norm of $TT^* = T$ seen as an operator from $L^1(M)$ to $L^\infty(M)$. This norm is equal to the $L^\infty(M \times M)$-norm of its kernel $K(x, y)$. By Cauchy-Schwartz inequality, we have

$$|K_\lambda(x, y)| \leq \big(K_\lambda(x, x) K_\lambda(y, y)\big)^{1/2} = \frac{N_\lambda}{\mathrm{Vol}(M)} \tag{2.3}$$

$$\Rightarrow \|K_\lambda\|_{L^\infty(M \times M)} = \|f_\lambda^2(x)\|_{L^\infty(M)} = \frac{N_\lambda}{\mathrm{Vol}(M)}$$

which proves (2.2). Notice that actually the proof also gives that there exists a (non trivial) function $u \in E_\lambda$ for which

$$\|u\|_{L^\infty(M)} = \Big(\frac{N_\lambda}{\mathrm{Vol}(M)}\Big)^{\frac{1}{2}} \|u\|_{L^2(M)}.$$

□

2.2. Almost sure L^q estimates. In this section, we prove the main estimate on the L^q norms. The proof is very much inspired from Shiffman-Zelditch [**13**]. The main step in our proof is the proof of the estimates (1.8).

According to Theorem 1, with $F_q(u) = \|u\|_{L^q(\mathbb{S}^d)}$,

$$\begin{aligned}(2.4)\quad |F_q(u) - F_q(v)| &= |\|u\|_{L^q} - \|v\|_{L^q}| \leq \|u - v\|_{L^q} \\ &\leq C N_k^{\frac{1}{2}-\gamma(q))} \|u - v\|_{L^2} \leq C N_k^{\frac{1}{2}-\gamma(q))} \mathrm{dist}(u, v),\end{aligned}$$

with

$$\gamma(q) = \begin{cases} \frac{d}{(d-1)q} & \text{if } \frac{2(d+1}{d-1} \leq q < +\infty \\ \frac{1}{4} + \frac{1}{2q} & \text{if } 2 \leq q \leq \frac{2(d+1)}{d-1} \end{cases}$$

We deduce

$$\|F\|_{\mathrm{Lips}} \leq C N_k^{\frac{1}{2}-\gamma(q)}$$

and consequently, according to the concentration of measure property (see Proposition 5.2) applied to the function $F_q(u) = \|u\|_{L^q}$ on the N_k-1 dimensional sphere $\mathbb{S}_k$ we get

$$(2.5)\qquad P_k(|F_q(u) - \mathcal{M}_{q,k}| > r) \leq 2e^{-cN_k^{2\gamma(q)} r^2}$$

where $\mathcal{M}_{q,k}$ is the median value of the function F on $\mathbb{S}_k$. This implies (1.8). To get (1.7), we need to estimate from below and above $\mathcal{M}_{q,k}$. According to (2.1), (5.1) and Lemma 2.1 (and the rotation invariance of the uniform measure on the sphere S_k), we have

$$P_k(|u(x)| > \lambda) = \frac{\Gamma\big(\frac{N_k}{2}\big)}{\Gamma\big(\frac{N_k-1}{2}\big)\Gamma\big(\frac{1}{2}\big)} \mathbf{1}_{\lambda \in \big(0, \frac{N_k^{1/2}}{\mathrm{Vol}(\mathbb{S}^d)^{1/2}}\big)} \int_0^\theta \sin^{N_k-2}(\varphi) d\varphi,$$

where the value of θ in the expression above is fied by the relation $\lambda = \frac{N_k^{1/2}}{\mathrm{Vol}(\mathbb{S}^2)^{1/2}} \cos(\theta)$.

Using

$$\mathbb{E}_k(|g|^q) = q \int_0^{+\infty} \lambda^{q-1} P_k(|g| > \lambda) d\lambda$$

we obtain

$$(2.6)\quad \mathcal{A}_{q,k}^q = \mathbb{E}_k(\|u\|_{L^q}^q) = \int_{S_k}\int_{\mathbb{S}^d} |u(x)|^q dx dp_k = q\int_{\mathbb{S}^d}\int_0^\infty \lambda^{q-1} P_k(|u(x)| > \lambda) d\lambda$$

$$= 2\frac{\Gamma(\frac{N_k}{2})}{\Gamma(\frac{N_k-1}{2})\Gamma(\frac{1}{2})}\Big(\frac{N_k}{\mathrm{Vol}(\mathbb{S}^d)}\Big)^{q/2} q\int_{\mathbb{S}^d}\int_0^{\frac{\pi}{2}} \cos(\theta)^{q-1}\sin(\theta)\int_0^\theta \sin^{N_k-2}(\varphi)d\varphi d\theta dx$$

$$= 2\frac{\Gamma(\frac{N_k}{2})}{\Gamma(\frac{N_k-1}{2})\Gamma(\frac{1}{2})}\Big(\frac{N_k}{\mathrm{Vol}(\mathbb{S}^d)}\Big)^{q/2}\mathrm{Vol}(\mathbb{S}^d)\int_0^{\frac{\pi}{2}} \sin^{N_k-2}(\varphi)\cos(\varphi)^q d\varphi$$

$$= \frac{(N_k)^{\frac{q}{2}}}{(\mathrm{Vol}(\mathbb{S}^d))^{\frac{q}{2}-1}}\frac{\Gamma(\frac{N_k}{2})\Gamma(\frac{q+1}{2})}{\Gamma(\frac{N_k+q}{2})\Gamma(\frac{1}{2})}.$$

We deduce

$$(2.7)\quad \mathcal{A}_{q,k} = \frac{(N_k)^{\frac{1}{2}}}{(\mathrm{Vol}(\mathbb{S}^d))^{\frac{1}{2}-\frac{1}{q}}}\Big(\frac{\Gamma(\frac{N_k}{2})\Gamma(\frac{q+1}{2})}{\Gamma(\frac{N_k+q}{2})\Gamma(\frac{1}{2})}\Big)^{\frac{1}{q}}$$

$$= \frac{\sqrt{2e}}{(\mathrm{Vol}(\mathbb{S}^d))^{\frac{1}{2}-\frac{1}{q}}}\Big(\frac{\Gamma(\frac{q+1}{2})}{\Gamma(\frac{1}{2})}\Big)^{\frac{1}{q}} e^{-e}\Big(1+\mathcal{O}(\frac{q}{N_k})\Big)$$

$$= \frac{e^{-e}}{(\mathrm{Vol}(\mathbb{S}^d))^{\frac{1}{2}}}\sqrt{q}\Big(1+\mathcal{O}(\frac{q}{N_k}) + o(1)_{q\to+\infty}\Big)$$

We also have

$$(2.8)\quad |\mathcal{A}_{q,k} - \mathcal{M}_{q,k}|^q = \big|\|F\|_{L^q(\mathbb{S}_k)} - \|\mathcal{M}_{q,k}\|_{L^q(\mathbb{S}_k)}\big|^q$$

$$\le \|F-\mathcal{M}_{q,k}\|_{L^q(\mathbb{S}_k)}^q = q\int_0^{+\infty}\lambda^{q-1}P_k(|F-\mathcal{M}_{q,k}| > \lambda)d\lambda$$

$$\le q\int_0^{+\infty}\lambda^{q-1}e^{-c_1 N_k^{2\gamma(q)}\lambda^2}d\lambda = \frac{q}{2N_k^{q\gamma(q)}c_1^{q/2}}\Gamma(q/2).$$

We deduce

$$(2.9)\qquad |\mathcal{A}_{q,k} - \mathcal{M}_{q,k}| \le \frac{C}{N_k^{\gamma(q)}}\sqrt{q}$$

which, according to (2.7), implies (1.7).

We can now deduce an estimation of the L^∞ norm

THEOREM 2. *Denote by $\mathcal{M}_{\infty,k}$ the median value of the function $F_\infty(u) = \|u\|_{L^\infty}$ on the unit sphere S_k (endowed with its uniform probability measure). Then there exists $c_0, c_1 > 0$ for any $k\in\mathbb{N}^*$ we have*

$$(2.10)\qquad \mathcal{M}_{\infty,k} \in [c_0\sqrt{\log(k)}, c_1\sqrt{\log(k)}]$$

and consequently according to (2.4) *and Proposition* 5.2,

$$(2.11)\qquad \begin{aligned} &P_k(u\in S_k; \Big| \|u\|_{L^\infty} > c_1\sqrt{\log(k)}+\Lambda) \le 2e^{-c\Lambda^2},\\ &P_k(u\in S_k; \Big| \|u\|_{L^\infty} < c_0\sqrt{\log(k)}-\Lambda) \le 2e^{-c\Lambda^2}\end{aligned}$$

PROOF. According to Sobolev embeddings, there exists $C_0 > 0$ such that for any k, any $q > 2$, and any $u \in E_k$

$$\|u\|_q \leq \|u\|_{L^\infty} \leq C_0 k^{-d/q} \|u\|_q$$

and choosing $q_k = a\varepsilon_0 \log(k)$, we obtain with $C_1 = C_0 e^{d/a\varepsilon_0}$ independent of k

$$\mathbb{E}_k(\|u\|_{q_k}) \leq \mathbb{E}_k(\|u\|_{L^\infty}) \leq C_1 \mathbb{E}_k(\|u\|_{q_k}) \tag{2.12}$$

On the other hand for any $q \in [2, \infty]$, by Proposition 5.2 we have
(2.13)

$$\begin{aligned}|\mathbb{E}_k(\|u\|_q) - \mathcal{M}_{q,h}| &\leq \int |\|u\|_q - \mathcal{M}_{q,h}| dP = \int_0^\infty P_k(|\|u\|_q - \mathcal{M}_{q,h}| > \lambda) d\lambda \\ &\leq 2\int_0^\infty e^{-c_1\lambda^2} = \sqrt{\pi/c_1}\end{aligned}$$

Since $q_k \leq \varepsilon_0 \log(N_k)$, (1.7) imply $\mathcal{M}_{q_k,h} \simeq \sqrt{q_k}$. We deduce consequently from (2.13) $\mathbb{E}_k(\|u\|_{q_k}) \simeq \sqrt{q_k}$, and also according to (2.12), $\mathbb{E}_k(\|u\|_{L^\infty}) \simeq \sqrt{q_k}$, hence, again according to (2.13) $\mathcal{M}_{\infty,k} \simeq \sqrt{q_k}$. □

Let us now remark that in the proof above, we used only Lemma 2.1 and the L^p estimates on eigenfunctions given by Theorem 1. However, in the case of a riemanian manifold with a group of isometries acting transitively, we can use Lemma 2.2 and interpolation to obtain that there exists $C > 0$ such that for any $\lambda \in \sigma(\sqrt{-\mathbf{\Delta}})$ and any $u \in E_\lambda$,

$$\|u\|_{L^p(M)} \leq C N_\lambda^{\frac{1}{2}-\frac{1}{p}} \|u\|_{L^2(M)}$$

Plugging this estimate instead of Theorem 1 in the proof above gives

THEOREM 3. *Consider (M, g) a riemanian manifold on which the group of isometries act transitively. for any $\lambda \in \sigma(\sqrt{-\mathbf{\Delta}}$, denote by E_λ the eigenspace of the Laplace operator of dimension $N_\lambda > 0$, endowed with the $L^2(M)$ norm and $\mathbb{S}_\lambda$ its unit sphere endowed with the uniform probability measure P_λ. Then for any $2 \leq q < +\infty$, we have*
(2.14)

$$\mathcal{A}_{q,\lambda}^q = \mathbb{E}(\|u\|_{L^q(\mathbb{S}^d)}^q) = \int_{u\in\mathbb{S}_\lambda} \|u\|_{L^q(\mathbb{S}^d)}^q dP_\lambda = \frac{N_\lambda^{q/2}}{Vol(\mathbb{S}^d)^{\frac{q}{2}-1}} \Big(\frac{\Gamma(\frac{q+1}{2})\Gamma(\frac{N_\lambda}{2})}{\Gamma(\frac{1}{2})\Gamma(\frac{N_\lambda+q}{2})}\Big),$$

$$\mathcal{A}_{q,\lambda} = \frac{1}{Vol(\mathbb{S}^d)^{\frac{1}{2}}} \sqrt{\frac{q}{e}} \Big(1 + \mathcal{O}(\frac{q}{N_\lambda}) + \mathcal{O}(\frac{1}{q})\Big)$$

and there exists $\mathcal{M}_{q,\lambda} > 0$ such that

$$\Big|\mathcal{M}_{q,\lambda} - \mathcal{A}_{q,\lambda}\Big| \leq C\frac{\sqrt{q}}{N_\lambda^{1/q}}, \tag{2.15}$$

$$P_\lambda(u \in S_\lambda;\ \|u\|_{L^q} - \mathcal{M}_{q,\lambda}\Big| > \Lambda) \leq 2e^{-N_\lambda^{\frac{2}{q}} r^2} \tag{2.16}$$

.

3. Eigenbases

We come back to the case of the sphere and spherical harmonic. We can identify the space of orthonormal basis of E_k (endowed with the L^2-norm) with the orthogonal group, $O(N_k)$ and endow this space with its Haar measure, ν_k. Let us recall that

$$\oplus_{k\in\mathbb{N}} E_k \tag{3.1}$$

is dense in $L^2(\mathbb{S}^d)$. The set of Hilbert bases of L^2 compatible with the decomposition (3.1) is

$$\mathcal{B} = \times_{k\in\mathbb{N}} O(N_k))$$

and is endowed with the product probability measure

$$\nu = \otimes_k \nu_k.$$

Now, we can proceed to prove (1.9) and (1.10). According to (1.8), we get

PROPOSITION 3.1. *There exists $C, c > 0$ such that for any $k \in \mathbb{N}$,*

$$\nu_k(\{B=(b_l)_{l=1}^{N_k} \in SU(N_k); \exists 1\leq l\leq N_k; \Big|\|b_l\|_{L^q}-\mathcal{M}_{q,k}\Big| > \Lambda) \leq c_0 e^{-c_1 k^{2(d-1)\gamma(q)}\Lambda^2} k^{d-1}. \tag{3.2}$$

Indeed, the map

$$B = (b_l)_{l=1}^{N_k} \in SU(N_k) \mapsto b_1 \in S_k$$

sends the measure ν_k to the measure P_k and consequently, according to (1.8) for any $1 \leq l_0 \leq N_k$

$$\nu_k(\{B = (b_l)_{l=1}^{N_k} \in SU(N_k); \Big|\|b_{l_0}\|_{L^q} - \mathcal{M}_{q,k}\Big| > \Lambda) \leq c_0 e^{-c_1 N_k^{-\gamma(q)}\Lambda^2}.$$

We deduce

$$\begin{aligned}\nu_k(\{B = (b_l)_{l=1}^{N_k} \in SU(N_k); \exists l_0 \in \{1,\dots,N_k\}; \Big|\|b_{l_0}\|_{L^q} - \mathcal{M}_{q,k}\Big| > \Lambda)\\ \leq c_0 e^{-c_1 k^{2(d-1)\gamma(q)q}\Lambda^2} k^{(d-1)},\end{aligned} \tag{3.3}$$

which implies (1.9). Let us define now

$$F_{k,r} = \{B = (b_k) \in SU(N_k); \forall l_0; \Big|\|b_{k,l_0}\|_{L^q} - \mathcal{M}_{q,k}\Big| \leq r\}$$

and

$$F_r = \cap_{k\geq 0} F_{r,k} = \{B = (b_{k,l}); k \in \mathbb{N}, l = 1, \cdots N_k; \forall k_0, l_0; \Big|\|b_{k_0,l_0}\|_{L^q} - \mathcal{M}_{q,k}\Big| \leq r\}$$

We have

$$\nu(F_r)^c \leq \sum_k \nu_k(F_{k,r}^c) \leq \sum_k c_0 e^{-c_1 k^{2(d-1)\gamma(q)} r^2} k^{d-1} \leq \sum_k c e^{-c' k^{2(d-1)\gamma(q)} r^2} \leq C e^{-c' r^2}. \tag{3.4}$$

This ends the proof of (1.9). To prove (1.10), it sufficies to follow the same strategy and use Theorem 2 instead of (1.8).

REMARK 3.2. *In the more general case of a riemanian manifold having a group of isometries acting transitively, the results above on Hilbert bases extend straightforwardly under the additional assumption that*

$$\liminf_{\lambda\in\sigma(\sqrt{-\Delta})} \log(N_\lambda) > 0.$$

4. Other manifolds

4.1. Tori. A natural class of examples for which the group of isometries acts transitively are the tori. As a consequence, as already mentionned, we deduce that Theorem 3 holds. Finally, Theorem 2 holds also with $\log(k)$ replaced by $\log(N_k)$, with the notable difference that the multiplicity N_k do not tend to $+\infty$ if $d \leq 4$. On the other hand, the proofs of (1.9) and (1.10) use the fact that the multiplicity tends to infinity at least polynomially, and it consequently holds on rational tori only if $d \geq 5$ (see Grosswald [**8**, Chapter 12])). For dimensions $d \leq 4$, we only obtain the weaker

$$\forall A > 0, \nu(\{B = (b_{k,l})_{k\in\mathbb{N}, l=1,\cdots N_k} \in \mathcal{B}; \forall k,l; \|b_{k,l}\|_{L^\infty(\mathbb{S}^d)} \leq A\}) = 0. \tag{4.1}$$

Finally, to complete our probabilistic upper bounds, we can also get lower bounds on tori: For $n \in \mathbb{N}$ Let us denote by $E_{n,d}$ the eigenspace of the Laplace operator associated with the eigenvalue λ and $r_k(\lambda)$ the number of presentations of the integer n as a sum of d squares of integers:

$$r_d(n) = \#E_{\lambda,d}, \qquad E_{\lambda,d} = \{(n_1, \dots, n_d) \in \mathbb{Z}^d; \sum_{j=1}^d n_j^2 = \lambda\}.$$

According to Weyl's formula,

$$\sum_{\frac{n}{2} < m \leq n} r_d(m) = Cn^{\frac{d}{2}} + O(n^{\frac{d-1}{2}})$$

From which we deduce that for any $d \geq 2$ there exists a sequence $\lambda_p \to +\infty$ such that

$$r_d(\lambda_p) \geq c\lambda_p^{\frac{d-2}{2}}$$

and by considering the following eigenfunction on $\mathbb{T}^d$ (having L^2-norm equal to 1),

$$u_p(x) = r_d(\lambda_p)^{-\frac{1}{2}} \sum_{N=(n_1,\dots,n_d)\in E_{\lambda_p,d}} e^{iN\cdot x}$$

we have according to Sobolev embeddings

$$c\lambda_p^{\frac{d-2}{4}} \sim \|u_p\|_{L^\infty} \leq C\lambda_p^{\frac{k}{2r}} \|u_p\|_{L^r}$$

Hence

$$c\lambda_p^{\frac{d-2}{4} - \frac{k}{2r}} \leq \|u_p\|_{L^r}$$

which is (1.4).

4.2. Other manifolds. We can extend our results on arbitrary manifolds, to the price of considering only approximate eigenfunctions. Here the Weyl asymptotics are playing a crucial role. Let us consider $0 < a_h < 1 < b_h \leq c$ two functions defined for $(0, h_0)$ such that

$$\lim_{h\to 0} b_h = 1 = \lim_{h\to 0} a_h \tag{4.2}$$

We assume that

$$b_h - a_h \geq Mh \tag{4.3}$$

for a constant M to be precised later Let E_h be the subspace of $L^2(M)$

$$E_h = \{u = \sum_{k\in I_h} z_k e_k(x),\ z_k \in \mathbb{C}\},\quad \widetilde{E}_h = \{u = \sum_{k\in I_h} z_k e_k(x),\ z_k \in \mathbb{R}\},$$
$$I_h = \{k \in \mathbb{N}; h\omega_k \in (a_h, b_h)\} \tag{4.4}$$

Let $N_h = dim(E_h)$. Let us recall that Weyl formula (with precise remainder) reads (see Hörmander [**10**])

$$N_h = (2\pi h)^{-d}\mathrm{Vol}(M)\mathrm{Vol}(S^{d-1}) \int_{(a,b)} \rho^{d-1} d\rho + O(h^{-d+1}) \tag{4.5}$$

and consequently, we have

$$\exists C > 0; \forall \lambda > 0 \left| \sharp\{k \in \mathbb{N}; \omega_k \leq \lambda\} - c_d \frac{\mathrm{Vol}\ (M)}{(2\pi)^d}\lambda^d \right| \leq \lambda^{d-1} \tag{4.6}$$

and

$$\left|\sharp\{k; \omega_k \in I_h\} - c_d \frac{\mathrm{Vol}\ (M)}{(2\pi)^d}\Big((h^{-1}b_h)^d - (h^{-1}a_h)^d\Big)\right| \leq C h^{-d+1}, \tag{4.7}$$

which implies

$$\begin{gathered}\exists M_0 > 0; b_h - a_h \geq M_0 h \Rightarrow \exists \beta > \alpha > 0;\\ \alpha h^{-d}(b_h - a_h) \leq N_h = \sharp\{k; \omega_k \in I_h\} \leq \beta h^{-d}(b_h - a_h)\end{gathered} \tag{4.8}$$

Let $(e_{h,j})_{j=1}^{N_h}$ be an orthonormal basis of the space E_h,

$$F_h(x) = \big(e_{h,j}(x)\big)_{j=1}^{N_h},$$

and $f_h(x) = \|F_h(x)\|_{l^2} = \big(\sum_{j=1}^{N_h} e_{h,j}^2(x)\big)^{1/2}$. We can identify a point $U = (u_j)_{j=1}^{N_h} \in \mathbb{S}_h$ with the function

$$u(x) = \sum_{j=1}^{N_h} u_j e_{h,j}(x) = U \cdot F_h(x). \tag{4.9}$$

We have the following other consequence of Weyl's formula (or rather a consequence of the proof by Hörmander [**10**])

LEMMA 4.1. *Let a_h, b_h as above. We assume that $b_h - a_h \geq M_0 h$ so that* (4.8) *is satisfied. We also assume if $a = b$ that their common value is positive. Then there exists $C_0 > 0$ such that for any $x \in M$ and any $h \in]0, 1]$ we have*

$$|e_{x,h}| \leq C_0 h^{-d}(a_h - b_h) \tag{4.10}$$

By decomposing $L^2(M)$ as a direct sum of spaces

$$L^2(M) = \oplus_k E_{h_k}$$

where E_{h_k} correspond to the choice $b_h - a_h = Mh$, we can prove the analogs of Theorems 1, 2. The only essential difference is that of course the elements of E_{h_k} are no more *exact* eigenfunctions, but rather *approximate* eigenfunctions:

$$-\mathbf{\Delta} e = h_k^{-2} e + \mathcal{O}(h_k^{-1})_{L^2(M)}$$

On Zoll manifolds (where all geodesics are periodic, the clustering of the eigenvalues (see [**1**]) allows to improve to

$$u \in E_{h_k} \Rightarrow -\mathbf{\Delta} u = h_k^{-2} u + O(1)_{L^2(M)}$$

5. Probability calculus on spheres

PROPOSITION 5.1. *Let us denote by p_N the uniform probability measure on the unit sphere $S(N)$ of $\mathbb{R}^N$. We have (see* [**5**, Appendix A]*)*

$$p_N(|x_1| > \cos(\theta)) = \mathbf{1}_{t\in[0,1[}D_N \int_0^\theta \sin^{N-2}(\varphi)d\varphi \tag{5.1}$$

with $D_N = \frac{c_1 c_{N-1}}{c_N}$ where c_k is the $k-1$-dimensional volume of the sphere $S(k)$.

Notice that

$$\begin{aligned} 1 = D_N \int_0^{\pi/2} \sin^{N-2}(\varphi)d\varphi &= D_N \int_0^1 (1-t^2)^{\frac{N-3}{2}} \\ &= \frac{D_N}{2}\int_0^1 (1-u)^{\frac{N-3}{2}} u^{-\frac{1}{2}} du = \frac{D_N}{2}\beta(\frac{N-1}{2}, \frac{1}{2}) \end{aligned} \tag{5.2}$$

where β is the beta function of Weierstrass,

$$\beta(x,y) = \frac{\Gamma(x)\Gamma(y)}{\Gamma(x+y)},$$

which implies

$$D_N = 2\frac{\Gamma(\frac{N}{2})}{\Gamma(\frac{N-1}{2})\Gamma(\frac{1}{2})}.$$

Finally let us recall the standard concentration of measure property (see e.g. Ledoux [**12**])

PROPOSITION 5.2. *Let us consider a Lipshitz function F on the sphere $\mathbb{S}^d = S(d+1)$ (endowed with its geodesi distance and with the uniform probability measure, μ_d). Let us define its median value $\mathcal{M}(F)$ by*

$$\mu(F \geq \mathcal{M}(F)) \geq \frac{1}{2}, \qquad \mu(F \leq \mathcal{M}(F)) \geq \frac{1}{2}.$$

Then for any $r > 0$, we have

$$\mu(|F - \mathcal{M}(F)| > r) \leq 2e^{-(d-1)\frac{r^2}{2\|F\|^2_{Lips}}} \tag{5.3}$$

References

[1] Arthur L. Besse, *Manifolds all of whose geodesics are closed*, Ergebnisse der Mathematik und ihrer Grenzgebiete [Results in Mathematics and Related Areas], vol. 93, Springer-Verlag, Berlin-New York, 1978. With appendices by D. B. A. Epstein, J.-P. Bourguignon, L. Bérard-Bergery, M. Berger and J. L. Kazdan. MR496885 (80c:53044)

[2] Jean Bourgain, *Eigenfunction bounds for the Laplacian on the n-torus*, Internat. Math. Res. Notices **3** (1993), 61–66, DOI 10.1155/S1073792893000066. MR1208826 (94f:58127)

[3] Jean Bourgain, *Moment inequalities for trigonometric polynomials with spectrum in curved hypersurfaces*, Israel J. Math. **193** (2013), no. 1, 441–458, DOI 10.1007/s11856-012-0077-1. MR3038558

[4] Jean Bourgain and Zeév Rudnick, *Restriction of toral eigenfunctions to hypersurfaces and nodal sets*, Geom. Funct. Anal. **22** (2012), no. 4, 878–937, DOI 10.1007/s00039-012-0186-3. MR2984120

[5] Nicolas Burq and Gilles Lebeau, *Injections de Sobolev probabilistes et applications* (French, with English and French summaries), Ann. Sci. Éc. Norm. Supér. (4) **46** (2013), no. 6, 917–962. MR3134684

[6] Nicolas Burq and Nikolay Tzvetkov, Probabilistic Well-posedness for the cubic wave equation. *Preprint, http://arxiv.org/abs/1103.2222, To appear Journal of the Eur. Math. Soc.* 2011.

[7] Pierre Deligne, *La conjecture de Weil. I* (French), Inst. Hautes Études Sci. Publ. Math. **43** (1974), 273–307. MR0340258 (49 #5013)
[8] Emil Grosswald, *Representations of integers as sums of squares*, Springer-Verlag, New York, 1985. MR803155 (87g:11002)
[9] Victor Guillemin, *Some classical theorems in spectral theory revisited*, Seminar on Singularities of Solutions of Linear Partial Differential Equations (Inst. Adv. Study, Princeton, N.J., 1977/78), Ann. of Math. Stud., vol. 91, Princeton Univ. Press, Princeton, N.J., 1979, pp. 219–259. MR547021 (81b:58045)
[10] Lars Hörmander, *The spectral function of an elliptic operator*, Acta Math. **121** (1968), 193–218. MR0609014 (58 #29418)
[11] Lars Hörmander, *The analysis of linear partial differential operators. IV*, Grundlehren der Mathematischen Wissenschaften [Fundamental Principles of Mathematical Sciences], vol. 275, Springer-Verlag, Berlin, 1985. Fourier integral operators. MR781537 (87d:35002b)
[12] Michel Ledoux, *The concentration of measure phenomenon*, Mathematical Surveys and Monographs, vol. 89, American Mathematical Society, Providence, RI, 2001. MR1849347 (2003k:28019)
[13] Bernard Shiffman and Steve Zelditch, *Random polynomials of high degree and Levy concentration of measure*, Asian J. Math. **7** (2003), no. 4, 627–646. MR2074895 (2005e:32037)
[14] Christopher D. Sogge, *Concerning the L^p norm of spectral clusters for second-order elliptic operators on compact manifolds*, J. Funct. Anal. **77** (1988), no. 1, 123–138, DOI 10.1016/0022-1236(88)90081-X. MR930395 (89d:35131)
[15] Jeffrey M. VanderKam, *L^∞ norms and quantum ergodicity on the sphere*, Internat. Math. Res. Notices **7** (1997), 329–347, DOI 10.1155/S1073792897000238. MR1440572 (99d:58175)
[16] Steve Zelditch, *A random matrix model for quantum mixing*, Internat. Math. Res. Notices **3** (1996), 115–137, DOI 10.1155/S1073792896000116. MR1383753 (97j:81085)

Laboratoire de Mathématique d'Orsay, Université Paris-Sud, UMR 8628 du CNRS, 91405 Orsay Cedex, France
E-mail address: nicolas.burq@u-psud.fr

Département de Mathématiques, Université de Nice Sophia-Antipolis Parc Valrose 06108 Nice Cedex 02, France et Institut Universitaire de France
E-mail address: lebeau@unice.fr

Contemporary Mathematics
Volume **630**, 2014
http://dx.doi.org/10.1090/conm/630/12673

Random Smooth Functions on Compact Riemannian Manifolds

Liviu I. Nicolaescu

CONTENTS

What follows is a survey of our investigation in [**17**, **18**, **20**] of the distribution of critical points and critical values of a random Morse function on a Riemann manifold defined as a Fourier-like eigen-series whose coefficients are independent Gaussian random variables.

1. The setup

Suppose that (M, g) is a smooth, compact, connected Riemann manifold of dimension $m > 1$. We denote by $|dV_g|$ the volume density on M induced by g. We assume that the metric is normalized so that

$$\operatorname{vol}_g(M) = 1. \tag{$*$}$$

For any $\boldsymbol{u}, \boldsymbol{v} \in C^\infty(M)$ we denote by $(\boldsymbol{u}, \boldsymbol{v})_g$ their L^2 inner product defined by the metric g. The L^2-norm of a smooth function $\boldsymbol{u}$ is denoted by $\|\boldsymbol{u}\|$.

Let $\Delta_g : C^\infty(M) \to C^\infty(M)$ denote the scalar Laplacian defined by the metric g. Fix an orthonormal Hilbert basis $(\Psi_k)_{k\geq 0}$ of $L^2(M)$ consisting of eigenfunctions of Δ_g,

$$\Delta_g \Psi_k = \lambda_k \Psi_k, \;\; \|\Psi_k\| = 1, \;\; k_0 < k_1 \Rightarrow \lambda_{k_0} \leq \lambda_{k_1}.$$

Fix an even Schwartz function $w : \mathbb{R} \to [0, \infty)$. For $\varepsilon > 0$ we set

$$w_\varepsilon(t) := w(\varepsilon t), \;\; \forall t \in \mathbb{R}.$$

2010 *Mathematics Subject Classification.* Primary 53B20, 57R45, 58C40, 58D20, 58J37, 58J50, 60B20, 60D05, 60G15.

This work was partially supported by the NSF grant, DMS-1005745.

Consider random functions on M of the form

$$\boldsymbol{u}_\varepsilon = \sum_{k\geq 0} u_k \Psi_k, \tag{1.1}$$

where the coefficients u_k are independent Gaussian random variables with

$$\boldsymbol{E}(u_k) = 0, \quad \boldsymbol{var}(u_k) = w_\varepsilon(\sqrt{\lambda_k}). \tag{1.2}$$

Note that

$$\Delta^N \boldsymbol{u}_\varepsilon = \sum_{k\geq 0} \lambda_k^N u_k \Psi_k, \quad \forall N > 0.$$

The fast decay of w, the Weyl asymptotic formula, [**6**, VI.4],

$$\lambda_k \sim const.k^{\frac{2}{m}}, \quad k \to \infty,$$

and the Chebyshev's inequality imply that

$$\sum_{k\geq 1} \boldsymbol{P}\Big(u_k \geq \frac{1}{k\lambda_k^N} \Big) \leq \sum_{k\geq 1} k^2 \lambda_k^{2N} w(\varepsilon\sqrt{\lambda_k}) < \infty.$$

Invoking the Borel-Cantelli Lemma we deduce that for any N the function $\Delta^N u_\varepsilon$ is almost surely (a.s.) in L^2. In particular, this shows that $\boldsymbol{u}_\varepsilon$ is a.s. smooth.

The covariance kernel of the Gaussian random function $\boldsymbol{u}_\varepsilon$ is given by the function

$$\mathscr{E}^\varepsilon : M \times M \to \mathbb{R},$$

$$\mathscr{E}^\varepsilon(\boldsymbol{p}, \boldsymbol{q}) = \boldsymbol{E}\big(\boldsymbol{u}_\varepsilon(p) \boldsymbol{u}_\varepsilon(q) \big) = \sum_{k\geq 0} w_\varepsilon(\sqrt{\lambda_k}) \Psi_k(\boldsymbol{p}) \Psi_k(\boldsymbol{q}).$$

The eigenfunctions Ψ_k satisfy the known pointwise estimates (see [**14**, Thm. 17.5.3] or [**19**, Thm 1.6.1]),

$$\|\Psi_k\|_{C^\nu(M)} = O\Big(\lambda_k^{\frac{m+\nu}{2}} \Big) \text{ as } k \to \infty, \quad \forall \nu \geq 0.$$

Since w_ε is rapidly decresing the above estimates imply that $\mathscr{E}^\varepsilon$ is a smooth function. More precisely, $\mathscr{E}^\varepsilon$ is the Schwartz kernel of the smoothing operator

$$w(\varepsilon\sqrt{\Delta}) : C^\infty(M) \to C^\infty(M).$$

Let us observe that if $w(0) = 1$, then as $\varepsilon \searrow 0$ the function w_ε converges uniformly on compacts to the constant function $w_0(t) \equiv 1$ and $w_\varepsilon(\sqrt{\Delta})$ converges weakly to the identity operator. The Schwartz kernel of this limiting operator is the δ-function on $M \times M$ supported along the diagonal. It defines a generalized random function in the sense of [**9**] usually known under the name of *white noise*. For this reason we will refer to the $\varepsilon \searrow 0$ limit as the *white noise limit*.

The asymptotic estimates in Proposition 1.2 show that the random field $\boldsymbol{du}_\varepsilon$ satisfies the hypotheses of [**1**, Cor. 11.2.2] for $\varepsilon \ll 1$. Invoking [**1**, Lemma 11.2.11] we obtain the following technical result.

PROPOSITION 1.1. *The random function $\boldsymbol{u}_\varepsilon$ is almost surely Morse if $\varepsilon \ll 1$.* □

For any $\boldsymbol{u} \in C^1(M)$ we denote by $\mathbf{Cr}(\boldsymbol{u}) \subset M$ the set of critical points of $\boldsymbol{u}$ and by $\boldsymbol{D}(\boldsymbol{u})$ the set of critical values[1] of $\boldsymbol{u}$.

To any Morse function $\boldsymbol{u}$ on M we can naturally associate three Borel measures.

[1]The set $\boldsymbol{D}(\boldsymbol{u})$ is sometime referred to as the *discriminant set* of $\boldsymbol{u}$.

- A positive Borel measure $\mu_{\boldsymbol{u}}$ on M defined by

$$\mu_{\boldsymbol{u}} := \sum_{\boldsymbol{p}\in\mathbf{Cr}(\boldsymbol{u})} \delta_{\boldsymbol{p}}.$$

- A *signed* Borel measure $\chi_{\boldsymbol{u}}$ on M defined by

$$\chi_{\boldsymbol{u}} := \sum_{\boldsymbol{p}\in\mathbf{Cr}(\boldsymbol{u})} (-1)^{m_{\boldsymbol{u}}(\boldsymbol{p})}\delta_{\boldsymbol{p}},$$

where $m_{\boldsymbol{u}}(\boldsymbol{p})$ de notes the Morse index of $\boldsymbol{u}$ at the critical point $\boldsymbol{p}$.
- A positive Borel measure $\boldsymbol{\sigma_u}$ on $\mathbb{R}$ defined by

$$\boldsymbol{\sigma_u} := \boldsymbol{u}_*(\mu_{\boldsymbol{u}}) = \sum_{t\in\mathbb{R}} \left|\, \boldsymbol{u}^{-1}(t)\cap\mathbf{Cr}(\boldsymbol{u})\,\right|\delta_t,$$

where $|S|$ denotes the cardinality of the set S.

Observe that

$$\operatorname{supp}\mu_{\boldsymbol{u}} = \operatorname{supp}\chi_{\boldsymbol{u}} = \mathbf{Cr}(\boldsymbol{u}),\ \ \operatorname{supp}\boldsymbol{\sigma_u} = \boldsymbol{D}(\boldsymbol{u}).$$

When $\boldsymbol{u}$ is not Morse, we set

$\mu_{\boldsymbol{u}} = \chi_{\boldsymbol{u}} := |dV_g|,\ \ \boldsymbol{\sigma_u} = \delta_0 =$ the Dirac measure on $\mathbb{R}$ concentrated at the origin.

Observe that for any Morse function $\boldsymbol{u}$ and any Borel subset $B\subset\mathbb{R}$ the number $\boldsymbol{\sigma_u}(B)$ is equal to the number of critical values of $\boldsymbol{u}$ in B counted with multiplicity.

To the random function $\boldsymbol{u}_\varepsilon$ we associate the random (or empirical) measures $\mu_{\boldsymbol{u}_\varepsilon}, \chi_{\boldsymbol{u}_\varepsilon}, \boldsymbol{\sigma}_{\boldsymbol{u}_\varepsilon}$ with expectations

$$\mu^\varepsilon := \boldsymbol{E}\big(\,\mu_{\boldsymbol{u}_\varepsilon}\big),\ \ \chi^\varepsilon := \boldsymbol{E}\big(\,\chi_{\boldsymbol{u}_\varepsilon}\big)\boldsymbol{\sigma}^\varepsilon := \boldsymbol{E}\big(\,\sigma_{\boldsymbol{u}_\varepsilon}\big)$$

defined in the obvious way. E.g., $\boldsymbol{\sigma}^\varepsilon$ is the measure on $\mathbb{R}$ uniquely determined by the equality

$$\int_{\mathbb{R}} f(t)\boldsymbol{\sigma}^\varepsilon(dt) = \boldsymbol{E}\left(\int_{\mathbb{R}} f(t)d\boldsymbol{\sigma}_{\boldsymbol{u}_\varepsilon}(dt)\right),$$

for any continuous and bounded function $f:\mathbb{R}\to\mathbb{R}$. We will refer to $\boldsymbol{\sigma}^\varepsilon$ as *the variational complexity* of the random function $\boldsymbol{u}_\varepsilon$, and to χ^ε as *the Morse complexity* of $\boldsymbol{u}^\varepsilon$. Observe that

$$N^\varepsilon = \int_M \mu^\varepsilon(d\boldsymbol{p}) = \int_{\mathbb{R}} \boldsymbol{\sigma}^\varepsilon(dt)$$

is the expected number of critical points of of the random function $\boldsymbol{u}_\varepsilon$. On the other hand, the Morse inequalities show that

$$\int_M \chi^\varepsilon(d\boldsymbol{p}) = \chi(M) = \text{the Euler characteristic of } M,\ \ \forall\varepsilon.$$

In the series of papers [**17, 18, 20**] we have investigated the following questions.

Q$_1$. *What happens to the quantities* $N^\varepsilon, \mu^\varepsilon, \chi^\varepsilon, \boldsymbol{\sigma}^\varepsilon$ *in the white noise limit ?*

Q$_2$. *What happens when the dimension of* M *is very large?*

Q$_3$. *How much of the geometry of* (M,g) *can we recover from the statistics of the random function* $\boldsymbol{u}_\varepsilon$*?*

In the remainder of the present paper we report on the progress we made in answering these questions.

On the surface, the first question seem to make little sense, because it is asking to find a pattern in the white-noise chaos. However, patterns do exist, and we would like to mention one that plays a central role in our investigation.

Fix a point $\boldsymbol{p}\in M$. This determines three Gaussian random vectors

$$\boldsymbol{u}_\varepsilon(\boldsymbol{p})\in\mathbb{R},\;\; d\boldsymbol{u}_\varepsilon(\boldsymbol{p})\in T^*_{\boldsymbol{p}}M,\;\; \operatorname{Hess}_{\boldsymbol{p}}(\boldsymbol{u}_\varepsilon)\in\operatorname{Sym}(T_{\boldsymbol{p}}M), \tag{RV}$$

where $\operatorname{Sym}(V)$ denotes the space of symmetric operators on the real Euclidean space V,

$$\operatorname{Hess}_{\boldsymbol{p}}(\boldsymbol{u}_\varepsilon): T_{\boldsymbol{p}}M\times T_{\boldsymbol{p}}M\to\mathbb{R},$$

is the Hessian of $\boldsymbol{u}_{\boldsymbol{\omega}}$ at $\boldsymbol{p}$ defined in terms of the Levi-Civita connection of g, and then identified with a symmetric endomorphism of $T_{\boldsymbol{p}}M$ using again the metric g. More concretely, if $(x^i)_{1\leq i\leq m}$ are g-normal coordinates at $\boldsymbol{p}$, then we can identify the random vector

$$\big(\,\boldsymbol{u}_\varepsilon(\boldsymbol{p}), d\boldsymbol{u}_\varepsilon(\boldsymbol{p}), \operatorname{Hess}_{\boldsymbol{p}}(\boldsymbol{u}_\varepsilon)\,\big)\in\mathbb{R}\oplus T^*_{\boldsymbol{p}}M\oplus\operatorname{Sym}(T_{\boldsymbol{p}}M).$$

with the centered Gaussian vector

$$\big(\,\boldsymbol{u}_\varepsilon(\boldsymbol{p}),\;(\partial_{x^i}\boldsymbol{u}_\varepsilon(p)\,)_{1\leq i\leq m},\;\partial^2_{x^ix^j}(\boldsymbol{u}_\varepsilon(\boldsymbol{p})\,)_{1\leq i,j\leq m}\,\big)\in\mathbb{R}\oplus\mathbb{R}^m\oplus\operatorname{Sym}_m,$$

where

$$\operatorname{Sym}_m:=\operatorname{Sym}(\mathbb{R}^m).$$

We define

$$s_m:=\frac{1}{(2\pi)^m}\int_{\mathbb{R}^m}w(|x|)dx,\;\; d_m:=\frac{1}{(2\pi)^m}\int_{\mathbb{R}^m}x_1^2w(|x|)dx, \tag{1.3a}$$

$$h_m:=\frac{1}{(2\pi)^m}\int_{\mathbb{R}^m}x_1^2x_2^2w(|x|)dx. \tag{1.3b}$$

We have the following key asymptotic estimates.

PROPOSITION 1.2. *For any $1\leq i,j,k,\ell\leq m$ we have the uniform in $\boldsymbol{p}$ asymptotic estimates as $\varepsilon\searrow 0$*

$$\boldsymbol{E}\big(\,\boldsymbol{u}_\varepsilon(p)^2\,\big)=s_m\varepsilon^{-m}\big(1+O(\varepsilon^2)\,\big), \tag{1.4a}$$

$$\boldsymbol{E}\big(\,\partial_{x^i}\boldsymbol{u}_\varepsilon(\boldsymbol{p})\partial_{x^j}\boldsymbol{u}_\varepsilon(\boldsymbol{p})\,\big)=d_m\varepsilon^{-(m+2)}\delta_{ij}\big(\,1+O(\varepsilon^2)\,\big), \tag{1.4b}$$

(1.4c)
$$\boldsymbol{E}\big(\,\partial^2_{x^ix^j}\boldsymbol{u}_\varepsilon(\boldsymbol{p})\partial^2_{x^kx^\ell}\boldsymbol{u}_\varepsilon(\boldsymbol{p})\,\big)=h_m\varepsilon^{-(m+4)}(\delta_{ij}\delta_{k\ell}+\delta_{ik}\delta_{j\ell}+\delta_{i\ell}\delta_{jk})\big(\,1+O(\varepsilon^2)\,\big),$$

$$\boldsymbol{E}\big(\,\boldsymbol{u}_\varepsilon(\boldsymbol{p})\partial^2_{x^ix^j}\boldsymbol{u}_\varepsilon(\boldsymbol{p})\,\big)=-d_m\varepsilon^{-(m+2)}\delta_{ij}\big(\,1+O(\varepsilon^2)\,\big), \tag{1.4d}$$

$$\boldsymbol{E}\big(\,\boldsymbol{u}_\varepsilon(p)\partial_{x^i}\boldsymbol{u}_\varepsilon(\boldsymbol{p})\,\big)=O(\varepsilon^{-m}),\;\;\boldsymbol{E}\big(\,\partial_{x^i}\boldsymbol{u}_\varepsilon(\boldsymbol{p})\partial^2_{x^jx^j}\boldsymbol{u}_\varepsilon(\boldsymbol{p})\,\big)=O(\varepsilon^{-(m+2)}). \tag{1.4e}$$

The above result shows that the statistics of the jets of order ≤ 2 of $\boldsymbol{u}_\varepsilon$ at a point in M have a well understood universal behavior in the white noise limit. The equality (1.4a) can be rewritten as

$$\int_M\mathscr{E}^\varepsilon(\boldsymbol{p},\boldsymbol{p})|dV_g(\boldsymbol{p})|=\operatorname{tr}w\big(\,\varepsilon\sqrt{\Delta}\,\big)\sim s_m\varepsilon^{-m},\;\;\varepsilon\searrow 0.$$

We can roughly interpret the above estimate as saying that the random function $\boldsymbol{u}_\varepsilon$ belongs to a Gaussian-Hilbert space $\boldsymbol{U}_\varepsilon\subset C^\infty(M)$ of dimension $\approx s_m\varepsilon^{-m}$.

We should say a few words about the proof the above asymptotic estimates. They are all estimates for the jets along the diagonal for the covariance kernel $\mathscr{E}^\varepsilon$ which, as mentioned earlier is the Schwartz kernel of $w(\varepsilon\sqrt{\Delta})$. These are eminently approachable using the technique pioneered by L. Hörmander [**12**]. Here are the nuts and bots of this technique.

The Fourier inversion formula implies that

$$w(\varepsilon\sqrt{\Delta}) = \frac{1}{2\pi}\int_{\mathbb{R}} \widehat{w}(t)e^{i\varepsilon t\sqrt{\Delta}}dt = \frac{1}{2\pi\varepsilon}\int_{\mathbb{R}} \widehat{w}\left(\frac{s}{\varepsilon}\right)e^{is\sqrt{\Delta}}ds, \tag{1.5}$$

where $\widehat{w}(s)$ is the Fourier transform of w,

$$\widehat{w}(s) = \int_{\mathbb{R}} e^{-isx}w(x)dx.$$

The second integral in the right-hand side of (1.5) shows that the $\varepsilon \to 0$ behavior of $w(\varepsilon\sqrt{\Delta})$ is dictated by the behavior of $e^{i\sqrt{\Delta}}$ on a short time interval $|t| < c$ since $\widehat{w}$ has extremely fast decay at ∞.

Techniques pioneered by J. Hadamard [**11**] can be used to provide extremely accurate approximations for the Schwartz kernel of $e^{i\sqrt{\Delta}}$ if $|t|$ is sufficiently small; see [**14, 19, 21**]. In conjunction with (1.5) these lead to the estimates in Proposition 1.2.

2. The white noise limit of the variational complexity

Let us point out a first difficulty in analyzing the variational complexity. Observe that if $\boldsymbol{u}$ is a fixed Morse function on M and c is a constant, then

$$\mathbf{Cr}(c+\boldsymbol{u}) = \mathbf{Cr}(\boldsymbol{u}),\ \ \mu_{c+\boldsymbol{u}} = \mu_{\boldsymbol{u}},\ \ \chi_{\boldsymbol{u}+c} = \chi_{\boldsymbol{u}},$$

but

$$\boldsymbol{D}(\boldsymbol{u}+c) = c + \boldsymbol{D}(\boldsymbol{u}),\ \ \boldsymbol{\sigma}_{\boldsymbol{u}+c} = \delta_c * \boldsymbol{\sigma}_{\boldsymbol{u}},$$

where $*$ denotes the convolution of two finite measures on $\mathbb{R}$. More generally, if X is a scalar random variable with probability distribution ν_X, then the expected variational complexity of the random function $X+\boldsymbol{u}$ is the measure

$$\boldsymbol{E}(\boldsymbol{\sigma}_{X+\boldsymbol{u}}) = \nu_X * \boldsymbol{\sigma}_u.$$

If $\boldsymbol{u}$ itself is a random function, and X is independent of $\boldsymbol{u}$, then the above equality can be rephrased as

$$\boldsymbol{E}(\boldsymbol{\sigma}_{X+\boldsymbol{u}}) = \nu_X * \boldsymbol{E}(\boldsymbol{\sigma}_u).$$

In particular, if the distribution ν_X is a Gaussian, then the measure $\boldsymbol{E}(\boldsymbol{\sigma}_{\boldsymbol{u}})$ is uniquely determined by the measure $\boldsymbol{E}(\boldsymbol{\sigma}_{X+\boldsymbol{u}})$ since the convolution with a Gaussian is an injective operation. It turns out that it is easier to understand the statistics of the variational complexity of a perturbation of $\boldsymbol{u}_\varepsilon$ with an independent Gaussian variable of cleverly chosen variance.

To explain this perturbation we need to introduce several quantities that will play a central part the sequel. We set

$$q_m := \frac{s_m h_m}{d_m^2}.$$

One can show that $q_m \geq \frac{m}{m+2}$. The sequence $(q_m)_{m\geq 1}$ can be interpreted as a measure of the tail of w, the heavier the tail, the faster the growth of q_m as $m \to \infty$. We set

$$r_n := \max(1, q_n),$$

and define $\boldsymbol{\omega}_m \geq 0$ via the equality

$$r_n = \frac{(s_m + \boldsymbol{\omega}_m)h_m}{d_m^2}. \tag{2.1}$$

Set $\check{s}_m := s_m + \boldsymbol{\omega}_m$ so that

$$r_m = \frac{\check{s}_m h_m}{d_m^2}. \tag{2.2}$$

Observe that

$$\boldsymbol{\omega}_m = 0 \Longleftrightarrow q_m = r_m \geq 1 \Longleftrightarrow \check{s}_m = s_m, \tag{2.3}$$

$$\lim_{m\to\infty} \frac{\boldsymbol{\omega}_m}{s_m} = 0, \quad \lim_{m\to\infty} \frac{r_m}{q_m} = \lim_{m\to\infty} \frac{\check{s}_m}{s_m} = 1. \tag{2.4}$$

Choose a scalar Gaussian random variable $X_{\boldsymbol{\omega}(\varepsilon)}$ with mean 0 and variance $\boldsymbol{\omega}(\varepsilon) := \boldsymbol{\omega}_m \varepsilon^{-m}$ independent of $\boldsymbol{u}_\varepsilon$. When $\boldsymbol{\omega}_m = 0$, then $X_{\boldsymbol{\omega}(\varepsilon)}$ is simply the deterministic constant 0. Form the new random function

$$\check{\boldsymbol{u}}_\varepsilon = X_{\boldsymbol{\omega}(\varepsilon)} + \boldsymbol{u}_\varepsilon.$$

We denote by $\check{\boldsymbol{\sigma}}^\varepsilon$ the expected variational complexity of $\check{\boldsymbol{u}}_\varepsilon$. We have the equality

$$\check{\boldsymbol{\sigma}}^\varepsilon = \gamma_{\boldsymbol{\omega}(\varepsilon)} * \boldsymbol{\sigma}^\varepsilon, \quad \boldsymbol{\omega}(\varepsilon) := \boldsymbol{\omega}_m \varepsilon^{-m}, \tag{2.5}$$

where $\boldsymbol{\gamma}_v(dt)$ denotes the centered Gaussian measure on $\mathbb{R}$ with variance v,

$$\boldsymbol{\gamma}_v(dt) = \frac{1}{\sqrt{2\pi v}} e^{-\frac{t^2}{2v}} dt.$$

Note that

$$\boldsymbol{N}^\varepsilon = \int_{\mathbb{R}} \check{\boldsymbol{\sigma}}^\varepsilon(dt) = \int_{\mathbb{R}} \boldsymbol{\sigma}^\varepsilon(dt).$$

The white noise behaviors of $\check{\boldsymbol{\sigma}}^\varepsilon$ and $\boldsymbol{\sigma}^\varepsilon$ are closely related to the distribution of eigenvalues of certain Gaussian ensembles of symmetric $(m+1) \times (m+1)$-matrices. To decribe this relationship we need to briefly recall some terminology from random matrix theory.

For $v \in (0, \infty)$ and N a positive integer we denote by GOE_N^v the space Sym_N of real, symmetric $N \times N$ matrices A equipped with a Gaussian measure such that the entries a_{ij} are independent, zero-mean, normal random variables with variances

$$\boldsymbol{var}(a_{ii}) = 2v, \quad \boldsymbol{var}(a_{ij}) = v, \quad \forall 1 \leq i < j \leq N.$$

Let $\rho_{N,v} : \mathbb{R} \to \mathbb{R}$ be the *normalized correlation* function of GOE_N^v. It is uniquely determined by the equality

$$\int_{\mathbb{R}} f(\lambda)\rho_{N,v}(\lambda)d\lambda = \frac{1}{N}\boldsymbol{E}_{\mathrm{GOE}_N^v}\big(\operatorname{tr} f(A)\big),$$

for any continuous bounded function $f : \mathbb{R} \to \mathbb{R}$. The function $\rho_{N,v}(\lambda)$ also has a probabilistic interpretation: for any Borel set $B \subset \mathbb{R}$ the expected number of eigenvalues in B of a random $A \in \mathrm{GOE}_N^v$ is equal to

$$N \int_B \rho_{N,v}(\lambda)d\lambda.$$

For any $t > 0$ we denote by $\mathcal{R}_t : \mathbb{R} \to \mathbb{R}$ the rescaling map $\mathbb{R} \ni x \mapsto tx \in \mathbb{R}$. If μ is a Borel measure on $\mathbb{R}$ we denote by $(\mathcal{R}_t)_*\mu$ its pushforward via the rescaling map

$\mathcal{R}_t$. The celebrated Wigner semicircle theorem, [**3, 16**], states that as $N \to \infty$ the rescaled probability measures

$$(\mathcal{R}_{\frac{1}{\sqrt{N}}})_*\big(\rho_{N,v}(\lambda)d\lambda\big)$$

converge weakly to the semicircle measure given by the density

$$\rho_{\infty,v}(\lambda) := \frac{1}{2\pi v} \times \begin{cases} \sqrt{4v-\lambda^2}, & |\lambda| \le \sqrt{4v} \\ 0, & |\lambda| > \sqrt{4v}. \end{cases}$$

THEOREM 2.1. *For $v > 0$ and $N \in \mathbb{Z}_{>0}$ we set*

$$\theta^{\pm}_{N,v}(x) := \rho_{N,v}(x)e^{\pm\frac{x^2}{4v}}.$$

(a) There exists a constant $C = C_m(w)$ that depends only on the dimension m and the weight w such that

$$\boldsymbol{N}^{\varepsilon} \sim C_m(w)\varepsilon^{-m}\big(1+O(\varepsilon)\big) \;\; \textit{as } \varepsilon \to 0. \tag{2.6}$$

More precisely

$$\begin{aligned} C_m(w) &= 2^{\frac{m+4}{2}} r_m^{\frac{1}{2}} \left(\frac{h_m}{2\pi d_m}\right)^{\frac{m}{2}} \Gamma\left(\frac{m+3}{2}\right) \int_{\mathbb{R}} (\gamma_{r_m-1} * \theta^+_{m+1,r_m})(y)\gamma_1(y)dy \\ &= 2^{\frac{m+4}{2}}\Gamma\left(\frac{m+3}{2}\right)\left(\frac{h_m}{2\pi d_m}\right)^{\frac{m}{2}} \int_{\mathbb{R}} \theta^-_{m+1,\frac{1}{r_m}} * \gamma_{\frac{r_m-1}{r_m}}(y)dy. \end{aligned} \tag{2.7}$$

(b) As $\varepsilon \searrow 0$ the rescaled probability measures

$$\frac{1}{\boldsymbol{N}^{\varepsilon}}\Big(\mathcal{R}_{\frac{1}{\sqrt{\check{s}_m\varepsilon^{-m}}}}\Big)_*\check{\boldsymbol{\sigma}}^{\varepsilon}$$

converge weakly to a probability measure $\check{\boldsymbol{\sigma}}_m$ on $\mathbb{R}$ uniquely determined by the proportionalities

$$\check{\boldsymbol{\sigma}}_m \propto \big(\gamma_{r_m-1} * \theta^+_{m+1,r_m}(y)\big)\gamma_1(y)dy \tag{2.8a}$$

$$\propto \theta^-_{m+1,\frac{1}{r_m}} * \gamma_{\frac{r_m-1}{r_m}}(y)dy. \tag{2.8b}$$

One immediate consequence of Theorem 2.1 is the following universality result.

COROLLARY 2.2 (**Universality**). *As $\varepsilon \to 0$ the rescaled probability measures*

$$\frac{1}{\boldsymbol{N}^{\varepsilon}}\Big(\mathcal{R}_{\frac{1}{\sqrt{\check{s}_m\varepsilon^{-m}}}}\Big)_*\boldsymbol{\sigma}^{\varepsilon}$$

converge weakly to a probability measure $\boldsymbol{\sigma}_m$ uniquely determined by the convolution equation

$$\gamma_{\frac{\omega_m}{\check{s}_m}} * \boldsymbol{\sigma}_m = \check{\boldsymbol{\sigma}}_m.$$

We can now formulate a satisfactory answer to question $\mathbf{Q}_3$.

THEOREM 2.3 (**Central limit theorem**). *Suppose that r_m has a limit $r \in [1,\infty]$ as $m \to \infty$. Then*

$$\lim_{m\to\infty} \boldsymbol{\sigma}_m = \gamma_{\frac{r+1}{r}}.$$

REMARK 2.4. The behavior of σ_m for m large depends more dramatically on the parameter r than the above theorem suggests. There are three qualitatively different regimes:

$$r = 1, \quad r \in (0, \infty), \quad r = \infty.$$

Each of these regimes is possible, depending on the weight w. For example, if $w = e^{-t^2}$ for $t \gg 0$ then $r = 1$. If $w(t) = e^{-(\log t)\log\log t}$ for $t \gg 0$ then $r = \infty$. Finally, if $w(t) = e^{-C(\log t)^2}$, then $r = e^{2/C}$ so, depending on C, r can have any value in $(1, \infty)$. □

Set

$$I_k(w) := \int_0^\infty w(r) r^k dr.$$

COROLLARY 2.5. *As $m \to \infty$ we have*

$$\begin{aligned} C_m(w) &\sim \frac{8}{\sqrt{\pi m}} \Gamma\left(\frac{m+3}{2}\right)\left(\frac{h_m}{\pi d_m}\right)^{\frac{m}{2}} \\ &\sim \frac{8}{\sqrt{\pi m}} \Gamma\left(\frac{m+3}{2}\right)\left(\frac{2I_{m+3}(w)}{\pi(m+2)I_{m+1}(w)}\right)^{\frac{m}{2}}. \end{aligned} \tag{2.9}$$

3. Statistical reconstruction of the geometry of M

In this section we deal with question $\mathbf{Q}_3$, namely how much of the geometry of (M, g) is detected by the random functions $\boldsymbol{u}_\varepsilon$. Recall that throughout we worked under the normalizing assumption $(*)$, i.e., $\mathrm{vol}_g(M) = 1$. Here is a first result.

THEOREM 3.1 (**Uniform distribution**). *As $\varepsilon \searrow 0$ the probablity measures $\frac{1}{N^\varepsilon}\mu^\varepsilon$ converge weakly to the Lebesgue measure $|dV_g|$ on M determined by the metric g.*

As its name suggests, the above results shows that in the white noise limit, we expect the critical points of the random function $\boldsymbol{u}_\varepsilon$ to be uniformly distributed on M with respect to the natural metric volume measure. In particular this shows that the metric measure on M can be determined from the statistical properties of $\boldsymbol{u}_\varepsilon$.

Following [**1**, §12.2] we define symmetric $(0,2)$-tensor g^ε on M given by

$$g^\varepsilon(X, Y) = \frac{\varepsilon^{m+2}}{d_m} \boldsymbol{E}\big(X\boldsymbol{u}_\varepsilon(\boldsymbol{p}), Y\boldsymbol{u}_\varepsilon(\boldsymbol{p})\big), \ \forall \boldsymbol{p} \in M, \ \ X, Y \in \mathrm{Vect}(M), \tag{3.1}$$

where Xu denotes the derivative of the smooth function u along the vector field X. The estimate (1.4b) implies that for ε sufficiently small the tensor g^ε is a Riemann metric and moreover g^ε converges in the C^0-topology to the metric g, a fact observed earlier by S. Zelditch [**24**]. We will refer to g^ε as the *stochastic metric* associated to the random function $\boldsymbol{u}_\varepsilon$.

The C^0-convergence is unsatisfactory because it does not guarantee that important metric invariants of g such as the Riemann curvature can be approximated by the corresponding invariants of g^ε. However, a much stronger result is true.

THEOREM 3.2 (**Approximation Theorem**). *The metrics g^ε converge in the C^∞-topology to the metric g.*

The construction of the metrics g^ε generalizes the construction in [**5**]. Note that for any $\varepsilon > 0$ we have a smooth map $\Xi_\varepsilon : M \to L^2(M,g)$

$$M \ni p \mapsto \Xi_\varepsilon(\boldsymbol{p}) := \left(\frac{\varepsilon^{m+2}}{d_m}\right)^{\frac{1}{2}} \sum_{k\geq 0} w_\varepsilon\big(\sqrt{\lambda_k}\,\big)^{\frac{1}{2}} \Psi_k(\boldsymbol{p})\Psi_k \in L^2(M,g).$$

Then g^ε is the pullback by Ξ_ε of the Euclidean metric on $L^2(M,g)$. Let us point out that [**5**, Thm. 5] is a special case of Theorem 3.2 corresponding to the weight $w(t) = e^{-t^2}$. However, the methods of [**5**] do not extend to this more general situation.

We should add a few words about the analytic result hiding behind Theorem 3.2. The key to this theorem is the convergence of the curvatures of g^ε to the curvature of g. The jets of the Schwartz kernel of $w_\varepsilon(\sqrt{\Delta})$ along the diagonal have a complete asymptotic expansion as $\varepsilon \searrow 0$, [**23**, Chap. XII]. The estimates in Proposition 1.2 are obtained from the leading terms of these asymptotic expansions. The curvature of g is hidden in the second order terms of this expansion.

4. A probabilistic interpretation of the Gauss-Bonnet theorem

To describe the Morse complexity, we need to make a brief detour and describe the *Lipschitz-Killing curvature measures* of a smooth Riemann metric h on M. For more details on this notion we refer to [**1**, Sec. 7.6] or [**7, 10**].

For any $j,k \leq m = \dim M$ we set denote by $\Omega^{j,k}(M)$ the space of smooth sections of

$$\Lambda^{j,k}T^*M := \Lambda^j T^*M \otimes \Lambda^k T^*M.$$

We will refer to such section as (j,k)-forms. Using the metric h we can identify the (j,k)-forms with bundle morphisms

$$\Lambda^j T^*M \to \Lambda^k T^*M.$$

In particular, we have a well defined trace

$$\operatorname{tr}_h : \Omega^{j,j}(M) \to C^\infty(M).$$

We also have a multiplication

$$\wedge : \Omega^{j_1,k_1}(M) \times \Omega^{j_2,k_2}(M) \to \Omega^{j_1+j_2,k_1+k_2}(M)$$

such that for any $\alpha_1 \in \Omega^{j_1}(M)$, $\alpha_2 \in \Omega^{j_2}(M)$, $\beta_1 \in \Omega^{k_1}(M)$ and $\beta_2 \in \Omega^{k_2}(M)$ we have

$$(\alpha_1 \otimes \beta_1) \wedge (\alpha_2 \otimes \beta_2) = (\alpha_1 \wedge \alpha_2) \otimes (\beta_1 \wedge \beta_2).$$

The Riemann tensor R_h of the metric h is a $(2,2)$-form and thus we can construct an exterior j-th power $R^{\wedge j} \in \Omega^{2j,2j}(M)$. The j-th *Lipschitz-Killing curvature measure* of the Riemann manifold (M,h) is then defined by

$$\mathcal{L}_h^j = \begin{cases} 0, & m-j \in 2\mathbb{Z}+1, \\ \dfrac{(-2\pi)^{\frac{m-j}{2}}}{\left(\frac{m-j}{2}\right)!} \operatorname{tr}_h\big(R_h^{\wedge \frac{m-j}{2}}\big)|dV_g|, & m-j \in 2\mathbb{Z}. \end{cases}$$

The 0-th Lipschitz-Killing measure is involved in the Gauss-Bonnet theorem, i.e.,

$$\int_M \mathcal{L}_h^0(d\boldsymbol{p}) = \chi(M). \tag{4.1}$$

We have the following result.

THEOREM 4.1 (**Morse complexity**).

$$\chi^{\varepsilon} = \mathcal{L}^0_{g^{\varepsilon}},$$

where g^{ε} is the stochastic metric defined in (3.1).

The above result yields a probabilistic proof for the Gauss-Bonnet equality (4.1). The above theorem implies

$$\int_M \mathcal{L}^0_{g^{\varepsilon}}(d\boldsymbol{p}) = \int_M \chi^{\varepsilon}(d\boldsymbol{p}) = \chi(M).$$

Letting $\varepsilon \searrow 0$ and invoking the Approximation Theorem we deduce

$$\int_M \mathcal{L}^0_g(d\boldsymbol{p}) = \lim_{\varepsilon\searrow 0}\int_M \chi^{\varepsilon}(d\boldsymbol{p}) = \chi(M).$$

5. Outline of proofs

5.1. Proof of Theorem 2.1 and Theorem 3.1. The key result behind Theorem 2.1 is a Kac-Rice type result. This result gives an explicit, yet quite complicated description of the measure $\check{\boldsymbol{\sigma}}^{\varepsilon}$. More precisely, for any Borel subset $B \subset \mathbb{R}$, the Kac-Rice formula provides an integral representation of $\check{\boldsymbol{\sigma}}^{\varepsilon}(B)$ of the form

$$\check{\boldsymbol{\sigma}}^{\varepsilon}(B) = \int_M f_{\varepsilon,B}(\boldsymbol{p})\,|dV_g(\boldsymbol{p})|,$$

for some integrable function $f_{\varepsilon,B} : M \to \mathbb{R}$. The core of the Kac-Rice formula is an explicit probabilistic description of the density $f_{\varepsilon,B}$.

THEOREM 5.1 (**Kac-Rice**). *Fix a Borel subset $B \subset \mathbb{R}$. For any $\boldsymbol{p} \in M$ define*

$$f_{\varepsilon,B}(\boldsymbol{p}) := \big(\det\big(2\pi\boldsymbol{S}(\,\boldsymbol{u}_{\varepsilon}(\boldsymbol{p})\,)\big)\big)^{-\frac{1}{2}}\,\boldsymbol{E}\Big(\,|\det \operatorname{Hess}_{\boldsymbol{p}}(\check{\boldsymbol{u}}_{\varepsilon})|\cdot \boldsymbol{I}_B(\,\check{\boldsymbol{u}}_{\varepsilon}(\boldsymbol{p})\,)\;\big|\;d\check{\boldsymbol{u}}_{\varepsilon}(\boldsymbol{p}) = 0\,\Big),$$

where

- $\boldsymbol{S}(\,\boldsymbol{u}_{\varepsilon}(\boldsymbol{p})\,) \in \operatorname{Sym}(T^*_{\boldsymbol{p}}M)$ *denotes the covariance form of the random vector $d\boldsymbol{u}_{\varepsilon}(\boldsymbol{p}) \in T^*_{\boldsymbol{p}}M$;*
- $\boldsymbol{I}_B$ *denotes the characteristic function of the Borel set B;*
- $\boldsymbol{E}(\,\textbf{var}\;|\;\textbf{cons}\,)$ *stands for the conditional expectation of the variable* **var** *given the constraint* **cons**.

Then

$$\check{\boldsymbol{\sigma}}^{\varepsilon}(B) = \int_M f_{\varepsilon,B}(\boldsymbol{p})\,|dV_g(\boldsymbol{p})|. \tag{5.1}$$

This theorem is a special case of a general result of Adler-Taylor, [**1**, Cor. 11.2.2]. Proposition 1.2 guarantees that the technical assumptions in [**1**, Cor. 11.2.2] are satisfied if $\varepsilon \ll 1$.

A immediate consequence of Proposition 1.2, and the key reason why the Kac-Rice formula can be applied succesfully to prove Theorem 2.1 is the following elementary observation.

For any $\boldsymbol{p} \in M$, we have the centered Gaussian random vector

$$\big(\,\check{\boldsymbol{u}}_{\varepsilon}(\boldsymbol{p}), d\check{\boldsymbol{u}}_{\varepsilon}(\boldsymbol{p}), \operatorname{Hess}_{\boldsymbol{p}}(\check{\boldsymbol{u}}_{\varepsilon})\,\big) \in \mathbb{R}\oplus T^*_{\boldsymbol{p}}M \oplus \operatorname{Sym}(T_{\boldsymbol{p}}M).$$

We fix normal coordinates $(x^i)_{1\leq i\leq m}$ at $\boldsymbol{p}$ and we can identify the above Gaussian vector with the centered Gaussian vector

$$\big(\,\check{\boldsymbol{u}}_\varepsilon(\boldsymbol{p}),\ (\partial_{x^i}\check{\boldsymbol{u}}_\varepsilon(p)\,)_{1\leq i\leq m},\ \partial^2_{x^ix^j}(\check{\boldsymbol{u}}_\varepsilon(\boldsymbol{p})\,)_{1\leq i,j\leq m}\,\big)\in\mathbb{R}\oplus\mathbb{R}^m\oplus\operatorname{Sym}_m.$$

Consider the rescaled random vector

$$(\check{s}^\varepsilon,v^\varepsilon,H^\varepsilon):=\big(\varepsilon^{\frac{m}{2}}\check{\boldsymbol{u}}_\varepsilon(\boldsymbol{p}),\varepsilon^{\frac{m+2}{2}}d\boldsymbol{u}_\varepsilon(p),\ \varepsilon^{\frac{m+4}{2}}\operatorname{Hess}_{\boldsymbol{p}}(\check{\boldsymbol{u}}_\varepsilon)\,\big).$$

From Proposition 1.2 we deduce the following (uniform in $\boldsymbol{p}$) estimates as $\varepsilon\searrow 0$.

$$\boldsymbol{E}(\,(\check{s}^\varepsilon)^2\,)=\check{s}_m\big(1+O(\varepsilon^2)\,\big),\tag{5.2a}$$

$$\boldsymbol{E}\big(\,v_i^\varepsilon v_j^\varepsilon\,\big)=d_m\delta_{ij}\big(\,1+O(\varepsilon^2)\,\big),\tag{5.2b}$$

$$\boldsymbol{E}\big(\,H^\varepsilon_{ij}H^\varepsilon_{kl}\,\big)=h_m(\delta_{ij}\delta_{k\ell}+\delta_{ik}\delta_{j\ell}+\delta_{i\ell}\delta_{jk})\big(\,1+O(\varepsilon^2)\,\big),\tag{5.2c}$$

$$\boldsymbol{E}\big(\,\check{s}^\varepsilon H^\varepsilon_{ij}\,\big)=-d_m\delta_{ij}\big(\,1+O(\varepsilon^2)\,\big),\tag{5.2d}$$

$$\boldsymbol{E}\big(\,\check{s}^\varepsilon v_i^\varepsilon\,\big)=O(\varepsilon),\ \ \boldsymbol{E}\big(\,v_i^\varepsilon H^\varepsilon_{jk}\,\big)=O(\varepsilon).\tag{5.2e}$$

These estimates show that the Gaussian vector $(\check{s}^\varepsilon,v^\varepsilon,H^\varepsilon)$ converges weakly to the centered Gaussian vector $(\check{s}^0,v^0,H^0)$ with the following statistical characteristics.

- s^0 is a Gaussian scalar random variable with mean zero and variance $\check{s}_m$
- v^0 is a $\mathbb{R}^m$-valued Gaussian vector with mean 0 and covariance $d_m\mathbb{1}_{\mathbb{R}^m}$, which is independent of both s^0 and H^0.
- H^0 is a random Gaussian symmetric $m\times m$ matrix which can be written as a sum
$$H^0=X\mathbb{1}_{\mathbb{R}^m}+A,$$
where A is a random metrics in the ensemble GOE_{m,h_m} and X is a scalar Gaussian random variable, independent of A, with mean zero and variance h_m.
- $\boldsymbol{E}(\check{s}^0,H^0_{ij})=-d_m\delta_{ij},\ \forall i,j=1,\dots,m.$

Theorem 2.1 is then a consequence of the Kac-Rice formula, aided by the above observation. The computations of the conditional expectation in the Kac-Rice formula is performed using the regression formula [**4**] and a trick of Fyodorov [**8**] involving random symmetric matrices in the ensemble $\mathrm{GOE}_{v,m}$. For details we refer to [**20**].

The proof of Theorem 3.1 also follows from the Kac-Rice formula coupled with the above elementary observation.

5.2. Proof of Corollary 2.2. According to (2.5) we have

$$\gamma_{\boldsymbol{\omega}_m\varepsilon^{-m}}*\boldsymbol{\sigma}^\varepsilon=\check{\boldsymbol{\sigma}}^\varepsilon.$$

Thus

$$\gamma_{\frac{\boldsymbol{\omega}_m}{\check{s}_m}}*\Big(\mathcal{R}_{\frac{1}{\sqrt{\check{s}_m\varepsilon^{-m}}}}\Big)_*\boldsymbol{\sigma}^\varepsilon=\Big(\mathcal{R}_{\frac{1}{\sqrt{\check{s}_m\varepsilon^{-m}}}}\Big)_*\check{\boldsymbol{\sigma}}^\varepsilon.$$

Hence

$$\lim_{\varepsilon\to 0}\frac{1}{\boldsymbol{N}^\varepsilon}\gamma_{\frac{\boldsymbol{\omega}_m}{\check{s}_m}}*\Big(\mathcal{R}_{\frac{1}{\sqrt{\check{s}_m\varepsilon^{-m}}}}\Big)_*\boldsymbol{\sigma}^\varepsilon=\check{\boldsymbol{\sigma}}_m.$$

We can now conclude by invoking Lévy's continuity theorem [**15**, Thm. 15.23(ii)]. □

5.3. Proof of Theorem 2.3. We have

$$\check{\boldsymbol{\sigma}}_m = \frac{1}{K_m}\theta^-_{m+1,\frac{1}{r_m}} * \gamma_{\frac{r_m-1}{r_m}} dy \tag{5.3}$$

where

$$\theta^-_{m+1,\frac{1}{r_m}}(\lambda) = \rho_{m+1,\frac{1}{r_m}}(\lambda) e^{-\frac{r_m\lambda^2}{4}},$$

and

$$K_m = \int_{\mathbb{R}} \theta^-_{m+1,\frac{1}{r_m}} * \gamma_{\frac{r_m-1}{r_m}}(y)dy = \int_{\mathbb{R}} \theta^-_{m+1,\frac{1}{r_m}}(\lambda)d\lambda = \int_{\mathbb{R}} \rho_{m+1,\frac{1}{r_m}}(\lambda)e^{-\frac{r_m\lambda^2}{4}}d\lambda.$$

We set

$$R_m(\lambda) := \rho_{m+1,\frac{1}{m}}(\lambda),\;\; R_\infty(x) = \frac{1}{2\pi}\boldsymbol{I}_{\{|x|\leq 2\}}\sqrt{4-x^2}.$$

Fix $c \in (0,2)$. In [**17**, §4.2] we proved that

$$\lim_{m\to\infty}\sup_{|x|\leq c} |\bar{R}_m(x) - R_\infty(x)| = 0, \tag{5.4a}$$

$$\sup_{|x|\geq c} |\bar{R}_m(x) - R_\infty(x)| = O(1) \;\text{ as } m\to\infty. \tag{5.4b}$$

Then

$$\rho_{m+1,\frac{1}{r_m}}(\lambda) = \sqrt{\frac{r_m}{m}} R_m\left(\sqrt{\frac{r_m}{m}}\lambda\right),\;\; \theta^-_{m+1,\frac{1}{r_m}}(\lambda) = \sqrt{\frac{r_m}{m}} R_m\left(\sqrt{\frac{r_m}{m}}\lambda\right) e^{-\frac{r_m\lambda^2}{4}}.$$

We now distinguish two cases.

Case 1. $r = \lim_{m\to\infty} r_m < \infty$. In particular, $r \in [1,\infty)$. In this case we have

$$K_m = \sqrt{\frac{r_m}{m}}\int_{\mathbb{R}} R_m\left(\sqrt{\frac{r_m}{m}}\lambda\right) e^{-\frac{r_m\lambda^2}{4}}d\lambda,$$

and using (5.4a)-(5.4b) we deduce

$$\lim_{m\to\infty}\int_{\mathbb{R}} R_m\left(\sqrt{\frac{r_m}{m}}\lambda\right) e^{-\frac{r_m\lambda^2}{4}}d\lambda = R_\infty(0)\int_{\mathbb{R}} e^{-\frac{r\lambda^2}{4}}dr = R_\infty(0)\sqrt{\frac{4\pi}{r}}.$$

Hence

$$K_m \sim K'_m = R_\infty(0)\sqrt{\frac{4\pi}{m}} \;\text{ as } m\to\infty. \tag{5.5}$$

Now observe that

$$\begin{aligned}\frac{1}{K'_m}\theta^-_{m+1,\frac{1}{r_m}}(\lambda)d\lambda &= \frac{1}{R_\infty(0)}R_m\left(\sqrt{\frac{r_m}{m}}\lambda\right)\frac{r_m}{\sqrt{4\pi}}e^{-\frac{r_m\lambda^2}{4}}d\lambda\\ &= \frac{1}{R_\infty(0)}R_m\left(\sqrt{\frac{r_m}{m}}\lambda\right)\gamma_{\frac{2}{r_m}}(d\lambda)\end{aligned}$$

Using (5.4a) and (5.4b) we conclude that the sequence of measures

$$\frac{1}{K'_m}\theta^-_{m+1,\frac{1}{r_m}}(\lambda)d\lambda$$

converges weakly to the Gaussian measure $\gamma_{\frac{2}{r}}$. Using this and the asymptotic equality (5.5) in (5.3) we deduce

$$\lim_{m\to\infty}\check{\boldsymbol{\sigma}}_m = \gamma_{\frac{2}{r}} * \gamma_{\frac{r-1}{r}} = \gamma_{\frac{r+1}{r}}.$$

This proves Theorem 2.3 in the case $r < \infty$ since

$$\gamma_{\frac{\omega}{\check{s}_m}} * \boldsymbol{\sigma}_m = \check{\boldsymbol{\sigma}}_m \text{ and } \lim_{m\to\infty} \frac{\boldsymbol{\omega}_m}{\check{s}_m} \stackrel{(2.4)}{=} 0.$$

Case 2. $\lim_{m\to\infty} r_m = \infty$. In this case we write

$$\theta^-_{m+1,\frac{1}{r_m}}(\lambda)d\lambda = \sqrt{\frac{4\pi}{m}} R_m\left(\sqrt{\frac{r_m}{m}}\lambda\right)\gamma_{\frac{2}{r_m}}(\lambda)d\lambda.$$

We have the following key fact whose proof can be found in [**20**].

LEMMA 5.2. *The sequence of measures*

$$R_m\left(\sqrt{\frac{r_m}{m}}\lambda\right)\gamma_{\frac{2}{r_m}}(\lambda)d\lambda$$

converges weakly to the measure $R_\infty(0)\delta_0$.

Lemma 5.2 shows that

$$K_m \sim K'_m = \sqrt{\frac{4\pi}{m}} R_\infty(0),$$

and

$$\lim_{m\to\infty} \frac{1}{K_m}\theta^-_{m+1,\frac{1}{r_m}}(\lambda)d\lambda = \delta_0.$$

On the other hand

$$\lim_{m\to\infty} \gamma_{\frac{r_m-1}{r_m}}(\lambda)d\lambda = \gamma_1(\lambda)d\lambda,$$

so that

$$\lim_{m\to\infty} \check{\boldsymbol{\sigma}}_m = \delta_0 * \gamma_1 = \gamma_1.$$

This completes the proof of Theorem 2.3. □

5.4. Proof of Corollary 2.5. Using (2.7) we deduce

$$\begin{aligned}
\varepsilon^m \boldsymbol{N}^\varepsilon &= 2^{\frac{m+4}{2}}\Gamma\left(\frac{m+3}{2}\right)\left(\frac{h_m}{2\pi d_m}\right)^{\frac{m}{2}} \int_{\mathbb{R}} \theta^-_{m+1,\frac{1}{r_m}} * \gamma_{\frac{r_m-1}{r_m}}(y)dy + O(\varepsilon)\\
&= 2^{\frac{m+4}{2}}\Gamma\left(\frac{m+3}{2}\right)\left(\frac{h_m}{2\pi d_m}\right)^{\frac{m}{2}} \int_{\mathbb{R}} \theta^-_{m+1,\frac{1}{r_m}}(\lambda)d\lambda + O(\varepsilon)\\
&= \underbrace{2^{\frac{m+4}{2}}\Gamma\left(\frac{m+3}{2}\right)\left(\frac{h_m}{2\pi d_m}\right)^{\frac{m}{2}} K_m}_{=C_m(w)} + O(\varepsilon).
\end{aligned}$$

Lemma 5.2 implies that as $m \to \infty$ we have

$$K_m \sim \sqrt{\frac{4\pi}{m}} R_\infty(0) = \frac{2}{\sqrt{\pi m}}.$$

We deduce that

$$C_m(w) \sim \frac{2^{\frac{m+6}{2}}}{\sqrt{\pi m}}\Gamma\left(\frac{m+3}{2}\right)\left(\frac{h_m}{2\pi d_m}\right)^{\frac{m}{2}} \quad \text{as } m \to \infty.$$

□

5.5. Proof of Theorem 3.2. Fix a point $\boldsymbol{p} \in M$ and normal coordinates (x^i) near $\boldsymbol{p}$. The equality (1.4b) shows that as $\varepsilon \to 0$ we have the following estimate, uniform in $\boldsymbol{p}$.

$$\boldsymbol{E}\big(\,\partial_{x^i}\check{\boldsymbol{u}}_\varepsilon(\boldsymbol{p})\partial_{x^j}\check{\boldsymbol{u}}_\varepsilon(\boldsymbol{p})\,\big) = d_m\varepsilon^{-(m+2)}\big(\delta_{ij} + O(\varepsilon)^2)\,\big).$$

Hence

$$g^\varepsilon(\partial_{x^i}, \partial_{x^j}) = \delta_{ij} + O(\varepsilon^2) = g_{\boldsymbol{p}}(\partial_{x^i}, \partial_{x^j}) + O(\varepsilon^2). \tag{5.6}$$

This proves (a) and (b) of Theorem 2.1.

With $\boldsymbol{p}$ and (x^i) as above we set

$$\mathscr{E}^\varepsilon_{i_1,\dots,i_a;j_1,\dots,j_b} := \frac{\partial^{a+b}\mathscr{E}^\varepsilon(x,y)}{\partial x^{i_1}\cdots\partial x^{i_a}\partial y^{j_1}\cdots\partial y^{j_b}}\big|_{x=y=0},$$

$$g^\varepsilon_{ij} := g^\varepsilon_{\boldsymbol{p}}(\partial_{x^i}, \partial_{x^j}),\ \ 1 \le i,j \le m.$$

We denote by K^ε_{ij} the sectional curvature of g^ε along the plane spanned by $\partial_{x^i}, \partial_{x^j}$. Using [**1**, Lemma 12.2.1] and that the sectional curvatures of a metric are inverse proportional to the metric we deduce as in [**17**, §3.3] that

$$K^\varepsilon_{ij} = \frac{d_m}{\varepsilon^{m+2}} \times \frac{\mathscr{E}^\varepsilon_{ii;jj} - \mathscr{E}^\varepsilon_{ij;ij}}{\mathscr{E}^\varepsilon_{i;i}\mathscr{E}^\varepsilon_{j,j} - (\mathscr{E}^\varepsilon_{i;j})^2}.$$

Observe that (1.4b) and (1.4c) imply that

$$K^\varepsilon_{ij} = O(\varepsilon^{-2})\ \text{ as } \varepsilon \searrow 0.$$

In fact, a much more precise result is true. In [**20**] we have shown that there exists a universal constant $\mathfrak{Z}_m$ that depends only on m and w such that

$$\mathscr{E}^\varepsilon_{ii;jj} - \mathscr{E}^\varepsilon_{ij;ij} = \varepsilon^{-(m+2)}\mathfrak{Z}_m K_{ij}(\boldsymbol{p})\big(\,1 + O(\varepsilon^2)\,\big),$$

where $K_{ij}(\boldsymbol{p})$ denotes the sectional curvature of g at $\boldsymbol{p}$. The estimate (1.4b) implies that

$$\mathscr{E}^\varepsilon_{i;i}\mathscr{E}^\varepsilon_{j,j} - (\mathscr{E}^\varepsilon_{i;j})^2 = d_m^2\varepsilon^{-2(m+2)}\big(\,1 + O(\varepsilon^2)\,\big).$$

Thus

$$K^\varepsilon_{ij} = \frac{\mathfrak{Z}_m}{d_m}K_{ij}(\boldsymbol{p})\big(\,1 + O(\varepsilon^2)\,\big).$$

To determine the constant $\frac{\mathfrak{Z}_m}{d_m}$ it suffices to compute it on a special manifold. Assume that M is the unit sphere S^m equipped with the round metric. This is is a homogeneous space equipped with an invariant metric g with positive sectional curvatures. The metrics g^ε are also invariant so there exists a constant $C_\varepsilon > 0$ such that $g^\varepsilon = C_\varepsilon g$. The estimate (5.6) implies that $C_\varepsilon = 1$ and thus $K^\varepsilon_{ij} = K_{ij}(\boldsymbol{p})$ so that $\frac{\mathfrak{Z}_m}{d_m} = 1$.

The convergence of sectional curvatures coupled with the technique of harmonic coordinates in [**2**, **22**] and the C^0-convergence of the metrics g^ε imply the C^∞-convergence of this family of metrics.

□

5.6. Proof of 4.1. The proof is based on another version of the Kac-Rice formula. Using the same notations as in Theorem 5.1 we define $\rho_\varepsilon : M \to \mathbb{R}$ by setting

$$\rho_\varepsilon(\boldsymbol{p}) = \big(\det\big(2\pi \boldsymbol{S}(\boldsymbol{u}_\varepsilon(\boldsymbol{p}))\big)\big)^{-\frac{1}{2}} \boldsymbol{E}\Big(\det \operatorname{Hess}_{\boldsymbol{p}}(\boldsymbol{u}_\varepsilon) \;\big|\; d\boldsymbol{u}_\varepsilon(\boldsymbol{p}) = 0\Big), \;\; \forall \boldsymbol{p} \in M.$$

Then, for any open subset $U \subset M$ we have

$$\chi^\varepsilon(d\boldsymbol{p}) = \rho_\varepsilon(\boldsymbol{p})|V_g(d\boldsymbol{p})|. \tag{5.7}$$

There is a more convenient version of this formula. Consider a new, rescaled random function

$$\boldsymbol{v}_\varepsilon := \left(\frac{\varepsilon^{m+2}}{d_m}\right)^{\frac{1}{2}} \boldsymbol{u}_\varepsilon.$$

Observe that $\boldsymbol{v}_\varepsilon$ and $\boldsymbol{u}_\varepsilon$ have the same critical points with the same indices so that

$$\chi_{\boldsymbol{v}_\varepsilon} = \chi_{\boldsymbol{u}_\varepsilon}.$$

Hence

$$\chi^\varepsilon = \tilde{\chi}^\varepsilon := \boldsymbol{E}(\chi_{\boldsymbol{v}_\varepsilon}).$$

Here is a first advantage of working with $\boldsymbol{v}_\varepsilon$: by construction, for any $\boldsymbol{p} \in M$, the covariance pairing of the random vector $d\boldsymbol{v}_\varepsilon(\boldsymbol{p})$ coincides with the metric g^ε on $T^*_{\boldsymbol{p}}M$. Using the metric g^ε we can identify this covariance paring with the symmetric operator

$$\boldsymbol{S}(d\boldsymbol{v}_\varepsilon(\boldsymbol{p})) = \mathbb{1}_{T^*_{\boldsymbol{p}}M}.$$

For any smooth function $f : M \to \mathbb{R}$ and any $\boldsymbol{p} \in M$ we denote by $\operatorname{Hess}^\varepsilon_{\boldsymbol{p}}(f)$ the Hessian of f at $\boldsymbol{p}$ defined in terms of the metric g^ε. More precisely

$$\operatorname{Hess}^\varepsilon_{\boldsymbol{p}}(f) = XYf(\boldsymbol{p}) - (\nabla^\varepsilon_X Y)f(\boldsymbol{p}), \;\; \forall X, Y \in \operatorname{Vect}(M),$$

where ∇^ε denotes the Levi-Civita connection of the metric g^ε. Using the metric g^ε we can identify this Hessian with a symmetric linear operator

$$\operatorname{Hess}^\varepsilon_{\boldsymbol{p}}(f) : (T_{\boldsymbol{p}}M, h^\varepsilon) \to (T_{\boldsymbol{p}}M, h^\varepsilon).$$

We then have another version of the Kac-Rice formula

$$\tilde{\chi}^\varepsilon(d\boldsymbol{p}) = \tilde{\rho}_\varepsilon(\boldsymbol{p}) dV_{g^\varepsilon}(d\boldsymbol{p}), \tag{5.8}$$

where

$$\begin{aligned}\tilde{\rho}_\varepsilon(\boldsymbol{p}) &= \frac{1}{\sqrt{\det 2\pi S(\boldsymbol{v}^\varepsilon(\boldsymbol{p}))}} \boldsymbol{E}\Big(\det \operatorname{Hess}^\varepsilon_{\boldsymbol{p}}(\boldsymbol{v}^\varepsilon) \,\big|\, d\boldsymbol{v}^\varepsilon(\boldsymbol{p}) = 0\Big)\\ &= (2\pi)^{-\frac{m}{2}} \boldsymbol{E}\Big(\det \operatorname{Hess}^\varepsilon_{\boldsymbol{p}}(\boldsymbol{v}^\varepsilon) \,\big|\, d\boldsymbol{v}^\varepsilon(\boldsymbol{p}) = 0\Big).\end{aligned}$$

A small miracle happens. As shown in [**1**, Eq. (12. 2.11)], the Gaussian random variables $\operatorname{Hess}^\varepsilon_{\boldsymbol{p}}(\boldsymbol{v}^\varepsilon)$ and $d\boldsymbol{v}^\varepsilon(\boldsymbol{p})$ are *independent* so that

$$\tilde{\rho}_\varepsilon(\boldsymbol{p}) = (2\pi)^{-\frac{m}{2}} \boldsymbol{E}\Big(\det \operatorname{Hess}^\varepsilon_{\boldsymbol{p}}(\boldsymbol{v}^\varepsilon)\Big).$$

Observe that Hessian $\operatorname{Hess}^\varepsilon(f)$ of a function f can also be viewed as a double form

$$\operatorname{Hess}^\varepsilon(f) \in \Omega^{1,1}(M).$$

In particular, $\operatorname{Hess}^\varepsilon(\boldsymbol{v}^\varepsilon)$ is a random $(1,1)$ double form and we have the following equality, [**1**, Lemma 12.2.1]

$$-2R_{g^\varepsilon} = \boldsymbol{E}\big(\operatorname{Hess}^\varepsilon(\boldsymbol{v}^\varepsilon)^{\bullet 2}\big), \tag{5.9}$$

where R_{g^ε} denotes the Riemann curvature tensor of the metric h^ε. On the other hand we have the equality [**1**, Eq. (12.3.1)]

$$\det \operatorname{Hess}^\varepsilon(\boldsymbol{v}^\varepsilon) = \frac{1}{m!} \operatorname{tr} \operatorname{Hess}^\varepsilon(\boldsymbol{v}^\varepsilon)^{\wedge m} \tag{5.10}$$

Using (5.9), (5.10) and the algebraic identities in [**1**, Lemma 12.3.1] we conclude that

$$\frac{1}{(2\pi)^{\frac{m}{2}}} \boldsymbol{E}\Big(\det \operatorname{Hess}^\varepsilon_{\boldsymbol{p}}(\boldsymbol{v}^\varepsilon) \Big) = \frac{1}{(2\pi)^{\frac{m}{2}} (\frac{m}{2})!} \operatorname{tr}\Big(-R_{h^\varepsilon}^{\wedge \frac{m}{2}} \Big).$$

Using this in (5.8) we deduce

$$\tilde{\chi}^\varepsilon = \mathcal{L}^0_{g^\varepsilon}.$$

□

References

[1] Robert J. Adler and Jonathan E. Taylor, *Random fields and geometry*, Springer Monographs in Mathematics, Springer, New York, 2007. MR2319516 (2008m:60090)

[2] Michael T. Anderson and Jeff Cheeger, *C^α-compactness for manifolds with Ricci curvature and injectivity radius bounded below*, J. Differential Geom. **35** (1992), no. 2, 265–281. MR1158336 (93c:53028)

[3] Greg W. Anderson, Alice Guionnet, and Ofer Zeitouni, *An introduction to random matrices*, Cambridge Studies in Advanced Mathematics, vol. 118, Cambridge University Press, Cambridge, 2010. MR2760897 (2011m:60016)

[4] Jean-Marc Azaïs and Mario Wschebor, *Level sets and extrema of random processes and fields*, John Wiley & Sons, Inc., Hoboken, NJ, 2009. MR2478201 (2010m:60003)

[5] P. Bérard, G. Besson, and S. Gallot, *Embedding Riemannian manifolds by their heat kernel*, Geom. Funct. Anal. **4** (1994), no. 4, 373–398, DOI 10.1007/BF01896401. MR1280119 (95g:58228)

[6] Isaac Chavel, *Eigenvalues in Riemannian geometry*, Pure and Applied Mathematics, vol. 115, Academic Press, Inc., Orlando, FL, 1984. Including a chapter by Burton Randol; With an appendix by Jozef Dodziuk. MR768584 (86g:58140)

[7] Herbert Federer, *Curvature measures*, Trans. Amer. Math. Soc. **93** (1959), 418–491. MR0110078 (22 #961)

[8] Yan V. Fyodorov, *Complexity of random energy landscapes, glass transition, and absolute value of the spectral determinant of random matrices*, Phys. Rev. Lett. **92** (2004), no. 24, 240601, 4, DOI 10.1103/PhysRevLett.92.240601. MR2115095 (2005h:82047)

[9] I. M. Gel′fand and N. Ya. Vilenkin, *Generalized functions. Vol. 4. Applications of harmonic analysis*, Academic Press [Harcourt Brace Jovanovich, Publishers], New York-London, 1964 [1977]. Translated from the Russian by Amiel Feinstein. MR0435834 (55 #8786d)

[10] Alfred Gray, *Tubes*, 2nd ed., Progress in Mathematics, vol. 221, Birkhäuser Verlag, Basel, 2004. With a preface by Vicente Miquel. MR2024928 (2004j:53001)

[11] J. Hadamard, *Lectures on Cauchy's Problem in Linear Partial Differential Equations*, Yale University Press, 1923.

[12] Lars Hörmander, *The spectral function of an elliptic operator*, Acta Math. **121** (1968), 193–218. MR0609014 (58 #29418)

[13] Lars Hörmander, *The analysis of linear partial differential operators. I. Distribution theory and Fourier analysis*, 2nd ed., Grundlehren der Mathematischen Wissenschaften [Fundamental Principles of Mathematical Sciences], vol. 256, Springer-Verlag, Berlin, 1990. MR1065993 (91m:35001a)

[14] Lars Hörmander, *The analysis of linear partial differential operators. III. Pseudo-differential operators*, Grundlehren der Mathematischen Wissenschaften [Fundamental Principles of Mathematical Sciences], vol. 274, Springer-Verlag, Berlin, 1994. Corrected reprint of the 1985 original. MR1313500 (95h:35255)

[15] Achim Klenke, *Probability theory. A comprehensive course*, Translation from the German edition, Universitext, Springer, London, 2014. MR3112259

[16] Madan Lal Mehta, *Random matrices*, 3rd ed., Pure and Applied Mathematics (Amsterdam), vol. 142, Elsevier/Academic Press, Amsterdam, 2004. MR2129906 (2006b:82001)

[17] L. I. Nicolaescu, *Critical sets of random smooth functions on compact manifolds*, arXiv: 1101.5990; to appear in Asian J. Math.
[18] L. I. Nicolaescu, *Complexity of random smooth functions on compact manifolds* , arXiv: 1201.4972, to appear Indiana U. Math. J.
[19] L. I. Nicolaescu, *The wave group and the spectral geometry of compact manifolds*, http://www.nd.edu/~lnicolae/Wave.pdf
[20] L. I. Nicolaescu, *Random Morse functions and spectral geometry*, arXiv: 1209.0639
[21] Marcel Riesz, *L'intégrale de Riemann-Liouville et le problème de Cauchy* (French), Acta Math. **81** (1949), 1–223. MR0030102 (10,713c)
[22] Stefan Peters, *Convergence of Riemannian manifolds*, Compositio Math. **62** (1987), no. 1, 3–16. MR892147 (88i:53076)
[23] Michael E. Taylor, *Pseudodifferential operators*, Princeton Mathematical Series, vol. 34, Princeton University Press, Princeton, N.J., 1981. MR618463 (82i:35172)
[24] Steve Zelditch, *Real and complex zeros of Riemannian random waves*, Spectral analysis in geometry and number theory, Contemp. Math., vol. 484, Amer. Math. Soc., Providence, RI, 2009, pp. 321–342, DOI 10.1090/conm/484/09482. MR1500155 (2010c:58040)

Department of Mathematics, University of Notre Dame, Notre Dame, Indiana 46556-4618

E-mail address: nicolaescu.1@nd.edu
URL: http://www.nd.edu/~lnicolae/

Contemporary Mathematics
Volume **630**, 2014
http://dx.doi.org/10.1090/conm/630/12674

On the Exponential Decay of Laplacian Eigenfunctions in Planar Domains with Branches

Binh T. Nguyen, Denis S. Grebenkov, and Andrey L. Delytsin

ABSTRACT. We consider the eigenvalue problem for the Laplace operator in a planar domain which can be decomposed into a bounded domain of arbitrary shape and elongated "branches" of variable cross-sectional profiles. When the eigenvalue is smaller than a prescribed threshold, the corresponding eigenfunction decays exponentially along each branch. We prove this behavior for Robin boundary condition and illustrate some related results by numerically computed eigenfunctions.

1. Introduction

The geometrical structure of Laplacian eigenfunctions has been thoroughly investigated (see the review [**1**] and references therein). When a domain can be seen as a union of two (or many) subdomains with narrow connections, some low-frequency eigenfunctions can be found localized (or trapped) in one subdomain and of small amplitude in other subdomains. Qualitatively, an eigenfunction cannot "squeeze" through a narrow connection when its typical wavelength is larger than the connection width. This qualitative picture has found many rigorous formulations for dumbbell shapes and classical and quantum waveguides [**2**–**12**]. Numerical and experimental evidences for localization in irregularly-shaped domains were also reported [**13**–**21**].

In a recent paper, we considered the Laplacian eigenvalue problem

$$\begin{aligned} \Delta u + \lambda u &= 0 \quad \text{in } \Omega, \\ u &= 0 \quad \text{on } \partial\Omega, \end{aligned} \tag{1.1}$$

for a large class of domains Ω in $\mathbb{R}^d$ ($d = 2, 3, ...$) which can be decomposed in a "basic" bounded domain V and a branch Q of a variable cross-sectional profile [**10**]. We proved that if the eigenvalue λ is smaller than the smallest eigenvalue μ among all cross-sections of the branch, then the associated eigenfunction u exponentially decays along that branch:

THEOREM 1.1. *Let $\Omega \subset \mathbb{R}^d$ ($d = 2, 3, ...$) be a bounded domain with a piecewise smooth boundary $\partial\Omega$ and let $Q(z) = \Omega \cap \{x \in \mathbb{R}^d \ : \ x_1 = z\}$ be the cross-section of Ω at $x_1 = z \in \mathbb{R}$ by a hyperplane perpendicular to the coordinate axis x_1 (Fig. 1).*

2010 *Mathematics Subject Classification.* 35J05, 35Pxx, 49Rxx, 51Pxx.
Key words and phrases. Localization, Laplace, Eigenfunctions, Exponential decay.

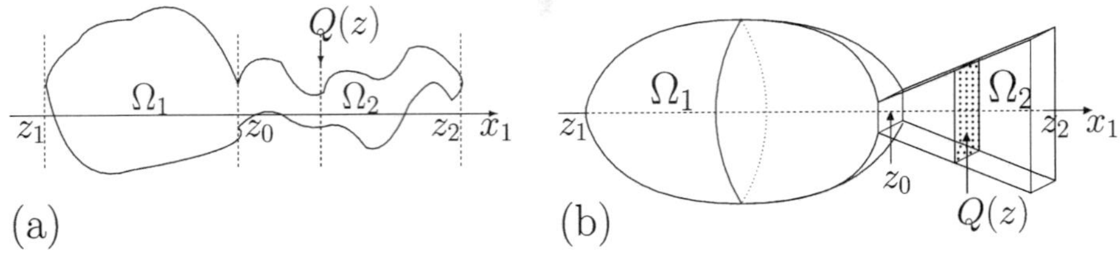

FIGURE 1. Two examples of a bounded domain $\Omega = \Omega_1 \cup \Omega_2$ with a branch Ω_2 of variable cross-sectional profile $Q(z)$. When the eigenvalue λ is smaller than the threshold μ, the associated eigenfunction exponentially decays in the branch Ω_2 and is thus mainly localized in Ω_1. Note that the branch itself may even be increasing.

Let

$$z_1 = \inf\{z \in \mathbb{R} \,:\, Q(z) \neq \emptyset\}, \qquad z_2 = \sup\{z \in \mathbb{R} \,:\, Q(z) \neq \emptyset\},$$

and we fix some z_0 such that $z_1 < z_0 < z_2$. Let $\mu(z)$ be the first eigenvalue of the Laplace operator in $Q(z)$, with Dirichlet boundary condition on $\partial Q(z)$, and $\mu = \inf\limits_{z\in(z_0,z_2)} \mu(z)$. Let u be a Dirichlet-Laplacian eigenfunction in Ω satisfying (1.1), and λ the associate eigenvalue. If $\lambda < \mu$, then

$$\|u\|_{L_2(Q(z))} \leq \|u\|_{L_2(Q(z_0))} \exp(-\beta\sqrt{\mu-\lambda}\,(z-z_0)) \quad (z \geq z_0), \tag{1.2}$$

with $\beta = 1/\sqrt{2}$. Moreover, if $(e_1 \cdot n(x)) \geq 0$ for all $x \in \partial\Omega$ with $x_1 > z_0$, where e_1 is the unit vector $(1, 0, ..., 0)$ in the direction x_1, and $n(x)$ is the normal vector at $x \in \partial\Omega$ directed outwards the domain, then the above inequality holds with $\beta = 1$.

In this theorem, a domain Ω is arbitrarily split into two subdomains, a "basic" domain Ω_1 (with $x_1 < z_0$) and a "branch" Ω_2 (with $x_1 > z_0$), by the hyperplane at $x_1 = z_0$ (the coordinate axis x_1 can be replaced by any straight line). Under the condition $\lambda < \mu$, the eigenfunction u exponentially decays along the branch Ω_2. Note that the choice of the splitting hyperplane (i.e., z_0) determines the threshold μ. Since μ is independent of the basic domain V, one can impose any boundary condition on $\partial\Omega_1$ (that still ensures the self-adjointness of the Laplace operator). In turn, the Dirichlet boundary condition on the boundary of the branch Ω_2 was relevant. Many numerical illustrations for this theorem were given in [**10**].

REMARK 1.2. It is worth stressing that the sufficient condition $\lambda < \mu$ involves purely spectral information: the eigenvalue λ in the whole domain and the smallest eigenvalue μ over all cross-sections. A simple geometrical condition on the basic domain Ω_1 can be formulated through the inradius ρ of Ω_1 (or Ω), i.e., the radius of the largest inscribed ball B_ρ. Since $B_\rho \subset \Omega$, the first Dirichlet eigenvalue λ is bounded as $\lambda \leq \lambda_1(B_\rho) = j^2_{\frac{d}{2}-1}/\rho^2$, where $j_{\frac{d}{2}-1}$ is the first positive zero of the Bessel function $J_{\frac{d}{2}-1}(z)$. A sufficient geometrical condition for getting exponentially decaying eigenfunction is then

$$\rho > j_{\frac{d}{2}-1}/\sqrt{\mu}. \tag{1.3}$$

For planar domains, the inequality yields $\rho/b > j_0/\pi$ where b is the largest width of the branch, and $j_0 \simeq 2.4048$. This inequality includes only the inradius of Ω_1

(or Ω) and the largest width of the branch, while the length of the branch can be varied arbitrarily. For instance, the localization in the basic domain Ω_1 may hold even when the area of the branch Ω_2 is arbitrarily large, as compared to the area of Ω_1.

REMARK 1.3. For higher dimensions ($d \geq 3$), the localization may sound even more striking, as the "branch" has to be "narrow" only in one direction (Fig. 1). For instance, if the branch Ω_2 has a constant width b in one direction, then the smallest eigenvalue μ in its cross-sections is greater than π^2/b^2. If the inradius ρ of Ω_1 is greater than $bj_{\frac{d}{2}-1}/\pi$ then the inequality (1.3) holds, and at least the first eigenfunction is localized in Ω_1. In the three-dimensional space, $j_{\frac{1}{2}} = \pi$ so that the inradius has to be just greater than b: $\rho > b$. A simple example is a domain decomposed into the unit cube $\Omega_1 = \{(x, y, z) \in \mathbb{R}^3 : 0 < x < 1,\ 0 < y < 1,\ 0 < z < 1\}$ and a parallelepiped $\Omega_2 = \{(x, y, z) \in \mathbb{R}^3 : 1 < x < L_x,\ -L_y < y < L_y,\ 0 < z < b\}$. When $b < 1/2$, the first eigenfunction is localized in the cube Ω_1, whatever the lateral spatial sizes L_x and L_y of the "branch" are.

In the remainder of the paper, we extend the above result to the Laplace operator in planar domains with Robin boundary condition. We also provide several numerical illustrations of localized eigenfunctions in planar domains in Sec. 3.

2. Extension for Robin boundary condition

We consider the eigenvalue problem for the Laplace operator in a planar domain Ω with Robin boundary condition on a piecewise smooth boundary $\partial\Omega$:

$$\begin{aligned} &\Delta u + \lambda u = 0 \quad \text{in } \Omega, \\ &\frac{\partial u}{\partial n} + hu = 0 \quad \text{on } \partial\Omega, \end{aligned} \tag{2.1}$$

where h is a nonnegative function, and $\partial/\partial n$ is the normal derivative directed outwards the domain. In that follows, we prove the following

THEOREM 2.1. *Let $\Omega = \Omega_1 \cup \Omega_2$, where $\Omega_1 \subset \mathbb{R}^2$ is a bounded domain, and*

$$\Omega_2 = \{(x, y) \in \mathbb{R}^2 \ :\ 0 < x < a,\ y_1(x) < y < y_2(x)\}, \tag{2.2}$$

is a branch of length $a > 0$ and of variable cross-sectional profile which is defined by two functions $y_1, y_2 \in C^1([0, a])$ such that $y_2(a) = y_1(a)$, $y_1'(x) \geq 0$ and $y_2'(x) \leq 0$. Let u and λ be an eigenfunction and eigenvalue of Ω satisfying the eigenvalue problem (2.1), with a nonnegative function h. We define $\mu = \inf_{x_0<x<a} \mu_1(x)$ where $\mu_1(x)$ is the first eigenvalue of the Laplace operator in the cross-sectional interval $[y_1(x), y_2(x)]$:

$$\begin{aligned} v''(y) + \mu_1(x)v(y) &= 0 \quad (y_1(x) < y < y_2(x)) \\ v'(y) - h_1(x)v(y) &= 0 \quad (y = y_1(x)) \\ v'(y) + h_2(x)v(y) &= 0 \quad (y = y_2(x)) \end{aligned} \tag{2.3}$$

where

$$h_i(x) \equiv h(y_i(x))\sqrt{1 + [y_i'(x)]^2} \qquad (i = 1, 2). \tag{2.4}$$

If $\lambda < \mu$, then

$$\|u\|_{L_2(\Omega(x))} \leq \|u\|_{L_2(\Omega(x_0))} \exp(-\beta\sqrt{\mu - \lambda}\ (x - x_0)) \quad (x \geq x_0), \tag{2.5}$$

where $\beta = 1/\sqrt{2}$ *and* $\Omega(x_0) = \{(x,y) \in \mathbb{R}^2 \ : \ x_0 < x < a, \ y_1(x) < y < y_2(x)\}$.

PROOF. The proof relies on Maslov's differential inequality and follows the scheme that we used in [**10**] for Dirichlet boundary condition. We consider the squared L_2-norm of the eigenfunction u in the "subbranch" $\Omega(x_0)$:

$$I(x_0) = \int\limits_{\Omega(x_0)} dxdy\ u^2 = \int\limits_{x_0}^{a} dx \int\limits_{y_1(x)}^{y_2(x)} dy\ u^2(x,y)$$

and derive the inequality for its second derivative:

$$I''(x_0) \geq 2(\mu - \lambda) I_0(x_0). \tag{2.6}$$

(i) From the first derivative

$$I'(x_0) = -\int\limits_{y_1(x_0)}^{y_2(x_0)} dy\ u^2(x_0, y),$$

we obtain

$$\begin{aligned} I''(x_0) &= -2 \int\limits_{y_1(x_0)}^{y_2(x_0)} dy\ u \frac{\partial u}{\partial x} - y_2'(x_0) u^2(x_0, y(x_0)) + y_1'(x_0) u^2(x_0, y_1(x_0)) \\ &\geq -2 \int\limits_{y_1(x_0)}^{y_2(x_0)} dy\ u \frac{\partial u}{\partial x}, \end{aligned}$$

where we used the conditions $y_2'(x) \leq 0$ and $y_1'(x) \geq 0$. Taking into account that

$$\begin{aligned} -\int\limits_{y_1(x_0)}^{y_2(x_0)} dy\ u \frac{\partial u}{\partial x} &= -\int\limits_{y_1(x_0)}^{y_2(x_0)} dy\ u \frac{\partial u}{\partial x} + \int\limits_{S(x_0)} dS\ u \frac{\partial u}{\partial n} + \int\limits_{S(x_0)} dS\ hu^2 \\ &= \int\limits_{\Omega(x_0)} dxdy\ \mathrm{div}(u \nabla u) + \int\limits_{S(x_0)} dS\ hu^2 \\ &= \int\limits_{\Omega(x_0)} dxdy\ (\nabla u, \nabla u) + \int\limits_{\Omega(x_0)} dxdy\ u \Delta u + \int\limits_{S(x_0)} dS\ hu^2 \\ &= \int\limits_{\Omega(x_0)} dxdy\ (\nabla u, \nabla u) - \lambda \int\limits_{\Omega(x_0)} dxdy\ u^2 + \int\limits_{S(x_0)} dS\ hu^2 \end{aligned}$$

where

$$S = S_1 \cup S_2, \qquad S_i = \{(x,y) \in \mathbb{R}^2 \ : \ 0 < x < a, \ y = y_i(x)\} \quad (i = 1, 2)$$

is the "lateral" boundary of Ω_2, we obtain

$$I''(x_0) \geq 2 \int\limits_{S(x_0)} dS\, hu^2 + 2 \int\limits_{\Omega(x_0)} dxdy \left(\frac{\partial u}{\partial y}\right)^2 - 2\lambda \int\limits_{\Omega(x_0)} dxdy\, u^2$$
$$= 2 \int\limits_{x_0}^{a} dx \left\{ h_2(x)\, u^2(x, y_2(x)) + h_1(x)\, u^2(x, y_1(x)) + \int\limits_{y_1(x)}^{y_2(x)} dy \left[\left(\frac{\partial u}{\partial y}\right)^2 - \lambda u^2 \right] \right\},$$

where $h_i(x)$ is defined by Eq. (2.4).

According to the Rayleigh principle, the first eigenvalue $\mu_1(x)$ of the eigenvalue problem (2.3) on the cross-sectional interval $[y_1(x), y_2(x)]$ can be written as
(2.7)

$$\mu_1(x) = \inf_{u \in H^1([y_1(x), y_2(x)])} \frac{\left[h_2(x)\, u^2(x, y_2(x)) + h_1(x)\, u^2(x, y_1(x)) \right] + \int\limits_{y_1(x)}^{y_2(x)} dy (\frac{\partial u}{\partial y})^2}{\int\limits_{y_1(x)}^{y_2(x)} dy\, u^2(x, y)},$$

from which we get

$$I''(x_0) \geq 2 \int\limits_{x_0}^{a} dx (\mu_1(x) - \lambda) \int\limits_{y_1(x)}^{y_2(x)} dy\, u^2(x, y)$$
$$\geq 2(\mu - \lambda) \int\limits_{x_0}^{a} dx \int\limits_{y_1(x)}^{y_2(x)} dy\, u^2(x, y) = 2(\mu - \lambda) I(x_0),$$

where $\mu = \inf\limits_{x_0 < x < a} \mu_1(x)$. That completes the first step.

(ii) We easily check the following relations:
(2.8)

$$I(a) = 0, \qquad I'(a) = 0, \qquad I(x_0) \neq 0 \quad (0 < x_0 < a), \qquad I'(x_0) < 0 \quad (0 < x_0 < a).$$

Note that the second relation relies on the assumption that $y_1(a) = y_2(a)$.

(iii) From the inequality (2.6) and relations (2.8), an elementary derivation implies the inequality (2.5). In fact, one multiplies (2.6) by $I'(x_0)$, integrates from x_0 to a, takes the square root and divides by $I(x_0)$ and integrates again from x_0 to x (see [**10**] for details). □

The statement of Theorem 2.1 for Robin boundary condition is weaker than that of Theorem 1.1 in several aspects:

- Theorem 2.1 employes an explicit parameterization of the branch through smooth height functions y_1 and y_2; in particular, the statement is limited to planar domains.
- The branch has to be non-increasing (conditions $y_1'(x) \geq 0$ and $y_2'(x) \leq 0$) and vanishing at the end (condition $y_1(a) = y_2(a)$).
- The inequality (2.5) characterizes the L_2-norm of the eigenfunction in the distant part of the branch, $\Omega(x)$, while the inequality (1.2) provided an estimate at the cross-section $Q(x)$.

- The decay rate in Eq. (2.5) involves the coefficient $\beta = 1/\sqrt{2}$ while the inequality (1.2) for non-increasing branches was proved for $\beta = 1$.

These remarks suggest that the statement of theorem 2.1 can be further extended while certain conditions may be relaxed.

We also note that the solution of the eigenvalue problem (2.3) has an explicit form

$$v(y) = c_1 \sin(\alpha y) + c_2 \cos(\alpha y), \tag{2.9}$$

with two constants c_1, c_2 and $\mu_1(x) = \alpha^2$, while the boundary conditions at the endpoints $y = y_1(x)$ and $y = y_2(x)$,

$$c_1[-h_1 \sin(\alpha y_1) + \alpha \cos(\alpha y_1)] + c_2[-h_1 \cos(\alpha y_1) - \alpha \sin(\alpha y_1)] = 0,$$
$$c_1[h_2 \sin(\alpha y_2) + \alpha \cos(\alpha y_2)] + c_2[h_2 \cos(\alpha y_2) - \alpha \sin(\alpha y_2)] = 0,$$

yield a closed equation on α:

$$(\alpha^2 + h_1 h_2) \sin \alpha(y_2 - y_1) + \alpha(h_1 + h_2) \cos \alpha(y_2 - y_1) = 0 \tag{2.10}$$

(here $h_{1,2}$ and $y_{1,2}$ depend on x). This equation has infinitely many solutions that can be found numerically. The first positive solution will determine $\mu_1(x)$.

3. Illustrations

In order to illustrate the geometrical structure of Laplacian eigenfunctions, we compute them for several simple domains. For all considered examples, we impose Dirichlet boundary condition for the sake of simplicity.

3.1. Sine-shaped branches. We consider the planar domain Ω composed of a basic domain Ω_1 (square of side L) and a branch Ω_2 of constant profile:

$$\Omega = \{(x, y) \in \mathbb{R}^2 \ : \ x \in (0, a), \ y \in (f(x), f(x) + b)\}. \tag{3.1}$$

For this example, we choose $f(x) = \sin(x)$, fix $b = 1$, $L = 1.54$ and take several values for the length a. Since the inradius of the square Ω_1 is greater than j_0/π, the first eigenvalue λ in these domains is smaller than $\mu = \pi^2$ for any length a so that the first eigenfunction should be localized in Ω_1 and exponentially decay along the branch Ω_2. This behavior is illustrated on Fig. 2.

3.2. Star-shaped domains. Figure 3 shows the first five Dirichlet-Laplacian eigenfunctions for a "star-shaped" domain which is formed by a disk with many elongated triangles. The inradius of this domain is much greater than the largest width of triangular branches that implies localization of the first eigenfunction. One can see that all the five eigenfunctions are localized in the disk and exponentially decay along the branches.

3.3. Elongated polygons. As we discussed at the beginning, the separation into a basic domain and a branch is conventional. We illustrate this point by showing the exponential decay of the first Dirichlet-Laplacian eigenfunction in elongated polygons for which the ratio between the diameter and the inradius is large enough. We start by considering a right triangle then extend the construction to general elongated polygons.

We consider a rectangle of sides a and b ($a \geq b$) on which a right triangle Ω with legs c and d is constructed as shown on Fig. 4. Note that the triangle is uniquely defined by one leg (e.g., d), while the other leg is $c = ad/(d - b)$. The

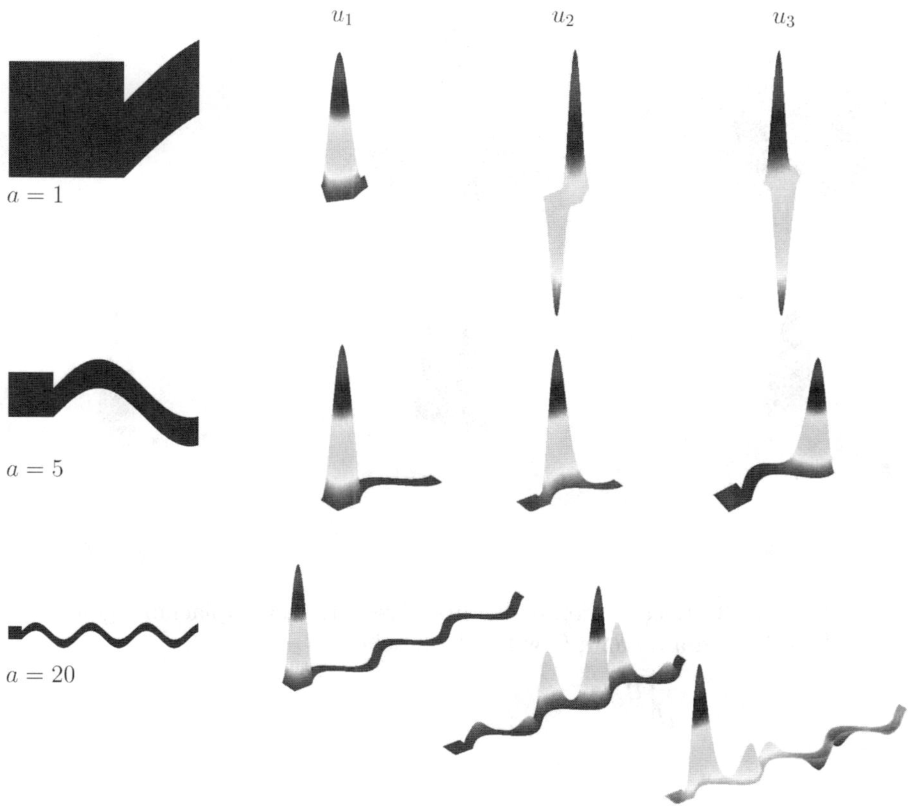

FIGURE 2. First three Dirichlet-Laplacian eigenfunctions for the planar domain Ω with sine-shaped branches, with $b = 1$, $L = 1.54$ and $a = 1$, $a = 5$ and $a = 20$. The first eigenfunction exponentially decays along the branch Ω_2, while the second and third eigenfunctions do not. The localization occurs in spite of the fact that the area of Ω_1 presents only 7.15% of the total area for the last domain with $a = 20$. Note that the third eigenfunction for $a = 5$ is localized at the end of the branch Ω_2.

vertical line at $x = a$ splits the triangle Ω into two subdomains: Ω_1 (a trapeze) and Ω_2 (a triangle). For fixed a and b, we are searching for a sufficient condition on d under which the eigenfunction u satisfying the eigenvalue problem (1.1), is localized in Ω_1 and exponentially decays along the subdomain Ω_2.

LEMMA 3.1. *Let Ω be a right triangle Ω defined by fixed $a, b, d > 0$ ($a \geq b$) (Fig. 4). Let $\zeta \equiv (d/b) - 1$, $P_A(\zeta)$ and $P_B(\zeta)$ be two explicit polynomials defined by Eq. (3.3), and $\zeta_0 \approx 0.0131$ is the zero of $P_A(\zeta)$. If both inequalities*

$$\zeta > \zeta_0, \qquad \frac{a^2}{b^2} > \frac{P_B(\zeta)}{P_A(\zeta)} \tag{3.2}$$

are fulfilled, then the first eigenfunction u of the Laplace operator in Ω with Dirichlet boundary condition exponentially decays in Ω_2.

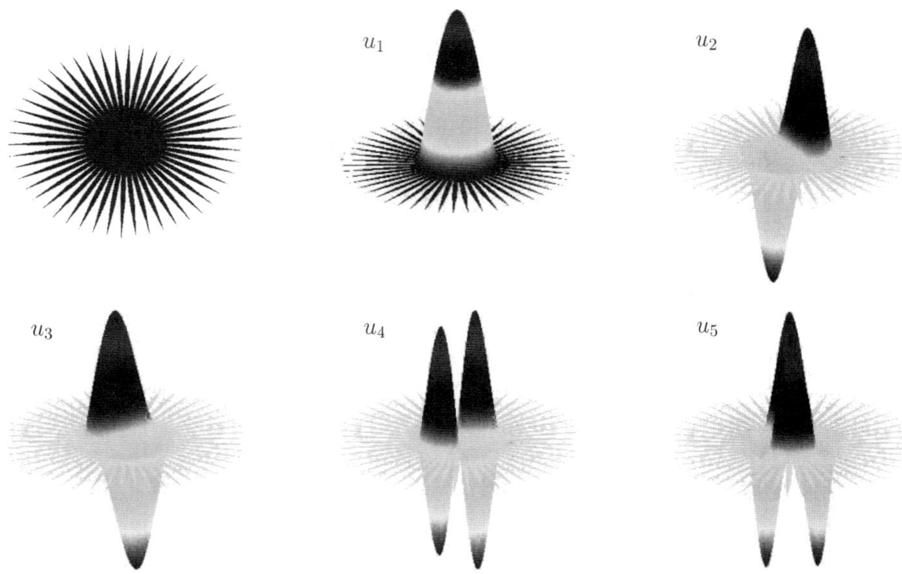

FIGURE 3. Localization of the first five Dirichlet-Laplacian eigenfunctions in a domain Ω with 51 branches.

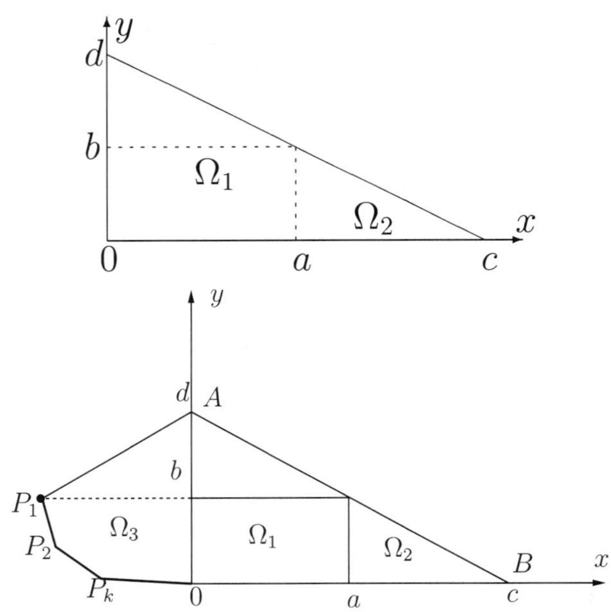

FIGURE 4. **Top:** A right triangle Ω is decomposed into a trapeze Ω_1 and a right triangle Ω_2. **Bottom:** An elongated polygon P with n vertices $\{0, B, A, P_1, P_2, \ldots, P_k\}$, where $k = n - 3$. Here, Ω_3 is the polygon including $n - 1$ vertices $\{0, A, P_1, P_2, \ldots, P_k\}$.

PROOF. The proof of the exponential decay relies on Theorem 1.1. For the "branch" Ω_2, the largest width is b so that the threshold $\mu = \pi^2/b^2$. Our goal is therefore to find a sufficient geometrical condition to ensure that the first eigenvalue λ_1 is smaller than μ. This condition can be replaced by a weaker condition $\gamma_1 < \pi^2/b^2$ for the first eigenvalue γ_1 of the Laplace operator in the trapeze Ω_1 with Dirichlet boundary condition.

The eigenvalue γ_1 can be found from the Rayleigh's principle as

$$\gamma_1 = \inf_{v \in H_0^1(\Omega_1)} \gamma(v), \qquad \gamma(v) \equiv \frac{(\nabla v, \nabla v)_{L^2(\Omega_1)}}{(v,v)_{L^2(\Omega_1)}}.$$

Taking a trial function

$$v(x,y) = y\left(y - d + \frac{d-b}{a}x\right)\sin(\pi x/a),$$

which satisfies Dirichlet boundary condition on the boundary of Ω_1, we look for such conditions that $\gamma(v) < \pi^2/b^2$, i.e.,

$$Q(\zeta) \equiv \frac{\pi^2}{b^2}(v,v)_{L^2(\Omega_1)} - (\nabla v, \nabla v)_{L^2(\Omega_1)} > 0.$$

The direct integration yields

$$Q(\zeta) = \frac{b^5}{720\pi^2 a}\Big(\kappa^2 P_A(\zeta) - P_B(\zeta)\Big),$$

where $\kappa = a/b$, and $P_A(\zeta)$, $P_B(\zeta)$ are two polynomials of the fifth degree:

$$P_A(\zeta) \equiv \sum_{j=0}^{5} A_j \zeta^j, \qquad P_B(\zeta) \equiv \sum_{j=0}^{5} B_j \zeta^j, \tag{3.3}$$

with the explicit coefficients:

$$\begin{array}{ll}
A_5 = 2\pi^4 - 15\pi^2 + 45 \approx 91.7741 & B_5 = 2\pi^4 + 15\pi^2 - 45 \approx 297.8622 \\
A_4 = 6(2\pi^4 - 10\pi^2 + 15) \approx 666.7328 & B_4 = 6(2\pi^4 + 10\pi^2 - 15) \approx 1671.0854 \\
A_3 = 30(\pi^4 - 4\pi^2 + 3) \approx 1827.9202 & B_3 = 30(\pi^4 + 3\pi^2) \approx 3810.5371 \\
A_2 = 20(2\pi^4 - 9\pi^2 + 9) \approx 2299.8348 & B_2 = 20(2\pi^4 + 3\pi^2) \approx 4488.5399 \\
A_1 = 30(\pi^4 - 6\pi^2) \approx 1145.7439 & B_1 = 30\pi^4 \approx 2922.2727 \\
A_0 = 12(\pi^4 - 10\pi^2) \approx -15.4434 & B_0 = 12\pi^4 \approx 1168.9091
\end{array}$$

Note that all $B_j > 0$ and $A_j > 0$ except for $A_0 < 0$. From the fact that $A_j < B_j$, one has $Q(\zeta) < 0$ for all $\zeta > 0$ when $\kappa = 1$ (i.e., $a = b$). We have therefore two parameters, ζ and κ, which determine the sign of Q and thus the exponential decay. Since $P_B(\zeta) > 0$ for all $\zeta \geq 0$, the condition $Q(\zeta) > 0$ is equivalent to two inequalities:

$$P_A(\zeta) > 0, \quad \kappa^2 > \frac{P_B(\zeta)}{P_A(\zeta)}. \tag{3.4}$$

One can check that $P_A(\zeta_0) = 0$ at $\zeta_0 \approx 0.0131$ and $P_A(\zeta) > 0$ if and only if $\zeta > \zeta_0$ that completes the proof. □

We remind that this condition is not necessary (as we deal with an estimate for the first eigenvalue). For given a and b (i.e., κ), the above inequalities determine

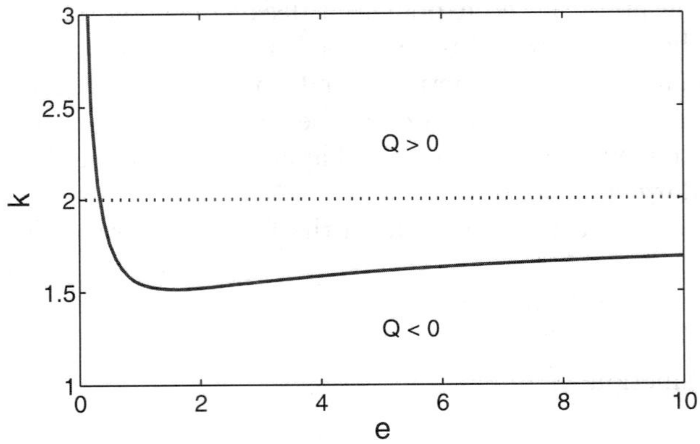

FIGURE 5. The diagram in the space of parameters ζ and κ for positive and negative signs of Q which correspond to localization and non-localization regions. For a given κ (e.g., $\kappa = 2$ shown by horizontal dotted line), one can determine the values of ζ, for which localization occurs.

the values of ζ (and thus the leg d) for which localization occurs. Alternatively, one can express a and b through the legs c and d (and parameter ζ) as

$$a = \frac{c\zeta}{\zeta + 1}, \qquad b = \frac{d}{\zeta + 1},$$

from which $\kappa = c\zeta/d$. For given c and d, one can vary ζ to get a family of inclosed rectangles (of sides a and b). The above inequalities can be reformulated as

$$\zeta > \zeta_0 \quad \Leftrightarrow \quad b < \frac{d}{\zeta_0 + 1},$$

$$k^2 > \frac{P_B(\zeta)}{P_A(\zeta)} \quad \Leftrightarrow \quad \frac{c}{d} > \frac{\sqrt{P_B(\zeta)}}{\sqrt{P_A(\zeta)}\,\zeta} \equiv f(\zeta).$$

The function $f(\zeta)$ can be checked to be monotonously decreasing so that the last inequality yields

$$\zeta > f^{-1}(c/d) \quad \Leftrightarrow \quad b < \frac{d}{f^{-1}(c/d) + 1}, \tag{3.5}$$

where f^{-1} denotes the inverse of the function $f(\zeta)$. This condition determines the choice of the inscribed rectangle (the size b) for a given triangle.

For the "worst" case $c = d$, for which a numerical computation yields $f^{-1}(1) \approx 1.515$, one gets

$$\frac{b}{d} < \frac{1}{2.515} \approx 0.3976.$$

This example shows that one can always inscribe a rectangle in such a way that $\lambda < \pi^2/b^2$. However, the "branch" Ω_2 in which an exponential decay of the eigenfunction is expected, may be small. Figure 6 illustrates these results.

REMARK 3.2. Any enlargement of the subdomain Ω_1 on Fig. 4 further diminishes the eigenvalue γ_1 and thus favors the exponential decay in Ω_2. In particular,

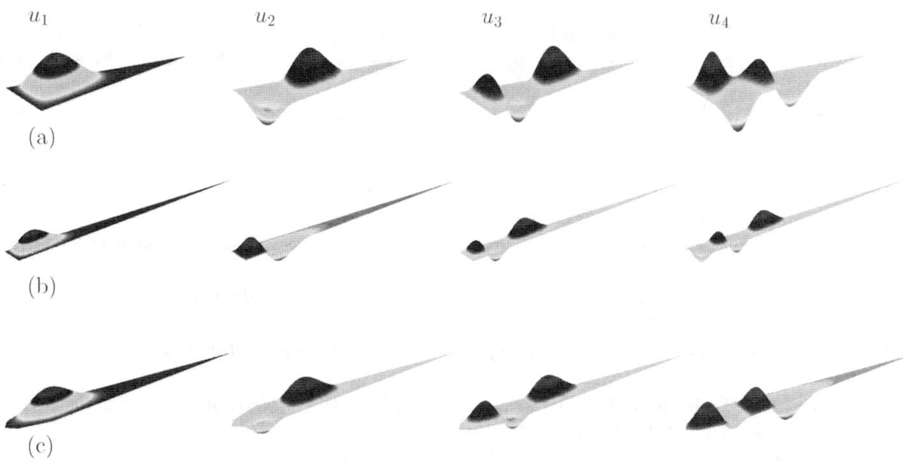

FIGURE 6. Several Dirichlet-Laplacian eigenfunctions in **(a)** the right triangle with $a = 2$, $b = 1$ and $\zeta \approx 0.32$ for which $c = 8.25$ and $d = 1.32$; **(b)** the right triangle with $a = 4$, $b = 1$ and $\zeta \approx 0.08$ for which $c = 61.14$ and $d = 1.07$; and **(c)** elongated hexagon. In all these cases, the first eigenvalue λ_1 is smaller than π^2, while the associated eigenfunction decays exponentially along the "branch" Ω_2. The other eigenfunctions are also concentrated in Ω_1.

for each positive integer n ($n \geq 3$), one can construct elongated polygons of n vertices for which the first Dirichlet-Laplacian eigenfunction is localized in Ω_1 (Fig. 4b). Figure 6c shows first eigenfunctions in elongated hexagons.

References

[1] D. S. Grebenkov and B.-T. Nguyen, *Geometrical structure of Laplacian eigenfunctions*, SIAM Rev. **55** (2013), no. 4, 601–667, DOI 10.1137/120880173. MR3124880

[2] Mark S. Ashbaugh and Pavel Exner, *Lower bounds to bound state energies in bent tubes*, Phys. Lett. A **150** (1990), no. 3-4, 183–186, DOI 10.1016/0375-9601(90)90118-8. MR1078396 (91h:35235)

[3] J. Goldstone and R. L. Jaffe, *Bound states in twisting tubes*, Phys. Rev. B **45** (1992), pp. 14100-14107.

[4] P. Duclos and P. Exner, *Curvature-induced bound states in quantum waveguides in two and three dimensions*, Rev. Math. Phys. **7** (1995), no. 1, 73–102, DOI 10.1142/S0129055X95000062. MR1310767 (95m:81239)

[5] P. Exner, P. Freitas, and D. Krejčiřík, *A lower bound to the spectral threshold in curved tubes*, Proc. R. Soc. Lond. Ser. A Math. Phys. Eng. Sci. **460** (2004), no. 2052, 3457–3467, DOI 10.1098/rspa.2004.1356. MR2166298 (2006d:81100)

[6] C. M. Linton and P. McIver, *Embedded trapped modes in water waves and acoustics*, Wave Motion **45** (2007), no. 1-2, 16–29, DOI 10.1016/j.wavemoti.2007.04.009. MR2441664 (2009e:35216)

[7] Shuichi Jimbo and Satoshi Kosugi, *Spectra of domains with partial degeneration*, J. Math. Sci. Univ. Tokyo **16** (2009), no. 3, 269–414. MR2597393 (2011d:35347)

[8] M. Filoche and S. Mayboroda, *Strong Localization Induced by One Clamped Point in Thin Plate Vibrations*, Phys. Rev. Lett. **103** (2009), 254301.

[9] O. Olendski and L. Mikhailovska, *Theory of a curved planar waveguide with Robin boundary conditions*, Phys. Rev. E **81** (2010), pp. 036606

[10] A. L. Delitsyn, B. T. Nguyen, and D. S. Grebenkov, *Exponential decay of Laplacian eigenfunctions in domains with branches of variable cross-sectional profiles*, Eur. Phys. J. B **85** (2012), 371.
[11] A. L. Delitsyn, B. T. Nguyen, and D. S. Grebenkov, *Trapped modes in finite quantum waveguides*, Eur. Phys. J. B **85** (2012), 176.
[12] Marcel Filoche and Svitlana Mayboroda, *Universal mechanism for Anderson and weak localization*, Proc. Natl. Acad. Sci. USA **109** (2012), no. 37, 14761–14766, DOI 10.1073/pnas.1120432109. MR2990982
[13] B. Sapoval, *Experimental observation of local modes of fractal drums*, Physica D **38** (1989), pp. 296-298.
[14] B. Sapoval, T. Gobron and A. Margolina, *Vibrations of fractal drums*, Phys. Rev. Lett. **67** (1991), 2974.
[15] B. Sapoval and T. Gobron, *Vibrations of strongly irregular or fractal resonators*, Phys. Rev. E **47** (1993), 3013.
[16] B. Sapoval, O. Haeberle and S. Russ, *Acoustical properties of irregular and fractal cavities*, J. Acoust. Soc. Am. **102** (1997), 2014.
[17] O. Haeberle, B. Sapoval, K. Menou and H. Vach, *Observation of vibrational modes of irregular drums*, Appl. Phys. Lett. **73** (1998), 3357.
[18] C. Even, S. Russ, V. Repain, P. Pieranski and B. Sapoval, *Localizations in Fractal Drums: An Experimental Study*, Phys. Rev. Lett. **83** (1999), 726.
[19] B. Hébert, B. Sapoval and S. Russ, *Experimental study of a fractal acoustical cavity*, J. Acoust. Soc. Am. **105** (1999) 1567.
[20] S. Felix, M. Asch, M. Filoche and B. Sapoval, *Localization and increased damping in irregular acoustic cavities*, J. Sound. Vibr. **299** (2007), 965.
[21] Steven M. Heilman and Robert S. Strichartz, *Localized eigenfunctions: here you see them, there you don't*, Notices Amer. Math. Soc. **57** (2010), no. 5, 624–629. MR2664041 (2011c:35101)

Laboratoire de Physique de la Matière Condensée, CNRS – Ecole Polytechnique, F-91128 Palaiseau, France
E-mail address: `binh-thanh.nguyen@polytechnique.edu`

Laboratoire de Physique de la Matière Condensée, CNRS – Ecole Polytechnique, F-91128 Palaiseau, France
E-mail address: `denis.grebenkov@polytechnique.edu`

Mathematical Department of the Faculty of Physics, Moscow State University, 119991 Moscow, Russia
E-mail address: `delitsyn@mail.ru`

Contemporary Mathematics
Volume **630**, 2014
http://dx.doi.org/10.1090/conm/630/12675

Bounds on the Principal Frequency of the p-Laplacian

Guillaume Poliquin

Abstract. This paper is concerned with the lower bounds for the principal frequency of the p-Laplacian on n-dimensional Euclidean domains. In particular, we extend the classical results involving the inner radius of a domain and the first eigenvalue of the Laplace operator to the case $p \neq 2$. As a by-product, we obtain a lower bound on the size of the nodal set of an eigenfunction of the p-Laplacian on planar domains.

1. Overview of the p-Laplacian

1.1. Physical models involving the p-Laplacian. Let Ω be a bounded open subset of $\mathbb{R}^n$. For $1 < p < \infty$, the p−Laplacian of a function f on Ω is defined by $\Delta_p f = \operatorname{div}(|\nabla f|^{p-2}\nabla f)$ for suitable f. The p-Laplacian can be used to model the flow of a fluid through porous media in turbulent regime (see for instance [**DHT**, **DTh**]) or the glacier's ice when treated as a non-Newtonian fluid with a nonlinear relationship between the rate deformation tensor and the deviatoric stress tensor (see [**GR**]). It is also used in the Hele-Shaw approximation, a moving boundary problem (see [**KMC**]).

Let us present a model, well known in the case of the Laplace operator, which remains very useful to understand the physical meaning behind some inequalities that we shall prove, in particular those involving the inner radius of Ω. The non-linearity of the p-Laplacian is often used to reflect the impact of non ideal material to the usual vibrating homogeneous elastic membrane, modeled by the Laplace operator. Thus, the following is used to describe a nonlinear elastic membrane under the load f,

$$
\begin{aligned}
-\Delta_p(u) &= f \qquad && \text{in } \Omega, \\
u &= 0 && \text{on } \partial\Omega.
\end{aligned} \tag{1.1}
$$

The solution u stands for the deformation of the membrane from the rest position (see [**CEP**, **Si**]). In that case, its deformation energy is given by $\int_\Omega |\nabla u|^p dx$. Therefore, a minimizer of the Rayleigh quotient,

$$
\frac{\int_\Omega |\nabla u|^p dx}{\int_\Omega |u|^p dx},
$$

2010 *Mathematics Subject Classification.* Primary 35P30; Secondary 58J50, 35P15.

Key words and phrases. Inradius, p-Laplacian, principal frequency, nodal set, capacity, interior capacity radius.

Research supported by a NSERC scholarship.

on $W_0^{1,p}(\Omega)$ satisfies $-\Delta_p(u) = \lambda_{1,p}|u|^{p-2}u$ in Ω. Here $\lambda_{1,p}$ is usually referred as the principal frequency of the vibrating non elastic membrane.

1.2. The eigenvalue problem for the p-Laplacian. For $1 < p < \infty$, we study the following eigenvalue problem:

$$\Delta_p u + \lambda|u|^{p-2}u = 0 \text{ in } \Omega, \tag{1.2}$$

where we impose the Dirichlet boundary condition and consider λ to be the real spectral parameter. We say that λ is an eigenvalue of $-\Delta_p$ if (1.2) has a nontrivial weak solution $u_\lambda \in W_0^{1,p}(\Omega)$. That is, for any $v \in C_0^\infty(\Omega)$,

$$\int_\Omega |\nabla u_\lambda|^{p-2}\nabla u_\lambda \cdot \nabla v - \lambda \int_\Omega |u_\lambda|^{p-2}u_\lambda v = 0. \tag{1.3}$$

The function u_λ is then called an eigenfunction of $-\Delta_p$ associated to the eigenvalue λ.

If $n \geq 2$ and $p = 2$, it is well known that one can obtain an increasing sequence of eigenvalues tending to $+\infty$ via the Rayleigh-Ritz method. Moreover, those are all the eigenvalues of Δ_2. Linearity is a crucial component in the argument.

For the general case, it is known that the first eigenvalue of the Dirichlet eigenvalue problem of the p-Laplace operator, denoted by $\lambda_{1,p}$, is characterized as,

$$\lambda_{1,p} = \min_{0 \neq u \in C_0^\infty(\Omega)} \left\{ \frac{\int_\Omega |\nabla u|^p dx}{\int_\Omega |u|^p dx} \right\}. \tag{1.4}$$

The infinimum is attained for a function $u_{1,p} \in W_0^{1,p}(\Omega)$. In addition, $\lambda_{1,p}$ is simple and isolated (there is no sequence of eigenvalues such that $\lambda_{k,p}$ tends to $\lambda_{1,p}$; see [**Lind2**]). Moreover, the eigenfunction u_1 associated to $\lambda_{1,p}$ does not change sign, and it is the only such eigenfunction (a proof can be found in [**Lind2**]). As for $\lambda_{2,p} > \lambda_{1,p}$, it allows a min-max characterization and every eigenfunction associated to $\lambda_{2,p}$ changes sign only once in Ω (it was first shown in [**ACFK**]; see also [**D1**]). It is not known if $\lambda_{2,p}$ is isolated. Via for instance Lyusternick-Schnirelmann maximum principle, it is possible to construct $\lambda_{k,p}$ for $k \geq 3$ and hence obtain an increasing sequence of eigenvalues of (1.2). There exist other variational characterizations of the Δ_p eigenvalues. However, no matter what variational characterization one chooses, it always remains to show that all the eigenvalues obtained exhaust the whole spectrum of Δ_p.

2. Introduction and main results

2.1. The principal frequency and the inradius. Using the domain monotonicity property, it is easy to obtain an upper bound for the principal frequency of the p-Laplacian. Indeed, for $B_r \subset \Omega$, we have that $\lambda_{1,p}(\Omega) \leq \lambda_{1,p}(B_r) = \frac{\lambda_{1,p}(B_1)}{r^p}$. Therefore, if we consider the largest ball that can be inscribed in Ω, we get

$$\rho_\Omega \leq \left(\frac{\lambda_{1,p}(B_1)}{\lambda_{1,p}(\Omega)} \right)^{\frac{1}{p}}, \tag{2.1}$$

where $\rho_\Omega := \sup\{r : \exists B_r \subset \Omega\}$. Note that unlike the case $p = 2$ corresponding to the Laplace operator, there are no explicit formulas for $\lambda_{1,p}$ of a ball.

Lower bounds involving the principal frequency are a greater challenge. Nevertheless, some results are known. Let us start by recalling that the classical Faber-Krahn inequality can be adapted to the p-Laplacian as noted in [**Lind1**, p. 224] and [**Hua**, p. 3353] (see also the rearrangement results in [**Ka**]) : among all domains of given n-dimensional volume, the ball minimizes $\lambda_{1,p}$. In other words, we have that $\lambda_{1,p}(\Omega) \geq \lambda_{1,p}(\Omega^*)$, where Ω^* stands for the n-dimensional ball of same volume than Ω.

For the Laplacian, lower bounds of the type

$$\lambda_{1,2}(\Omega) \geq \alpha_{n,2}\, \rho_\Omega^{-2}, \tag{2.2}$$

where $\alpha_{n,2} > 0$ is a positive constant, have been studied extensively. If $n = 2$, the first result proved in that direction is due to J. Hersch in [**He**], and states that for convex simply connected planar domains, the latter inequality holds with the constant $\alpha_{2,2} = \frac{\pi^2}{4}$. This result was later improved by E. Makai in [**Mak**]. For all simply connected domains, he obtained the constant $\alpha_{2,2} = \frac{1}{4}$. An adaptation of Makai's method for the pseudo p-Laplacian was studied in [**B**] and lead to a similar lower bound for simply-connected convex planar domains.

By a different approach, W. K. Hayman also obtained a bound for simply-connected domains with the constant, $\alpha_{2,2} = 1/900$. R. Osserman (see [**O1, O2, O3**]) later improved that result to $\alpha_{2,2} = 1/4$. R. Osserman also relaxed the assumption of simple connectedness of the domain by considering the connectivity k of a planar domain has a parameter. He obtained a similar lower bound for domains of connectivity $k \geq 2$, a result that was improved in [**Cr**].

In higher dimensions, it was first noted by W. K. Hayman in [**Hay**] that if A is a ball with many narrow inward pointing spikes removed from it, then $\lambda_{1,2}(A) = \lambda_{1,2}(\text{Ball})$, but in that case the inradius of A tends to 0. Therefore, bounds of the type,

$$\lambda_{1,2}(\Omega) \geq \alpha_{n,2}\, \rho_\Omega^{-2},$$

are generally not possible to obtain even if Ω is assumed to be simply connected. The higher dimensional case for Euclidean domains is discussed in Section 3. Similar bounds on manifolds are presented in Section 4.

In the next subsection, we extend some of these results to the case $p \neq 2$. The size nodal set of an eigenfunction of the p-Laplacian is also discussed. All proofs are given in Section 5.

2.2. First eigenvalue of the p-Laplacian and inradius of planar domains. We present some lower bounds for the principal frequency involving the inradius of a planar domain with an explicit constant,

$$\lambda_{1,p}(\Omega) \geq \alpha_{2,p}\, \rho_\Omega^{-p}.$$

We do so by adapting proofs obtained for the usual Laplacian. We need two main ingredients: a modified Cheeger-type inequality (see the original result for the Laplacian in [**Che**]; a generalized version for the p-Laplacian can be found in [**KF**]), and a geometric inequality relating the ratio of the length of the boundary of a domain and its area. Regarding the modified Cheeger inequality, it consists of an adapted version of a result in [**O4**]:

LEMMA 2.1. *Let (S, g) be a Riemannian surface, and let $D \subset S$ be a domain homeomorphic to a planar domain of finite connectivity k. Let F_k be the family of* *relatively compact subdomains of D with smooth boundary and with connectivity at most k. Let*

$$h_k(D) = \inf_{D' \in F_k} \frac{|\partial D'|}{|D'|}, \tag{2.3}$$

where $|D|$ is the area of D' and $|\partial D'|$ is the length of its boundary. Then,

$$\lambda_{1,p}(D) \geq \left(\frac{h_k(D)}{p}\right)^p. \tag{2.4}$$

The crucial point of Lemma 2.1 resides in the fact that the Cheeger constant is computed among subdomains that have a connectivity of at most the connectivity of the domain, which allows us to use geometric inequalities accordingly.

We start by proving the following extension of Osserman-Croke's result to the p-Laplacian. We actually generalize a stronger result that was implicit in [**Mak**] and [**O1**], but made explicit in [**G**]. Instead of considering the inner radius, we use the reduced inradius, which is defined by:

$$\tilde{\rho}_\Omega := \frac{\rho_\Omega}{1 + \frac{\pi \rho_\Omega^2}{|\Omega|}}.$$

Notice that $\frac{\rho_\Omega}{2} < \tilde{\rho}_\Omega < \rho_\Omega$.

The first main result consists of extending classical planar inradius bounds of the Laplace operator to the case of the p-Laplacian:

THEOREM 2.2. *Let Ω be a domain in $\mathbb{R}^2$. If Ω is simply connected, then for all $p > 1$, we have that*

$$\lambda_{1,p}(\Omega) \geq \left(\frac{1}{p\,\tilde{\rho}_\Omega}\right)^p. \tag{2.5}$$

If Ω is of connectivity $k \geq 2$, then for all $p > 1$, we have that

$$\lambda_{1,p}(\Omega) \geq \frac{2^{p/2}}{k^{p/2} p^p\, \rho_\Omega^p}. \tag{2.6}$$

It is hard to say whether these bounds are optimal for the case $p \neq 2$ since we can not compute eigenfunctions and eigenvalues explicitly on simple domains, unlike the case of the usual Laplacian. Also note that for the Laplace operator, instead of the constant $\frac{1}{4}$ given by (2.2), the better constant ≈ 0.6197 was found using probabilistic methods in [**BaC**].

REMARK 2.3. For a bounded domain Ω in $\mathbb{R}^n$, the ground state problem associated to the ∞-Laplacian is the following:

$$\min\{|\nabla u| - \lambda_{1,\infty} u; -\Delta_\infty u\} = 0, \tag{2.7}$$

where $\Delta_\infty u := \sum_{i,j=1}^n \frac{\partial u}{\partial x_i}\frac{\partial u}{\partial x_j}\frac{\partial^2 u}{\partial x_i \partial x_j}$. It is a notable feature that $\lambda_{1,\infty} = \frac{1}{\rho_\Omega}$, i.e. the value of $\lambda_{1,\infty}$ can immediately be read off the geometry of Ω, without any topological assumptions on Ω (see [**JL**] and reference therein for additional details).

2.3. The limiting case p=1. As $p \to 1$, the limit equation formally reads

$$(2.8)\qquad -\operatorname{div}\left(\frac{\nabla u}{|\nabla u|}\right) = \lambda_{1,1}(\Omega) \qquad \text{in } \Omega,$$
$$u = 0 \qquad \text{on } \partial\Omega,$$

where $\lambda_{1,1}(\Omega) := \lim_{p\to 1^+} \lambda_{1,p}(\Omega) = h(\Omega)$, where $h(\Omega) = \inf_{D\subset\Omega} \frac{|\partial\Omega|}{|\Omega|}$ with D varying over all non-empty sets $D \subset \Omega$ of finite perimeter (see [**KF**, **KS**] for instance). Here, Ω is assumed to be smooth enough. If we restrict the subdomains considered in the computation of $h(\Omega)$ to smooth simply connected ones, we get

PROPOSITION 2.4. *If Ω is a simply connected planar domain, then*

$$(2.9)\qquad \lambda_{1,1}(\Omega) \geq \frac{1}{\tilde{\rho}_\Omega}.$$

2.4. A bound on the size of the nodal set in the planar case. This subsection is devoted to the study of the size of the nodal set $\mathcal{Z}_\lambda = \{x \in \Omega : u_\lambda(x) = 0\}$ of an eigenfunction u_λ of (1.2). Yau's Conjecture (see [**Y**]) asserts that the size of the nodal set of an eigenfunction u_λ is comparable to $\lambda^{1/2}$.

Donnelly and Fefferman (see [**DF**]) proved Yau's Conjecture for real analytic manifolds. However, if one assumes only that (M, g) is smooth, Yau's Conjecture remains partially open. In the planar case, it is known that $\mathcal{H}^1(\mathcal{Z}_\lambda) \geq C_1\lambda^{1/2}$, where $\mathcal{H}^1$ stands for the 1-dimensional Hausdorff measure (see [**Bru**]). If $n \geq 3$, lower bounds were obtained (see recent works of Sogge-Zelditch in [**SZ**], Colding-Minicozzi in [**CM**], and Mangoubi in [**M**]).

We generalize the lower bound in the planar case for the p-Laplacian on a planar bounded domain. However, the situation is slightly different from the one for the Laplace operator mainly since it is still not known whether the interior of $\mathcal{Z}_\lambda$ is empty or not (see for instance [**D1**, **GM**] for a discussion on that matter). The result is the following:

THEOREM 2.5. *Let Ω be a planar bounded domain. There is a constant $M > 0$ such that for $\lambda > M$, there exists a positive constant C such that*

$$(2.10)\qquad \mathcal{H}^1(\mathcal{Z}_\lambda) \geq C\lambda^{1/p}.$$

In order to prove the previous theorem, we need to start proving the analog of a classical result of the Dirichlet Laplacian stating that every eigenfunction must vanish in a ball of radius comparable to the wavelength:

PROPOSITION 2.6. *Let $\Omega \subset \mathbb{R}^n$ be a bounded domain. Let $R = \left(\frac{C}{\lambda}\right)^{\frac{1}{p}}$, where $C > \lambda_{1,p}(B_1)$, the first eigenvalue of a ball of radius* 1. *Then, any eigenfunction u_λ of* (1.2) *vanishes in any ball of radius R.*

This result can be deduced from (2.1).

3. Lower bounds involving the inradius in higher dimensions

3.1. Lieb's and Maz'ya-Shubin's approaches to the inradius problem. Recall that if $p \leq n-1$, W. K. Hayman's counterexample holds. Therefore, it is not possible to get such lower bounds in that case even if Ω is simply connected or if $\partial\Omega$ connected. Nevertheless, E. Lieb obtained a similar lower bound by relaxing the condition that the ball has to be completely inscribed in Ω. By doing so, one can also relax some hypotheses on Ω. Instead, throughout this section, we shall only assume that Ω is an open subset of $\mathbb{R}^n$. No assumptions on the boundedness or on the smoothness of the boundary of Ω are required. We denote the bottom of the spectrum of $-\Delta_p$ by $\lambda_{1,p}(\Omega)$. In the case that Ω is a bounded domain, $\lambda_{1,p}(\Omega)$ corresponds to the lowest eigenvalue of $-\Delta_p$ with Dirichlet boundary condition as defined in (1.4). In the general case, we write

$$\lambda_{1,p}(\Omega) = \inf_{u\in C_0^\infty(\Omega)} \left\{ \frac{\int_\Omega |\nabla|^p dx}{\int_\Omega |u|^p dx} \right\}. \tag{3.1}$$

Lieb's method to get such a bound is to allow a fixed fraction $\alpha \in (0,1)$ of the Lebesgue measure of the ball to stick out of Ω. The result can be found in [**L**, p. 446]. It states that for a fixed $\alpha \in (0,1)$, if

$$\sigma_{n,p}(\alpha) = \lambda_{1,p}(B_r)\delta_n^{p/n}(\alpha^{-1/n} - 1)^p,$$

where $\delta_n = \frac{r^n}{|B_r|}$, then

$$\lambda_{1,p}(\Omega) \geq \frac{\sigma_{n,p}(\alpha)}{(r^L_{\Omega,\alpha})^p}. \tag{3.2}$$

Here $r^L_{\Omega,\alpha} = \sup\{r : \exists B_r \text{ such that } |B_r \setminus \Omega| \leq \alpha|B_r|\}$. Notice that the constant in the lower bound is not totally explicit since it depends on $\lambda_{1,p}(B_1)$, which is not known explicitly for $p \neq 2$.

Maz'ya and Shubin obtained a similar bound, but instead of using Lebesgue measure, they considered the Wiener capacity. The goal of this section is to generalize results of [**MS**] to the p-Laplacian to cover the case where $p < n$. To do so, we mainly use results stated in [**Maz**] and simplify the approach used in [**MS**], while losing the explicit constants in the bounds. Recall that p-capacity is defined for a compact set $F \subset \Omega$, where $n > 2$, as

$$\operatorname{cap}_p(F) = \inf_u \left\{ \int_{\mathbb{R}^n} |\nabla u|^p : u \in C_0^\infty(\Omega), u|_F \geq 1 \right\}. \tag{3.3}$$

Fix $\gamma \in (0,1)$. A compact set $F \subset B_r$ is said to be (p,γ)-negligible if

$$\operatorname{cap}_p(F) \leq \gamma \operatorname{cap}_p(\overline{B}_r). \tag{3.4}$$

We are ready to state the main result of this section and its corollaries:

THEOREM 3.1. *If $1 < p \leq n$, there exist two positive constants $K_1(\gamma,n,p)$ and $K_2(\gamma,n,p)$ that depend only on γ,n,p such that*

$$K_1(\gamma,n,p)r_{\Omega,\gamma}^{-p} \leq \lambda_{1,p}(\Omega) \leq K_2(\gamma,n,p)r_{\Omega,\gamma}^{-p}, \tag{3.5}$$

where $r_{\Omega,\gamma} = \sup\{r : \exists B_r, \overline{B}_r \setminus \Omega \text{ is } (p,\gamma)-\text{negligible}\}$ is the interior p-capacity radius.

A direct application of Theorem 3.1 is the following:

COROLLARY 3.2. *If $1 < p \leq n$, then $\lambda_{1,p}(\Omega) > 0 \iff r_{\Omega,\gamma} < +\infty$.*

Corollary 3.2 gives a necessary and sufficient condition of strict positivity of the operator $-\Delta_p$ with Dirichlet boundary conditions. For instance, let Ω be the complement of any Cartesian grid in $\mathbb{R}^{n\geq 3}$ (this example is adapted from [**Maz**, p. 789]. Clearly, for any $\gamma \in (0,1)$, $r_{\Omega,\gamma} = +\infty$; thus, $\lambda_{1,p}(\Omega) = 0$. However, if Ω is a narrow strip, $r_{\Omega,\gamma} < +\infty$, implying that $\lambda_{1,p}(\Omega) > 0$.

Note that since $\lambda_{1,p}(\Omega) > 0$ does not depend on γ, we immediately get the following:

COROLLARY 3.3. *If $1 < p \leq n$, the conditions $r_{\Omega,\gamma} < +\infty$ for different γ's are all equivalent.*

Also, one can show that for $1 < p \leq n$, the lower bound given by Theorem 3.1 implies the lower bound obtained earlier by Lieb, (3.2). To do so, one needs to use an isocapacity inequality that can be found in [**Maz**, Section 2.2.3], stating that if F be a compact subset of $\mathbb{R}^n$, then

$$\operatorname{cap}_p(F) \geq \omega^{p/n} n^{(n-p)/n} \left(\frac{n-p}{p-1}\right) |F|^{(n-p)/n}, \tag{3.6}$$

where equality occurs if and only if F is a ball.

PROPOSITION 3.4. *If $\alpha = \gamma^{n/(n-p)}$ and $1 < p \leq n$, then $r^L_{\Omega,\alpha} \geq r_{\Omega,\gamma}$. In particular, Theorem 3.1 implies (3.2).*

PROOF. Let $C = \omega^{p/n} n^{(n-p)/n} \left(\frac{n-p}{p-1}\right)$ and fix $\gamma \in]0,1[$. Suppose that

$$\operatorname{cap}_p(\overline{B}_r \setminus \Omega) \leq \gamma \operatorname{cap}_p(\overline{B}_r);$$

therefore, using (3.6), we get that

$$\begin{aligned} |\overline{B}_r \setminus \Omega| &\leq C^{-\frac{n}{n-p}} \operatorname{cap}_p(\overline{B}_r \setminus \Omega) \\ &\leq C^{-\frac{n}{n-p}} \gamma^{\frac{n}{n-p}} \operatorname{cap}_p(\overline{B}_r)^{\frac{n}{n-p}} \\ &= \alpha |\overline{B}_r|, \end{aligned}$$

yielding the desired result. □

3.2. Convex domains in $\mathbb{R}^n$. Another way to avoid such difficulty is to consider convex domains in $\mathbb{R}^n$. Doing so, we can prove the following:

PROPOSITION 3.5. *If Ω be a convex body in $\mathbb{R}^n$, then the following inequality*

$$\lambda_{1,p}(\Omega) \geq \left(\frac{1}{p\rho_\Omega}\right)^p, \tag{3.7}$$

holds for all $p > 1$:

The proof is based on two key facts (see [**O3**, p. 26]), namely that the inequality,

$$|\partial\Omega| \geq h(\Omega)|\Omega|,$$

is required to be true only for subdomains bounded by level surfaces of the first eigenfunction of Ω (see the proof of Lemma 2.1), and that if Ω is convex, then those subdomains Ω' are also convex (see [**BL**, Theorem 1.13]). Recalling that $\frac{1}{\rho_{\Omega'}} \geq \frac{1}{\rho_\Omega}$, the proposition then follows easily.

3.3. Further discussion. A striking difference between the usual Laplace operator and the p-Laplacian is that it is possible to obtain bounds of the type (2.2) in higher dimensions, as long as p is "large enough" compared to the dimension. Indeed, W.K. Hayman's observation remains valid in the case $p \leq n-1$ since for such p, every curve has a trivial p-capacity. On the other hand, for $p > n$, even a single point has a positive p-capacity (see [**Maz**, Chapter 13, Proposition 5 and its corollary]). Consequently, W. K. Hayman's counterexample no longer holds since removing a single point has an impact on the eigenvalues as it can be seen from (1.4). Taking the latter observation into account, it is possible to prove the following theorem:

THEOREM 3.6. *Let $p > n$ and let Ω be a bounded domain. Then there exists a positive constant $C_{n,p}$ such that*

$$\lambda_{1,p}(\Omega) \geq \frac{C_{n,p}}{\rho_{\Omega}^{p}}. \tag{3.8}$$

It is known that if $p > n-1$, every curve has a positive p-capacity. Therefore, W. K. Hayman's counter example does not work in that case, leading to the following result:

THEOREM 3.7. *For $p > n-1$, suppose that Ω is a bounded domain such that $\partial\Omega$ is connected, then there exists $C_{n,p} > 0$ such that*

$$\lambda_{1,p}(\Omega) \geq \frac{C_{n,p}}{\rho_{\Omega}^{p}}.$$

This result can be viewed as a generalization of classical results for the Laplacian on simply connected planar domains ($p = n = 2$) discussed earlier however, without the explicit constant. Also notice that the result can not hold if $p = n - 1$ as noted by W. K. Hayman for the case $p = 2, n = 3$.

Theorems 3.6 and 3.7 were suggested by D. Bucur and their proofs can be found in [**Poli**].

4. Lower bounds involving eigenvalues of the p-Laplace operator on manifolds

4.1. A two-sided inradius bound for nodal domains on closed Riemannian surfaces. Some results concerning the inner radius of nodal domains on manifolds were obtained in [**M1, M2**]. As suggested by D. Mangoubi, it is possible to extend a two sided estimate valid for closed surfaces, [**M1**, Theorem 1.2] to the case of the p-Laplacian. Let U_λ denote the λ-nodal domain associated to the eigenfunction $u_{p,\lambda}$. The result is the following:

THEOREM 4.1. *Let (S, g) be a closed surface. Then, there exists two positive constants c, C such that*

$$c\lambda^{-1/p} \leq \rho_{U_\lambda} \leq C\lambda^{-1/p}.$$

The proof of the case $p = 2$ relies on two main tools, namely Faber-Krahn, which still holds for the p-Laplace operator, and a Poincaré inequality, [**M1**, Theorem 2.4]. The latter can also be generalized to our setting. We have the following result:

LEMMA 4.2. *Let $Q \subseteq \mathbb{R}^2$ be a cube whose edge is of length a. Fix $\alpha \in (0, 1)$ and let u be in $C_0^\infty(Q)$ such that u vanishes on a curve whose projection on one of the edges of Q has size $\geq \alpha a$. Then, there exists a constant $C(\alpha)$ such that*

$$\int_Q |u|^p dx \leq C(\alpha) a^p \int_Q |\nabla u|^p dx.$$

Since the proof of Theorem 4.1 and Lemma 4.2 are direct extensions of what is done in [**M1**], we omit details.

4.2. Lower bounds involving the inradius on surfaces. We can adapt some inradius results of [**O1**] valid for simply connected domains of surfaces (S, g) with controlled Gaussian curvature to the case of the p-Laplacian:

PROPOSITION 4.3. *Let S be a surface. Let $D \subset S$ be a simply connected domain. Denote by K its Gaussian curvature and by β the quantity $\int_D K^+$, where $K^+ = \max\{K, 0\}$. If $\beta \leq 2\pi$ the inequality,*

$$\lambda_{1,p}(D) \geq \left(\frac{1}{p\ \rho_D}\right)^p, \tag{4.1}$$

holds.

If the surface has a negative Gaussian curvature, then we have the following:

PROPOSITION 4.4. *Let S be a simply connected surface. Let $D \subset S$ be a simply connected domain such that $K \leq -\alpha^2$ on D, $\alpha > 0$, where K stands for its Gaussian curvature. The stronger inequality*

$$\lambda_{1,p}(D) \geq \left(\frac{\alpha \coth(\alpha \rho_D)}{p}\right)^p \tag{4.2}$$

holds.
Furthermore, if S is a complete surface, then one has the following,

$$\lambda_{1,p}(D) \geq \left(\frac{\alpha \coth(\alpha R_D)}{p}\right)^p, \tag{4.3}$$

where R_D stands for the circumradius of D.

4.3. Negatively curved manifolds. The first result, called McKean's theorem (see [**McK**]) in the case of the Laplace operator, concerns manifolds of negative sectional curvature:

PROPOSITION 4.5. *Let (M, g) be a complete and simply connected Riemannian manifold. Let $D \subset M$ be a domain such that its sectional curvature is bounded by $\leq -\alpha^2, \alpha > 0$, then*

$$\lambda_{1,p}(D) \geq \frac{(n-1)^p \alpha^p}{p^p}. \tag{4.4}$$

The next result is valid on an arbitrary surface (without additional assumptions, such as being simply connected), but it is only valid for doubly-connected domains.

PROPOSITION 4.6. *Let (S, g) be a surface. Let $D \subset S$ be a doubly-connected domain such that $K \leq -1$ where K stands for its Gaussian curvature, then*

$$\lambda_{1,p}(D) \geq \frac{1}{p^p}. \tag{4.5}$$

4.4. Minimal submanifolds in $\mathbb{R}^n$. It is also possible to prove the following using similar arguments:

PROPOSITION 4.7. *Let D be a domain on a m-dimensional minimal submanifold in $\mathbb{R}^n$. If D lies in a ball of radius R, then*

$$\lambda_{1,p}(D) \geq \left(\frac{m}{p}\right)^p. \tag{4.6}$$

5. Proofs

5.1. Proof of Lemma 2.1. By (2.3), it follows that if one proves (2.4) for all domains in a regular exhaustion of D, then (2.4) will also hold for D. Knowing that every finitely-connected domain has a regular exhaustion by domains of the same connectivity, we may assume that D has a smooth boundary. Hence, the p-Laplacian Dirichlet eigenvalue problem admits a solution u_1 corresponding to $\lambda_{1,p}$ and may be chosen without loss of generality such that $u_1 \geq 0$.

Let $g = u_1^p$. Then, Hölder's inequality implies that

$$\begin{aligned}\int_D |\nabla g(x)| dx &= p\int_D |u_1(x)|^{p-1}|\nabla u_1(x)| dx \\ &\leq p||u_1||_p^{p-1}||\nabla u_1||_p.\end{aligned}$$

Dividing by $||u_1||_p^p$, one gets

$$\frac{1}{p}\frac{\int_D |\nabla g(x)| dx}{\int_D |g(x)| dx} \leq \frac{||\nabla u_1||_p}{||u_1||_p} = \lambda_{1,p}(D)^{1/p}. \tag{5.1}$$

For regular values t of the function g, we define the set

$$D_t = \{y \in D : g(y) > t\}.$$

We want to show that the connectivity of D_t is at most k, since it will imply by (2.3) that

$$L(t) \geq h_k(D)A(t), \tag{5.2}$$

for all regular values of t. Here, $L(t)$ is the length of the boundary of D_t and $A(t)$ the area of D_t.

Since t is regular, the boundary ∂D_t consists of a finite number, say m, of smooth curves $C_1, C_2, ..., C_m$ along which $\nabla g \neq 0$. Let D_t' be any connected component of D_t. If the connectivity of D_t' were greater than k, then the complement of D_t would contain a component lying completely in D of boundary say, C_l. Since $u_1 \in C(D) \cap W^{1,p}(D)$ and

$$\int_D |\nabla u_1|^{p-2}\nabla u_1 \cdot \nabla v \, dx = \lambda_{1,p}\int_D |u_1|^{p-2}u_1 v \, dx \geq 0,$$

for all non negative v in $C_0^\infty(D)$, u_1 is p-superharmonic (see [**Lind3**, Theorem 5.2]). By definition of p-superharmonicity (again [**Lind3**, Definition 5.1]), the comparison principle holds. Therefore, since $u_1 = \sqrt[p]{t}$ on C_l, it follows that $u_1^p \geq t$ in the internal region of C_l, which contradicts the fact that the internal region lies in the complement of D_t', so that the function g has to be such that $g < t$.

Now, since the set of singular values of g is a closed set of measure zero by Sard's theorem, its complement is a countable union of open intervals I_n. Thus, we can define the following

$$E_n = \{p \in D : g(p) \in I_n\}.$$

Then, by the coarea formula,

$$\begin{aligned}\int_D |\nabla g(x)| \, dx &\geq \sum_{n=1}^{\infty}\int_{E_n} |\nabla g(x)| \, dx = \sum_{n=1}^{\infty}\int_{I_n} L(t) \, dt \\ &\geq h_k(D)\sum_{n=1}^{\infty}\int_{I_n} A(t) \, dt = h_k(D)\int_D g(x) \, dx. \end{aligned} \tag{5.3}$$

Combining (5.1) with (5.3) completes the proof.

5.2. Proof of Theorem 2.2, of Propositions 4.3, and 4.4. We use the following lemma which was proved in [**G**]:

LEMMA 5.1. *If* $\Omega' \subset \Omega$, *then*

$$\tilde{\rho}_{\Omega'} \leq \tilde{\rho}_{\Omega}.$$

PROOF OF THEOREM 2.2. Combining Lemma 2.1 with Lemma 5.1 and Bonnesen's inequality (see [**Bon**]) then yields the desired result.

In order to adapt the argument to planar domains of connectivity k, one must use a generalized geometric inequality that can be found in [**Cr**, Theorem 1], stating that

$$\frac{|\partial\Omega|}{|\Omega|} \geq \frac{\sqrt{2}}{\sqrt{k\rho}}.$$

Combining Lemma 2.1, [**Cr**, Theorem 1], and the fact that $\rho_{\Omega'} \leq \rho_{\Omega}$ provided that $\Omega' \subset \Omega$ yields the desired result. □

PROOF OF PROPOSITION 4.3. We start by using Burago and Zalgaller's inequality that can be found in [**BZ**], and that states that

$$\rho_D|\partial D| \geq |D| + \left(\pi - \frac{1}{2}\beta\right)\rho_D^2 \iff \frac{|\partial D|}{|D|} \geq \frac{1 + \frac{(\pi - \frac{1}{2}\beta)\rho_D^2}{|D|}}{\rho_D} \geq \frac{1}{\rho_D}. \tag{5.4}$$

By definition of $h_1(D)$, together with (5.4), one gets the desired result.

□

PROOF OF PROPOSITION 4.4. Since S is simply connected, one can use [**I**, Theorem 1] or [**O3**, Theorem 8(c)] to get that

$$|\partial D| \geq \frac{\alpha|D|}{\tanh \alpha\rho_D} + \frac{2\pi}{\alpha}\tanh\frac{\alpha\rho_D}{2} \geq \alpha|D|\coth(\alpha\rho_D).$$

Combining the previous inequality with Lemma 2.1 and the fact that $\rho_{\Omega'} \leq \rho_{\Omega}$ provided that $\Omega' \subset \Omega$ proves the first part of Theorem 4.4. To conclude the proof, one must use [**O3**, Theorem 8], which states that

$$|\partial D| \geq \alpha|D|\coth(\alpha R_D).$$

Combining the latter inequality with Lemma 2.1 and the fact that for any subdomain D', its circumradius satisfies $R' \leq R, \coth(\alpha R') \geq \coth(\alpha R)$ yields the desired result. □

5.3. Proof of Proposition 2.6. Consider a ball B_R and suppose that $u_\lambda \neq 0$ inside B_R. Since $u_\lambda \neq 0$ in B_R, there exists a nodal domain A of u_λ such that $B_R \subset A$. By (2.1), we have that

$$R \leq \rho_A \leq \left(\frac{\lambda_{1,p}(B_1)}{\lambda_{1,p}(A)}\right)^{\frac{1}{p}}.$$

Since the restriction of u_λ corresponds to the first eigenfunction on A, $\lambda_{1,p}(A) = \lambda$. Thus, we get that

$$R \leq \left(\frac{\lambda_{1,p}(B_1)}{\lambda}\right)^{\frac{1}{p}},$$

leading to a contradiction.

5.4. Proof of Theorem 2.5. Notice that for the p-Laplacian, it is still unclear whether $\text{int}(\mathcal{Z}_\lambda)$ is empty or not. Nevertheless, if $\text{int}(\mathcal{Z}_\lambda) \neq \emptyset$, then $\mathcal{H}^1(\mathcal{Z}_\lambda) = +\infty$.

Suppose that $\text{int}(\mathcal{Z}_\lambda) = \emptyset$. We need to show that $\exists C > 0$ such that $\mathcal{H}^1(\mathcal{Z}_\lambda) \leq C\lambda^{\frac{1}{p}}$ for λ large enough.

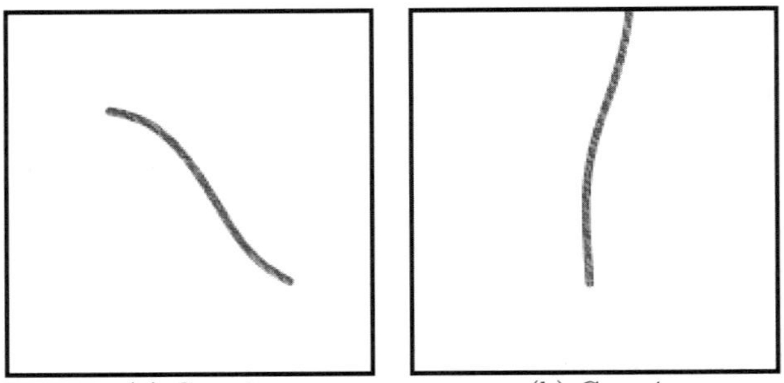

(a) Case 3 (b) Case 4

FIGURE 1. Nodal lines in a square.

By Proposition 2.6, Ω can be split into squares S_c of side $c = \text{Area}(\Omega)\lambda^{-1/p}$ such that each square contains a zero of u_λ. Take λ large enough to allow that the center of each square corresponds to a zero of u_λ. We represent the various cases of nodal lines in a square in Figure 1 and Figure 2. Recall the following Harnack inequality:

THEOREM 5.2 (Theorem 1.1 of [**Tr**]). *Let $K = K(3\rho) \subset \Omega$ be a cube of length 3ρ. Let u_λ be a solution of* (1.2) *associated to the eigenvalue λ such that $u_\lambda(x) \geq 0$ for all $x \in K$, then*

$$\max_{K(\rho)} u_\lambda(x) \leq C \min_{K(\rho)} u_\lambda(x),$$

where C is a positive constant that depends on n, p and on λ.

Notice that in order for Theorem 5.2 to fail, we know that every neighborhood of the boundary of the nodal domain must contain points such that u_λ changes sign. To do so, the nodal line must be closed. Therefore, nodal lines depicted in Figure 1 can not occur.

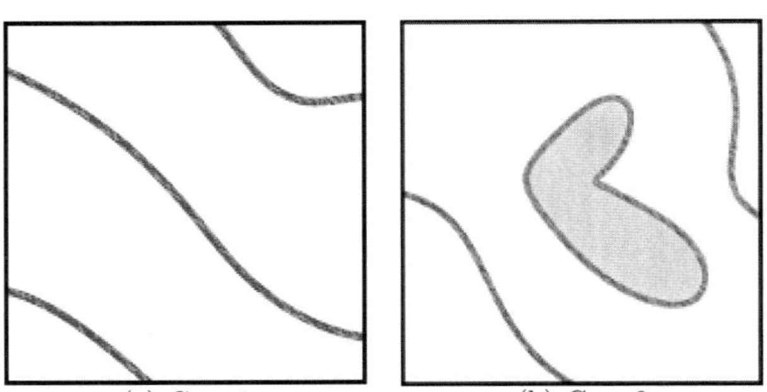

(a) Case 1 (b) Case 2

FIGURE 2. Closed nodal lines in a square.

Suppose that there is a closed nodal line inside a square. If it were the case, it would mean that there exists a nodal domain A included the square. Since the eigenfunction restricted to A would then correspond to the first one, domain monotonicity would yield a contradiction. Indeed, we would have the following

$$\lambda_{1,p}(S_1)\lambda\,\mathrm{Area}(\Omega)^{-p} = \lambda_{1,p}(S_c) \le \lambda_{1,p}(A) = \lambda,$$

yielding that

$$\lambda_{1,p}(S_1) \le \mathrm{Area}(\Omega)^p,$$

a contradiction (simply rescale Ω if necessary).

Therefore, any nodal line inside any such square must be at least of length greater than or equal to $C_1\lambda^{-1/p}$. Since $\mathrm{Area}(\Omega) = \mathrm{Area}(S_c)\lambda^{2/p}$, there are roughly $\lambda^{2/p}$ squares covering Ω, implying that

$$\mathcal{H}^1(\mathcal{Z}_\lambda) \ge C_2\lambda^{2/p}\lambda^{-1/p} = C_2\lambda^{1/p}.$$

5.5. Proof of Theorem 3.1. Using [**Maz**, Theorem 14.1.2], we get the following lemma:

LEMMA 5.3. *Let F be a compact subset of $\overline{B}_r$.*
1. If $1 < p \le n$, for all $u \in C^\infty(\overline{B}_r)$ such that $u \equiv 0$ on F, there exists a positive constant $C_1(n,p)$ depending only on n and p such that

$$\mathrm{cap}_p(F) \le \frac{C_1(n,p)}{r^{-n}} \frac{\int_{\overline{B}_r} |\nabla u|^p}{\int_{\overline{B}_r} |u|^p}. \tag{5.5}$$

2. If $1 < p \le n$, for all $u \in C^\infty(\overline{B}_r)$ such that $u \equiv 0$ on F, where F is a negligible subset of $\overline{B}_r$, and

$$||u||_{L^p(\overline{B}_{r/2})} \le C||\nabla u||_{L^p(\overline{B}_r)}, \tag{5.6}$$

then there exists a positive constant $C_2(n,p)$ depending only on n and p such that

$$\mathrm{cap}_p(F) \ge \frac{C_2(n,p)}{r^{-n}} \frac{\int_{\overline{B}_r} |\nabla u|^p}{\int_{\overline{B}_r} |u|^p}. \tag{5.7}$$

Let us also recall the following two properties of the p-capacity (which are proved in [**Maz**]):

- The p-capacity is monotone: $F_1 \subset F_2 \implies \mathrm{cap}_p(F_1) \le \mathrm{cap}_p(F_2)$.
- The p-capacity of a closed ball of radius r can be computed explicitly:

$$\mathrm{cap}_p(\overline{B}_r) = r^{n-p}\,\mathrm{cap}_p(\overline{B}_1) = r^{n-p}\omega_n \left(\frac{|n-p|}{p-1}\right)^{p-1}, \tag{5.8}$$

where ω_n is the area of the unit sphere $S^{n-1} \subset \mathbb{R}^n$ and $p \neq n$.

We are ready to begin the proof of the lower bound. The ideas used in the following proof come from [**MS**].

LOWER BOUND OF THEOREM 3.1. Fix $\gamma \in (0,1)$ and choose any $r > r_{\Omega,\gamma}$. Then, any ball $\overline{B}_r$ is of non-negligible intersection, i.e.

$$\mathrm{cap}_p(\overline{B}_r \setminus \Omega) \ge \gamma\,\mathrm{cap}_p(\overline{B}_r).$$

Since any $u \in C_0^\infty(\Omega)$ vanishes on $\overline{B}_r \setminus \Omega$, one can use Lemma 5.3, part 1. Using (5.9) and the explicit value of the p-capacity of a closed ball, one gets the following:

$$\begin{aligned} \int_{\overline{B}_r} |u|^p dx &\le \frac{C_1(n,p)}{r^{-n}\operatorname{cap}_p(\overline{B}_r \setminus \Omega)} \int_{\overline{B}_r} |\nabla u|^p dx \\ &\le \frac{C_1(n,p)}{r^{-n}\gamma \operatorname{cap}_p(\overline{B}_r)} \int_{\overline{B}_r} |\nabla u|^p dx \\ &\le \frac{C_1(n,p)}{r^{-p}\gamma \operatorname{cap}_p(\overline{B}_1)} \int_{\overline{B}_r} |\nabla u|^p dx. \end{aligned} \tag{5.9}$$

Choose a covering of $\mathbb{R}^n$ by balls $\overline{B}_r = \overline{B}_r^{(k)}$, $k = 1, 2, ...$, so that the multiplicity of this covering is at most $N = N(n)$, which is bounded since for example, for $n \geq 2$, the following estimate is valid (see [**R**, Theorem 3.2]):

$$N(n) \le n\log(n) + n\log(\log(n)) + 5n.$$

Sum up (5.9) to get the following:

$$\begin{aligned} \int_{\mathbb{R}^n} |u|^p dx &\le \sum_k \int_{\overline{B}_r^{(k)}} |u|^p dx \\ &\le \frac{C_1(n,p)}{r^{-p}\gamma \operatorname{cap}_p(\overline{B}_1)} \sum_k \int_{\overline{B}_r^{(k)}} |\nabla u|^p dx \\ &\le \frac{C_1(n,p)N(n)}{r^{-p}\gamma \operatorname{cap}_p(\overline{B}_1)} \int_{\mathbb{R}^n} |\nabla u|^p dx. \end{aligned}$$

Since for all $u \in C_0^\infty(\Omega)$, we have that

$$\frac{\gamma \operatorname{cap}_p(\overline{B}_1) r^{-p}}{C_n N(n)} \le \frac{\int_\Omega |\nabla u|^p dx}{\int_\Omega |u|^p dx},$$

we get that

$$\lambda_{1,p}(\Omega) \ge K_1(\gamma, n, p) r^{-p} = \frac{\gamma \operatorname{cap}_p(\overline{B_1})}{C_1(n,p)N(n)} r^{-p}. \tag{5.10}$$

Taking the limit of (5.10) as $r \searrow r_{\Omega,\gamma}$ yields the desired result. □

The proof of the upper bound is very similar to the last one, but uses the second part of Lemma 5.3. However, this proof is different from the one given in [**MS**], but has the disadvantage of not yielding an explicit constant. Nevertheless, no such constant are known in the case of the p-Laplacian (recall that Lieb's constant for the lower bound and that the upper bound given in (2.1) are not totally explicit since they both depend on $\lambda_{1,p}(B)$).

UPPER BOUND OF THEOREM 3.1. Fix $\gamma \in (0,1)$. Consider $r_{\Omega,\gamma}$. By definition, we know that

$$\operatorname{cap}_p(\overline{B}_{r_{\Omega,\gamma}} \setminus \Omega) \le \gamma \operatorname{cap}_p(\overline{B}_{r_{\Omega,\gamma}}).$$

We know want to use Lemma 5.3, part 2. Let $F = \overline{B}_{r_{\Omega,\gamma}} \setminus \Omega$. Clearly, F is a negligible subset of $\overline{B}_{r_{\Omega,\gamma}}$. It is also clear that any test function $u \in C_0^\infty(\Omega)$ will vanish identically on F. Therefore, for any such function, using Poincaré inequality, we get

$$||u||_{L^p(\overline{B}_{r_{\Omega,\gamma}/2})} \le ||u||_{L^p(\overline{B}_{r_{\Omega,\gamma}})} \le C||\nabla u||_{L^p(\overline{B}_{r_{\Omega,\gamma}})}.$$

Therefore, one can use Lemma 5.3 part 2 and get :

$$\begin{aligned}\int_{\overline{B}_{r_{\Omega,\gamma}}} |u|^p dx &\geq \frac{C_2(n,p)}{r_{\Omega,\gamma}^{-n}\operatorname{cap}_p(\overline{B}_{r_{\Omega,\gamma}}\setminus\Omega)}\int_{\overline{B}_{r_{\Omega,\gamma}}} |\nabla u|^p dx \\ &\geq \frac{C_2(n,p)}{r_{\Omega,\gamma}^{-n}\gamma\operatorname{cap}_p(\overline{B}_{r_{\Omega,\gamma}})}\int_{\overline{B}_{r_{\Omega,\gamma}}} |\nabla u|^p dx \\ &\geq \frac{C_2(n,p)}{r_{\Omega,\gamma}^{-p}\gamma\operatorname{cap}_p(\overline{B}_1)}\int_{\overline{B}_{r_{\Omega,\gamma}}} |\nabla u|^p dx.\end{aligned}$$

Choose a covering of $\mathbb{R}^n$ by balls $\overline{B}_{r_{\Omega,\gamma}} = \overline{B}^{(k)}_{r_{\Omega,\gamma}}$, $k = 1,2,...$, so that the multiplicity of this covering is at most $N = N(n)$, and get

$$\begin{aligned}\int_{\mathbb{R}^n} |\nabla u|^p dx &\leq \sum_k \int_{\overline{B}^{(k)}_{r_{\Omega,\gamma}}} |\nabla u|^p dx \\ &\leq \frac{r_{\Omega,\gamma}^{-p}\gamma\operatorname{cap}_p(\overline{B}_1)}{C_2(n,p)}\sum_k \int_{\overline{B}^{(k)}_{r_{\Omega,\gamma}}} |u|^p dx \\ &\leq \frac{r_{\Omega,\gamma}^{-p}\gamma\operatorname{cap}_p(\overline{B}_1)N(n)}{C_2(n,p)}\int_{\mathbb{R}^n} |u|^p dx.\end{aligned}$$

For such $u \in C_0^\infty(\Omega)$, we have that

$$\lambda_{1,p}(\Omega) \leq \frac{\int_{\mathbb{R}^n} |\nabla u|^p dx}{\int_{\mathbb{R}^n} |u|^p dx} \leq \frac{r_{\Omega,\gamma}^{-p}\gamma\operatorname{cap}_p(\overline{B}_1)N(n)}{C_2(n,p)};$$

thus, yielding the desired result with $K_2(\gamma,n.p) = \frac{\gamma\operatorname{cap}_p(\overline{B}_1)N(n)}{C_2(n,p)}$. □

5.6. Proofs of Propositions 4.5, 4.6, and 4.7. The proofs are very straightforward and consist of combining Lemma 2.1 with use [**O3**, p.26, eq. (121)-(122)], [**O2**, p. 1208, Eq. (4.31)], or [**O4**, Eq. (14)] respectively.

Acknowledgements

This paper is part of the author's Ph.D. thesis under the supervision of Iosif Polterovich. The author is thankful to Iosif Polterovich and to Dima Jakobson for suggesting the problem. Moreover, he wishes to thank Dorin Bucur, Yaiza Canzani, Daniel Grieser, Dan Mangoubi, Pavel Drabek, and Robert G. Owens for useful discussions and references. Finally, the author thanks Gabrielle Poliquin for her help with the figures.

References

[ACFK] Aomar Anane, Omar Chakrone, Mohamed Filali, and Belhadj Karim, *Nodal domains for the p-Laplacian*, Adv. Dyn. Syst. Appl. **2** (2007), no. 2, 135–140. MR2489043 (2010d:35135)

[B] Gabriella Bognár, *A lower bound for the smallest eigenvalue of some nonlinear eigenvalue problems on convex domains in two dimensions*, Appl. Anal. **51** (1993), no. 1-4, 277–288, DOI 10.1080/00036819308840217. MR1279006 (95c:35184)

[BaC] Rodrigo Bañuelos and Tom Carroll, *Brownian motion and the fundamental frequency of a drum*, Duke Math. J. **75** (1994), no. 3, 575–602, DOI 10.1215/S0012-7094-94-07517-0. MR1291697 (96m:31003)

[BL] Brascamp, H.J. and Lieb, E. H., *Some inequalities for Gaussian measures and the long-rage order of one-dimensional plasma*, Functional Integration and Its Applications (A.M. Arthurs, ed.), Clarendon Press, Oxford, (1975), 1-14.

[Bon] T. Bonnesen, *Über eine Verschärfung der isoperimetrischen Ungleichheit des Kreises in der Ebene und auf der Kugeloberfläche nebst einer Anwendung auf eine Minkowskische Ungleichheit für konvexe Körper* (German), Math. Ann. **84** (1921), no. 3-4, 216–227, DOI 10.1007/BF01459405. MR1512031

[Bru] Jochen Brüning, *Über Knoten von Eigenfunktionen des Laplace-Beltrami-Operators* (German), Math. Z. **158** (1978), no. 1, 15–21. MR0478247 (57 #17732)

[BZ] Ju. D. Burago and V. A. Zalgaller, *The isoperimetric problem in the case of limitation of the width of a domain on a surface* (Russian), Trudy Mat. Inst. Steklov **76** (1965), 81–87. MR0202098 (34 #1972)

[CEP] Fabrizio Cuccu, Behrouz Emamizadeh, and Giovanni Porru, *Nonlinear elastic membranes involving the p-Laplacian operator*, Electron. J. Differential Equations (2006), No. 49, 10 pp. (electronic). MR2226922 (2006m:35090)

[Che] Jeff Cheeger, *A lower bound for the smallest eigenvalue of the Laplacian*, Problems in analysis (Papers dedicated to Salomon Bochner, 1969), Princeton Univ. Press, Princeton, N. J., 1970, pp. 195–199. MR0402831 (53 #6645)

[CM] Tobias H. Colding and William P. Minicozzi II, *Lower bounds for nodal sets of eigenfunctions*, Comm. Math. Phys. **306** (2011), no. 3, 777–784, DOI 10.1007/s00220-011-1225-x. MR2825508

[Cr] Christopher B. Croke, *The first eigenvalue of the Laplacian for plane domains*, Proc. Amer. Math. Soc. **81** (1981), no. 2, 304–305, DOI 10.2307/2044213. MR593476 (82e:35061)

[D1] Pavel Drábek and Stephen B. Robinson, *On the generalization of the Courant nodal domain theorem*, J. Differential Equations **181** (2002), no. 1, 58–71, DOI 10.1006/jdeq.2001.4070. MR1900460 (2003g:35170)

[DF] Harold Donnelly and Charles Fefferman, *Nodal sets of eigenfunctions on Riemannian manifolds*, Invent. Math. **93** (1988), no. 1, 161–183, DOI 10.1007/BF01393691. MR943927 (89m:58207)

[DHT] J. I. Díaz, J. Hernández, and L. Tello, *On the multiplicity of equilibrium solutions to a nonlinear diffusion equation on a manifold arising in climatology*, J. Math. Anal. Appl. **216** (1997), no. 2, 593–613, DOI 10.1006/jmaa.1997.5691. MR1489600 (98i:86005)

[DTh] Jesus Ildefonso Diaz and François de Thélin, *On a nonlinear parabolic problem arising in some models related to turbulent flows*, SIAM J. Math. Anal. **25** (1994), no. 4, 1085–1111, DOI 10.1137/S0036141091217731. MR1278892 (96a:35076)

[G] Daniel Grieser, *The first eigenvalue of the Laplacian, isoperimetric constants, and the max flow min cut theorem*, Arch. Math. (Basel) **87** (2006), no. 1, 75–85, DOI 10.1007/s00013-005-1623-4. MR2246409 (2007e:35213)

[GM] S. Granlund and N. Marola, *Vanishing properties of sign changing solutions to the p-laplace type equations in the plane*, Preprint, 2012.

[GR] Roland Glowinski and Jacques Rappaz, *Approximation of a nonlinear elliptic problem arising in a non-Newtonian fluid flow model in glaciology*, M2AN Math. Model. Numer. Anal. **37** (2003), no. 1, 175–186, DOI 10.1051/m2an:2003012. MR1972657 (2004e:86022)

[Hay] W. K. Hayman, *Some bounds for principal frequency*, Applicable Anal. **7** (1977/78), no. 3, 247–254. MR0492339 (58 #11468)

[He] Joseph Hersch, *Sur la fréquence fondamentale d'une membrane vibrante: évaluations par défaut et principe de maximum* (French, with English summary), Z. Angew. Math. Phys. **11** (1960), 387–413. MR0125319 (23 #A2622)

[Hua] Yin Xi Huang, *On the eigenvalues of the p-Laplacian with varying p*, Proc. Amer. Math. Soc. **125** (1997), no. 11, 3347-3354, DOI 10.1090/S0002-9939-97-03961-0. MR1403133 (97m:35081)

[I] V. K. Ionin, *Isoperimetric and certain other inequalities for a manifold of bounded curvature* (Russian), Sibirsk. Mat. Ž. **10** (1969), 329–342. MR0240753 (39 #2098)

[JL] Petri Juutinen and Peter Lindqvist, *On the higher eigenvalues for the ∞-eigenvalue problem*, Calc. Var. Partial Differential Equations **23** (2005), no. 2, 169–192, DOI 10.1007/s00526-004-0295-4. MR2138081 (2006a:35226)

[Ka] Bernhard Kawohl, *Rearrangements and convexity of level sets in PDE*, Lecture Notes in Mathematics, vol. 1150, Springer-Verlag, Berlin, 1985. MR810619 (87a:35001)

[KF] B. Kawohl and V. Fridman, *Isoperimetric estimates for the first eigenvalue of the p-Laplace operator and the Cheeger constant*, Comment. Math. Univ. Carolin. **44** (2003), no. 4, 659–667. MR2062882 (2005g:35053)

[KMC] John R. King and Scott W. McCue, *Quadrature domains and p-Laplacian growth*, Complex Anal. Oper. Theory **3** (2009), no. 2, 453–469, DOI 10.1007/s11785-008-0103-9. MR2504764 (2010c:35218)

[KS] Bernd Kawohl and Friedemann Schuricht, *Dirichlet problems for the 1-Laplace operator, including the eigenvalue problem*, Commun. Contemp. Math. **9** (2007), no. 4, 515–543, DOI 10.1142/S0219199707002514. MR2348842 (2009b:35116)

[L] Elliott H. Lieb, *On the lowest eigenvalue of the Laplacian for the intersection of two domains*, Invent. Math. **74** (1983), no. 3, 441–448, DOI 10.1007/BF01394245. MR724014 (85e:35090)

[Lind1] Peter Lindqvist, *A note on the nonlinear Rayleigh quotient*, Analysis, algebra, and computers in mathematical research (Luleå, 1992), Lecture Notes in Pure and Appl. Math., vol. 156, Dekker, New York, 1994, pp. 223–231. MR1280948 (95f:35189)

[Lind2] Peter Lindqvist, *On the equation* $\operatorname{div}(|\nabla u|^{p-2}\nabla u) + \lambda|u|^{p-2}u = 0$, Proc. Amer. Math. Soc. **109** (1990), no. 1, 157–164, DOI 10.2307/2048375. MR1007505 (90h:35088)

[Lind3] Peter Lindqvist, *Notes on the p-Laplace equation*, Report. University of Jyväskylä Department of Mathematics and Statistics, vol. 102, University of Jyväskylä, Jyväskylä, 2006. MR2242021 (2007h:35102)

[M] Dan Mangoubi, *A remark on recent lower bounds for nodal sets*, Comm. Partial Differential Equations **36** (2011), no. 12, 2208–2212, DOI 10.1080/03605302.2011.558556. MR2852075

[M1] Dan Mangoubi, *On the inner radius of a nodal domain*, Canad. Math. Bull. **51** (2008), no. 2, 249–260, DOI 10.4153/CMB-2008-026-2. MR2414212 (2010c:58037)

[M2] Dan Mangoubi, *Local asymmetry and the inner radius of nodal domains*, Comm. Partial Differential Equations **33** (2008), no. 7-9, 1611–1621, DOI 10.1080/03605300802038577. MR2450173 (2009g:35040)

[Mak] E. Makai, *A lower estimation of the principal frequencies of simply connected membranes* (English, with Russian summary), Acta Math. Acad. Sci. Hungar. **16** (1965), 319–323. MR0185263 (32 #2732)

[Maz] Vladimir Maz'ya, *Sobolev spaces with applications to elliptic partial differential equations*, Second, revised and augmented edition, Grundlehren der Mathematischen Wissenschaften [Fundamental Principles of Mathematical Sciences], vol. 342, Springer, Heidelberg, 2011. MR2777530 (2012a:46056)

[McK] H. P. McKean, *An upper bound to the spectrum of* Δ *on a manifold of negative curvature*, J. Differential Geometry **4** (1970), 359–366. MR0266100 (42 #1009)

[MS] Vladimir Maz′ya and Mikhail Shubin, *Can one see the fundamental frequency of a drum?*, Lett. Math. Phys. **74** (2005), no. 2, 135–151, DOI 10.1007/s11005-005-0010-1. MR2191951 (2006m:58050)

[O1] Robert Osserman, *A note on Hayman's theorem on the bass note of a drum*, Comment. Math. Helv. **52** (1977), no. 4, 545–555. MR0459099 (56 #17297)

[O2] Robert Osserman, *The isoperimetric inequality*, Bull. Amer. Math. Soc. **84** (1978), no. 6, 1182–1238. MR0500557 (58 #18161)

[O3] Robert Osserman, *Bonnesen-style isoperimetric inequalities*, Amer. Math. Monthly **86** (1979), no. 1, 1–29, DOI 10.2307/2320297. MR519520 (80h:52013)

[O4] Robert Osserman, *Isoperimetric Inequalities and Eigenvalues of the Laplacian*, Proceedings of the International Congress of Mathematics, Helsinki, (1978).

[Poli] G. Poliquin, *First eigenvalue of the p-Laplacian, p-capacity estimates, and the inradius of Euclidean domains*, in preparation.

[R] C. A. Rogers, *Packing and covering*, Cambridge Tracts in Mathematics and Mathematical Physics, No. 54, Cambridge University Press, New York, 1964. MR0172183 (30 #2405)

[Si] M. A. J. Silva, *On a viscoelastic plate equation with history setting and perturbation of p-Laplacian type*, IMA Journal of Applied Mathematics Advance Access, (2012), 1-17.

[SZ] Christopher D. Sogge and Steve Zelditch, *Lower bounds on the Hausdorff measure of nodal sets*, Math. Res. Lett. **18** (2011), no. 1, 25–37, DOI 10.4310/MRL.2011.v18.n1.a3. MR2770580 (2012c:58055)

[Tr] Neil S. Trudinger, *On Harnack type inequalities and their application to quasilinear elliptic equations*, Comm. Pure Appl. Math. **20** (1967), 721–747. MR0226198 (37 #1788)

[Y] Shing Tung Yau, *Problem section*, Seminar on Differential Geometry, Ann. of Math. Stud., vol. 102, Princeton Univ. Press, Princeton, N.J., 1982, pp. 669–706. MR645762 (83e:53029)

Département de mathématiques et de statistique, Université de Montréal, CP 6128 succ. Centre-Ville, Montréal, H3C 3J7, Canada

E-mail address: gpoliquin@dms.umontreal.ca